CW00553925

ESTATES GAZETTE LAW REPORTS

Estates Gazette Law Reports

1976

Volume 1

Series edited by
Barry Denyer-Green LLM PhD FRICS
of the Middle Temple, barrister

Cases edited by
J Muir Watt OBE MA
of the Inner Temple, barrister

Audrey Boyle BA
Publisher's editor

A member of Reed Business Publishing

THE ESTATES GAZETTE LIMITED
151 WARDOUR STREET, LONDON W1V 4BN

First published 1995
ISBN for complete set of 2 volumes: 0 7282 0218 2
ISBN for this volume: 0 7282 0219 0

ISSN 0951-9289

Origination by Keyword Publishing Services
Printed and bound by Bell and Bain Ltd, Glasgow

CONTENTS

TABLE OF CASES

Titles of cases shown in bold type
Names of parties reversed shown in ordinary type

INDEX OF SUBJECT-MATTER

COMPENSATION FOR ACQUISITION OF LAND — *continued*
Valuation — *continued*

Gas main
Land with planning permission for development sterilised by construction of gas main — Claim for £51,000 on basis of near agreement discounted for risks — Held, award of £27,000 reflected value of land ripe for development.
***Vaynor Developments Ltd* v *Wales Gas Board (LT)*, 196**

Injurious affection
Land taken for construction of underground railway tunnel — Claim for injurious affection for noise and vibration — Held, reduction in gross value persuasive evidence in making small award for injurious affection.
***Bradd* v *London Transport Executive (LT)*, 179**

Pointe Gourde principle
Claim for development value of acquired land — Reasonable expectation of plot density in no-scheme world — Held, *Pointe Gourde* principle applied and diminution in value caused by council's scheme disregarded — Residual valuations accepted by tribunal as only evidence.
***St Pier Ltd* v *Lambeth London Borough (LT)*, 189**

Vacant possession or subject to tenancy
Houses subject to controlled tenancies — Tenants rehoused after notice of entry — Houses padlocked by council without any other overt act of possession — Whether owner entitled to compensation on vacant possession basis — Held, owner entitled to vacant possession value because right to possession retained by owner until locking of houses.
***Buckingham Street Investments Ltd* v *Greater London Council (LT)*, 178**

COMPULSORY PURCHASE
Compulsory purchase order
Application to quash
CPO for road widening purposes — Alleged secret agreement affecting excess failure to challenge order within six-week time-limit — Held, failure to challenge within time-limit excludes courts jurisdiction to consider application.
***R* v *Secretary of State for the Environment*, ex parte *Ostler (CA)*, 11**
CPO quashed because of breach of rules of natural justice — Held, inspector and Secretary of State had taken into account matters not in issue at inquiry — Furthermore, inspector's recommendation was based on observations made at his site inspection, after the inquiry, which the appellant had no possibility to comment on.
***Fairmount Investments Ltd* v *Secretary of State for the Environment (CA)*, 7**

CONTRACT
Accord and satisfaction
Extent of liability
Programme of building works based on assumption that road would be closed by certain date — Work begun but disrupted by delay in road closure — Employers by letter recognise liability for consequential loss — Held, on its true construction employers accepted liability for losses incurred after date of letter.
***Bacal Construction (Midlands) Ltd* v *Northampton Development Corporation (CA)*, 127**

Implied term
Geological conditions
Whether statement in tender documents as to geological make-up of site implied term or warranty — No express term in contract to entitle contractors to remuneration for extra work caused by misdescribed ground conditions — Held, strong commercial reasons for implying such a term.
***Bacal Construction (Midlands) Ltd* v *Northampton Development Corporation (CA)*, 127**

"Subject to contract"
Sale by tender
Land to be sold by tender — Tender documents set out all the terms of the contract and required to be attached to offer — On letter of acceptance of offer "Subject to contract" mistakenly typed — Words held to be meaningless.
***Michael Richards Properties Ltd* v *Corporation of Wardens of St Saviour's Parish, Southwark (Ch)*, 163**

DEROGATION FROM GRANT
Lease
Unlawful use
Unlawful change of use by underlessee of land leased by Crown Estate Commissioners — Enforcement notice served with consent of Commissioners, as required by statute — Held, argument that by giving consent Commissioners were in derogation of their grant rejected — Doctrine cannot be prayed in aid of something unlawful — Cannot be applied to fetter discretionary power exercised for public good.
***Molton Builders Ltd* v *Westminster City Council (CA)*, 150**

DWELLING HOUSE
Meaning of — See **Town and country planning** (Dwellinghouse)

EMPLOYMENT
Right to possession
Job for life
Contract of service creating "job for life" — Occupation of house ancillary to service and rent-free — Held, no licence for life and no tenancy for life under Settled Land Act 1925 created.
***Ivory* v *Palmer (CA)*, 41**

ESTATE AGENTS
Deposit
Engagement of person as estate agent — Whether he had actual or ostensible authority to receive pre-contract deposit from would-be purchaser as agent of vendor — Whether vendor liable to purchaser for acts or default of estate agent in dealing with deposit — Held, vendor has no claim, right or control over deposit — Purchaser entitled to return of deposit on demand.
***Sorrell* v *Finch (HL)*, 20**

Evidence
Plaintiff had contractual right to be appointed sole selling agent, "at any rate in the first instance", of all houses built on any land found by him for developer — Further, developer had no right to terminate, irrespective of any length of notice — Held, on evidence, there was a contract of agency but different in terms — Determinable on reasonable notice — Developers entitled to appoint other agents to act solely or jointly with plaintiff — Plaintiff in repudiatory breach.
***Knight* v *Calderlodge Developments (QB)*, 13**

Commission
Quantum meruit
No express contract for plaintiffs to act as agents for sale of defendant's home — Incomplete agreement as parties did not establish sale price — Held, claim for *quantum meruit* damages failed — No contract to be implied from defendant's awareness of plaintiff's continued and successful efforts to find purchaser — On the facts, acted as agents to purchasers.
***Rosenthal* v *Allan (QB)*, 27**

Theft — See **Theft** (Sentencing — Estate agents)

FAIR RENT — See **Landlord and tenant** (Rent Acts — Rent assessment)

FEES — See **Compensation for acquisition of land** (Disturbance — Surveyors fees); **Estate agents** (Commission)

FIRE DAMAGE — See **Landlord and tenant** (Landlord and Tenant Act 1954 Part II — New tenancy application)

FORFEITURE — See **Landlord and tenant** (Forfeiture)

GAS MAIN — See **Compensation for acquisition of land** (Valuation)

HIGHWAYS
Dedication
Evidence
Appeal against magistrates finding that lane not public highway — Documentary evidence of dedication and oral evidence of user inconclusive — Held, on that basis court could not rule decision unreasonable or perverse.
***Hall* v *H B Howlett & Partners Ltd (QB)*, 35**

Highway authority
Repairs
Whether three footpaths were "out of repair" within the meaning of Highways Act 1959 s59 — Held, out of repair denotes becoming unsound or impaired by neglect or use — To be contrasted with "obstructed" which requires positive act — Therefore, one footpath with barbed wire fence across it was not out of repair, but footpaths with vegetation naturally growing on them were.
***Hereford and Worcester County Council* v *Newman (CA)*, 32**

Obstruction
Lawful excuse
Proprietor erects permanent bollard on footway outside shop to protect his property and reduce risk of injury to persons — Held, no evidence to justify finding of "lawful excuse".
***Atfield* v *Ipswich Borough Council (QB)*, 37**

HOUSING
Rehabilitation order
Reasons for refusal
Applications for rehabilitation orders in respect of properties subject to clearance and compulsory purchase orders refused — Held, in deciding to refuse council entitled to take account of evidence and findings of compulsory purchase inquiry — Statutory duty to provide reasons for refusal discharged by letters stating council policy to demolish existing and erect new housing.
***Elliott* v *Southwark London Borough Council (CA)*, 4**

LAND REGISTRATION
Option
New lease
Whether option to renew underlease void against purchaser of the registered freehold land — Not registered as land charge before purchase complete — However, notice of underlease lodged against freehold on land register — Effect of s23(1) of Land Charges Act 1972 — Held, option protected by notice of underlease, even though no mention of option.
***Kitney* v *MEPC Ltd (Ch)*, 49**

LANDLORD AND TENANT
Assignment — See **Vendor and purchaser** (Assignment of lease)
Business tenancies — See **Landlord and tenant** (Landlord and Tenant Act 1954 Part II)
Forfeiture
Adjournment
Forfeiture for non-payment of rent — Held, judge wrong to adjourn case for a second time and not make possession order after plaintiff had made out his case — No need for *mandamus* as judge likely to determine case very soon.
***Wathen* v *White (QB)*, 47**
Harassment
Civil action
Landlord cuts off electricity and water — Whether civil remedy for harassment — Held, tenant entitled to damages for breach of implied term of tenancy including quiet enjoyment — No liability under Rent Act 1965 s30(2).
***McCall* v *Abelesz (CA)*, 82**
Landlord and Tenant Act 1954 Part II
New tenancy application
Business premises rendered unfit for occupation by fire — Whether tenants had remained in occupation, so as to exercise right to renew lease — Held, on the evidence they had — Premises, although damaged, still in existence after fire — Same rights of occupation before fire as after — Demolition of premises by landlord did not affect tenants' rights of occupation.
***Morrison Holdings Ltd* v *Manders Property (Wolverhampton) Ltd (CA)*, 70**
Tenants then commenced proceedings in High Court claiming, specific performance of agreement for 50-year lease — Held, Court of Appeal upheld refusal at first instance to grant stay of county court proceedings, pending determination of High Court action — Argument rejected that county court was debarred from deciding issues which were relevant to both cases and which were normally relevant to granting of relief which it had no jurisdiction to grant — S43A of 1954 Act.
***Rushton* v *Smith (CA)*, 62**
Lease
Unilateral mistake
Landlord unilaterally mistaken over effect of court order — Upon this basis grants tenant new lease on favourable terms — Held, lease not to be set aside — No termination effected — Landlord's possession claim failed.
***ITC Pension Fund Ltd* v *Pinto (Ch)*, 67**
Lease or licence
Business premises
Whether agreement lease or licence constituting "management agreement" — Occupant did not devote whole time to business — Held, treated by both parties as tenancy — Therefore, lease — True subtenancy protected by statute.
***Wang* v *Wei (QB)*, 66**
Leasehold enfranchisement
Price of freehold
Enfranchisement of detached house with 59 years of lease unexpired — Whether tenant's bid to be included — Division of marriage value — Housing Act 1974 s118 — Held, bargaining power of landlord and tenant equal — Valuation accordingly.
***Norfolk* v *Trinity College, Cambridge (LT)*, 215**
Semi-detached house
Claimed to enfranchisement — Whether occupied as "house" within s2 Leasehold Reform Act 1962 — Held, test to be applied is what the reasonable man would conclude — A "house" despite that access existed from it through party wall to adjoining property and door to that property bricked-up — Additional strip of land not enfranchised as not let with house.
***Gaidowski* v *Gonville & Caius College, Cambridge (CA)*, 72**
Option
New lease
Protected tenancy with rent controlled by statute — Option for a new lease at commercial yearly rack-rent at which the demised premises might reasonably be expected to be let in "open market" — Held, action for specific performance not defeated because of impossibility of performance — Value of the rack-rent equivalent to controlled rent.
***Newman* v *Dorrington Developments Ltd (Ch)*, 76**

Rent Acts
Rent assessment
Fair rent — Allegations of improper basis of determination — Held, powers of court restricted to considering errors of law — Committee entitled to base decision on own experience and knowledge — Valuation can be arrived at by reference to comparables outside immediate area.
***Meredith* v *Stevens (QB)*, 78**
Fair rent — Furnished premises — Held, assessment panel not bound to determine first what fair rent would be if property was unfurnished and then make additions for furniture — Taking into account general proposition that furnished tenancies are let for shorter periods did not constitute taking landlord's personal circumstances into account.
***Campbell* v *Gardner (QB)*, 81**
Whether tenants' application to rent assessment committee can be withdrawn — Held, an application under Sch 6 to Rent Act 1968, cannot be withdrawn once matter has been referred to committee.
***Hanson* v *London Rent Assessment Committee (QB)*, 79**

Rent review clause
Construction: general
Clause failed to make provision for position if landlord and tenant were in default of agreement on new rent — Held, that in absence of agreement tenant should pay rent on use-and-occupation basis — This sum to be calculated in accordance with those terms of rent review clause upon which agreed sum would have been based.
***Beer* v *Bowden (Ch)*, 83**
Delivery of letter addressed to Secretary of RICS summarising effect of rent review clause and containing express request not to appoint valuer for time being — Whether or not sufficient to appoint valuer — Held, an "application" within meaning of clause — Meaning of stipulation requiring application to be made "not less than six months before March 25 1974".
***Rank City Wall Ltd* v *Gallaher & Johnston Allen Ltd (QB)*, 98**
Time-limits
Building lease — Whether or not time-limits of the essence — Held, distinction between "option clauses" and "obligation clauses" false and unhelpful dichotomy — Held, correct approach to construction is to ask whether parties are to be imputed with intention that limits should be strictly adhered to — Commercially important consequences for landlord and tenant on operation of rent review clause — On the basis of business efficacy, courts should readily impute intention that time-limits mandatory.
***United Scientific Holdings Ltd* v *Burnley Corporation (CA)*, 89**
Clause providing for dispute to be referred to arbitration by lessor — Failure of landlords to refer within time provided by lease — Clause to be properly construed as an option rent review clause and not an obligation rent review clause — Time-limit mandatory — Too late for landlords to refer matter to arbitration.
***Mount Charlotte Investments Ltd* v *Leek, Westbourne & Eastern Counties Building Society (Ch)*, 83**
Complicated clause in underlease finishing with term that an arbitrator is to be appointed on application by lessor within specified time-limit — Held, term for protection of tenant to be construed accordingly — Time-limit mandatory.
***Stylo Shoes Ltd* v *Wetherall Bond Street W1 Ltd (CA)*, 86**
Rent review clause in underlease modelled on Form 8, 1 in *Encyclopedia of Forms and Precedents* (1966) vol 12 — According to terms landlords some 18 months late with their notice demanding revised rent — Held, on true construction time-limit did not need to be strictly observed — Delay not unreasonable, but borderline case.
***Accuba Ltd* v *Allied Shoe Repairs Ltd (Ch)*, 88**

Repairs
Implied covenant
Reservation of commons parts in residential multi-storey building to local authority landlord — Tenants dependent upon lifts common stairs and corridors for enjoyment of their leases — Held, implication of term by necessity — Obligation upon landlords to take reasonable care to keep lifts, staircases, corridors and associated lighting in reasonable repair — Additional repairing covenant implied by Housing Act 1961 s32.
***Liverpool City Council* v *Irwin (HL)*, 53**
Service charges
Determination
Lease provided for service charges to be "ascertained and certified by the lessor's managing agents acting as experts and not as arbitrators" — Held, proper construction that managing agents are to be somebody different from the lessor — Lessor and managing agent company same person — Therefore, no valid certificate — Term to be implied that service charges to be fair and reasonable.
***Finchbourne Ltd* v *Rodrigues (CA)*, 51**
Surrender of lease
Valuation
Valuers jointly instructed by landlord and tenant to assess surrender value of lease — Proceedings by landlord to upset valuation disclosed

VAT
Taxable supply
Concession agreement under which company were to sell certain goods during specified hours from shops at Heathrow Airport — Held, true nature of agreement was to confer right to occupy land — No taxable supply.
***British Airports Authority* v *Commissioners of Customs and Excise (QB),* 134**

THEFT
Sentencing
Estate agent
Appeal against sentence — Conviction on guilty plea of nine counts of theft of deposits — Sentenced to 15 months imprisonment concurrently on each count — Held, proper sentence.
***R* v *Greenfield (CA),* 25**

TOWN AND COUNTRY PLANNING
Blight notice
Counternotice
Blight notice served in respect of "white" land — Counternotice objections that no proposal to acquire and no reasonable endeavours to sell — On professional advice land not offered for sale — Held, objection to blight notice upheld on ground that no reasonable endeavours to sell.
***Perkins* v *West Wiltshire District Council (LT),* 182**
Blight notice served in respect of dwellinghouse — Line of proposed road cuts across accessway — Counternotice that no part of hereditament to be acquired — Held, rights appurtenant to land not part of hereditament — Objection upheld.
***Ley* v *Kent County Council (LT),* 183**
Conservation area
Direction to cease demolition
Form of direction did not comply with express words of statute — Contained no reference to it being expedient that it should take immediate effect — However, implicit from wording of direction that this was the case — Held, conviction upheld.
***R* v *Endersby Properties Ltd (CA),* 155**
Dwellinghouse
Meaning of
Top floor of premises occupied by owner as home — Other two lower floors used by owner to carry on estate agent's business — Held, Secretary of State correct to decide that premises had dual residential and business use — Not "dwellinghouse" within meaning of General Development Order 1973.
***Scurlock* v *Secretary of State for Wales (QB),* 149**
Enforcement notice
Appeal to Secretary of State
Refusal to grant permission for cash-and-carry warehouse — Whether in reaching conclusion that warehouse would adversely affect local authority's project for central shopping centre minister required evidence — Held, minister was entitled to rely on experience and opinion — Proceedings were not akin to criminal case.
***Wholesale Male Order Supplies Ltd* v *Secretary of State for the Environment (QB),* 137**
Three enforcement notices served prohibiting use of three adjacent pieces of land for sale of caravans — Use for that purpose established on two areas — Two notices quashed, other upheld — Appellant argued that minister should have treated all three pieces as single planning unit — However, had abandoned at inquiry relevant appeal ground that matters in notice did not allege breach of planning control — Minister not obliged to find out for himself possible grounds of appeal.
***West Cheshire Caravan Co Ltd* v *Ellesmere Port Borough Council (QB),* 143**
Two enforcement notices issued in respect of site — One properly required extensions to building to be pulled down — The other, applicable to entire site, required cessation of a change of use — Second notice amended by minister so as to apply to areas where extensions stood and where two cottages had once stood, even though rest of site could take advantage of use classes order — Held, remitted for reconsideration to minister — Wrong and oppressive to view site of extensions as without same established use — Whole site, including where cottages stood, to be treated as single planning unit.
***Joyce Shopfitters Ltd* v *Secretary of State for the Environment (QB),* 144**
Whether inspector erred in deciding change of use commenced after January 1 1964 — Held, inspector was entitled to reach decision he made — Question of fact.
***Catton* v *Secretary of State for the Environment (QB),* 150**
Validity
Material change of use in part of planning unit — Enforcement notice can be served either in respect of whole or relevant part of planning unit — Held, minister can correct errors in notice provided no injustice.
***Morris* v *Secretary of State for the Environment (QB),* 146**

Planning permission
Compensation
Planning permission for development of restaurant, residential club, flat and garages revoked — Existing building listed — Proper basis of valuation — Whether interest recoverable — Held, value of right to develop assessed at below offer by third party — Interest not allowed.
***Burlin* v *Manchester City Council (LT),* 205**
Conditions
Planning permission made subject to condition requiring existing office accommodation on site only to be used as ancillary to non-office uses — Promoted authority's policy for controlling and restricting office accommodation — Held, condition reated fairly and reasonably to development site — Policy intended to apply both outside and inside area covered by the permission.
***AI&P (Stratford) Ltd* v *London Borough of Tower Hamlets (QB),* 139**
Planning permission subject to condition that no vehicles should be sold or displayed for sale on forecourt — Definition of "forecourt" — Held, no objection that effect of condition would be to cancel former established use
***Leighton & Newman Car Sales Ltd* v *Secretary of State for the Environment (QB),* 154**
Material change of use
Previous use of site by building firm for storage of materials — Sale of building materials becomes predominant use — Held, no material change of use — Means of subsequent disposal of materials when they leave the site of no importance in land-use terms.
***Snook* v *Secretary of State for the Environment (QB),* 148**
Outline permission
Deemed refusals of planning authority to grant permission to further develop site with outline planning permission — *Mandamus* sought to compel minister to hear appeals — Argument as to whether development had begun so as to preserve outline permission — Held, outline permission single permission and not severable — Previous development had preserved permission.
***Percy Bilton Industrial Properties Ltd* v *Secretary of State for the Environment (QB),* 141**
Use classes
Enforcement notice served on applicant to discontinue use of barn for servicing and maintenance of vehicles — Previously used by sculptor as studio — At that time it was equipped with anvils, benches and metals — Held, that Secretary of State was entitled to decide that use as sculptor's studio is *sui generis* and reject applicant's claim that barn had acquired character of "general industrial building".
***Tessier* v *Secretary of State for the Environment (QB),* 137**

TREE ROOTS DAMAGE — See **Nuisance** (Tree)

TRUST FOR SALE
Beneficial interest
Order for sale
Mother, protected tenant, bought reversion — Property conveyed jointly to mother, daughter and son-in-law — Latter two paid mortgage and cost of modernisation and repairs — Held, no order for sale — Mother persuaded to buy only on the basis that she would always keep her home.
***Charlton* v *Lester (QB),* 131**

TRUSTS
Beneficial interest
Joint tenants
Wife responsible for marriage breakdown — Husband provided all purchase moneys and willing to pay compensation — Held, inequitable to make order for sale — Where husband puts up purchase price and land conveyed as beneficial joint tenants, purpose to provide home for husband had not failed.
***Hayward* v *Hayward (Ch),* 46**
Resulting trust
Presumption
House purchased by mother but conveyed to mother, daughter and son-in-law as beneficial tenants in common — Held, undisturbed finding that conveyed "simply for convenience" — Negatived any application of a presumption of advancement — Plea of estoppel failed *in limine* — Held, presumption of resulting trust in favour of mother applied.
***Gross* v *French (CA),* 129**

VALUATION
Compensation — See **Compensation for acquisition of land** (Valuation)
Lease, surrender of — See **Landlord and tenant** (Surrender of lease)

VALUE ADDED TAX — See **Taxation** (VAT)

VENDOR AND PURCHASER
Assignment of lease
Illegality
Illegal clause, contrary to 1968 Rent Act, providing for purchase at premium of assignor's furniture — Held, clause capable of severance from legal elements of agreement to assign lease — Specific perform-

VENDOR AND PURCHASER — *continued*
Assignment of lease — *continued*
Illegality — *continued*

ance of contract to assign, with payment of reasonable consideration for furniture.

Ailion v ***Spiekermann (Ch), 48***

Completion of sale
Notice to complete

Contract of sale of land incorporating Law Society's conditions, including condition permitting completion on certain dates after service of notice — Presence of this condition not inconsistent with further contractual obligation to complete by a fixed date — Obligation continues to complete by date fixed or within reasonable time thereafter — Service of notice not prerequisite to enforcement of sale.

Woods v ***Mackenzie Hill Ltd (Ch), 162***

Misrepresentation
Damages

Contract of sale of land signed by one joint tenant without the authority of other — Signatory held to have misrepresented to purchaser that he owned house and was in position to sell — No specific performance — Effect of Misrepresentation Act 1967 on rule in *Bain* v *Fothergill* — Held, damages not limited to expenses incurred — Entitled to recover for loss of bargain.

Watts v ***Spence (Ch), 158***

Option
Purchase of freehold interest

Addition of new term in option to purchase above landowner's signature — Offer in old agreement discharged by parties accepting alteration — Held, new agreement unenforceable — No sufficient writing to meet s40 of Law of Property Act 1925 — No part performance by co-operation in applying for planning permission.

New Hart Builders Ltd v ***Brindley (Ch), 165***

Option to purchase exercisable by notice in writing to be sent "by registered or recorded delivery post" — Notice in fact sent by ordinary post — Held, option validly exercised — Rule of construction same as for statutory instruments — With no express term to make the mode of acceptance mandatory provision was only permissive.

Yates Building Company Ltd v ***R J Pulleyn & Sons (York) Ltd (CA), 157***

Right of pre-emption
Registration

Right of pre-emption is registrable as Class C (iv) land charge, if taken by a council vendor under s104 of the Housing Act 1957, whatever the position may be in case of private sale.

First National Securities Ltd v ***Chiltern District Council (Ch), 39***

WATER
Pollution
Cause

Pollution of streams traced to breaches in walls of lagoons containing effluent — Held, appellant did not "cause" pollution simply by entering into contract and allowing adjoining landowners to directly discharge effluent into lagoons — "Causing" requires some active operation — Conviction quashed — River (Prevention of Pollution) Act 1951 s2(1)

Price v ***Cromack (QB), 119***

AUCTIONS

A

QUEEN'S BENCH DIVISIONAL COURT

February 20 1975

(Before Lord WIDGERY CJ, Mr Justice ASHWORTH and Mr Justice BRIDGE)

LOMAS v RYDEHEARD

Estates Gazette March 13 1976

(1975) 237 EG 801

Mock Auctions Act 1961—Meaning of "sale of goods by way of competitive bidding"—Meaning of "highest bid" in the context of a Dutch auction process—General point of constitution of proceedings—Separate incidents add up to "conducting a mock auction"

B

This was an appeal by Mr Francis John Lomas, of Wolsey Road, Lytham Road, Blackpool, Lancashire, from the dismissal by Preston Crown Court on May 2 1974 of his appeal against his conviction by Blackpool justices on seven charges of offences against the Mock Auctions Act 1961 brought by the respondent, Mr Alan Rydeheard, the chief superintendent of police, Blackpool.

Mr I McCulloch (instructed by Simpson, Silvertown & Co, agents for D Betesh & Co, of Manchester) appeared for the appellant, and Mr W R Wickham (instructed by Norton, Rose & Co, agents for A C Brewer, of Lancashire Police Authority, Blackpool) represented the respondent.

C

Giving the first judgment, BRIDGE J said: This is an appeal by case stated from a decision of the Crown Court at Preston given on May 2 1974, affirming the conviction by justices of the petty sessions at Blackpool of the present appellant of seven offences in contravention of the provisions of the Mock Auctions Act 1961. That statute is unfamiliar territory in this court, and I turn at once to the provisions which create the relevant offence. Section 1 (1) provides:

> "It shall be an offence to promote or conduct, or to assist in the conduct of, a mock auction at which one or more lots to which this Act applies are offered for sale."

D

Subsection (2) prescribes penalties for the offence, and then by subsection (3) it is provided:

> "Subject to the following provisions of this section, for the purposes of this Act a sale of goods by way of competitive bidding shall be taken to be a mock auction if, but only if, during the course of the sale (a) any lot to which this Act applies is sold to a person bidding for it, and either it is sold to him at a price lower than the amount of his highest bid for that lot, or part of the price at which it is sold to him is repaid or credited to him or is stated to be so repaid or credited, or," and then (b) is immaterial for present purposes, "or (c) any articles are given away or offered as gifts."

E

The circumstances out of which the seven informations on which the appellant was convicted arose occurred at premises known as the Bonnie Street Sale Rooms on two successive days, July 29 and 30, as long ago as the year 1972. The evidence of a number of police officers who visited those sale rooms from time to time on those two dates is set out at some length in the case stated. It should be said at once that there is no dispute that if the activities going on in those sale rooms amounted to the holding of a mock auction or mock auctions, the appellant was a person who was either conducting or assisting in the conduct of those mock auctions. Again there is no doubt that the articles which were being offered for sale and sold were within the phrase "lots to which this Act applies." The primary issue canvassed in this court as a ground on which it is said that the convictions should be quashed relates to the question whether the relevant activities fell within the phrase in section 1 (3), "a sale of goods by way of competitive bidding." That phrase is the subject of a more elaborate definition in section 3 (1) of this Act, which is in these terms:

F

G

> "In this Act 'sale of goods by way of competitive bidding' means any sale of goods at which the persons present, or some of them, are invited to buy articles by way of competitive bidding, and 'competitive bidding' includes any mode of sale whereby prospective purchasers may be enabled to compete for the purchase of articles whether by way of increasing bids or by the offer of articles to be bid for at successively decreasing prices or otherwise."

It is perhaps a convenient shorthand to refer to the two different kinds of competitive bidding which that definition contemplates as an English auction and a Dutch auction, the latter phrase being one which is well understood in practice.

H

With that definition in mind it is necessary to turn and look at the seven instances in the evidence of goods being sold or offered for sale in circumstances which led the Crown Court to the conclusion that there were in this case sales of goods by way of competitive bidding. The first instance is spoken to by a police officer who visited the sale rooms on July 29 and whose evidence is summarised in the case. Mr Velleman was the auctioneer or salesman, and he said to the assembled company in the sale room "Right, I'll make this a good lot," and he held up a cigarette lighter, two transistor radios and two alarm clocks. He said "I'm not asking £10, £9 or £8: hands up who will give me £7," whereupon several persons present put their hands up, and one of them was selected to receive the lot by Mr Velleman. She did. Some ballpoint pens were then distributed free among members of the public in attendance, and having given to the lady the lot which she had offered to buy at £7 by putting her hand up, Mr Velleman said, "There, I don't want your £7, lady. Give it her back, Walter. Let her have them for £1, no, 14 bob." In the second instance to which I refer, which occurred the following morning, again Mr Velleman was the salesman. The articles being offered for sale in this case were umbrellas. Mr Velleman said, "These umbrellas must be worth £2.50. I don't ask £2.50, £2 not even 30 bob. Today's price £1. Hands up. Blue one over there. Red one over there. Lady over there. What colour blue one?" A number of persons who put their hands up received umbrellas of their chosen colour. The third instance relates to a later time on July 30 when, after distributing some boxes free which turned out to contain cuff links, Mr Velleman said, facetiously, obviously, in the context of what had gone before, to his assembled prospective customers, shaking the boxes, "Can you hear anything? No? Well, that's because the box is empty. Who's going to buy an empty box? Shall I give it to you for 25 bob? No. Who wants an empty box

J

K

for 10 bob?" Thereupon there were a number of takers at 10 bob of the boxes, which were not empty but contained, like the ones that had earlier been distributed, in each case a pen and a pair of cuff links.

The learned circuit judge who presided at the Crown Court helpfully in the case sets out the judgment he gave, explaining the court's reasons for arriving at a conclusion that there were here mock auctions being conducted. The way it is put in the judgment in one succinct paragraph is as follows: "The issue the court is concerned with is what would have happened if nobody had put their hands up. It is clear that a further inducement would have been given to people to put their hands up by putting the price down further. The idea was to get members of the public in who were looking for a bargain. If nobody put their hands up, then the goods would be reduced to the lower price. This was a mock auction: there was competitive bidding." Mr McCulloch, on behalf of the appellant, has attacked that reasoning and said that there was no basis in the evidence for the inference which is drawn in that paragraph by the court that if no one had put their hands up at a certain price, for example, at the figure of £1 in relation to the umbrellas, the umbrellas would then have been offered at a lower price. Mr McCulloch submits that the court could not legitimately conclude that this evidence established sales of goods by way of competitive bidding within the meaning of section 3 (1) of the statute, because it was not legitimate to look at what would have happened if hands had not been put up at a particular point; all the court could look at was what did happen. For my part, I am wholly unable to accept that argument. The question the court had to decide was whether persons at the sale were being invited to buy articles by way of competitive bidding, and that requirement was satisfied by an offer of articles to be bid for at successively decreasing prices in the context of this Dutch auction situation. It seems to me that the evidence abundantly justified the inference drawn by the Crown Court as indicated in the passage from the judgment which I have read, and that the drawing of that inference established as clearly as could be, and perfectly legitimately, that the business at these sale rooms was being carried on by way of offers of articles to be bid for at successively decreasing prices. Accordingly, in my judgment the main contention on which this appeal has been argued fails.

But there are a number of subsidiary points to which I must make reference. The offending activities which contravened section 1 (3) of the statute setting out the seven offences of which the appellant was convicted were in six cases the giving away of free gifts contrary to section 1 (3). But in one case, the case of the lady who bought the mixed lot of transistors and clocks, what was alleged was that that lot had been sold to her at a price lower than the amount of her highest bid for that lot in contravention of section 1 (3) (a). Mr McCulloch submits that her bid of £7 for the lot, which it will be remembered she was eventually allowed to purchase for 14 shillings, could not properly be termed her highest bid because she had never made any lower bid. That is, as I understand, the argument, but it seems to me that in the statutory context where one form of competitive bidding within the explicit definition of section 3 (1) contemplates a Dutch auction as opposed to an English auction, the phrase "his highest bid" in section 1 (3) (a) must be construed as applicable to the first effective bid made by a taker as the seller progressively reduces his offer price. Another point which Mr McCulloch argued before us, although not strongly pressed, was to the effect that when the free gifts were given away, which brought the proceedings within the definition of a mock auction, because that is one of the activities which turns an otherwise respectable auction into a mock auction under section 1 (3), the gifts in question were given after the relevant sales by way of competitive bidding were completed. He refers to the provisions of the Sale of Goods Act 1893 as to when a sale must be considered as complete. But the whole answer to this argument, it seems to me, is clearly provided by an express provision again in the interpretation section of the Act, section 3 (5):

> "For the purposes of this Act anything done in or about the place where a sale of goods by way of competitive bidding is held, if done in connection with the sale, shall be taken to be done during the course of the sale, whether it is done at the time when any articles are being sold or offered for sale by way of competitive bidding or before or after any such time."

In the light of that provision, Mr McCulloch's point on the time of the gift in relation to the time of the relevant sale is clearly unarguable.

Finally, it is said on the appellant's behalf that the way in which the prosecution here sought to distinguish between the seven different occasions when seven different offences were committed by the appellant was inaccurate and did not accord with the statute, in that each separate information related to a separate instance of an offending activity under section 1 (3); and it is said that because you may find different offending activities under subsection (3) which bring the proceedings within the definition of a mock auction, that does not necessarily mean that you find that there are separate mock auctions being conducted on the different occasions when these activities are carried on. It is said that what constitutes the offence here is the conducting of a mock auction. Speaking for myself, I think there is considerable force in this point, and I am inclined to think it is right that a single offence is committed by one who conducts or assists in the conduct of a single mock auction, and there may be a single mock auction at which a variety of the offending activities, as I have called them, under subsection (3) are carried on. But the difficulty in the way of the appellant so far as this point is concerned is it was never taken in the court below, nor is anything said about it in the case. The question of what amounts to a single mock auction as opposed to a series of distinct and separate mock auctions must be a question of fact. I should imagine that in a sale room like these Bonnie Street Sale Rooms in Blackpool it is perfectly clear that there is a separate auction each day. It may well be that there is a separate auction each morning and each afternoon, or given appropriate evidence, it might be apt to say that there were half a dozen auctions or more which could be seen to be separate and distinct carried on within the space of a single day. But all those questions must be questions of fact, and since the point was never raised below, it does not seem to me possible for us to give the appellant any relief in respect of it, subject only to this. The prosecution alleged seven distinct offences by reference to the distinct offending activities which were established by the evidence on seven different occasions. They must be able in the evidence to point to a clear connection between a particular offending activity and a relevant sale by way of competitive bidding, and if, as Mr McCulloch cogently pointed out, one looks at the facts on which the last of the seven convictions was founded, when plastic combs were given away free, and therefore in contravention (if there was a mock auction going on at the time) of section 1 (3) (a), it is quite impossible to find anything in the evidence to which that distribution of free plastic combs can be sufficiently proximately related to say that that evidence justified a conviction for that seventh offence. Accordingly I would allow the appeal on that ground alone, and to the strictly limited extent of saying that the conviction of the appellant on the seventh of the seven cases of which he was convicted should be quashed. Subject to that, I would dismiss the appeal and affirm the other six convictions.

ASHWORTH J: I agree.

LORD WIDGERY: I agree.

The respondent was awarded four-fifths of the costs of the appeal.

COMPULSORY PURCHASE AND COMPENSATION

COURT OF APPEAL

July 18 1975

(Before Lord Justice BUCKLEY, Lord Justice ORR and Sir GORDON WILLMER)

BROMILOW v GREATER MANCHESTER COUNCIL

Estates Gazette March 13 1976

(1975) 237 EG 799

Compensation—Site beside bone-works—Injunction issued against owner of works before compensation hearing—Tribunal entitled to take into account possibility of works continuing to be used despite injunction—No error in law, tribunal's assessment affirmed

This was an appeal by Mr John Bromilow, owner of a site at 77 Manchester Road, Westhoughton, near Manchester, from a decision of the Lands Tribunal dated July 3 1974 determining at £6,500 the compensation payable on acquisition of the land by the respondents, the Greater Manchester Council. The decision of the Lands Tribunal was reported at (1974) 232 EG 1232.

Mr A A Rumbelow (instructed by Stevensons, agents for Berry & Berry, Cocker Smith & Co, of Bolton) appeared for the appellant, and Mr M H Spence (instructed by Turner, Peacock, agents for the solicitor to the council) represented the respondents.

Giving judgment, BUCKLEY LJ said: This is an appeal from a decision of the Lands Tribunal on June 6 1974 relating to a site at Westhoughton, near Manchester, known as 77 Manchester Road. At the time of the decision the site lay within the urban district of Westhoughton, but it now lies within the metropolitan district of Greater Manchester. This is a site of probable growth, and the prospective plans for its development include the introduction of new access roadways which would considerably increase its accessibility from Manchester and no doubt from other areas as well. In September 1971 the appellant obtained a certificate under the Land Compensation Act 1971, section 17, to the effect that this property would have been given planning permission for development as an office block were it not for the fact that it lies upon the route of a proposed new road called route 225. On October 26 1971 the appellant served notice on the local authority under the Town and Country Planning Act 1962, section 139, requiring the authority to buy the land upon the basis that it had been adversely affected by the planning arrangements, and that notice was accepted by the local authority. Consequently, the value of the land had to be determined, and that matter came before the Lands Tribunal, when the decision was reached from which the present appeal is brought. The difference between the two sides, that is to say the appellant and the acquiring authority, was that whereas the appellant's valuer, a Mr Derek Johnson, valued the site at £30,833 upon the basis that it would be a site that would attract a developer who would wish to develop it for office use, the district valuer took the view that it would command no market for development in that way, and upon that basis he valued it at £6,500, a very substantial difference.

In close proximity to the site, within 100 yds of it, there is a building which was in use as a bone-works, where animal waste was processed, and that was an activity which was liable to occasion offensive smells. In October 1971 the local authority obtained an injunction in the Chancery Court of the County Palatine of Lancaster restraining the owners of the bone-works perpetually from so carrying on their business there as to cause a nuisance by stenches. We do not know what subsequently occurred, if anything, in relation to that injunction. We do know that the bone-works continued in operation until October 1974, and that on July 15 1974 the owners of the bone-works circularised their customers stating their intention then to discontinue the operation of the bone-works. They did not, however, in fact discontinue until a somewhat later date, namely October 7 1974. The proceedings came before the Lands Tribunal for determination in June 1974, and Mr Johnson and the district valuer both gave evidence. It appears from a note of that evidence which has been read to us—it was not before the Lands Tribunal, because this is evidence which relates to the evidence which was given before the Lands Tribunal—that Mr Johnson in the course of the re-examination said that the public health authority had served a notice of closure on the works, referring to the bone-works. It has not been made clear to us by counsel on either side precisely what he meant by "a notice of closure." It also emerges from the evidence which has been put before this court that the district valuer in the course of his evidence referred to the fact that an injunction had been obtained, and he went on to say that the Westhoughton council had been striving for 10 years to get something done about the bone-works, but without success.

That being the state of the evidence, the member of the Lands Tribunal who heard the case said this:

"The district valuer accepted that although the 1971 master plan and written statement had no statutory significance, it was envisaged that Westhoughton would increase in size. He did not accept, however, that the locality of the site itself could be regarded as included in the potential growth area. In his opinion there would be no demand at all for the site for an office-block development. After inspection, I find myself in agreement with the district valuer as to this lack of demand. This finding does not rest on a rejection of Mr Johnson's evidence regarding the influence of the motorway network on development trends generally. It rests on the location of the site itself. As has been indicated, the site is in a locality where it is the industrial use that predominates. Opposite the site there is a paint-works and a plant-hire depot. Less than 50 yds away there is a large engineering works fabricating steel plates. And within 100 yds there is a bone-works. This last factor would, I think, be decisive in causing any potential office-block developer to lose completely any interest he might otherwise have shown in the site. The bone-works are physically unattractive. More importantly, they give forth a noisome smell. The district valuer's evidence was that in his capacity as valuation officer he had received a series of proposals from local residents for reductions of assessment (both on the 1963 valuation list and on the 1973 list) because of the presence of these works; the local authority had been trying for years to get the bone-works closed; and although an enforcement notice had been served, this had been ineffective."

Then the member of the Lands Tribunal went on to say that as he did not consider that there would be any demand for a site for development as an office block, the section 17 certificate that I have mentioned was not of significance.

We are asked to disturb that decision of the Lands Tribunal, which accepted the district valuer's valuation in preference to the valuation of the appellant's valuer on the grounds contained in the passage which I have just read. There is an appeal to this court from the Lands Tribunal in a case of this nature only on questions of law, and therefore, in order to succeed in this appeal, the appellant has to show that in some way or other the Lands Tribunal erred in law. The way in which Mr Rumbelow has put his case is this. He says that there was unchallenged evidence before the court that the closure notice had been served, and that the Lands Tribunal rejected or disregarded that evidence, and so (I think Mr Rumbelow would go as far as to say this) the Lands Tribunal reached a conclusion that it could not reasonably reach in that state of the evidence. He has asked us to entertain the evidence about what has since occurred in relation to the closure of the bone-works as indicating that the member of the Lands Tribunal was mistaken in attributing to the existence of the bone-works the decisive influence upon the position which the member considered that it had.

I have looked with care at the evidence which had a bearing upon this point which was available to the Lands Tribunal when the matter was decided. It was not very voluminous; it consists of the various passages to which I have made reference. It seems to me impossible to say on the state of that evidence that there were not grounds in the evidence upon which the tribunal could reasonably arrive at the view that it did. The evidence of Mr Bell that the local authority had been striving for 10 years to get something done about the bone-works without success seems to me to have been ample evidence to justify the member in arriving at the conclusion that the bone-works must be treated as something that was likely to go on, notwithstanding that the injunction had been granted, and that the existence of the bone-works was a baleful influence upon the value of the appellant's site which would in truth have dissuaded any developer who might consider buying the site for development for offices from going on with any such project. For this reason, it seems to me that there is no ground for saying that the Lands Tribunal erred in law, and in those circumstances I do not think there is any assistance that we can give to the appellant. I appreciate that he probably feels that his case was dealt with on a view of the facts that was unduly pessimistic from his point of view, but it was a view which I think the member of the Lands Tribunal was perfectly entitled to take upon the evidence before him. Accordingly, in my judgment, this appeal must fail.

ORR LJ: I agree.

SIR GORDON WILLMER: I also agree.

The appeal was dismissed with costs.

COURT OF APPEAL

February 24 1976

(Before Lord Justice MEGAW, Lord Justice JAMES and Lord Justice GEOFFREY LANE)

ELLIOTT AND OTHERS v SOUTHWARK LONDON BOROUGH COUNCIL

Estates Gazette April 3 1976
(1976) 238 EG 41

Development area—Council's CPO followed by owners' applications for rehabilitation orders—Council entitled, in refusing rehabilitation, to say merely that their policy was to demolish existing properties and put up new housing accommodation on the sites—"More than a mere statement of the council's conclusion"—"Did state the salient reason why the houses could not be rehabilitated"

This was an appeal by Mr Alpheus Elliott and three other owners of houses in the Selborne Road area of Camberwell, London SE5, against the refusal of Willis J in the Queen's Bench Division on February 2 1976 to declare that the respondents, Southwark London Borough Council, had failed to carry out their statutory duty under the Housing Act 1974 to consider, determine and give reasons for their refusal to make rehabilitation orders in respect of the appellants' houses, all of which were subject to the Selborne Road (No 1) Compulsory Purchase Order 1973.

Mr P Boydell QC, Mr D Keane and Mr P Crawford (instructed by Jeffrey Gordon & Co) appeared for the appellants, and Mr G Moriarty QC and Viscount Culross (instructed by the London Borough of Southwark) represented the respondents.

Giving the judgment of the court, JAMES LJ said that the appeal had been expedited by order of the court, and the judgment of Willis J had been given only on February 2 1976 on the hearing of an originating summons taken out on January 22 1976. By that summons ten plaintiffs claimed declarations and injunctions against the London Borough of Southwark, and four of those plaintiffs were the appellants in the present appeal.

The background to the case was the introduction by legislation in 1974 of the concept of rehabilitation of houses as an alternative to demolition, clearance and redevelopment. The relevant statutory provisions were the Housing Act 1974, section 114, and the tenth schedule to that Act as amended by the Housing Rents and Subsidies Act 1975. Section 114 applied to houses that were comprised in a clearance area under Part III of the Housing Act 1957 and fell into any one of three categories which were likely to include many houses comprised in clearance areas. Subsection (2) stated:

> Where any house to which this section applies (a) was included in the clearance area by reason of its being unfit for human habitation, and (b) in the opinion of the local authority is capable of being, and ought to be, improved to the full standard, the local authority may make and submit to the Secretary of State a rehabilitation order in relation to that house.

Paragraph 3 (2) of schedule 10 read:

> Where the owner of a house to which section 114 . . . applies and which was included in the clearance area by reason of its being unfit for human habitation requests the local authority to make a rehabilitation order in respect of the house, and the local authority refuse to make an order, they shall give him in writing their reasons for so refusing.

In April 1968 the council declared a development area known as "Selborne Road development area." In March 1973 they made a clearance order under Part III of the Housing Act 1957 in respect of that area, and in May 1973 they resolved to effect the clearance by exercise of compulsory purchase powers. On October 12 they submitted a compulsory purchase order to the Secretary of State for confirmation. At that date there were 157 dwelling-houses in the area, of which 72 had already been purchased by the council by agreement with the owners. Of the remaining 85 houses, 70 were included in the proposed order on the ground that they were unfit for human habitation. The four appellants were owner-occupiers of four of those 70 houses. They and many other owner-occupiers to which the order applied objected to the making of the order. Some disputed the alleged unfitness of their houses. Many urged that the defects could be remedied and improvements made so that the houses would conform to the required standard. Many had done work of maintenance and repair. None wanted to leave. All wanted to preserve the existing community. Whatever compensation was payable, many people suffered considerable distress when their homes were taken from them by compulsory purchase

A and they were forced to leave familiar surroundings and start afresh in a strange and sometimes less convenient area. This was expressly recognised in the White Paper, *Better Homes: The Next Priorities,* presented to Parliament in 1973.

The court felt great sympathy with persons who found themselves in the position of the appellants and the other occupiers of houses within the order. However, the Secretary of State ordered a local public inquiry which was held during two days in January 1974, and the inspector reported to the Secretary of State on April 22 1974 recommending that subject to certain modifications the order be confirmed. The Secretary of State wrote his letter of decision confirming the B order, known as the London Borough of Southwark (Selborne Road) (No 1) Compulsory Purchase Order 1973, on December 2 1974, the same day that section 114 of the 1974 Act came into effect. On March 17 1975 the council served notices to treat. On May 8 and 9, many of those who had objected to the making of the order and who had advanced their objections at the public inquiry wrote to the council making application for "a rehabilitation order to be made in respect of houses which are the subject of the abovementioned order," in other words, the order in question in the present appeal. The letters were all in the same terms, and expressly suggested that the appropriate reply to the C application was "either to accept it or to give reasons for refusing it." Each letter referred to the property or properties in which the signatory had the necessary interest as owner. The applications were considered by a subcommittee of the housing committee, and subsequently by the housing committee of the council. The recommendation of these committees was accepted by the council, who resolved on July 16 1975 to refuse the application. On July 21 the deputy town clerk wrote to each applicant saying that the application was refused for the reason that the policy of the council was to demolish the existing properties and erect new housing D accommodation on the sites.

Demolition began on January 12 1976. Discussions and negotiations had not succeeded in bringing about any compromise or modification, except that a number of occupiers had been rehoused. Those who remained were constant in their desire to save their homes and the existing community, while the council remained unshaken in the opinion that their duty lay in the implementation of the order and the clearance of the area by demolition in order to facilitate redevelopment. So the appellants and six others joined as plaintiffs in the originating summons seeking declarations that the council had failed to carry out their statutory duties E under the 1974 Act and that their decision refusing to make rehabilitation orders was invalid, with injunctions ordering the council to reconsider and determine their requests according to law and restraining the council from demolishing any of the houses until they had dealt with the requests for rehabilitation orders according to law. On January 17 they obtained an interim injunction. On February 2 the matter was heard by Willis J, who refused to grant the relief sought. It was from his judgment that four of the plaintiffs were appealing. They were said to represent the interests of all the owners of houses affected by the decision of the learned judge.

F It was desirable to begin by saying something as to the function of the court in a case such as the present one. It was not for the court to determine issues which by statute fell to be decided by a body such as a local housing authority. Thus, for example, it was no part of the court's function to arrive at a view, or a decision, even if it had the material necessary to do so, on such questions as whether or not a particular house, or the houses in a particular area, should be rehabilitated or should be demolished in order that new housing might be built. Its function was not to substitute its own view or opinion or decision on matters which Parliament had left to the judgment of the local housing authority without provision for appeal. Its function was, where such G issues were raised in proceedings of this nature, to decide whether the local authority, in reaching its decision, had acted in accordance with the statutory provisions by which Parliament had defined the authority's powers, duties and procedure. The appellants invited the court to say that the council did not comply with their statutory duty under section 114 (2) of the 1974 Act and paragraph 3 (2) of the tenth schedule. They said that the council were under a duty to consider the applications and to consider them individually, and having discharged that duty, were under the further duty to give to each appellant in writing the reasons for refusing. They said that the material before the court, showed, at least H by way of inference, that the council did not apply their minds to those considerations which were relevant to the applications, either individually or at all, and that the letter of July 21 did not give any or any sufficient reasons for the refusal. The council's case was a denial of these assertions.

The learned judge had said little in his judgment upon the question whether the council had given consideration to the applications. In the state of the evidence, there was no need for him to say more than he did. There was no evidence of any impropriety. There was evidence that the application had been first before a subcommittee of the housing committee J and then before the housing committee itself, and that finally it had been the subject-matter of a resolution of the council. In those circumstances the learned judge accepted the submission of counsel that to hold that proper consideration had not been given would be "an outrageous thing for a court to do." Upon the question whether the council had properly discharged the duty to give reasons for the refusal, the judge said that paragraph 3 (2) of schedule 10 had to be looked at in the context of the legislation; that in the particular circumstances of the case it must appear to the satisfaction of the court asked to review the reasons that the recipient should fairly understand why the housing authority was not K able to accede to the request; and that the question here was whether those reasons did pass the test. He (Willis J) construed the letter of July 21 as stating "for the reasons that the whole of the area has got to be demolished and the sites used for housing purposes." He thought that that statement was a sufficiently clear indication by the local authority of reasons which could not be said to be either inadequate or insufficient in the circumstances of the case, "bearing in mind what had gone before," and that the statement was intelligible to the recipient of the refusal.

The words "bearing in mind what had gone before" were one of a number of references which the judge made to the L fact that the plaintiffs' contentions had been advanced at the local public inquiry in January 1974. Mr Boydell, for the appellants, emphasised the prominence given to this feature in the judgment, and argued that the judge was in error in his interpretation of what had taken place at the inquiry. The judge used the expressions, "plainly canvassed and dealt with in evidence on both sides," "fully ventilated and dealt with in the report by the inspector," and "thoroughly thrashed out," in relation to the issue of rehabilitation in so far as it was raised at the inquiry. Their Lordships were inclined to the view that the words "fully ventilated" and "thoroughly thrashed out" were something of an overstate- M ment, but this was a matter of minor criticism of the judgment. Then Mr Boydell advanced a number of other propositions. Much attention had been devoted to his submission that section 114 not only conveyed a power but imposed a duty upon the local authority to consider making a rehabilitation order, irrespective of a request to do so made by an owner of property which qualified for consideration. Mr Moriarty, on behalf of the council, disputed the proposition and supported his argument by a detailed analysis of the relevant statutory provisions, but their Lordships found it unnecessary to come to any conclusion on this issue, because it was common ground between the parties that once a request had been made pursuant to schedule 10, paragraph 3 (2), the local authority were under a duty to consider the matters relevant to the application and to give reasons for a refusal.

Mr Boydell's next proposition was that in the discharge of the duty, the local authority should pay regard to relevant and disregard irrelevant matters. Mr Moriarty did not challenge that. Then Mr Boydell said that if a statute imposing a duty to consider some matter expressly defined the relevant factors to be considered, the person on whom the duty lay must consider those factors; if and in so far as the statute did not define the relevant considerations, on the other hand, it was for the court to construe the statute and determine what relevant considerations were to be implied. In relation to a request for the making of a rehabilitation order, the local authority must consider two initial questions which were prescribed in the Act: (a) whether the particular house was capable of being improved to the full standard, and (b) if so, whether it ought to be so improved. Consideration of whether the house ought to be improved, said counsel, involved the consideration of other factors which the court should imply as necessary to proper consideration by the local authority in order to comply with the spirit and to fulfil the purpose of the Act. He (Mr Boydell) expressed these compendiously as the matters involved in balancing the difficulty and expense of improving a particuar home and the alternative of demolishing that house and redeveloping the site. Again there was no challenge to Mr Boydell's proposition that the words of the Act required consideration of the two initial factors, whether a particular house was capable of being improved, and if so, whether it ought to be improved. Also, it was common ground that when considering "ought" the local authority could and should take into consideration the area in which the house was situated and the neighbouring properties.

It was clear that the matters which the local authority should consider when deciding whether a house, capable of improvement to the full standard, ought to be improved under the provisions of a rehabilitation order rather than be demolished, and the weight which should be given to one or another factor, would vary from case to case. It was not for the court to prescribe a list of matters which must always be considered or to prescribe which factors should be given more weight than others. It was worth repeating that the function of the court, where such issues were raised, was not to substitute its own opinion or decision on matters which Parliament had left to the judgment of the local authority, but to decide whether the council in reaching their decision had acted in accordance with the statutory provisions. The complaint that matters were wrongly considered rested upon the contention that the council relied upon the evidence and arguments at the public local inquiry and the findings of the inspector. The complaint that the council failed to consider relevant matters was that the council took no account, or not sufficient account, of changes since January 1974 in respect of the law, housing policy, economic conditions and costings. In their Lordships' judgment the council were entitled to take into consideration what had transpired at the public local inquiry. If they had chosen to do so, they could have included in the reasons for refusal express references to the evidence given at the inquiry and to the findings of the inspector: see *Givaudan & Co v Minister of Housing and Local Government* [1966] 3 All ER 696 at 699. Although it was true that the evidence and arguments adduced and advanced at the inquiry were not, and could not be, directed to the making of a rehabilitation order, for at that date no such order could be made, there could be no doubt that the objectors at the inquiry were contending for the same result which could now be achieved by a rehabilitation order. It was clear from the inspector's report that the arguments ranged round the merits of demolition as against rehabilitation. Thus there was material fit to be taken into consideration when the requests for rehabilitation orders were received. Their Lordships could find no evidence indicating that the local authority had failed to consider what, if any, changes, had taken place since January 1974. It was suggested that the inadequacy of the purported reasons for refusal gave rise to the inference of such failure, but their Lordships shared the view of the judge that in the face of the evidence of the elaborate committee procedure followed by the council it was impossible to infer that they did not take the relevant factors into consideration.

The only other complaint was of a failure to give reasons or adequate reasons. At one stage, it was suggested on behalf of the appellants that the giving of reasons had relevance to the basis upon which compensation for compulsory purchase was payable. In the end it became clear that the difference, if any, in respect of the basis of compensation was not a material factor, and the point was not pressed. On the general issue, Mr Boydell said that where a statute expressed a duty to give reasons, the court should imply a condition that reasons given in discharge of that duty should be adequate and intelligible. Lord Parker said in *Mountview Court Properties Ltd v Devlin* (1970) 21 P&CR 689 that what reasons were sufficient depended upon the facts of the case. He (Lord Parker) cited from the judgment of Megaw J in *Re Poyser and Mills's Arbitration* [1964] 2 QB 467 at 478: "Parliament provided that reasons shall be given, and in my view that must be read as meaning that proper, adequate reasons must be given. The reasons that are set out must be reasons which will not only be intelligible but which deal with the substantial points that have been raised." Mr Boydell argued that the reasons given by the council here did not pass that test. He said that if a body under a statutory duty to give reasons failed to do so, then (a) (which did not apply in this case) if the relevant statute prescribed the sanction of quashing the order or decision, the court should quash the order or decision, and (b) in other cases, if the failure to give reasons was sufficiently serious in the circumstances of the case, the court should quash the order or decision, but if the court did not quash the order or decision, the court should make an order requiring reasons to be given. He (counsel) submitted that the purported reasons were inadequate and did not convey to those who requested rehabilitation why the local authority was refusing the requests.

It was argued for the council, their Lordships thought rightly, that there was no evidence that the council took into account irrelevant matters or failed to take relevant matters into account. The duty imposed by the statute was a duty to act in an administrative, not a judicial, capacity, and it was open to the council to arrive at a policy decision. It was further submitted for the council that the purported reason was adequate and intelligible. Against the background of the inquiry, the inspector's report and the Secretary of State's decision letter, the appellants and those owners of houses to whom similar letters were sent were fully informed by the terms of the letter that the request for a rehabilitation order was refused because it could not be granted consistently with the policy of the housing authority to demolish the existing and erect new housing accommodation. Their Lordships thought that this submission was correct. The letter was more than a mere statement of the council's conclusion. It did state the salient reason why the houses in question could not be rehabilitated. Their Lordships did not think it was necessary to include in the reason given any reference to the cost of demolition as against the cost of improvement or any reference to the practicability of the houses being improved to full standard. There was certainly no need to refer to the various matters of detail which their Lordships assumed, in the absence of evidence to the contrary, had been taken

A into account when considering whether the houses ought to be improved to the full standard. To those who received the letters, the reason for the refusal was intelligible and sufficient without the inclusion of these details.

It followed that the appeals failed, but this decision did not govern what might be different circumstances in other cases. The wording used for the giving of reasons in this case passed the test; the same wording in another case might fail to do so. The duty to give reasons pursuant to statute was a responsible one and could not be discharged by the use of vague general words which were not sufficient to bring to the mind of the recipient a clear understanding of why B his request for a rehabilitation was being refused.

The appeal was dismissed. Nor order was made as to costs. The appellants were refused leave to appeal to the House of Lords.

COURT OF APPEAL

November 5 1975

(Before Lord Justice CAIRNS, Lord Justice JAMES and Lord Justice SHAW)

C FAIRMOUNT INVESTMENTS LTD v SECRETARY OF STATE FOR THE ENVIRONMENT AND ANOTHER

Estates Gazette May 1 1976

(1975) 238 EG 337

Owners of Southwark houses object to CPO—Inquiry followed by inspection of the properties—Financial viability of owners' proposals for rehabilitation commented on by inspector in report—Held not to have been in issue at the inquiry—Inspector's reasons included extent of settlement of foundations, attributed by him to expansion and contraction of clay base—This matter also held not to have D been in issue at the inquiry—Inspector's reliance on "telltale" observed during his inspection held improper—CPO quashed on ground of breach of rules of natural justice

This was an appeal by Fairmount Investments Ltd, from a judgment of Cusack J in the Queen's Bench Division on February 25 1975 dismissing their application for an order to quash the London Borough of Southwark (Aldbridge Street) (No 1) Compulsory Purchase Order 1973, made by the second respondents, the London Borough of Southwark, on January 9 1973 and confirmed by the first respondent, the Secretary of State for the Environment, on April 9 1974.

Mr M Mann QC and Mr D G Robins (instructed by E Laytons & Charles Ingham, Clegg & Crowther, agents for John Gorna & Co, of Manchester) appeared for the appellants, and Mr W J Glover QC and Mr H K Woolf (instructed by the Treasury Solicitor) represented the first respondent. The second respondents took no part in the proceedings.

Giving judgment, CAIRNS LJ said: This is an appeal from a decision of Cusack J given on February 25 of this year on an application under the Housing Act 1957 to quash a compulsory purchase order made under that Act by the second respondents, the London Borough of Southwark, and confirmed by the first respondent, the Secretary of State for the Environment. The order was made on January 9 1973 F under section 43 of the Act. It is called the London Borough of Southwark (Aldbridge Street) (No 1) Compulsory Purchase Order 1973. It covered 68 houses. The appellant company, Fairmount Investments Ltd, is the owner of 64 of them. The area in which these houses were situated had been declared under section 42 of the Act to be a clearance area, it being the wish of the local authority to clear it for the purpose of providing a public open space. The compulsory purchase order required the houses to be demolished as being unfit for human habitation. The owners objected to the order on the ground that the houses were not unfit for human habitation, alternatively that if they were, they were suitable for rehabilitation rather than for demolition. The G Secretary of State ordered a public local inquiry under the third schedule to the Act, paragraph 3 (3) of which provides as follows:

> If any objection duly made is not withdrawn, the Minister shall, before confirming the order, either cause a public local inquiry to be held or afford to any person by whom an objection has been duly made as aforesaid and not withdrawn an opportunity of appearing before and being heard by the person appointed by the Minister for the purpose, and, after considering any objection not withdrawn and the report of the person who held the inquiry or of the person appointed as aforesaid, may, subject to the provisions of this part of this schedule, confirm H the order with or without modification.

The Secretary of State, of course, fulfils the functions which were conferred upon the Minister by that subparagraph. The inquiry was held on September 18 and 19 1973, and after hearing the evidence for the local authority and for the owners, after hearing the submissions of advocates on both sides, and after inspecting the premises, the inspector made a report recommending that the order be confirmed. The Secretary of State accepted this recommendation and confirmed the order. The owners applied to the High Court, as they were entitled to do under paragraph 2 of the fourth schedule to the Act, for the order to be quashed. The J application came before Cusack J, who dismissed it, and the owners appeal. It is provided in paragraph 2 (2) of the fourth schedule to the Act that such an order may be quashed "if it is not within the powers of the Act or if the interests of the applicant have been substantially prejudiced by any requirement of the Act not being complied with." The owners made an application under each of those heads, contending that the order was not within the powers of the Act, because the rules of natural justice had not been observed, in that the inspector and the Secretary of State took into account matters not raised at the inquiry and paid attention to something which the inspector observed in the K course of his inspection which the owners had no opportunity to deal with; and they contended, for the same reasons, that the Secretary of State failed to comply with the requirements of the Act and that they were thereby substantially prejudiced.

Section 4 (1) of the Act provides: "In determining for any of the purposes of this Act whether a house is unfit for human habitation, regard shall be had to its condition in respect of the following matters, that is to say (a) repair; (b) stability; (c) freedom from damp; (cc) internal arrangement; (d) natural lighting; (e) ventilation; (f) water supply; (g) drainage and sanitary conveniences; (h) facilities for L storage, preparation and cooking of food and for the disposal of water; and the house shall be deemed to be unfit for human habitation if and only if it is so far defective in one or more of the said matters that it is not reasonably suitable for occupation in that condition." The case presented by the council was that the houses were defective in a number of respects, and the inspector reported in paragraph 78 of his report that there were defects under (a), (b), (c), (g) and (h) in varying degrees which rendered the properties unfit for human habitation by reason of the sum of those defects. The finding of unfitness is not under attack M by the owners. But a finding in paragraph 78 of the report which is attacked by them is that under (b), which is headed "Stability" and is in these terms:

> There is evidence of settlement in all of these properties which is evidenced in many humped and sloping ground floors, sloping first floors, the majority of door heads are out of alignment varying from about ¼ in to 1½ in in the worst cases. Many side back-addition walls are bulged; some of these walls have been rebuilt owing to war damage, but some others appear to have been rebuilt pre-war; this is apparent by materials and workmanship. These rebuilt walls are generally sound. Most back walls to the back additions appear to be somewhat bulged outwards.

A This is evidenced by excessive mortar filling between the sash window and the wall in the external window reveal. The back walls of the three-storeyed blocks appear to be bulged outwards at second-floor level and this is evidenced by long vertical cracks in the staircase partition walls at or near their junctions with the back walls. These cracks extend through the partition walls from staircase to bedrooms.

I would underline the next sentence:

A "tell-tale" has been secured across a typical crack in reference P62, and this has sheared through. A number of the lean-to scullery blocks are pulling away from the main building at the roof intersection level with the main back addition wall and fractured brickwork is frequently evidenced at this point.

B Then in paragraph 81 (c) of the report the inspector expresses this opinion:

The settlement which is evident in all the houses in clearance area no 1 would appear to be due to the foundations not having been taken deeply enough into the clay so as to avoid that area which is subject to seasonal expansion and contraction. Because of this and other previously stated defects I am of the opinion that satisfactory rehabilitation would not be a financially viable proposition and that the award of discretionary grants would be unlikely.

In paragraph 102 he said:

C I am satisfied that [with an exception to which it is not necessary to refer] the best method of dealing with the conditions is the demolition of all the buildings in the clearance areas. The buildings are old and of an out-moded type, but they are of a character and general location which might warrant consideration for rehabilitation, were they of satisfactory general structural condition. The properties in the order are largely held by one owner who wishes to rehabilitate and improve his holding. Since for reasons already stated, I do not consider this to be a financially viable proposition, the council are justified in seeking to acquire the properties in order to consolidate the total of three ownerships into one controlling interest so that the clearance area can be redeveloped for approved purposes.

D Then in paragraph 104 he formally recommended that "the London Borough of Southwark (Aldbridge Street) (No 1) Compulsory Purchase Order 1973 be confirmed," with certain minor qualifications. It is contended by the owners that the financial viability of rehabilitation was not an issue raised at the inquiry; nor was it suggested in evidence that the foundations had not been taken deep enough into the clay; and in connection with settlement, it is complained that the owners had no opportunity of dealing with the "tell-tale," this being something which the inspector observed after the close of the inquiry when he went to inspect the premises. In these respects it is contended that the inspector E in reaching his conclusions, and the Secretary of State in acting upon the inspector's recommendation, did not comply with the rules of natural justice, and therefore that the inquiry was a nullity, so that the confirmation of the order was not within the powers of the Act, or alternatively a requirement of the Act was not complied with and the owners have been substantially prejudiced thereby.

There is no reported case in which the issues correspond exactly with those raised on this appeal, but very valuable guidance is to be found in the decision of Browne J (as he then was) in *Hibernian Property Co Ltd v Secretary of State for the Environment* (1973) 27 P & CR 197. That was F a case relating to an inquiry under the Housing Act 1957 where the inspector after the inquiry had visited the premises and had conversations with the occupiers which were taken into account in the report to the Secretary of State, and Browne J held that this was not in accordance with natural justice. The judgment contains ample citation of authorities in this court and in the House of Lords, and contains, among other things, these propositions: at p 207, that the functions of the inspector and the Secretary of State are quasi-judicial and that each is bound to observe the rules of natural justice; at p 210, citing from the judgments of Denning and Parker LJJ in *Steele v Minister of Housing and Local Government and Another* (1956) 6 P & CR 386, that G the inspector must not take into consideration information coming to him in the absence of one party, and must not pay attention to material unless the parties have had the opportunity of commenting upon it; and at p 212, that the test as to prejudice is not whether the party is proved to have been prejudiced, but whether there was a risk of prejudice. I respectfully agree with all of those propositions.

So far as the actual decision is concerned, it is contended on behalf of the Secretary of State in this appeal that the observation of the "tell-tale" is something very different from obtaining oral information from people seen at the time of the inspection. For my part, I can see no relevant difference H between the two. It is conceded that what the inspector sees in the course of his inspection is by way of evidence in the course of the inquiry. Whatever may be the position in relation to seeing, for instance, cracks and other physical features of the houses of that kind, it appears to me that a "tell-tale" is, as its name very aptly shows, something which is informing a person who looks at it of something just as would a piece of information given orally or a notice placed in the premises. The "tell-tale" by having sheared is saying, "At some stage there was some settlement which caused this crack to widen." Another decision from which I derive J assistance is that of the Divisional Court in *R v Paddington & St Marylebone Rent Tribunal* [1949] 1 KB 666, where the decision of a rent tribunal was based in part on their having found on inspection of the premises that the height of some rooms in some flats was low. This matter was not referred to at the hearing, and this was one of the grounds upon which the court granted an order of certiorari and quashed the decision.

It was contended on behalf of the Secretary of State that the issue of stability was clearly raised at the hearing of this inquiry; that settlement was referred to in that connection, and that settlement would only be relevant to stability if it K was settlement which was continuing or was liable to continue; that the inspector was therefore entitled to take into account anything he observed in the course of his inspection which tended to show that the settlement had not ceased; and further, that he was entitled to form a view based on his professional experience (the inspector is a Fellow of the Royal Institute of British Architects) as to the cause of the settlement. With regard to financial viability, it is said that the owners had put before the inspector a valuation directed to showing that the cost of carrying out conversion and improvement on the premises in a way that they were willing to do was such that the value of the houses after repair would be L £1,450 more than their original value, plus the cost of such conversion and improvement, minus an improvement grant of £2,400 which they expected to receive, and it was contended that the inspector was entitled to conclude from his professional knowledge that rehabilitation in the sense in which it was found to be required here by the inspector, that is to say, rehabilitation involving major structural work, would cost much more and would not be financially viable.

I cannot accept these contentions. It appears to me from the report that the matter of stability or instability was dealt with at the hearing only as part of the defects constituting unfitness. In this connection I quote from paragraph 14 of M the report, under the heading "The case of the council": "*On unfitness.* The houses in the clearance areas suffered from external and internal disrepair, there was evidence of settlement, rising and penetrating dampness, there were poorly defined changes of level inside the houses, the external wcs were not readily accessible, facilities for the preparation and cooking of food and for the disposal of waste water were inadequate." Then paragraph 15: "The council were satisfied that the dwellings were unfit for human habitation and that the most satisfactory method of dealing with the conditions in the clearance areas was by the demolition of all the

A buildings in the clearance areas." Then paragraph 37 (b): "The properties were unfit for the reasons already stated. The objectors' proposed unit of rehabilitation was not favoured by the council; in the council's opinion a maisonnette over a ground-floor flat was unsuitable on social grounds. It created a fire risk, and a pram belonging to the upper floor tenant would have to be kept in the common entrance passage which was about 3 ft wide (3 ft 3 in where measured by the inspector)." It will be observed that the council were there saying that the properties were unfit "for reasons already stated," and the objection to rehabilitation was put forward on the basis that "a maisonnette over a B ground-floor flat was unsuitable on social grounds." In paragraph 52, in the case for the objectors, the matter of unfitness was put as follows: "The grounds of unfitness of each of these properties were examined in great detail. In general terms it was submitted that all the alleged defects listed in the principal grounds, under items (a) to (h) of section 4 of the Housing Act 1957, were, upon examination, trivial or exaggerated. Any disrepair that existed in the particular house or its curtilage did not constitute a threat to the health of or cause any serious inconvenience to the occupants. There was no evidence of instability which would constitute a threat to the occupants of any house. Indeed, the local C authority had, under the heading of stability, mentioned only external settlement or internal settlement. The existence of settlement did not indicate the probability, nor indeed the slightest possibility, of further movement, and this was no ground for unfitness." Again it is "unfitness" that is being referred to.

The grounds on which the local authority were contending that rehabilitation was unsuitable were, therefore, that the owners' proposals for conversion and improvement would not create satisfactory units; and further, it appears from other parts of their contentions that they were saying that D in any case they wanted to turn the area into a public open space, a matter which is not relevant on this appeal. Whether the inspector's opinion that settlement was due to the foundations not having been taken deeply enough was an inference from any evidence given at the inquiry or was inferred from what he observed on the inspection is not apparent, but there is nothing to indicate that the owners' counsel or witnesses ever had the opportunity of dealing with it. If it had been raised they might have been able to counter it, which would have a major bearing on the practicability of rehabilitation. With regard to financial viability, if it is indeed correct (and there is nothing in the report to show E whether it is so or not) that the inspector drew some inference about the cost of rehabilitation from the valuation put forward for conversion, this was not an inference that should have been drawn without the owners having the opportunity of challenging it. Finally, the "tell-tale," though it may seem a small matter, was so far as appears the one piece of evidence which pointed towards progressive settlement. It was not very cogent evidence without something to show how long it had been there and when it had sheared, but it seems that the inspector placed some reliance upon it, and in my view it was as wrong for him to do so as it was for the rent tribunal in the *Paddington* case to place reliance on the height of the F rooms observed in the absence of the parties. I regard anything observed by an inspector on an examination of property as evidence just as much as anything that is put before him in writing or told to him by witnesses, and indeed, that is accepted on behalf of the respondents to this appeal. In so far as a view does no more than help him to a clearer understanding of evidence given at the hearing, it can properly be taken into account without the need for any further opportunity of comment. Nor would such an opportunity be necessary if witnesses on one side said that a certain defect was present, and witnesses on the other side denied it, and observation indicated directly which side was speaking the G truth. But if some feature is observed which nobody has so far mentioned, it is not in accordance with natural justice for it to be used in support of a conclusion unless the parties are told of it and are enabled to present argument, and it may be evidence, upon it.

Nothing inconsistent with the views that I have expressed is, in my judgment, to be found in the case to which we were referred by Mr Glover, *Salsbury v Woodland* [1970] 1 QB 324. There the judge who had tried a road accident case went after the court hearing to have a view of the road where the accident had occurred. The grounds of appeal were that he should not have had a view, and that on the whole of the H evidence, including what he had seen on the view, the decision was against the weight of evidence. Both of these arguments were rejected. In the present case there is no suggestion that the inspector should not have had a view. Everybody agrees that it was a proper thing for him to do. But the question is what he should observe, and what use should he make of what he does observe. In the *Salsbury* case, at p 343, Widgery LJ (as he then was) in the course of his judgment said: "The expression 'view' is used indifferently to describe two very different things. Sometimes it refers to what Denning LJ spoke of as a judge going to see some public place, where all that is involved is the presence J of the judge using his eyes to see in three dimensions and true colour something which had previously been represented to him in plan and photograph." That was the type of view which the Court of Appeal found that the judge had had in that case and which they held not to be objectionable. Similarly Sachs LJ, at p 350 of the report, said: "Knowing how plans and photographs may give an incomplete impression of a place, it may, indeed, often be wise to go and have a look in order to get a first hand impression of the locality as a whole—to obtain a clearer and three-dimensional picture, so that, in effect, the evidence falls into place. It must be remembered that all he is doing is to appreciate the evidence K already given in the light of a static background." It was a different use, it appears to me, which was made by the inspector in this case of the view which he had. Another case that was mentioned was a very recent one of *Weatherall v Harrison,* reported in *The Times* on November 1. There it was decided that a professionally-qualified magistrate can give effect as part of the background of a case to his expert knowledge. Well, of course, so can an inspector at an inquiry of this kind. But that is very different from basing his decision on matters which are not raised at the inquiry or on a piece of evidence obtained in the absence of the parties. L

Mr Woolf drew our attention to the Town and Country Planning (Inquiry Procedure) Rules 1974. In rule 11 of those rules provision is made for a site inspection, and in rule 12 (2) there is provision for the Secretary of State to notify the parties if new evidence or a new issue has come to light after the close of an inquiry, and to give an opportunity for representations to be made or an opportunity to ask for the reopening of the inquiry. It was pointed out that no reference appears in those rules to the reopening of an inquiry by an inspector. Admittedly the rules do not apply to this particular inquiry with which we are concerned, but it is said that if in any circumstances reopening by an inspector were con- M sidered appropriate, that would be something which would be dealt with in these rules, which were made after consultation with the Council on Tribunals. I cannot accept that because there is no reference in the rules to an inquiry ever being reopened by an inspector there could be no circumstances in which it would be proper and necessary for him to reopen an inquiry. It is self-evident that in some cases reopening would be needed: if, for instance, an important change of circumstance took place between the time when he concluded the hearing and the time when he made his report, or if a vital and completely new fact came to light during that period.

A The only question, to my mind, is in what circumstances such as reopening is necessary.

Next Mr Woolf says that under the third schedule, paragraph 3 (3) (which I have read), there need not be a public inquiry at all. What would be natural justice if there were no inquiry, but the alternative procedure provided by that sub-paragraph were adopted, it is unnecessary for us on this appeal to go into. But Mr Woolf goes on to say that in the case of *Local Government Board v Arlidge* [1915] AC 120 the House of Lords held that it was a mistake to suppose that an inquiry made in circumstances parallel to these is B subject to all the rules applicable to a trial in a court of justice. The actual decision in the case was that there is no need for a party to be shown a copy of the report made by the inspector to the Minister, and there is no right to obtain a copy of such a report. But it will be noted that in the course of the speeches in the House of Lords it was made abundantly clear that there are certain rules of natural justice which are required to be observed in connection with such an inquiry as well as in a court case. Lord Haldane LC, at p 133, referred to the well-known case of *Board of Education v Rice* [1911] AC 179 in that connection, and referring to some of the observations of Lord Loreburn in C a speech in that case, said: "But he went on to say that he did not think it was bound to treat such a question as though it were a trial. The board had no power to administer an oath, and need not examine witnesses. It could, he thought, obtain information in any way it thought best, always giving a fair opportunity to those who were parties to the controversy to correct or contradict any relevant statement prejudicial to their view." At p 141, in the speech of Lord Parmoor near the foot of the page, is an actual citation from *Board of Education v Rice* on the same point.

One question of importance here is whether there was some new issue of fact raised after the inquiry. Mr Woolf D contends that there was not; that the issues at the inquiry were unfitness and whether rehabilitation was appropriate, and that both of these were fully gone into. But if the broad issues are dealt with, as they were here, by considering certain sub-issues at the hearing, and then in the report some quite different sub-issue is considered and a conclusion upon it is reached by the inspector, that is indeed, to my mind, a new issue. Financial feasibility of such rehabilitation as might be needed to remedy defects in the foundation was never considered at the inquiry. It is said that it is inconsistent for the owners to attack the decision on rehabilitation while accepting the decision on unfitness. I do not for a moment E accede to that. Acceptance of the decision on unfitness does not involve acceptance of all the grounds upon which that decision was based. It was put to us that it was for the owners to present their case of financial feasibility on the basis of structural rehabilitation. Again, I cannot accept that. Until it was raised that that form of rehabilitation was not financially feasible it was not something that the objectors could be expected to deal with.

Finally, there is the contention that the reopening of an inquiry of this kind would be a matter of considerable expense to a great many people, and delay, both of which are matters to be avoided. They could be avoided if the inspector found F it possible to put out of his mind something which he had observed and on which he ought not to base his report. That, of course, relates only to the matter of the "tell-tale." So far as the other matters are concerned, the issues which were not dealt with at the hearing, but which were dealt with as part of the inspector's opinion and on the basis of which he made his recommendations, I can only say that it seems to me unfortunate that if he were going to deal with it in that way he should not himself have raised it before the hearing came to an end, which might have avoided any calling together again of the parties for the purpose of continuing the inquiry. For the reasons that I have given, I am of the opinion that in this case the rules of natural justice were not G adhered to, that in consequence the confirmation of the order was not within the powers of the Act, and also that the owners were substantially prejudiced by the requirement of the Act for an inquiry, which must obviously be a proper inquiry, not being complied with. Accordingly, I would allow the appeal and quash the order.

Agreeing, JAMES LJ said that as the court was differing from the learned judge below, and out of courtesy to the careful arguments of Mr Glover and his learned junior Mr Woolf, he ought perhaps to deliver a separate judgment, but he did not propose to review all the factors or all the points which had been covered in the judgment of Cairns LJ. The H inspector had based his conclusion that rehabilitation was not a financially viable proposition on unfitness generally and (in particular) on the houses not having been taken deeply enough into the clay to avoid the area subject to seasonal expansion and contraction. That opinion appeared to be based expressly upon that which was evident in all the houses, and referred to what the inspector saw on his inspection; and it was because of the feature that he found, that it was apparent that the foundations had not been taken deeply enough into the clay, that the inspector concluded that the rehabilitation was not financially viable. The Secretary of J State for the Environment by letter dated April 9 1974 reported to the borough and the objectors his decision to confirm the compulsory purchase order. The letter stated that the inspector's findings, conclusions and recommendations had been generally accepted, and it made specific reference as a matter of recital to the matters of the foundations and the apparent cause for the evident settlement. Paragraph 8 of the letter read: "Careful consideration has been given to the objectors' arguments that the dwellings were not unfit and that they were capable of being converted and rehabilitated at a reasonable cost. While recognising that it is government policy to secure the rehabilitation of older houses where this K is practicable, it is noted that the inspector found that the dwellings had serious defects, including evidence of settlement, and that their general structural condition was not satisfactory." It appeared to him (his Lordship) that the finding of the inspector on his inspection of the premises as to the probable cause of the evident settlement played some part in his ultimate conclusion, and also, naturally enough, was a feature taken into account by the Secretary of State.

The argument for the respondents was that neither the Secretary of State nor the inspector had failed to exercise their functions in accordance with the rules of natural justice. The inspector was entitled to make an inspection; L indeed he was invited to make an inspection, and was invited to arrive at conclusions based on what he saw at the inspection. The issue "rehabilitation or demolition" only arose for determination, said the respondents, if the inspector reached, as in fact he did reach, the conclusion that the properties were unfit, and therefore was an issue to be determined by him on the basis, already decided, that the properties were unfit by reason of instability, or, using the words of the Act, "a condition in respect of stability." Upon that basis, the condition of the premises in respect of stability had already, said the respondents, been fully canvassed, and it was a factor which the inspector would have to consider M when he went on to the next issue, namely, that of possible rehabilitation. The respondents said that upon that issue the appellants decided to rest their case upon the evidence that they put before the inspector as to their feasibility project and the figures they advanced as to the value of the houses then as they were, the cost of immediate repairs, the cost of renovation and alteration to provide for a ground-floor flat and a maisonnette on the upper floor, in which calculations they did not include any element at all for the cost of the work of putting right any structural defects due to settlement. The respondents said that the appellants in this regard were in

A that difficulty which confronts every man who tries to ride two horses at once. If they had included calculations of the cost of work based on work being necessary to rectify defects of instability, it would go to undermine their case that the houses were not unfit, and if they chose to rest their case without incorporating such material, then, said the respondents, the inspector was fully entitled to consider the evidence that was before him, and if he rejected that evidence put forward by the appellants, then he was entitled to reach the conclusion that the case for rehabilitation had not been made out. So the respondents contended that Cusack J was right in the way he approached that aspect of the matter and dealt B with at page 9 of his judgment.

He (his Lordship) heeded the warning to be found in the case of *Local Government Board v Arlidge* and did not seek to apply principles applicable to the trial of a suit between parties in a court of law to the conduct of an inspector conducting an inquiry which he had been appointed by the Secretary of State to conduct. But he did find, having warned himself in that way, that although the inspector was entitled to go about the inspection in the way he did, was entitled to take note of what he had found, was entitled to reach conclusions based upon what his senses told him, still, whatever senses he used and applied, he did do something which he C ought not to have done, and that was to take into account certainly his opinion that the condition as to stability was affected by the foundations not having been taken deeply enough into the clay, and possibly the " tell-tale," without giving the opportunity to the objectors at the inquiry of dealing with these matters in respect of the financial feasibility of rehabilitation. He (James LJ), for his part, was not very much impressed with the argument based upon the finding of the " tell-tale." It was an argument put forward by Mr Mann in a very restricted way, and one could see why; indeed, the respondents saw why. The reference to the " tell-tale " in paragraph 78 of the report was a plain statement of D fact: " A ' tell-tale ' has been secured across a typical crack in reference P62 and this has sheared through." The inspector who made that report was a professional gentleman, an architect with high qualifications, and one had to assume that he would not be misled into drawing conclusions from that bare fact. Nevertheless, it was a possible argument of which the appellants could avail themselves that there was something there in the report which might possibly have reference to the later opinion as to the cause of the settlement. But as it stood by itself, he (his Lordship) would not regard the reference to the " tell-tale " as being any indication E that the inspector took something into account that he ought not to have done without giving the appellants notice of the fact or giving them the opportunity to deal with it either by evidence or argument.

But on the other aspect, the causation of the cracks evidencing settlement, he was quite satisfied that that was a matter which the inspector gleaned from his inspection and which did go towards his conclusions. This was not a new major issue, but was a new sub-issue on rehabilitation affecting the heart of the appellants' case. It introduced a matter with which they would have to deal specifically if they were going to meet it by evidence or by argument, because it was F a matter which, on the face of it, falsified their feasibility project, being inconsistent with it. They were not given the necessary opportunity, and they were not given the information that they should have been given. He (his Lordship) thought that that was a breach of the rules of natural justice, and for that reason the position was that the appellants had not had a fair inquiry. In saying that, he did not for one moment wish to imply that the inspector thought he was behaving unfairly or thought otherwise than that he was behaving fairly. It was not a case of attributing any moral blame at all, but there was here a breach of the rules of natural justice applicable to the conduct of the inquiry. For those reasons, he thought the appeal should be allowed and G the order quashed.

SHAW LJ: As I agree fully with both the judgments which have been delivered, I content myself by adding only a few sentences. This court is not called upon to apply any other than familiar and well-established principles to the particular facts. The only difficulty has been in seeing what were the facts upon which the inspector ultimately rested his conclusion as to the financial feasibility of the project of rehabilitation. The difficulty arose because he took account of a material factor which came to his notice on an inspection after the hearing was concluded and which had not been canvassed before him. It was urged on behalf of the respon- H dents that the general question of stability was raised and investigated at the hearing. So it was; but the impact on the decision of that question of evidence not led or tested at the inquiry was never considered during the hearing. There was neither discussion nor argument as to a state of things which in the end appears to have been regarded by the inspector as weighty, even if not of itself decisive. This was simply that the foundations were defective. There can be no doubt that that was a matter which affected his mind (and the objectors could not have known it) in deciding whether the rehabilitation contended for by the appellants was financially practicable or not. The " tell-tale " is perhaps of J little consequence, except that it is clear from the way in which the inspector put the matter that it was something that reinforced his view as to the non-feasibility of rehabilitation. That being so, it seems to me that there was a failure to recognise what natural justice required, namely, that a proper opportunity should have been afforded to the objectors of dealing with this new aspect of the problem. I would allow this appeal.

The appeal was allowed, with costs in the Court of Appeal and below. Leave to appeal to the House of Lords was given on an undertaking that the Secretary of State would K not seek to alter the order for costs made by the Court of Appeal and would not apply for costs of the appeal to the House of Lords.

COURT OF APPEAL

March 16 1976

(Before Lord DENNING MR, Lord Justice GOFF and Lord Justice SHAW)

R v SECRETARY OF STATE FOR THE ENVIRONMENT EXPARTE OSTLER L

Estates Gazette June 26 1976

(1976) 238 EG 971

Compulsory acquisition for road-widening purposes—Alleged secret agreement by Department's officer with objectors about access—Neighbouring owner affected by agreement did not know about it, did not object and did not attack order within six weeks—Owner held to be caught by rule in Smith v East Elloe Rural District Council—Unable to challenge order—Smith's case unaffected by the Anisminic appeal

This was an appeal by the Secretary of State for the M Environment from an order of the Queen's Bench Divisional Court granting Mr Sydney Ostler, of Boston, Lincolnshire, leave to move out of time to quash two orders made by the appellant in March and May 1974 for compulsory acquisition of land for the purpose of constructing an inner relief road near Boston town centre.

Mr H K Woolf (instructed by the Treasury Solicitor) appeared for the appellant, and Mr B A Payton (instructed by Eland, Hore & Paterson, agents for Jebb & Tunnard, of Boston) represented the respondent.

Giving judgment, LORD DENNING said that the case raised

A the question whether the decision of the House of Lords in *Smith v East Elloe Rural District Council* [1956] AC 736 had been overruled or varied by the later decision in the House of Lords in *Anisminic Ltd v Foreign Compensation Commission* [1969] 2 AC 147. The dispute arose out of a very important scheme for an inner relief road at Boston in Lincolnshire. In 1972 proposals were made for the acquisition of land for the new road, which was designed to take traffic away from the middle of the town. Objections had to be lodged by October 1972, and among those objections were some by firms who objected that the proposed access roads would affect their premises. The present respondent, Mr B Ostler, claimed that as a result of representations made by them, those firms were assured by an officer of the Department of the Environment that they need not worry, because access was going to be by way of Crowthorne Lane, a road which adjoined the respondent's own premises. The respondent now claimed that that was a secret agreement of which he had had no knowledge, and that as he had had no reason to object to the proposals at the time, he had failed to attend the inquiry into the scheme which was held in September 1973.

The proposed roads which were the subject of the 1973 C inquiry were approved in May 1974. Later that year a supplementary order was made whereby it was proposed that Crowthorne Lane should be widened and that this should be done in such a way that part of the respondent's premises would be affected. There was an inquiry into those proposals in December 1974, which the respondent attended. At that inquiry he was told by the inspector appointed to conduct the inquiry that he could not go into the former scheme which had been the subject of the earlier inquiry. The inspector made his report after the inquiry, and his recommendations were approved in July 1975. That decision was not challenged by the respondent at that time. In December D 1975 the respondent applied to the Divisional Court for an order of certiorari to quash, not the latest order, but the earlier orders, claiming that they were invalid. He claimed there was bad faith and a want of natural justice. The Divisional Court decided that the hearing should go forward, and from that decision the Secretary of State for the Environment now appealed, relying upon *Smith* and claiming that further inquiry into the proposals should not be permitted.

The Highways Act 1959 gave the Secretary of State ample power to lay out roads, make new ones, stop up side roads and do all other ancillary works. The court had been told that 80 per cent of the land needed for the Boston scheme E had been acquired and 90 per cent of the demolition work done for the new road. Was the scheme to be held up or set aside by reason of Mr Ostler's application? The relevant provisions, which were in much the same terms as those considered in *Smith v East Elloe*, were in Schedule 2 to the Act. Paragraph 2 said, "If a person aggrieved by a scheme or order . . . desires to question the validity thereof, or of any provision contained therein . . . he may, within six weeks from the date on which the notice [of the confirmation of the order] . . . is first published, make an application for the purpose to the High Court." Paragraph 4 said, "Subject to the provisions of the last foregoing paragraph, a scheme or F order . . . shall not, either before or after it has been made or confirmed, be questioned in any legal proceedings whatever. . . ." Those were strong words, "shall not be questioned in any legal proceedings whatever," and as he (his Lordship) read *Smith*, the majority of the House held that a clause in those terms, in a case where a woman had charged the clerk to the council with fraud, was of such effect that it barred an application to quash the order, even though it was on the ground of fraud. *Smith v East Elloe* had stood for 12 or 13 years; but it had been considered in the G *Anisminic* case, where the House held that the provision of the Foreign Compensation Act 1950 that "the determination by the commission of any application made to them under this Act shall not be called into question in any court of law" applied to a determination properly so called. It did not apply to a purported determination, so that where there had been no proper determination the decision could be called in question.

There were some observations in *Anisminic* which appeared to throw some doubt upon *Smith v East Elloe*, but he (Lord Denning) saw grounds for distinguishing between H the two decisions. *Anisminic* dealt with a complete ouster of the jurisdiction of the court, whereas in *Smith*, as in the present case, the court had ample powers to inquire into matters so long as the applicant came within six weeks. It was more like a limitation period than a complete ouster. Then the *Anisminic* case dealt with a judicial body, the Foreign Compensation Commission, whereas the case with which the present appeal was concerned concerned a quasi-judicial tribunal. There was a world of difference between the two cases, because the public interest was very much in the picture when an administrative decision was in issue. In the case of an inquiry into roads, the decision was very much J in the nature of an administrative matter. Although an administrative decision could be questioned by certiorari, there was a distinction between a judicial decision and an administrative decision. The policy of the legislature appeared to be that when road inquiries had been held and orders made the citizen had a remedy if he went to court within six weeks. Once the six weeks had expired the normal consequences of the making of an order ensued, and as a matter of public policy, could not be challenged weeks or years later. The 1959 Act was designed to bring to an early end all the objections that could be made to a scheme, for the good reason that in the public interest, once a scheme had K been approved and a compulsory purchase order made much had to be done to implement it, and objections could not be left on one side to be dealt with perhaps years later. If a scheme was to be set aside it ought to be done at once. Mr Ostler was seeking to challenge the result of the 1973 inquiry when much had already been done under it. The court was bound by *Smith v East Elloe*, and the appeal should be allowed.

Agreeing, GOFF LJ said that there were two grounds on which *Anisminic* was distinguishable from *East Elloe*. First a judicial decision and an administrative or executive decision might be different. Though it had been said that a Minister L acted in a quasi-judicial capacity, he was nevertheless conducting an administrative or executive inquiry where questions of policy came in and must influence his decision. Secondly, the ratio in *Anisminic* was on a question of jurisdiction, where the tribunal, however eminent, had misconceived the effect of the Act and acted outside its jurisdiction; whereas in the present case, one was dealing with an actual decision made within the jurisdiction but attacked on the ground of fraud or *mala fides*. It could not be gainsaid that some of the speeches in *Anisminic* did appear to cast doubt upon *East Elloe*, but it was certainly not expressly overruled, nor did any of their Lordships say that it was wrong. So it M stood as a decision directly to the point, and the appeal should be allowed.

Also agreeing, SHAW LJ said that the case fell squarely within *Smith v East Elloe*. If that decision had survived *Anisminic*, which he thought it had, it followed that any challenge to the 1973 inquiry whether by certiorari or otherwise, was precluded, and the appeal should be allowed.

Leave to appeal to the House of Lords was refused.

ESTATE AGENTS

A

QUEEN'S BENCH DIVISION

February 18 1976

(Before Judge WINGATE QC, sitting as a deputy judge of the Division)

KNIGHT v CALDERLODGE DEVELOPMENTS AND OTHERS

Estates Gazette April 10 1976 and April 17 1976

(1976) 238 EG 117 and 189

Estate agent claims right to be appointed sole selling agent, "at any rate in the first instance," for houses erected on sites found by him for developer—Developer said to have no right to terminate this arrangement by any length of notice—No such agreement, written or oral, express or implied, made out on the evidence—"Would in any case be subject to so many exceptions as to be uncertain"—Further point on agent's conduct—Judgment for developer

B

This was a claim by Mr William Knight FSVA FRVA FIArb, sole principal of the firm of Knight, Benjamin & Co, estate agents, of Cathedral Chambers, Tombland, Norwich, against Calderlodge Developments Ltd, Mr Lloyd Brian Summers, property developer, of 415 Unthank Road, Norwich, and Norfolk Garden Estates Ltd, for damages for breach of an agency contract or contracts relating to sale of houses on sites developed by the defendants.

C

The plaintiff appeared in person. Mr J Wood QC and Mr G Draycott (instructed by Mills & Reeve, of Norwich) represented the defendants.

Beginning his reserved judgment, JUDGE WINGATE said: This action concerns a dispute between an estate agent and a property developer. The plaintiff, Mr Knight, has carried on the business of an estate agent under the name of Knight, Benjamin & Co in the city of Norwich for a number of years. Mr Knight is an energetic man, a city councillor with many contacts and interests, and has been successful in his business. Mr Summers, the second defendant, to whom I have referred as a property developer, is a solicitor who ceased to practise as such in 1962 and since that time has built up a property company group. It is sufficient to say that by virtue of either of his own holdings or of his wife's, he directly or indirectly controls both the defendant companies, Calderlodge Developments Ltd and Norfolk Garden Estates Ltd, and has been the moving spirit in all their affairs. He has been very successful in the acquisition, sale and development of land through these companies, which have operated for the most part in the Norwich area. Each of these men had much to gain from their association in business, which, as far as this case is concerned, started in March 1967, when Mr Summers, on behalf of Norfolk Garden Estates, wrote a letter on the 14th of that month to the plaintiff appointing him as sole agent for all sales of housing units arising out of the development by Norfolk Garden Estates of the Blithewood estate on the terms set out in the letter. The relationship of principal and agent was expanded into other fields and continued until 1973, when it terminated in very unhappy circumstances giving rise to these proceedings. During the six years and more of their association the plaintiff acted for the defendant in a number of different transactions involving the acquisition of land, the sale of land and the sale of housing units on land developed by one or other of the defendant companies, and also on occasions in the obtaining of finance to further the operations of the defendant companies.

D

E

F

The way in which the plaintiff has put his case against each of the defendants in these consolidated actions has made it necessary to go into a number of matters, some of which in the final event can be seen to be irrelevant to the issues which I have to decide. My task of eliminating irrelevant matter during the trial has not been made any easier by the fact that the plaintiff decided to withdraw his instructions to his solicitors and counsel soon after the commencement of the proceedings and has conducted his case in person, which he did with ability. The plaintiff is a very intelligent person and has considerable appreciation of what is and is not material to his case. I understand that he is interested in law and is a student of one of the Inns of Court. But none the less, in a case of some complexity, such as this, the judge trying the case must be careful not to exclude matters from his consideration as irrelevant before he has a clear picture of the issues in the case. Let me say at once that it has not been easy to ascertain and state with precision the various ways in which the plaintiff has sought to put his case against each of the defendants. There are two actions by the plaintiff, one against Calderlodge Developments, Mr Summers and Norfolk Garden Estates, and the other against Norfolk Garden Estates and Mr Summers. Stated in simplest terms, the issue I have to consider is whether at the material time, that is to say in January of 1973, there was in existence a contract of agency or sole agency between the plaintiff and the defendants, or one or other of them, and if so, what were its terms. The plaintiff's case is that the agreement between himself and the defendants advanced far beyond the terms of the original agreement stated in the letter of March 14 1967, which stated the usual terms on which an estate agent is employed to sell properties on behalf of his principal.

G

H

J

Some examination of the pleadings becomes necessary at this stage to see how the case on paper has been developed by the plaintiff, making due allowance for the fact that, save in the early stages, he settled the pleadings himself without experience in such matters. He first began proceedings in December 1973 in the Norwich County Court against Calderlodge Developments Ltd only. It was pleaded by his solicitors as a simple claim for fees, calculated at 2 per cent, due to the plaintiff as sole selling agent for the defendant company in respect of the sale of five housing units at a site named Templemere, otherwise known as the stadium, which was being developed by Calderlodge. The defence denied, *inter alia,* the plaintiff's appointment as sole selling agent for the sale of the housing units, and counterclaimed a declaration that the plaintiff was never at any material time appointed as sole selling agent. In answer to a request for further and better particulars of the agreement, the plaintiff alleged that it was oral and written and to be implied from the conduct of the parties. The oral agreement with Calder-

K

lodge was alleged to have been made on October 6 1970 between the plaintiff and Mr Summers, acting on behalf of Calderlodge. It may be noted in passing that Calderlodge was not incorporated until October 1971. Reliance was also placed on a document called the heads of agreement, dated February 10 1969, which was made between the plaintiff, Norfolk Garden Estates and Mr Summers, to which agreement Calderlodge was not and could not have been made a party, and on a letter dated January 28 1972 from Bird & Partners, accountants, written on behalf of Calderlodge, which stated that it was the intention of that company that the plaintiff should act as sole agent in respect of an estate at Martineau Lane and the stadium, the latter being the site known, when developed, as Templemere. Finally it was alleged that the plaintiff would rely on the course of conduct between the plaintiff, Mr Summers and Norfolk Garden Estates from about 1965, and that Calderlodge by conduct approved the agreements made between the plaintiff, Mr Summers and Norfolk Garden Estates.

Soon after the delivery of these particulars the action was transferred, on March 5 1974, to the High Court. On April 23 1974 the plaintiff dispensed with the services of his solicitors. On April 26 1974 the plaintiff delivered an amended statement of claim in which he added Norfolk Garden Estates and Mr Summers as defendants, alleging that the selling agency agreement was with all the defendants. On June 10 1974 particulars were delivered alleging an additional oral agreement, on September 12 1972, between the plaintiff and one Meadows, a director of Calderlodge. I think I should mention here that in the course of the trial for the first time the plaintiff alleged a further oral agreement to have been made with him by Mr Meadows on September 18 1972. In further particulars dated July 29 1974 reliance was placed by the plaintiff, in relation to the agreement alleged on September 12 1972 with Mr Meadows, on the production by Mr Meadows of a brochure on which the name of the plaintiff's firm was printed and shown as selling agents. In a yet further set of particulars delivered by the plaintiff on October 14 1974 the plaintiff stated that Mr Summers at all times exercised absolute authority on behalf of the two companies, Calderlodge and Norfolk Garden Estates, and that when Calderlodge was formed Mr Summers assured the plaintiff that this would not affect the plaintiff in his agency agreements. The plaintiff also alleged that Mr Hart, the signatory of the letter of January 28 1972 on behalf of Calderlodge, in various conversations prior to January 28 1972 made it clear that the plaintiff would act and continue to act as agent for Calderlodge.

Other allegations are made by the plaintiff in the pleadings in this action against Calderlodge, Mr Summers and Norfolk Garden Estates, but I think I have stated those which are directly relevant to the issues of fact relating to the alleged existence of an agency agreement between the plaintiff and the defendants. On November 25 1974 the plaintiff began a second action against Norfolk Garden Estates and Mr Summers. Taking together the allegations made in the statement of claim and in particulars which were delivered on February 14 1975, the plaintiff states that he was appointed selling agent to act *in futuro* in respect of about 535 houses to be erected on a site referred to in these proceedings as Eaton, at a fee of 2 per cent on the selling price; that this agreement was made jointly and severally with Norfolk Garden Estates and Mr Summers; that it was oral and in writing, and that in so far as it was oral, it was made on divers dates between December 1967 and December 1972 between the plaintiff and Mr Summers at the latter's home, 415 Unthank Road, Norwich, and in so far as it was in writing, it was contained in the heads of agreement dated February 10 1969. When asked to particularise further the oral part of the agreement, a number of meetings with Mr Summers are relied on in the particulars dated March 12 1975, namely, meetings on December 29 and December 30 1967, May 26, June 10 and November 11 1970, March 30 1971 and April 18 1972.

It is clear from these pleadings—and this became even clearer in the course of the trial—that it is impossible to spell out from the documents referred to the contracts of agency alleged by the plaintiff, and that if he is to succeed, it must be by the evidence relating to the alleged oral agreements and the conduct of the parties. I should perhaps explain here that in referring in some detail to the way in which the plaintiff's case has grown and been substantially added to during the interlocutory stages, I have in mind the defendants' submission that this has been on account of the absence of any clear evidence of the agreement alleged by the plaintiff, so that he found it necessary to reconstruct and add to his case as it progressed. Furthermore, it is clear from the pleadings, and from his presentation of his case before me, that the agency contract or contracts alleged by the plaintiff confers or confer rights and obligations which go far beyond those arising from the usual form of estate agency agreement, such as that created by the letter of March 14 1967. However, the exact terms of the agreements relied on by the plaintiff do not clearly emerge from the pleadings, nor did the plaintiff find it easy to elucidate them in the course of the case.

As I have said, the agency agreements alleged by the plaintiff differ in important respects from what one might call the norm. First of all, it is said that whenever the plaintiff found a suitable site for development and successfully introduced it to the defendants, he had a vested right to be appointed as the sole selling agent, at any rate in the first instance, for the sale of all units erected on that site in the course of development by one or other of the defendants. Such a term would really have to come into existence at the time of the employment of the plaintiff to find a site and then be incorporated as a term of an agency agreement once the site had been introduced to the defendants and acquired by them. The plaintiff alleges that by the time the relationship between the parties came to an end in 1973 the stage had been reached in the case of the relevant sites, which were Eaton, Martineau Lane and the stadium, when his right to act as agent had vested in him and had been admitted and confirmed by the defendants. Again, it is said that this agency was to be performed in the future and over a considerable period of time. The plaintiff conceded that in the event of a site being sold in whole or in part, as happened in several instances, instead of being developed by the defendants, he would have no cause for complaint. In other words, his vested right to commission as selling agent on the sale of housing units on the sites could properly be defeated by the defendants' decision to sell the land instead of developing it. Again, the performance of the contract of agency might have to be postponed indefinitely for want of finance for development, or want of demand for houses. The right to be appointed as *sole* selling agent has also to be considered in this context, because, as the history of the association between the plaintiff and the defendants shows, there were occasions when other agents were employed by the defendants to sell houses on their behalf along with the plaintiff without protest from the plaintiff that it was in breach of his vested right to act as sole agent. It is also the plaintiff's case that the defendants had no right to terminate his contract of agency by any length of notice, as would normally be their right in the case of the usual estate agency contract. So any termination by them of his agency without just cause would be a breach of contract, entitling him to damages.

So far I have been concerned to outline the plaintiff's case as to the contracts of agency alleged by him, and in doing so I have not distinguished at this stage the case against one defendant from that alleged against another, although of course I bear such distinctions in mind. I must now mention an important issue which was raised by the defence in answer

A to the plaintiff's claim that his agreement was wrongfully breached by the defendants in January of 1973. The plaintiff alleges that the defendants in January 1973 committed a breach of his agency agreement, and in particular he claims that on January 24 1973 at an interview with Mr Summers, then acting on his own behalf and as agent for Calderlodge and Norfolk Garden Estates, Mr Summers informed the plaintiff that his commission of 2 per cent was to be reduced to 1 per cent. He (the plaintiff) contends that this purported reduction was a breach of a term which went to the root of the contract of agency, entitling him to treat the defendants' conduct as a repudiation of the agency agreements; alternatively, that the defendants wrongfully thereafter terminated the agency. The defendants allege that if, which they deny, they committed any breach of a term of the agency agreement in 1973, the plaintiff so conducted himself at the time, and thereafter, that they were entitled to and did terminate his appointment as selling agent. The conduct complained of may be described shortly as making repeated threats to disclose information about Mr Summers to the press, to the Law Society and to RMC Ltd, a company which had control of Norfolk Garden Estates, and to other persons, and to ruin the second defendant, Mr Summers, unless certain financial demands, which the plaintiff was making upon the defendants, were met. These are serious allegations, of which further particulars were given in the course of the interlocutory proceedings. The plaintiff's reply to these allegations in the course of the trial I can only describe as qualified, that is to say, it fell well short of an unqualified denial that he made statements of the kind complained of, but he denied that the word "threat" was justified. His case is that such statements as he made in this respect were to achieve a meeting with Mr Summers with a view to a settlement, and not to exert pressure to gain a financial advantage.

As already indicated, in the course of the hearing of this consolidated action it has been necessary to go into a large number of matters. I say "necessary" because I found no reasonable way of avoiding doing so, although in the final event many of these matters turned out to be irrelevant to the real issues in the case. I am sure that the plaintiff, however, would have felt very dissatisfied had I prevented him from bringing these matters to my attention, and it may well be that they have played a part in explaining the history of the relationship between the parties and its final breakdown, and have helped me to decide where the truth lies. I do not, however, propose to deal with all the various side issues in this judgment. I now come to the history of this dispute as I find it to be upon the evidence in the case, dealing with the salient facts only. When referring to the main events of the years 1967-73, I do not propose to deal at this stage with the various meetings between the plaintiff, Mr Summers and Mr Meadows at which it is alleged oral agreements as to the plaintiff's agency were made. I shall deal with these all together after dealing with the history of the parties' relationship and its final demise.

The story starts, as I have already indicated, with the letter of March 14 1967 signed by Mr Summers on behalf of Norfolk Garden Estates and addressed to the plaintiff's firm, appointing him as sole agent for all sales at the Blithewood estate. This conforms with the usual pattern of such agreements, except that by clause 1 it is terminable by either side at any time without any period of notice being required. Commission of 2 per cent is payable in the event of the plaintiff introducing a purchaser who actually completes. In 1967 the question of a public flotation of Norfolk Garden Estates was considered, and the plaintiff carried out inquiries and negotiations on the instructions of Mr Summers with a view to obtaining a quotation on the Stock Exchange. In particular, in 1968 the plaintiff made contact with G F Nash Securities, of Kettering, with a view to that firm giving advice and assistance to enable a public flotation to take place late in 1968. It is sufficient to say that nothing came of this, but that was through no fault of the plaintiff, who at the time in question was a director of Norfolk Garden Estates and as it turned out spent time and effort with no direct reward. I say "no direct reward," because the plaintiff's association with the defendants over the years yielded very substantial rewards, and had it continued would no doubt have yielded even further rewards. However, although the plaintiff at the time accepted what might be called the luck of the draw, namely, work without pay where the project was unsuccessful, in the circumstances of the final breach he undoubtedly felt a deep sense of grievance that in some cases, of which the negotiations with Nash are an example, his work went unrewarded either directly or indirectly, and finally that this association with the defendants was terminated unlawfully, thus robbing him of the opportunity of earning further commission. I shall, again, have to refer later to this aspect of the plaintiff's complaints of the defendants' treatment of him. It is convenient to mention here that in July 1968 the plaintiff became a director of Norfolk Garden Estates and was entitled to an honorarium of £500 a year as director's remuneration.

Towards the end of 1968 Norfolk Garden Estates experienced a cash crisis, and the plaintiff played a part in obtaining some temporary finance for the company from a company called Mesco. There was a disagreement between the evidence of the plaintiff and Mr Summers as to the importance of the part played by the plaintiff, and particularly with regard to an important meeting which took place on December 24 1968 between Mr Summers, the plaintiff and the chairman of Mesco, following which further time was obtained for the repayment of a loan which Mesco had made to Norfolk Garden Estates of £100,000. While Mr Summers may have underrated the contribution made by the plaintiff in obtaining valuable time for repayment from Mesco—and in saying this I have regard to the report by Mr Summers outlining the history of Norfolk Garden Estates in November 1968, which is document 402/1—I think the plaintiff has exaggerated his part, but I do not find it necessary to say more than that. This is one of a number of peripheral matters which have been the subject of evidence and argument. At about the same time, that is to say late 1968, approaches were made to the Gulf Oil Company to obtain finance for Norfolk Garden Estates, and again the plaintiff played a part in this, the extent of which is in dispute. In January 1969 Gulf agreed to make a loan of £250,000 to Norfolk Garden Estates which greatly assisted to relieve the financial embarrassment, but the directors of Norfolk Garden Estates were required to guarantee the loan. The plaintiff was unwilling to enter into such an agreement, and meetings took place between the plaintiff and Mr Summers on January 18 and January 19 1969. I find upon the evidence that the plaintiff made it a condition of his entering into such a guarantee that Mr Summers should transfer to the plaintiff 5 per cent of the company's equity by transferring shares to the plaintiff from his (Mr Summers's) own shareholding.

The evidence of the plaintiff as to what passed at this interview differed from that of Mr Summers, but I prefer the latter's evidence. Mr Summers was able to refer to a contemporaneous note to refresh his memory as to what passed between him and the plaintiff on Saturday January 18 and Sunday January 19, and following this Mr Summers wrote a letter of January 19 1969 to the plaintiff which I find correctly summarises the points which were at issue between them. The plaintiff asserted that it was not justifiable to ask the directors to give a guarantee and that Gulf were only asking shareholder directors to give guarantees, and he (the plaintiff) tried unsuccessfully to persuade other directors to support him in this attitude. I prefer the evidence of the defendants that Gulf required all directors to give guarantees, whether shareholders or not. The plaintiff's refusal put Mr

A Summers and the company in a very difficult situation at a time when the company desperately needed the Gulf loan, and I think the plaintiff was using the situation to obtain a share in the equity of Norfolk Garden Estates. I should perhaps mention that the defendants' evidence was—and I accept it—that before accepting directorships the plaintiff and others were told that they might have to give guarantees of this kind. The letter of January 19 1969, written by Mr Summers, sets out his account of this disagreement. The disagreement was eventually resolved, and after negotiations the parties entered into the heads of agreement dated February 10 1969. This provided for the resignation of the B plaintiff as a director, a continuation of the business association between Norfolk Garden Estates, Mr Summers and the plaintiff, and the payment of a proportion of the fees claimed by the plaintiff to be due to him. So far as it affects the issues in this case, the significant clause in the heads of agreement is clause 9. The words "It is hoped" make it clear that no binding obligation to appoint the plaintiff as agent is entered into under this agreement, but it is provided that in the event of further work being undertaken by the plaintiff's firm, the scale of fees contained in the fourth schedule is to apply. This provides for 2 per cent commission.

C On February 24 1969 the plaintiff wrote a letter to Mr Summers regarding a shop site called Flower Pot Lane for acquisition by Norfolk Garden Estates, the last sentence of which is directly relevant to the contention of the plaintiff that he had a right to be appointed as agent for the letting or sale of units erected on a site introduced by him to the defendants. The reply by Mr Summers, writing on behalf of Norfolk Garden Estates, makes it clear that the practice of the company was not to enter into any binding obligation in this respect, and the plaintiff made no reply, nor any kind of challenge, to this clear statement. On June 3 1969 the plaintiff wrote to Mr Summers with regard to a proposed D acquisition of land by Norfolk Garden Estates, and in the last sentence assumes that in the event of an acquisition and development taking place, he (the plaintiff) will have the opportunity of reselling the finished units in the usual way. In the light of the whole of the evidence, I find this to be an expression of hope rather than an assertion of a legal right. On July 18 1969 the plaintiff wrote a letter setting out the services he claimed to have rendered to Mr Summers, and the matter appears to have been resolved by an agreement set out in a letter of July 31 1969 from Mr Summers.

On September 24 1969 Mr Summers informed the plaintiff of the appointment of another agent, Mr Alan Ebbage, to E act as selling agent in the sale of the remaining units at Eaton Village, and informed the plaintiff that he could, if he wished, continue as selling agent in competition with Mr Ebbage. The plaintiff's reply on September 29 1969 regrets the appointment of another agent, but in no way suggests that this is in breach of his own right to continue to act as sole agent or at all. A similar situation arose at Tithe Barn, another site being developed by the defendants, as appears from a letter of October 21 1969 from Mr Summers to the plaintiff, and by a letter of March 24 1970 Mr Summers notified the plaintiff of his intention to instruct another firm of agents, Alan Ebbage, to act as agent at the F Tithe Barn estate, so that the plaintiff and Alan Ebbage would be selling in competition with each other. Furthermore, in the penultimate paragraph Mr Summers contemplates one or other of the two firms being appointed as sole agents. The plaintiff's reply on March 25 1970 accepts this situation, stating that he had no alternative but to do so. It may be that these examples are not exact parallels of the situation which existed in January 1973, but in my judgment they are significant.

For some years the possibility of Norfolk Garden Estates acquiring Norwich City Stadium for development had been under consideration by both the plaintiff and Mr Summers. G This project involved the purchase of the shareholding in the company owning the stadium. Letters of October 19, November 25 and December 2 1970 from Mr Knight show that matters were progressing towards a final stage. On April 26 1971 the plaintiff wrote a letter to Mr Summers at the latter's request, describing the site and giving a valuation figure of £74,000. As appears from the plaintiff's letter of November 25 1970, Mr Summers had not decided which of his companies was going to buy the stadium, and indeed it was contemplated that a new company might be formed for this purpose. The plaintiff stated in evidence that he assumed in the summer of 1971 that it would be acquired by Norfolk H Garden Estates, and a letter of July 22 1971 shows that the plaintiff was looking to that company for his fees on the acquisition. The plaintiff was then informed by Mr Summers that the stadium would be acquired by a company controlled by Mr Summers' wife, and in response to a request by Mr Summers the plaintiff wrote the letter of August 4 1971 to Mrs Summers. In the event, the stadium was acquired by Calderlodge, which was a wholly-owned subsidiary of a company that was controlled by Mrs Summers but for all practical purposes operated by Mr Summers. I should mention here that the shareholding in Norfolk Garden Estates had been taken over by the Poynter Group in 1969, which company J was in turn taken over by RMC, a public company, in April 1972. RMC remained in control of Norfolk Garden Estates until January 3 1973, when Mr Summers again acquired the shareholding. There was a service agreement between Mr Summers and the Poynter Group (later RMC). The plaintiff considered that the acquisition of the stadium by Calderlodge had deprived Norfolk Garden Estates of a valuable asset and that this deal was carried out by Mr Summers in breach of his obligations to Norfolk Garden Estates and to its parent company, and also in breach of the service agreement between Mr Summers and the Poynter Group (later RMC). The plaintiff suggested that Mr Summers had thus involved himself K in a conflict-of-interest situation.

On January 28 1972 Mr Hart wrote a letter to the plaintiff on behalf of Calderlodge Developments Ltd. This letter dealt with the plaintiff's entitlement to remuneration for his part in the acquisition by Calderlodge of an industrial site at Martineau Lane, the stadium company and site, and a site near Swansea known as Llansamlet. The plaintiff was paid a sum of £2,000 in full settlement of his fees in connection with all three acquisitions. The letter sets out what the payment is intended to cover. This payment was accepted by the plaintiff, although he claimed that the sum was far less than L he should have been paid. Two points arise here for comment. First of all, it is to be noted that the plaintiff is being paid a fee for acting in the acquisition of sites, as indeed he was in a number of other instances where he was instrumental in the acquisition of land, but he was not always paid a fee on such acquisitions. However, it seems to me to be relevant, as a matter of probability, to the question whether there was an agreement between the parties that whenever the plaintiff had been instrumental in the acquisition of a site, he should later be appointed sole selling agent if and when it was developed by the defendants, that the plaintiff was in a large number of cases—and the case of the stadium is immediately M in point here—paid fees for his work in the acquisition of the site. It seems to me that it cannot be argued convincingly, as the plaintiff attempted to do, that without the alleged agreement the plaintiff had insufficient consideration for what he did for the defendants. In this regard the schedule of payments put in by the defendants, showing what was paid to the plaintiff over the years between 1965 to 1973, totalling a sum of £67,500, is in point. Some of this was paid on acquisitions; more on the sale of houses. The second point for comment is this. The letter from Mr Hart makes it clear that no further agency is going to be offered in respect of the Welsh site, and having regard to its distance from Norwich

A the plaintiff accepts this without argument. However, in the last sentence of the letter, Mr Hart, on behalf of Calderlodge, states that it is the intention of the company that the plaintiff should act as sole agents in the development of the estates at Martineau Lane and the stadium. The plaintiff did not comment on this in his reply. It seems to me that here again we find evidence of intention on the part of the defendants but not of any binding agreement. However, both sides certainly conducted themselves on the basis that the plaintiff would be the selling agent of housing units built on the site, and a letter of June 6 1972 from Mr Meadows to Mr Bennett is consistent with this, though in that letter it is made clear that formal B instructions had not yet been given.

Throughout 1972 the plaintiff continued to act as the defendants' agent in respect of various matters, of which the acquisition of nos 10 and 11 Tombland by Norfolk Garden Estates and the subsequent sale of the property is an example. At this time, however, there was little or no selling of new houses on the defendants' estates, as few were available for sale. Indeed, at the time of the break between the parties in January 1973, of the three sites in respect of which the plaintiff claims to be sole selling agent, that is, Eaton, the stadium and Martineau Lane, the last one, an industrial site, was the C only one which had any immediate prospect of sale. On September 12 1972, according to the plaintiff, there was a meeting between the plaintiff and Mr Meadows of Calderlodge, and there was a discussion about the development of the site of the stadium, to be known as Templemere. According to the plaintiff, he informed Mr Meadows that in view of the fact that Mr Summers had a service agreement with RMC, and of the conflict of interest which had arisen, it would be inappropriate for his firm to place housing units on the open market until this had been cleared up. He also stated that Mr Meadows appointed him to act as selling agent for Templemere and confirmed this later on September 18 in an oral conversation, when he (Mr Meadows) handed him (the plaintiff) a brochure for Templemere which carried the plaintiff's name as selling agent. Mr Meadows denied handing the brochure to the plaintiff, and denied giving the plaintiff instructions to act as agent on these occasions. I accept the evidence of Mr Meadows in preference to that of the plaintiff. Though I have no doubt that up until January 1973 both Mr Meadows and Mr Summers intended to give the plaintiff the opportunity to act as selling agent for Templemere, I find that there was nothing said at these meetings which carried the matter any further. The mention in the brochure of the plaintiff's name as selling agent is consistent E with the then admitted intention of the parties, although, in the light of the plaintiff's evidence of his reservations because of the conflict-of-interest situation, it would seem that any firm offer which might have been made to appoint him as agent would not have been accepted by him. It is also to be noted that the plaintiff's name never appeared as selling agent on the board at the site, although again it is clear from a letter of January 12 1973, signed by Mr Meadows' secretary, that it was believed that the plaintiff would be the selling agent for the flats to be offered for sale in a few weeks at the stadium site.

At the beginning of January 1973 Mr Summers acquired F the shareholding in Norfolk Garden Estates. This came as an unwelcome surprise to the plaintiff, who had been in touch with two possible purchasers of Norfolk Garden Estates, namely, Wilson Connelly Ltd and Francis Parker Ltd. The plaintiff said that when he had discussed the conflict-of-interest question with Mr Summers in the previous autumn, they considered that a way out of the difficulty would be for Mr Summers to obtain a purchaser for Norfolk Garden Estates and thus obtain a release of Mr Summers from his service agreement. The plaintiff said that it was clear from this conversation that Mr Summers had discussed with RMC the purchase from RMC of their interest in Norfolk Garden Estates. As I have said, on or about January G 3 1973 the plaintiff learned from RMC that Mr Summers had purchased the equity of Norfolk Garden Estates. On January 5 the plaintiff saw Mr Summers. According to the plaintiff, Mr Summers said that he was now the owner of Norfolk Garden Estates but that this would make no difference to the plaintiff. On January 24 the plaintiff called at Mr Summers' house at 415 Unthank Road, Norwich. The plaintiff's account of this important interview differed in material respects from that of Mr Summers, whose evidence I prefer. In particular, the plaintiff denied making any threats, and said that Mr Summers stated that the company had decided to pay only 1 per cent to all agents, and that H when he referred to the heads of agreement of February 10 1969 Mr Summers said, "You can refer to that at your peril."

The plaintiff has made it part of his case that Mr Summers committed a breach of contract on behalf of himself and the two defendant companies at this interview on January 24 1973 by a unilateral reduction of the plaintiff's commission from 2 per cent to 1 per cent at a time when his contract of agency was on foot. Mr Summers denied that. He denied that he said anything which could be construed as a final decision by the defendants to cut the commission to 1 per cent at that stage. He said he informed the plaintiff that he J was considering such a reduction, and I prefer his evidence to that of the plaintiff, although it is quite true that within a matter of a day or so, as appears from the correspondence, Mr Summers had decided on behalf of the companies that some reduction would have to be made. It is to be noted here that it is clear from his subsequent actions that the plaintiff thought that the amount of future commission was negotiable. The importance of this finding arises from the contention relied on by the plaintiff that on January 24 1973 the defendants, by their purported reduction of commission, committed a breach of the contract of agency which he was entitled to treat as a breach going to the root of the contract, K and that nothing that he may be found to have done thereafter can affect his claim to damages.

As I have said, I prefer the evidence of Mr Summers as to what passed between these two men at this interview. Mr Summers said that on January 24 the plaintiff called on him at about 8 o'clock in the evening. He said that he mentioned to the plaintiff as sympathetically as he could that the Swansea contracts had been exchanged, and that this was a big disappointment to the plaintiff, as Mr Summers appreciated it would be, because, as this sale had been arranged through other agents, the plaintiff would not get the substantial commission he was hoping to do as a result of the efforts L he had made to find a purchaser. The plaintiff became excited and said he would take his file and papers to Feltham —the offices of RMC—if Mr Summers did not pay him £36,000. Subsequently the sum of £35,000 was mentioned by the plaintiff. Mr Summers became upset. He understood what the plaintiff meant by his reference to Feltham, and realised he might be embarrassed by the revelations the plaintiff might seek to make as to the profits on the sale of Swansea. Mr Summers said that the plaintiff said he would ruin and destroy Mr Summers' businesses, and that he spoke of pulling them apart. Mr Summers said that at some stage M the rate of agents' commission was raised and that he (Mr Summers) told the plaintiff that they were considering alterations to the level of fees they were paying. Mr Summers said that they were prepared to discuss this question with the plaintiff, but that they had agreed 1 per cent with another firm as the basic level, with a minimum of £100 per housing unit. Mr Summers made some notes of this conversation and that which followed, and was able to refer to them to refresh his memory. There was another interview, probably on January 26, when Mr Summers said the plaintiff soon became excited and made the same threats and the same demands and mentioned the press. The plaintiff referred to

A the *Sunday Times*, and said he would take a brown paper parcel to the editor's waiting room. He again made it clear that he wanted £35,000. On January 27 there was a further meeting between the plaintiff and Mr Summers at about noon. The plaintiff made the same sort of threats and then referred to legal proceedings. The plaintiff again demanded £35,000 and said that if Mr Summers did not agree, he would go immediately to RMC and the *Sunday Times*.

As I have already said, I accept the evidence of Mr Summers with regard to his account of what passed between him and the plaintiff on these occasions, and I need not refer in further detail to his account of what the plaintiff said, except B to note that for this interview on January 27 Mr Summers had prepared a statement referring to blackmail. He said to the plaintiff, "What you are doing is blackmail in the full criminal law sense, a threat to disclose information and a demand for money coupled with the threat to disclose." The plaintiff said, "Maybe you are right, but there is nothing you can do about it; you have got to do what I say." Mr Summers refused and requested the plaintiff to leave. He said he would not leave, and Mr Summers left him in the lounge for about a quarter of an hour. Mr Summers then returned, and the plaintiff left soon thereafter. Mr Summers said that C the plaintiff referred to his services to himself and the companies, and that the plaintiff was clearly upset and different from his normal self. I accept this account of what occurred. Both men then on January 29 1973 wrote letters which crossed. In his letter, which is the first statement in writing of his claim, the plaintiff placed reliance on the heads of agreement of February 1969 and suggested that the reduction of fees from 2 per cent to 1 per cent was a breach of that agreement. He also was clearly very upset about the outcome of the sale of the Swansea site, with which much of the letter was concerned, and in effect he charged Mr Summers with bad faith. Mr Summers' letter of January 29 1973 is D in my opinion absolutely consistent with his account of the interview which had taken place between the plaintiff and himself. After references to the threats made by the plaintiff, he makes a very fair and indeed generous offer to the plaintiff to mend the breach in their relationship.

A number of letters followed, which speak for themselves. I am satisfied that the plaintiff continued to make threats against Mr Summers and that these became more specific in character as time went on. The letter of February 26 1973 from Mr Summers to the plaintiff's solicitors is a sufficient indication of how the plaintiff's conduct was making any kind of future relationship impossible. The plaintiff's soli- E citors denied these allegations, but I am satisfied that the plaintiff never told them the truth. The evidence that the plaintiff was engaged over a period of time from January 24 to the end of June 1973 in a campaign of threats against Mr Summers is overwhelming. In addition to the evidence of Mr Summers and what appears in the correspondence, I had the evidence of Mr Hart, Mr Meadows, Mr Chaplin, Mr Warren and Mr Pooley that on many different occasions the plaintiff repeated his threats to these witnesses to injure Mr Summers and his companies one way or another. I do not propose to refer in detail to the evidence of these gentlemen, some of which is supported by the evidence of tape recordings. F It is sufficient to say that I found the witnesses to be reliable and in no way prejudiced against the plaintiff, as clearly appeared from the way in which they gave their evidence. Like Mr Summers, they regretted very much that the plaintiff had resorted to the tactics of threats. The plaintiff in his evidence admitted uttering most of the statements attributed to him, but denied that they amounted to "threats," a word he took exception to, and insisted that by making such statements he was only trying to bring pressure to bear on Mr Summers in order to achieve a meeting with him when a settlement could be arrived at, and not to extract money or a financial bargain of benefit to himself. I find this interpretation of his actions impossible to accept. In my opinion G his threats—and I think the word is abundantly justified—were made to force Mr Summers and his companies to come to an agreement with the plaintiff under which the plaintiff would be entitled to considerable financial benefits.

Before stating my conclusions as to what follows from this finding of fact, I must deal with the allegation that on various dates from 1967 onwards the plaintiff had entered into oral agreements with Mr Summers, Mr Meadows and the companies represented by them whereby he was appointed selling agent for the sites being developed, or to be developed, by one or other of the defendants, and also that from conduct such agreements are to be implied. In the final result, my H conclusion as to this may be of little practical effect, having regard to what I find must be the consequences of the plaintiff's conduct on and after January 24 1973, but since it has been an issue throughout the long trial, I feel I should deal with it. I find that there was no agreement that the plaintiff should have a vested right to be appointed agent where the site in question was one which he had introduced to the defendants. That this was the policy of the defendants was admitted by Mr Summers, but in my judgment there is no evidence, save the plaintiff's, which I reject, to support the existence of an express or implied agreement to that effect. J Furthermore, as I have already pointed out, such an agreement would be subject to so many exceptions, for example, where the defendants resold the site without development, as to be too uncertain in its operation to be defined and enforced.

The plaintiff sought to support his contention as to the oral agreements alleged on the various dates by reference to entries in his diaries for the years in question. The existence of such diary entries appears to have been first raised by the plaintiff in a letter of March 14 1975 addressed to the defendants' solicitors. These diaries were submitted for inspection by the defendants in May 1975, and by a letter dated June K 3 1975 the defendants' solicitors stated their wish to check the claim by the plaintiff that these entries were made contemporaneously with the events they recorded. By a letter of June 11 1975 the plaintiff gave detailed information as to his practice in making such entries, and under paragraph (d) of his letter he stated that they were used for making essential notes of meetings normally within a very few days after the event. Under paragraph (e) he referred to his practice of transferring items of importance from his pocket diary and local-authority diary at convenient intervals, and at the end of each year in particular. Finally, under paragraph (f) he stated that all other entries would be found to have been L made contemporaneously. The defendants submitted these diaries to a handwriting expert, and in the light of his opinion they were not prepared to accept that the entries were in any sense contemporaneous records. In a letter of July 9 1975 the plaintiff began to resile from the position he had undoubtedly taken up with regard to the importance of these entries as giving valuable support to his assertion of oral agreements on the various dates alleged in the pleadings. In the course of his evidence, while denying that all the relevant entries—which are conveniently set out in the first report of Mr Wallace, the handwriting expert—were made in 1974, he agreed that they could not be regarded as a contemporaneous M record, but stated that they were not handed over as such. He further admitted that the entry for December 29-30 1967 was probably put in by him two or three years later and that it was possible that these entries were entered at the same time as the entries made for 1972, that is to say, five years after the events recorded under 1967.

Mr Wallace said that in his opinion the characteristics and qualities of the relevant entries in the diaries for December 1967 and December 1972 satisfied him that they were written under the same conditions, the same circumstances and the same environment, and he thought the writer was in the same mood and occupied the same position while both sets of entries were being written. I accept his opinion. With regard to the other entries, Mr Wallace was not prepared to go so far, but was of the opinion that they were all written by the same pen. The plaintiff agreed that this was probable, but contended that during these years he had kept the same Parker biro pen and only used it occasionally, for reasons which he explained, and that the fact that the same pen might have been used to make all these entries did not show that they were all entered up at the same time. I do not think that that evidence enables me to reach a final conclusion on this point, nor is it necessary for me so to do, although it is to be noted also that the entries made by this same pen are all in respect of matters material to this action. I think it is sufficient for me to say that I am unable to attach any importance whatever to the entries as supporting the plaintiff's oral evidence. I think that there are very good reasons why the plaintiff should now regret that he put these entries forward to buttress his case. Clearly, the only probative effect they could have would be on the basis that the entries were contemporaneous, or nearly so, and this was the basis, I think, on which they were put forward. I think the intention was to impress the defendants with the strength of the plaintiff's case. The plaintiff, in my judgment, withdrew from the brink just in time after he appreciated the effect of the handwriting expert's opinion.

The plaintiff's conduct in relation to these entries has not enhanced his claim to be treated as a credible witness. On the dates, or some of the dates, in question there were probably discussions between the plaintiff and Mr Summers about business matters, and these may have included the prospects of selling houses to be built on the defendants' sites, but I accept Mr Summers' evidence—and indeed that of Mr Meadows—that on none of these dates were any specific instructions given to the plaintiff to act as selling agent on behalf of the defendants. In my judgment such business discussions as took place from time to time over the years between the plaintiff, Mr Summers and Mr Meadows added little or nothing to the relationship recognised by the heads of agreement, that is to say, that so long as it suited both parties, the plaintiff would act as selling agent for the defendants' companies as and when they acquired sites and developed them, but that this was at all times subject to the right of either party to give reasonable notice to terminate the arrangement and to the right of the defendants to vary its terms by appointing other agents to act jointly with the plaintiff, or altering his rate of commission. If the plaintiff was unwilling to act as agent on the terms proposed by the defendants then he would, of course, be at liberty to terminate his agency. I find that this was the situation in January 1973 and that the agency then in existence with the two companies was on the terms I have outlined. I find, however, that at no material time was there any contract of agency between the plaintiff and the second defendant.

I have said that the agency was determinable by reasonable notice, the length of which would depend on the relevant circumstances at the time it was given: see the case of *Martin Baker Aircraft Co Ltd v Murison* [1955] 2 QB 556. I have not overlooked the fact that in the letter of March 14 1967, by which the plaintiff was appointed to act as sole agent by Norfolk Garden Estates in respect of Blithewood, it was expressly provided that no period of notice was required on either side, but there was no such express provision in the heads of agreement on February 10 1969 or subsequently agreed, and I think it is reasonable to imply that to determine the agency which existed in January 1973 a reasonable notice was necessary on either side. If it had become necessary—which I do not think it is—for me to decide what length of notice would have been reasonable in January 1973, I should have had regard to the business actually done by the plaintiff for the defendants at the material time, and to other relevant facts. As already mentioned, no sales were actually proceeding in respect of the Eaton or stadium sites, though preparatory work was being done, such as the collection of names of interested persons in the case of stadium, so that Martineau Lane, an industrial site, was the only one in respect of which the plaintiff was active in trying to find a purchaser. From the available information, which was considerable and not in dispute, I consider that a notice specifying weeks rather than months would have been an adequate period of time to determine this contract of agency. However, my conclusion as to this issue is of little practical importance in view of the course of events in 1973.

Mr Summers refused to negotiate a settlement of the dispute with the plaintiff unless the latter withdrew his threats and allegations. This was made clear in Mr Summers' letter of February 26 1973, and was repeated again and again in the correspondence between the parties and their respective solicitors. The plaintiff, far from withdrawing his threats, took action in respect of them. For example, in June 1973 he reported some of his complaints to his professional body, the Incorporated Society of Valuers and Auctioneers, with a serious innuendo as to the conduct of Mr Summers. Inevitably the Law Society, to whom the matter was referred, refused to take it up. Relations were finally terminated in July 1973 by a letter of July 23 from the defendants' solicitors. This letter, and the one to which it is a reply, speak for themselves. In my judgment, the defendants were abundantly justified in bringing the relationship of principal and agent to an end. Indeed, I think they would have been so justified at any time on and after January 24 1973. Mr Summers said that he regarded the plaintiff's demand for £35,000, coupled with the threats, as an outrage. In my view, Mr Summers' reaction was fully justified. The plaintiff thereafter had conducted a campaign of threats and allegations of bad faith against Mr Summers, and by his actions undermined the basis for the relationship of principal and agent. I think the defendants, and Mr Summers in particular, showed considerable patience and forbearance in giving the plaintiff ample opportunity to withdraw his threats and allegations following January 24 1973. However, the plaintiff made it clear beyond any doubt whatever that the relationship which had existed between them for so long, and which he had destroyed, could not be rebuilt. By his actions he repudiated the contract.

The plaintiff argued that by their agent, Mr Summers, the companies committed a breach going to the root of the contract by a unilateral variation of his right to commission at 2 per cent by a reduction to 1 per cent, and that the court is, therefore, not concerned with what happened thereafter, because the plaintiff was entitled to treat this purported reduction as a breach going to the root of the contract and as entitling him to damages in the action. This submission ignores my findings of fact and also, in my judgment, the principles of law applicable in this situation. In this context I was referred to the case of *The Mihalis Angelos* [1971] 1 QB 164 and particularly to the passage in the judgment of Edmund Davies LJ, as he then was, which is to be found at the bottom of p 202 and the top of p 203. In my judgment the principle there expressed is applicable here. I have found that at the first interview between Mr Summers and the plaintiff on January 24 1973 the plaintiff's hostile attitude arose from his learning from Mr Summers that the Swansea site was sold without any regard to him. I have also found that a reduction of his commission from 2 per cent to 1 per cent was put forward for consideration only in the first instance. I do not find that this amounted to a breach by the defendants of the contract of agency or one going to the root of the contract. In any event, the conduct of the plaintiff in this very first interview, and persisted in thereafter, entitled the defendants in my judgment to treat the plaintiff as having repudiated the contract, although it is true that the defendants did not at once break off all relations with the plaintiff but gave him the opportunity to mend the relationship. In the same way, the plaintiff did not treat the alleged breach by the defendants, that is to say, the alleged reduction of his commission to 1 per cent, as entitling him to treat the contract as at an end, but on the contrary made counter-proposals, as appears from the correspondence, on the basis of a continuing relationship.

I should add here that it was submitted to me that in any event I ought not to hold that a unilateral reduction of the plaintiff's commission from 2 per cent to 1 per cent was a breach which went to the root of the contract. Having regard to the evidence, which I accept, that other agents at the material time were prepared to act as agents for the defendants for 1 per cent commission, and that at this time it was not difficult to sell houses, it may well be that it was not such a breach as would have entitled the plaintiff to treat the contract as repudiated. However, the plaintiff did not treat the contract as such, and in the light of my findings it seems to me unnecessary for me to state my conclusion as to whether the proposed reduction of the rate of commission would have been a breach going to the root of the contract.

If, contrary to my findings, there was any breach of contract by the defendants or any of them such as was alleged by the plaintiff, he failed to prove that any damages flowed from such breach. On the contrary, the evidence satisfied me beyond any doubt that any damages which the plaintiff has suffered as the result of the termination of his contract of agency with the defendants are the result of his own conduct, which was of such a grave and serious nature that in my opinion the defendants had no alternative but to bring their business association with the plaintiff to an end. This was, in my judgment, the inevitable consequence of the conduct of the plaintiff. It may seem surprising that an association which had for so long been profitable to the plaintiff should have been so recklessly put at risk by him. The explanation, I think, lies in an obsessional sense of grievance against Mr Summers which the plaintiff has nursed for a long time, as appeared from the evidence. Although an intelligent and able person, the plaintiff appeared to be blinded by envy and dislike of Mr Summers, so that he persisted to the end in a course which was ruinous to his own interests, despite the advice of third parties such as Mr Hart, who had a regard for the plaintiff and for whom the plaintiff professed to have a regard. The plaintiff in 1969 had unsuccessfully attempted to bring pressure on Mr Summers to obtain a share in the equity of Norfolk Garden Estates, and throughout the years which followed it is clear that he was dissatisfied with his rewards. In 1973 he employed somewhat similar but rougher tactics to those employed in 1969 to obtain a larger share of the profits which the companies were making, and which he thought he deserved, having regard to his work over the years in helping to build up Mr Summers' position. These proceedings are a final attempt by the plaintiff to obtain what I am sure he has convinced himself is his due. The plaintiff has had every opportunity, in a hearing which has lasted many days, to put his case forward, and by evidence and argument to persuade me that he has a claim for substantial damages against the defendants. I hope I have indicated sufficiently in this judgment the reasons for my finding here that the plaintiff's claim fails, and that there must be judgment for the defendants.

The action was dismissed with costs.

HOUSE OF LORDS

May 12 1976

(Before Lord WILBERFORCE, Lord SALMON, Lord EDMUND-DAVIES, Lord FRASER OF TULLYBELTON and Lord RUSSELL OF KILLOWEN)

SORRELL AND ANOTHER v FINCH

Estates Gazette May 29 1976

(1976) 238 EG 639

Pre-contract deposit with estate agent—Vendor does not impliedly or ostensibly authorise estate agent to recieve such a deposit, whether as agent for the vendor, "stakeholder," or in any other capacity—A prospective purchaser who pays such a deposit can look only to the estate agent for its return, and loses it if the estate agent fails to repay it—Rayner v Paskell & Cann (1948) 152 Estates Gazette 270 cited and approved by the House

This was an appeal by Mr David John Finch, of Acton Lane, Watford, from a decision of the Court of Appeal affirming, by a majority, a judgment of Judge MacNair at Watford County Court awarding the respondents, Mr and Mrs Malcolm John Sorrell, of Exmouth Road, Watford, £550 on their claim for the return of a pre-contract deposit paid to an estate agent in 1971.

Mr David Kemp QC and Mr J H Vallat (instructed by Hart, Fortgang & Co, of Edgware) appeared for the appellant, and Mr C French QC and Mr G M Freeman (instructed by Enever, Freeman & Co, of Ruislip) represented the respondents.

LORD WILBERFORCE said that he had had the advantage of reading in advance the text of the speech to be delivered by Lord Edmund-Davies. He agreed with it and would allow the appeal.

LORD SALMON said that for the reasons given in the speech prepared by Lord Edmund-Davies, he too would allow the appeal.

LORD EDMUND-DAVIES: When an owner of property asks an estate agent to find a purchaser for his house, and the estate agent receives a deposit from a person who has it in mind to buy the house, but at a time when the negotiations are still "subject to contract," is the vendor liable to repay the depositor in the event of the estate agent misappropriating the sum deposited? This question, which has arisen with a certain measure of frequency in the courts over the last 30 years and has been variously answered, has at last reached your Lordships' House for determination. The particular facts giving rise to the question are always important, and I begin with those of the present case. For such purpose I cannot do better than to quote from the unreported judgment of Lord Denning MR:

> Mr Finch was the owner of 11 Cardiff Road, Watford. I will call him "the vendor." In 1971, he decided to sell it and buy another house. He called on Emberdene Estates. Mr Levy of that firm ("the estate agent") came round and said that the house would fetch about £5,250. Later he put it up to £5,500. About 15 young couples came to see the house. Five of them—Mr Smythe, Mr Maynard, Mr Bence, Mr Barry and Mr Farrell—all paid deposits. But the vendor did not know of this except that Mr Smythe and Mr Maynard both said that they had paid a deposit to Emberdene Estates. Eventually the vendor got tired of people coming round. He told the estate agent that if Mr Barry was prepared to buy, there was no reason for showing others round. Then there came Mr and Mrs Sorrell. I will call them "the purchasers." The vendor showed them round. They said they liked the house and would contact Emberdene Estates. On November 22 1971 they went to the offices of Emberdene Estates. They saw Mr Levy, the estate agent, who told them that the price was £5,500 and that a deposit of 10 per cent would be required, that is, £550. He said that the money would have to be transferred to the building society with whom he dealt, and that this would be enough to get a 90 per cent mortgage, ie £4,950. The purchasers said: "We were under the full impression at this stage that we were to have first option on the property." The

A purchasers raised the amount in two instalments. The first was for £100, on which the estate agent said: "Make it £112 to cover the surveyor's fees." So on November 22 1971 they gave him a cheque for £112, for which he gave them a receipt:

> November 22 1971.
>
> Received from Mr M J Sorrell £112, being preliminary deposit re 11 Cardiff Road, Watford. Price £5,500 freehold, subject to contract.

The estate agent produced a mortgage proposal form to the building society and the purchasers completed some of the details. On December 2 1971 the purchasers went to pay the balance of £450. The estate agent offered to lend them £200 towards it, "as B the building society likes to be sure it is your own money, and not borrowed money." This was another representation that the deposit was going to the building society. But the purchasers declined the loan. They said they had enough to pay the balance. So they drew a cheque for £450, and the estate agent gave them this receipt:

> December 2 1971.
>
> Received from Mr and Mrs Sorrell the sum of £450, being balance of 10 per cent deposit re 11 Cardiff Road, Watford, Herts. Price £5,500, subject to contract.

The purchasers filled in the forms and did all that was necessary for the building society, but later the agent disappeared. The purchasers saw the vendor (Mr Finch) and told him about the C deposit. The vendor said that it was not his problem. The purchasers said: "I thought we would get the deposit back from Levy (the agent) or Finch (the vendor), but primarily from Levy"—that is, primarily from the agent. Such are the facts.

In his clear and careful judgment the learned county court judge came to the conclusion that the decided cases, and in particular *Burt v Claude Cousins & Co* [1971] 2 QB 426, compelled him to hold in favour of the Sorrells, who had sued Mr Finch for the return of their £550. In the Court of Appeal Browne LJ and Sir John Pennycuick affirmed that judgment, but Lord Denning MR dissented, saying, "Any apparent authority is negatived by the estate agent's repre- D sentation that he was receiving it for the building society. So the vendor is not liable. On this ground I would allow the appeal."

The claim was originally based on contract alone, but by amendment was extended to assert the vendor's liability in tort for "the fraudulent misrepresentation of the said representative that he intended at the time the said deposit was paid over to hold the same as a deposit whereas at the material time he intended to convert the same to his own use" and for the fraudulent conversion which later followed. Dealing first with the claim in contract, it is common ground that liability in the vendor is dependent upon the estate agent E having acted on his behalf and as his agent in receiving the £550 claimed. The *fons et origo* of asserting that such was the case is *Ryan v Pilkington* [1959] 1 WLR 403, where, unknown to the prospective vendor, the estate agent obtained two sums by way of deposit from the prospective purchaser, receipting the first payment as "agent for" the vendors and the second simply as "agent." The sale having gone off, the depositor failed to recover from the estate agent the sums paid, and thereupon sued both him and the property-owner. The trial judge found that the estate agent had acted in the matter as agent for the owner and was therefore not liable, but that the owner himself was. Dismissing his appeal, F Hodson LJ cited an observation of Lord Russell of Killowen in *Luxor* (*Eastbourne*) *Ltd v Cooper* [1941] AC 108 at 124, that in circumstances such as those of the instant case "No obligation is imposed on the agent to do anything," and continued at p 409:

> It is *prima facie* in the interests of the vendor that the proposed purchaser should give an earnest of his expressed intention to become owner of the property in question. In my judgment, looking at the matter through the eyes of a reasonable prospective purchaser, the act of the agent in taking a deposit is within the scope of the apparent authority which is given to the agent by his principal.

G **Morris and Willmer LJJ concurred, holding that the estate agent had acted within his ostensible authority, and each attaching importance to the fact that to the plaintiff's knowledge, in return for the payments he had made, he was given acknowledgements that the sums had been received merely as "agent," a fact quite inconsistent with the depositor's later assertion that there was a personal liability in the agent to repay. As I later observed in *Maloney v Hardy & Moorshead* (1970, unreported), the case was therefore authority for saying that the estate agent had the vendor's implied authority to receive on her behalf a deposit from a potential purchaser. But in the light of later developments in H this branch of the law, it is indeed unfortunate that the Court of Appeal did not then have before it the decision of Lord Greene MR delivered 11 years earlier in *Rayner v Paskell & Cann* and reported only in (1948) 152 EG 270, which became more widely known only through the researches of his successor, Lord Denning MR, when printed in 1971 as an appendix to his judgment in *Burt v Claude Cousins & Co* (*ante*, at p 439). Dismissing an estate agent's appeal against a decision that he was obliged to repay a deposit paid to him "subject to contract," and for which he had given a receipt with no indication of the capacity in which he was acting, Lord Greene MR said:**

J

> **The respondent was entitled to expect the appellants to hold the money *as stakeholders* until he entered into a binding contract, and . . . as this was the usual practice in the case of estate agents who took deposits subject to contract, the appellants were stakeholders and had wrongfully parted with his money . . . it must be taken that the appellants received the money on the terms customary in the profession, whether the respondent knew of them or not. . . . On the findings of fact the judge could not do otherwise than hold that the appellants were under the duty of *stakeholders*.**

Another decision unfortunately not referred to when the Court of Appeal was considering *Ryan v Pilkington* was that K delivered some months earlier in the Court of Criminal Appeal by Lord Goddard CJ in *R v Pilkington* (1958) 42 Cr App R 233. There, in relation to circumstances indistinguishable from those of the instant case, he held "that the inference was irresistible that the appellant was receiving the money as stakeholder," and added:

> **When these persons paid their money to the appellant . . . they meant that, if no contract was concluded, they were to have their money back. Who else was to have it? The vendor would have no right to the money. . . . There was no engagement on either side and the money had to be refunded.**

There was, indeed, no novelty about these observations, L and the law had long been similar in Australia and New Zealand; see, for example, *Wells v Birtchnell* (1893) 19 VLR 473, *Whinfield v Lovell* [1926] VLR 185, *Fischer v Parry* [1963] VLR 97, *Egan v Ross* [1928] 29 NSWR 382, and *Richards v Hill* [1920] NZLR 724, where, holding liable the land agents who had parted with a deposit received by them, Salmond J said at p 728:

> **The defendants had no right so to part with the deposit. It was held by them for and on account of the plaintiff until and unless a complete contract was effected between her and the vendor. No such contract was ever effected, and the defendants must account for the deposit accordingly to the plaintiff.**

M

It occasions no surprise that these Commonwealth decisions were not cited to Sachs J (as he then was) before he delivered what was to prove his seminal judgment in *Goding v Frazer* [1967] 1 WLR 286, which has powerfully influenced this branch of the law ever since. Unfortunately, neither was *Rayner v Paskell* (*ante*), though on a later occasion the learned judge was to make his comments on it. In that case, like the present, there was no mention of the terms on which the deposit was paid "subject to contract." Observing (p 293G) that "there is no authority that binds a judge of first instance on the issue," and holding the house-owner liable in an action

claiming the sum deposited as money paid by the plaintiff to the defendant's use, the learned judge said (p 293C):

> To my mind, . . . in the absence of some express provisions, the vendor must remain responsible for its return whether or not it has remained throughout in the hands of the estate agent. So long, however, as it remains in the hands of the estate agent, the latter is, on the happening of any event entitling the purchaser to its return (and that includes a simple demand at any time before some contract is made affecting the allocation of the deposit), in law personally liable to return it. That is not to my mind because an estate agent technically agrees to be a stakeholder but because he must be deemed to be authorised by the vendor to pay it over immediately it became due to the purchaser and accordingly the deposit then becomes money had and received to the use of the purchaser. If and in so far as it is the practice of estate agents to retain in their hands these deposits, the practical result is the same as if they were stakeholders, the risk of whose solvency is borne by the vendor. In essence, however, my opinion is that estate agents take the deposit as agents of the vendor who throughout is liable for its return; with the purchaser having also in appropriate circumstances a remedy against any estate agent who, in fact, retains that deposit. (It is perhaps as well to add that if, contrary to my above view, estate agents do in law receive such deposits as stakeholders, I would have held that the vendor must all the same bear the risk of their insolvency.)

In the unreported Court of Appeal decision in *Maloney v Hardy & Moorshead* (February 12 1970) which turned on whether an estate agent had authority to engage a subagent, Russell LJ expressed misgivings over the use of the phrase "as stakeholder" in relation to the receipt of pre-contract deposits, adding: "Whether the money stays with the agent is entirely in the hands of the purchaser, and the vendor has no say in it or control over it. In such circumstances I would not hold the vendor responsible." I, too, pointed out:

> The essence of stakeholding in vendor and purchaser cases is that a *binding* contract of sale has been entered into, and the intending purchaser deposits with a third party a sum to be held pending completion; meanwhile the third party holding their deposit may part with it to neither contracting party without the consent of the other, and if competing demands arise he can interplead. But . . . in the present case the plaintiff was entitled to demand the return of his £450 and none could lawfully gainsay him. The [vendor] thus lacking any ability to prevent the estate agent from complying with such a demand, it seems difficult to see how the latter can be said ever to have been holding the money as stakeholders for anyone.

And so we come to *Burt v Claude Cousins & Co* [1971] 2 QB 426, where judicial battle was well and truly joined. There, as in the instant case, the prospective purchaser was given a simple acknowledgement of receipt of the deposit paid to the estate agent, who later defaulted. Dissenting from the conclusion of Sachs and Megaw LJJ that the property-owner was liable to make good that default, Lord Denning MR held that only when the estate agent, being duly authorised to do so, received a deposit "as agent for the vendor" is the latter liable; if he received it "as stakeholder," he is under a duty to hold in *in medio* pending the outcome of a future event.

> He does not hold it as agent for the vendor, nor as agent for the purchaser. He holds it as trustee for both to await the event: see *Skinner v Trustee of the Property of Reed (a Bankrupt)* [1967] Ch 1194, 1200 *per* Cross J. Until the event is known, it is his duty to keep it in his own hands; or to put in on deposit at the bank. If the purchaser should become entitled to the return of his deposit, he must sue the estate agent or solicitor for it: see *Eltham v Kingsman* (1818) 1 B & Ald 683; *Hampden v Walsh* (1876) 1 QBD 189. He cannot sue the vendor, because the vendor has never received it, or become entitled to receive it.

At p 436C the Master of the Rolls continued:

> I cannot believe that he receives it as agent for the vendor; for if that were so, the estate agent would be bound to pay it it over to the vendor forthwith; and the vendor alone would be answerable for its return. That cannot be right. Seeing that no contract has been made, the vendor is not entitled to a penny piece. If the estate agent should pay it over to the vendor, he does wrong; and if the vendor goes bankrupt, the estate agent is answerable for it . . . *Rayner v Paskell & Cann*. Seeing that the estate agent must not, before a contract is made, hand the deposit over to the vendor, what is he to do with it? Clearly he must keep it in his own hands until a contract is made, or the purchaser asks for it back. And what is he then but a stakeholder? . . . If no contract is made, the estate agent must return the deposit to the purchaser and can be sued if he does not. . . . The proper inference is that he receives the money as stakeholder and not as agent for the vendor. I cannot agree, therefore, with the decision of Sachs J in *Goding v Frazer*.

Sachs LJ, who adhered to the view he had expressed in the earlier case, summarised his "conclusion as to the terms upon which an estate agent is authorised by a vendor to receive and hold a deposit (subject, of course, to any contrary agreement between them)" in the following passage (p 447G), into which I have taken the liberty to insert letters to demarcate the different points involved:

> (*a*) He is authorised, or perhaps it would be better to say instructed, to hold the deposit in his own possession unless and until an event occurs upon which he is authorised to dispose of it.
>
> (*b*) In the event of the purchaser demanding its return before any contract is concluded, ie during the "pre-contract" period, he has to return the deposit to him.
>
> (*c*) In the event of a contract being concluded, it is to be disposed of in accordance with the terms of that contract, be they express or implied.
>
> (*d*) The instruction to hold the deposit in his, the estate agent's, possession is one which during the pre-contract period precludes him in the absence of the consent of both the depositor and the vendor from handing it over to the vendor or any person the latter may nominate; but, of course, entitles him to place it in his, the estate agent's, account at a bank of repute.
>
> (e) The above terms correspond with the practice of estate agents as set out in the evidence in the instant case and also in other cases which have come before the courts. Indeed, it has now become, to my mind, a matter of judicial knowledge that it is the practice of estate agents to receive deposits on the above-mentioned standard terms.

Of the foregoing conclusions, Mr French, counsel for the respondents, founded himself upon (*b*) and (*c*). But I understood him to say that he respectfully regarded (*a*) and (*d*) as too emphatically expressed, and that the outcome of the decided cases was not unequivocal enough for any taking of judicial notice to arise. Megaw LJ, concurring in the conclusion arrived at by Sachs LJ (452B), basically founded himself on the proposition that:

> In the absence of special circumstances justice requires that the prospective vendor rather than the prospective purchaser should bear the loss. It is the prospective vendor who has chosen the estate agent; who has clothed him with the capacity of agent; and who has enabled him to ask for and receive a deposit in connection with the business to which the agency relates.

Such agency arose, in the view of Megaw LJ, from the indisputable proposition that "every agent has implied authority to do whatever is necessary for, or ordinarily incidental to, the effective execution of his express authority in the usual way" (*Bowstead on Agency*, 13th ed art 28); and further, from the evidence of the purchaser's wife in that case "that she knew that it was not uncommon for estate agents to ask for and get deposits when, as here, the price was agreed; she agreed it was a usual thing." That decision was considered a few months later in *Barrington v Lee* [1972] 1 QB 326, where the estate agent expressly received a pre-contract deposit "as stakeholder." The facts differed in an important respect from the instant case, for the depositor had there obtained judgment against the estate agent, and the Court of Appeal unanimously held that it must follow that he could not also get judgment against the prospective vendor. But, even had that event not happened, Lord Denning MR would nevertheless have held in favour of the vendor, while Stephenson LJ and I took the view that in those circumstances we would have been obliged to follow *Burt v Claude Cousins & Co* (*ante*) and hold the vendor liable. I expressed the view that "were the judicial slate clean" I, in common with the Master of the Rolls, should have exculpated the vendor, whereas Stephenson LJ (344CD and 346FH) regarded the suggested relationship of estate agent and vendor as "a surprising sort of agency" in rendering both agent and principal liable in certain circumstances, but not unique on that account. And so, at long last, to the instant case, where Lord Denning MR summarised the effect of the foregoing decisions in the following propositions (with all of which he again expressed disagreement):

(i) If the agent signs the receipt expressly as agent for the vendor, the proposed purchaser can recover the deposit from the vendor: *Ryan v Pilkington* (1959).

(ii) If the agent signs the receipt in his own name without qualification, not saying that he signs it on behalf of the vendor or anyone else, the proposed purchaser can again recover the deposit from the vendor: *Burt v Cousins* (1971).

(iii) If the agent signs the receipt expressly as stakeholder, the proposed purchaser can still recover the deposit from the vendor: *Barrington v Lee* (1972).

Browne LJ while acknowledging the great force of the dissenting views of the Master of the Rolls held himself bound by the authorities to support propositions (ii) and (iii), and added that they also established that "(iv) when someone who wants to sell his house puts it into the hands of an estate agent, he gives the estate agent authority to receive a pre-contract deposit on his behalf." Browne LJ accordingly felt bound to hold that the vendor was liable. So also did Sir John Pennycuick, who echoed the hope expressed by Browne LJ that the whole topic would soon be considered by your Lordships' House.

The primary basis of the vendor's liability seemingly is that to engage an estate agent "to find a purchaser" is to confer upon him authority to receive money as agent for his principal (*per* Hodson LJ, *Ryan v Pilkington, ante,* at p 409); it seems, therefore, to be a case of actual, though implied, authority, save that at an earlier stage in his judgment Hodson LJ said (p 407), "the vital question in this case is whether he was acting within the scope of his *apparent* authority," and answered it by saying (p 409): "It is surely relevant to remember that ability to purchase is one of the qualities which one looks for in a purchaser. Accordingly, the requirement for the payment of a sum of money by way of deposit would seem to me to be permissible as evidence of some ability to purchase, some indication that the proposed purchaser was not a man of straw." But the validity of such a conclusion depends upon the extent to which the taking of a pre-contract deposit *on such terms as to make the prospective vendor liable therefor* can be regarded as reasonably incidental to the simple engaging of an estate agent to find a ready, willing and able purchaser. Some limits on his authority to commit the owner must, and have been, imposed; for example, the estate agent cannot enter into a binding contract of sale (*Keen v Mear* [1920] 2 Ch 574, 579) nor one of letting (*Navarro v Moregrand Ltd* [1951] WN 335). And, as I ventured to point out in *Barrington v Lee* (*ante*, p 339D), "It is too simplistic to say that, because it was the would-be vendor who appointed the estate agent in the first place and so put him in a position to receive a deposit, the justice of the case demands that the vendor should be held liable to recompense the purchaser if loss results," The same could be said of many persons undoubtedly appointed as agents, certain of whose activities nevertheless fall outside their authority; see the monitory observations of Lord Halsbury LC in *Farquharson Bros & Co v King & Co* [1902] AC 325 at 331-2. In such circumstances, as in the instant case, where the issue raised is which of two innocent parties is to suffer for the default of a third, the matter has to be determined in accordance with strict legal principle and with nothing else. As Dixon CJ said in *National Insurance Co of New Zealand v Espagne* [1961] 105 CLR 569:

Intuitive feelings for justice seem a poor substitute for a rule of law antecedently known, more particularly where all do not have the same intuitions.

For my part, I remain of the view I expressed in *Barrington v Lee* (*ante,* at 339E) that:

. . . the just solution of the problem of which of two innocent parties should suffer should depend very largely (and possibly conclusively) upon what rights could have been asserted by each of them in respect of the money in the agent's hands at all material times.

It being common ground that before a contract was concluded neither vendor nor estate agent could gainsay the purchaser's demand for its return, I fail to see that the justice of the case demands that the vendor, however personally innocent, should be held liable to repay the depositor in the event of the agent defaulting. It is not open to the prospective purchaser to deny knowledge of his unfettered legal right to get his money back at that stage, and if, with that actual or imputed knowledge, he chooses to pay a deposit and leave it in the estate agent's hands, while one must naturally have sympathy with him, such intuitions of justice as I possess do not demand that he should be recouped by a vendor who shares his innocence and differs from him only in engaging someone to find a purchaser for his house. In this I am glad to find myself in the company of Sir John Pennycuick, who in the present case said: "I do not myself find much force in the consideration of relative hardship."

It has to be said respectfully that the hitherto prevailing majority view as to the authority of an estate agent in the pre-contract stage is an extremely odd one. It involves the inference that he possesses the authority of the vendor to receive it as his agent; but although the deposit is received in that representative capacity, the recipient must nevertheless return it to the depositor at his request, and the principal has no control over it, save that, in the words of Sachs J in *Goding v Frazer* (*ante*, p 293A):

There seems to be nothing in law to preclude the vendor from making such bargain as he would wish with the estate agent as to how and by whom any deposit should be held, and in particular (if he so preferred) from arranging that the deposit be paid into the hands of his solicitor or of himself.

I know of no such right in the vendor nor of any support for it in the authorities, and in my judgment it is non-existent. It is right to add that the learned judge seemed to have later resiled from that view; see *Burt v Claude Cousins & Co* (*ante,* at 447G to 448B). It is the fact that in this alleged relationship of agency the vendor has no control over property alleged to have been received on his behalf, which makes it so unlikely and so wide a departure from the ordinary law (see *Edgell v Day* (1865) LR 1 CP 80) as to be unacceptable. Equally odd, with respect, is the view expressed by Sachs LJ in *Burt v Claude Cousins & Co* (at p 449) that "a claim manifestly does lie against the estate agent whatever the answer to the question as to status or capacity," whereas it is a truism that where an agent makes a contract solely in his capacity as agent he is not liable to the third party thereon, even if sums paid to him as agent by a third party still remain in his possession (*Ellis v Goulton* [1893] 1 QB 350 (CA) and *Bowstead on Agency,* 13th ed 373). For my part, the difficulty is not cleared up by adverting, as Megaw LJ did in *Burt v Claude Cousins & Co* (as p 455CG), to the fact that an agent may undertake personal liability to the third party in the course of making a contract on behalf of his principal, and that it would not be foreign to the concept of agency for the estate agent to have authority from the prospective vendor to take the deposit as his agent and to take it subject to the obligation that he must repay it to the depositor on demand. In none of the reported cases did the estate agent expressly enter into any such arrangement with the vendor, and I see no ground for implying one. On the contrary, the better view appears to me to be that his liability to account to the depositor rests solely upon an implied term of the transaction between him and the depositor and springs in no way, direct or indirect, from the relationship between him and the vendor.

In my respectful judgment, these cases have unfortunately been "off course" ever since the extempore judgments in *Ryan v Pilkington* (*ante*) were delivered, despite the repeated efforts of Lord Denning MR to put matters right. It may well be that the prevailing error sprang originally from a two-fold source which was itself attributable to confusing terminology. First, the styling of the intermediary as an "estate agent" may, however subconsciously, have induced the notion that whatever he did in relation to the vendor's property he did on his behalf; as my noble and learned friend, Lord Salmon, observed in the course of submissions, such confusion might never have arisen had the intermediary been styled a "realtor" or by some such other description. And secondly, the consistent assertion right back to *Rayner v Paskell & Cann* (*ante*) that he received a pre-contract deposit as a "stakeholder" may well have led to blurring the difference between his position *vis-à-vis* the depositor according as to whether the deposit was paid before or after the contract of sale had been concluded. Whether or not such factors contributed in some measure to the decision in *Ryan v Pilkington,* the critical comments of Mr David Kemp strike me as very cogent. He points out that the claim there not exceeding £200, no appeal lay on a question of fact, and the county court judge had found (p 404) that the deposit "was paid to the first defendant as agent for the second defendant and not as a stakeholder." The appeal on law must therefore have been based on the trial judge having misdirected himself or on the complete absence of evidence to support such a finding. But "Pilkington had gone into the witness box saying that he was at all times agent for a disclosed principal, Gem, and that the plaintiff ought to sue Gem and not him, Pilkington" (see Willmer LJ, p 413). The next witness called was none other than Gem himself, and he never denied having given authority to Pilkington to receive a deposit on his behalf, nor did he challenge Pilkington's authority when Ryan first informed him of the deposit he had paid. On the totality of the evidence it would indeed have been impossible to exculpate Gem, for there was the uncontradicted evidence of Pilkington of express authority; and Morris LJ stressed (p 412) that the case depended on its own facts. Hodson LJ made clear (pp 408-9) that no expert evidence was called concerning the practice of estate agents and that "the question of law whether the action of the agent here was within the scope of his authority is not directly covered by authority, and one has to apply first principles." In my respectful judgment, it is not in accordance with first principles to hold that the estate agent in such circumstances as the present was authorised to receive *on the vendor's behalf* a pre-contract deposit in the absence of express or impled authority so to do, and in neither way was such authority here given. Nor does the prospective vendor's knowledge that a deposit had been received of itself impose any liability upon him to repay it.

For the foregoing reasons, had the Court of Appeal not been bound by the earlier authorities to which I have referred, I hold that they should have allowed the vendor's appeal from the adverse judgment of the learned county court judge. This House being free to consider the matter afresh, I accordingly would reverse the majority decision of the Court of Appeal. It only remains for me to deal shortly with the further ground of claim, introduced by amendment. It will be recalled that this was based on the allegation that Levy fraudulently misrepresented to the plaintiffs that he intended to retain as a deposit the sum paid to him, whereas in reality he intended to convert it to his own use, and thereafter proceeded to do so. Assuming such allegations to be established, there would still be no liability in the defendant unless Levy was acting in the course of his employment by the defendant in making such a fraudulent misrepresentation, or his action was authorised or ratified by the defendant, or can be regarded as having been done within his actual or apparent authority; see *Bowstead* 13th ed 329 and the many cases there cited. Accordingly, holding as I do that Levy had no authority from the defendant to receive any deposit as his agent, and there being no suggestion of the latter's complicity in any fraudulent misrepresentation, it follows that nothing Levy said in order to induce the payment of that deposit can implicate the vendor. The claim under this alternative head must accordingly also fail. In thus holding that the appeal should be allowed, I must make clear that I share the deep misgivings of your Lordships that the innocent plaintiffs should, through absolutely no fault of theirs, find themselves in their present unfortunate plight. Any steps which can be taken by the appropriate professional body to prevent the recurrence of such cases should be promptly taken, for they are not uncommon, they arouse great concern, and they tend to give the whole profession a bad name, despite the high integrity of the vast majority of its members.

LORD FRASER OF TULLYBELTON said that he had had the advantage of reading in draft the speeches of Lord Edmund-Davies and Lord Russell of Killowen. He agreed with them, and he too would allow the appeal.

LORD RUSSELL OF KILLOWEN: This appeal raises a question of general importance in cases in which A, wishing to sell his house, goes to an estate agent to find a possible purchaser; the estate agent obtains in the course of negotiations a deposit or deposits from a possible purchaser or possible purchasers; negotiation does not reach the stage of contract; and the deposit or deposits cannot be recovered from the estate agent, who has spent them for his own purposes and is insolvent or has otherwise made off with them. Is the vendor then to be held liable to repay the depositors? In the Court of Appeal counsel for the vendor found himself obliged by previous decisions of that court, unpalatable as they were to Lord Denning MR, to accept the following propositions:

(i) If the agent signs the receipt expressly as agent for the vendor, the proposed purchaser can recover the deposit from the vendor: *Ryan v Pilkington* [1959] 1 WLR 403.

(ii) If the agent signs the receipt in his own name without qualification, not saying that he signs it on behalf of vendor or anyone else, the proposed purchaser can recover the deposit from the vendor: *Burt v Claude Cousins & Co* [1971] 2 QB 427.

(iii) If the agent signs the receipt expressly as stakeholder, the proposed purchaser can still recover the deposit from the vendor: *Barrington v Lee* [1972] 1 QB 326.

Counsel for the vendor therefore had to rely upon a particular feature of the present case as a distinction. Lord Denning MR upheld that distinction but his brethren found themselves unable so to do. In your Lordships' House the previous decisions are of course open to review, and of them consideration is first required, after a summary of the facts in the instant case.

The estate agent was, so far as is known, a man with no professional qualifications, and a member of no professional organisation. He was, unknown to either the vendor appellant or to any of the possible purchasers concerned, a dishonest rogue and an undischarged bankrupt. He traded under the fairly high-sounding style of Emberdene Estates. The vendor, having decided to sell his house, 11 Cardiff Road, Watford, called on the estate agent in order to put his house on the market. The estate agent took particulars of the house for this purpose, the idea being that it should be offered for negotiation at £5,500. Nothing was said about any deposits being taken by the estate agent, nor, naturally, about the capacity in which he was to receive such deposits. In fact no mention was made of commission on introduction of a willing purchaser. As a result of the activities of the estate agent several possible purchasers came to see the house. Five of them (apart from the respondents) paid deposits to the estate agent, presumably the receipts given by the estate agent being in like form to those given to the respondents. When the respondents (who may conveniently be called the purchasers) came to see the property, the vendor knew that two possible purchasers had paid a deposit to the estate agent, but nothing was said between vendor and purchasers as to this. Subsequently the purchasers returned to the estate agent, who told them that a deposit of 10 per cent of the purchase price of £5,500 would be required. They paid this in two sums to the estate agent by cheque made out to the estate agent in his trade name. He gave them two receipts made out on Emberdene Estates' writing-paper, signed with no definition of the capacity in which he took the money: that is to say, he did not in terms say either that he received as the agent of the vendor or that he received as stakeholder. In each case the house and price were identified, and it stated that the receipt was for a sum being preliminary deposit or balance of deposit re 11 Cardiff Road, but no reference was made to any vendor, who might for all that appeared on the document have been trustees or a mortgagee. Both documents contained the words " subject to contract." Accordingly we have a situation in which (i) the vendor did not expressly authorise the estate agent to receive as agent for the vendor pre-contract deposits from possible purchasers, and (ii) the estate agent did not purport to receive any deposit as agent for whoever might be the vendor of the property.

An estate agent, despite the style, is an independent person, engaged ordinarily on a commission basis, to find and introduce a willing purchaser. He is not the agent of the vendor to contract on his behalf; his actions are attributable to the vendor only in a strictly limited sense, as for example the making of representations as to the condition of the property. In my opinion an estate agent has neither actual (impled) nor ostensible (apparent) authority to ask for or receive a pre-contract deposit as agent for the vendor. It is true that it has become quite common for estate agents to receive pre-contract deposits, and that in one sense this earnest of genuine interest is beneficial to the vendor; but it is also beneficial to the purchaser who hopes thereby to get his foot in the door, and indeed, in time of acute housing shortage it is I believe not unknown for a would-be purchaser to press a deposit on the estate agent for that very reason. In my opinion the decision in *Ryan v Pilkington* [1959] 1 WLR 403 (CA), in so far as it was based upon the footing that an estate agent has implied actual authority or ostensible apparent authority to receive a pre-contract deposit as agent for the vendor, is erroneous. The second of the three cases in the Court of Appeal which counsel for the vendor felt obliged to accept below was *Burt v Claude Cousins & Co* [1971] 2 QB 426. The decision in that case (where the estate agent did not purport to receive the pre-contract deposit as agent for the vendor) was based upon in the decision in *Ryan v Pilkington* that the estate agent had either implied or ostensible authority to receive a pre-contract deposit as agent for the vendor, and accordingly must also in my opinion be regarded as erroneous. I find it impossible to accept a theory of authority from a vendor to the estate agent to accept a deposit *as agent for the vendor* on terms that so long as all remains in negotiation the estate agent must on no account pay the deposit to the vendor or to the vendor's order, but on the contrary must on demand of the purchaser pay it back to the purchaser without reference to the vendor: see per Sachs LJ [1971] 2 QB at pp 447-448. That appears to me with all respect to be juridically unacceptable. The third case mentioned above is *Barrington v Lee* [1972] 1 QB 326, in which the estate agents had received the deposit " as stakeholders" The majority felt themselves constrained by the decision in *Ryan v Pilkington* and the reasoning of the majority in *Burt v Claude Cousins & Co* to hold that the vendor was *prima facie* liable to refund the deposit to the purchaser, on the footing that the estate agents had implied authority from the vendor: though since the purchaser had sued the estate agents to judgment, that operated as a bar on the facts of the case to an action against the vendor.

The subject matter of these decisions is of course at large in this House. In my opinion, (i) in cases such as the present it is wrong to say that by the engagement of an estate agent there is conferred upon the estate agent either implied or ostensible authority to receive a deposit from a would-be purchaser as agent of the vendor, and (ii) if the estate agent receives a deposit either without other definition of his character or in terms " as stakeholder," and the estate agent goes bankrupt or otherwise defaults, the vendor is not liable to the purchaser. (As to " stakeholder " I adhere to the view expressed by me obiter in *Maloney v Hardy & Moorshead* noted at [1971] 2 QB at p 442.) The crucial point in my opinion is that at all times until contract the purchaser is the only person with any claim or right to the return of the deposit moneys, and his right is a right on demand: whereas the vendor has no such claim or right and no control over the deposit moneys. It has been said that " in justice " the one of two innocent people to suffer should be the vendor who chose the estate agent. But this seems to me too loose an approach to the problem, which should be solved by analysis of legal rights and relationships. A would-be purchaser is not obliged to pay a pre-contract deposit, and can in any event require that it be paid into joint names. The vendor, on the other hand, if the line of cases mentioned is upheld, may (as here) find himself liable to repay the whole string of deposits —worth perhaps more than the house—without being able to avoid or control the situation: for the suggestion that he might insist that the cyclostyled particulars should bear in bold lettering the warning that any pre-contract deposits be not paid to the estate agent's sole name is hardly practical. My opinion is not of course intended to cast any doubt on the liability of the vendor for the default of a stakeholder auctioneer: in such cases the deposit is paid to the auctioneer on contract, and the purchaser is required by the vendor to make the payment: he has no option. As to the alternative ground based upon fraud, I have nothing to add to the reasoning of my noble and learned friend Lord Edmund-Davies. In the circumstances I need not enlarge upon the particular feature of this case which Lord Denning MR considered to be justification for distinguishing the previous decisions. For these reasons, in my opinion this appeal should be allowed.

COURT OF APPEAL (CRIMINAL DIVISION)

March 26 1976

(Before Lord Justice ROSKILL, Lord Justice LAWTON and Mr Justice PAIN)

R v GREENFIELD

Estates Gazette June 12 1976

(1976) 238 EG 795

Theft of deposits by estate agent—Payment of deposits into overdrawn business account—Fifteen months a light sentence—" loud and clear " warning by court

This was an appeal by Keith Howard Greenfield, of Ty Gwyn, Beddau, Llantrisant, Cardiff, against nine sentences of 15 months' imprisonment (to run concurrently) imposed on him at Cardiff Crown Court on October 2 1975 upon his plea of guilty to nine counts of theft.

Mr J Griffith Williams (instructed by the registrar of criminal appeals) appeared for the appellant. The Crown was not represented.

Giving the judgment of the court, PAIN J said that the

A applicant pleaded guilty on October 1 1975 at Cardiff Crown Court to nine counts of theft. He was sentenced to 15 months' imprisonment concurrently on each count, making a 15-month sentence in all. By leave of the single judge, he now appealed against sentence. He started business as an estate agent at Barry in Wales in 1974. At that time he was 23 years old, and his only training had been as an assistant in another estate agent's business. When he set up on his own, he had personal and business bank accounts, with overdraft facilities on each for £1,500. As he received deposits for houses he paid them into his business account, not keeping them separate from other moneys. In doing that, he was doing B nothing unlawful; indeed one cause of his downfall might be that it was not unlawful to do this very thing. The court regarded the case as a classic example of how undesirable it was for young men with no proper training not only to be able to set themselves up in business as estate agents, but to be in a position to accept large sums of money from the public. The only course the prudent and honest estate agent could take to protect himself and his clients was to put deposits into a separate account which had to remain untouched, come what may. In the present case the appellant drew on sums from his business account, which included clients' deposits, for the general purpose of running his C business. An overdraft was run up. It reached a stage where D any cheque issued on the business account would almost certainly be returned unless there were sufficient funds to meet it. Despite that, the appellant continued paying deposits into that account and using it for the general purposes of the business. In this way he was stealing deposits paid by clients.

So far as estate agency frauds were concerned, 15 months was a fairly light sentence. No doubt the trial judge had taken pity on Greenfield because of his youth and inexperience. The present court had given every weight to the plea put forward for Greenfield and to letters received from members of his family. One difficulty was that there still E seemed to remain a doubt in the family as to whether what the appellant did was in any way dishonest. Let it be said loud and clear that any use by an estate agent, for his own purposes, of deposits placed with him by clients was dishonest. It was quite purposeless for him to say he hoped to be able to repay the money in one way or another. He was in a position of trust. It was not his money, and he should not make use of it for his own purposes. It was because that principle was not clearly understood that the appellant fell into the grievous position he was in. Having paid due attention to all the circumstances of the case, the court did not regard the sentence as one which could be altered. F

FEES AND COMMISSIONS

A

QUEEN'S BENCH DIVISION

March 24 1975

(Before Mr Justice O'CONNOR)

ROSENTHAL v ALLAN

Estates Gazette June 12 1976

(1975) 238 EG 797

Estate agents' commission—No express instructions to sell house—Would-be agent has no quantum meruit claim based on continued efforts to sell property—No implication of an agreement from vendor's supposed "acceptance" of continued intervention by insurance broker

B

This was a claim by Mr and Mrs Herman Rosenthal, of Chanctonbury Way, Woodside Park, London N12, against Mrs K M Allan, of Wyre Grove, Edgware, London, for commission of £1,093 on the sale of Chatsworth, Austell Gardens, Mill Hill, London NW7, in 1972.

Mr P H Latham (instructed by John Wood & Co) appeared for the plaintiffs, and Mr I McCulloch (instructed by A E Hamlin & Co) represented the defendant.

Giving judgment, O'CONNOR J said: Early in 1971 the Allan family were living in a house named Chatsworth, Austell Gardens, Mill Hill, and the freehold was in Mrs Allan's name. In the summer of 1971 the marriage broke down and Mrs Allan put her affairs in the hands of her solicitor, Mr Cowan. In September of 1971 Mr Cowan advised her that the house should be put up for sale, and he instructed Hamptons, well-known estate agents, to act as agents to find a purchaser for the house. Among other things they advertised the house in *The Sunday Times* in October 1971, and among the people who saw the advertisement and did something about it were a Mr and Mrs Milton. Mr Milton is a baker in a substantial way of business, and he and his wife were looking for a new house in the Hampstead or Totteridge or Mill Hill area. Along with particulars of other houses, the particulars of this house were sent to them by Hamptons in October 1971. Apparently they drove past it, did not like the look of it and forgot about it. Hamptons rapidly busied themselves with finding a purchaser for the house. At some stage during the winter of 1971-72 Mr Allan through his solicitors put a caution on the title, and negotiations had to come to a halt. It is quite plain that the matrimonial proceedings were of considerable complexity and very anxious-making for Mrs Allan. She had got children whose schooling was in dispute, and I need not say more than that I am quite satisfied that the issues of children and money were occupying her time to a degree which is understandable, and she relied entirely on Mr Cowan with whom she was in daily communication, including the weekends, about her affairs. At some stage it must have become plain that the hearing of the divorce proceedings which were to be contested had been fixed for, I think, May 3 1972. It is obvious that in preparing for the hearing of those proceedings Mrs Allan would be subjected to great personal pressures during the run-up period to the hearing. The Miltons in the meantime had been negotiating for one or two houses, and this was a time when there was a seller's market and if no contract was signed the would-be buyers found that the offers that they had made subject to contract had been topped by somebody else. Nevertheless, as I have said, the Miltons were negotiating for one or two houses which they liked. As far as Mrs Allan's house was concerned, Hamptons had produced some people called Lavey who were interested in buying it and had made an offer subject to contract which was thought to be a reasonably good one. But as I have said, nothing could be done about it until the dispute between Mr and Mrs Allan had been resolved.

The plaintiffs in this case claim that they are entitled to be paid the estate agent's commission normally paid by a vendor by Mrs Allan, because they say they were engaged as her agents and effectively introduced the purchaser who bought the house. Mr Rosenthal is an insurance broker, having his own business in central London, and his wife helps him in the business. They have a small office at home. They both told me, and I accept their evidence about it, that from time to time their business put them in touch with people who were either wishing to buy or to sell houses, and from time to time they went into what they called the estate agent's business and acted as agent and collected commission. They said that they had done this over the years a few times, about 10 times. There is no reason why they should not have done so, and I accept their evidence that they have done so. They had known Mr Allan's family for a number of years, because Mr Rosenthal had been his insurance broker. I have not been told what Mr Allan's business was, but there was that relationship. Equally Mr Rosenthal had been insurance broker to Mr Milton and his businesses, and there was evidence that Mr Milton owned some 14 patisserie shops in various parts of London. The Rosenthals learned from Mr Allan that the matrimonial home was to be sold. Mrs Rosenthal had never been to it; Mr Rosenthal had. They came to the conclusion that the house, which they knew to be a new one, might be just the thing which their clients the Miltons were looking for. They decided that it might be a good idea to see whether a purchase by the Miltons could be effected. It is not clear from their evidence whether they had any discussions with the Miltons about it beforehand, but they certainly decided that they would see whether they could put the two parties together. The first thing to do was to go to see Mrs Allan and the house. They achieved that some time in April. According to them it was on April 10; according to Mrs Allan it was much nearer her divorce, she thought it was the Sunday before the divorce, that would be April 30. It does not matter, because there is no dispute that a meeting did take place. It was not by any prior arrangement. Mr and Mrs Rosenthal got into their motor car, drove over to Mrs Allan's house and found her at home. She was very suspicious of their arrival, because she knew that Mr Rosenthal was her husband's insurance broker, and she thought that he was introduced as a spy into the camp preparatory to the divorce suit which was about to be heard.

There is a dispute of evidence as to what took place at that meeting. Ultimately they all went in the house, and according to Mrs Rosenthal she told Mrs Allan, "I think we have got a purchaser for the house who would be very interested in it." After she had seen round the house she was enthusiastic, thinking it was just the thing that the Miltons were looking for. According to her, in the course of conversation she said to Mrs Allan, "If we find a purchaser who will buy the house, will you pay us the ordinary estate agent's fee?" According to her, Mrs Allan said, "Yes." Mr Rosenthal's account of the crucial conversation was a bit different. According to him, it was he who did the talking. For my part, having heard both Mr and Mrs Rosenthal, the probability is that it was Mrs Rosenthal who did the talking. However, Mr Rosenthal's evidence about this was that he told Mrs Allan that he had a potential purchaser and asked whether he could bring him to see the house; that he said in the course of the conversation, "In the event of my clients being willing and able to purchase the house, I would look to you for commission," and that Mrs Allan said, "Certainly." He said, "I think I explained I would require the normal estate agent's commission." So it is the plaintiffs' case that there was an express agreement made on that visit, whether it be April 10 or April 30 matters not, whereby Mrs Allan agreed that they should act as her agents to try to sell the house for her. Mrs Allan says it did not happen that way at all. She denies that there was an express agreement to pay commission or that commission was mentioned. According to her, the Rosenthals came, and they were enthusiastic on behalf of their friends and clients, the Miltons; they wanted to see the house, and she told them that she could do nothing without referring to Mr Cowan. She allowed Mrs Rosenthal to look at the house, and according to her there was no agreement on the payment of commission or that the Rosenthals should act as her agents.

As to precisely what took place at that meeting, the burden of proving an agreement is on the plaintiffs. In my judgment, for reasons which will appear quite soon, I have come to the clear conclusion that they have not discharged that burden. Mrs Allan was in no condition to make an agreement with anybody, and I am quite satisfied that she would never have dreamed of hiring estate agents for the sale of a house at this stage without conferring with Mr Cowan. If the question of selling the house was discussed in any commercial terms, I am satisfied that all that she said to the Rosenthals was, "Well, you will have to get in touch with Mr Cowan," And indeed, that is their evidence. They accept that at that meeting she said, "You will have to get in touch with Mr Cowan." Now the Rosenthals' account of the matter, given particularly by Mrs Rosenthal, is that an attempt was made—and I think this was agreed by all parties to be on a Sunday afternoon—to get in touch with Mr Cowan without success, and according to Mrs Rosenthal, Mrs Allan was then persuaded to ring up Mr Milton then and there and make an appointment to view the house. Mrs Allan stout-heartedly denies that that occurred, and I prefer her evidence on that topic. I am satisfied that all that happened at this meeting was that Mrs Allan told the Rosenthals that if they wanted to busy themselves with introducing the Miltons to the house they should get in touch with Mr Cowan. She may or may not have said "Hamptons." If that be right, then there was no express agreement made at that first meeting between the parties that the Rosenthals should act as the defendant's agents in the sale of the house. I am fortified in coming to that conclusion because it is common ground between Mr and Mrs Rosenthal and Mrs Allan that at that meeting nobody said, "What are you asking for the house? What is the price?" It is quite plain that the sort of agency which Mr Rosenthal speaks to cannot be concluded, namely to introduce a ready and willing purchaser to the house, until you know what it is the purchaser has got to pay. It is common ground that they never asked and did not know what the asking price of the house was. I have no doubt that being shrewd business people they could put it in a bracket, but there it is: at that meeting they did not ask.

To my mind it is an absolutely crucial factor in setting up the business relationship of a vendor and agent that either the vendor should say, "I have been advised that the house is worth so much and it is that sort of figure that I am looking for or better," or alternatively should say to somebody who is a would-be estate agent, "Will you please advise me as to what I ought to ask for the house?"; and it is not suggested by the Rosenthals that they fitted into either of those positions. On their own evidence it is quite plain that no concluded agreement was come to. I am quite sure that Mrs Rosenthal in particular has now convinced herself, as a result of what happened, that she did mention commission and that Mrs Allan agreed to pay it. I can well understand how she has come to that conviction. It was a result of what took place subsequently. But I am quite satisfied that it did not happen that way, and that there was no agreement by Mrs Allan to pay estate agent's commission or any commission, nor was there any agreement by her to engage the services of the Rosenthals as her agents to sell her house at that time. She looked upon them, and I think rightly, as people who were interested on behalf of their client, Mr Milton, and if she thought about it at all, which is unlikely, she assumed that any financial reward that might come to the Rosenthals would result from their relationship with the Miltons. She said in her evidence that she realised that the Rosenthals were shrewd business people and they were not in this for nothing, and she assumed that they were going to make some money out of insurances which they would place for the Miltons if they got the house.

Now that being the position, the claim based on an express agreement must fail. It is said in the alternative that the court ought to imply a binding agreement at a subsequent stage between the parties as a result of which the court ought to order the defendant to pay a remuneration to the Rosenthals on what is known as a *quantum meruit*. The case is put in this way. It is said that after the initial meeting it must have become apparent to Mrs Allan that the Rosenthals would busy themselves mightily in trying to effect a sale of her house to the Miltons and a purchase by the Miltons of her house. At some stage the court ought to imply an agreement on her part in continuing to accept their intervention as imposing a duty upon her to pay them some remuneration. Well, what happened? There has been a good deal of dispute about the form the negotiations took at the end of April and in particular during the early part of May. What is clear is that on May 17 Mr Milton signed a contract to buy the house for £76,250. At a later stage that price was adjusted because there was a dispute over some of the boundaries. That fact is without doubt clear. How did it come about? The Rosenthals having seen the house undoubtedly persuaded the Miltons that it was just the house for them, and it became necessary to get them to see it. Mr Rosenthal's evidence about it is that he was told he had to arrange this through Mr Cowan, and that he did telephone to Mr Cowan and ask him whether it would be all right to view the house. Mr Cowan was not overenthusiastic, because the state of affairs at this stage was that the divorce suit had been settled, and it had been a term of the settlement that the caution should be lifted and that Mrs Allan should effect a sale of the house within three months, and that the money should be applied in certain agreed ways under the settlement, one of which was that part of it be placed in trust for the children of the marriage. It was therefore well appreciated by Mr Cowan that it was his duty to get the highest price that he could for the house.

Mr Cowan has got no recollection of this telephone call. I am satisfied that there was such a call to him. I am not

A in the least surprised that there is no note of it. He busied himself with Mrs Allan's affairs to a very great degree, and I am sure that he acted perfectly properly and successfully. I am satisfied that there was a call to him in which he agreed that the Rosenthals could take the Miltons to see the house. At that stage the Laveys were making further offers, and they were standing at about £65,000. It was Mr Rosenthal's evidence that he asked Mr Cowan how much the house was going for, what was the asking price, and was told £70,000. I am satisfied that his recollection about that conversation is wrong. I do not think he was told the price; I do not think he asked it. I believe that at that stage, whether it was Mr B or Mrs Rosenthal who made the phone call (and it matters not), they were only concerned in getting an appointment for their clients Mr and Mrs Milton to view the house. In due course that was arranged, and on May 14, which was a Sunday, Mr and Mrs Rosenthal and their son and Mr and Mrs Milton went to see the house by appointment. The fact that I am satisfied Mr Rosenthal had not inquired the price emerges from the evidence of Mr Milton. After he and his wife had looked round the house—I may say that they recognised from the outset that it was a house on which they had some particulars which had long since been thrown away and forgotten back in October of 1971; they naturally C kept quiet about that, and nobody is to criticise them for so doing—having looked over the house, Mrs Milton fell for it and was extremely enthusiastic and said it was just the house that she wanted. Mr Milton gave evidence to say that he said to Mrs Allan, "Well, how much do you want for the house?," and got the answer, "£70,000."

Now I do not believe, if the Rosenthals had known that that was the asking price, that it is credible that the Miltons would not have been told so by them. Mr Milton's evidence is clear that he did not know what was being asked until he asked Mrs Allan, and she agreed that he had asked her and D that she had said £70,000, and I accept that evidence about it, and it throws a lot of light on the true nature of the relationship in this case, because it follows that as at that meeting when £70,000 was asked, Mr Milton said on the spot, "I will pay it." Of course he was told, "You will have to go and see my solicitor Mr Cowan on the Monday morning." It is quite apparent, if that situation is accepted, that Mr and Mrs Rosenthal, who had been busying themselves about this matter, had been doing so, if anything, as agents for the purchaser, for the Miltons. The evidence, indeed, of what took place on May 14 confirms that, because after Mrs Allan had said, "I want £70,000 for the house," according E to her (and I accept her evidence about it) Mr Milton and Mr Rosenthal conferred about it. They were talking about mortgages and what money could be raised on it and so forth. I am satisfied that that took place, and it all points to the true position that the Rosenthals regarded themselves as acting for the Miltons and not for Mrs Allan, because if they had been really acting for Mrs Allan it would be an absolute prerequisite that they should find out what sum of money was being asked for the house before they went round looking for purchasers, and they did not do so. The true relationship at that stage was that they were advising Mr and Mrs Milton.

F The next two days were extremely busy, because I may say that on the periphery of the case Mrs Milton had telephoned Hamptons. She had done that after Mrs Rosenthal had learnt either from Mrs Allan or Mr Cowan, it does not matter, that Hamptons had got the agency for the house. She had communicated with Hamptons, spoken to Mr Sharpe, said that she had got a possible purchaser for this property, but she was not prepared to reveal the name of the purchaser, and asked if Hamptons would pay a split commission. That they refused to do. That again points to the fact that she did not think that she had got a direct agreement with Mrs Allan for the full commission, because there would be no point in asking Hamptons to split a commission if she had G got a firm agreement with the vendor to pay the full commission if they produced a purchaser, and there would be no need for that conversation which admittedly took place. Mrs Milton telephoned Mr Sharpe and she was told the house was sold, because indeed a firm offer from the Laveys had been made through Hamptons at £65,000 or thereabouts, and that happened apparently on May 8. Mrs Milton, who had not seen Chatsworth at that stage, said, "Well it is only a standby in case the negotiations for the house in Totteridge for which we are bargaining fall through," and she was not overinterested in it. As far as Hamptons were concerned, H nothing more happened. On Monday May 15 Mr Milton's solicitor went along to see Mr Cowan and to try to pay him £7,000 deposit. Mr Cowan refused it; he said no, that the house would have to go to the highest bidder. There was a firm offer from other would-be purchasers, and he would have to inquire whether they were prepared to pay more than £70,000 before he could advise his client to sign a contract. During Monday and Tuesday the house went up. The Laveys made an increased offer. Mr Milton made an increased offer, and Mr Cowan then came to the sensible conclusion of saying: "Well, I have got two people who are ready and willing purchasers, both with the money, both J bidding for the house. The sensible way of resolving this is for each of them to put a sealed bid in an envelope and the one which is greatest wins." That was to take place on Wednesday morning, May 17.

According to Mrs Rosenthal, at some stage on the Monday or the Tuesday, it matters not, Mrs Allan telephoned her and said to her effectively: "If Mr Milton will put an extra £1,250 in his envelope, he will have the house. Don't tell Mr Cowan I am taking sides." I am quite satisfied that this conversation in this shape never took place at all. Whether or not there was some conversation between Mrs Rosenthal and Mrs Allan about the sale of the house, I am quite K satisfied that that was not said to Mrs Rosenthal. I do not know where she got it from. It may be that it was her brain-child to say, "The way in which Mr Milton will win, because he has already offered £75,000 or even £76,000, will be just to top it up a little bit to see if we can beat the Laveys." I am satisfied that Mrs Allan did not say that. It is really not credible that she would say, "Don't tell Mr Cowan about it," because her view had been (and I accept her evidence about this) that she had been negotiating with the Laveys for a long period and that she was anxious for them to have the house, and that she had been told by Mr Cowan that the house must go to the highest bidder. If the L Miltons turned out to be the highest, they were to have it. There was no earthly reason for her to take sides in the matter at that stage, and I do not believe she did. Therefore, I do not accept the evidence that she intervened in that fashion. Once that finding has been made, then the basic ground on which an agreement can be implied that she agreed to accept the services of the Rosenthals as her agents for the sale of her house disappears. They are left where they always were, as agents for Mr and Mrs Milton. The fact is that Mrs Milton said that she received a phone call from Mrs Rosenthal saying that she had been told this by M Mrs Allan. I have no reason to doubt that there was such a conversation. It may be that it was just to provoke the Miltons into putting up a little bit more, because I am quite satisfied that the Rosenthals wanted the Miltons to have the house. Mrs Milton was desperately anxious to have it, and it may be that that is how it came about. I am quite satisfied that nothing took place at that stage which altered the relationship between the Rosenthals and Mrs Allan so as to make them her agents for the sale of a house requiring her to remunerate them.

Mr Milton put his bid in his envelope. The Laveys' representative arrived, and he tried to impose a further term on

Mr Cowan, saying: "What happens if the bids are level? Will my man have it?" Mr Cowan would not agree, whereupon Mr Lavey's representative walked out and his envelope never was opened. That left the Miltons with the field clear. £76,250 was opened and accepted and the contract signed. On May 18 Hamptons sent in a bill for commission for selling the house. Mr Cowan obviously told Mrs Allan about it, to which she said: "Good gracious me, they have not sold the house. The Rosenthals introduced the Miltons." The correspondence shows that there was then a dispute, but at that stage it is noticeable that Mrs Rosenthal does not suggest either orally or in writing that she ever said to Mr Cowan: "We are entitled to the commission because Mrs Allan agreed that we should act as her agents for the normal commission. We have done so, we have sold the house, we have succeeded in getting £11,000 more than Hamptons were trying to get rid of it for, and we are entitled to the full commission." What she said is that she had a conversation in which there was a consideration as to whether she should be paid half the commission, and that Mr Cowan was saying, "Yes, that seems fair," as I have no doubt from the correspondence, because he replied to Hamptons having received their letter asking for £1,200. I will say no more about the question as to whether they were entitled to any commission over this transaction, about which, as I have said, they had done nothing. What Mr Cowan wrote to them was: "Thank you for your letter. Upon referring your account to Mrs Allan I am instructed that Mr Milton who has agreed to purchase the property was not introduced through your firm, that you did not negotiate with him, and although your firm's account refers to discussions with a Mr and Mrs Milton, would you please let me know if this is the case, and if so upon what basis you consider that commission is due to your firm. In fact a claim for the commission has been made to Mrs Allan from the people who did introduce Mr Milton," and that was the Rosenthals. In the end, for whatever reason, the house ultimately went to completion on October 11 1972, and in due course, for whatever reason, Mr Cowan decided that it was Hamptons who should be paid commission on the sale and they were to be paid.

Now what about the Rosenthals? They had written on this matter a couple of times. In June Mrs Rosenthal wrote to Mr Cowan: "Referring to our conversation over the phone of some weeks ago when it was agreed that you would write to me confirming the position following my introduction of the purchaser of the above property, the contracts have now been exchanged, and I should be obliged to have your early reply regarding the commission due to me for this transaction." Mr Cowan replied: "Thank you for your letter. While it is not admitted there is any commission due to you from Mrs Allan, we are pursuing the matter with Hamptons and will write to you again in due course." That is June 28. Nothing happened. There was some question during the summer as to whether the deal was going to go on, because there was some difficulty about making title to parts of the property, and indeed, as I have already said, an adjustment in the purchase price was ultimately negotiated and agreed. In September Mrs Rosenthal wrote to Mr Cowan: "Thank you for your letter of June 28. I have been informed by both vendor and purchaser that completion will take place on September 29, and I am looking forward to hearing favourably from you in early course regarding the commission due for effectively transacting the sale of the above property." Mr Cowan replied a month later: "We thank you for your letter of 14th. As we have stated previously in correspondence we do not agree that there is any commission due to you on the sale of Chatsworth."

When that letter was received, the Rosenthals put the matter in the hands of their solicitors, Alexander Fine & Co. The partner dealing with it at the time, on their evidence, was a sick man, and he may or may not have got the correct part of the story. Mr Fine apparently was ill, and in due course dropped out of the matter. But the fact is that on October 20 on behalf of the Rosenthals he sent off a letter to Mrs Allan saying: "Mr Rosenthal informed you that he had somebody who was looking for a house, and that if your house was suitable this person would buy your house, and in these circumstances Mr Rosenthal stated that he and his wife wanted the estate agent's commission." So it will be seen that in October the first letter trying to set out an express agreement is written. The Rosenthals' story is that it was Mr Rosenthal who did the talking and made the agreement. As I have already pointed out, on Mr Rosenthal's recollection of what he thinks he said, no agreement was possible. The solicitor then gave Mrs Allan seven days to pay £1,093 or to be subject to a High Court action for its recovery. No doubt because he was ill, he sent the letter to the wrong address, and the people who received it hunted round. They thought they knew some people called Allan somewhere nearby, because Mrs Allan had moved by now to an address in Edgware. They hunted round and they found that they did know somebody called Allan. They assumed it must have been those people, whoever they were. I think it was a taxi-driver. At all events the letter was passed to him. He opened it to see whether it was anything to do with him. He found that it was not. He knew where the real Mrs Allan lived, and he took it round to her. She was very reasonably extremely angry and upset that a letter of this nature from the Rosenthals should reach her out of the blue. Having tried to get hold of Mr Cowan, because again it was a Sunday, she telephoned to the Rosenthals.

There is a dispute as to what took place on that occasion. According to Mrs Allan, Mr Rosenthal answered the telephone, and he remained on the phone together with his wife on an extension. Mrs Rosenthal did a great deal of talking; there was discussion about commission, and so forth, and about the letter. She said: "Why did you send this letter to me? What was all this about, giving me only seven days and sending it to the wrong address?" and all the rest of it, and Mrs Rosenthal asked a number of questions, the details of which Mrs Allan cannot remember, but Mrs Rosenthal talked 19 to the dozen—and having heard Mrs Rosenthal, and this is no criticism of her, I am satisfied that she did talk 19 to the dozen because it is her way to do so; I am not blaming her for it; it is just the way in which she is made. According to Mrs Rosenthal, she made a detailed note after the conversation was over, recording her recollection of it, which is set out in the agreed bundle of correspondence. According to her, Mrs Allan having telephoned her said, "Fay," because they used Christian names, "are you going to take me to court?" According to Mrs Rosenthal, she said, "Yes." So Mrs Allan said, "Well, why me?" and got the answer, "Because you gave me permission to sell your house." And she said, "I do not want to go to court." She said, "I do not want to lose the commission." Now it is noted, "You gave me permission to sell your house" is quite different to saying, "You agreed to pay me commission if I sold your house." According to Mrs Rosenthal's note of this matter: "I reminded her that she had told me that she had spoken to Mr Cowan about the commission, and he thought I was a shrewd businesswoman and he would try to get the full commission for me." Then she records that she put the following questions: "Did you give me permission to sell your house?" Answer: "Yes." "Did you say that you would give us the commission?" Answer: "Yes." "Did I get you £11,250 more?" Answer: "Yes." "Did I in fact sell your house?" Answer: "Yes." "Do you want me to have the commission?" Answer: "I know I have got to pay commission, and I want you to have it." Then according to Mrs Rosenthal the conversation went on, saying that Mrs Rosenthal said that it was her impression that Mr Cowan was not going to pay commission to anybody, Hamptons or her, and the money was all to go into a trust fund. She got the answer from Mrs Allan, "If that is what happens, I will give you half the commission."

This was, according to her, a conversation which Mrs Rosenthal said took place at a social meeting when she and her husband went to call on Mrs Allan some time in September. The precise date is not known. It may be that it was September 14 or thereabouts, but it does not matter, at some stage late in September, perhaps early in October. The parties are agreed that there was an occasion when this occurred, because a Mrs Downs was present arranging a party for some children. I accept Mrs Allan's recollection about that meeting that she never offered to pay £500 to Mrs Rosenthal. I am quite satisfied she would not have dreamed of doing any such thing without consulting Mr Cowan. She is not that sort of woman. I suspect that Mrs Rosenthal has convinced herself of this conversation because throughout this period she, not unnaturally, felt that she and her husband were entitled to something. They had brought on the sale, they had introduced the purchaser, they had got another £10,000 odd for the vendors, and I can well understand that they thought that they should have some remuneration. But I am quite satisfied that Mrs Allan did not say this at the time, and even if she had it would not affect the issue. The purpose of this was to try to persuade me that Mrs Allan had made an admission that she had engaged the Rosenthals as her agents to sell the house. In so far as the answers suggest that, I reject them. I am quite satisfied that Mrs Allan was not intending to, nor did she, make any admission that she had engaged the services of the Rosenthals to act as her agents for the sale of the house. If anything is wanted to clinch that view, it comes from Mrs Rosenthal's recollection as to how the phone conversation of October 22 ended. According to her, she said there was little more discussion appertaining to the sale of the house to Mr Milton. "Mrs Allan agreed that everything that I said was right, and she said to me 'What can I do?' I said to her, 'Your expenses are going to be very high.' She said, 'You are telling me; I know, but I cannot change my solicitor in midstream; what should I do?'" Mrs Rosenthal said: "I said to her, get in touch with your solicitor and tell him that it is you who is paying the piper and you will call the tune. She said she would phone her solicitor at 9.30 on Monday morning and phone me during the week." Having seen Mrs Allan in the witness-box, I am quite satisfied that at that stage the very last thing that she would have suggested doing would be sacking Mr Cowan as her solicitor. It is wholly out of character, and I am quite satisfied that she said nothing of the kind. If anything was said about changing solicitors, it came from Mrs Rosenthal and not from Mrs Allan, and if so it certainly was not understood by Mrs Allan in a plethora of conversation, I can well understand that too.

If the whole purpose so far as the plaintiffs' case is concerned is seeking to set up this conversation as an admission of an agreement to engage the services of the Rosenthals, it gets nowhere near it. I reject the plaintiffs' account of that conversation. The Rosenthals say that Mr Rosenthal was not even in the house, that he had gone off to hunt for this letter which had been wrongly delivered. I prefer Mrs Allan's recollection about what took place on that occasion. In the result, the plaintiffs have not succeeded in making out any case which entitles them to succeed in this action. It need only be said that I can well understand that they feel sore about it, that they did introduce the purchaser, they did as I have said get considerably more money than the agents who were handling the matter before. But they are not entitled to recover that commission from the defendant unless they can show that there was an express agreement to pay them the commission, or alternatively that there was an agreement, either express or one which was to be implied, that they were engaged to act as the defendant's agents in the sale of her house, and she would then remunerate them on a *quantum meruit*. They have not succeeded in doing that. The truth is that throughout the relevant part of this case, which is up to the moment when contracts were signed, they were acting as the Miltons' advisers and agents, and they were never retained by the defendant. It follows that although I have got considerable sympathy for Mr and Mrs Rosenthal in the way in which this transaction has gone, they are not entitled to succeed in these proceedings, and there must be judgment for the defendant.

The defendant was awarded costs.

HIGHWAYS

A

COURT OF APPEAL

February 21 1975

(Before Lord Justice CAIRNS, Lord Justice LAWTON and Mr Justice MACKENNA)

HEREFORD AND WORCESTER COUNTY COUNCIL v NEWMAN

Estates Gazette January 10 1976

(1975) 237 EG 111

Growth of vegetation makes a highway "out of repair" only in so far as it interferes with the surface of the way—Obstruction of the way by a barbed-wire fence does not put it into disrepair—Majority allows Worcestershire footpath appeal in part only

B

This was an appeal by Hereford and Worcester County Council from a decision of the Queen's Bench Divisional Court on April 24 1974 dismissing the council's appeal from orders made by justices sitting at Redditch at the instance of the respondent, Mr Peter John Newman, now of Prior House, Cleeve Prior, Evesham, Worcestershire, directing the then Worcestershire County Council to put in proper repair three footpaths at Inkberrow, Worcestershire.

C

Mr P Freeman QC and Mr K Schiemann (instructed by Sharpe, Pritchard & Co, agents for the clerk to the council) appeared for the appellants, and Mr R Tucker QC and Mr H Wolton (instructed by Cartwright & Lewis, of Edgbaston, Birmingham) represented the respondent.

Giving the first of the reserved judgments, MacKenna J said that the facts were few and simple. The first of the footpaths in question, no 19, had a 7-ft-high hawthorn hedge growing in the middle, and passage along it was impossible. The second, no 20, had a barbed-wire fence and thick undergrowth across it, so that a passage could be effected only by "negotiating" the fence and forcing a way through the undergrowth. The third, described as the junction of footpaths 73 and 75, was crossed by a barbed-wire fence of several strands, and passage at this junction could be effected only by climbing over or under the wire. Each of the three paths was admittedly a highway for whose repair the appellants were made responsible by section 44 of the Highways Act 1959. That section provided: "The authority who are for the time being the highway authority for a highway maintainable at the public expense shall, subject to the following subsection, be under a duty to maintain the highway." Section 295 provided that "maintenance" included "repair," and that "maintain" and "maintainable" should be construed accordingly. Section 59 provided machinery for the enforcement of the duty imposed by section 44. If a person alleged that a highway "maintainable" at the public expense was "out of repair" he might serve a notice on the highway authority whom he alleged to be liable to "maintain" it, and if the authority admitted those allegations, he might apply to the justices for an order requiring them, if the justices found that the highway was out of repair, to put it in proper repair within such reasonable period as might be specified in the order. Other provisions of the Act dealt with the protection of public rights. Section 116 (3), in its original form, provided that it should be the duty of the council of a district "to prevent, as far as possible, the stopping-up or obstruction of highways in their district." By an amendment, that provision now applied to county councils. Section 129 provided that if an obstruction arose in a highway from an accumulation of snow, or from the falling down of banks on the side, or from any other cause, the highway authority should cause the obstruction to be moved from time to time and in any case within 24 hours of service of a notice from a justice of the peace. Lastly, section 299 provided that no provision of the Act relating to obstruction of or other interference with highways should be taken as affecting any right of a highway authority, under any other enactment or rule of law, to remove any obstruction from a highway or otherwise abate a nuisance or other interference with a highway.

Were the footpaths in question "out of repair" within the meaning of section 59? On behalf of the appellants, it was contended that a highway was "out of repair" only if it was impassable or difficult to use because of the condition of its surface. For the respondent, it was submitted that a highway was out of repair not only where the surface was decayed but where by reason of obstructions, whether caused by nature or by the acts of men, it was no longer possible to use it or it could be used only with difficulty. He (his Lordship) was of opinion that the latter argument was correct. An obstructed highway, pending the removal of the obstruction, could be said to be "out of repair" and to remain out of repair until the obstruction was removed. The removal of the obstruction was a work of repair. That was not using the words "repair" and "out of repair" in any unnatural sense. As Lord Porter had said in *London and North Eastern Railway Co v Berriman* [1946] AC 278 at 307, "The exact meaning of repair is perhaps not easy to define, but it contains, I think, some suggestion of putting right that which has gone wrong." If a footpath had become blocked, either by growth of vegetation or by erection of a fence across it, something had gone wrong, and the man who cut down the vegetation or laid low the fence across the highway put right that which had gone wrong. Du Parcq J had given much the same meaning to "repair" in *Bishop v Consolidated Properties Ltd* (1933) 148 LT 407, one of the cases relied upon by Lord Widgery CJ in the court below. There a downfall pipe, which it was the landlord's duty to repair, had become blocked by a dead pigeon. The pipe had overflowed and the tenant's property was damaged. The landlord, who had failed to remove the pigeon, was held to be in breach of his repairing covenant. At p 410 du Parcq J said:

> I have to consider . . . whether it can be properly said that in this case the water system was out of repair; in other words, whether there was a breach of the covenant to keep in good repair. I have come to the conclusion that a pipe which is choked and not able to do its duty as a pipe is out of repair. . . . I think that one has to remember that "to repair" after all merely means to prepare or make fit again to perform its function; it means to put in order, and, I think, no more.

Similarly, the Suez Canal was "out of repair" when it was blocked by sunken ships and for that reason could not be

A used as a waterway. In contending for a narrower meaning of "repair," the appellants had relied on two arguments arising out of the provisions of the Act of 1959. They pointed first to the use of the word "maintain" in section 44 (1) and to the definition of "maintain" in section 295, where it was said to *include* "repair." This, they said, meant that "maintain" was wider than "repair," and therefore "repair" had a meaning which excluded such other matters as were maintenance only. He (his Lordship) was uncertain whether the draftsman of the 1959 Act really meant the duty of maintenance to include matters other than repair, but even if he did, so that "maintain" was wider than "repair," the question would still remain whether the removal of obstructions B was maintenance but not repair, and he (his Lordship) would answer that it was repair. The second argument put forward by the appellants was that as section 116 imposed a duty to prevent obstructions, a meaning should be given to the word "repair" in section 59 which did not include the removal of obstructions. He (his Lordship) did not read the Act in that way. He thought that the duty of repair imposed by earlier sections included the duty of removing obstructions, and that the duty of preventing obstructions imposed by section 116 was a "cumulative" one, to borrow an expression used in *R v Heath* (1865) 6 B & S 578, 12 LT 492. That was C a strong authority in the respondent's favour. To understand it, it was necessary to look at some of the provisions of the Highways Acts of 1835 and 1862. Section 6 of the 1835 Act provided:

The inhabitants of every parish maintaining its own highways . . . shall proceed to the election of one or more persons to serve the office of surveyor in the said parish for the year then next ensuing . . . which surveyor shall repair and keep in repair the several highways in the said parish for which he is appointed, and which are now or hereafter may become liable to be repaired by the said parish.

D There one met the two words "maintaining" and "repairing" in conjunction. The section referred to the old duty of the inhabitants to "maintain" and to the new duty of the surveyor to "repair." He (his Lordship) would not expect the new duty to be less extensive than the old, so as to omit matters of maintenance, not being repair, for which the surveyor was not to be responsible. Section 20 imposed penalties on the surveyor for neglect of his duty, and section 27 empowered him to make a rate in order to raise money "for carrying the several purposes of this Act into execution." As in the 1959 Act, there were sections dealing with the removal of obstructions. Section 26 provided that if there should be E an obstruction from snow, or from the falling down of banks, or from any other cause, the surveyor should, from time to time, and within 24 hours if ordered by a justice, cause it to be removed. Sections 63 to 69 dealt with the removal by the surveyor of particular kinds of obstructions including, in section 69, encroachment by buildings. Then there was the 1862 Act, which supplemented the 1835 provisions by enabling different parishes to join together in setting up highway boards to do the work for all the parishes which had formerly been done by a surveyor for each of them separately. Section 17 was in these terms:

F The highway board shall maintain in good repair the highways within their district, and shall . . . as respects the highways in each parish within their district, perform the same duties, have the same powers, and be liable to the same legal proceedings as the surveyor of such parish would have performed, had, and been liable to if this Act had not passed. . . .

Section 20 provided for the highway board charging the expenses they had incurred for the common use of the several parishes to a district fund, to be contributed by the parishes according to a prescribed scheme. The section continued:

. . . But the expenses of maintaining and keeping in repair the highways of each parish within the district, and all other expenses in relation to such highways, except such expenses as are in this G Act authorised to be charged to the district fund, shall be a separate charge on each parish.

In *R v Heath* a highway board was seeking to recover from one of its parishes part of its costs of indicting one T Burrows for obstructing a highway. Burrows had erected houses partly on his own land and partly on a paved footway which adjoined his land and formed a public highway. He was indicted for a nuisance, tried at the assizes, found guilty and ordered to pay the taxed costs of the prosecution. He paid the costs and at once removed the obstruction, but the taxed costs were not equal to the whole of the prosecutor's expenses, and the board took proceedings against the parish to recover H the £60 difference. The parish disputed liability, and the case was tried at Quarter Sessions, which stated a case for the opinion of the Court of Queen's Bench. The board put its case in two ways, the first of which involved the following three propositions: (i) if a surveyor had himself removed Burrows's obstructions, this would have been work of repair within the meaning of section 6 of the Act of 1835, and he would have been entitled to make a rate to cover the cost under section 27; (ii) if he had proceeded by indictment, as the board had in fact done, this would equally have been work of repair, and a rate could have been made to recover the cost of the litigation; (iii) by section 17 of the Act of J 1862, the board had the same power that a surveyor would have had. The second argument was that the costs of the litigation were recoverable as an expense of "maintaining and keeping in repair the highways" of the township under section 20 of the Act of 1862, alternatively as "other expenses in relation to such highways" under the same section. Counsel for the parish tried to meet the board's argument on section 6 of the 1835 Act by pointing out that that section dealt with repair while sections 26 and 29 dealt with the removal of obstructions, saying, in effect, that the powers and duties of removal in the later sections excluded such powers and duties from section 6. Here it was that Crompton J observed that K the sections were "cumulative." When counsel attempted to distinguish between the costs of removing an obstruction and the costs of litigation, Cockburn CJ answered:

If the surveyor is entitled to charge the expenses of removing a nuisance by manual or mechanical labour, why is he not entitled to charge the expenses of doing it by legal proceedings? . . . Litigation leads to the same end.

Giving judgment, the Chief Justice supported the board's first argument, as follows:

If this had been the case of a prosecution by the surveyor under the Act of 1835 for removing an obstruction on a highway I L should have been disposed to hold that he had the power to include the expenses of it in a highway rate; for by section 27 he was directed to make a rate in order to raise money for carrying out the several purposes of that Act into execution. The main purpose of the Act was to repair the highways and keep them in proper condition, but the existence of an obstruction on a highway amounting to a nuisance is inconsistent with that condition. And therefore, according to a wise and liberal construction of the Act, the expenses of such a prosecution might have been fairly and legitimately included in the highway rate.

The closing sentences were expressed in a slightly different way in the *Law Times Report,* but the meaning was the same. Having thus dealt with the 1835 Act, the Chief Justice pointed M to the words of section 20 of the 1862 Act about "all other expenses in relation to such highways" and said that they made the section a wider one than section 17 of the earlier Act. He concluded that the expenses claimed against the parish fell within these wider provisions. Crompton J said the same about both the earlier Act and the later:

Mr McIntyre (counsel for the highway board) rested his case on short and simple propositions. The highway board are bound to maintain the highways in good repair; it is impossible to do that without removing obstructions; and in many cases obstructions cannot be removed without having recourse to law. And he said that the costs of such legal proceedings are expenses which may be charged upon that parish under statutes 5 & 6 Will. 4, c 50 [the Act of 1835] and 25 & 26 Vict. c 61 [the Act of 1862]. I have not heard anything to meet that argument and I think his propositions are correct.

And a little later he said:

> Supposing 5 & 6 Will. 4 c 50 did not include such an expense as that in question, though I think it does, it is clearly an expense in relation to highways within 25 & 26 Vict. c 61, section 20.

Blackburn J thought that section 20 of the later Act was no wider than section 27 of the earlier Act, that the expenses of litigation would have been recoverable by a rate under section 27 as being incurred for the purpose of repair, and that they were equally recoverable under section 20. Those judgments of authoritative judges clearly supported the conclusion that the removal of an obstruction which was not merely temporary in its nature and which made the whole or part of a highway impassable, or at least very difficult to use, was work of repair, and that the highway, until that work was done, was out of repair. Mr Freeman objected, very truly, that *Heath's* case was an old authority, more than 100 years old, and asked why the construction of a modern statute like the 1959 Act should be governed by it. He pointed out, accurately, that the decision, not being one of the Court of Exchequer Chamber, was not binding on this court. The answer to that submission lay partly at least in the connecttion between the 1959 Act and the old highways law. That Act was in direct succession to the 1862 Act and the 1835 Act. It was not a slavish adherence to authority which would preserve a meaning given to the word "repair" in those earlier Acts by judges such as Blackburn J, even when sitting in the Queen's Bench, and never subsequently challenged. For the rest, the answer to the submission was that what the judges said in *Heath's* case was good sense. It was not surprising that the Divisional Court based its judgment on that case.

Before ending, he (his Lordship) would briefly and respectfully comment on two passages in the judgment of the Lord Chief Justice in the present case. The first was at page 944 of the report in the *Weekly Law Reports,* where Lord Widgery said that if an obstacle was one which ought to have been prevented or remedied in the course of "normal routine maintenance" its existence caused the highway to be out of repair. He then went on to give an example of an obstruction for which proceedings would not lie under section 59: "If, however, a builder chose to dump tons of rubble on a footpath thus rendering it impassable it would, I think, be an abuse of language to say that the highway authority had allowed the footpath to become out of repair." That, with respect, was a mistaken view. If the obstruction to the use of the highway was such that it could fairly be said to put it out of repair, the duty to remove the obstruction, or cause it to be removed, existed however it might have got there. The authority's liability was not based on its having wilfully "allowed" the obstruction to be erected, nor was its duty of removal limited to what might be achieved in the course of "normal routine maintenance." The second passage was at 945 of the report where the Lord Chief Justice dealt with a fourth footpath which was sent back to the justices for further consideration. As neither party in the present proceedings had appealed against that part of the Divisional Court's order, his (MacKenna J's) observations on that passage were *obiter*. The case was one of a footpath which had become flooded with effluent from a cesspit, and the Lord Chief Justice said: "If the effluent was originally carried under the footpath in a culvert which has collapsed or become obstructed, this may well be a failure in maintenance. If, however, someone has built a cesspit in such a way that the effluent floods the footpath I do not think that the result is to put the footpath out of repair." If, as appeared to be the case, the flooding made the footpath impassable, the path was out of repair. The cause of that condition was relevant only in considering the time which it was reasonable to give the local authority for removing it or causing it to be removed. The appeal should be dismissed.

Cairns LJ said that he did not feel able to give the word "repair" the weight which MacKenna J had done. In his opinion the word had to be considered in its ordinary meaning and the context in which it was used. Not every act of putting a thing right was a repair. Whatever the answer with regard to the correctness of the actual decision in *Bishop's* case, du Parcq J appeared to have given an unduly extended meaning to "repair." To wind up a watch and set the figures to the correct time was to make it fit again for its normal function, but no one could describe such an operation as "repair." In relation to a highway, no one, so he (his Lordship) ventured to suggest, would describe the removal of an obstruction as a "repair." It was, in his judgment, striking that in all the Highways Acts "repair" and "removal of obstructions" were separately dealt with. Despite *R v Heath,* he did not feel constrained to give to the word "repair," as used in the Highways Act, any meaning other than the natural and normal meaning. Highways could only be "out of repair" when the surface was in some way defective or disturbed. In the present case he could not say that the footpath obstructed by the wire was "out of repair." He would accordingly allow the appeal in respect of footpath 73 and 75, but dismiss the appeal in respect of paths 19 and 20 on the basis that the growth of vegetation had interfered with their surfaces.

Lawton LJ said that section 44 of the Highways Act 1959 imposed upon highway authorities the duty to maintain highways. Keeping a highway in repair was only one aspect of maintaining it. The words "repair" and "maintain" were not, in his opinion, synonymous. Since the 18th century, and probably long before, justices had had a duty to see that the King's subjects could pass freely on the highway. Blackstone in his *Commentaries,* in a section dealing with public nuisances, said: "Of this nature are annoyances on highways, bridges and public rivers, by rendering them inconvenient or dangerous to pass; either positively by actual obstruction or negatively, by want of reparation" (23rd edition, vol 4, p 205). It was only the inhabitants at large who were responsible for want of reparation; individuals were responsible for obstruction and encroachments. That distinction was reflected in the General Highways Act 1773. It was also to be found in the Highways Act 1959, but before showing that he (his Lordship) had to consider the phrase "out of repair" in section 59. It was an adjectival phrase, one in ordinary usage, and in his opinion, when used in relation to highways or bridges it connoted the restoration to a sound or unimpaired condition of that which had become unsound or impaired by neglect or use. If *Hawkins' Pleas of the Crown* was a reliable guide, in at least one case the indictment charging want of repairs referred to the highway being "in decay" (see 8th edition vol 1 pp 698-704). That fitted in with two of the meanings of repair given in *Murray's Oxford Dictionary*. The 19th century precedents of indictments in the 2nd and 15th editions of *Archbold's Pleadings and Evidence in Criminal Cases* reflected the same concept of decay or neglect.

A highway got out of repair because the highway authority, over a long period, had not done its duty. It was therefore just that the law should make mandatory orders for repairs. That was what section 59 did. A highway, however, could become obstructed or encroached upon overnight. If the obstruction had been caused by weather, for example a landslide, the highway authority could not reasonably be expected to remove the obstruction quickly. It might be impossible to do so. The highway authority had a duty to prevent, so far as possible, the obstruction of highways (section 116), and was given powers to enable it to do so (sections 124, 125, 131 and 133). Whether a highway was out of repair was a question of fact for justices. In his (Lawton J's) judgment,

A Parliament, by confining the magisterial remedy to cases in which a way or bridge was out of repair, had intended to draw the distinction between positive and negative conduct causing nuisances to highways which the law had recognised for over 200 years. In coming to this conclusion, he had not derived much help from the meaning given to the word "repair" in leases or other contexts. What mattered in the present case was the historical background. He did not share MacKenna J's opinion that the decision in *R v Heath* was a valuable guide. The problem in that case was the mundane one as to which authority should pay the costs of prosecuting someone who had obstructed the highway. He was doubtful B whether the Court of Queen's Bench directed its attention specifically to the question in the present case, namely whether a highway which had been obstructed, or had become obstructed, could be said to be "out of repair."

Turning to the facts in the present case, there was a hawthorn hedge in the middle of footpath 19 making passage along it impossible. That finding by the justices was expressed in few words, and more detail would have been helpful, because if the hedge had been planted then it would seem, on the authority of *Hawkins* (*supra*), that it should be regarded as an encroachment or a positive act of obstruction. C He inferred, however, that the justices found that it had grown up where it had because the highway authority had done nothing to the path for a long period. Nature did tend to take over when footpaths were neglected. He would adjudge that footpath out of repair. By the same reasoning footpath 20 was out of repair because of the undergrowth which had grown up. That was due to neglect. The barbed-wire fence, however, had got where it had because of positive action of someone. He therefore adjudged the junction of footpaths 73 and 75 not to be out of repair, but to be obstructed by the barbed-wire fence. He would therefore allow the appeal in respect of the junction of footpaths 73 and 75, but D dismiss it in respect of footpaths 19 and 20.

The appeal was accordingly dismissed in so far as it related to the orders made concerning footpaths 19 and 20, and allowed only in respect of the junction of footpaths 73 and 75. The appellants were ordered to pay the costs. They were granted leave to appeal to the House of Lords on terms that they did not ask for costs against the respondent in the House of Lords.

E

QUEEN'S BENCH DIVISIONAL COURT

March 4 1975

(Before Lord WIDGERY CJ, Mr Justice BRIDGE and Mr Justice EVELEIGH)

HALL v H B HOWLETT & PARTNERS LTD AND OTHERS

Estates Gazette March 20 1976

(1975) 237 EG 875

Obstruction of footpath—Whether lane a public highway—Statements in conveyances of property bounded on one side by lane held not to be conclusive—Points on rights-of- F **way granted or not granted in conveyance of nearby properties also held inconclusive—Absence of evidence of user of lane by through traffic "noticeable"—Lane not made out to be a public highway, obstruction charges correctly dismissed**

This was an appeal by Mr Christopher Myles Hall, director of the Council for the Protection of Rural England, against the dismissal by justices sitting at Watlington, Oxfordshire, of three informations preferred by him under the Highways Act 1959 alleging that the respondents, H B Howlett & Partners Ltd, of Great Holcombe Farm, Watlington, and Mr Adrian Nixey and Mr Reginald Nixey, both of Langley Hall Farm, Watlington, unlawfully obstructed with fencing Holcombe G Lane, Watlington, being a public highway.

Mr P J Crawford (instructed by Blyth, Dutton, Robins, Hay, agents for Marshall & Galpin, of Oxford) appeared for the appellant, and Mr N B Primost (instructed by Cole & Cole, of Oxford, and Parrott & Coales, of Aylesbury) represented the respondents.

Giving judgment, LORD WIDGERY said: This is an appeal by case stated by justices for the county of Oxford in respect of their adjudication as a magistrates' court at Watlington, when they dismissed three informations laid by the present appellant against the present respondents, each information H alleging that the respondent in question did without lawful authority or excuse wilfully obstruct the free passage along a lane known as Holcombe Lane, which lane was alleged by the prosecutor to be a public road, the form of the obstruction being by barbed wire, posts and the like. There was, as I understand it, never any question but that the lane in question had been obstructed to the extent alleged, and the issue below, and indeed the issue in this court, was really restricted to the question of whether the lane was a public highway, proof of which of course was an essential feature of this charge. I have some sympathy, I confess, with the appellant in this case. It is evident that it is a matter of con- J sequence to him; the industry and trouble taken with regard to presenting the case below is self-testimony to that, and one can well understand that, having had a hearing on that scale before the magistrates, and this being the only court to which an appeal can lie, it would be understandable enough if a layman expected a review in this court to be rather more comprehensive than it can be. The fact is, as is well known to lawyers, that the only appeal in this court can lie on a point of law, and Parliament seems deliberately to have left many of these vexed questions of fact to justices, possibly because of their local knowledge and relying upon common-sense. In order to succeed in these proceedings, the appellant K has got to show either that the magistrates followed some wrong principle, or that they reached a conclusion of fact for which there was no evidence, or that they reached a conclusion which was perverse, or to put it another way, one which no reasonable bench could reach on the material before them, and it is obvious that that is a heavy burden which the appellant necessarily assumes in this court, and for reasons which I shall give in greater detail in a moment I certainly have come to the conclusion that that onus has not been discharged.

The road in question, conveniently called Holcombe Lane, is of a length of about one mile and a half, and it begins in L the village of Newington at a point near the Stag public house. It takes a loop to the north and east, passing and no doubt serving Great Holcombe Farm, and perhaps a number of buildings in the vicinity. It then turns south and east and takes that general direction until it comes to Little Holcombe Farm. Proceeding on beyond there it turns again south east and comes on what is indisputably a main public highway. But Mr Crawford, on behalf of the appellant, says that it does not stop at the main public highway but continues beyond, where there is a further path continuing in a south-easterly direction. As I say, there is no dispute that the obstructions complained of were created by the respective M respondents, and it is perhaps worth noting that of those obstructions the first, second and third took place in a part of the lane north and west of Little Holcombe Farm. The fourth obstruction took place quite close to Little Holcombe Farm. The relevance of that is this: it came as a great surprise to each member of the court to find that when this matter was investigated before the justices no one put before the justices any information based upon the definitive map prepared under the National Parks and Access to the Countryside Act 1949. We have been told in this court that the definitive map does not show a public way through Holcombe

A Lane, but it shows a public footpath starting in Newington Village, somewhat south of Holcombe Lane, running (generally speaking) in a southerly and easterly direction, and linking up with Holcombe Lane some few hundred yards to the west of Little Holcombe Farm. Thereafter, the footpath defined on the definitive map follows the line of Holcombe Lane down to the main road to which I have referred. Accordingly, if and so far as the contents of the definitive map were relevant in this case—and as they were not before the justices I hesitate to comment upon them—three of the obstructions would have taken place in that part of Holcombe Lane not included on the definitive map, but the fourth would B have been on that later length of the lane. The fourth obstruction left a stile for pedestrians and therefore presumably was not an obstruction of the footpath. I think, as I have said, that we should not be affected by matters which were not before the justices, but I think it really is necessary to say that the failure to assert, as I understand it, any footpath over the rest of Holcombe Lane during the inquiries under the Act of 1949 is to say the least a difficulty for those who allege that there was here evidence upon which dedication of a public highway could be inferred.

The appellant below produced a number of old documents C and also called a number of witnesses as to actual user, and I must deal briefly with these two broad heads under which this evidence called resolved itself. We were shown first of all, as were the magistrates, a large number of maps going back to the end of the 18th century, which demonstrate that there was a road of some kind, probably along the line of Holcombe Lane, from those days. The justices received certain expert evidence which was called to enable them to interpret whether the road was at that time a public highway or not, but it seems that the justices were not able to regard that information as being in any sense conclusive, and we have not been asked to look at it again. Accordingly, the D maps as such demonstrate only that there was a road and they do not, I think, provide any significant assistance on the question before us, namely whether that road was or was not a public highway. In addition to the maps, we have been shown a number of miscellaneous documents, many of them documents of title, upon which reliance of some kind or another is made. In particular, there is at the northern and western end of Holcombe Lane a property known as Cobshall Cottage, and we have been shown two deeds of conveyance affecting Cobshall Cottage. The first, which is dated December 23 1879, in the parcels describing Cobshall Cottage used this phrase: "All those two cottages or tene- E ments formerly one cottage or tenement with the gardens outbuildings and appurtenances to the same respectively belonging situate and being at Holcombe in the Parish of Newington in the County of Oxford formerly in the occupation of," and then names are given, and the description follows that the property thereby conveyed is "bounded on the east by a public lane." Now the relevance of that information, as tortuously expressed from the deed, is that the lane in question must be part of Holcombe Lane, albeit at its almost extreme northern and western end, and as Mr Crawford rightly says, there is in that document an indication that Holcombe Lane was reputed to be at all events a public F lane. The same point is repeated in a further indenture dealing with the same property on October 2 1905, where the parcels have been slightly varied and the point to which I am now referring is made in these words: "bounded towards the east or front by a public lane called Holcombe Lane." Again the point is thus seen that the draftsman thought that Holcombe Lane was a public way at that time.

I do not for a moment seek to disparage the value of such entries, but of course one has to give each document of this kind the weight which its own circumstances justify, and if Holcombe Lane was not a public lane at that time all that it means is that one draftsman has introduced an error in the terms of the conveyance. It has some value; its weight is a G matter ultimately for the tribunal of fact. Another point which was made, although not persisted in, by Mr Crawford was that in the case of Cobshall Cottage the only access to an undisputed public way was down Holcombe Lane, and he said that the absence of any grant of a right-of-way over that section of Holcombe Lane is again indicative of the fact that everybody thought this was public and therefore no such grant was required. I am totally unimpressed by that argument. The overwhelming probability was that this cottage would already have had some right over Holcombe Lane, and if that was the case there was no need for the conveyance to provide for it afresh. A similar comment is made in H respect of Starveall Cottage which is situated somewhat to the south of Holcombe Lane and much further east than the property to which I have just referred. In order to obtain access to Starveall Cottage it is necessary to go down a track from the cottage to Holcombe Lane and then to find one's way out via Holcombe Lane itself. We have been shown a conveyance of Starveall Cottage which contains an express grant of a right-of-way from the cottage to Holcombe Lane but no express grant of a right-of-way thereafter. This, says Mr Crawford, is indicative that there is a public right-of-way in Holcombe Lane and this is why no further grant is required. Again, it seems to me a much simpler explanation J that no private right was necessary initially when the cottage and the surrounding land was in single ownership, and the occupier could walk over his own land to Holcombe Lane, but when the property was sold the occupier of Starveall Cottage did require such a right, and accordingly it was granted to him. I can find no substance and weight to be attached to those documents, so far as this matter is concerned.

Then we were shown what on any view must be an important matter, namely an inclosure award, and this is, as inclosure awards go, somewhat difficult to interpret, K because it contains a great deal of unpunctuated English, but under the side-note "Private Roads" the inclosure award makes provision for the laying out of "one other private carriage road and driftway of the like breadth of twenty feet leading out of the Chalgrove Road at the west end of Chalgrove Lane in a south-west direction," and further details are appended. We were told yesterday that that description is related to Holcombe Lane, in other words, that which is being described is Holcombe Lane. If that is so, and it is not disputed, I find that this is a point of great significance, because when inclosures were made the prime purpose was to set out new boundaries, boundaries which L had not existed, and I should have thought that if the commissioners set out a new private road in an inclosure award it is almost conclusive that the commissioners did not think that there was already a public highway there, because there is no basis to establish and lay out a new private road over existing public highway. I think this is a point of considerable weight to go into the scales when those scales are operated by the tribunal of fact concerned with this matter.

Then we were referred to one or two incidental references to Holcombe Lane which appeared in public documents during the last century. The first one to which I would wish M to refer is a board minute of the Watlington highways board on June 3 1868. This minute has a side-note "Holcombe Lane," and obviously referring to that side-note the minute reads: "Mr Hamp Haywarden Newington stated Mr Shrubb of Holcombe had complained of the state of Holcombe Lane. Mr Hamp was informed by the chairman of the board that Mr Shrubb should make his complaint in writing in order that the matter might be formally before the board." So one sees the record there of Mr Hamp complaining of lack of repair, and the comment is fairly made that it is unlikely that he would be complaining of lack of repair except on

A the basis that the road was a public way, but the matter, as has been seen, was dealt with inconclusively on June 3 1868, and was deferred to the next meeting of the board of July 1 1868. Against the side-note "Holcombe Lane" we find: "Mr Hamp having failed to make his complaint in writing as suggested by the chairman at the last meeting the matter was allowed to drop." So really no conclusion was reached at all, except that Mr Hamp for good reason or ill seems to have been trying to obtain a statement of fact that the road was a public road, but the inconclusive nature of the inquiry seems to me to make it impossible to attach very much weight to that. There is another similar reference to B trees obstructing Holcombe Lane. This is an annual meeting of the parish held in Newington called on March 30 1915, and the minutes disclose that "Mr Belcher complained of a fallen tree across Little Holcombe Lane and proposed that the clerk be instructed to write to the owner and ask him to remove it. This was seconded by Mr Moores and carried unanimously." Again it may be only a straw in the wind, but it seems to indicate that somebody thought that some part of Holcombe Lane was public, but it may well be for all we know, and for all the justices knew, that that part to which reference was made as the carriageway was the part of the lane which is not now disputed to be public highway C by virtue of the Act of 1949. I hope Mr Crawford and those instructing him will not think that an inadequate review of the effect of the documentary evidence. In its result, it seems to me to show very little weight one side or the other. It may be that the justices could have been influenced for or against the appellant by these documents, but I cannot believe that they would have been heavily influenced either way because of the nature of the documents on which I have already commented.

Then the second class of evidence relied upon by the appellant below was the evidence of user, and a number of D witnesses were called to say what had been happening in the way of access to and user of this road. Unfortunately, because it adds to the appellant's difficulties, the road has on any view become heavily overgrown in the last 20 or 30 years and clearly has not been used in that sort of period to the extent to which it may have been used 100 years ago when it was of normal carriage width, but we have nevertheless had the advantage of the justices' clerk's notes of evidence, evidence given by witnesses as to user. Of course, it would be a great disadvantage to us if we had to try the issue of fact in this case, because we did not see the witnesses. It would equally be a great disadvantage because E we have not got a verbatim transcript, only a good summary note taken by, I think, the clerk to the justices. But we have not got to try the issue of fact, and as I have already endeavoured to explain, one must only look at the general weight of the evidence given before the justices to see whether it could be said that their decision was perverse or otherwise open to inquiry by us. Here again, I must attempt a summary of the evidence, because it would be intolerable to go through it all, and unprofitable as well. Most of the evidence as to use called below seems, from the notes, to have been use by local people, for their private affairs. One witness went courting in Holcombe Lane; motor cyclists had F used it for a scramble; someone had gone there looking for stray sheep; the hunt had gone down there; and other people had gone down there as a pleasant recreational walk from their home in or around the village. All that is a matter upon which an inference of dedication can be drawn, but what strikes me as being noticeably absent from the evidence called in this case is evidence of user of the lane as a through lane. This may be because it has been overgrown at the relevant time, and has not been used as a through lane, but since the inquiry before the justices was whether the evidence justified an inference that the public highway had been dedicated, the fact that there was noticeably little evidence G of user as a through way by members of the public seems to me to weaken to a very considerable degree the extent to which this evidence might have importance attached to it. I do not for a moment suggest that proof of user as a through way is vital to the inference of a public highway, but the effect of it is bound to be more impressive, and in its absence here I do not get the impression, looking at the matter before us, that the evidence of actual user was particularly impressive.

Having come to that conclusion, both on the documents and on the oral evidence, there is only, I fear, one conclusion in this case, that is, that one cannot possibly say that the H justices' decision was perverse or that this was a decision which no reasonable bench might have reached. Whether we would have come to the same conclusion is another matter, but I find it impossible to say that the case for the appellant is made out, and I would accordingly dismiss this appeal.

BRIDGE J: I agree.

EVELEIGH J: I agree.

The appeal was dismissed with costs.

J

QUEEN'S BENCH DIVISIONAL COURT

June 11 1975

(Before Lord WIDGERY CJ, Mr Justice KILNER BROWN and Mr Justice WALLER)

ATFIELD v IPSWICH BOROUGH COUNCIL

Estates Gazette May 8 1976

(1975) 238 EG 415

Obstruction of highway—Intention to protect property, shop-customers, etc, does not justify shop-owner in placing K bollard in footway—Lawful excuse in case of temporary obstruction cannot be extended to facts where obstruction intended to be permanent

This was an appeal by the Ipswich Borough Council against the dismissal by Ipswich justices of a summons alleging that the respondent, Mr Douglas Atfield, of 17 St Stephen's Lane, Ipswich, did on September 16 1974 obstruct the free passage of St Stephen's Lane without lawful authority or excuse, contrary to section 121 (1) of the Highways Act 1959.

Mr A Fletcher (instructed by Sharpe, Pritchard & Co, L agents for E K Dixon, of Ipswich) appeared for the appellants, and Mr P Collins (instructed by Turner, Martin & Symes, of Ipswich) represented the respondent.

Giving judgment, KILNER BROWN J said that the justices had undoubtedly approached the matter with a degree of sympathy for the defendant and done their best to find a way round the obligations of the Act of 1959 because of their sympathy for him. What emerged was that the defendant had been carrying on a business as an antique dealer since May 1967 in a shop at 17 St Stephen's Lane, Ipswich. The shop was an historic building and no less than 400 years old. It was in a side street which was not a major public highway, M a one-way street little used by vehicular traffic except as a means of access to nearby buildings. Unhappily for the defendant, in or about 1969 a large building was erected on land close to his shop. Two large shops were opened in January or February 1970, and in consequence a number of supply vehicles passed and repassed the defendant's shop many times a day. These vehicles frequently mounted the footway; damage had been caused to the footway, and the justices also thought there was danger to pedestrians. The defendant sought the appellant council's help, asking them to put up a bollard outside his premises, but the council would not agree, considering that such a bollard would cause an obstruc-

A tion. They offered to strengthen the highway, and so on, but this was not satisfactory to the defendant, who on September 16 1974 took matters into his own hands by erecting a metal bollard on the footway outside his shop. The justices found that he did so with the intention that the bollard should remain in position indefinitely. It was a substantial bollard, seven inches from the edge of the kerb, leaving a passageway two feet three inches wide along the footway. The justices reached the conclusion that in all the circumstances the defendant had acted " with lawful excuse, in order to protect his property and reduce the risk of injury to his customers and to other persons using the footpath." Accordingly, they B dismissed the information.

Mr Collins, for the defendant, had done his best to maintain that decision. He had said in broad, general terms that the justices had used the expression " lawful excuse " quite properly, because it should be interpreted broadly; that the law had always adopted a reasonable approach, and the C courts had tried to apply a test of reasonableness; and that, recognising that most of the cases dealt with temporary user, where safeguarding property and so on was accepted as a defence, that same result should follow where a permanent fixture was concerned. He (his Lordship) found himself unable to accept those contentions. In his view the answer to the question posed for consideration was that the justices in the present case did not come to a correct determination and decision in point of law. The bollard was plainly an obstruction, and there was no evidence before the court to justify the finding of " lawful excuse." There had been a gallant attempt by the justices to transmit sympathy into terms of D evidence, but the evidence was not there. The case should be remitted with a direction to convict.

LORD WIDGERY and WALLER J agreed, and an order was made accordingly. The appellants were awarded costs.

HOUSING ACTS

A

CHANCERY DIVISION

February 20 1975

(Before Mr Justice GOULDING)

FIRST NATIONAL SECURITIES LTD v CHILTERN DISTRICT COUNCIL

Estates Gazette April 17 1976

(1975) 238 EG 187

Right of pre-emption registrable as an estate contract if taken by a council vendor under power in Housing Act 1957, whatever the position may be in the case of a private sale

B

This was a claim by First National Securities Ltd, second mortgagees of 39 Hildreth Road, Prestwood, Buckinghamshire, for an order vacating a class C (iv) land charge registered against the property by predecessors of the defendants, the Chiltern District Council.

Mr P Cowell (instructed by Davis & Co, of Harrow) appeared for the plaintiffs, and Mr G Lightman (instructed by Sharpe, Pritchard & Co) represented the defendants.

Giving a reserved judgment, GOULDING J said that the question for decision concerned a right of first refusal or pre-emption of a house, 39 Hildreth Road, Prestwood, Buckinghamshire. The right was given to the defendant local

C

authority by covenant in a deed of conveyance of the property, and stood registered as an estate contract under the Land Charges Act 1972, the registration having originally been made under the Land Charges Act 1925. It was not suggested that these Acts exhibited any relevant difference.

By a conveyance dated July 21 1969 Amersham Rural District Council, predecessors of the defendants, conveyed 39 Hildreth Road to Mr Roy Kingsnorth. Among the covenants entered into by the purchaser for himself and his successors was paragraph 13 of the fourth schedule, which read:

D

> Not at any time during a period of five years next following the date of this conveyance to (i) let the said property at a rent which would be in excess of any rent charged by the council for houses of the same type belonging to the council on the Prestwood Estate (ii) enter into any contract of sale in respect of the said property either operative at the date thereof of any later date at a price greater than the purchase price paid hereunder by the purchaser plus such increase for improvements (if any) effected by the purchaser as may be agreed between the purchaser and the council . . . (iii) convey his interest or any part of such interest in the said property or any part or parts thereof . . . without first giving to the council notice in writing of such his desire and in such event the council shall have the option which shall be
>
> E
>
> exercised by the council within one calendar month from the date when such notice shall be received by the council of repurchasing the property at the original purchase price recited in this conveyance together with such amount as represents the value of any improvements effected by the purchaser but less such amount as represents depreciation. . . .

Shortly afterwards, the council registered the right of pre-emption in paragraph 13 as a land charge, class C (iv). The purchaser, who had mortgaged the house to the council, on January 16 1973 executed a second mortgage of the property in favour of the plaintiffs, First Securities Ltd. Within the

F

five-year period of the right of pre-emption he wrote a letter dated April 18 1974 to the council, and the council replied. The letters were not in evidence, but the plaintiffs conceded for the purposes of the action that by virtue of the exchange of the letters the council had exercised its right of pre-emption, so that it was to be assumed that there was a subsisting contract as between Mr Kingsnorth and the council for sale of the property to the council. The plaintiffs had, however, gone into possession of the property as mortgagees, and wished to sell it in the exercise of their power of sale. They claimed to be free to do so on the basis that their mortgage was prior to the contract of sale between Mr

G

Kingsnorth and the council. The council said that its own claim had priority, because the right of pre-emption was acquired and registered before the mortgage to the plaintiffs, so that the plaintiffs' loan was made with notice of the right of pre-emption. Therefore, said the council, the plaintiffs were in equity bound by the right and by the contract between Mr Kingsnorth and the council which flowed from it.

Having regard to the general scheme of the Acts of 1925 and 1972, the relevant provisions of which did not appear to be different in meaning, he (his Lordship) thought that it was only where an obligation was inherently capable of enforcement in suitable circumstances against the contracting landowner's successors in title that it came within the pur-

H

view of the Act of 1972 at all. Such an obligation brought into being an equitable interest in land: *London & South Western Railway Co v Gomm* (1881) 20 Ch D 562 at 580, 581. A good example was an ordinary contract for the sale of land. All lawyers knew that the purchaser got an equitable interest in the land as soon as the contract was signed. An option to buy land was a different sort of contract. The landowner was only bound to sell if and when the grantee of the option called on him to do so. None the less, the grantee of the option had an interest in the land even before he exercised his right: *Gomm*'s case, in the passage already

J

referred to. A right of pre-emption in the sense of a right of first refusal was a step further away from an actual sale. The grantee of the right of pre-emption would only become a purchaser if the landowner wished to sell and he (the grantee) then chose to buy. The plaintiffs here argued that unlike an actual purchase or an option to purchase, a mere contractual right of pre-emption conferred no interest in land. It was never enforceable against a successor in title of the grantor, even one with notice of the right, but only personally against the grantor or his personal representatives. Therefore, said the plaintiffs, it was not an obligation affecting land within the Land Charges Act, and it ought not to be on the

K

register. The argument was based largely on the decision of the Court of Appeal in *Manchester Ship Canal Co v Manchester Racecourse Co* [1901] 2 Ch 37 at 50, where in the reserved judgment of the court it was stated plainly that a right of first refusal did not create an interest in land in the sense of *Gomm*'s case, and the contrary view of Farwell J below was rejected. The council, on the other hand, said that a right of pre-emption did confer an interest in land, because it belonged to the same genus as an option to purchase or an actual purchase, though to a different species of that genus.

A Mr Lightman, taking his stand on certain recent cases in appellate courts, said that the Court of Appeal's statement in the *Manchester Ship Canal* case ought not now to be regarded as a current statement of the law. Alternatively, counsel said that the law had been altered by necessary implication from the terms of the Law of Property Act 1925 and the Land Charges Act 1925, two Acts of Parliament which came into operation at the same time. It was submitted that sections in those Acts, and later statements or silences in cases in the House of Lords and Court of Appeal, were really inexplicable if the law still was that a right of pre-emption did not confer an interest in land.

B He (his Lordship) thought it very highly probable from the recitals and the operative terms of the conveyance of July 1969 that such conveyance had been executed by the Amersham Rural District Council in exercise of the powers conferred by section 104 of the Housing Act 1957, and that the particular clause he had to consider in the relevant conveyance was inserted pursuant to section 104. He had offered Mr Cowell an opportunity to investigate the facts to see whether there was any other power under which the Amersham council might have sold the house, and he (counsel) had declined, seeking a decision on the matter as it stood. That being so, he (Goulding J) proceeded on the footing that the C terms of section 104 were indeed relevant to the case. The important subsection for present purposes was subsection (3), which allowed a local authority selling a house to impose conditions:

(a) limiting the price at which the house may be sold during any period not exceeding five years from the completion of the sale;

(b) limiting the rent at which the house may be let to the limit imposed by section 20 of the Rent Act 1957 during that period; and

(c) precluding the purchaser (including any successor in title of his and any person deriving title under him or any such successor) from selling or letting the house during any such period unless he has notified the authority of the proposed sale or letting and D offered to resell or sell the house to them and the authority have refused the offer or have failed to accept it within one month after it is made, and prescribing or providing for the determination E of the price to be paid in the event of the acceptance of such an offer.

Subsection (5) provided that where any condition such as was mentioned in paragraph (a), paragraph (b) or paragraph (c) was imposed on sale of a house, it should be registered in the register of local land charges by the proper officer in such manner as should be specified by rules. Mr Lightman had continued his argument by pointing to the express indication in section 104 (3) (c) that the condition a local authority might impose was to preclude not only the purchaser but "any successor in title of his and any person deriving title under him or any such successor" from disposing of the house F without giving the local authority first refusal. This, said counsel, meant that if Mr Cowell was right, and under the general law a mere contractual right of pre-emption did not give rise to an interest in land binding successors in title of the grantor, then Parliament had erected a special class of such agreements. Mr Cowell, on the other hand, submitted that if Parliament, for its own purposes, authorised the making of certain agreements, it was not to be inferred that such agreements, when made on the invitation of Parliament, would have any higher validity than they would have had under the general law had the authorising Act not been G passed.

He (his Lordship) accepted Mr Lightman's argument on this point. Whatever might be the true explanation of the *Manchester Ship Canal* case, he felt no doubt that agreements made in pursuance of section 104 of the Act of 1957 were valid as contemplated by Parliament. It was thus not necessary for him to express any opinion on the present state of the law relating to mere private contracts of first refusal. The registration ought to remain, and the motion must be dismissed. That in fact meant the dismissal of the action, the parties having agreed to treat the hearing of the motion as the trial of the cause. The plaintiffs must pay the council's H costs.

LANDLORD AND TENANT
GENERAL

A

COURT OF APPEAL

March 6 1975

(Before Lord Justice CAIRNS, Lord Justice ROSKILL and Lord Justice BROWNE)

IVORY AND ANOTHER v PALMER

Estates Gazette February 7 1976

(1975) 237 EG 411

Foreman carpenter's employment described in negotiations as "a job and a house for life"—Order for possession of house made despite wrongful termination of employment by estate—No licence for life, no tenancy for life under Settled Land Act 1925

B

This was an appeal by Mr Eric Palmer, of 22 Tavistock Road, Roborough, Plymouth, from a decision of Judge Chope at Plymouth County Court on August 5 1974 granting possession of 22 Tavistock Road to Mr Eric James Ivory, Mr James Gilbert Sydney Gammell and Mr Charles Matthew Farrer, trustees of the Maristow Estate, Devon.

Mr G M Godfrey QC and Mr S A B Parish (instructed by Lucien A Isaacs & Co, agents for Arthur Goldberg, of Plymouth) appeared for the appellant, and Mr J A S Hall QC and Mr J G Hull (instructed by Wolferstans, of Plymouth) represented the respondents.

C

Giving judgment, CAIRNS LJ said: This is an appeal from a decision of His Honour Judge Chope, sitting at the Plymouth County Court on August 5 of last year. The plaintiffs in the action are the trustees of a settlement. The defendant is a man of 53 or 54, who had been employed by the plaintiffs as a foreman carpenter. While so employed he had the occupation of a house belonging to the plaintiffs. The time came when he was dismissed from his employment, but he remained in occupation of the house. The plaintiffs brought an action for possession against him. He defended and counterclaimed. He contended that by an oral agreement between himself and a representative of the plaintiffs he had been employed for life, subject to his doing his work properly, and had been given a rent-free tenancy of this house for the joint lives of himself and his wife. He counterclaimed a declaration to that effect; alternatively, that he was a protected tenant under the Rent Act. Then, as an alternative, if he were held not to be entitled to remain in possession, he claimed £500 in respect of expenditure which he said he had incurred on improving the house, damages for loss of the right to occupy it, and damages for wrongful dismissal. The judge held that he was employed for life subject to working properly, and that he had a licence to occupy the cottage so long as he continued in that employment. The judge found that the employment was in fact terminated by the plaintiffs without good reason and without proper notice to the defendant; that the defendant had suffered no loss of earnings, at any rate that he had proved no loss of earnings, but that he was entitled to some damages based on the cost of removal to another residence and the loss of the rent-free residence with which he had been provided, and he assessed the damages at £200. The defendant appeals, contending that the claim for possession should have been dismissed, that a declaration should have been made in accordance with his counterclaim, and that damages should be awarded to him on the basis of that claim. The plaintiffs have served a cross-notice, contending that the judge should have held that the employment was not for life but was subject to reasonable notice, and that the right to occupy the house was a licence subject to revocation upon reasonable notice.

D

E

F

It is convenient to deal first with the issue raised by the plaintiffs' cross-notice, and this involves giving a rather fuller account of the facts. In September 1973 the plaintiffs advertised for an estate carpenter. The defendant applied for the job. He was interviewed by Mrs Hood, who had authority on behalf of the plaintiffs to engage staff. There was a discussion between them, in the course of which it was agreed that the defendant should be employed as foreman carpenter and that he should have a house which he and his wife could occupy free of rent. The defendant's evidence was that Mrs Hood said that he would have a job for life and security for life and a home for his widow. Mrs Hood, though her answers were not always consistent with each other, said in chief that she had told the defendant that if he worked satisfactorily, it was a job for life. The judge's finding on the matter was this: "I am driven to the conclusion that what the plaintiffs offered was a job for life as long as Mr Palmer worked satisfactorily." The salary was later fixed at £1,500 a year. On September 20 1973 a letter was written by the plaintiffs, addressed to the defendant, setting out terms of employment, making no reference to its being employment for life, and indeed, on the contrary, providing for two weeks' notice on either side. The defendant said that he never received that letter. The judge accepted that he never received it, and it disappears from the picture altogether. The question is: what was the contract resulting from the discussion between the defendant and Mrs Hood, together with the later fixing of the salary at £1,500 a year and the taking up by the defendant of his work and his going to live with his wife in the house, which he very shortly did? All went well, apparently, until the early part of 1974, when the plaintiffs formed the opinion that the defendant was not working satisfactorily. He was interviewed on February 11. There was a dispute on the evidence as to the effect of the interview, but the judge held, in favour of the defendant, that he was told then that he was going to be dismissed and that he protested against that. On February 28 he was dismissed summarily and called upon to give up possession of the house. He then left the employment of the plaintiffs and took up some other work. According to his evidence, he first took up work as an area manager for a certain company, and then later he left that and began to work on his own account. The judge held that the plaintiffs had no good ground for saying that the defendant's work was unsatisfactory, and accordingly found that he had been wrongfully dismissed.

G

H

J

K

The plaintiffs' contention in this court as to the terms of employment is that the defendant was employed subject to reasonable notice on either side; that what was said about a job for life was no more than an expression of expectation that, if all went well, the job would in fact be a permanent one. Alternatively, it was argued that if there was any contractual term about a job for life, it did not have the effect contended for by the defendant. There is no doubt that a person may be employed on a contract of service for life, even though there is no undertaking on his part to continue in the employment of the employer: see *Salt v Power Plant Co Ltd* [1936] 3 All ER 322, a decision of the Court of Appeal, and *McClelland v Northern Ireland General Health Services Board* [1957] 1 WLR 594, a decision of the House of Lords. These cases show that the courts will lean against such a construction and that clear words are needed to bring it about, though in both of those authorities the word "permanent," taken with other factors, was held to connote employment for life. Here the very words "a job for life" were used, according to evidence which was accepted by the learned judge, and nothing else was said as to the duration of the employment or as to any circumstances in which it could be terminated. In my opinion, the judge was entitled to find on the evidence here that what was said was intended to be contractual in effect, or, putting it more accurately, according to a passage which has been shown to us by Mr Godfrey, on behalf of the defendant:

> "The conduct of each party and the language which was used was such as to show that each was reasonably entitled to conclude from the attitude of the other that that was the effect of the conversation."

The passage in question is in the speech of Lord Reid in *McCutcheon v David MacBrayne Ltd* [1964] 1 WLR 125 at the foot of p 128. I cannot say that the judge was wrong in his conclusion on the facts as to this matter, and in so far as the question is one of law, I cannot say that he was wrong to hold that there was a contract to the effect of the defendant being employed for life. The exact effect of that expression may be open to some doubt, but for my part I cannot see any other possible interpretation than that it meant either "a job for the whole of your natural life," or "a job for the whole of your working life." The latter would seem to be so much the more reasonable construction that I would so interpret it in this context. The judge did not go into the question of the exact meaning of it.

What then is the effect of the contract so far as the right to occupy the house is concerned? The judge held that the defendant had a licence to occupy the house ancillary to his contract of employment, and that when the employment was terminated, although wrongfully, the licence came to an end. Mr Godfrey contends that this is wrong, and that the defendant occupied as a licensee for life on a licence which could not be revoked, or as a tenant for life under the Settled Land Act 1925, or as a tenant under a lease for life which, by reason of section 149 of the Law of Property Act 1925, would take effect as a lease for 90 years. The defendant's original case of a tenancy for the joint lives of himself and his wife has not been pursued on appeal, and Mr Godfrey did not seek to establish an alternative case under the Rent Act. The contention which he developed most fully was that in favour of a licence for life. There is no doubt that a licence for life can be granted. If the licensor purports to revoke such a licence, is the effect that the licence is revoked, leaving the licensee to claim in damages for breach of contract, or is the licence irrevocable during the life, entitling the licensee to remain in occupation? In *Foster v Robinson* [1951] 1 KB 149, where a farm worker had been told that he could live in a cottage rent-free until he died, Lord Evershed MR said at p 156:

> Since the recent decision in *Winter Garden Theatre (London) Ltd v Millennium Productions Ltd* [1948] AC 173 I think that, although a licence of that kind may, apart from the terms of the contract, be revoked, it may now be taken that if the landlord, having made that arrangement, sought to revoke it, he would be restrained by the court from doing so. Thus the result is arrived at that the tenant was entitled as licensee to occupy the premises without charge for the rest of his days, and he did so.

Singleton LJ expressed himself to a similar effect at pp 160 and 161. The actual claim in that case had been a claim for possession against the widow of the farm worker. An order for possession had been made in the county court, and that order was affirmed by the Court of Appeal. However, there have been several later cases in which the present Master of the Rolls has referred to *Foster v Robinson* as authority for the proposition that a licensor will not be allowed to eject a licensee in breach of contract: see *Combe v Combe* [1951] 2 KB 215 at 219; *Errington v Errington* [1952] 1 KB 290 at 298; and *Binions v Evans* [1972] Ch 359 at 367. Mr Godfrey contends that a dictum of Lord Upjohn in *National Provincial Bank v Ainsworth* [1965] AC 1175 at 1239 is to be understood in a similar sense. I am not altogether satisfied that that is a correct reading of that sentence in Lord Upjohn's speech.

It is unnecessary in this case to decide what is the correct rule in respect of a licence which is a licence for life *simpliciter,* because the contract in the present case was a contract for service and for a licence. The judge held that the licence was ancillary to the service, and in my opinion that is plainly right. The occupation of the house was rent-free, and it cannot have been intended that the defendant should continue to occupy the house rent-free if his service came to an end, for whatever reason. Mr Godfrey says that the defendant would be prepared to pay a fair rent for the premises, but that would be making a fresh bargain for the parties. Counsel also contends that despite the normal rule that a contract of service cannot be specifically enforced, an injunction can be obtained in exceptional cases to restrain an employer from dismissing a servant, and he cites *Hill v C A Parsons & Co* [1972] Ch 305, and especially this passage from the judgment of the Master of the Rolls at p 314:

> Suppose that a senior servant has a service agreement with a company under which he is employed for five years certain—and, in return, so long as he is in the service, he is entitled to a free house and coal—and at the end to a pension from a pension fund to which he and his employers have contributed. Now, suppose that, when there is only six months to go, the company, without any justification or excuse, gives him notice to terminate his service at the end of three months. I think it plain that the court would grant an injunction restraining the company from treating the notice as terminating his service.

I do not believe that the Master of the Rolls would have said the same of a contract for service for life which had been running only for a few months when the dismissal took place. Even assuming that this contract could in effect be specifically enforced by an injunction (which I do not for a moment accept), there has been no attempt so to enforce it. The defendant has accepted the dismissal from the employment, and he has given up his employment. He clearly cannot then claim to retain the benefit of the licence to occupy the house. It is contended that the plaintiffs are not entitled to rely upon their own wrong in dismissing him in breach of their contract with him. But, of course, in a sense, whenever there is a wrongful dismissal, the employer is taking advantage of his own wrong. He is entitled, although he was wrong in dismissing the servant, to refuse to have him working for him any more. If, for instance, the servant was entitled to the use of a motor car during the period of his service, the employer would be entitled to refuse to let him have it any longer, and in that sense to take advantage of his own wrong. I cannot see that what is claimed by the plaintiffs in this case in respect of the possession of this house is any more taking advantage of their own wrong than those circumstances which I have mentioned.

I therefore reach the conclusion that there was not here an irrevocable licence for the life of the defendant. Then did he become a tenant for life under the Settled Land Act?

A The contention that he did is based on the decision in *Binions v Evans*, where the majority of the Court of Appeal held that where the trustees of an estate agreed to permit a woman "to reside in and occupy" a cottage "free of rent for the remainder of her life or until determined as hereinafter provided," she became a tenant for life under the Settled Land Act. That case, in my view, stretched to the very limit the application of the Settled Land Act. The Master of the Rolls at p 366 of the report, says:

> But it was suggested here that the defendant was a tenant for life under the Settled Land Act 1925, with some support from *Bannister v Bannister* [1948] 2 All ER 133. I cannot think this B can be right. A tenant for life under that Act has power to sell the property, and to lease it (and to treat himself or herself as the owner of it): see sections 38 and 72 of the Settled Land Act 1925. No one would expect the defendant here to be able to sell the property or to lease it. It would be so entirely contrary to the true intent of the parties that it cannot be right.

I cannot accept that a right of occupation expressed to be granted for the period of a job, which in its turn is described as a job for life, could carry with it a Settled Land Act tenancy. As to an ordinary lease for life, the short answer is that there was here no rent fixed. There are no materials which would enable a rent to be quantified, and in C my view there are no circumstances which would justify the court in holding that there was a lease rather than a licence.

I therefore reject all the contentions directed to establishing that the defendant is entitled to remain in possession of the premises. His remedy is in damages. Then how are the damages to be quantified? For the loss of the employment the damages would be the difference, if any, between his salary under the contract with the plaintiffs and his earnings after dismissal, multiplied by an appropriate multiplier. Here, however, there was no evidence that his earnings after dismissal were any less than his contractual salary. So that element of damages does not arise. For the loss of the D house, the damages would be the value of any improvements which he had made to the house itself, as distinct from furniture, which he could take away; the expense of removal, and a figure based on the extra cost of a new home to him. No evidence was given of the loss in any of these three respects, and, as was pointed out in *Sunley (B) & Co Ltd v Cunard White Star Ltd* [1940] 1 KB 740 at p 745 by the judgment of the Court of Appeal, consisting of Scott, Mackinnon and Clauson LJJ,

> "It was the business of the plaintiffs to establish, if disputed, that the wrong had been done, so also it was the business of the E plaintiffs to prove the amount of pecuniary loss occasioned to them."

It might be argued that a new house would be likely to cost something, whereas the one he has lost cost him nothing, but this is to overlook the fact that the salary that he was paid by the plaintiffs must have taken into account that he had a free house, and there is nothing to show that the amount of his later earnings did not fully compensate him for the loss of the free house. We were referred to a decision in *McClelland v Crowther*, an unreported case heard by my brother Browne LJ as a judge of first instance, where he assessed the figure of £7,000 for the loss of a cheap house F in a personal injuries case, but there the plaintiff was incapable or almost incapable of earning anything in future, and it was plainly necessary to add to the figure based on the loss of earnings the value of the house, as to which the evidence was that its rental value was about £15 a week more than he was charged for it. I do not find that that decision is of any assistance to the court in assessing damages in this case. It was pointed out to us that in a 19th-century case, *In re English Joint Stock Bank* (1867) LR 4 Eq 350, Page-Wood V-C found, as a matter of principle, that the plaintiff was entitled to damages in respect of the loss of the use of a house, and the matter was referred for inquiry as to the damages in accordance with the common Chancery practice. G The practice in the county court is that the damages are assessed at the time of trial. Although application was made on behalf of the defendant for damages to be referred to the registrar, the learned judge refused to refer them. He was perfectly entitled so to refuse, and there has been no appeal against that interlocutory decision of his.

In those circumstances, it was necessary for the defendant to establish the amount of his damages at the trial. Mr Godfrey says: "Well, there are some materials upon which damages could be assessed. Mrs Hood agrees that in the course of her conversation with the defendant she said that he had in this house the equivalent of £10,000 if he worked H satisfactorily." I dare say that the value of the house was in the neighbourhood of £10,000, but that is really of no assistance in reaching a conclusion as to how much worse off the defendant would be in having to go to some other house the value of which is completely unknown. The other figure that appears in the evidence is the figure of £5.25, which had been assessed by a rent tribunal as the fair rent of this particular house; but I make the same observation about that, that there are no materials to make a comparison between the value of that house and the cost of any other house that the defendant may have to acquire, taking into account, as it would be necessary to do also, any change J in the rate of money earnings. The conclusion I have reached is that the defendant has simply failed to establish any sum of damages. The learned judge, taking a very broad view of the matter admittedly—I think he used the expression "taking a figure out of the air"—assessed the damages at £200. The plaintiffs have not appealed against that to say it is excessive, and I cannot say that it is insufficient. I would dismiss the appeal.

ROSKILL LJ said that he agreed that the appeal should be dismissed, though on the first point he would have been content, had it been necessary, to have upheld the decision K below on the ground that there was not on the facts of the case the contract of employment between the plaintiffs and the defendant which the judge found had been concluded. He (his Lordship) could not help thinking that if one looked at the negotiations between Mrs Hood and the defendant through the eyes of Mackinnon LJ's famous "officious bystander," and asked what these two people would have replied if he had said to them when they thought they had reached agreement, "Have you decided that the man is to have this job and the cottage literally for the rest of his life, and his widow the cottage thereafter," Mrs Hood would have turned to him and said, "Good heavens, no. but that is a L likely result if he continues to give us good service over the years," and the defendant would not have disagreed. [Continuing, his Lordship said:] Now let me assume, as I am most readily prepared to do, since my Lord takes the opposite view, that I am wrong in what I have just said, and Mr Godfrey has a correct finding of fact and a correct conclusion of law as to the nature of the contract—that it was a contract of employment for life. The argument then is that the plaintiff thereby acquired a contractual licence to occupy the house for life, even though his service contract came to an end, since that termination of his service contract came to an end through its wrongful repudiation by the plaintiffs. M The starting-point for the consideration of this argument is an elementary proposition of contract law. It was discussed in one of the cases referred to in *Hill v Parsons*, to which my Lord has already referred, *Vine v National Dock Labour Board*, reported in this court in [1956] 1 QB 658, and in the House of Lords, in [1957] AC 488. I can take the relevant quotations from the judgments of the Master of the Rolls and Sachs LJ in *Hill v Parsons*. In *Vine's* case in this court Jenkins LJ (as he then was) said:

> "But in the ordinary case of master and servant the repudiation or the wrongful dismissal puts an end to the contract, and the

contract having been wrongfully put an end to a claim for damages arises. It is necessarily a claim for damages and nothing more. The nature of the bargain is such that it can be nothing more."

In the House of Lords this characteristically lucid statement was expressly adopted by Viscount Kilmuir LC ([1957] AC 488 at 500). In reference to what he called the ordinary master and servant case the Lord Chancellor said, "If a master wrongfully dismisses the servant, either summarily or by giving insufficient notice, the employment is effectively terminated, albeit in breach of contract."

I am not going to say anything about *Hill v Parsons* (a majority decision of this court on an interlocutory appeal), beyond that it obviously was a very special case in which the plaintiff was seeking to assert the continuity of his contract of employment by inviting the court to restrain the employer from giving effect to his purported dismissal. The present is not such a case. Once the defendant—the present appellant—was dismissed, under what was an ordinary service agreement between employer and employee, he had, as a matter of the ordinary contract law of master and servant, no alternative but to accept that dismissal as an accomplished fact, whether it was right or wrong. If it was right, he had no remedy; if it was wrong, his remedy, as was pointed out in *Vine's* case, lay in damages. But he cannot, in one and the same breath, as I ventured to point out to Mr Godfrey yesterday, accept that his service agreement is at an end, and yet assert that his right to the continued occupation of the cottage remains unaffected. This is not a severable contract. It is a single contract of employment between employer and employee. Once that contract goes, it goes as a whole, and that, with the greatest respect to Mr Godfrey's argument, is the short answer to this appeal.

Mr Godfrey sought to argue that, notwithstanding what I have just said, none the less there was in this case a form of contractual licence which equity would protect. Over the years there has been much discussion in this court and in the House of Lords about a married woman's supposed equity. I say nothing about that, nor about the cases upon which it was supposedly based, nor about the present statutory position. So far as I am aware, whatever the rights and wrongs of the different views that were expressed at different times in this court and elsewhere upon that subject, those cases have no application whatever here. I find it quite impossible to see how this man can have acquired any interest in this cottage which at any time in legal history equity would have enforced or protected by injunction or otherwise. There was no estate in land here. At the most he had a contractual licence to occupy that cottage for so long, but only for so long, as he remained in the estate's employment. Once his service agreement went, then, in my judgment, it went as a whole. So, on the second point, I find myself in entire agreement with my Lord. I venture to think, on the basis that the learned judge reached the right conclusion on the first point and that I am wrong, that the learned judge, on the second point, on which he decided this case against the defendant, clearly reached the right conclusion.

So far as the rather strange argument under the Settled Land Act is concerned, I have nothing to add to what my Lord has said. On damages all I would say is this. A plaintiff suing for damages for breach of contract (for that is what the defendant was doing by reason of his counterclaim) must prove his case. *Sunley v Cunard* is the clearest possible authority for that proposition, if authority is wanted for anything so elementary. With the greatest respect to those who represented him below, his pleading did not adequately plead the loss he claimed to have suffered. There may have been a very good reason for that omission; indeed, there probably was, because I suspect he may not have suffered any loss, or at least could not prove he had suffered any loss. But whether he could have proved it or not, he did not plead or prove it. I would not myself wish to take a pleading point against him in a county court case if in the event he had proved the loss. But not only did he not plead it; he never proved it. Mr Godfrey has made a valiant effort to make bricks without straw, but in my judgment there simply is no evidence which would justify an increase in the damages. For the reasons my Lord has given, £200 may indeed have been too much. It certainly is not shown to have been too little. In the result, therefore, I would dismiss the appeal.

BROWNE LJ said that he also agreed that the appeal should be dismissed. He was bound to say that he thought the contract alleged by Mr Palmer and the rather different contract found by the judge were so improbable that it was almost incredible that anybody on behalf of the plaintiffs should ever have made such a contract. The plaintiffs knew little about the defendant or his work at the time when the contract was made, and evidently knew nothing about his capacity as a supervisor, which was what he was going to be. There was no mutuality between the parties, in that the defendant could leave at any time, whereas the estate was bound for his life if he chose to stay with them. It seemed to him (his Lordship) that all the probabilities were that what was said about a job for life, and so on, was merely the expression of a hope or expectation of what would happen if all went well, rather than a contractual term. Nevertheless, after much hesitation, he (Browne LJ) had come to the conclusion, in agreement with that of Cairns LJ, that the court should not interfere with the decision below on this point.

[Continuing, his Lordship said:] On the rest of the case I agree that, for the reasons already given by both my Lords, Mr Godfrey's most interesting arguments should be rejected on all points. So far as the Settled Land Act point is concerned, I add very little. In the case to which my Lord has already referred, *Binions v Evans* [1972] Ch 359, the majority of this court, as I understand them, based their decision on the previous decision of this court in *Bannister v Bannister* [1948] 2 All ER 133, but as I read the decision in *Bannister*, the court did not in fact in that case base its decision on the Settled Land Act 1925, but on more general principles of the law of trusts. It is true that at p 137, between letters B and C, the court varied the form of the declaration made by the county court and said this: "A trust in this form has the effect of making the beneficiary a tenant for life within the meaning of the Settled Land Act 1925, and consequently there is very little practical difference between such a trust and a trust for life *simpliciter*." As I understand it, however, the actual decision in the case was not based on the Settled Land Act.

So far as damages are concerned, I entirely agree with my Lords that Mr Palmer entirely failed to call any evidence whatever which would have justified the county court judge in awarding him any damages, certainly any damages in excess of the £200 which were in fact awarded on the counterclaim. Accordingly, as I say, I agree that this appeal fails and must be dismissed.

The appeal was dismissed with costs, the order not to be enforced without further leave of the court. Leave to appeal to the House of Lords was refused.

COURT OF APPEAL

May 21 1975

(Before Lord Justice CAIRNS and Lord Justice ORMROD)

HUTCHINGS v HUTCHINGS

Estates Gazette February 21 1976

(1975) 237 EG 571

"Transfer of property" order may properly require transfer of a council house tenancy despite the existence of a

A **covenant against assignment, provided the council's consent to the transfer is or will be given—Observations on the part the court plays in determination of such issues**

These were appeals by Mrs Caroline Elizabeth Hutchings, of 17 Loder Gardens, Worthing, from orders made by Judge White at the Worthing County Court on January 9 1975 rejecting her application for an order under section 24 of the Matrimonial Causes Act 1973 requiring the respondent, her former husband, Mr Frederick Thomas Hutchings, also of 17 Loder Gardens, to transfer the tenancy of the house to her, and granting the respondent an order under section 17 of the Married Women's Property Act for possession of B the house within eight weeks. The respondent entered a cross-notice appealing against the judge's finding that on the merits the appellant had a good case for a transfer of the tenancy.

Mr R J Seabrook (instructed by Ward, Bowie & Co, agents for Davies, Thomas & Cheale, of Worthing) appeared for the appellant, and Mr A C B Hunter (instructed by Arbeid & Co, of Worthing) represented the respondent.

Giving judgment, CAIRNS LJ said that the parties were married in August 1940, when the husband was 23 and the wife 38. They had two children who were now grown up. C In 1957 the husband took the tenancy of 17 Loder Gardens, which was a council house. In May 1972 the wife petitioned for a divorce on the grounds of the husband's unreasonable behaviour. After some vacillation on his part the suit became undefended, and a decree nisi was pronounced on February 13 1973. It was made absolute on March 28 1973. Neither party had so far remarried, but the husband was contemplating remarriage. The house was still occupied by the parties despite the dissolution of the marriage.

The case raised questions whether a council tenancy, which by its terms was subject to a covenant not to assign, was D such an interest as could be the subject of a transfer of property order under section 24, and if so, whether the court in its discretion should ever make such an order. The county court judge found that the wife had a strong claim because of her past contributions to the house and her future needs, but because of the condition of the tenancy against assignment he (Judge White) thought it doubtful whether the husband had any estate which he could transfer. In expressing that opinion, the judge was accepting the authority of a passage in the judgment of Dunn J in *Brent v Brent* [1975] Fam 1. He went on to say that if there was jurisdiction to make the order he considered that it should, as a E matter of discretion, be refused, both because the question whether a transfer should be allowed was a matter of policy for the local authority and because the making of an order would have required the husband to break his contract with the authority. A further reason the judge gave for his decision was that if the transfer were effected the wife's right to retain the house was so tenuous that an order ought not to be made. The wife appealed, contending that the judge was wrong in law; the husband by his cross-notice contended, as an alternative to the grounds on which the judge based his decision, that the judge was wrong in finding that on the merits the wife had a good claim for a transfer.

F So far as the law was concerned, it had been further elucidated by two decisions of the Court of Appeal. One was the case of *Thompson v Thompson* [1975] 2 WLR 868, in which it was held, contrary to what had been suggested by Dunn J in what was in fact an obiter passage in his judgment, that the tenancy of a council house was property to which a transfer of property order could apply. It was pointed out in argument in the present case that in *Thompson v Thompson* there was no covenant against assignment, whereas here there was. The other case was *Hale v Hale*, so far reported only in *The Times* for February 25 1975 [see now [1975] 1 WLR 931—Ed]. That was a case of a contractual weekly tenancy where again it was held that G that constituted property which could be transferred under section 24. Those cases indicated in terms, particularly *Hale v Hale*, that if there was a covenant against assignment then that meant that a transfer of property order could not be made. It was said that here was a covenant against assignment. But that must always be subject to there being no obligation on the part of the local authority as landlords to insist upon that covenant. The tenancy could not be assigned without their consent. But if they had consented, then it became an assignable tenancy.

In his (Cairns LJ's) view, the local authority had quite H clearly consented to the transfer of the property, having regard to an exchange of letters which took place in June 1974. A letter from the council which contained the words "if your client succeeds in her application it would be the intention of this department to transfer her to smaller premises" could only mean that if an order was made which had the effect of transferring the tenancy to the wife, then the council would regard her as the tenant, and as a tenant who could be given four weeks' notice to quit and then rehoused in a smaller residence. That being the position, and it being clear from the later correspondence that unless a transfer of property order was made the council would J not be willing to find such an alternative residence for the wife, he (his Lordship) was clearly of opinion that there was jurisdiction on the part of the court to make the order. He further took the view that it was proper for the court to exercise its jurisdiction in favour of making such an order.

Mr Hunter had referred the court to section 7 of the Matrimonial Homes Act 1967, which dealt with the transfer of a protected or statutory tenancy from one spouse to another, and pointed out that under that section notice had to be given to the landlord, whereas there was no such provision in relation to section 24. He (his Lordship) could not see the relevance of that. It being now accepted that, generally K speaking, a weekly tenancy, including a council tenancy, was property which would fall within section 24, he failed to see how it could be said that any analogy with section 7 of the Matrimonial Homes Act could properly be drawn. In the end the question here was whether, on the merits, the wife ought to have the order made in her favour. The county court judge was clearly of opinion that on the merits she was so entitled. He (Cairns LJ) agreed with him. It was a case of a long marriage in which it could be taken that the wife's contribution to the upkeep of the home entitled her as between herself and her husband to an equal right to be awarded a share of the home if it were something L which could be divided. In these circumstances, this home not being something which was divisible, it was appropriate to look to the future and see which partner would suffer the greater hardship by having to leave. Having regard to the husband's much greater earning capacity than that of the wife, and to his being much younger, the judge was right in his view that on the merits the transfer of property order ought to be made. He (his Lordship) would allow the appeal and make such an order.

Agreeing, ORMROD LJ said that the case illustrated the role which the court could properly play in these difficult questions of council tenancies. It would be obviously M invidious to expect the housing manager to decide these difficult questions of justice as between the former spouses. This was peculiarly the sphere of the court, which had facilities for investigating such matters and adjudicating upon them. Equally it would be quite improper for the court to interfere with the housing policy of the local authority. It followed that the court and the housing authority must work together. The court in these cases would always need to know the attitude of the housing authority to the problem. The solicitors for the wife in this case very properly took the matter up with the local authority at an early stage, and as a result the court knew the authority's attitude. He (his Lordship) did not regard the provision against assignment as any obstacle to the exercise of the court's powers under section 24, and he entirely agreed with what had been said with regard to the merits. The former wife in this case required all the protection and support which the court could give her.

The court allowed the appeal, granted the order sought and set aside the order for possession which had been made in favour of the respondent.

CHANCERY DIVISION

November 1 1974

(Before Mr Justice WALTON)

HAYWARD v HAYWARD

Estates Gazette February 21 1976

(1974) 237 EG 577

Matrimonial home vested by husband in self and wife as joint tenants beneficially—Wife leaves and seeks order for sale—Wife held responsible for break—Inequitable for her to be allowed to turn husband out of his home, order for sale refused—Important question raised by Walton J—Does a joint beneficial tenancy created by one party who puts up the money imply a purpose by which each party may remain in residence for life?

By this summons, Mrs Winifred Emily Hayward, of Trowbridge, Wilts, claimed an order for sale of 57 Somerset Road, Frome, Somerset, formerly the matrimonial home of herself and the respondent, her husband, Mr William Robert John Hayward.

Mr D J M Campion (instructed by Turner, Peacock, agents for Daniel & Cruttwell, of Frome) appeared for the applicant, and Mr W H Goodhart (instructed by Knapp-Fishers) represented the respondent.

Giving judgment, WALTON J said that the parties were married in 1932. The husband was now 68 and the wife 65. Their first two matrimonial homes were purchased entirely out of moneys provided by the husband. Their third home, 57 Somerset Road, Frome, with which this summons was concerned, was also, for all practical purposes, acquired with moneys all of which traced their origin back to the husband, and not surprisingly, since the wife did not work or have any independent income. The property was in joint names, there being an express trust for the two parties as joint tenants beneficially. The marriage was stormy in its latter years, and at the beginning of 1973 the wife walked out with her unmarried son, Graham, who had been living at home. Graham had now bought a house in which his mother resided. If he got married, she would have to leave. Mutual notices severing the joint beneficial tenancy had been given, and the property was now held upon trusts under which husband and wife were beneficial tenants in common in equal shares. The wife now applied for an order for sale with vacant possession and division of the proceeds. There were no prospects either of reconciliation or of divorce. He (his Lordship) thought both parties were to some extent acting unreasonably: the husband in wishing to live in a house which was too large for him and would be a burden to maintain, and the wife in insisting on a sale now, when she had no immediate need of her share of the proceeds of sale and when that money was soundly invested in property which would almost certainly increase in value. But he was not really concerned with the reasonableness of the parties' attitudes. He was troubled as to whether the £6,000 which each party would be likely to receive from a sale would be of any use in providing alternative accommodation. There was no evidence that it would. But he had not felt this was a circumstance he ought to take into account, however convinced he was that an order for sale would simply result in the local authority having to rehouse the husband now and the wife at some time in the future.

It was accepted by both sides that he (Walton J) had a discretion as to whether to order a sale. The dispute was mainly in relation to the strength which should be attributed to the various established criteria and their applicability to the present case. A number of cases had been cited: *Jones v Challenger* [1961] 1 QB 176; *Rawlings v Rawlings* [1964] P 398; *Appleton v Appleton* [1965] 1 WLR 25; *Bedson v Bedson* [1965] 2 QB 666; *Re Solomon* [1967] Ch 573; *Jackson v Jackson* [1971] 1 WLR 1539; and *Burke v Burke* [1974] 1 WLR 1063. From these authorities, Mr Campion, for the wife, drew the following principles, with which Mr Goodhart did not seriously quarrel:

(1) Where there was a trust for sale, the operation of that trust was suspended while the underlying purpose of that trust—in the present case, the provision of a matrimonial home—could be achieved.

(2) As soon as that purpose ceased to be capable of being achieved—in other words, as soon as the marriage was at an end in law or in fact—the trust for sale came into operation, but the court had a discretion, and would exercise that discretion by ordering a sale unless it was inequitable to do so; the test was not whether it was reasonable, but whether it was equitable.

(3) The fact that a wife had left her husband without just cause was not an absolute bar to the making of an order at her request.

(4) The fact that the party applying had invested money in the property was a fact which helped the court to come to a conclusion that there should be a sale, but there was no principle that the mere fact that the applying party had not invested money would be a bar to a sale.

(5) Where there was a real possibility that the beneficial interests might be varied in divorce proceedings, that might be a ground for withholding an order for sale.

Mr Goodhart said that in the present case there were a number of significant factors which, in conjunction, should result in the court refusing an order:

(1) Almost all the money was provided by the husband, as in *Bedson v Bedson*.

(2) The wife walked out without just cause.

(3) The husband had offered to pay rent, showing both his attachment to the property and his desire to play fair with his wife, since the sum offered (£5 per week) was obviously the maximum he could possibly afford.

(4) The wife did not at the moment require the money.

He (his Lordship) thought there was only one question of fact for real decision, and that was who was responsible for the wife having walked out. Having seen both parties in the witness box, he had no doubt that all the trouble in the marriage was due to the wife's attitude. Her husband spoke with the voice of truth and humility as against the voice of his wife, which was of stubborn pride. She was unable to see that she was a selfish domineering woman. Having decided that issue in favour of the husband, he (Walton J) felt that it would be inequitable for the husband to be turned out of the house for which he had provided all the money at the suit of his wife, who bore the responsibility for the breakdown of their marriage, when he was willing to pay her what he could by way of compensation.

That was sufficient to dispose of the matter, but he (his Lordship) would add that in considering the authorities, he had been puzzled by the stress laid upon the provision of the original purchase price by one or both parties: at first sight, this had nothing to do with the trusts with which one was dealing, whether those trusts had been declared at the instance of one party who provided the whole of the purchase money, or of both parties who had contributed thereto in whatever shares and proportions. However, it had occurred to him that in a case where one party—usually the husband—provided the whole of the purchase money, and arranged that the trusts should be either *simpliciter* those arising out of a conveyance to the husband and wife as joint tenants at law or, as here, to husband and wife as joint tenants beneficially, it might not be the whole of the purpose of the purchase merely to provide a matrimonial home for the parties. Spelling out what would be the legal position if equity had not interfered by enabling the joint tenancy to be so easily split, both would have occupied the property jointly during their joint lives, and the survivor then for the remainder of his or her life, thus ensuring that both of them had a roof over their heads for the term of his or her life. In other words, he (Walton J) thought that from a conveyance in this form, where only one person had provided the purchase money, one ought to infer that individual as well as collective provision of a home was intended. This led, of course, to the result that if one of the parties walked out of the home, well and good, but that party could not possibly ask for execution of the trusts so long as the other wished to remain in residence, as this would run counter to the implied purposes, or one of the implied purposes, for which the property was acquired. *Per contra,* if the trusts were for the parties as beneficial tenants in common, the situation would be completely different, since the survivor would have no right to the whole of the house, which obviously could not in a normal case be intended to be shared with the devisee of the first to die. Thus the purpose was restricted to that of providing a matrimonial home only.

It was right to say that this explanation had never before, so far as he (his Lordship) could find from the reported cases, been put forward. But he could not think that where a husband purchased a house as a residence for himself and his wife as joint tenants beneficially, he (the husband) did not contemplate and intend that he should be able to reside there as long as he wished, even if the marriage broke down. Lord Denning MR had come near to suggesting this kind of reasoning in *Bedson v Bedson* at p 675. He (Walton J) therefore thought it ought to be permissible to reach the conclusion he had reached by a very much shorter route: namely, to hold that as the husband provided the whole of the purchase price (or perhaps more accurately, that there was no element of investment by the wife of moneys in the house), and as the conveyance was taken into the names of the parties as joint tenants beneficially, the purposes for which the property was acquired had not failed, since the husband wished to continue to live there, and therefore no sale should be ordered.

His Lordship refused to order a sale. He ordered that the husband, so long as he remained in possession, should pay the wife £5 a week. No order was made as to costs.

QUEEN'S BENCH DIVISIONAL COURT

January 28 1976

(Before Lord WIDGERY CJ, Mr Justice KILNER BROWN and Mr Justice WATKINS)

WATHEN AND OTHERS v WHITE

Estates Gazette April 3 1976
(1976) 238 EG 45

Forfeiture for non-payment of rent—Judge right to adjourn case for further evidence as to tenant's means, but wrong to grant subsequent adjournment on tenant producing virtually no basis for relief—Mandamus nevertheless refused—Every reason to assume landlord would shortly obtain order at resumed hearing

In these proceedings Mr Mark William Gerard Wathen and others, the trustees of certain property in Norfolk, applied for an order of mandamus directed to His Honour Judge Adrian Head requiring him to hear and determine according to law their action against Sir Christopher Robert Meadows White, of Hill House, Northrepps, Norfolk, for possession of Hill House and arrears of rent.

Mr L Marshall (instructed by Sharpe, Pritchard & Co, agents for Mills & Reeve, of Norwich) represented the applicants, and Mr H K Woolf (instructed by the Treasury Solicitor) appeared as *amicus curiae*. The respondent was neither present nor represented.

Giving the first judgment, WATKINS J said that in April 1971 the applicants let Hill House to the respondent for a term of 25 years. The lease contained the usual proviso for forfeiture on non-payment of rent. Unhappily the respondent, having gone into occupation, defaulted in his payment of rent, with the result that he forfeited his lease. The applicants then applied to the county court for possession and arrears of rent. The matter came before His Honour Judge Head sitting at Norwich on October 2 1975. The applicants were represented by a solicitor; the respondent was not represented by solicitor or counsel, but was present in person. He admitted he was in arrears with the rent. Neither the applicants nor the respondent sought an adjournment at that hearing, but the judge ordered an adjournment, as he was entitled to do provided he acted judicially. In adjourning the matter the judge invited the solicitor acting for the applicants to contact solicitors for the respondent. The purpose of that was to gain information as to whether or not the respondent would be in a position to pay the arrears. The solicitor did as he was asked by the judge, and received information that the respondent might be in a position to pay the arrears in the near future. That information was given to the judge on the resumed hearing in December 1975. Once again the judge did not make a decision on the case. What he did was to further adjourn the matter until February 6 1976. He also made an order, the purport of which was to recite that the respondent had given the court an undertaking to assign over to the applicants so much as was received by him from other trustees as would cover the arrears of rent.

The judge was entitled to adjourn the hearing on his own motion, and he (his Lordship) thought that there could be no criticism of the first order for an adjournment, since the court was not at the time in a position to be able to judge with any degree of certainty the period which should be given the respondent to enable him to comply with the terms of the lease and pay off the rent arrears. That left the matter of the second adjournment. At any time after the first hearing it had been open to the solicitors for the applicants to seek a further hearing before the judge. That was done, and when the matter was restored to the judge's list evidence was tendered to the effect that there was some possibility of moneys coming to the respondent from a trust but no indication of when it would arrive. A judge had considerable power to assist any person brought before him for forfeiture based upon non-payment of rent, but what he was called upon to do was to make the necessary order once it had been established the plaintiff had made out his case. Although the length of time to be allowed to a defendant was a discretionary matter, that in no way diminished the duty of a judge to proceed to judgment at the instance of the plaintiff once the case had been made out. He (his Lordship) had reached the conclusion that the applicants had succeeded in establishing their case at the adjourned hearing, and that accordingly an order for possession should have been made in December.

There remained the question whether the discretionary remedy of mandamus should issue. There was much to be said for allowing an order to go, but whether so Draconian a step was necessary was a matter not free from question. There was every reason to suppose that in a few days' time, on February 6, the judge would carry out his duty, and in the circumstances it appeared that there was no need for a prerogative order.

Agreeing, LORD WIDGERY said that the judge had been motivated by no desire other than to produce justice and fairness between the parties, but that had resulted in his being unfair to the landlords. The first mistake he had made was not to remind himself there was a duty upon him to make the order. He had not wrongly exercised his discretion in adjourning the case in October, when there were vague possibilities that the tenant might come into funds. By the time December was reached, however, it was really obvious that no moneys were available. The judge should then have gone on and given judgment for the applicants.

Also agreeing, KILNER BROWN J said that he was confident that the judge would exercise his discretion properly when the hearing was resumed in February.

The application was dismissed, no order being made as to costs.

CHANCERY DIVISION

December 8 1975

(Before Mr Justice TEMPLEMAN)

AILION v SPIEKERMANN AND ANOTHER

Estates Gazette April 3 1976
(1975) 238 EG 48

Premium element in contract for assignment of protected tenancy and purchase of assignor's chattels—Purchasers allowed into occupation before completion held entitled to ignore notice to complete on the illegal basis and sue for specific performance of the contract's legal elements—"Does not follow that every purchaser of a protected tenancy at a premium is entitled to specific performance"

This was a claim by Mr Jack Albert Ailion, of Penywern Road, Earl's Court, London SW5, against Mr Eric Spiekermann and his wife, Mrs Joan Spiekermann, of 59 Oakwood Court, Kensington, London W14, for possession of those premises and for a declaration that a contract of April 8 1974 for assignment of a leasehold term thereof had been validly rescinded. The defendants counterclaimed specific performance of the contract to the extent that it was legal.

Mr R Pryor (instructed by Alfille & Co) appeared for the plaintiff, and Mr L Hoffmann (instructed by Davenport, Lyons & Co) represented the defendants.

Giving judgment, TEMPLEMAN J said that the contract of April 8 1974 provided for assignment by Mr Ailion to Mr and Mrs Spiekermann of the balance of a lease of 59 Oakwood Court expiring on December 24 1976 at a yearly rent of £850. There was the usual consideration consisting of covenants to indemnify the vendor, but the contract also required the purchasers to pay £3,750 for certain chattels subsequently valued at £604.75. This valuation had not been proved or accepted, but it was common ground that the chattels were worth much less than £3,750. It was also common ground that at the date of the contract the vendor and the purchasers were aware that the chattels were not worth £3,750 and that it was not lawful for the vendor to require or receive more than their value. The vendor wanted the money, and the purchasers needed the flat. The illegality in question arose from the fact that the lease of the flat constituted a protected tenancy within the Rent Act 1968. Section 86 (1) provided that any person who as a condition of the assignment of a protected tenancy required the payment of any premium was guilty of an offence. By section 88, where the purchase of any furniture had been required as a condition of the assignment of a protected tenancy, then, if the price exceeded the reasonable price of the furniture, the excess would be treated as if it were a premium required to be paid as a condition of assignment. Offering furniture at a price the vendor knew, or ought to know, was unreasonably high was an offence under section 89 (1). After contract in the present case the purchasers were allowed into possession of the flat as licensees on the terms of a letter of April 26 requiring them to "vacate the premises on demand in the event of them failing to complete the purchase in accordance with the contract." The date for completion was May 6, but the purchasers had trouble raising the money, and on May 16 the vendor served notice to complete; he also revoked the purchasers' licence to occupy the flat and demanded possession by the following day. The present litigation followed, in which the vendor claimed possession and rescission, while the purchasers counterclaimed specific performance of the contract on payment of a reasonable value for the chattels.

For the present purpose it did not matter whether the vendor committed all the offences alleged. It was admitted, and it was sufficient, that the contract originated in an illegal offer by the vendor under section 89 and could not be completed without an illegal receipt by the vendor under section 86 (2). The vendor had fallen foul of the statutory control of rented residential accommodation; the court was not concerned with the merits or demerits of the matter, and merely observed that one of the objects of the Rent Act 1968 was plainly to protect a purchaser requiring accommodation from a vendor seeking to exploit the financial value of the controlled rent and security of tenure established by Parliament. Thus, for example, section 90 of the Act provided that an excess premium should be recoverable even after it was paid. If the purchasers here had kept quiet and paid the £3,750, they could after taking the assignment, have sued successfully for the recovery of the amount by which the £3,750 exceeded the reasonable value of the chattels. Their argument in these proceedings was that they were only seeking to achieve, without further breach of the law, the result which section 90 was designed to produce, an assignment without the illegal premium. The argument for the vendor was that the contract, being illegal, could not be performed according to its terms, and that the court had no power to inflict a different bargain on the parties. Severance of the good and bad portions of the contract was, it was said, impossible or unprecedented.

He (his Lordship) thought that the effect of the legislation was to divide the contract into three separate elements. The vendor was contractually and legally bound to assign the lease in return for a covenant of indemnity; he was bound to transfer the chattels in return for their reasonable value; and he was contractually but illegally entitled to a premium higher than the value of the chattels. Parliament having effected this clear and distinct division between the illegal elements of the contract, the court was not powerless to remove the gilt and leave the gingerbread. The absence of precedent did not worry him (Templeman J) as long as there was no lack of principle. Where there were legal and illegal elements in a contract which were capable of severance, the jurisdiction to enforce the legal elements would only be exercised in a proper case. The fact that the purchasers knew of the illegality at the outset was a powerful reason why they should not obtain any relief, but the Rent Act was designed to protect persons in the position of the purchasers. They could not insist on the elimination of the illegal premium from the draft contract without losing the flat. They had no choice if they needed somewhere to live. They committed no offence. Again, if the purchasers, by obtaining specific performance without payment of the illegal premium were able to put themselves in a better position than if they had completed the contract according to its terms, that would in general be a good reason for the court declining assistance;

A but section 90 made all the difference. Whether a purchaser knew of the illegality or not, he could recover his illegal payment.

It did not follow that every purchaser of a protected tenancy at a premium was entitled to specific performance. The remedy was discretionary. If the vendor was ignorant of the facts or the law, or if the purchaser tempted the vendor with a cheque book, or if the vendor changed his mind before the purchaser altered his position in reliance on the contract, the court might decline to make a specific performance order. It all depended on the circumstances. In the present case the initiative came wholly from the vendor, and he now wanted B to recoil from the contract because he could not obtain the illegal payment on which, until the illegality became public, he insisted. Moreover, the purchasers had been allowed to go into the occupation of the flat. If they were now evicted it would appear that they had been turned out because they failed to raise and pay an illegal premium and to keep quiet about the illegality until after the contract had been completed by an assignment. That appearance was not consistent with the purpose of the Act or the administration of justice. If he (his Lordship) made no order either for possession or for specific performance, washing his hands of a contract containing an illegal element, the vendor would remain liable C for the rent and covenants under the lease and the position of both the vendor and the purchasers would be unsatisfactory. In these circumstances there would be an order for specific performance of the contract to assign the lease and to transfer the chattels in return for the consideration to which the vendor had been confined by Parliament.

CHANCERY DIVISION

February 24 1976
(Before Mr Justice WHITFORD)

D

KITNEY v MEPC LTD AND ANOTHER

Estates Gazette May 8 1976
(1976) 238 EG 417

Option to renew underlease—Option not registered as a land charge, void as against a purchaser—Head lease assigned and title to it shortly afterwards registered—Assignee of head lease subsequently acquires freehold reversion—Freehold merges with head lease and absolute title registered, subject however to underlease—Option provision in underlease held to affect freehold—Lessee entitled to extension

E

This was a preliminary point upon an originating summons by Mr Alan William Kitney, of Oak Knoll, Furzefield Road, Beaconsfield, Buckinghamshire, for a declaration that he was entitled to an extension of a lease of two shops in Harrow, Middlesex, for a further term of 21 years from December 25 1974. The defendants were MEPC Ltd, of Brooke House, 113 Park Lane, London W1, in whom the reversion expectant upon determination of the lease was now vested, and Greater London Properties Ltd, of 22 Conduit Street, London W1, the original lessors.

F Mr M Browne QC and Mr I L B Romer (instructed by Dale & Newbery) appeared for the plaintiff, and Mr P M F Horsfield (instructed by Clifford-Turner & Co) represented the first defendants. The second defendants took no part in the proceedings.

Giving judgment, WHITFORD J said: I have to consider a preliminary question in a proceeding in which the plaintiff is a Mr Kitney and the defendants are two companies, MEPC and Greater London Properties Ltd. Only MEPC are, however, concerned with the preliminary point. The facts relevant to this point are not really in dispute. In December 1932 the second defendants were granted a 99-year building lease of a property in Harrow, Middlesex. Hereunder in April 1933 the second defendants granted a lease, which I shall G call "the underlease," to a company, GB Outfitters Ltd. It was an underlease of two shops and living accommodation over them. It was granted for a term of 42 years at a yearly rental of £700. The only relevant provision for present purposes is to be found in the proviso to clause 5 (2). The proviso, which is in substance an option to renew, is in these terms:

> Provided always and the lessors hereby agree that if the lessee shall be desirous of extending this lease for a further term of 21 years from the expiration of the term hereby granted, and of such desire shall give to the lessor six months previous notice in writing, then and in such case, provided the lessee shall have H punctually paid the rents hereby reserved and observed and performed all the covenants and conditions on the lessee's part herein contained, this lease shall be continued for such further term of 21 years at the same rents as are hereinbefore reserved, and subject to the same terms covering the conditions as are herein contained except this proviso.

A memorial of this lease was recorded in the Middlesex Deeds Department of the Land Registry in June 1933. In October 1935 the underlease was assigned to Peters Outfitters Ltd. In March 1946 it was assigned to the plaintiff and his sister-in-law, Mrs E S Kitney, jointly. They occupied the J shop premises, where they carried on the business of gentlemen's outfitters under the name of A E Kitney, also collecting the rents of the residential flats, until December 5 1949, when the underlease was assigned to a company, W & G Kitney Ltd. This latter company changed their name to Alan Kitney Ltd in 1965. The underlease was assigned to the plaintiff in October 1966. The plaintiff duly and understandably gave notice of his desire to renew the underlease, and it is accepted that the notice was properly given, but his right to enforce the option is contested.

I must turn briefly to the changes of interest and title on the lessor's side. The second defendants assigned their lease K to a company, Monument Property Trust Ltd, in January 1934, and that company in their turn assigned it to the first defendants on July 25 1947. In October 1947 this leasehold title was registered at the Land Registry. In April 1962 there was a conveyance of the freehold, and the freehold title was registered for the first time on April 16 1962. In March 1969 the freehold was transferred to the first defendants; the freehold and the leasehold merged, and in May 1969 the first defendants were registered as proprietors with title absolute. The first defendants assert that although the plaintiff served his notice under and in accordance with the terms agreed in the underlease, this is an option which he is not entitled L to exercise because it is a charge registrable under the Land Charges Act which has never in fact been registered. The plaintiff accepts that in this court, in the light of the decision of Buckley J in *Beesly v Hallwood Estates Ltd* [1960] 1 WLR 549, it is not open to him to challenge the assertion that this option to renew is a registrable charge. He also accepts that it was not registered as a land charge prior to registration of the leasehold and freehold titles. It being agreed that this is a class C (iv) charge, the effect of section 13 of the Land Charges Act falls next to be considered. Section 13 (1) provides that certain class A charges shall be void as against M purchasers unless the charge has been registered in the register of land charges before completion of the purchase. Section 13 (2) deals with the effect of failure to register, *inter alia,* class C charges, and is accordingly the relevant subsection to be considered. Here again these charges are declared void as against purchasers unless the charge has been registered. But there is a difference, because the requirement is that the charge shall be registered in the "appropriate" register before completion. There was a certain amount of debate on the issues as to whether the Middlesex Deeds Department of the Land Registry was in 1934 an "appropriate" register for the purposes of section 13 (2),

A for prior to January 1934, when the lease was transferred to Monument Property, and to July 1947, when it was acquired by the first defendants, nothing was registered anywhere else. What was registered in the Middlesex Deeds Department was a memorial of the lease, and the option was not specifically entered upon that register.

The first defendants at this stage stand in this position: they assert that as against them the charge is void. They say that it is a registrable charge and was never registered as such in any registry before their purchase was completed. I must now turn to section 23 (1) of the Land Charges Act, which is (omitting the words which are not of relevance to the B particular issue which I have to consider) in substance in these terms: "As respects . . . land charges . . . required to be registered . . . after the commencement of this Act, this Act shall not apply thereto, if and so far as they affect registered land, and can be protected under the Land Registration Act 1925 by lodging or registering a creditor's notice, restriction, caution, inhibition or other notice." So if the land be registered land and the interests can be protected by some appropriate notice, then the Land Charges Act is of no effect. The plaintiff says that this is registered land and that his option can be protected, and has indeed been protected, C by the lodging of a notice of his lease against a registration of the freehold title. That there is a note of the lease on the land register stands admitted. What the position was when the leasehold title was registered is not known and apparently cannot now be ascertained. But against the registration of the defendants' title absolute is found on the charges register under head 2, "Lease dated April 2 1933 of the land tinted pink on the filed plan to GB Outfitters Ltd for 42 years from December 25 1932 at the rent of £700 and insurance rent." How the entry got there no one apparently knows, nor is it known whether a corresponding entry was made against the registration of the leasehold title. At all events, the plaintiff D says, "This is registered land. The lease upon which I base my case is registered, and that registration has protected my option." Section 23 of the Land Charges Act provides that the Act shall not apply to land charges affecting registered land, which by definition means land registered under the Land Registration Act 1925 or any enactment replaced by that Act.

By the provisions of section 5 of the Land Registration Act, the first defendants' registration of their title took effect subject to encumbrances appearing on the register—in this case the underlease—and subject to over-riding interests. Section 20 of the Act provides that dispositions of land regis- E tered in absolute title shall confer an estate subject only to encumbrances on the register and over-riding interests. Although under section 48 the plaintiff, or his predecessors as underlessees, could have applied to register their underlease under and in accordance with the terms of the Act, they never in fact did so. Section 49 omits the separate registration of various other interests including land charges. At the date the underlease was entered into, there was no registration under the Land Charges Act. A memorial of the lease was entered upon the Middlesex Deeds Register, but in 1937 Middlesex became a compulsory area and that old register was closed. The entries upon the register can be seen by the F land registrar but by nobody else. The first certain entry on the land register is the first defendants' leasehold title in 1947. This title was registered 30 years ago. When the freehold title was first registered in 1962, as when the first defendants' title was registered in 1969, the position is as I have already stated it. The first defendants' interest has been registered for nearly 30 years, and the freehold title has been registered from the dates stated. The first defendants say that this is all irrelevant, for they acquired their interest at a time when the land was not registered land and when the charge was not registered. At the time of their acquisition, section 13 (1) —so they suggest—must necessarily have applied, and, having once applied it, it is their argument that they are left free G thereafter, for they purchased the land before the land was registered land and completed before any registration was effected. The exception in respect of registered land is not, they say, to be regarded as applying to land which was not registered at the date of any relevant transaction, though it may subsequently have become registered land. Further, they say that in any event the charge on the register relates only to the underlease, and there is no mention of the option, and in any event, neither actual nor constructive notice of the option has any relevance to the point at issue.

I am plainly considering legislation in which the intention H is that on one register or another there should be sufficient notice of interests touching land to enable prospective purchasers and other interested persons to know how far the land in question may be encumbered. In the case of registered land, it is plainly considered that the appropriate register is the land register, and it is no doubt for that reason that there is an express exception so far as the Land Charges Act is concerned against any necessity to register under that Act if the land be registered land. At the date when the first defendants acquired the interest on which they now rely, the land was registered land. The option was not only a charge which could be protected, it was in my judgment protected by J notice of the lease. I am of the opinion that counsel for the plaintiff was right when he submitted that this option is effective in the circumstances which I have above outlined, and that the matter stands concluded by the entry on the register, for this reason: that the first defendants are in fact suing in respect of their possession of the freehold interest, and by the date that they acquired that the land in question had become registered land. There were two other matters on the plaintiff's side to which I should make brief reference. It was suggested that the entry in the Middlesex Registry would in itself be adequate to protect their position, but I do not think that this is so, having regard to the particular pro- K visions of section 13 of the Land Charges Act. I do not think it was an "appropriate registry." It was not a register appropriate to the registration of an interest by way of land charges on the material to which I have been referred. Secondly, it was suggested as an alternative line of defence to the argument of the first defendants that it was open to the plaintiff to assert that he had an over-riding interest which was protected by reason, as I understood it, of the occupation of the premises by himself and his sister-in-law and the fact that they were collecting the rents for the residential flats over the period to which I have already briefly referred. This argument, bringing into consideration, as it does, other L provisions under the Land Registration Act, is essentially dependent upon the land being registered land as at the relevant date, and as at that date the only registration that was noted was the entry on the Middlesex Deeds Register. That was not, on the information available to me and upon the submissions made by counsel with reference to the relevant Acts, an Act which was in effect replaced by the Land Registration Act, and therefore I am not of the opinion that this point either is good. It was said by counsel for the defendants that if it were held that the entry on the land register defeated his clients' claim, he would seek to rectify the register. It was agreed during the hearing that it would M be premature at this stage to consider any question of rectification, for the necessary information is not available.

Counsel for the plaintiffs seeks an order for the execution of a new lease in terms to be settled by the court in the absence of agreement. In the absence of agreement, in my view, the consideration of the terms upon which a new lease should be granted ought to be deferred, and I accordingly propose to declare that the plaintiffs are entitled to a new lease in terms to be settled by the court, if not agreed.

A declaration was made accordingly, subject to any claim

A for rectification of the register which the first defendants might wish to pursue. His Lordship said that if leave to appeal was necessary, he would grant it.

COURT OF APPEAL

April 13 1976

(Before Lord Justice CAIRNS, Lord Justice ORR and Lord Justice BROWNE)

FINCHBOURNE LTD v RODRIGUES

Estates Gazette June 5 1976

B **(1976) 238 EG 717**

Service charges—Amount of lessee's contribution to be "ascertained and certified by the lessors' managing agents acting as experts and not as arbitrators"—Lessors subsequently discovered to be the managing agents themselves—Agents' certificate invalid in such circumstances—"Managing agents must, according to the terms of the lease, be somebody different from the lessors"—Further point on charges to be included in a valid certificate—An implied term must be read into the lease by which the charges are to be "fair and reasonable"

C This was an appeal by Finchbourne Ltd, landlords of a flat at 11 Holland Park Avenue, Holland Park, London W11, from a judgment of Judge McIntyre at West London County Court on July 1 1975 dismissing their claim against the tenant, Mrs Leila Rodrigues, for a contribution towards service charges amounting to £183.

Mr A L Price QC and Mr J J Davis (instructed by Lieberman, Leigh & Co) appeared for the appellants, and Mr D Wood (instructed by Isadore Goldman & Son) represented the respondent.

Giving judgment without calling on counsel for the respondent, Cairns LJ said: This is an appeal from an order of D His Honour Judge McIntyre, sitting in the West London County Court, deciding five preliminary points which arose in proceedings between plaintiff landlords and a defendant tenant and, on the basis of his decision on those points, dismissing the plaintiffs' claim. The plaintiffs claimed as landlords under a lease of a flat money alleged to be due by way of so-called contribution, a contribution constituting a service charge. There were two actions, one for an interim payment and the other for a balance payment. The actions were consolidated, and no separate issue arises in either action. The total of the claim was only £183.70, but the matter is of E importance to the plaintiffs because the flat is in a block of flats which they own, and if they are not entitled to recover this sum they may be in difficulty in recovering similar sums from a number of other tenants.

The lease was dated March 17 1967, and was granted by Josun Ltd, the predecessors of the plaintiff company, to the defendant. It was a lease of flat no 25 in Linton House, 11 Holland Park Avenue, London W11. The term was 125 years. A premium of £9,250 was paid. The rent was £6 a year, but there was also payable this contribution, which is, putting it shortly, a contribution to the expenses of maintaining the block of flats as a whole. It is necessary to read a number F of the clauses in the lease, and I start with one or two of the definitions contained in clause 1. Subclause (e) provides: "'The reserved property' shall mean that part of the property not included in the flats including the property more particularly described in the second schedule hereto"—in effect, the hall, staircase and so forth. Then "(h) 'The surveyor' shall mean the surveyor for the time being of the lessors," and "(i) 'Contribution' shall mean the yearly sum defined in the eighth schedule hereto." Clause 4: "The lessors hereby covenant with the lessee that the lessors will from time to time and at all times during the said term observe and perform each and every of the obligations on the lessors' part set out in the seventh schedule hereto." Then going G straight to the seventh schedule, that provides, "(1) Keep the reserved property in good order repair and condition and will in particular in a good and workmanlike manner (a) as to the external parts," and the internal parts, and there are particulars as to painting and so on. Paragraph (2), "Keep the property insured in the full amount of the current cost of rebuilding thereof," the cost to be determined by the surveyor, whose decision shall be binding on the parties; then, "(b) To pay all premiums necessary for effecting and maintaining such insurance. . . ." Paragraph (3), "To pay the charges incurred in meeting all outgoings paid and payable by the lessors in respect of the property including in particular," and H then there are set out water rates, electricity and gas charges, and the cost of renting a public telephone; and subparagraph (iv), which is important, says:

The reasonable and proper fees payable by the lessors to their managing agents for the time being for carrying out the general management and administration of the property including (but without prejudice to the generality of the foregoing) all fees payable to such agents in connection with the collection of the rents and contribution and the interim sums payable by the respective owners of the flats and all other outgoings payable by the lessors in respect of the property or any part thereof and the preparation of all accounts in connection with the calculation and J assessment of the contribution and interim sums and arrangements for the supervision of any works which may be carried out pursuant to paragraph (1) of this schedule and (if undertaken by such agents) preparation of specifications in connection therewith.

Paragraph (4) of the schedule provides, "To cause to be kept clean and maintain in proper order and replace where necessary the part of the reserved property firstly described in the second schedule to this lease including the furnishings thereof and dustbins therein." Paragraph (5), "To use their best endeavours adequately to light all the reserved property," and further details are set out as to that. I need not read paragraphs (6) or (7). Paragraph (8), "To use their K best endeavours to employ suitable staff to maintain an adequate porterage," and so on. Paragraph (9) provides for accommodation for the porters, paragraph (10) the supply of hot water, and paragraph (11), "Repair and keep in running order the lifts installed in the property. . . ." Paragraph (12), "To maintain repair and renew if necessary the equipment (including boilers pipes and radiators) not used exclusively by any one lessee for the provision of hot water and central heating." Then the eighth schedule is the one which defines the contribution, and it is of importance in this case. It is defined in this way:

L "A yearly sum equal to 4 per centum of the amount which the lessors shall from time to time have expended during the year immediately preceding the date hereunder mentioned in (a) meeting the outgoings costs expenses and liabilities incurred by them in carrying out their obligations under the provisions of the seventh schedule hereto (except paragraphs 6 and 7 thereof) and (b) paying from time to time the costs and expenses of and incidental to making repairing maintaining amending and cleansing all or any ways roads pavements gutters sewers channels drains pipes wires cables water-courses walls party walls party structures fences and works and other apparatus matters and conveniences which shall belong to or be used for the premises in common with any other part or parts of the property or any neighbouring or adjoining M premises or which shall form part of the reserved property such contribution to be payable on the first day of April in every year. The amount of such contribution shall be ascertained and certified by the lessors' managing agents acting as experts and not as arbitrators once each year throughout the term on the first day of April in each year (or if such ascertainment shall not take place on the said first day of April then the said amount shall be ascertained as soon thereafter as may be possible as though such amount had been ascertained on the aforesaid first day of April) commencing on the first day of April 1967 and such certificate shall contain a fair summary of the lessors' expenses and a copy thereof shall be supplied to the lessee (but not more frequently than once in every yearly period computed from the first day of

April in any year to the 31st day of March in the then next following year). Provided always and it is hereby agreed (1) that the lessee shall pay such a sum on account of the contributions payable by the lessee under this schedule as the lessors' managing agents shall certify as being a reasonable interim sum in order to meet such of the lessors' current liabilities as will be properly attributable to the contribution (hereinafter referred to as 'interim sum' or in the plural as 'interim sums') to be paid on account of the contribution: such interim sums shall be paid on the first day of April and the first day of October in every year in advance commencing on the half-yearly date next following the date of this lease: (2) that the contribution payable by the lessee hereunder (or such balance as shall remain after giving credit for any interim sums) shall be paid by the lessee or any proper balance found to be repayable to the lessee shall be so repaid to him within 14 days after the issue of the certificate of the lessors' managing agents of the amount of the contribution.

I need not read the proviso which follows, as the first period of the lease is no longer relevant, nor need I read the final paragraph of the schedule. The beneficial ownership of the flat was alleged by the defendant to be in Mr David Pinto. The certificates on which the actions were based were made by a so-called firm, D Pinto & Co, purporting to be the managing agents. In fact, Mr Pinto was the sole proprietor of that business. The five preliminary questions which the learned judge was called upon to decide are set out at the end of his judgment on page 44 of the bundle. He said:

My answers to the preliminary points are therefore these: (1) Were the plaintiffs mere nominees of D Pinto & Co as regards (a) the reversion, and (b) the management of the flats, when the certificates sued on were issued? My answer is yes without hesitation. Mr Pinto was the principal and the company was his mere nominee.

That finding was challenged in the notice of appeal. I very much doubt that that challenge could have been continued, since the claim here was of a limit below that on which an appeal on issues of fact is permissible: but whether that be so or not, it was not pursued.

(2) Was Mr David Pinto thereby disqualified from issuing the certificates? My answer is also yes. You cannot have someone charged with the duty of managing agent under this lease when that person has everything to gain and nothing to lose by charging the tenants whatever he chooses. Managing agents cannot carry out their function as experts if they are devoid of accountability to the tenants and so much bound up in this matter as D Pinto & Co were.

The first argument presented by Mr Price on behalf of the plaintiffs on this appeal was that the judge's answer to that question was wrong.

(3) Is the issue of a valid certificate a condition precedent to the recovery of service charges under the eighth schedule to the lease?

The same point arises on this issue. Of course the answer is yes. Otherwise the tenants could not know what sums they have to pay.

That was not challenged on the appeal.

(4) Apart from affirmative answers to issue (1) and issue (2) above, can the plaintiffs recover costs and outgoings other than fair and reasonable costs and outgoings? The answer is academic, but yes. You cannot properly assess the sums which are due unless they are shown to be fair and reasonable.

That was the second point which was argued on this appeal.

(5) In the eighth schedule, do the words "costs expended" cover only moneys actually paid, or can they also cover amounts incurred? The fifth issue depends upon the wording of the eighth schedule. In my opinion the defendant is right here, and the plaintiffs can only recover moneys which they show that they have laid out prior to certification.

That was not challenged on the appeal.

The judgment had to deal with evidence called on matters which are no longer in dispute, and the learned judge delivered a robust judgment, making animadversions on the status of the plaintiff company, defects in its record, its relationship to Mr Pinto, the failure to call him as a witness, and other matters. We, however, have only had to consider two questions, which are both matters of construction of the lease: (1) Could a certificate by Mr Pinto in his capacity as proprietor of D Pinto & Co be a sufficient certificate for the purposes of the eighth schedule? (2) Can the plaintiffs recover only fair and reasonable outgoings? On the first point Mr Price contends that the only function of the certifier is to calculate, not to judge, whether any costs were properly incurred. Mr Price points out that the certificate is not stated to be "final and binding" and that the implication from the requirement that a summary of the expenses is to be provided for the lessee is that he is entitled to challenge them. Counsel relies on a passage in the judgment of Devlin J in *Minster Trust Ltd v Traps Tractors Ltd* [1954] 1 WLR 963 distinguishing between a certifier and an arbitrator: see pp 973 to 975 of the report. He also cites the judgment of Roskill J (as he then was) in *Frank H Wright (Constructions) Ltd and others v Frodoor Ltd and another* [1967] 1 All ER 433 as to the circumstances in which a certificate of accountants acting as experts and not as arbitrators with regard to the value of certain liabilities, could be challenged in court: see pp 453 to 454 of the report. And I quote from p 455 a citation from the judgment of Denning LJ (as he then was) in *Dean v Prince* [1954] Ch 409 at 427 in these words:

For instance, if the expert added up his figures wrongly; or took something into account which he ought not to have taken into account, or conversely; or interpreted the agreement wrongly; or proceeded on some erroneous principle. In all these cases the court will interfere.

Mr Davis, following on behalf of the appellants, referred to a decision of this court in *Arenson v Arenson* [1973] Ch 346, reading only one short passage from the judgment of Buckley LJ at the foot of p 371:

In my judgment a clear distinction is to be drawn between the position of a third party who is required to adjudicate in such a way as this and one to whom the parties delegate the function of ascertaining some matter of fact. If a vendor and a purchaser of goods sold by weight delegate to a third party the function of ascertaining the weight of a particular parcel of such goods, they are not seeking an adjudication: they are delegating a ministerial function which with the necessary skill and equipment they could perform themselves and about which, if they did it accurately, no dispute could arise, for the weight of the parcel would be a pure question of fact allowing no room for difference of opinion.

It is fair to counsel to say that neither of them claimed that any of these authorities was directly in point. Undoubtedly under this lease the managing agents were to act as experts and not as arbitrators. Whether the certificate of managing agents properly appointed could be challenged does not arise on this appeal, though the passage I have read from *Dean v Prince* suggests that in some circumstances it could be challenged. Even if it be a fact that the function of certifying could have been equally well done by the lessors, that does not arise here, because Buckley LJ was certainly not saying that if the contract provided for the making of the assessment by a third party it would be a fulfilment of the contract for it to be made by one of the parties to the contract. I express no opinion as to whether the function of the managing agents here went no further than to ascertain whether the sum said to have been spent had actually been spent. Even assuming that that is so, the managing agents were not a mere calculating machine. I note that they were to "certify as experts." Though the certificate, in contrast to the report of surveyors referred to elsewhere, was not declared to be binding, the tenant is entitled to assume that an expert mind has been applied to the figures, and that unless there is something glaringly wrong with them he can rely on the certificate. It is said that if the lessor were a company the managing agents might be a subsidiary of that company. However that may be, I am quite satisfied that the managing agents are intended to be somebody different from the lessor.

Once it is found that the real lessor here was Mr Pinto, it follows that the managing agent must, according to the

A terms of the lease, be somebody different from Mr Pinto. Indeed, I would go further and say that it was a misdescription to describe Pinto & Co as agents at all. That so-called firm was merely another name for Mr Pinto. If that is so, then there has been no valid certificate within the meaning of the eighth schedule, and that is an end of the plaintiffs' case. However, I will express my opinion briefly on the second point. Is there an implication that the costs claimed are to be "fair and reasonable"? It is contended that no such implication is necessary to give business efficacy to the contract. Passages from the speeches in the House of Lords in the recently-decided case of *Liverpool City Council v Irwin* B [1976] 2 WLR 562 are referred to as the most recent statement of the principles on which terms can be implied. Taking the strictest of tests on that matter, I am of opinion that such an implication must be made here. It cannot be supposed that the lessors were entitled to be as extravagant as they chose in the standards of repair, the appointment of porters, etc. Mr Price says that there would come a point without any implied term where the costs might be so outlandish as not to come within the description of the seventh schedule at all. In my opinion, the parties cannot have intended that the landlords should have an unfettered discretion to adopt the highest conceivable standard and to charge the tenant with C it. Stress is laid on the provision that, in so far as the fees payable to the managing agents were included, it was expressly provided that it was to be a "reasonable and proper fee." There was a special reason for drawing attention to this, seeing that the agents themselves were to be the certifiers. It is not, in my view, a case where the expression of one amounts to an exclusion of the other. In my opinion, the learned judge arrived at the right conclusion on both points. I would dismiss the appeal.

ORR LJ: I agree that this appeal should be dismissed for the reasons given by my Lord, and I would only add a few D words as to the second point. The eighth schedule to the lease provides that the amount of the lessee's contribution "shall be ascertained and certified by the lessors' managing agents acting as experts and not as arbitrators." Whereas here the managing agent is an estate agent, two matters very much within his expertise are what work is reasonably required to be done in order to comply with the terms of the lessors' obligations under the lease, and what is the reasonable cost of such work, but merely to add up the figures of expenses incurred on the work done does not call for expertise. Mr Price's answer to this point, as I understood it, was that the wording of the eighth schedule, "acting as experts and not as arbi- E trators," which I agree with him is a phrase commonly used in legal documents, is designed to do no more than to make clear that the agent is not to act as an arbitrator, but this ignores the words "acting as experts" and I decline to construe the passage as if it read, "acting neither as experts nor as arbitrators." I too would dismiss the appeal.

BROWNE LJ: I agree that this appeal should be dismissed, and I agree entirely with the reasons given by my Lords. I only add a few words. Under the lease the tenant's obligation under clause 4 (2) (b) is "to pay the contribution," and "contribution" is defined in clause (1) (i) as "the yearly sum defined in the eighth schedule hereto." The eighth schedule F provides that "The amount of such contribution shall be ascertained and certified by the lessors' managing agents acting as experts and not as arbitrators. . . ." It seems to me that this provision can only mean that the amount of the contribution, as my Lord has said, is to be ascertained and certified by a third party, other than the landlord himself, acting as an expert. The intention clearly was that the tenant should be entitled to rely on the expertise of such a third person. On the judge's findings of fact here, which are not challenged, the managing agents and the lessor were in fact the same person. Accordingly, I agree with my Lord that there is no valid certificate for the purposes of the eighth schedule in this case and that this is the end of the case. G I also agree, for the reasons given by my Lord, that a term must be implied in the lease that the costs should be "fair and reasonable." I agree that the appeal should be dismissed.

The appeal was dismissed with costs, less the costs of preparation of certain documents.

HOUSE OF LORDS

March 31 1976

H (Before Lord WILBERFORCE, Lord CROSS OF CHELSEA, Lord SALMON, Lord EDMUND-DAVIES and Lord FRASER OF TULLYBELTON)

LIVERPOOL CITY COUNCIL v IRWIN AND ANOTHER

Estates Gazette June 19 1976 and June 26 1976

(1976) 238 EG 879 and 963

Premises let to tenant subject to no express obligation on part of landlord—Some such obligation necessarily to be implied—In the case of multi-storey premises, an obligation must be implied by which the landlord assumes responsi- J bility for repair and maintenance of lifts and other common parts, rubbish chutes, etc—De Meza v Ve-Ri-Best Manufacturing Co Ltd (1952) 160 Estates Gazette 364 cited in support by Lord Wilberforce and Lord Edmund-Davies—Further point on the statutory repairing obligation in section 32 (1) (b), Housing Act 1961—"Keep" in repair covers a case in which the design of a fitting is defective from the outset

This was an appeal by Mr Leslie Irwin and his wife, Mrs Maureen Irwin, of 50 Haigh Heights, Liverpool 3, from a decision of the Court of Appeal granting their landlords, Liverpool City Council, an order for possession of their K maisonnette and dismissing their counterclaim for damages and an injunction based on an allegation that the council were in breach of an implied obligation with regard to repair of the maisonnette and of certain common parts.

Mr G Godfrey QC and Mr D M Evans (instructed by Kingsford, Dorman & Co, agents for the Vauxhall Community Law Centre, of Liverpool) appeared for the appellants, and Mr H E Francis QC and Mr J Boggis (instructed by Howlett & Clarke Cree & Co, agents for the solicitor to the council) represented the respondents.

In his speech, LORD WILBERFORCE said: This case is of general importance, since it concerns the obligations of local L authority, and indeed other, landlords as regards high-rise or multi-storey dwellings towards the tenants of these dwellings. This is a comparatively recent problem, though there have been some harbingers of it in previous cases. No 50 Haigh Heights, Liverpool, is one of several recently-erected tower blocks in the district of Everton. It has some 70 dwelling units in it. It was erected 10 years ago following a slum clearance programme at considerable cost, and was then, no doubt, thought to mark an advance in housing standards. Unfortunately, it has since turned out that effective slum clearance depends upon more than expenditure upon steel and concrete. There are human factors involved too, and it M is these which seem to have failed. The defendants moved into one of the units in this building in July 1966: this was a maisonnette of two floors, corresponding to the ninth and tenth floors of the block. Access to it was provided by a staircase and by two electrically-operated lifts. Another facility provided was an internal chute into which tenants in the block could discharge rubbish or garbage for collection at the ground level. There has been a consistent history of trouble in this block, due in part to vandalism, in part to non-co-operation by tenants, in part, it is said, to neglect by the corporation. The defendants, with other tenants, stopped payment of rent, so that in May 1973 the corporation had to start proceedings for possession. The defendants put in a counterclaim for damages and for an injunction, alleging that the corporation was in breach of its implied covenant for quiet enjoyment, that it was in breach of the statutory covenant implied by section 32 of the Housing Act 1961, and that it was in breach of an obligation implied by law to keep the "common parts" in repair.

The case came for trial in the Liverpool County Court before His Honour Judge T A Cunliffe. A good deal of evidence was submitted, both orally and in the form of reports. The judge himself visited the block and inspected the premises: he said in his judgment that he was appalled by the general condition of the property. On April 10 1974 he gave a detailed and careful judgment granting possession to the corporation on the claim, and on the counterclaim judgment for the defendants for £10 nominal damages. He found that the defects alleged by the defendants were established. These can be summarised as consisting of (i) a number of defects in the maisonnette itself—these were significant but not perhaps of major importance; (ii) defects in the common parts, which may be summarised as continual failure of the lifts, sometimes of both at one time, lack of lighting on the stairs, dangerous condition of the staircase with unguarded holes giving access to the rubbish chutes, and frequent blockage of the chutes. He found that these had existed or been repeated with considerable frequency throughout the tenancy, had gone from bad to worse, and that while some defects in the common parts could be attributed to vandalism, not all could be so attributed. No doubt also some defects, particularly the blocking of the rubbish chutes, were due to irresponsible action by the tenants themselves. The learned judge decided that there was to be implied a covenant by the corporation to keep the common parts in repair and properly lighted, and that the corporation was in breach of this implied covenant, of the covenant for quiet enjoyment and of the repairing covenant implied by the Housing Act 1961, section 32. The corporation appealed to the Court of Appeal, which allowed the corporation's appeal against the judgment on the counterclaim. While agreeing in the result, the members of that court differed as to their grounds. Roskill and Ormrod LJJ held that no covenant to repair the common parts ought to be implied. Lord Denning MR held that there should be implied a covenant to take reasonable care, not only to keep the lifts and stairs reasonably safe, but also to keep them reasonably fit for use by the tenant and his family and visitors. He held, however, that there was no evidence of any breach of this duty. The court was agreed in holding that there was no breach of the covenant implied under section 32 of the Housing Act 1961; the tenants did not seek to uphold the judge's decision on the covenant for quiet enjoyment, and have not done so in the House.

I consider first the tenants' claim in so far as it is based on contract. The first step must be to ascertain what the contract is. This may look elementary, even naïve, but it seems to me to be the essential step and to involve, from the start, an approach different, if simpler, from that taken by the members of the Court of Appeal. We look first at documentary material. As is common with council lettings there is no formal demise, or lease or tenancy agreement. There is a document headed "Liverpool Corporation, Liverpool City Housing Dept" and described as "Conditions of Tenancy." This contains a list of obligations upon the tenant—he shall do this, he shall not do that, or he shall not do that without the corporation's consent. This is an amalgam of obligations added to from time to time, no doubt, to meet complaints, emerging situations, or problems as they appear to the council's officers. In particular there have been added special provisions relating to multi-storey flats which are supposed to make the conditions suitable to such dwellings. We may note under "Further special notes" some obligations not to obstruct staircases and passages, and not to permit children under 10 to operate any lifts. I mention these as a recognition of the existence and relevance of these facilities. At the end there is a form for signature by the tenant stating that he accepts the tenancy. On the landlords' side there is nothing, no signature, no demise, no covenant: the contract takes effect as soon as the tenants sign the form and are let into possession. We have, then, a contract which is partly, but not wholly, stated in writing. In order to complete it, in particular to give it a bilateral character, it is necessary to take account of the actions of the parties and the circumstances. As actions of the parties, we must note the granting of possession by the landlords and reservation by them of the "common parts"—stairs, lifts, chutes, etc. As circumstances we must include the nature of the premises, viz a maisonnette for family use on the ninth floor of a high block, one which is occupied by a large number of other tenants, all using the common parts and dependent upon them, none of them having any expressed obligation to maintain or repair them.

To say that the construction of a complete contract out of these elements involves a process of "implication" may be correct: it would be so if implication means the supplying of what is not expressed. But there are varieties of implications which the courts think fit to make, and they do not necessarily involve the same process. Where there is, on the face of it, a complete bilateral contract, the courts are sometimes willing to add terms to it, as implied terms: this is very common in mercantile contracts where there is an established usage: in that case the courts are spelling out what both parties know, and would, if asked, unhesitatingly agree, to be part of the bargain. In other cases, where there is an apparently complete bargain, the courts are willing to add a term on the ground that without it the contract will not work—this is the case, if not of *The Moorcock* (1889) 14 PD 64 itself on its facts, at least of the doctrine of *The Moorcock* as usually applied. This is, as was pointed out by the majority in the Court of Appeal, a strict test, though the degree of strictness seems to vary with the current legal trend, and I think that the majority were right not to accept it as applicable here. There is a third variety of implication, that which I think Lord Denning MR favours, or at least did favour in this case, and that is the implication of reasonable terms. But though I agree with many of his instances, which in fact fall under one or other of the preceding heads, I cannot go so far as to endorse his principle: indeed, it seems to me, with respect, to extend a long, and undesirable, way beyond sound authority. The present case, in my opinion, represents a fourth category, or I would rather say a fourth shade on a continuous spectrum. The court here is simply concerned to establish what the contract is, the parties not having themselves fully stated the terms. In this sense the court is searching for what must be implied.

What then should this contract be held to be? There must first be implied a letting, ie a grant of the right of exclusive possession to the tenants. With this there must, I would suppose, be implied a covenant for quiet enjoyment, as a necessary incident of the letting. The difficulty begins when we consider the common parts. We start with the fact that the demise is useless unless access is obtained by the staircase: we can add that, having regard to the height of the block, and the family nature of the dwellings, the demise would be useless without a lift service: we can continue that there being rubbish chutes built into the structures and no other means of disposing of light rubbish there must be a right to use the chutes. The question to be answered—and it is the only question in this case—is what is to be the legal relationship between landlord and tenant as regards these matters. There can be no doubt that there must be implied (i) an easement for the tenants and their licensees to use the stairs, (ii) a right in the nature of an easement to use the lifts, (iii) an easement to use the rubbish chutes. But are these easements to be accompanied by any obligation upon the landlord, and what obligation? There seem to be two alternatives. The first, for which the council contends, is for an easement coupled with no legal obligation, except such as may arise under the Occupiers' Liability Act 1957 as regards the safety of those using the facilities, and possibly such other liability as might exist under the ordinary law of tort. The alternative is for easements coupled with some obligation on the part of the landlords as regards the maintenance of the subject of them, so that they are available for use. In order to be able to choose between these, it is necessary to define what test is to be applied, and I do not find this difficult. In my opinion such obligation should be read into the contract as the nature of the contract itself implicitly requires, no more, no less: a test, in other words, of necessity. The relationship accepted by the corporation is that of landlord and tenant: the tenant accepts obligations accordingly, in relation *inter alia* to the stairs, the lifts and the chutes. All these are not just facilities, or conveniences provided at discretion: they are essentials of the tenancy without which life in the dwellings, as a tenant, is not possible. To leave the landlord free of contractual obligation as regards these matters, and subject only to administrative or political pressure, is in my opinion inconsistent totally with the nature of this relationship. The subject-matter of the lease (high-rise blocks) and the relationship created by the tenancy demand, of their nature, some contractual obligation on the landlord.

I do not think that this approach involves any innovation as regards the law of contract. The necessity to have regard to the inherent nature of a contract and of the relationship thereby established was stated in this house in *Lister v Romford Ice & Cold Storage Co Ltd* [1957] AC 555. That was a case between master and servant and of a search for an implied term. Viscount Simons makes a clear distinction between a search for an implied term such as might be necessary to give "business efficacy" to the particular contract and a search, based on wider considerations, for such a term as the nature of the contract might call for, or as a legal incident of this kind of contract (p 579). If the search were for the former, he says, "I should lose myself in the attempt to formulate it with the necessary precision." We see an echo of this in the present case, when the majority in the Court of Appeal, considering a "business efficacy" term—ie a "*Moorcock*" term—found themselves faced with five alternative terms and therefore rejected all of them. But that is not, in my opinion, the end, or indeed the object, of the search. We have some guidance in authority for the kind of term which this typical relationship (of landlord and tenant in multi-occupational dwelling) requires in *Miller v Hancock* [1893] 2 QB 177. There Bowen LJ said at p 180:

> The tenants could only use their flats by using the staircase. The defendant, therefore, when he let the flats, impliedly granted to the tenants an easement over the staircase, which he retained in his own occupation, for the purpose of the enjoyment of the flats so let. Under those circumstances, what is the law as to the repairs of the staircase? It was contended by the defendant's counsel that, according to the common law, the person in enjoyment of an easement is bound to do the necessary repairs himself. That may be true with regard to easements in general, but it is subject to the qualification that the grantor of the easement may undertake to do the repairs either in express terms or by necessary implication. This is not the mere case of a grant of an easement without special circumstances. It appears to me obvious, when one considers what a flat of this kind is, and the only way in which it can be enjoyed, that the parties to the demise of it must have intended by necessary implication, as a basis without which the whole transaction would be futile, that the landlord should maintain the staircase, which is essential to the enjoyment of the premises demised, and should keep it reasonably safe for the use of the tenants, and also of those persons who would necessarily go up and down the stairs in the ordinary course of business with the tenants; because, of course, a landlord must know when he lets a flat that tradesmen and other persons having business with the tenant must have access to it. It seems to me that it would render the whole transaction inefficacious and absurd if an implied undertaking were not assumed on the part of the landlord to maintain the staircase so far as might be necessary for the reasonable enjoyment of the demised premises.

Certainly that case, as a decision concerning a claim by a visitor, has been overruled (*Fairman v Perpetual Investment Building Society* [1923] AC 74). But I cite the passage for its commonsense as between landlord and tenant, and you cannot overrule commonsense. There are other passages in which the same thought has been expressed. *De Meza v Ve-Ri-Best Manufacturing Co Ltd* (1952) 160 EG 364 was a case of failure to maintain a lift in which Lord Evershed MR, sitting with Denning and Romer LJJ, held the landlords liable in damages for breach of an implied obligation to provide a working lift. The agreement was more explicit than the present agreement, in that there was an express demise of the flat "together with the use of the lift," but I think there is no doubt that the same demise or grant must be implied here, and if so can lead to the same result. In *Penn v Gatenex Co Ltd* [1958] 2 QB 210, a case about a refrigerator in a flat, Sellers LJ said this at p 227:

> If an agreement gives a tenant the use of something wholly in the occupation and control of the landlord, for example, a lift, it would, I think, be accepted that the landlord would be required to maintain the lift, especially if it were the only means of access to the demised premises. I recognise that a lift might vary in age and efficiency, but in order to give meaning to the words "the use of" and to fulfil them, it should at least be maintained so that it would take a tenant up and down, subject to temporary breakdown and reasonable stoppages for maintenance and repairs.

That was a dissenting judgment, but Lord Evershed MR (p 220) makes a similar observation as to lifts. These are all reflections of what necessarily arises whenever a landlord lets portions of a building for multiple occupation, retaining essential means of access. I accept, of course, the argument that a mere grant of an easement does not carry with it any obligation on the part of the servient owner to maintain the subject-matter. The dominant owner must spend the necessary money, eg in repairing a drive leading to his house. And the same principle may apply when a landlord lets an upper floor with access by a staircase: responsibility for maintenance may well rest on the tenant. But there is a difference between that case and the case where there is an essential means of access, retained in the landlord's occupation, to units in a building of multi-occupation: for unless the obligation to maintain is, in a defined manner, placed upon the tenants, individually or collectively, the nature of the contract, and the circumstances, require that it be placed on the landlord.

It remains to define the standard. My Lords, if, as I think, the test of the existence of the term is necessity, the standard must surely not exceed what is necessary having regard to the circumstances. To imply an absolute obligation to repair would go beyond what is a necessary legal incident and would indeed be unreasonable. An obligation to take reasonable care to keep in reasonable repair and usability is what fits the requirements of the case. Such a definition involves—and I think rightly—recognition that the tenants themselves have their responsibilities. What it is reasonable to expect of a landlord has a clear relation to what a reasonable set of tenants should do for themselves. I add one word as to lighting. In general I would accept that a grant of an easement of passage does not carry with it an obligation on the grantor to light the way. The grantee must take the way accompanied by the primeval separation of darkness from light, and if he passes during the former must bring his own illumination. I think that the case of *Huggett v Miers* [1908] 2 KB 278 was decided on this principle, and possibly also *Devine v London Housing Society Ltd* [1950] 2 All ER 1173. But the case may be different when the means of passage are constructed, and when natural light is either absent or insufficient. In such a case, to the extent that the easement is useless without some artificial light being provided, the grant should carry with it an obligation to take reasonable care to maintain adequate lighting—comparable to the obligation as regards the lifts. To impose an absolute obligation would be unreasonable; to impose some might be necessary. We have not sufficient material before us to see whether the present case on its facts meets these conditions.

I would hold, therefore, that the landlords' obligation is as I have described. And in agreement, I believe, with your Lordships, I would hold that it has not been shown in this case that there was any breach of that obligation. On the main point, therefore, I would hold that the appeal fails. It will be seen that I have reached exactly the same conclusion as that of Lord Denning MR, with most of whose thinking I respectfully agree. I must only differ from the passage in which, more adventurously, he suggests that the courts have power to introduce into contracts any terms they think reasonable or to anticipate legislative recommendations of the Law Commission. A just result can be reached, if I am right, by a less dangerous route. As regards the obligation under the Housing Act 1961, section 32, again I am in general agreement with Lord Denning MR. The only possible item which might fall within the covenant implied by this section is that of defective cisterns in the maisonnette giving rise to flooding or, if this is prevented, to insufficient flushing. I do not disagree with those of your Lordships who would hold that a breach of the statutory covenant was committed in respect of the matter for which a small sum of damages may be awarded. I would allow the appeal as to this matter and dismiss it for the rest.

Lord Cross: I have had the advantage of reading the speeches of my noble and learned friends Lord Wilberforce, Lord Salmon and Lord Edmund-Davies. I agree with them that on the main point—the liability of the respondent council to pay damages to the appellants for failure to keep the staircases and chutes in repair and the lifts in working order—this appeal should be dismissed; but that it should be allowed so far as concerns the claim under section 32 of the Housing Act 1961 relating to the lavatory cistern inside the maisonnette. I do not wish to add anything with regard to the latter claim, but in view of its general importance and because I am—with respect to him—unable to agree with a passage in the judgment of the Master of the Rolls I will add a few words of my own on the main point.

When it implies a term in a contract the court is sometimes laying down a general rule that in all contracts of a certain type—sale of goods, master and servant, landlord and tenant, and so on—some provision is to be implied unless the parties have expressly excluded it. In deciding whether or not to lay down such a *prima facie* rule the court will naturally ask itself whether in the general run of such cases the term in question would be one which it would be reasonable to insert. Sometimes, however, there is no question of laying down any *prima facie* rule applicable to all cases of a defined type, but what the court is being in effect asked to do is to rectify a particular—often a very detailed—contract by inserting in it a term which the parties have not expressed. Here it is not enough for the court to say that the suggested term is a reasonable one the presence of which would make the contract a better or fairer one; it must be able to say that the insertion of the term is necessary to give—as it is put—"business efficacy" to the contract, and that if its absence had been pointed out at the time both parties—assuming them to have been reasonable men—would have agreed without hesitation to its insertion. The distinction between the two types of case was pointed out by Lord Simonds and Lord Tucker in their speeches in *Lister v Romford Ice & Cold Storage Co Ltd* [1957] AC 555 at pp 579 and 594, but I think that Lord Denning in proceeding—albeit with some trepidation—to "kill off" Lord Justice Mackinnon's "officious bystander" must have overlooked it. Counsel for the appellant did not in fact rely on this passage in the speech of the Master of the Rolls. His main argument was that when a landlord lets a number of flats or offices to a number of different tenants, giving all of them rights to use the staircases, corridors and lifts, there is to be implied, in the absence of any provision to the contrary, an obligation on the landlord to keep the "common parts" in repair and the lifts in working order. But for good measure, counsel also submitted that he could succeed on the "officious bystander" tests.

I have no hesitation in rejecting this alternative submission. We are not here dealing with an ordinary commercial contract by which a property company is letting one of its flats for profit. The respondent council is a public body charged by law with the duty of providing housing for members of the public, selected because of their need for it, at rents which are subsidised by the general body of ratepayers. Moreover, the officials in the council's housing department would know very well that some of the tenants in any given block might subject the chutes and lifts to rough treatment and that there was an ever-present danger of deliberate damage by young "vandals," some of whom might in fact be children of the tenants in that or neighbouring blocks. In these circumstances, if at the time when the respondents were granted their tenancy one of them had said to the council's representative, "I suppose that the council will be under a legal liability to us to keep the chutes and the lifts in working order and the staircases properly lighted," the answer might well have been—indeed I think, as Roskill LJ thought, in all probability would have been—"Certainly not." The official might have added in explanation, "Of course we do not expect our tenants to keep them in repair themselves, though we do expect them to use them with care and to co-operate in combating vandalism. The council is a responsible body conscious of its duty both to its tenants and to the general body of ratepayers, and we will always do our best in what may be difficult circumstances to keep the staircases lighted and the lifts and chutes working; but we cannot be expected to subject ourselves to a liability to be sued by any tenant for defects which may be directly or indirectly due to the negligence of some of the other tenants in the very block in question." Some people might think that it would have been, on balance, wrong for the council to adopt such an attitude; but no one could possibly describe such an attitude as irrational or perverse.

I turn, therefore, to consider the main argument advanced by the appellants. One starts with the general principle that the law does not impose on a servient owner any liability to keep the servient property in repair for the benefit of the owner of an easement. If I let you a house on my land with a right-of-way to it over my property, and the surface of the way in need of repair, you cannot call on me to repair it if I have not expressly agreed to do so. I can say, "I do not use that road much myself and so the fact that it is out of repair does not trouble me. If it troubles you, you can repair it yourself." I see no reason why the same principle should not be applicable when the owner of a house lets part of an upper storey in it to a single tenant. The landlord would, no doubt, be subject to the liability of an occupier under the Occupiers' Liability Act 1957; but as a matter of contract between himself and his tenant he could, I think, say, "I agree that the staircase is dark and somewhat in need of repair; but I am content with it as it is. If you are not content with it you can repair it and light it yourself." But must it follow that the same principle must be applied to the case where a landlord lets off parts of his property to a number of different tenants, retaining in his ownership "common parts"—halls, staircases, corridors and so on—which are used by all the tenants? I think that it would be contrary to commonsense to press the general principle so far. In such a case I think that the implication should be the other way, and that instead of the landlord being under no obligation to keep the common parts in repair and such facilities as lifts and chutes in working order unless he has expressly contracted to do so, he should—at all events in the case of ordinary commercial lettings—be under some obligation to keep the common parts in repair and the facilities in working order unless he has expressly excluded any such obligation. This was the view taken by the Court of Appeal in *Miller v Hancock* [1893] 2 QB 177, and though the actual decision in that case, which gave a visitor the right to sue on the implied obligation, was wrong and was later overruled by this House, I think that so far as concerns the position as between landlord and tenant the Court of Appeal was right. I agree, however, with your Lordships that the obligation to be implied in such cases is not an absolute one, but only a duty to use reasonable care to keep the common parts and facilities in a state of reasonable repair and efficiency.

So far as concerns ordinary commercial lettings, I do not suppose that the acceptance by this House of the correctness of the view expressed by the Court of Appeal in *Miller v Hancock* as to the implied obligations of a landlord in such cases is of much importance, for normally the tenancy agreement contains detailed provisions as to the extent of the landlord's liability for the repair, cleaning and lighting of the common parts and the maintenance of the lifts, coupled often with provisions for the payment by the tenants of a service charge separate from the basic rent to cover the cost incurred by the landlord in such repair and maintenance. But the question remains—and it is the only question in this appeal over which I have felt any doubt—whether the considerations which make it to my mind impossible to apply the " officious bystander " test in this case ought not to lead to the drawing of a distinction between ordinary commercial lettings and the grant of tenancies of council flats, so that while in the former class the landlord should be under an obligation unless he has expressly excluded it, the general rule as to the repair of easements should apply in the latter class and the council only be under an obligation if it has expressly assumed it. But on reflection I do not think that the differences between the letting of council flats and of privately-owned flats are great enough or clear cut enough to justify the drawing of such a distinction. Nowadays most tenants pay less than the full economic rent for their accommodation, though in the case of privately owned properties the subsidy is at the expense of the landlord and not of the local community, and it is not only council flats that suffer from " vandalism." If local authorities wish to avoid any contractual liability to their tenants with regard to the repair and lighting of the common parts and maintenance of the lifts and chutes then they must expressly exclude it. But to succeed in their claim the appellants had to prove negligence on the part of the respondent council, and I agree with all the judges in the Court of Appeal and with your Lordships that they did not prove it in this case.

LORD SALMON: On July 11 1966 Liverpool City Council accepted the defendant and his wife as their tenants of a council maisonnette consisting of three bedrooms, together with a sitting room, kitchen, bathroom, wc and an outside balcony. This dwelling was on the 9th floor of a 15-storey block known as Haigh Heights which comprised some 70 similar dwellings. The only access to any of the 15 storeys was by two lifts and a staircase. These lifts and this staircase remained in the possession and control of the council. The original rent was £3 1s 2d a week inclusive of rates. This was undoubtedly a low rent; and it still is, although I understand that it has now been increased by about three times that amount. The tenancy was terminable by either party giving to the other four weeks' notice in writing ending on any Monday. After a rent strike by the tenants (including the defendant), caused by what the county court judge found to be appalling conditions to which the tenants in this multi-storey block were all subjected, a notice to quit was served upon the defendant in 1973. Subsequently the council brought an action to evict him, and on June 18 1973 the defendant filed a defence and a counterclaim for £10 nominal damages. The county court judge found, rightly, that there was no defence to the action, and made an order for possession. He awarded the defendant the £10 damages counterclaimed. The judgment on the counterclaim was based chiefly on the ground that the council, in breach of its obligations, continuously failed to keep the lifts in operation and left the staircase in complete darkness, and also on the ground that the council was in breach of section 32 (1) (b) of the Housing Act 1961, to which I shall presently refer. Without challenging any of the facts, the council appealed from that judgment, and the appeal was allowed, the majority of the Court of Appeal, Roskill and Ormrod LJJ, deciding that the council was under no obligation of any kind to keep the lifts or the staircases in repair, and Lord Denning MR deciding that although the council was under a duty to use reasonable care to keep the lifts working and the staircase lit and in repair, they had not failed in that duty. The court of Appeal unanimously concluded that the council was not in breach of the Housing Act 1961. The defendant now appeals from that judgment.

This appeal turns chiefly upon whether the council was under any, and if so what, contractual obligation to their tenants. The printed conditions of tenancy dated July 11 1966 imposed a great many express obligations upon the tenants but did not expressly impose any obligations of any kind upon the council. It has been argued that the council should not be taken to have accepted any legal obligations of any kind. After all, this was a distinguished city council which expected its tenants happily to rely on it to treat them reasonably without having the temerity to expect the council to undertake any legal obligations to do so. I confess that I find this argument and similar arguments which I have often heard advanced on behalf of other organisations singularly unconvincing. Clearly, there was a contractual relationship between the tenants and the council with legal obligations on both sides. Those of the tenants are meticulously spelt out in the council's printed form, which mentions none of the council's obligations. But legal obligations can be implied as well as expressed. In order to discover what, if any, are the council's implied obligations, all the surrounding circumstances must be taken into account. Among the more important surrounding circumstances are the following. This was a block 15 storeys high which was built to be let to parents with young children. The lifts and staircases were obviously provided by the council as being necessary amenities for their tenants which they impliedly gave the tenants and their families and visitors a licence to use. As Bowen LJ said in *Miller v Hancock* [1893] 2 QB 177 at 181:

> This is not the mere case of a grant of a [licence] without special circumstances. It appears to me obvious, when one considers what [a block] of this kind is, and the only way in which it can be enjoyed, that the parties to the demise of it must have intended by necessary implication, as a basis without which the whole transaction would be futile, that the [council] should maintain [the lifts and staircases, which are] essential to the enjoyment of the premises demised, [and that the council] should keep [them] reasonably safe. . . . It seems to me that it would render the whole transaction inefficacious and absurd if an implied undertaking were not assumed on the part of the [council] to maintain the [lifts and staircases] so far as might be necessary for the reasonable enjoyment of the demised premises.

Could it in reality have been contemplated by the council or their tenants that the council undertook no responsibility to take, at any rate, reasonable care to keep the lifts in order and the staircases lit? No doubt the tenants also owed a

A duty to use the lifts and staircases reasonably; indeed, so much was clearly implied in the printed terms of the tenancy. Can a pregnant woman accompanied by a young child be expected to walk up 15, or for that matter nine, storeys in the pitch dark to reach her home? Unless the law, in circumstances such as these, imposes an obligation upon the council at least to use reasonable care to keep the lifts working properly and the staircase lit, the whole transaction becomes inefficacious, futile and absurd. I cannot go so far as Lord Denning MR and hold that the courts have any power to imply a term into a contract merely because it seems reasonable to do so. Indeed, I think that such a proposition is B contrary to all authority. To say, as Lord Reid said in *Young & Marten Ltd v McManus Childs Ltd* [1969] 1 AC 454 at 465 that "no warranty ought to be implied in a contract unless it is in all the circumstances reasonable" is in my view quite different from saying that any warranty or term which is in all the circumstances reasonable ought to be implied in a contract. I am confident that Lord Reid meant no more than that unless a warranty or term is in all the circumstances reasonable there can be no question of implying it into a contract, but before it is implied much else besides is necessary, eg that without it the contract would be inefficacious, futile and absurd.

C The decision in *Miller v Hancock* [1893] 2 QB 177 to the effect that a visitor to demised premises who met with an injury could take advantage of the implied contractual terms between the landlord and tenant and accordingly sue the landlord for injuries which the visitor suffered as a result of the breach of those terms was naturally overruled in *Fairman v Perpetual Investment Building Society* [1923] AC 74. But the general propositions of Bowen LJ to which I have referred have never been overruled. It has, however, been made clear that those propositions were not intended to impose an absolute obligation to maintain, but only an obligation to D take all reasonable care to maintain the lifts and staircase for the reasonable enjoyment of the demised premises (*Dunster v Hollis* [1918] 2 KB 795 at 803). For my part, I do not think that the propositions laid down by Bowen LJ as modified are in any way weakened—indeed, I think they are supported—by two of the authorities cited by Roskill LJ, namely, *In re Comptoir Commercial Anversois v Power, Son & Co* [1920] 1 KB 868 and *R v Paddington & St Marylebone Rent Tribunal ex parte Bedrock Investments Ltd* [1947] KB 984. In the first of these cases, Scrutton LJ said at pp 899-900 "[the court] ought not to imply a term merely because it would be a reasonable term to include if the parties had E thought about the matter, or because one party, if he had thought about the matter, would not have made the contract unless the term was included; it must be such a necessary term that both parties must have intended that it should be a term of the contract and have only not expressed it because its necessity was so obvious that it was taken for granted." In the second authority, Lord Goddard CJ said at p 990: "No covenant ought ever to be implied unless there is such a necessary implication that the court can have no doubt what covenant or undertaking they ought to write into the agreement." I find it difficult to think of any term which it could be more necessary to imply than one without which F the whole transaction would become futile, inefficacious and absurd, as it would do if in a 15-storey block of flats or maisonnettes, such as the present, the landlords were under no legal duty to take reasonable care to keep the lifts in working order and the staircases lit.

It may be that further codification of the law of landlord and tenant is desirable. The recommendations of the Law Commission referred to in the Court of Appeal may be translated into statutes sooner or later—perhaps much later. I respectfully agree with Lord Denning MR that, in the meantime, the law should not be condemned to sterility and that the judges should take care not to abdicate their traditional role of developing the law to meet even the advent of G tower blocks. The next point for decision is whether the defendant has proved that the continuous failure of the lifts and of the lights on the staircases was due to the council's failure to take reasonable care. The difficulty is that in his pleadings the defendant never alleged that the council owed a duty to take reasonable care. He alleged that they were under an absolute duty to maintain the lifts and keep the staircases lit. If any such absolute duty rested on the council —which, as I have indicated, I cannot accept—the defendant, on the judge's findings of fact, would clearly be entitled to succeed. I of course recognise that in the county court pleadings are apt not to be so strictly regarded as in the H High Court. I am also conscious that most of the evidence on both sides seems to have been directed to the issue of whether or not the failure of the lifts and the lights were to be attributed to the council's fault. Nevertheless, had failure to take reasonable care (which was the council's only obligation under the contract) been pleaded, as it should have been, it may be that the council would have armed themselves with more convincing evidence that they had done everything which could reasonably be expected of them. I have, with some reluctance and doubt, come to the conclusion that it would not be fair to find against them on an issue which has never been pleaded against them, or indeed J expressly raised before the county court judge. My doubts are certainly not diminished by the impression I have drawn from the judge's notes that if failure to take reasonable care had been pleaded the judge might well have found it proved. The lifts were out of action inordinately often, and only a little more than half the time on account of vandalism. Moreover, Mr Tyrer, the council's district housing manager, conceded that a lot of the damage was not done by the children in the blocks—Haigh Heights being one of a group of three blocks. Since, however, only an absolute obligation was pleaded against the council to which they had a complete K answer, I do not think it would be right for the reasons I have already given to find against them on the ground that they failed to take reasonable care. I would accordingly dismiss the appeal in so far as it relates to the lifts and staircase.

It remains to consider whether the council were in breach of their obligations under section 32(1)(b) of the Housing Act 1961, which admittedly applies to the tenancy in question. It reads, so far as relevant, as follows: "In any lease of a dwelling-house, being a lease to which this section applies, there shall be implied a covenant by the lessor . . . (b) to keep in repair and proper working order the installations in the L dwelling-house—(i) for the supply of water . . . for sanitation (including . . . sanitary conveniences . . .)." The judge found that every time a water closet was used, the water overflowed and was apt to flood the floor and escape on to the landing, where it lay without any means of draining away. Whether the ball-cock as fitted caused this tiresome fault, or whether it was due to the design of the sanitary convenience, is not clear, nor in my view does it matter. Some tenants tried using pails to catch the overflow. Others attempted to bend the ball-cock down, which stopped the overflow but did not allow sufficient water to flush the water closet efficiently. For my part I do not understand how on any acceptable M construction of the section, it can be held that in the circumstances I have recited the council complied with their statutory obligations to keep the sanitary conveniences in proper working order. I can well understand that sanitary conveniences may be in proper working order even if they are too small or there are too few of them, but how they can be said to be in proper working order if every time they are used they may swamp the floor passes my comprehension. I would accordingly allow the appeal in relation to that part of the counterclaim based on the council's breach of the Housing Act 1961, and reduce the damages awarded from £10 to £5.

A LORD EDMUND-DAVIES: The questions to which this appeal gives rise fall into two parts: (1) Were the respondents, the Liverpool City Council, under any obligation to the plaintiffs as ninth-floor tenants of Haigh Heights, Everton, in respect of the common parts of that tower block which they, as landlords, retained in their possession and control? If so, what was the nature of their obligation, and were they in breach of it? (2) Were the respondents in breach of section 32(1) of the Housing Act 1961?

The conditions of tenancy signed by the appellants imposed no express duty on their landlords. They were drafted mainly B with houses in mind, and towards the end there were added "Further special notes for multi-storey dwellings." These imposed on the tenants certain prohibitions in respect of hallways, staircases and lifts, but again imposed no express obligations on the landlords. Yet, to all save tenants occupying ground-floor maisonnettes, the tenancies were useless unless adequate means of ascent were provided. Even so, the finding of the majority of the Court of Appeal was that there was no sort of obligation on the landlords to keep available such access, without which the premises were not worth even the extremely low weekly rent of £3 1s 2d fixed in July 1966. Such a conclusion is explicable only on the C basis that the members of the Court of Appeal adopted what I respectfully regard as an initially wrong approach to the novel problem presented by the facts. The case for the tenants was founded upon *Miller v Hancock* [1893] 2 QB 177, where a landlord was held liable to compensate his tenant's visitor for personal injuries sustained while descending stairs leading from the tenant's second-floor premises owing to the worn and defective condition of one of the stairs. *Woodfall* (27th ed p 577) cites that decision as authority for the proposition that "where the landlord of a building let out in flats or offices retains the possession or control of a staircase, there is an implied agreement by him with his D tenants to keep the staircase in repair." That set the Court of Appeal off on considering in what circumstances a contractual term could be implied, and that understandably but unfortunately led them to *The Moorcock* (1889) 14 PD 64, CA. It had not been cited in *Miller v Hancock* (*ante*), but the Court of Appeal considered that it enshrined the only possible basis for implying such a term as that contended for by the tenants. It is right to say, furthermore, that such was the only basis advanced on behalf of the tenants themselves at that time. The Court of Appeal accordingly proceeded to consider whether, in the light of *The Moorcock* (*ante*), such a term could be implied in the tenancy agreement. E Roskill LJ (with whom Ormrod LJ agreed) said ([1975] 3 WLR 677):

> I cannot agree . . . that it is open to us in the court at the present day to imply a term because subjectively or objectively we as individual judges think it would be reasonable so to do. It must be *necessary* in order to make the contract work, as well as reasonable so to do, before the court can write into a contract as a matter of implication some term which the parties have themselves, assumedly deliberately, omitted to do.

Lord Denning MR, on the other hand, "with some trepida- F tion" (which was understandable), took a different view, and, after referring to some out of the "stacks" of relevant cases, said (*ibid* 670B):

> In none of them did the court ask: what did both parties intend? If asked, each party would have said he never gave it a thought: or the one would have intended something different from the other. Nor did the court ask: Is it necessary to give business efficacy to the transaction? If asked, the answer would have been: "It is reasonable, but it is not necessary." The judgments in all those cases show that the courts implied a term according to whether or not it was reasonable in all the circumstances to do so. . . . This is to be decided as a matter of law, not as a matter of fact.

I have respectfully to say that I prefer the views of the G majority in the Court of Appeal. Bowen LJ said in the well-known passage in *The Moorcock* (*ante*, at p 68):

> In business transactions such as this, what the law desires to effect by the implication is to give such business efficacy to the transaction as must have been intended at all events by both parties who are businessmen; . . . to make each party promise in law as much, at all events, as it must have been in the contemplation of both parties that he should be responsible for. . . .

That is not to say, of course, that consideration of what is reasonable plays no part in determining whether or not a term should be implied. Thus, in *Hamlyn & Co v Wood & Co* [1891] 2 QB 488 at 491, decided only two years after *The* H *Moorcock* (to which he had been a party) Lord Esher said:

> The court has no right to imply in a written contract any such stipulation unless, on considering the terms of the contract in a reasonable and business manner, an implication necessarily arises that the parties must have intended that the suggested stipulation should exist. It is not enough to say that it would be a reasonable thing to make such an implication. It must be a necessary implication in the sense that I have mentioned.

Bowen and Kay LJJ, who had also been members of *The Moorcock* court, delivered similar judgments. The touchstone is always *necessity* and not merely *reasonableness*: see, J for example, the judgment of Scrutton LJ in *Reigate v Union Manufacturing Co* [1918] 1 KB 592 at 605, and in the case cited below by Roskill LJ, In *re Comptoir Commercial Anversois v Power, Son & Co* [1920] 1 KB 868 at 899. But be the test that of necessity (as I think, in common with Roskill and Ormrod LJJ) or reasonableness (as Lord Denning MR thought), the exercise involved is that of ascertaining the presumed intention of the parties. Whichever of these two tests one applies to the facts of the instant case, in my judgment the outcome would be the same, for, in the words of Roskill LJ (*ibid* at 677H):

> K I find it absolutely impossible to believe that the Liverpool City Council, if asked whether it was their intention as well as that of their tenants of these flats that any of the implied terms contended for by Mr Godfrey should be written into the contract, would have given an affirmative answer. Their answers would clearly have been "No."

It follows that, had such continued to be the case presented on the appellants' behalf to your Lordships' House, for my part I should have rejected it. But it was not, for Mr Godfrey adopted before your Lordships a previously unheralded and more attractive approach, which was very properly not objected to by Mr Francis despite its late appearance on the L scene. As an alternative to his argument based on *The Moorcock* (*ante*), Mr Godfrey submitted before this House that an obligation is placed upon the landlords in all such lettings of multi-storey premises as are involved in this appeal by the general law, as a legal incident of this kind of contract, which the landlords must be assumed to know about as well as anyone else. This new approach was based largely upon *Lister v Romford Ice & Cold Storage Co Ltd* [1957] AC 555, a case concerning the incidents of a contract of service between master and servant, in which Viscount Simonds said (at p 576):

> M For the real question becomes, not what terms can be implied in a contract between two individuals who are assumed to be making a bargain in regard to a particular transaction or course of business; we have to take a wider view, for we are concerned with a general question, which, if not correctly described as a question of status, yet can only be answered by considering the relation in which the drivers of motor-vehicles and their employers generally stand to each other. Just as the duty of care, rightly regarded as a contractual obligation, is imposed on the servant, or the duty not to disclose confidential information (see *Robb v Green*), or the duty not to betray secret processes (see *Amber Size & Chemical Co Ltd v Menzel*), just as the duty is imposed on the master not to require his servant to do any illegal act, just so the question must be asked

and answered whether in the world in which we live today it is a necessary condition of the relation of master and man that the master should, to use a broad colloquialism, look after the whole matter of insurance. If I were to try to apply the familiar tests where the question is whether a term should be implied in a particular contract in order to give it what is called business efficacy, I should lose myself in the attempt to formulate it with the necessary precision. The necessarily vague evidence given by the parties and the fact that the action is brought without the assent of the employers shows at least *ex post facto* how they regarded the position. But this is not conclusive; for, as I have said, the solution of the problem does not rest on the implication of a term in a particular contract of service but upon more general considerations.

From this basis one reverts to *Miller v Hancock* itself, where Lord Esher MR said (at p 179):

> What . . . are the rights of the tenants and the duties of the landlord towards them? Their only mode of access to their tenements was . . . by this staircase. This may be called an easement, but it was in my opinion, under the circumstances, such an easement as the landlord was bound to keep so as to afford a reasonably safe entrance and exit to the tenants. It seems to me that there is an implied obligation on the part of the landlord to the tenants to that effect, or else he is letting to the tenants that which will be of no value to them.

Bowen LJ (p 180) and Kay LJ (182) expressed similar views. Mr Godfrey submitted (and it is important to stress that Mr Francis did not challenge) that in lettings of the kind here under consideration, the general law confers upon the tenants easements of access by the staircases and lifts provided such as would give them a legal remedy were the landlord to prevent the tenants from enjoying them by, for example, locking the lifts or erecting a barrier across the stairs. But the question is whether the general law imposes any duty upon the landlord save the duty of non-interference, and above all, whether it obliges the landlord to repair such means of access. Mr Francis denies that any such extended obligation exists and relies upon the well-established principle that the law imposes, for example, no duty of repair on the servient owner in respect of a right-of-way over his land, leaving it to the dominant owner to effect such repair as he finds necessary for the proper enjoyment of his easement. Accordingly, so ran his argument, to hold that any obligation of repair rests upon the Liverpool Corporation would be a radical and impermissible departure from well-established law, and only Parliament can impose such an obligation. But there appears to be no technical difficulty in making an *express* grant of an easement coupled with an undertaking by the servient owner to maintain it. That being so, there seems to be no reason why the easement arising in the present case should not by implication carry with it a similar burden on the grantor. As Bowen LJ said in *Miller v Hancock* (*ante*, at p 181):

> It was contended by the defendant's counsel that, according to the common law, the person in enjoyment of an easement is bound to do the necessary repairs himself. That may be true with regard to easements in general, but it is subject to the qualification that the grantor of the easement may undertake to do the repairs either in express terms or by necessary implication. This is not the mere case of a grant of an easement without special circumstances. It appears to me obvious, when one considers what a flat of this kind is, and the only way in which it can be enjoyed, that the parties to the demise of it must have intended by necessary implication, as a basis without which the whole transaction would be futile, that the landlord should maintain the staircase, which is essential to the enjoyment of the premises demised. . . .

There is modern support for such a view. Thus in *De Meza v Ve-Ri-Best Manufacturing Co Ltd* (1952) 160 EG 364, where a fourth-floor flat had been demised "together with the use of the lift" and the lift had been out of order for three years, the tenants were held entitled to damages for the landlords' failure to maintain it in working order. Lord Evershed MR (with whom Denning and Romer LJJ concurred) dealt with the submission advanced on the landlords' behalf that there was no express covenant to repair in the tenancy agreement, and that none could be implied, by saying that the terms of the agreement imposed upon the landlords the obligation to maintain a working lift. The tenants are afforded a further measure of support by the observation made *obiter* by Sellers LJ in his dissenting judgment in *Penn v Gatenex Co Ltd* [1958] 2 QB 210 at p 227 that:

> If an agreement gives a tenant the use of something wholly in the occupation and control of the landlord, for example, a lift, it would, I think, be accepted that the landlord would be required to maintain the lift, especially if it were the only means of access to the demised premises. I recognise that a lift might vary in age and efficiency, but in order to give meaning to the words "the use of" and to fulfil them, it should at least be maintained so that it would take a tenant up and down, subject to temporary breakdown and reasonable stoppages for maintenance and repair.

I therefore conclude that the city council were under an obligation to the tenant in relation to the maintenance of stairs and lifts in Haigh Heights in such a condition as to enable them to be used as means of access to and from their maisonnettes. This also involved the maintenance of reasonably adequate lighting of the staircases at such times and in such places as artificial lighting was called for.

The next question that arises is: what is the nature and extent of such obligation? In other words, is it absolute or qualified? If the former, any failure to maintain (save of a wholly minimal kind) would involve a breach of the landlord's obligation, and in *Hart v Rogers* [1916] 1 KB 646 at 650 Scrutton J considered that such was the view taken by the court in *Miller v Hancock* (*ante*). But later decisions such as *Dunster v Hollis* [1918] 2 KB 795 and *Cockburn v Smith* [1924] 2 KB 119 treat the duty only as one of reasonable care, and such is the conclusion I have come to also. To impose an absolute duty upon the landlords in the case of buildings in multiple occupation would, I think, involve such a wide departure from the ordinary law relating to easements that it ought not to be held to exist unless expressly undertaken and should not be implied. Then, adopting the standard of reasonable care, were the landlords shown to have been in breach? The county court judge made no such finding, and this for the good reason that no such breach had been alleged in the counterclaim of the tenants, who were then asserting that the landlords owed an absolute duty. In these circumstances, and for the reasons appearing in the judgments of Lord Denning MR and Roskill LJ, it would, I think, be wrong now to hold on such evidence as was adduced that lack of reasonable care had been established. I therefore concur with my Lords in dismissing this part of the appeal.

There remains the Housing Act question. It is clear that section 32 (1) (*b*) of the Act of 1961 imposes an *absolute* duty upon the landlord "to keep in repair and proper working order the installations in the dwelling-house. . . ." It could be said that the opening words ("to keep . . .") apparently limit the landlord's obligation to preserving the *existing* plant in its original state and create no obligation to improve plant which was, by its very design, at all times defective and inefficient. But the phrase has to be read as a whole and, as I think, it presupposes that at the inception of the letting the installation was "in proper working order," and that if its design was such that it did not work properly the landlord is in breach. Bathroom equipment which floods when it ought merely to flush is clearly not in "working order," leave alone "proper" working order (if, indeed, the adjective adds anything). To say that such whimsical behaviour is attributable solely to faulty design is to advance an explanation that affords no excuse for the clear failure

A " to keep . . . in proper working order." Just as badly-designed apparatus has been held not of " good construction " (*Smith v A Davies & Co (Shopfitters) Ltd* (1968) 5 KIR 320, per Cooke J), so in my judgment the landlords here were in breach of section 32 (1) (*b*) by supplying bathroom equipment which, due to bad design, throughout behaved as badly as did the Irwins' cistern. I do not, however, find established any of the other statutory breaches alleged. In the result, while otherwise dismissing the appeal, I hold that the appellants are entitled to succeed in the one respect indicated, and I concur in the award of damages of £5 in respect thereof.

LORD FRASER: I have had the advantage of reading in B print the speech of my noble and learned friend on the Woolsack. I agree with him that there is to be implied, as a legal incident of the kind of contract between these landlords and these tenants, an obligation on the landlords to take reasonable care to maintain the common stairs, the lifts and the C lighting on the common stairs. I agree also that the landlords have not been shown to be in breach of that obligation. With regard to the second point, I am of opinion that the landlords were in breach of their statutory obligation under section 32 (1) (*b*) (i) of the Housing Act 1961 in respect of the cisterns, but not in any other respects. The cisterns, apparently because of their faulty design, were so inefficient that tenants had either to bend the ball-cock arm so that the cistern did not fill completely, with the result it did not flush the lavatory properly, or to leave the ball-cock in the designed position, with the result that the cistern overflowed and caused flooding. Such a cistern was clearly not in " proper D working order " and in my opinion the landlords failed to " keep " it in proper working order. I would allow the appeal on the second point and reduce the damages to £5.

LANDLORD AND TENANT
BUSINESS TENANCIES

A

COURT OF APPEAL

March 25 1975

(Before Lord Justice MEGAW, Lord Justice STEPHENSON and Sir JOHN PENNYCUICK)

RUSHTON AND OTHERS v SMITH AND ANOTHER

Estates Gazette January 24 1976

(1975) 237 EG 259

Business tenancy—Notice to terminate followed by tenant's application to county court—Tenant also commences Chancery action claiming specific performance of agreement for 50-year lease—County court judge's decision not to grant adjournment of application for new lease under Act upheld on appeal—In proceedings under the 1954 Act the county court has jurisdiction to determine equity questions to do with relevant matters such as the existence of a tenancy between the parties

B

This was an appeal by Mr Roy Frederic Rushton, Mr Peter Harry Knight, Mr Peter Halsworth Field Phillips and Mr Anthony Sherwood Brooks New, carrying on business at 39, 40, 41 and 42a Cloth Fair, London EC1, as the Seely & Paget Partnership, architects, from an order of Judge Leonard in the Mayor's and City of London Court on January 17 1975 refusing to adjourn or stay an application by the partnership in that court for a new tenancy of the premises occupied by them, pending the hearing of an action by the partnership in the Chancery Division against Mr Paul Edward Paget, and against the present respondents, Mr John Lindsay Eric Smith and his wife, Mrs Christian Margaret Smith, sued as trustees of the Landmark Trust, for specific performance of an agreement for the grant of a 50-year lease of the premises.

C

Mr G M Godfrey QC and Miss E Gloster (instructed by Stone, Odell & Frankson, of Banstead) appeared for the appellants, and Mr N T Hague (instructed by Stephenson, Harwood & Tatham) represented the respondents. Mr J D Waite (instructed by Field, Fisher & Martineau) held a watching brief on behalf of Mr Paget.

D

Giving judgment, MEGAW LJ said: This appeal arises out of proceedings under Part II of the Landlord and Tenant Act 1954. The appellants are Messrs Rushton, Knight, Phillips and New, who carry on business as partners in an architects' partnership under the name Seely & Paget Partnership. I shall call the appellants, collectively, "the tenants." The respondents in the appeal are Mr Smith and Mrs Smith, who are concerned in this litigation in their capacity as trustees of the Landmark Trust. I shall call them "the landlords." The Landmark Trust is concerned in the preservation of small buildings of architectural or historic importance or interest. The premises in question are 39 (in part), 40, 41 and 42a Cloth Fair, in the City of London. Since 1925 a partnership (or, it may be, successive partnerships) had been carried on in these premises under the name Seely & Paget. In 1963 Mr Paul Paget, who had, I think, been one of the original partners, became the sole partner, or proprietor, of the firm. Mr Paget also became the freehold owner of the premises in question in

E

F

Cloth Fair. On October 8 1963 Mr Paget took into partnership the four gentlemen whom I have named above and whom I am calling "the tenants." This was done by a partnership agreement of that date, Mr Paget being described therein as "the senior partner." In that partnership agreement, which appears at p 33 of the bundle, clause 2 provides as follows: "Period. Subject to the senior partner being entitled to retire at any time from the firm, the period of the partnership will be of three years' duration from April 1 1963." Clause 4 is marginally described as "Business Accommodation." Subclause (1) reads: "The firm will continue for business purposes its occupation of the accommodation in 39, 40 and 42a Cloth Fair together with permissive use of the reception room and offices in 41 Cloth Fair as hitherto, all as shown on the schedule forming part of this agreement." Clause 4, subclause (2) reads:

G

> The premises to which reference is made in clause 4 (1) of these heads will remain the sole property of the senior partner and will be leased to the firm at a rent of £1,800 per annum, inclusive of all fixtures, fittings and office furniture taken over at April 1 1963, for the term of 50 years from April 1 1963, determinable by the partnership on three months' notice at any time and by the senior partner or his representatives on three months' notice at any time after the dissolution of the partnership.

H

On May 20 1966 a supplemental agreement was made between the same parties, extending the partnership agreement for three years to April 1 1969. On that latter date Mr Paget retired from the partnership. The tenants continued as partners. Nothing that I say is to be construed as indicating any view whether the partnership was or was not dissolved; whether the existing partnership continued without Mr Paget; or whether a new partnership was created. That is not an issue before us. It may be an issue hereafter. I express no views on it. Mr Paget, having retired, sold the premises in Cloth Fair to Mr and Mrs Smith, the landlords, in their capacity as trustees of the Landmark Trust. The contract of sale was dated August 19 1970. Completion date was shown to be September 29 1970. Mr Paget sold as beneficial owner. The only relevant part of the contract which I need cite is special condition of sale K (c). That is as follows:

J

> K. The purchasers hereby agree with the vendor so that the provisions of this clause shall remain in full force and effect notwithstanding completion of the sale . . . (c) to grant to Messrs Seely & Paget an initial term of three years of parts of 39, 40, 41 and 42a Cloth Fair comprising the whole of the office premises now occupied by them at an initial exclusive rent of £1,800 per annum together with . . .

K

—and then I need not read the specific provisions that follow. The contract contained no reference to the term of the partnership agreement regarding the 50-year lease from April 1 1963, and the lease for an initial term of three years referred to in special condition K (c) was not executed.

It would seem—and I do not think that there is controversy about this—that the tenants continued after Mr Paget's retirement, and after the sale of the premises, to occupy

A the premises as before, paying the rent, though after the sale to the defendants, the landlords, the rent was presumably paid to them, the landlords, and not to Mr Paget. However, on July 18 1974 the landlords, through their solicitors, served on the tenants a notice purporting to be under section 25 of the Landlord and Tenant Act 1954, terminating the tenants' tenancy on January 25 1975. The notice indicated that the landlords would not oppose an application to the court under Part II of the 1954 Act for the grant of a new tenancy. A further notice, also under section 25, was served by the landlords on November 22 1974. That further notice gave notice of termination on B September 30 1975. That notice was served to cover a possible different view of the date of the expiry of the tenancy. The first notice, that of July 18 1974, purporting to terminate the tenancy, did not come out of the blue, nor as a shock to the tenants. Further, the correspondence between the parties before the issue of that notice in some minor degree, and after November 1974 more specifically, indicated to the landlords that the tenants were, or might be, asserting that the tenancy by reference to which the landlords were giving their notice was not accepted by the tenants as being the tenancy which they possessed.

C The rateable value of the premises is recorded as being £2,880. The county court jurisdiction, as laid down in section 63 (2) (a) of the 1954 Act as amended by the Administration of Justice Act 1973, is subject to the limit of a rateable value of £5,000. So this is well within the county court jurisdiction so far as concerns the Landlord and Tenant Act proceedings. The tenants' first response to the landlords' section 25 notice of July 18 1974 was to submit an application (which appears at p 1), in proper time under the Act and the rules, in the Mayor's and City of London Court, on November 18 1974, applying for the grant of a new tenancy. On the face of it, that application may appear somewhat odd in relation to the D issues which are being raised: for by it the tenants appear to accept that their tenancy was, indeed, a term of three years, whereas they now seek to assert that no question of the grant of a new tenancy under the 1954 Act arises, because they (the tenants) are entitled to a tenancy of 50 years from April 1 1963, so that Part II of the Landlord and Tenant Act 1954 (if it should so long survive) would not become relevant to the premises until the year 2013. However, there is no suggestion by the landlords that any waiver or estoppel arises from the tenants' application having been expressed in that form. It is accepted that by the relevant time, the landlords were aware of the tenants' primary contention that E the tenancy under which they held had not expired. For their part, the tenants made this application when they did in order to prevent themselves from being defeated by the stringent time provisions of the Act and the rules in the event of their failing to be able to rely, broadly speaking, on their alleged 50-year lease. The landlords on December 2 1974 submitted their answer proposing a new lease for seven years with a rent of £15,000 pa, and other terms into which it is unnecessary to go. Then on January 3 1975 the landlords, seeking to take advantage of section 24A of the Landlord and Tenant Act 1954, which had been introduced by section 3 of the Law of Property Act 1969, submitted an application to the F Mayor's and City of London Court asking for the determination of an interim rent.

In January 1975, the tenants sought to make a fairly drastic amendment of their application originally submitted on November 18 1974. I do not think that the precise date is material for present purposes, but we were told, as I understood it, that the application to make this amendment was in fact made to the judge in the Mayor's and City of London Court at a hearing on January 17 1975, and was then allowed by him. The amendment of the formal application raised a point, which I do not think is relevant at the present stage of the proceedings, as to the validity of the landlords' notice of termination served on July 18. Paragraph 2 of the G amended application read thus:

> In the alternative to (1) above, if (which is denied) the said notice was served on us on a correct date and the said date of termination is correct, and if (which is denied) the respondents are right in their contention that we hold the premises under a lease or an agreement for a lease for a term of three years from September 29 1970, then we apply to the court for the grant of a new tenancy pursuant to Part II of the Landlord and Tenant Act 1954.

Then further particulars are set out. That paragraph 2 was intended, I think, indirectly, and no doubt deliberately H indirectly, to indicate the tenants' contention that the landlords' assumption of a tenancy of three years commencing on September 29 1970 was not accepted by the tenants. I say that it was indirectly indicated and deliberately so. That is because the tenants were minded to seek to have the issue of the tenancy determined, not in the Mayor's and City of London Court, but in the Chancery Division of the High Court. The landlords had been, some time earlier, at any rate by the middle of November 1974, apprised of the general intention of the tenants to adopt some such course. Hence the tenants presumably were not anxious to appear themselves to raise directly, in the Mayor's and City of London Court J proceedings, the issue or issues which they were going to contend could only be decided, or should properly be decided, in a different court. They (the tenants) then launched proceedings in the Chancery Division of the High Court by writ issued on January 16 1975 and by a statement of claim delivered on the same day. In those proceedings, Mr Paget was made the first defendant and the landlords were made the second defendants. The essence of the relief claimed was a declaration against both the first and second defendants that the tenants (who were the plaintiffs in the Chancery action) were entitled to an order for specific performance of the agreement contained in the partnership agreement of K October 8 1963 for the grant to the tenants of a 50-year lease running from April 1 1963 at a rent of £1,800 pa. Alternatively, against the first defendant, Mr Paget, there was a claim for damages; and there were various claims for other relief.

The writ and statement of claim in the Chancery proceedings were part of the material put before the learned judge in the Mayor's and City of London Court in the proceedings on January 17 1975, to which I shall come shortly. The defence in the Chancery Division proceedings had not then been delivered. It has been delivered since. By paragraph 5 of that defence, which was delivered, coupled with a counter- L claim, on March 11 of this year, the second defendants—that is the landlords—contended that the sale agreement "created an equitable tenancy for the term of three years from September 29 1970," and that that equitable tenancy had been "duly determined" by the notice given on July 18 1974 under section 25 of the Landlord and Tenant Act. Paragraph 7 of the defence asserted that "prior to the date of the contract for sale . . . the plaintiffs"—that is, the gentlemen whom I have called "the tenants"—"or one or more of them were fully aware that the purchase price payable" under that contract of sale by the landlords to Mr Paget "was negotiated on the footing that the plaintiffs' interest in the M premises was only the agreement to grant a term of three years from the date of the transfer." It was denied that the tenants were entitled to a 50-year term, or to any term; and it was contended that the plaintiffs, the tenants, had expressly or impliedly represented to the landlords that they, the tenants, had no rights of occupation of the premises, or that, after completion of the purchase by the landlords, the plaintiffs would occupy the premises solely by virtue of the agreement for a three-year term, and that the landlords relied on those representations. Particulars were given of that assertion. By paragraph 8 of the defence it was further asserted

A that the agreement by the first defendant, Mr Paget, to grant to the plaintiffs, the tenants, a 50-year term in the partnership agreement was terminated by mutual agreement between them prior to the date of the contract of sale. By paragraph 9 it was asserted that the agreement and rights in relation to a 50-year term were not overriding interests within section 70 (1) of the Land Registration Act 1925, and further or alternatively that the plaintiffs, the tenants, were estopped by their conduct from claiming any rights under that agreement against the second defendants, the landlords. By paragraph 10 it was asserted that if the 50-year term remained valid and subsisting, the landlords were entitled to determine B that term by three months' notice at any time after the partnership between Mr Paget and the plaintiffs, the tenants, had been dissolved.

Pursuant to the proceedings in the Chancery Division, the landlords gave notice under the Rules of the Supreme Court, order 16, rule 8, addressed to the first defendant in those proceedings, Mr Paget, claiming that they were entitled to damages or an indemnity in the event of the plaintiffs' claim in that Chancery action succeeding against the landlords. I return to what happened in relation to the proceedings in the Mayor's and City of London Court. The tenants applied in C that court for an adjournment or stay of the proceedings pending the determination of the action which they had started in the Chancery Division of the High Court. On January 17 1975 that application was heard and refused by Judge Leonard. He did, however, grant an adjournment of the proceedings under the Landlord and Tenant Act in order to enable an appeal to be taken to this court should the tenants be so minded. It is that appeal which is before us now. The essence of the submissions made on behalf of the tenants is that the Mayor's and City of London Court, having the jurisdiction of a county court, does not have jurisdiction to determine an issue which the tenants wish to raise in this D case: that is, the issue that they, the tenants, are entitled to a decree of specific performance of the provision of the partnership agreement under which the agreement was made that they should have a 50-year lease of the premises. The tenants also contend that if they are wrong on that submission, and if there was indeed jurisdiction in the Mayor's and City of London Court, nevertheless that jurisdiction is discretionary, and the judge was wrong in exercising his discretion so as to prevent the tenants from having the opportunity to have determined in the Chancery Division the proceedings which they have started in that Division before the proceedings in the Mayor's and City of London Court should be E further heard and decided.

The value of the premises with which we are here concerned is accepted to be very substantially over the limit of £5,000 which is prescribed in section 52 (1) (d) of the County Courts Act 1959. That section, which relates to the jurisdiction of the county court in equity proceedings, provides by subsection (1), " A county court shall have all the jurisdiction of the High Court to hear and determine any of the following proceedings, that is to say . . . (d) proceedings for the specific performance . . . of any agreement for the sale, purchase or lease of any property, where, in the case of a sale or purchase, the purchase money, or in the case of a lease, F the value of the property, does not exceed the sum of £5,000." So in the ordinary way the county court, while it has full jurisdiction to deal with proceedings for specific performance where the value of the property is not more than £5,000, has no such jurisdiction where the value of the property is greater than £5,000. Here, as I have said, it is not disputed that the value of the property is greater than £5,000. However, the matter does not stop there, because there are the specific statutory provisions of the Landlord and Tenant Act 1954. By section 63 (2) of that Act it is provided that: " Any jurisdiction conferred on the court by any provision of Part II of this Act or conferred on the tribunal by Part I of the Landlord and Tenant Act 1927 shall, subject to the G provisions of this section, be exercised (a) where the rateable value of the holding does not exceed £5,000, by the county court." Then there are further provisions in later subsections of that section whereby the parties can, by consent, agree to the transfer of what I may call a county court jurisdiction case to the High Court, and vice versa; and there is power also in the court itself so to do, on the application of either of the parties. I should refer also at this stage to the definition of " tenancy," which is set out in section 69 (1) of the same Act. " Tenancy " is there defined as follows: " ' Tenancy ' means a tenancy created either immediately or derivatively out of the freehold, whether by a lease or under- H lease, by an agreement for a lease or underlease or by a tenancy agreement or in pursuance of any enactment. . . . " I need not read the rest of it. While I am referring to the provisions of the Act, I should also cite section 43A, which was introduced into the 1954 Act by section 13 of the Law of Property Act 1969. It provides as follows:

> Where the rateable value of the holding is such that the jurisdiction conferred on the court by any other provision of this Part of this Act is, by virtue of section 63 of this Act, exercisable by the county court, the county court shall have jurisdiction (but without prejudice to the jurisdiction of the High Court) to make any declaration as to any matter arising under this Part of this J Act, whether or not any other relief is sought in the proceedings.

In this case the tenants are seeking a declaration that they are entitled to a decree of specific performance of what they say is the agreement between them and Mr Paget initially in the partnership agreement, whereby it was agreed that they should be granted a lease of the premises for 50 years; and, the tenants would seek to say, when the freehold of the premises was sold by Mr Paget to the landlords, the tenants' equitable right which they had by virtue of that agreement with Mr Paget survived and the burden of it became binding upon the landlords. The statement of claim which has been delivered in the Chancery Division is concerned to set out K the history of the matter directed towards building up such a case. It is a part of the tenants' submissions (and this is not, I think, in dispute as a matter of fact) that at all material times they remained in occupation of the premises throughout this period, even though no lease for 50 years was ever executed in pursuance of the agreement: " So," say the tenants,

> " We, under the doctrine of *Walsh* v *Lonsdale* (1882) 21 Ch D 9, are entitled to specific performance of that agreement as against the landlords: therefore we have a tenancy which runs for 50 years from April 1963, and therefore these proceedings by the landlords terminating the tenancy are wholly abortive L as a matter of law because our right is specific performance, creating a bar to the existence of any other tenancy than that which would run for 50 years."

Those, say the tenants, are proceedings which, not merely should, but can only, be determined in the Chancery Division of the High Court: they cannot be determined in the county court because of the provision of section 52 (1) (d) of the County Courts Act which I have read, with its limitation of value of £5,000 on the county court's jurisdiction. The landlords, on the other hand, say that the county court has jurisdiction because of the provisions of the Landlord and Tenant Act 1954 itself. Mr Hague on their behalf submits M that by reference to the definition of " tenancy " which I have already read from section 69 of the Act, and the provisions of section 63 (2), the county court has jurisdiction. It must have jurisdiction, it is said, to determine any matter which is relevant in relation to the question whether or not there is here a tenancy such as the tenants claim that there is. In those circumstances, the county court has jurisdiction to deal with the question of specific performance: not, indeed, so as to grant a declaration or a decree of specific performance, but so as to decide whether or not the tenants would be entitled to such a declaration or to such a decree. For if

A the county court judge should hold that they are not, then what may be called the defence to the landlords' proceedings under the 1954 Act fails. It is an issue which has to be decided as an integral part of the 1954 Act proceedings.

It is said on behalf of the tenants that the question of the county court's jurisdiction in this case is similar to that which was dealt with by the Court of Appeal in *Foster v Reeves* [1892] 2 QB 255, and that that decision is conclusive in favour of the tenants. In that case the defendant had entered on premises under an executory agreement for a lease. He subsequently gave six months' notice to quit, as if he were on a yearly tenancy, and he left. An action was brought B against him in the county court for a quarter's rent accruing due after the time when he had given up possession. The value of the premises exceeded £500, which at that stage was the limit of county court jurisdiction in equity matters; and so the judge had no jurisdiction to decree specific performance of the agreement. But the county court judge was of opinion that it was a case in which specific performance would be decreed and that he was, therefore, bound to treat the defendant as a tenant under the terms of the agreement; and so he gave judgment for the plaintiff for the rent claimed. On appeal to the Queen's Bench Division, that judgment was C upset; and in this court the judgment of the Queen's Bench Division was upheld. It was held "that the equitable doctrine that a person who enters under an executory agreement for a lease is to be treated as in under the terms of the agreement, can only be applied where the court in which the action is brought has concurrent jurisdiction in law and equity, and that the plaintiff could not recover in the action"—because there the county court had jurisdiction in law but it did not have jurisdiction in equity. Lord Esher, in his judgment in that case, said that it was "a puzzling point." However, there is a subsequent decision of this court, *Cornish v Brook Green Laundry Ltd* [1959] 1 QB 394, which in my judgment is of D vital importance on the question of jurisdiction. There the question arose under the provisions of the Landlord and Tenant Act 1954. I do not propose to go into the rather complicated facts of the case. It is, I think, sufficient for this purpose to read that part of the headnote which relates to the problem with which we are here concerned, and a passage from the judgment of Romer LJ delivering the judgment of the court. Holding no 3 in the headnote (at p 395) is in these terms:

> That the county court judge was right in exercising jurisdiction under section 63 (2) of the Act of 1954, notwithstanding the provisions of the County Courts Acts, although the value of the E premises exceeded £500, as he was not required to enforce any equitable right, but merely to decide whether or not such a right existed.

Foster v Reeves [1892] 2 QB 255 was distinguished. Though that is recorded in the headnote as being a "holding" of the court, it appears that it was strictly speaking *obiter dictum*, because of the earlier holdings. They made it unnecessary for the court to express a view on that particular point. But the court did express a view, and that view, though not binding on us, is, in my judgment, persuasive guidance which we ought to follow. The passage from the judgment of the F court to which I wish to refer begins at p 412:

> In the present case the value of 23 Lower Belgrave Street is in excess of £500 and therefore a county court judge could not decree specific performance of an agreement to grant a lease of it. The difficulty which the judge felt in the present case was whether, having regard to the decision and judgments in *Foster v Reeves*, he had power to inquire whether, as the applicant alleged, a *Walsh v Lonsdale* equity had been created between the trustees and Brook Green, even though he was not being asked to enforce it if he came to the conclusion that it had. In our judgment the judge was quite right in concluding that he had jurisdiction to entertain the question. By section 62 (2) of the Act it is provided that. . . .

Then the learned Lord Justice quotes the words of that G subsection as it then stood, and he goes on:

> One of the matters which necessarily arise in applications under Part II [of the Landlord and Tenant Act 1954] (as it arose, indeed, in the present case) is whether the relationship of landlord and tenant exists between the respective parties, and unless a county court judge has unfettered jurisdiction to determine that question, it is difficult to see how he can properly exercise the powers which are vested in him by the section. So long as the rateable value of the holding is not in excess of £500, he can, in our judgment, and must, inquire into the existence of a tenancy affecting that holding, whether such tenancy is said to have been created at law or in equity. In our opinion the H judge expressed the position correctly in his judgment when he said: "I am not now asked, as was the county court judge in *Foster v Reeves*, to enforce any equitable right, whether to money, specific performance, rectification or anything else. I am merely asked to determine the question as to the nature of the right under which Brook Green hold its premises, and answer that question I must if I am to discharge my duty under section 63 (2) (a).

Romer LJ goes on, "We agree with those observations of the judge." That, in my view, is directly relevant to the issue of jurisdiction with which we are concerned. That case does not appear to have been cited to the learned county court J judge; but, as I understand it, from the reasons he gave for his judgment he had arrived at his conclusion essentially upon that self-same reasoning.

It was submitted to us that there was, or might be, some degree of inconsistency between the views expressed in the judgment that I have just read and certain passages in the judgment of Hodson LJ in *Airport Restaurants Ltd v Southend Corporation* [1960] 1 WLR 880. In particular, our attention was called to a passage at p 881, where Hodson LJ said this:

> "The position is that an application for a new lease was made, the landlord having given notice of termination of the tenancy— K it has been conceded for the purposes of this application that there may be either one tenancy or two tenancies of the premises here in question—and the tenants at a late stage decided to challenge the validity of the notice after they had themselves made their application for a new lease. Being unable to do that in the county court proceedings, as soon as their attention was drawn to this point by counsel they issued a writ in the High Court claiming a declaration"

—and so forth. It is said that the implication from the use by Hodson LJ of the words "Being unable to do that in the county court proceedings" is that the learned Lord Justice was expressing the view that the county court would not L have had jurisdiction. I do not see anything in either of the other two judgments in that case which would support that implication; and I do not think that Hodson LJ had in mind in any shape or form the sort of question with which we are here concerned. Even if one were to take the view, which I do not myself think is a correct view, that anything said in the *Airport Restaurants* case would otherwise be relevant to the question of jurisdiction in the present case, it is to be observed, first, that that case was decided before section 43A of the Landlord and Tenant Act 1954, which I have read earlier, was enacted; and secondly, that *Cornish v Brook Green Laundry* was not, it would seem, cited. Following M the reasoning of the court in *Cornish's* case, I would hold without hesitation that the learned judge has jurisdiction to deal with the issue of specific performance so far as it was necessary to deal with it for the purposes of the proceedings under the Landlord and Tenant Act 1954. He is not obliged in those proceedings to issue a declaration or to grant a decree of specific performance, and it may well be that he has not jurisdiction to do so. But he is not debarred from deciding the issues which are relevant merely because in other circumstances the decision of those issues might normally be followed by the granting of reliefs or the making of

A orders which he does not have jurisdiction to grant or make.

I come, then, to the second argument put forward on behalf of the tenants. Assuming that the learned judge did have jurisdiction to deal with this matter, they contend that nevertheless he ought, in his discretion, in the circumstances here prevailing, to have stayed the proceedings before him in order to enable what I may call the specific performance issue to be determined in the Chancery Division of the High Court. It is submitted on behalf of the tenants that that would have been the appropriate course for the hearing and determination of what, it is suggested, may be complex and difficult issues of fact and law relating to specific performance. B Whether or not the proceedings under the Landlord and Tenant Act 1954 would thereafter be resumed in the Mayor's and City of London Court would depend upon the outcome of the Chancery Division proceedings. I do not think that it would be desirable, in the circumstances of this appeal, since there are issues which may have to be determined hereafter, to say anything which could be regarded as indicating a view on the merits of the matter. I will content myself by saying simply this: I think that, in all the circumstances of this case, the learned judge exercised his discretion correctly. Accordingly, I would dismiss the appeal.

C STEPHENSON LJ: Can the judge go on with these proceedings in his court now that the other proceedings have been started in the Chancery Division of the High Court? I agree with my Lord that he can. On jurisdiction, I find the reasoning of the county court judge and the Court of Appeal which approved it in *Cornish's* case completely convincing. The observations at the end of this court's judgment in that case, which my Lord has quoted, were admittedly *obiter dicta,* but I cannot doubt that if they had been cited to the Court of Appeal in the *Airport Restaurants* case Hodson LJ would not have there made the observations cited by my Lord on which Mr Godfrey has fastened. And I agree in effect with D the judge of the Mayor's and City of London Court, whose decision we are asked to reverse, that section 43A of the Act of 1954 made that observation obsolete in 1969 and supports his own view that he has jurisdiction to determine the question, raised also in the Chancery Division proceedings, as to the nature of the right under which the tenants hold these premises and whether it amounted to a tenancy created in equity by an agreement for a lease. Then should the judge go on with the proceedings if, as I agree, he can? Again I agree with the judge. Here the *Airport Restaurants* case has some relevance, but not much, since each case of the exercise of judicial discretion depends on its own facts. At one time E I thought that the absence of Mr Paget, the first defendant in the High Court proceedings, from the proceedings in the Mayor's and City of London Court might be a strong reason, which the judge did not seem to have considered, for staying the proceedings until the High Court proceedings had been concluded, notwithstanding the delay that that would inevitably cause. But I am satisfied, on all the material that has been put before us, that the judge was right to refuse any stay or adjournment for this purpose; and like my Lord, without saying any more I too would dismiss the appeal.

SIR JOHN PENNYCUICK: I agree with both the judgments which have been delivered. Once satisfied on the issue of F jurisdiction, as I am satisfied by the judgment of the Court of Appeal in *Cornish v Brook Green Laundry Ltd,* I am far from being persuaded that the learned judge here exercised his discretion wrongly by refusing a stay of the county court proceedings pending determination of the action in the Chancery Division. On the contrary, on the particular facts of the present case, his decision seems to me to be eminently fair and sensible. So far as I can see, no such technicalities of equity law are involved in the case as make it particularly appropriate for hearing by the Chancery Division. I would dismiss this appeal.

The appeal was dismissed with costs.

G

QUEEN'S BENCH DIVISION

July 4 1975

(Before Judge Edgar FAY, sitting as a deputy judge of the division)

WANG v WEI AND ANOTHER

Estates Gazette February 28 1976

(1975) 237 EG 657

Business premises—Occupant under purported management agreement did not in fact manage the business full-time, but started another of his own and behaved generally as a tenant—Agreement a sham, manager entitled to grant subtenancy—Subtenant accordingly protected by statute H

This was an originating summons by Mrs Yuan Wee Wang against Mr and Mrs William Wei, of 135 Cromwell Road, South Kensington, London SW7, for possession of business premises known as the Orchid House Chinese restaurant, 134 Cromwell Road, together with the flat occupied by the defendants in the adjacent premises, no 135. The plaintiff also sought damages for trespass or alternatively arrears of rent.

Mr D T A Davies (instructed by Cripps, Harries, Willis & Carter) appeared for the plaintiff, and Mr C J Whybrow (instructed by C L MacDougall & Co) represented the defendants. J

Giving judgment, JUDGE FAY said that Mrs Yuan Wee Wang acquired a leasehold interest in both the restaurant and the flat in February 1963, and until 1971 she herself ran the restaurant business. In that year she fell ill, found herself unable to continue with the business, and handed it over to Mr Martin Wang. A written agreement, due to expire on November 28 1974, was drawn up between them. Mr Wang operated the business until 1974. Then, by an agreement expressed to operate as a subtenancy from May 20 to November 30 1974, he handed it over to Mrs Wei. She and Mr William Wei remained in occupation after the end of K November 1974, claiming the protection of Part II of the Landlord and Tenant Act 1954. The issue turned on whether Mrs Wei had a lease or a licence in the premises, and that in turn depended on whether Mr Wang had a lease or a licence.

The agreement between Mrs Wang and Mr Wang had been drawn up by solicitors. On its face, it was a management agreement saying nothing about Mr Wang's status on the premises. It was implicit that his presence there was as an employee. There was nothing in the agreement about the flat. That was Mrs Wang's residence. In 1972, however, she went to Hong Kong for medical treatment. She retained a key to the flat and left personal belongings locked in it, and L it remained fully furnished with her possessions. Mr Wang then used the flat in connection with the business and paid an extra £20 a week to Mrs Wang. Mrs Wei was employed at the restaurant in 1973 by Mr Wang, and her husband moved into a room in the flat. Other rooms were occupied by other members of staff. Subsequently Mr Wang entered into the transaction with Mrs Wei which he (Judge Fay) had already outlined, with the result that from May 1974 Mr and Mrs Wei occupied both the restaurant and flat and ran the business on their own account.

If Mr Wang had power to grant a leasehold interest, Mrs Wei, having a business tenancy for over six months, would M enjoy the protection of Part II of the Landlord and Tenant Act 1954. Under the agreement entered into by Mr Wang with Mrs Wang, he was described as "manager." It was provided that he was to act on the owner's behalf, and that he must devote his whole time to the business. The financial provisions of the agreement, on the other hand, were such as were ordinarily found in a lease rather than a contract of managership. Mr Wang had, for example, injected capital of his own into the business. There was no rule of law which prevented the parties from agreeing that Mr Wang should have merely a managership, with its consequential

A status of licensee, and cogent among the items of evidence must be the terms the parties had set their hands to: *Shell-Mex v Manchester Garages* [1971] 1 WLR 612. The agreement, however, was equivocal. If the parties had treated it as a managership agreement it could have conferred no more than a licence upon Mr Wang. But in fact it had not been treated as such. In particular, Mr Wang had paid no attention to the provision requiring him to devote his whole time to the business. He started another business of his own, and acted as though he were tenant of the Orchid House business and premises. Mrs Wang did not object, although she knew of the presence of the Weis.

B Where an agreement was a sham, the reality of the situation had to be examined, and as both parties had treated the agreement as a tenancy agreement, this conferred a tenancy upon Mr Wang. It followed that Mr Wang had an estate in the land and could grant an assignment or sublease. Accordingly Mrs Wei was a subtenant. For all practical purposes, the flat also became part of the business premises and of the business letting. It was accepted by counsel that in those circumstances both premises were protected under Part II of the Act, so that there must be judgment for Mr and Mrs Wei in respect of both the restaurant and the flat on the claim for possession. That C meant that the defendants had since December 1 1974 been tenants of the plaintiff under the provisions of the Act of 1954 relating to continuation of business tenancies, and the plaintiff would have judgment for £3,208 as arrears of rent.

CHANCERY DIVISION

June 18 1975

(Before PLOWMAN V-C)

D

ITC PENSION FUND LTD v PINTO

Estates Gazette March 6 1976

(1975) 237 EG 725

Mistake in landlord's legal department as to effect of order in Landlord and Tenant Act proceedings—Tenant offered advantageous terms for temporary tenancy—Mistake quite one-sided, tenant entitled to take the benefit—Landlord's possession claim fails

This was a claim by ITC Pension Fund Ltd against Mr David Pinto, estate agent, surveyor and valuer, of 15 Dover E Street, London W1, for possession of the fourth floor of 14 Dover Street. The defendant counterclaimed a declaration that he was entitled to retain possession by virtue of a tenancy which had not determined.

Mr P R R Sinclair (instructed by Trower, Still & Keeling) appeared for the plaintiffs, and Mr J M Chadwick (instructed by Lieberman, Leigh & Co) represented the defendant.

Giving judgment, PLOWMAN V-C said: In this action the plaintiffs, as landlords, claim possession of the fourth floor of 14 Dover Street, W1, and damages for trespass. The defendant claims to be entitled to retain possession by virtue of a tenancy which has not determined, and counterclaims F a declaration to that effect. The defendant carries on business as an estate agent, surveyor and valuer in the adjoining property, 15 Dover Street, from which there is access to the fourth floor of No 14, and for a number of years he has occupied the only room on that floor in conjunction with his premises at No 15 for the purposes of his business. He claims to be entitled to continue that occupation. No 14 is a building consisting of a basement, ground floor and five upper floors. Immediately before January 3 1969, the freehold of the building was vested in London Property Inc, from whom the plaintiffs bought it as an investment and with a view to redevelopment in 1970. The building was subject to two head leases. The first of these was a lease of the G ground floor and basement for a term expiring on September 29 1973. The second was a lease of the upper floors for a term expiring on December 24 1972 at a rent of £2,270 per annum. The upper floors were separately underleased for terms expiring a day or two before December 24 1972. Among those underleases was one of the fourth floor, dated April 2 1964, for a term expiring on December 23 1972. This underlease (which I will call "the 1964 underlease") had become vested in the defendant in 1968 by assignment. The rent payable under it was £650 per annum.

On January 3 1969 two things happened. First, the head H lease of the upper floors was assigned to the Trustees of the Manifold Charitable Trust, and secondly, those trustees granted to the defendant an underlease (which I will call "the 1969 underlease") of the upper floors (including the fourth floor) for a term expiring on December 23 1972 at a rent of £4,300 per annum. I have heard much argument on the question whether the 1964 underlease thereupon merged in the 1969 underlease, a matter to which I will revert later. On December 28 1971 the defendant served on the plaintiffs a notice under section 26 of the Landlord and Tenant Act 1954 requesting a new tenancy of all five upper floors of No 14 commencing on December 25 1972. That notice is J referable only to the 1969 underlease. On March 2 1972 the plaintiffs served on the defendant a notice under section 40 (1) of the Act requiring information about the occupation and subtenancies of the upper floors. I refer to that subsection, which is as follows:

> Where any person having an interest in any business premises, being an interest in reversion expectant (whether immediately or not) on a tenancy of those premises, serves on the tenant a notice in the prescribed form requiring him to do so, it shall be the duty of the tenant to notify that person in writing within one month of the service of the notice:
>
> K (a) whether he occupies the premises or any part thereof wholly or partly for the purposes of a business carried on by him, and
>
> (b) whether his tenancy has effect subject to any subtenancy on which his tenancy is immediately expectant and, if so, what premises are comprised in the subtenancy, for what term it has effect (or, if it is terminable by notice, by what notice it can be terminated), what is the rent payable thereunder, who is the subtenant, and (to the best of his knowledge and belief) whether the subtenant is in occupation of the premises or of part of the premises comprised in the subtenancy and, if not, what is the subtenant's address.

On March 21 1972 the defendant, in reply to the section 40 notice, stated that D Pinto & Co occupied the fourth floor L office for the purposes of the business carried on by the defendant and gave particulars of the subletting of the other floors, but did not specifically mention either the 1964 or the 1969 underlease. On March 29 1972 the plaintiffs served on the defendant a notice that they would oppose an application to the court under Part II of the Act on the ground set out in section 30 (1) (f) of the Act, namely

> "that on the termination of the current tenancy the landlord intends to demolish or reconstruct the premises comprised in the holding or a substantial part of those premises or to carry out substantial work of construction on the holding or part thereof and that he could not reasonably do so without obtaining M possession of the holding."

On April 24 1972 the defendant issued an originating application to the Westminster County Court for a new lease of the upper floors.

On May 9 1972 the plaintiffs put in their answer, stating that they opposed the grant of a new tenancy in accordance with section 30 (1) (f). They then instructed their surveyors, Richard Ellis & Son, to prepare a scheme for the reconstruction of No 14. This was done, and on June 15 1972 the plaintiffs applied for planning permission. The application for a new lease was heard by Judge Stockdale on June 28

1972 while the application for planning permission was still pending. The judge reserved judgment, but indicated that he proposed to refuse the application for a new tenancy on the ground set out in section 30 (1) (f), but that since the plaintiffs would be unable to proceed with their scheme until the lease of the ground floor and basement fell in on September 29 1973, he would exercise his power under section 31 (2) of the Act and substitute September 29 1973 for December 25 1972 in the defendant's section 26 notice. Section 31 (2) is as follows:

Where in a case not falling within the last foregoing subsection the landlord opposes an application under the said subsection (1) on one or more of the grounds specified in paragraphs (d), (e) and (f) of subsection (1) of the last foregoing section but establishes none of those grounds to the satisfaction of the court, then if the court would have been satisfied of any of those grounds if the date of termination specified in the landlord's notice or, as the case may be, the date specified in the tenant's request for a new tenancy as the date from which the new tenancy is to begin, had been such later date as the court may determine, being a date not more than one year later than the date so specified:

(a) the court shall make a declaration to that effect, stating of which the said grounds the court would have been satisfied as aforesaid and specifying the date determined by the court as aforesaid, but shall not make an order for the grant of a new tenancy;

(b) if, within 14 days after the making of the declaration, the tenant so requires the court shall make an order substituting the said date for the date specified in the said landlord's notice or tenant's request, and thereupon that notice or request shall have effect accordingly.

On July 21 1972 Judge Stockdale made a consent order which was in the following terms:

It is hereby declared that the respondents have not established any ground on which they are entitled under section 30 of the Landlord and Tenant Act 1954 to oppose the application of the applicant for a new tenancy of the premises known as 14 Dover Street, London W1, but that the court would have been satisfied on the following ground, namely that specified in section 30 (1) (f) of the said Act, that the respondents would have an intention to demolish the said premises if the date specified in the applicant's request for a new tenancy under section 26 of the said Act as the date from which the new tenancy is to begin had been September 29 1973. It is ordered that the last-mentioned date be substituted for the date specified in the said tenant's request.

The effect of this was to continue the 1969 underlease until September 29 1973, but the defendant was hoping that by then a scheme for the redevelopment of 13, 14 and 15 Dover Street as a whole, about which he was in negotiation with Richard Ellis & Son, would have been worked out and that he would be given office accommodation in the redeveloped block.

The legal aspects of the matters to which I have been referring were being dealt with on behalf of the plaintiffs by Mr W A Tacey, a solicitor in the legal department of the Imperial Group at Bristol, of which the plaintiffs form part. He got into a complete muddle as to the effect of the judge's order, and appears to have thought that it operated on the 1964 underlease rather than on the 1969 underlease. He said in evidence: "I mistakenly believed that the essence of the court order was to keep the 1969 underlease in existence until December 1972 and then in some way to resurrect the tenant's rights under the 1964 underlease for nine months." It is not easy to see why he went wrong, because on July 19 1972 Richard Ellis & Son had written a letter to his department stating the effect of the anticipated order quite accurately. They said this:

Thank you for your letter of June 29, the contents of which are appreciated and have been recorded by our estate records department.

We have assumed from the information you have given us we should demand rent from Mr David Pinto in respect of the periods from December 23 1972 to September 9 1973 at the rent reserved in his underlease dated January 3 1969 granted to him by the trustees of the Manifold Charitable Trust, under which we imagine he would continue to hold the premises by virtue of the county court order you have secured.

In this connection we are wondering if you would be kind enough to confirm that the rent we should demand with effect from December 23 1972 should be at the rate of £4.300 per annum exclusive, reserved by the above-quoted underlease, and not the £2,270 exclusive reserved in the head lease to the Manifold Charitable Trust.

Mr Tacey, however, saw matters differently. On July 21 he replied as follows:

Thank you for your letter of July 19. I confirm that you should demand rent from Mr David Pinto for the period December 25 1972 to September 29 1973. The Westminster county court judge is in fact today delivering his reserved verdict, but it is understood that the effect of this will be to extend Pinto's lease to the September quarter day 1973.

Up to December 25 1972 you should continue to collect rent from the Manifold Charitable Trust at the existing rate. After December 25 1972 rent will be payable by Pinto at the current rate payable under his underlease of the fourth floor, namely £650 per annum, since the essence of the court order is an extension of his occupation lease, not his lease of the upper floors.

At about this time the defendant became aware that the plaintiffs were busy acquiring the interests of the other subtenants of the upper floors of No 14 with a view to obtaining vacant possession. He was under no illusions about the effect of the court order, and viewed with some apprehension the prospect of having to pay £4,300 per annum under the 1969 underlease from December 1972 to September 1973 while upper floors (except his own fourth floor) were vacant and producing no rent. He was therefore relieved to get the following letter, dated October 19 1972, from Mr Tacey:

Dear Sirs,

Fourth Floor, 14 Dover Street, W1

As you are aware, with effect from December 25 1972 you will become direct tenants of my clients in respect of the fourth-floor premises until September 1973. My clients do not have a copy of the lease under which you hold the fourth floor and I should be most obliged if you could let me have a copy of your lease.

Pausing there, that is clearly a reference to the 1964 underlease, because the plaintiffs already had a copy of the 1969 underlease. The letter goes on:

I shall, of course, be responsible for your copying charges in this matter.

The impression that that letter made on the defendant's mind was that the plaintiffs were making the reasonable suggestion that as from Christmas 1972 he should stay on until September 29 1973, not under the 1969 underlease, but under the 1964 underlease at a rent of £650 per annum. Accordingly, when on November 17 1972 Richard Ellis & Son wrote to him on the question of renewing a contract for servicing the lift at No 14 (the maintenance of which was the tenant's obligation under the 1969 underlease) he replied on November 20 1972 as follows:

We thank you for your letter of November 17.

Our lease on the whole of the upper parts of 14 Dover Street terminates on December 25 1972 and at that date we cease to be responsible for the servicing and maintaining of the lift, and indeed the cleaning and general maintenance of the common parts. We assume that from that date all these services will be taken over by your clients for the benefit of the tenants who remain in occupation of the various premises in the upper parts.

We shall, of course, be continuing in occupation of the fourth-floor room only after that date and naturally we shall expect the services to the common parts to be continued by your client.

On November 22 1974 the plaintiffs renewed their request for a copy of the 1964 underlease. They wrote to the defendant as follows:

A Re 4th Floor, 14 Dover Street, W1.

You will recall that I wrote to you on October 19 asking for a copy of the lease of the fourth floor under which you will continue to hold these premises until September 1973. I shall be most obliged if you would let me have a copy of your lease.

A copy of your letter of November 20 to my clients' managing agents has been passed to me, and of course my clients are unable to deal with the various queries in your letter until such time as we have a copy of the lease, since it is necessary to examine the various responsibilities of the landlords under that lease, which responsibilities will be taken over by my clients with effect from the December quarter day.

On December 6 1972 the defendant replied as follows:

B Dear Sirs,

4th Floor, 14 Dover Street

I am sorry for the delay in dealing with your request for a copy of my lease under which I hold the fourth-floor premises and this is now enclosed.

I pause to say that that was the 1964 underlease.

Kindly let us have a remittance for our copying charges of 60p.

Then there is a postscript:

I take the opportunity whilst writing of enclosing my firm's C cheque for the quarter's rent due on December 25 in the sum of £162.50.

Under the 1964 underlease the rent was payable quarterly in advance. This letter was acknowledged on December 8 1972 by the plaintiffs, who wrote as follows:

Thank you for your letter of December 6 enclosing a copy of the lease under which you hold the fourth-floor premises. I enclose herewith a cheque in the sum of 60p in respect of your copying charges. I am obliged for your cheque in the sum of £162.50 in respect of the December quarter's rent.

Similar payments in advance were made in respect of the rent due on March 25 and June 24 1973, but when the D defendant sent a cheque for the Michaelmas rent, Richard Ellis & Son returned it, saying:

As you know, the rent in respect of the above premises is payable quarterly in advance, and accordingly we return your cheque for £162.50 as we envisage you will be vacating the premises on September 29.

On October 1 1973 the plaintiffs were notified that their application for planning permission was refused. In March 1974 they gave notice of appeal to the Secretary of State, and on November 4 1974 the Secretary of State dismissed the appeal. The plaintiffs are therefore no longer in a E position to avail themselves of section 30 (1) (f) of the Act, and instead of reconstructing the property have undertaken a substantial programme of internal refurbishment not involving planning consent, and in order to assist to that end the defendant has moved out of the fourth floor as a temporary measure and without prejudice to his claim in this action that he is entitled to remain in occupation under a tenancy which has never been determined.

The plaintiffs issued their writ on November 15 1973. Their case can be summarised as follows: (1) On the granting of the 1969 underlease, the 1964 underlease merged in it; (2) alternatively, the defendant is estopped from contending that F it did not merge; (3) by virtue of the county court judge's order, the defendant's tenancy under the 1969 underlease came to an end on September 29 1973; (4) alternatively, that order estops the defendant from asserting the continued existence of the 1964 underlease; (5) ergo, the plaintiffs are entitled to possession and damages for trespass. The defendant's case can be summarised as follows: (1) The 1964 underlease did not merge in the 1969 underlease; (2) the defendant is not estopped from so contending; (3) the plaintiffs themselves are estopped from asserting that it did merge; (4) the county court judge's order did not affect the defendant's tenancy of the fourth floor under the 1964 underlease, and accordingly at all times since December 25 1972 the defendant has held the fourth floor under a tenancy from the G plaintiffs upon the terms of the 1964 underlease and is entitled to possession of that floor by virtue of that tenancy; (5) alternatively the defendant is entitled to possession of the fourth floor by virtue of a tenancy for a term of years commencing on December 25 1972 and ending on September 28 1973, the terms of which are the terms of the 1964 underlease (so far as the same are applicable to and not inconsistent with the said term of years); (6) this tenancy has not terminated. In reply, the plaintiffs say that if (which they deny) they did grant or agree to grant the defendant the tenancy referred to in paragraph 5 of my summary of the defendant's case, H they did so under a mistake, which they plead as follows:

(a) In granting or agreeing to grant such lease the plaintiffs and their agents were under a mistake (i) as to the true effect of the order of His Honour Judge Stockdale referred to in paragraph 9 of the statement of claim in that they erroneously believed that the consequence of the said order was that the defendant became entitled to occupy the fourth floor from December 25 1972 until September 29 1973 upon the terms of the 1964 sublease and (ii) as to their contractual rights and obligations *vis-à-vis* the defendant in the circumstances which arose after the said order had been made in that the plaintiffs and their agents erroneously believed that the 1969 sublease determined on J December 25 1972 and that thereafter until September 29 1973 such contractual rights and obligations arose under the 1964 sublease rather than under the 1969 sublease.

(b) At all material times or at any rate by December 6 1972 the defendant knew of the said mistakes of the plaintiffs.

Particulars of knowledge: The plaintiffs will rely on the terms of a letter dated December 6 1972 from the defendant to the plaintiffs.

(I have read that letter.)

(c) In the further alternative, if the defendant did not know of the said mistakes of the plaintiffs then such mistakes were common to the plaintiffs and the defendant in that they shared K the same erroneous beliefs referred to in subparagraph 5 (a) hereof.

In my judgment, it matters not whether the 1964 underlease did or did not merge in the 1969 underlease, nor, in my judgment, is it material to determine whether the 1964 underlease survived the county court judge's order. I am prepared to assume, without so deciding, that the plaintiffs are right about both these matters. But the crux of the matter is that after Christmas 1972 the defendant remained in possession of the fourth floor with the consent of the plaintiffs and on the understanding that he should be entitled to continue in occupation until September 20 1973, paying a rent of £650 L per annum and otherwise on the terms of the 1964 underlease. In my judgment, the inference is irresistible, first, that a new tenancy was thereby created and, secondly, that the arrangement under which the defendant remained in possession operated as a surrender of the 1969 underlease.

That leaves only the question of mistake, and in my judgment there is nothing in it. The only mistake proved was the unilateral mistake of Mr Tacey. The defendant was under no misapprehension as to the effect of the county court judge's order, nor was he at any material time aware that the plaintiffs were mistaken as to its effect. I was referred to the equitable rule which was stated by Denning LJ (as he M then was) in *Solle v Butcher* [1950] 1 KB 671, 693 in these terms:

"A contract is also liable in equity to be set aside if the parties were under a common misapprehension either as to facts or as to their relative and respective rights, provided that the misapprehension was fundamental and that the party seeking to set it aside was not himself at fault."

In the present case there was no common misapprehension, and the fault was all on the side of the plaintiffs, in the sense that there was no reasonable excuse for the mistake, the only explanation for it being that Mr Tacey got into a muddle.

A Since the defendant's present tenancy has never been terminated in accordance with the provisions of the Landlord and Tenant Act, it still continues. In the circumstances I dismiss the action, and on the counterclaim I will make the alternative declaration sought, namely, a declaration that the defendant holds the fourth floor under a tenancy for a term of years commencing on December 25 1972 and ending on September 28 1973, the terms of which are now the terms of the 1964 sublease (so far as the same are applicable to and not inconsistent with the said term of years), which tenancy has not terminated.

The action was dismissed with costs of both claim and B counterclaim.

COURT OF APPEAL

December 17 1975

(Before Lord Justice STEPHENSON, Lord Justice SCARMAN and Sir Gordon WILLMER)

MORRISON HOLDINGS LTD v MANDERS PROPERTY (WOLVERHAMPTON) LTD

Estates Gazette June 5 1976

(1975) 238 EG 715

C **Wolverhampton shop premises damaged by fire—Clause in lease allowing landlords to determine in such event—Tenants remove stock, etc, and landlords serve notice purporting to exercise power in clause—Tenants held to have evinced an intention to exert their right of occupancy to an extent entitling them to apply for a new lease under the Landlord and Tenant Act 1954**

This was an appeal by Morrison Holdings Ltd against a decision of Judge Davison at Wolverhampton County Court of February 6 1975 holding that they had no *locus standi* to pursue an application for the grant of a new tenancy of D business premises at 31 Dudley Street, Wolverhampton, from the respondents, Manders Property (Wolverhampton) Ltd.

Mr G Godfrey QC and Mr M Rich (instructed by Paisner & Co) appeared for the appellants, and Mr J Mills QC and Mr D A McConvill (instructed by Manby & Steward, of Wolverhampton) represented the respondents.

Giving the first judgment, SCARMAN LJ said that the appeal raised a short question which might be put in this way: did these applicants for a new tenancy have the *locus standi* to make their application under Part II of the 1954 Act? The appellants applied for a new tenancy by an application dated October 17 1974. The respondents filed an answer in which E they first set out one of the statutory grounds for opposing the grant of a new tenancy, namely, that they intended to demolish and reconstruct the premises. That issue had not yet been considered by the county court judge, for the reason that at the invitation of the parties he directed his attention to a preliminary point also raised in the landlords' answer, the argument that the appellants were not entitled to the grant of a new tenancy, since prior to the issue of their application they had vacated the whole of the premises and were no longer in occupation of the whole or any part. Other matters alleged in the answer did not fall for consideration in the present appeal. The proceedings had taken a strange F and unusual course before the judge, who dealt with two preliminary points. The first of these was dealt with on January 31 1975, when he held that the landlords were within their rights in serving a notice under clause 6 (4) of the lease. That point was no longer the subject of litigation. The second preliminary point was the point on which the appellants had come to the Court of Appeal, since the judge found that they were no longer in occupation and therefore had no *locus standi* to apply for a new tenancy under the 1954 Act.

The appellants were tenants pursuant to an underlease dated July 6 1972, under which they held a tenancy running from March 25 1972 until December 24 1977 at a rent of £4,750 a year. The premises consisted of a shop on the G ground floor and a basement. They were part of a larger building known as the Central Arcade. That was a large structure consisting of shops, four of them facing Dudley Street and some 11 facing the arcade, which lay to the west of Dudley Street and ran from Dudley Street towards the rear of the landlords' other property known as the Manders Centre. The tenancy was a business tenancy in that the underlease contemplated the tenants carrying on a retail business of clothiers. The landlords owned a large area—the court was told some four acres—in the centre of Wolverhampton, including the arcade. They had in mind a large redevelopment scheme, had redeveloped a good deal H of the area behind the arcade and were working towards Dudley Street. It was undoubtedly because the landlords had in mind their redevelopment programme, and because of their desire at some time to recover possession of their Dudley Street frontage, including no 31, that the lease was framed in the way it was. It included a break clause which enabled the landlords to give notice determining the tenancy on December 25 1975 or at any time thereafter. It also contained some not very unusual clauses in regard to insurance against certain risks, including fire, but clause 6 (4), already mentioned, which was one not frequently found, was clearly dictated by the particular situation in which the landlords J found themselves. It had figured largely in the present appeal. By its terms, the rent due under the lease was to be suspended, if the premises were destroyed or damaged by fire or any other insured risk, until they were made fit for occupation once more; and the clause continued:

Provided that in the event of the said Central Arcade being so substantially damaged by fire or other insured risks as to render it in the opinion of the lessors' architects economically incapable of rebuilding or reinstatement in its existing form and layout the lessors may at any time following such damage give to the lessees notice in writing forthwith terminating this demise whereupon this lease shall immediately be cancelled and all obligations by K either party brought to an end.

The landlords by May 19 1974 had got their demolition contractors right up to the back of these premises and were intending to demolish the area immediately in the rear, but not of course 31 Dudley Street, which was then in the occupation of the appellants for business purposes. On May 20 a fire occurred. The county court judge described it as devastating; Mr Mills (for the landlords) as catastrophic and disastrous. The fire burned down the arcade itself. It did not wholly destroy 31 Dudley Street, but it did considerable damage to the roof and to the upper storeys, which were L not, of course, in possession of the appellants and not included in the demise. The judge said that he was satisfied that no actual part occupied by the appellants was actually burned by fire, but that he was satisfied, nevertheless, that the structure of the whole building was affected. The walls were affected by water. The premises were also affected by smoke. The judge found that the building was rendered wholly and completely unfit for occupation as business premises on May 20. It was, in his finding, quite impossible for the tenants to carry on the business in the premises at that date. He added as a further factor that all the services, ie the drains and the electricity supply, had been destroyed in the fire. M

The landlords had to consider the situation. So did the tenants. The attitude of the latter was that they were insisting on retaining their right of occupation, recognising for the time being that physical occupation for business purposes was not possible and inviting the landlords to act under the lease to reinstate the premises so that they could resume trading. Their subsequent actions were consistent with that interpretation of their letter to the landlords on May 21. The judge found that the tenants never gave up possession of the keys, and that they left some fixtures and fittings even after the bulk of the stock had been salvaged; but that they then left the premises and did not return again, certainly at no time before June 17, the date on which, on one view of the case, the landlords took the law into their own hands and demolished what was left of no 31. The landlords received an honest and competent report from their architects, on which they acted perfectly bona fide under clause 6 (4), giving an immediate notice terminating the contract under that clause. As it happens, the tenants' letter of May 21 and the landlords' clause 6 (4) notice, which they gave on May 22, crossed in the post and this led to some confusion. In the course of the confusion each party in fact reiterated and emphasised their respective attitudes. During this period the premises were boarded up and made reasonably safe so far as passers-by were concerned. On June 17 the landlords went into no 31 with their bulldozers and knocked the premises down. Since that date 31 Dudley Street, even to the extent that it was left by the fire, had disappeared save for the land on which the premises were built. The court was told that there had been rebuilding on the site to a plan quite different from the 31 Dudley Street which was the subject-matter of the demise. The judge never did get so far as to consider the merits of the substantive plea by the landlords that they required the premises for demolition and reconstruction and that therefore a new tenancy should not be granted. That issue remained, with all its difficulties, to be considered if the tenants were successful in this appeal and were found to have the *locus standi* to make their application.

The point now before the court turned on the construction of sections 23 and 24 of the 1954 Act and on the proper inference to be drawn as to the relationship of the tenants to the property after the devastating fire on May 20. Mr Mills, for the landlords, submitted that from the date of the fire these applicants were not tenants under a tenancy to which the Act applied, because they no longer occupied these premises, and certainly no longer occupied them for the purposes of a business carried on by them. This question had been considered by the courts, and in particular by Cross J in *I & H Caplan Ltd v Caplan (no 2)* [1963] 1 WLR 1247. To be able to apply for a new tenancy under the Act, a tenant had to show either that he was continuing in occupation of the premises for the purposes of a business carried on by him or, if events over which he had no control had led him to absent himself from the premises, that he continued to exert and claim his right of occupancy. In the present case the absence of the tenants from the premises after the fire was not their choice but was brought about by the state in which the fire had left the premises—a state which was none of their own making. Nevertheless, they exhibited immediately after the fire and continued to exhibit an intention to return and to claim their right of occupancy, and reminded the landlords from time to time of what they thought were the landlords' obligations to reinstate.

The judge was so impressed with the devastating nature of the fire that he was not prepared to draw the inference that the absence of the tenants from the property was temporary. He though the reality of the situation was that the tenants had gone for ever. Mr Mills submitted that that was an inference which indeed he was right to draw because of the catastrophic nature of the fire. He (his Lordship) thought that when events such as he had detailed arose, and a tenant was faced with difficulties of occupation such as confronted these tenants, it must be a question of fact whether the tenant intended to cease occupation or whether he was (as the judge found these tenants were) cherishing the hope of return but making it quite clear that he regarded himself as possessing a right of occupancy and calling upon the landlord to reinstate. Mr Mills said that the whole of that was quite unrealistic because these landlords had, immediately following the fire, given a notice under clause 6 (4) indicating that the Central Arcade was economically incapable of being reinstated to the old design and terminating the tenancy. But one had to look at the matter as the tenants were then looking at it. They had premises which themselves had not been destroyed. The judge expressly found that the part held by the appellants still stood. They (the appellants) also must be assumed to have appreciated their position under the Landlord and Tenant Act 1954. That was this, that the landlords could not bring the contractual tenancy to an end save by taking steps detailed in Part II of the Act. In particular, the landlords could not bring the tenancy to an end merely by serving a notice under clause 6 (4) if the tenants were still in occupation or claiming their right of occupation at the time it was served, because in terms it was not a notice as required by the Act. It was to be observed, moreover, that these difficulties created no technical problem for the landlords in this case, because at a later date, on July 1 1974, realising that there might be argument as to the effectiveness of the clause 6 (4) notice, they did serve an appropriate notice under section 25.

That led to a further question arising on this appeal. As Cross J recognised in *I & H Caplan Ltd v Caplan*, the applicant for a new tenancy had to continue in occupation, or maintain his right to occupy, right up to the date of the order made by the court. Were these applicants still occupiers after the demolition in June and at the time, some two weeks later, when the section 25 notice was served? He (his Lordship) would find it surprising if action which, on the view of the case he had formed, was action not lawfully open to the landlords, but was nevetheless taken by them, could be said to destroy the appellants' *locus standi* to apply under this part of the Act. He did not think the court was driven to such an unjust conclusion. It was plain that the cesser of occupation in this case was not to be regarded as an abandonment of occupation by the tenants. They had at all times indicated not only their hope but their intention to go back as soon as the premises were fit for occupation. Nothing he (Scarman LJ) had said bore upon the substantive ground in the landlords' answer, namely, that they required the premises for reconstruction. It might be that they had a strong case. All that the court was concerned to determine was whether the appellants had a right to apply to the court. For the reasons he had given, he (his Lordship) could find nothing in the facts of the case and nothing in the 1954 Act to debar them from pursuing their application. He would allow the appeal.

Sir Gordon Willmer said that he had come to the same conclusion. The question was whether the appellants had shown such continuing occupation as to give them a *locus standi* to assert any rights under the 1954 Act. This was a mixed question of fact and law. After the fire the appellants as tenants enjoyed the same right to occupy the premises as they had enjoyed before. The premises, although damaged, were still in existence. The appellants had, during the first day or two following the fire, exercised their rights by going into the premises in order to save as much stores and equipment as they could. It seemed to him (his Lordship) that the tenants, who had been in continuous occupation up to the fire and immediately after the fire and who retained the intention to occupy, remained both in fact and in law the occupiers of the premises at the relevant time. They retained the keys; they were thus in a position to exclude the public or to open the door and invite the public to come in. If some person had been invited to come in, he (Sir Gordon Willmer) apprehended that the tenants would have had some difficulty in escaping their liabilities under the Occupiers' Liability Act 1957. He did not see how the subsequent demolition of the building at the instance of the landlords could possibly have affected one way or the other such rights as the tenants had.

Stephenson LJ agreed, and the appeal was allowed. The appellants were awarded costs above and below.

LANDLORD AND TENANT
LEASEHOLD REFORM

A

COURT OF APPEAL
March 26 1975
(Before Sir GEORGE BAKER P, Lord Justice ORMROD and Sir GORDON WILLMER)
GAIDOWSKI v GONVILLE & CAIUS COLLEGE, CAMBRIDGE

Estates Gazette April 24 1976
(1975) 238 EG 259

Leasehold Reform Act 1967—Premises a "house" within Act though within the five-year occupation period they were combined with a room taken from the adjoining property—Strip of land running behind houses not however "let with" relevant premises—Tenant gets freehold of house but not of strip

B

This was an appeal by Mr Joseph Gaidowski, of 6 Harvey Road, Cambridge, from a judgment of Judge Connolly Gage in the Cambridge County Court on April 29 1974 dismissing his application for a declaration that he was entitled, under the terms of the Leasehold Reform Act 1967, to require the respondents, the Master, Fellows and Scholars of Gonville & Caius College, Cambridge, to convey to him the freehold of 6 Harvey Road. By a cross-notice, the respondents appealed against the judge's conclusion that a strip of land running behind the gardens of 5 and 6 Harvey Road was let to the appellant with no 6.

C

Mr M Barnes (instructed by Seaton Taylor & Co, agents for Vintners, of Cambridge) appeared for the appellant, and Mr N Hague (instructed by Francis & Co, of Cambridge) represented the respondents.

Giving the first of the reserved judgments, ORMROD LJ said: This is an appeal from a judgment of His Honour Judge Connolly Gage, sitting at Cambridge County Court on April 29 1974, dismissing an originating application by Mr Gaidowski, the appellant, for a declaration that he was entitled, under the Leasehold Reform Act 1967, to require the freeholders, the Master and Fellows of Gonville & Caius College, to convey to him the freehold of 6 Harvey Road, Cambridge, together with a strip of garden running parallel to Harvey Road and lying at the foot of the back gardens of the houses in that road. No 6 is one of a pair of semi-detached Victorian houses, the other being no 5, which were built about 1880 by an architect, Mr Morley, on land belonging to the college. In 1883 the college granted a lease to the late Mr J N Keynes, a former Fellow of Pembroke College, for a term of 99 years from December 25 1881, of nos 5 and 6 together at an annual rent of £30. Mr Keynes and his family occupied no 6, but except in one respect, which is highly material to this case, there was no evidence as to the occupation or use of no 5. The important fact was that at some date prior to 1958 an opening was made in the wall separating the adjoining ground-floor front rooms of nos 5 and 6, and the doorway leading from the hall of no 5 into this room was bricked up. This extra room appears to have been used as a library by Mr and Mrs Keynes. After the death of Mr and Mrs Keynes the unexpired portion of the lease of nos 5 and 6 was transferred to Mr Gaidowski by their personal representative, Sir Geoffrey

D

E

Keynes, by an assignment dated December 5 1958. Mr Gaidowski has lived in no 6 ever since, but in 1972 he filled up the opening in the wall between nos 5 and 6 and reopened the doorway into the hall of no 5, replacing the original door which had been stored in the cellar. Nos 5 and 6 were, therefore, restored to their original condition.

F

By a notice of application dated January 12 1973 Mr Gaidowski claimed the freehold of "All that house garage and garden premises as shown on the attached plan," that is of 6 Harvey Road and the garden strip already referred to. On June 7 1973 the college gave a notice in reply disputing Mr Gaidowski's claim on the grounds "that during part of your five-year period of residence you have not occupied (in whole or in part) a 'house' as defined by the Act, in that the relevant building included the ground-floor front room of no 5 and was not a 'house' by reason of section 2 (2); and that, in respect of the garden strip, this was not 'let with' no 6 and so was not part of the premises as defined by section 2 (3)." It was common ground in this court and in the court below that the appellant fulfilled all the other criteria prescribed by the Leasehold Reform Act. He was at the date of his notice of application the tenant of a leasehold house on a "long" lease and at a "low" rent, as defined in the Act, and the rateable value of no 6 was £182 per annum, that is, within the prescribed limit of £200. The sole issue on this part of the case was, and is, whether he can fulfil the remaining requirement, namely, that he had occupied the house as his only or main residence for at least five years immediately preceding the date of his notice. Mr Hague, for the college, concedes that if no 6 had never been altered by the taking in of the front room of no 5, there would have been no answer to Mr Gaidowski's claim. He also concedes that after the expiry of five years from the time of what he calls the "re-conversion" or reinstatement of nos 5 and 6, the college could not resist Mr Gaidowski's claim to enfranchise no 6. The real question in this case, therefore, is the effect of alterations to the premises during the five-year period, or, put in another way, how does the Act apply to premises which have been altered, as these premises have been altered, during the crucial five-year period.

G

H

J

Mr Hague's main submission, both here and below, is that so long as no 6 consisted of the original house plus the additional room taken in by the Keynes's from no 5, the house was not a "house" within the Act, because it was caught by section 2 (2), and that therefore Mr Gaidowski could not prove that he had occupied a "house" which was within the Act during the requisite period of five years, the alterations having been done in 1972, barely a year before the date of the notice of application. Mr Barnes' contention on behalf of the appellant is, and was, that no 6 could at all times reasonably be called a "house" with the use of an additional room which did not form or could reasonably be regarded as not forming a part of no 6. If that is so, it would not be caught by section 2 (2). That subsection reads as follows:

K

> References in this Part of this Act to a house do not apply to a house which is not structurally detached and of which a material part lies above or below a part of the structure not comprised in the house.

A The additional room certainly lay above and below a part of the structure not comprised in no 6, and therefore, if it is to be treated as an integral part of no 6 before the reinstatement, no 6 would be excluded by the subsection from the operation of the Act so long as this situation remained unchanged. The learned county court judge rejected Mr Barnes' submission that the additional room was not part of the house and his alternative submission that it was not a material part, and accordingly dismissed the originating application. The issue therefore turned on the meaning to be given to the word "house" for the purposes of the Act. Section 2 (1) contains what Mr Barnes called the basic defini-
B tion of a house and reads as follows:

For the purposes of this Part of this Act, "house" includes any building designed or adapted for living in and reasonably so called, notwithstanding that it is not structurally detached. . . .

This is a fairly elastic definition. All that has to be shown is that the premises might reasonably be called a house. The fact that they might equally well be called something else will not take them out of the Act (*Lake v Bennett* [1970] 1 QB 663). Mr Hague's contention leads to an anomaly which ought, if possible, to be avoided. It was agreed by both counsel, who have great experience in this field, that the only
C practical reason for the conclusion in the Act of subsection (2) of section 2 is a conveyancing one. "Flying freeholds," or "freehold in the air," give rise to peculiarly difficult conveyancing problems. These are obviated by this subsection, which simply excludes from the provisions of the Act any premises the enfranchisement of which would lead to difficulties of this kind But no such difficulties can arise in this case, because the notice of application is confined to no 6 in its original form. No attempt is being made here to secure the freehold of the front room of no 5, and the only result of accepting Mr Hague's submission would be to postpone the
D enfranchisement of no 6 until 1977, at some additional expense to the tenant. Mr Barnes' contention also is not free from difficulty, in that it makes his client's claim to the freehold of no 6 depend on what must be an impressionistic conclusion as to what can or cannot reasonably be called a "house."

With all respect to the two very experienced counsel who argued this case with the utmost care and lucidity, I cannot suppress my doubt whether on the facts of this case this problem actually arises at all. If one begins with section 1, the Act confers on the tenant the right to acquire the freehold of "the house and premises." I ask myself, what house and
E premises? The answer must be, the house and premises identified in the notice of application, that it, 6 Harvey Road in its condition at the date of the notice. This house is a leasehold house, let on a long lease, at a low rent, and of a rateable value below £200. Now comes the crucial part, that is, paragraph (b) of subsection (1), which reads: "at the relevant time . . . [the applicant] has been . . . occupying it as his residence for the last five years. . . ." I find it difficult to understand why "it" should not refer to the leasehold house identified in the notice of application, that is no 6 in its condition at that date. If this is correct, the fact that for some part of the five-year period no 6 was connected to the front
F room of no 5, which was cut off from no 5, becomes irrelevant. I very much doubt whether the draftsman of the Act could have intended subsection (2) of section 2 to apply to the house referred to as "it." If it does, it leads to the curious result that by altering the property at any time before making his application the tenant can reduce the rateable value to a figure within the £200 and so bring it within the Act, but cannot eliminate the effect of section 2 (2), which has ceased to have any practical effect. However, this point was not taken in the court below, and this court must therefore deal with the case on the basis of the arguments addressed to the learned judge.

G But for authority, the learned judge would have accepted Mr Barnes' contention that before the reinstatement what was occupied was the house, no 6, together with the extra room. I too think that even with the extra room attached, references to the "house" in this case could reasonably be taken to mean no 6 by itself. The matter can be tested by asking how a reasonable man looking over the property would describe it. If he attached more importance to structure than to use, I think he would say that the extra room was obviously the front room of no 5 and that no 6 was the house. If he were thinking of purchasing no 6, he would probably at least inquire whether the extra room was to be included in the
H sale of no 6. If he were more impressed by use, and he found the extra room carpeted and curtained to match the adjoining room in no 6, and fully furnished, he might describe the whole as no 6 into which an extra room had been incorporated. If he found it undecorated, unfurnished, and used simply as a storeroom for unwanted furniture, he might have his doubts. To some observers, much might depend on how easy or otherwise it would be to restore these two houses to their original condition. Had the occupant of no 6 been content to lock the door into no 5 and stand a bookcase against it, the mere fact of making an opening in the wall between the two houses could not be said to have made this room part
J of no 6. In the present case the door frame leading into the hall of no 5 was still in place and the door itself was still available, so that all that was required by way of reinstatement was to remove the bricks from the doorway, close the opening between the houses and rehang the door. Quite a different situation may arise where two or more cottages are converted into a single house. Party walls may be removed, front doors eliminated, one or more staircases may be taken out, so that the identity of the original cottages completely disappears. My conclusion, therefore, is that the "house" in this case was always no 6 as it was originally built, and that
K it could reasonably be called a house whether or not the communication with the front room of no 5 was opened or bricked up. It is accordingly unnecessary to consider Mr Barnes' second point, that in any event the additional room was not a "material part" of no 6, although I would be inclined to think that if it was a part of no 6, it was a material part. Certainly if it was to be included in the conveyance of the freehold to Mr Gaidowski, the conveyancing difficulties which section 2 (2) seems designed to avoid would arise.

Turning now to the two cases referred to by the learned judge, *Peck v Anicar Properties Ltd* [1971] 1 All ER 517 and *Wolf v Crutchley* [1971] 1 All ER 520, I think, contrary to
L the judge's view, that *Peck's* case helps rather than hinders Mr Barnes. In that case, as in the present case, the application was limited to that part of the premises which did not lie above or below a part of the structure not comprised in the house. The only question was whether the "house" must be taken to include the ground floor of the adjoining house which had been joined to the ground floor of the premises sought to be enfranchised, forming together a single shop. It was held that the word "house" could reasonably be applied to the building consisting of one half of the shop and the living accommodation above. It is to be observed that consequently no conveyancing problem arose in that case,
M because the application was restricted to that part of the property which was affected by section 2 (2). I do not think that the fact that the ground floor was used as a shop rather than as part of a residence is sufficient to distinguish that case on its facts from the the present. We are concerned with a reasonable description of a building, and if the structure in *Peck's* case could reasonably be called a house, I think that the present case is, if anything, stronger. No 6 is to all appearances a "house." The extra room does not, in my opinion, make it inappropriate to refer to no 6 by itself as a house. In *Wolf v Crutchley* the question was whether 5 Gliddon Road could reasonably be called a house in itself, notwith-

A standing that there was a connecting door to the adjoining premises which were used by the applicant for letting out rooms. Cairns LJ attached importance to the fact that no part of the adjoining house was used as part of the applicant's residence, which was wholly in no 5. Lord Denning MR made the same point, but also emphasised the fact that no 5 was "structurally a separate house." In the present case no 6 is and was structurally a separate house, and as I have said, the extra room could reasonably be described, in structural terms, as the front room of no 5. (Interestingly, it was so described by the respondents themselves in their notice in reply.) Domestically, for reasons which I have given, this B extra room was used by Mr Gaidowski only as a place to store some surplus furniture. For these reasons, I do not think that there is anything in these two cases which should have caused the judge below to have departed from his primary impression. I would accordingly allow this appeal in so far as the application relates to 6 Harvey Road as described in the original lease.

I now turn to the second part of the case, which concerns the strip of garden lying across the end of the gardens of these houses, which is the subject-matter of the respondents' notice in this court. This strip of land was not included in the original lease to the late Mr J N Keynes, and came into Mr C Gaidowski's possession as the result of an entirely separate series of transactions. To succeed in his application to acquire the freehold of this strip Mr Gaidowski must rely on the terms of section 2 (3), which are as follows:

> Subject to the following provision of this section, where in relation to a house let to and occupied by a tenant reference is made in this Part of the Act to the house and premises, the reference to premises is to be taken as referring to any garage, outhouse, garden, yard and appurtenances which at the relevant time are let to him with the house and are occupied with and used for the purpose of the house or any part of it by him or by another occupant.

D The history of this strip is as follows. The freehold of it is, and at all material times was, in the college, and it was leased to Mr Morley for a term of 99 years at about the same time as the lease of nos 5 and 6 was granted to Mr Keynes, possibly with a view to providing access to the rear of a house which Mr Morley was building in Harvey Road. It never was so used, and seems to have remained in Mr Morley's possession until his death. In 1938 his executors, Lloyds Bank Ltd, granted an underlease of it for the residue of the term of the original lease, less one day, to the late Mrs Keynes. She appears to have used it as a garden, connected by a gate E or in some other way with the garden of no 6, until her death, and it was included in the assignment of nos 5 and 6 to Mr Gaidowski by Sir Geoffrey Keynes, although described in a separate schedule. So at that time it was let by Lloyds Bank or their successors in title to Mr Gaidowski. It was certainly not then "let with" nos 5 and 6, since the lessors were different persons. Eventually, in 1964, the college took an assignment of the residue of the lease from two ladies in whom it had become vested subject to the underlease which had been assigned to Mr Gaidowski. Accordingly, the college became the lessor of the strip to Mr Gaidowski as the lessee.

The question, therefore, is whether as a result of this series F of transactions the strip could properly be said to be let at the relevant time, that is, the date of the notice of application, to Mr Gaidowski "with the house." All that can be said is that at the relevant time the lessors and the lessee of no 6, and the lessors and the lessee of the strip, had become the same persons. In my judgment, this is not enough to establish that the strip is "let with" no 6. There is no direct authority on the point, but Mr Barnes referred us to a passage in Megarry J's book on the Rent Acts (10 ed, p 99 and following), and to cases there cited. For my part, I doubt whether these cases are of any assistance in the present case because the purposes of the Act with which we are concerned are quite different from those of the Rent Acts. Cairns LJ in G *Wolf's* case rejected a similar submission on this ground, and I respectfully agree with the view which he expressed. The matter may be tested in this way. Suppose the college had granted a weekly tenancy of the strip as a vegetable garden to Mr Gaidowski a week or so before the notice of application had been served. Could it possibly be argued that the strip was "let with" the house in such circumstances? If the question were to be answered in the affirmative, Mr Gaidowski would be entitled to acquire a potentially valuable piece of freehold land merely because he had become a weekly tenant of the same landlord. In my judgment, "let with" implies some reasonably close connection between the transactions H of letting the house and letting the strip. The learned judge expressed the opposite view, but with respect I cannot agree with him. In my view the respondents therefore succeed on the contention put forward in their notice, and I would accordingly exclude the strip from the declaration to which the appellant is entitled in respect of the house.

SIR GORDON WILLMER: The appellant, Mr Gaidowski, is the tenant of a leasehold house at no 6 Harvey Road, Cambridge, which he occupies as his residence. He claims to be entitled under the Leasehold Reform Act 1967 to acquire the freehold of the house. It is not in dispute that the house is let to him on a long tenancy at a low rent, and J that the rateable value of the house is within the limits specified in the Act. But the claim is resisted on the ground that during part of the period of five years preceding the date when he gave notice of his claim—and, indeed, at all material times up to 1972—he also occupied a room within no 5. The room in question is the ground-floor front room of no 5. Prior to 1972 this room was connected to the front of no 6 by an access doorway. Moreover, until 1972 this room in no 5 was separate from the rest of the house, the connecting doorway having been blocked up. It was not in dispute that this front room of no 5 lay below a part of the structure not K comprised in no 6, in that it lay below the upper part of no 5. It is therefore contended on behalf of the respondents that Mr Gaidowski's claim is excluded by section 2 (2) of the Act. Mr Gaidowski gave evidence, which was not challenged, that prior to 1972, when he bricked up the connecting doorway between no 6 and the room in no 5 and reopened the doorway from that room into the rest of no 5, he used the room in question solely as a storeroom. Can it be said, therefore, that he occupied this room as part of his residence?

Bearing in mind the words of Lord Wilberforce in *Parsons v Trustees of Henry Smith's Charity* [1974] 1 WLR 435 at 440C, that in this jurisdiction the issue is one "which must L be largely factual and one of commonsense," I would say, as a matter of commonsense, that he did not. He certainly never resided in the room in question. Had he used the room for the purpose of carrying on a business or profession, or had he simply left it empty, it could hardly be said that he occupied it as part of his residence. It seems to me that the same applies to a mere storeroom. The importance of occupation as a residence was emphasised by Lord Denning MR in *Peck v Anicar Properties Ltd* [1971] 1 All ER 517. He indicated three matters to be looked at in determining what is the "house" in any given case, namely, (1) the lease M itself, (2) the portion occupied as a residence, and (3) the physical condition of the structure. In the present case the lease itself does not help, for it is a lease of both houses, nos 5 and 6. As to the physical condition of the structure, this was at the material time as I have described it. But as to the second matter, if one looks at the portion occupied as a residence, one can only come to the conclusion that the portion so occupied was no 6 itself without the addition of the room in no 5. In *Peck's* case the Court of Appeal had no difficulty in concluding that the tenant claiming the benefit of the Act was not defeated by the mere fact that on the ground floor under his residence there was a shop which

A extended underneath the adjoining residence. I can see no valid distinction between that case and this. In my view, it is conclusive in favour of Mr Gaidowski in the present case. I think that the learned judge came to a wrong conclusion, and that the appeal should be allowed so far as concerns the house.

There remains for consideration the subsidiary question whether Mr Gaidowski's right to acquire the freehold extends also to the garden strip at the back of the houses. In order to bring himself within section 2 (3) of the Act Mr Gaidowski must show that at the date when he gave notice of his claim this garden strip was (a) " let to him with the house," and (b)
B " occupied with and used for the purpose of the house." Mr Gaidowski no doubt satisfies the second of these requirements, but I find difficulty in saying that the garden strip was " let to him with the house." The conveyancing history which led to his becoming the tenant of the college in relation to the garden strip was quite different from that in relation to the house, no 6. The mere fact that he now occupies both properties as tenant of the same landlord is clearly not sufficient of itself. Effect must be given to the requirement that the garden strip must also be " let to him with the house." I do not go so far as to say that the properties must both
C be let to him as part of the same transaction. But there must, in my view, be at least some connecting link between the letting of the one property and the letting of the other. No such link exists in the present case. On the contrary, the only link between the letting of the garden strip and any house is with no 8. In the circumstances, I do not think that Mr Gaidowski brings himself within section 2 (3) in relation to the garden strip, and I do not agree with the view expressed obiter by the learned judge in relation thereto. It follows that in my judgment, while Mr Gaidowski has the right to acquire the freehold of the house, No 6, he does not enjoy the same right in relation to the garden strip.

D BAKER P: It is perhaps worth adding a word about 5 Harvey Road. The only evidence about its occupation is that a Professor Tilly lived there for two or three years, but it must be an inevitable inference that in Cambridge no 5 did not remain empty either before or after Mr and Mrs Keynes

E bricked up the doorway from the hall of no 5 to the disputed room. The availability of purchasers or sub-tenants, the requirements of individuals at any particular time, the price or rent obtainable, the length of time for which a prospective occupier wanted the premises and other factors could each and all have a bearing on whether Mr and Mrs Keynes, or later Mr Gaidowski, would, in the light of their own need for or use of the room, want to part with the whole or only the truncated portion of no 5 if it became vacant. For example, a family with children would, I should think, inevitably want the front room (that is, the disputed room), which measures 23 ft 4 in by approximately 15 ft. All this,
F coupled with the known facts, and especially that the door frame was in place and the door kept in the cellar, leads me to the conclusion that the disputed room always remained an integral part of no 5, and that no 6 as it was originally built was a " house."

By the underlease on sale dated September 19 1938, Lloyds Bank, as executors of Mr Morley, disposed of the residue of the lease of the " garden strip," less one day, to Mrs Keynes for the sum of £300 at the rent of one shilling a year. I think this nominal rent may have tended to confuse the issue. Suppose the rent to be substantial, and that Mr Gaidowski had failed to pay, any proceedings by the college for payment
G or forfeiture of the garden strip would have to be separate from and irrespective of their rights in nos 5 and 6 Harvey Road. I would prefer, like Sir Gordon Willmer, to reserve my opinion on whether properties must both be let as part of the same transaction to fall within section 2 (3) of the Act, but I am satisfied that on the facts and history of this case the garden strip was not at the relevant time let to Mr Gaidowski with the house. Having had the opportunity of reading the judgments which have just been delivered, I would add that I entirely agree with all that my Lords have said.

The appeal was allowed to the extent of a declaration that
H the appellant was entitled to acquire the freehold of the house, but was not entitled to acquire the freehold of the garden strip. The appellant was awarded three-quarters of his costs both in the Court of Appeal and below. Leave to appeal to the House of Lords was refused.

LANDLORD AND TENANT

RENT ACTS

A

CHANCERY DIVISION

May 15 1975

(Before Mr Justice BRIGHTMAN)

NEWMAN v DORRINGTON DEVELOPMENTS LTD

Estates Gazette January 31 1976

(1975) 237 EG 335

Protected tenancy—Option for further lease at "commercial yearly rack-rent at which demised premises might reasonably be expected to be let in the open market"—Option not incapable of performance though rack-rent substantially more than the registered rent—Order for specific performance of contract constituted by exercise of option

B

This was a claim by Mr Frederick Allen Newman, of 7 Dudley House, Westmoreland Street, Marylebone, London W1, against Dorrington Developments Ltd, for specific performance of a contract for a further lease of his flat constituted by exercise of an option contained in a lease of June 24 1970.

Mr T R F Jennings (instructed by Nabarro, Nathanson) appeared for the plaintiff, and Mr E J Prince (instructed by Philip Ross, Elliston & Bieber) respresented the defendants.

Giving judgment, BRIGHTMAN J said: This is an action by a tenant to compel landlords to grant a new lease in accordance with an option for renewal which the tenant claims to have exercised. The landlords submit that the option has become incapable of performance by reason of the fact that the rent is restricted by the Rent Act 1968. By a lease dated June 24 1970 the then landlord let a flat, 7 Dudley House, Marylebone, to the then tenant for a term expiring on June 24 1973 at the yearly rent of £550, plus a service charge which I can ignore. Clause 5 (iv), so far as relevant for present purposes, provided as follows:

C

(a) If the lessee shall be desirous of taking a lease of the demised premises for a further term of three years from the expiration of the term hereby granted at the rent and on the terms and conditions hereinafter mentioned and shall not more than 12 nor less than six months before the expiration of the term hereby granted (time being deemed to be of the essence) give to the lessors notice in writing of such his desire and if he shall have paid the rent hereby reserved (hereinafter called "the current rent") and shall have performed and observed the covenants and stipulations herein contained and on his part to be performed and observed up to the termination of the tenancy hereby created then the lessors will let the demised premises to the lessee for the further term of three years from the twenty-fourth day of June One thousand nine hundred and Seventy-three at a rent to be determined as hereinafter provided and subject in all other respects to the same stipulations as are hereinbefore contained except this clause for renewal. . . .

D

E

(b) The rent for the said further term (hereinafter called "the new rent") shall be such annual sum as shall be agreed in writing between the lessors and the lessee or their respective surveyors but if within two months after the date of the lessee's notice (time being deemed to be of the essence) of his desire of taking the said lease for a further term of three years agreement shall not have been reached on the new rent then the new rent shall be determined in accordance with the Arbitration Act 1950 by a single arbitrator to be appointed by the President for the time being of the Royal Institution of Chartered Surveyors as being the commercial yearly rack-rent at which the demised premises might reasonably be expected to be let in the open market for an unbroken term commencing on the same day and of the same duration as the said further term by a willing lessor to a willing lessee on the same terms covenants and conditions (except as to rent) as are herein contained provided that notwithstanding a determination lower than the current rent the new rent shall in no circumstances be less than the current rent.

F

In 1972 the term became vested in the plaintiff and the reversion became vested in the defendants. On June 24 1972 the option first became exercisable and would continue to be exercisable until December 24 1972. By notice dated June 30 1972 the plaintiff purported to exercise the option by requiring the defendant to let the demised premises to him for a further term of three years from June 24 1973 at a rent to be determined in accordance with clause 5 (iv) (b) of the lease.

G

It is common ground that on the date when the tenancy was created, ie June 24 1970, it was (and still is) a protected tenancy within the meaning of section 1 of the Rent Act 1968, and a regulated tenancy within the meaning of section 7. As a result, the rent of the demised premises became subject under section 20 to a limit which is called by the Act "the contractual rent limit." If a rent has been registered under Part IV of the Act the contractual rent limit is basically the registered rent. Where no rent has been registered, then the contractual rent limit is basically to be determined as follows: (a) if not more than three years before the regulated tenancy began the dwelling-house was subject to another regulated tenancy, the contractual rent limit would be the rent payable under that other tenancy; (b) otherwise, the contractual rent limit is the rent payable under the terms of the lease creating the tenancy. Section 20 provides that where the rent payable for any contractual period of a regulated tenancy of a dwelling-house would exceed the contractual rent limit, the amount of the excess shall be irrecoverable. Section 46 provides, in effect, that in determining the fair rent, which is the rent to be registered, regard shall be had to all the circumstances and, in particular, to the age, character and locality of the dwelling-house and its condition, but it shall be assumed that the "number of persons seeking to become tenants of similar dwelling-houses in the locality" is not substantially greater than the number of available houses. In other words, scarcity value is to be disregarded. The contractual period of this tenancy expired on June 24 1973, leaving aside for this purpose the option to renew. I understand that there was no registered rent on that day, but that an application for the registration of a rent was made to the rent officer on July 6. On November 29 the fair rent was determined to be £890 per annum. This accordingly became the registered rent and *prima facie* took effect from the date of the application. It follows that if the defendants had granted a new tenancy to anyone on June 24 1973 a yearly rent in excess of £550 would not have been recoverable. The landlords could, however, have granted a new tenancy to anyone after July 6 1973 at a recoverable rent not exceeding £890. It is common ground that the commercial yearly

H

J

K

rack-rent in the open market (to use the phraseology in the lease) would have exceeded £890 if the Act had not existed.

As I have already said, the lease provides that in default of agreement (and none exists in the present case) the rent shall be "the commercial yearly rack-rent at which the demised premises might reasonably be expected to be let in the open market . . . by a willing lessor to a willing lessee." That, say the defendants, is an impossibility in the present case, because an open market commercial rent presupposes an unrestricted market. In the present case the market is not unrestricted; section 46 prevents scarcity value from being taken into account. A "rack-rent" was defined by William Blackstone as "a rent of the full value of the tenement or near it"; see *Commentaries on the Laws of England*, vol 2 p 43. In support of his proposition, counsel for the landlords referred me to *Rees v Marquis of Bute* [1916] 2 Ch 64. In that case, a freeholder of a number of cottage properties let at weekly rents wished to give his tenants the opportunity of buying long leases which would be almost equivalent to freeholds. An auction sale was held. One tenant of a cottage, let at a weekly rent of about 4s, bought for £55 a 99-year lease of his cottage at an annual rent of £1. A month before the sale the Rent Act of 1915 had been passed, but this was not appreciated by the vendor or purchaser. Both sides wished to honour the bargain, but the vendor was unwilling to do so until it had been ascertained that it could lawfully be carried into effect. The house in question was one to which the Act applied. The Act provided that the payment of a premium for a tenancy could not be required and if paid could be recovered. Consequently, the agreement to pay the premium in that case was an illegal agreement, and as it had not been performed, the defendant ought, it was decided, to be relieved from it. I do not think that this case supports the defendants' proposition. In the *Bute* case the payment of the purchase price or premium was illegal, and if paid could be recovered. The contract was inevitably one of which the court would not order specific performance because one term of the contract was not a lawful term.

I was next referred to *Hollies' Stores Ltd v Timmis* [1921] 2 Ch 202. That was a case where there was an option for the renewal of a lease, but it was a term of the lease that the payment of the rent should be guaranteed by three named guarantors. One of the guarantors had died before the exercise of the option. The lessees, upon the purported exercise of the option, naturally could not produce all three guarantors, but they offered instead to pay, in advance, the entire rent under the renewed lease or to secure it by a deposit of government securities sufficient for that purpose. The judge held that the contract was impossible of performance, because one of the specified parties could not concur in the guarantee. The only purpose of the guarantee being to secure the payment of the rent, and the tenants being able and willing to secure the rent beyond a peradventure, I am not certain that I would myself have been willing to decide the case against the tenants. But however that may be, the decision seems to me a long way from that with which I am concerned, and it does not give me any assistance.

In *Brilliant v Michaels* (1945) 114 LJ Ch 5 the defendant agreed to let a flat from a future date at a rent in excess of the standard rent payable under the Rent Restriction Acts. The plaintiff sought specific performance. It was held on the facts that there was no consensus ad idem and therefore no binding agreement. However, the learned judge added this on the enforceability of the agreement had there been any agreement to enforce (I read from p 8):

> There is one other matter to which I think perhaps I ought to make some reference, since the point is a novel one on which there has been some argument. As I have already indicated, the rent, which was eventually settled in the final of these receipts, was a rent which appears to have been substantially in excess of what is called the standard rent applicable to these premises by virtue of the current rent restriction legislation. It is quite clear that, if two parties agree that one is to pay and the other is to receive a rent in excess of the so-called standard rent, neither that part of the bargain nor the whole contract is thereby rendered illegal. The fact is that in so far as the rent inserted or agreed upon is in excess of the standard rent, the excess is not recoverable at law. Had the plaintiff succeeded, the question would have arisen whether, in circumstances such as appear here, the court would have made an order for specific performance of an agreement, a material term of which was not wholly enforceable at law. In the circumstances which have happened, it is unnecessary for me to reach a final conclusion, but for what it is worth, I will say that I am not yet satisfied that the court ought to grant specific performance, having regard to the form of decree common in these cases, of an agreement one term of which to the knowledge of the court is unenforceable in whole or in part, particularly where, as in this case, the subject-matter, the rent and the limitation of the amount of rent, which can be charged, is a matter of public policy. I therefore say, without expressing it any more affirmatively than that, that I am not satisfied that it would be right in such a case to grant a decree of specific performance with such a term as to rent in the agreement.

Lastly, counsel for the landlords, referred me to *Maurey v Durley Chine (Investments) Ltd* [1953] 2 QB 433. In that case there was a furnished tenancy for a term expiring on June 24 1953 at a rent of £525. On March 30 1949 the tenant was granted for good consideration an option for the immediate grant of a new lease for the further term of seven years from June 24 1953 at the like rent of £525. The tenancy due to end in June 1953 unless the option were exercised was within the Furnished Houses (Rent Control) Act 1946. In 1950, the rent was fixed under the Act at the reduced rate of £425 per annum. In December 1951 the tenant purported to exercise the option. At that date the 1946 Act, originally due to expire in 1947, had been extended to March 1953. Later on, it was extended to March 1954. The position therefore was that at the date when the option was exercised (December 1951) and the new lease ought to have been forthwith executed, the new lease would not have been subject to any legislative rent restriction at the moment it commenced, although by the time the term of the new lease fell into possession such a restriction had come into existence. The option was held to have been validly exercised. However, Jenkins LJ, reading the judgment of the court, added this:

> If at the time when the new lease ought to have been granted (that is, forthwith after the exercise of the option) the duration of the Act of 1946 as then fixed had extended beyond the date of the commencement of the new term (namely June 24 1953), we would, as at present advised, have been disposed to hold the defendants relieved on the ground that the new lease could not lawfully have been granted at a recoverable rent of £525 per annum, and that the agreement under which it was to have been granted had accordingly become impossible of performance because it could not be legally performed.

The cases which I have read seem to come to this, that if a contract is made, whether by ordinary offer and acceptance or by the grant and exercise of an option, for a sale or lease at a price or rent which is greater than that which can be lawfully recovered at the time when the contract is due to be performed, the court will not force the bargain upon the parties, because it is not a bargain which can lawfully be implemented according to its terms. These authorities are not, in my view, decisive of the present case. Here the option is not, in terms, an option for the grant of a lease at a rent which exceeds the permitted rent. It is an option for the grant of a lease at a rent which is determined to be "the commercial yearly rack-rent at which the demised premises might reasonably be expected to be let in the open market." A somewhat similar formula was considered in *Rawlance v Croydon Corporation* [1952] 2 QB 803. The case arose under section 9 of the Housing Act 1936. This section provided that where a local authority was satisfied that a house was unfit for human habitation they might serve a notice to repair upon

A the person having control of the house, and that was defined as the person who received the rack-rent of the house or would receive it if the house were let at a rack-rent. "Rack-rent" was defined as rent which was not less than two-thirds of the full net annual value of the house. A notice was served on the respondent in respect of a house let by him at a yearly rent of £45. This was the permitted rent under the Rent Restriction Acts. It was admittedly less than two-thirds of the annual value of the house on the supposition that the house was outside the ambit of the Acts. Nevertheless it was held that the respondent was in receipt of a rack-rent, because he was receiving the full rent which he was capable of receiving B by law. I quote this passage from the judgment of Romer LJ:

> When the legislation was first introduced into our economic and social system, and it may be for some considerable time thereafter, landlords still received a profit rental although they were in the main precluded from increasing it. The general trend of the legislation has long since resulted in landlords only receiving rentals which are usually a great deal less than those received by owners of equivalent but uncontrolled properties. Nevertheless this change cannot alter the fact that landlords who are affected by the legislation are undoubtedly receiving the full rent which their properties are capable of yielding, in the sense that they are receiving the maximum which is permitted by the law. In other C words, they are receiving the "rack-rents" of their premises. The argument to the contrary overlooks the fact that value is not an absolute but a relative conception. The value of any particular thing can only be ascertained in the light of circumstances which affect or control its disposability. For example, the apparent owner of property might have some defect in his title which would lessen its selling value below that of another similar property the title to which was flawless. Similarly in the present case the rack-rent or annual value of the property in question is conditioned by the fact that its owner cannot increase its annual yield beyond the permitted maximum. In my opinion the legislature, in section 9, was applying itself to a factual and not to a hypothetical position. If the standard rent is the greatest rent that is obtainable in respect of any particular premises then it is D the full rent of those premises, the rack-rent, notwithstanding that (and indeed because) the owner is restricted from receiving the higher rent which the premises, if uncontrolled, would command.

The *Rawlance* case was recently applied by the Court of Appeal in *Gidlow-Jackson v Middlegate Properties Ltd* [1974] 1 QB 361 when considering the meaning of "the letting value of the property" in section 4 of the Leasehold Reform Act 1967.

In the result, I do not think that there is any impossibility involved in an agreement by a landlord to grant a lease at "the commercial yearly rack-rent at which the demised E premises might reasonably be expected to be let in the open market" notwithstanding that the rent is controlled by statute. The case is not in essence different from any other case in which the quantum of rent is reduced by some form of legislative control. Take this example by way of illustration. A dwelling-house may be lettable at a low rent because it lies in a district zoned for residential occupation, but if it were available for office use it might be let at a much higher rent. Nevertheless the lower rent is truly a rack-rent. No doubt there are other cases in which the hand of the legislature precludes the realisation of the highest rent potential of a property. Value, as Romer LJ said, can only be ascertained in the light F of factors which affect or control its disposability. In the result, I decide that the plaintiff, having exercised the option, is entitled to an order for specific performance. I will hear argument as to the form of order appropriate to the circumstances. Strictly speaking, the rent is a matter for arbitration, but proceedings for arbitration should have been commenced in September 1972. Perhaps the parties can agree the rent so as to avoid the formalities and the expense of an arbitration, if that is still the appropriate mode of determination in the absence of agreement.

G **QUEEN'S BENCH DIVISIONAL COURT**

December 2 1974

(Before Lord WIDGERY CJ, Mr Justice MELFORD STEVENSON and Mr Justice WATKINS)

MEREDITH v STEVENS

Estates Gazette February 21 1976

(1974) 237 EG 573

Fair rent of Clwyd bungalow—Committee entitled to consider comparables drawn from what they considered a relevant area—No force in remaining criticisms advanced —Landlord's appeal dismissed

H This was an appeal by Mr Kenneth Meredith, solicitor, of Middleton, Manchester, from a decision of the Denbighshire Rent Assessment Committee fixing a fair rent of £26 per month in respect of a bungalow owned by the appellant at 11 Pen Lan, Towyn, Abergele, North Wales.

Mr A R C Kirsten (instructed by K Meredith) represented the appellant, and Mr H K Woolf (instructed by the Treasury Solicitor) appeared as *amicus curiae*.

Giving judgment, LORD WIDGERY said: This is an appeal under section 13 of the Tribunals and Inquiries Act 1971 brought by the landlord, Mr Meredith, against a decision of the Denbighshire Rent Assessment Committee given on J March 27 1974 determining the fair rent of a dwelling at 11 Pen Lan, Towyn, Abergele, in the county of Clwyd, at £26 per month. The brief history of the matter is that this dwelling, which was I think a bungalow, one of a development of 22 undertaken by the same landlord, had been let under a lease for the contractual rent of £29.25 a month. The matter was referred to the rent officer for determination of a fair rent, and the landlord being dissatisfied with the rent officer's determination appealed to the rent assessment committee, as he was entitled to do under the terms of the Rent Act 1968. The rent assessment committee entered upon K consideration of the rent, and after their deliberations produced as a fair rent for these premises the sum of £26 a month. I am not going to read the reasons given by the committee in detail. It has always to be remembered, as has been pointed out more than once before, that these committees are not staffed by Chancery draftsmen and one must not be unduly particular about the language which they have used. If the basis of their reasoning is clear, that is all that can be expected in such circumstances. These reasons, if I may pass a compliment to the committee in passing, are good ones which have been well and intelligently set out, but nevertheless there is some criticism.

L The first criticism raised in the notice of motion, though not very hotly pursued by Mr Kirsten, was that the committee held that there was dampness in the front bedroom and that this was contrary to the evidence. A surveyor had said there was no damp in the room. The committee had examined the premises and they had seen no sign of rising damp, but they had seen signs of mildew which in their wisdom they decided was due to some dampness in the premises, and in their reasons they indicate that they were satisfied that there was dampness, although of course they do not attempt to put a figure on the dampness element. A further complaint is that the committee took into account M the detrimental effects of an adjoining poultry farm when in fact the poultry farm had been closed down before the inquiry held by the committee. There is nothing in the notes of evidence of the committee about the poultry farm. It is however said by Mr Kirsten that when the inspection took place the poultry farm was closed down, and he contends strongly that the committee have wrongly depreciated the rent on account of the existence of the farm, which was no longer existing at the relevant time. Actually all that the committee have done is to record that the tenant contended that there were unpleasant smells from the poultry farm, but the committee point out that the tenant agreed that the public health inspector had not considered that this was sufficient to amount to a statutory nuisance, and the only fair reading of that paragraph is that the committee were not impressed with the argument of the poultry farm anyway, regardless of whether this was operating at the date of the inquiry. They go on in their reasons to review other contentions. In (5) they say: "The evidence of the landlord was that the bungalow had been built in 1970 at a cost of £4,050 and that the rent of £351 per annum exclusive of rates had been calculated to give a return of 8 per cent on the capital." In other words, the landlord was inviting the committee to at least have regard to the contractor's theory and consider fixing a fair rent by reference to the capital value of the property, and the committee clearly have done as bidden by the Act, and they have considered the evidence. It is equally clear that they did not regard this as conclusive, and indeed they say so in their reasons. In paragraph (7)—and this is the real gist of the committee's decision—they say they were "unable to find any comparable property in the immediate area where a fair rent had been fixed by a rent assessment committee, but the committee have experience of market rentals of properties along the Denbighshire and Flintshire coastal area." If they did not find suitable comparables in the area, they were entirely within their rights in deciding in what area comparable market rents might help, and that is all they say they have done in paragraph (7). Finally, they say that they considered the calculation based on the contractor's theory, to which I have already referred, but did not find it conclusive. That again is within their rights; they are perfectly entitled to say that evidence of that kind is not regarded by them as either helpful or conclusive, or whatever other phrase they may choose. They go on to say that although the market rental is now above the contractual rent under the lease, after taking into account the scarcity provision of the Act they think it is a fair rent.

I can find absolutely nothing in those reasons to indicate an error of law. We get a lot of these cases of course—considerable money is often involved—and landlords very often come here with a sense of grievance, but our powers are restricted to considering an error in law, and once one realises that the committee are entitled to use their own experience and knowledge and they are not bound to follow any specific valuation evidence which may be tendered to them, it must become quite obvious that all the matters in dispute here are valuation matters and not matters of law at all. I would dismiss this appeal.

MELFORD STEVENSON J: I agree.

WATKINS J: I agree.

The appeal was accordingly dismissed.

QUEEN'S BENCH DIVISIONAL COURT

November 11 1975

(Before Lord WIDGERY CJ, Mr Justice PARK and Mr Justice MAY)

HANSON v LONDON RENT ASSESSMENT COMMITTEE AND ANOTHER; R v LONDON RENT ASSESSMENT COMMITTEE EX PARTE HANSON

Estates Gazette February 28 1976

(1975) 237 EG 651

An application for a fair rent to be fixed cannot be withdrawn once the matter has been referred to the rent assessment committee

This was an application by Mr John Hanson, of 21 Bramerton Street, London SW3, for an order of certiorari to bring up and quash a decision of the first respondents, a London Rent Assessment Committee of the London Rent Assessment Panel, fixing a fair rent for 21 Bramerton Street, owned by the second respondents, the Church Commissioners for England, at £900 per annum exclusive of rates. The applicant also appealed against the committee's decision.

Mr G F Hastings (instructed by Lorenz & Jones) appeared for the applicant; Mr H K Woolf (instructed by the Treasury Solicitor) for the first respondents; and Mr R Moshi (instructed by Radcliffes & Co) for the second respondents.

Giving judgment, LORD WIDGERY said: In these proceedings Mr Hastings moves for an order of certiorari to bring up into this court with a view to its being quashed a decision made by a London Rent Assessment Committee on September 7 1974 whereby it fixed as a fair rent for premises known as 21 Bramerton Street, Chelsea, a rent of £900 per annum exclusive of rates. The applicant today is the tenant of 21 Bramerton Street, and his name is John Hanson. It is he who moves for the order of certiorari to quash the decision of the rent assessment committee. The landlords of the premises are the Church Commissioners, and they of course are concerned with the proceedings as the recipients of whatever rent is in due course determined as the fair rent.

The facts of this case, so far as relevant, are extremely limited in scope and quantity. The Rent Act 1968, amongst other activities, provides for the fixing of a fair rent of residential property in certain circumstances, and the method which is adopted in the relevant part of the Act of 1968 is the appointment of officials known as rent officers, who, on application from landlords or tenants, or now the local authority, fix what they regard as a fair rent for the premises in question. If either the landlord or the tenant, or any other interested party, is dissatisfied with the rent so fixed, the Act provides for the issue to be referred to a body called a rent assessment committee whose duty it is officially to determine the fair rent. All that had happened in this case. The premises, as I have said, were let to Mr Hanson by the Church Commissioners. The Church Commissioners approached the rent officer asking him to determine the fair rent for the premises, and the rent officer determined a fair rent of £800 a year. The tenant (the present applicant for certiorari), being dissatisfied with that determination, objected to it, and as a result required the matter to be determined by a rent assessment committee, as I have endeavoured to explain.

It is appropriate at this point to look a little more closely at the provisions of the Act which bring about that consequence, and they are conveniently collected in Schedule 6 of the Rent Act 1968, where one has a simple and clear code of conduct which is laid down for persons concerned with the provisions to which I have referred. I do not take time by looking at the procedure of applications to a rent officer which occupy the first four paragraphs of Schedule 6. One comes to the action which has to be taken by persons dissatisfied wtih the rent officer's conclusion when paragraph 5 of Schedule 6 is reached. That provides: "After considering, in accordance with paragraph 4 above, what rent ought to be registered or, as the case may be, whether a different rent ought to be registered"—and then there is an important provision—he is required to "notify the landlord and the tenant accordingly by a notice stating that if, within twenty-eight days of the service of the notice or such longer period as he or a rent assessment committee may allow, an objection in writing is received by the rent officer from the landlord or the tenant the matter will be referred to a rent assessment committee." So, following the decision of the rent officer, both parties are told what that decision is and told that unless objection is made within twenty-eight days that rent will be confirmed. The schedule goes on in paragraph 6 to describe what would happen if objection is taken. It is laid down there that: "If such an objection as is mentioned in paragraph 5 above is received, then (a) if it is received within the period of twenty-eight days specified in that paragraph or a rent assessment committee so direct,

A the rent officer shall refer the matter to a rent assessment committee"—an important sentence, in my view, because it is clearly mandatory and it says in the plainest terms that if either of the parties to the contract makes objection within twenty-eight days, the rent officer *shall* refer the matter to the rent assessment committee. When we get on to paragraph 7 of the schedule we are told what happens when the matter is so referred. Paragraph 7 requires the rent assessment committee to give the sort of notice which one would expect in cases of this kind, and if either of the parties wants an oral hearing, it is provided in paragraph 8 that the rent assessment committee shall set up an oral B hearing accordingly. When all that has been done, and when the rent assessment committee has made such inquiries as they think fit, their duty is to be found in paragraph 9 of the schedule in these terms: "The committee shall make such inquiry, if any, as they think fit and consider any information supplied or representation made to them in pursuance of paragraph 7 or paragraph 8 above, and (a) if it appears to them that the rent registered or confirmed by the rent officer is a fair rent, they shall confirm that rent; (b) if it does not appear to them that that rent is a fair rent, they shall determine a fair rent for the dwelling-house." Again there is no room for choice, options or doubt. The duties C of the committee are clearly laid down once the matter has been referred to them. They must conduct such inquiries as the schedule provides for, and then come up with a fair rent, either by confirmation of the rent fixed by the rent officer or by the exercise of their own judgment.

As I have already indicated, in this case the landlords approached the rent officer and obtained an assessment of the fair rent from him. The parties, including the present applicant, who was the tenant, were notified of what had happened, and within twenty-eight days an objection was raised by the tenant to the rent assessed by the rent officer, which I remind myself again was £800 per year. Notice was D given to the rent assessment committee that an oral hearing was required. A date was fixed for the oral hearing, August 15 1974. Everything was prepared for the committee to hear the parties and then proceed in accordance with its duty to fix a fair rent. But a few days before August 15 1974 the present applicant, Mr Hanson, evidently began to doubt the wisdom of the course which he had adopted. He was advised by well-known surveyors in this matter, and reading between the lines, one is driven to the conclusion that, if one may use the vernacular, he was getting cold feet and was not at all sure that it was in his interest to go on with this matter. E So eventually he told his surveyors on August 8 to make arrangements to withdraw his objection. On August 8 a letter was written by the advisers of the tenant to the rent assessment panel in these terms:

> Dear Sirs,
> No 21 Bramerton Street: Objection to rent assessment.
> We write following our telephone conversation of this morning. Acting on behalf of Mr Hanson of the above address, we confirm that we shall not be placing our objection before the London Rent Assessment Panel on Thursday August 15.

There is evidence which indicates how that letter came to F be written. The evidence, which I think we should accept at this stage, because it is not contradicted, is to the effect that the surveyors acting for the applicant telephoned to the rent assessment committee to ask what the procedure would be if the applicant wished to withdraw his objection to the rent officer's figure. The evidence is that they were told that a letter of withdrawal would be sufficient, and in some way the phraseology similar to that in the letter which I have read seems to have been mentioned and approved. I say "in some way," because I am far from satisfied that this letter was actually composed in the office of the rent assessment panel. But that a clerical officer employed in that office gave some assistance to the compilation of the letter is not G seriously in dispute, and so the letter went off.

It will be rembered that August 15 was the date fixed for the hearing of this matter before the panel. August 15 duly arrived, and the panel were informed that the office had received the letter of August 8 which I have read. The panel took the view that the phraseology was ambiguous; it was not altogether clear whether the applicant was abandoning all interest in the matter or whether he was merely saying he would not come in person and hoped that the committee would look after his interests in his absence, or something like that. To cut a long story short, the committee decided that they should not act on the letter of August 8 but should H hear the matter. Since the tenant was not present, hearing the matter on his side did not include hearing any evidence from him. After the hearing stage in the morning of August 15, the committee moved on, as is their practice, to look at the house and inspect it. When they got to the house they could not get in because they were told that Mr Hanson's objection had been withdrawn and there was no further matter which concerned them about this house. Some confusion followed as a result of that, and eventually the committee went away. They did however fix a fair rent, and they fixed a fair rent at £900 a year, and it is the fact that that exceeded what the rent officer himself fixed, which rent J the applicant reluctantly might have been prepared to pay, and that is the cause of our being here today dealing with this particular dispute. The way in which it is put is two-fold. It is submitted by Mr Hastings on behalf of the applicant that there is power for someone who has objected and set the wheels of Schedule 6 in motion to withdraw his objection and thus resile from the battle, as it were, leaving the matter as though he had never objected at all. If that is the law, as Mr Hastings submits it is, then it would follow, he says, that upon the writing of the letter of August 8 1974 Mr Hanson's objection was withdrawn, his liability to have his rent increased if the panel so thought was also nullified, K and he was left in a position as though he had never made any objection at all. That is the argument on the applicant's side, and it is said that if that is right, there are two consequences which follow. The first is that the panel lacked jurisdiction when it eventually proceeded to fix the rent as it did. Alternatively, I think it is said that there was a denial of natural justice, in that Mr Hanson did not in the event have an opportunity of making oral representations to the panel.

The central and vital fact in this dispute is whether there is a right for an objector to withdraw his objection, and if L so, what the consequences of that withdrawal may be. I approach this problem on the footing that in general, where you have a statutory procedure of this kind which involves the making of objections or applications, in general a person who makes an objection or an application should have the right to withdraw if he can do so without prejudicing other interested parties. I think that considerable support for that general proposition is to be found in the case of *Boal Quay Wharfingers Ltd v King's Lynn Conservancy Board* [1971] 1 WLR 1558. The passage which illustrates the point that I am now seeking to make is to be found in the judgment M of Salmon LJ at page 1569 where he referred to the Rent Act, although to a different part of it than that with which we are concerned, and said this: "The Rent Act 1968 contains an example of such a statutory provision"—that is to say a provision for actual withdrawal. He continued: "The Landlord and Tenant (Rent Control) Act of 1949, however, contained no such provision; and in *R v Hampstead & St Pancras Rent Tribunal, ex parte Goodman* [1951] 1 KB 541 the court held that since there was nothing in that Act to prevent an application being withdrawn, it could be withdrawn at any time. That I think is an authority for the view I have expressed." I respectfully agree with what

Salmon LJ says, and I start with the assumption that if the existence of a right of withdrawal can march in double harness with the protection of other parties concerned, then the right of withdrawal ought to be there. But I do not think in this case that it is possible to say that a withdrawal, at all events once the matter is before the committee, can be made wihout a possible prejudice to other parties.

I am impressed by the fact that other parts of the Rent Act 1968, such as those referred to by Salmon LJ, do specifically contemplate withdrawal of an application in certain circumstances. The part we are concerned with contains no such provision at all. Furthermore, the issue which arises when application is made to a rent officer to fix a fair rent is not a matter which is simply *inter partes* and simply concerns the landlord and tenant at the moment when the issue arises. The effect of the fixing of a fair rent is that the rent is fixed *in rem,* as Lord Parker put it in one of his judgments, for others who come as landlord or tenant thereafter. It is not, I think, consistent with the general policy of the Act, or indeed with the purpose enshrined in Schedule 6 which I have read, that there should be any kind of unlimited right for a person, who has by objection set the machinery in motion, to decide to withdraw his objection and thus as it were reverse the machinery to a position which it formerly occupied. I think it is just possible that there may be a right of withdrawal for an objector if he withdraws before the matter has been referred to the rent assessment committee. He has to be very quick if he is going to withdraw before that happens. I would not decide it today because the point does not arise directly here, but it may very well be that there is an unlimited right of withdrawal of objection up to the point when the matter is actually referred to the rent assessment committee, but thereafter in my judgment there is no general right of withdrawal at all; that is to say there is no means whereby on a so-called withdrawal the party can extricate himself and put himself in the position as though he had never lodged an objection at all. Again, as this branch of the law develops, it may be that other consequences will be found to flow from a purported withdrawal. It may be that it will be right to say—and here again, another day will be the day to say it—that a purported withdrawal excuses the rent assessment committee from providing an oral hearing for which the withdrawer had previously opted, and there may be many requirements of that kind whereby following a purported withdrawal the procedure can be somewhat streamlined and made the more effective, and a decision obtained more quickly. But what I am quite satisfied about, speaking for myself, is that once the matter has been referred to the rent assessment committee the rent assessment committee must produce an answer in the terms of a fair rent, and no attempted withdrawal or any other manoeuvres on the part of the parties can avoid the obligation on the rent assessment committee to come up with a fair rent, doing its own assessment of what is a fair rent for the premises.

It is in my view, therefore, on the facts of this case, whatever construction one puts on the letter of August 8 1974, that in so far as it was intended to be or purported to be a withdrawal of the objection, it was ineffective for that purpose. It was much too late for any such withdrawal to be of effect at all, and therefore the committee were entirely within their jurisdiction in hearing the application on the morning of August 15, as in fact they did. Furthermore, the fact that they had jurisdiction, and the fact that the withdrawal was ineffectual, as I have described, means that no error of law can be found in the committee's findings, and that is relevant because, although I am not sure whether I have mentioned it in the course of this judgment heretofore, apart from the application for certiorari, there is an appeal against the rent assessment panel's decision brought under the Tribunals and Inquiries Act. That appeal in my judgment must be dismissed, because for the reasons which I hope I have made clear there is in my view no error of judgment discernible in the decision of the committee. There remains of course the application for certiorari, and all I find it necessary to say about that is that it was brought far too late to justify consideration in this case at all. The decision of the rent assessment committee was dated September 7 1974, and the motion for leave to quash it by certiorari is dated July 25 1975, or something like nine months later, which, as the Court of Appeal emphasised only last week, is much too long for any ordinary certiorari application and would be fatal in this case, in my judgment, to the application of the prerogative order. For those reasons I would refuse these applications.

PARK J: I agree.

MAY J: I also agree.

QUEEN'S BENCH DIVISIONAL COURT

February 13 1976

(Before Lord WIDGERY CJ, Mr Justice KILNER BROWN and Mr Justice WATKINS)

CAMPBELL v GARDNER

Estates Gazette April 10 1976

(1976) 238 EG 115

Fair rent of furnished premises—No rule binding committee to start by finding unfurnished rent—General propositions about typical periods of furnished tenancies do not represent a reference to the personal circumstances of the landlord—Committee's assessment confirmed

This was an appeal by Mr Thomas McDonald Campbell, of 33 Rosetta Road, Nottingham, from a decision of the East Midlands Rent Assessment Panel dated June 18 1975 fixing the fair rent of the premises let to him at that address by the respondent, Mr W A Gardner, at £8.50 per week inclusive of furniture but exclusive of rates.

Mr C Anderson (instructed by Turner Peacock, agents for Fraser, Brown, White & Pears, of Nottingham) appeared for the appellant. The respondent appeared in person.

Giving the first judgment, WATKINS J said that the appellant was a student and in July 1974 became a tenant of the respondent in premises at 33 Rosetta Road, Basford, Nottingham. The premises were in a residential area about three miles from the town centre, and had been built in 1910. It was a house built for the artisan class. By the contract he entered into with the respondent, the appellant agreed to pay a weekly rental of £13 inclusive of rates and furniture. On January 20 1975 the appellant made application to a rent officer for a fair rent. On February 11 1975 the rent officer registered a fair rent of £5.50 per week exclusive of rates, and indicated that 72 pence per week was attributable to furniture. The respondent, in turn, appealed to the rent assessment panel, which on June 18 1975 resolved that the rent to be registered should be £8.50 per week exclusive of rates. It was from that decision that the appellant now appealed.

The effect, in general terms, of the 1974 Rent Act, so far as furnished lettings were concerned, was to bring them into the same category of control as unfurnished lettings under the 1968 Act and to allow not only the same procedures for assessment of a fair rent to apply but also for the same criteria to be used for assessing a fair rent. Neither of the Acts defined "fair rent," but a number of criteria were set out in section 46 of the 1968 Act. A rent officer was entitled to have regard to the capital return to the landlord, the area in which the premises were situated, etc. By section 46 (3) no regard was to be paid in assessing a fair rent to disrepair

A or defects due to a tenant's failure to do improvements which he should do, or to the personal circumstances of either landlord or tenant. What a committee had to do in assessing what a fair rent should be was generally to be as fair as possible both to landlord and tenant and see that the rules of natural justice were obeyed.

Counsel for the appellant had criticised the approach of the tribunal here. He had contended that a panel should first determine what a fair rent would be if the premises were unfurnished, and then add on something for furniture. In the present case the panel had made no secret of the fact that they had not adopted that approach. They had viewed B the premises as a whole and then had asked themselves what would be a fair rent. There was nothing wrong with that approach. Then counsel had contended that the panel had erred in law in taking into account the fact that in general, tenants of furnished tenancies remained for shorter periods than tenants of unfurnished tenancies and that therefore the landlords' expenses were greater. That, contended counsel, represented taking the landlord's personal circumstances into account, which was forbidden. Further, it was argued, the panel had concerned itself with the personal circumstances of those landlords who let premises to students. There were C no valid grounds for such criticism. At no time had the panel paid special regard to the personal circumstances of the landlord. The panel had expressed itself in general terms, and he (his lordship) thought that the appeal should be dismissed.

Lord Widgery and Kilner Brown J agreed. The appeal was dismissed with no order as to costs.

D

COURT OF APPEAL

December 19 1975

(Before Lord DENNING MR, Lord Justice ORMROD and Lord Justice SHAW)

McCALL v ABELESZ AND ANOTHER

Estates Gazette May 1 1976

(1975) 238 EG 335

A civil action for harassment lies at a tenant's suit for breach of a term of the tenancy, but not for breach of statutory obligation under section 30 (2) of the Rent Act 1965

This was an appeal by Mr Leonard McCall, of Claremont E Road, Cricklewood, London, against a judgment of Judge Counsell at Willesden County Court on December 18 1974 awarding him only £75 on his claim against the respondents, Erno Abelesz and Jacob Ostreicher, trading as Riverside Property Services, for damages for harassment in breach of section 30 (2) of the Rent Act 1965. By a cross-notice, the respondents appealed against the award of any damages at all.

Mr J P Singer (instructed by Alexander & Partners) appeared for the appellant. The first respondent appeared in person, and Mr H Carlisle (instructed by the Treasury Solicitor) as *amicus curiae*.

F Giving judgment, Lord Denning said that Mr McCall, who came from the West Indies, became the tenant of a furnished room in 3 Claremont Road, Cricklewood, London, in March 1968. He had to put coins into meters for gas and electricity supplied by his landlords, who were in turn supplied by the gas and electricity boards. The defendants bought 3 Claremont Road in May 1973, and left everything to their manager, a Mr Arran. Mr McCall paid his rent of £3.50 a week regularly by post to the defendants. A few months later, the defendants got a bill for £435 for gas, most G of it supplied before they bought the property. They did not pay the bill, and in October 1973 the gas board cut off the gas. In December 1973 the public health authorities drew the landlord's attention to the condition of the house, and Mr Arran went to look at it for the first time. He found that in Mr McCall's room the meters were intact and the money there, but in two other rooms the meter locks were broken and the money taken. The gas board refused to give the tenants a direct supply of gas, and in March 1974 the two other tenants left owing about £200 in rent which they never paid. About this time the electricity and water were also cut off. The landlords offered Mr McCall alternative H accommodation in a different area and at a higher rent, but he refused it. The landlords finally had the gas, electricity and water restored, but Mr McCall took proceedings in the county court. It was suggested at the first hearing that his plea might be breach of contract, but after consideration his counsel pleaded it as damages for breach of section 30 (2) of the Rent Act 1965. The judge ruled that it was not a case for aggravated damages, and awarded £75. Mr McCall wanted more, and appealed; the landlords cross-appealed, maintaining that no civil claim lay under the section.

Mr Singer, for Mr McCall, cited tort books which suggested J that there was a civil remedy for harassment under section 30 (2). He said that there had been several cases in which county courts had awarded such damages, but in most of them the point had been assumed without argument. On the face of it, the section did not give rise to a civil remedy, but it was still open to the court, on an examination of the whole Act, to hold that there was a remedy in damages. An example was *Groves v Lord Wimborne* [1898] 2 QB 402; on the other hand, it was held that there was no remedy in damages in *Cutler v Wandsworth Stadium Ltd* [1949] AC 398. Lord du Parcq there suggested that Parliament should explicitly state whether it was intended that there should be K a civil remedy or not in any particular case, and it appeared that section 30 (4) was intended to comply with Lord du Parcq's suggestion. It only provided, however, that "nothing in this section shall be taken to prejudice any liability or remedy to which a person guilty of an offence thereunder may be subject in civil proceedings." In some statutes Parliament had stated that breach of a particular obligation should give rise to a civil remedy, but no statute had been found in which it was expressly stated that a criminal offence should not give rise to a civil remedy in damages.

He (his Lordship) had come to the conclusion that there L was no need for any new civil remedy for harassment, for which there was already a perfectly good civil action for damages. The tenant here could sue for breach of an implied term of the tenancy, and possibly also for breach of the covenant for quiet enjoyment. That covenant extended to any conduct of the landlord calculated to interfere with the tenant's peace or comfort, including any mental upset or distress caused the tenant. On the evidence in the present case, he (his Lordship) doubted whether it was right to find the defendant landlords guilty of harassment under the section, and on this point he agreed with what Ormrod LJ was about to say. But it was enough to decide that the M section did not give rise to a civil action for damages. There should be judgment for the landlords on the appeal and cross-appeal.

Agreeing, Ormrod LJ added that in his view the tenant had failed to prove that the landlords withheld supplies with intent to cause him to give up the premises; there was only a drifting on their part.

Shaw LJ agreed with both judgments, and an order was made in the terms proposed by Lord Denning.

LANDLORD AND TENANT

RENT REVIEW

A

CHANCERY DIVISION

June 12 1975

(Before Mr Justice FOSTER)

BEER AND OTHERS v BOWDEN

Estates Gazette January 3 1976

(1975) 237 EG 41

Rent review clause provides for " such rent as shall thereupon be agreed between the landlord and the tenant," nothing being said as to the position in default of agreement—Lease is not void, tenant must pay on use-and-occupation basis

B

This was an originating summons by Mr Anthony Wilders Beer, of Barnet, London, Mr William Paul Elliott De Beer, of Richmond, London, and Mr Michael Wills De Beer, of Ontario, Canada, asking what, on the true construction of a clause in a lease of 54-56 Torbay Road, Paignton, Devon, to the defendant, Mr Harold Herbert Bowden, was the basis for calculation of the rent of the demised premises for the period from March 25 1973 to March 24 1978.

Mrs E B Solomons (instructed by Scott, Son & Chitty) appeared for the plaintiffs, and Mr G W Jaques (instructed by Boxall & Boxall, agents for R Hancock & Son, of Callington) represented the defendant.

C

Giving judgment, FOSTER J said that the summons arose out of a rather curious question in regard to what was supposed to be a rent review clause. By a lease dated July 17 1968 the plaintiffs demised to the defendant premises known as 54-56 Torbay Road, Paignton, Devon, for 10 years from March 25 1968. In May 1971 a memorandum was made by which the parties agreed to extend the term to a period of 14 years. In the original lease the reddendum was expressed:

> Until the twenty fourth day of March one thousand nine hundred and seventy three (yearly and proportionately for any fraction of a year) the rent of £1,250 per annum and from March 25 1973 such rent as shall thereupon be agreed between the landlords and the tenant but no account shall be taken of any improvements carried out by the tenant in computing the amount of increase if any and in any case not less than the yearly rental payable hereunder such rent to be paid in advance by four equal quarterly payments on the four usual quarter days the first of which payments shall be made on March 25 1968. . . .

D

The memorandum of 1971 provided:

> In consideration of the covenants on the part of the tenant contained in the within written lease the landlords agree that the term of years . . . shall be read and construed as if the term of fourteen years were substituted in the place of ten years and the reference . . . to a rent review in respect of the rent to be charged for the said premises from March 25 1973 shall be read and construed as if there were also inserted reference to a rent review for the rent to be charged for the said premises from March 25 1978. . . .

E

The parties had failed to agree what the rent should be on the first review date, from March 25 1973. The plaintiffs had obtained a valuation to the effect that a proper rent for the period in question would be £2,850, and they were in fact prepared to take £2,650, but the defendant did not agree with that or any other figure. On the question raised by the summons, the plaintiffs submitted that he (his Lordship) should read into the reddendum of the lease, after the words " such rent as shall . . . be agreed between the landlords and the tenant," words such as, " and if there shall be no agreement the rent payable shall be a proper and reasonable rental having regard to the market value thereof to be fixed by the court." For the tenant, it was suggested that if one found that a new rent had not been agreed the rent of £1,250 was to continue, and reliance was placed on the words " in any case not less than the yearly rental payable hereunder."

F

He (his Lordship) thought that it was perfectly clear that the original rent of £1,250 was to be paid only up to March 24 1973 and was not to be paid thereafter. The words relied on for the tenant, " in any case not less than the yearly rental payable hereunder," in truth quite plainly assumed that some other rent was going to be paid after March 24 1973, and not the original rent; they were to be applied only if that rent happened to be less than £1,250. In his (Foster J's) judgment, there was a complete hiatus as to the rent to be paid after March 24 1973 and again, because of the memorandum, after March 24 1978, but there was in contemplation a rent different from the original rent. It had been suggested on behalf of the plaintiffs that if no rent was in fact agreed or provided for in a lease the term was void, and that accordingly the lease now in question was void from the date specified for the first review, March 25 1973. He (his Lordship) considered, however, that where one found a hiatus such as this it was incumbent on the tenant to pay the landlord a sum in respect of the use and occupation of the premises while the tenancy was in existence. The amount the tenant must pay for use and occupation would be subject, in the present case, to no account being taken of any improvements he had made, and would in any case be not less than £1,250. The summons should accordingly be sent back to the master for him to decide what was the proper sum the defendant should pay during the currency of the lease for use and occupation. This would be a matter of evidence before him with experts on each side. After March 25 1973, therefore, and in the absence of agreement again in 1978, the proper amount which the tenant should pay was the amount referable to his use and occupation during the currency of the tenancy. When counsel drew up an agreed minute they could use the phrase " for use and occupation " or " what the premises were worth "; the phrases meant the same thing.

G

H

J

K

CHANCERY DIVISION

June 20 1975

(Before Mr Justice TEMPLEMAN)

MOUNT CHARLOTTE INVESTMENTS LTD v LEEK, WESTBOURNE & EASTERN COUNTIES BUILDING SOCIETY

Estates Gazette January 31 1976

(1975) 237 EG 339

Rent review clause—Arbitrator to be nominated on application by the lessors, application to be made before a

specified date—Clause held to be in the nature of an option, and therefore strictly construed against landlords—C Richards & Son Ltd v Karenita (1971) 221 EG 25 and United Scientific Holdings Ltd v Burnley Corporation (1974) 231 EG 1543 cited and applied

This was a claim by Mount Charlotte Investments Ltd, of City Road, London EC, against Leek, Westbourne & Eastern Counties Building Society, of Leek, Staffordshire, for a declaration that the plaintiffs had validly applied, under a rent review clause in a lease of April 8 1968, to the President of the Royal Institution of Chartered Surveyors for appointment of an arbitrator to determine a revised rent of office premises at 89 Queen Street, Cardiff, for a period of six years beginning with March 25 1975.

Mr J S Colyer (instructed by Baileys, Shaw & Gillett) appeared for the plaintiffs, and Mr C J Slade QC and Mr J A Moncaster (instructed by Gouldens, agents for Knight & Sons, of Newcastle-under-Lyme) represented the defendants.

Giving judgment, TEMPLEMAN J said: This is the latest (but I fear not the last) dispute over a rent review clause in a lease. The landlords failed to refer the rent to arbitration within the time provided by the lease, and the tenants claim that the landlords have lost their chances to increase the rent. The lease is dated April 8 1968. It was made between the landlords' predecessors in title and the tenants. It was a lease of a shop and premises, 89 Queen Street, Cardiff, for a term of twenty-one years from March 25 1967, and for use as offices of a building society or any other trade or business which the landlords might permit. The lease provided that the tenants should hold the premises "yielding and paying therefor during the said term and so in proportion for any less time than a year the respective rents following: (a) during the first eight years of the term" £2,750 per annum, which was defined as "the basic rent"; then "(b) during the next six years of the said term," from March 25 1975 to March 25 1981, "whichever of the two following yearly rents is the greater that is to say (i) the basic rent or (ii) such amount as may be agreed between the lessors and the lessees before the expiration of six calendar months before the end of the eighth year of the said term or (in the absence of such agreement by such last mentioned date) such amount as may be determined by an arbitrator to be nominated by the President for the time being of the Royal Institution of Chartered Surveyors on the application of the lessors to be made within fourteen days after the date six calendar months before the end of the eighth year of the said term and so that in case of such arbitration the amount to be determined by the arbitrator shall be such as in his opinion shall represent the fair yearly rent for the demised premises at the end of the eighth year of the said term having regard to rental values of property then current"; and then "(c) during the last seven years of the said term . . . whichever of the two following rents is the greater that is to say: (i) the rent reserved for the previous six years or (ii) the rent equivalent to a fair yearly rent as at the end of the fourteenth year of the said term to be agreed or calculated on the same basis *mutatis mutandis* as set out above in respect of the eighth year." There followed a provision that the rent should be paid on the usual quarter days, in arrears. The lease thus provided for negotiations between the landlords and the tenants before September 25 1974, and if those negotiations failed to reach an agreement, then a reference to arbitration by the landlords before October 9 1974, with the obvious objective of seeing that the new rent should be determined before June 24 1975 if possible, before the new rent became payable for the first time. The landlords opened negotiations with the tenants by a letter dated September 20 1974 and negotiations proceeded through surveyors. The date for an application to the President of the Royal Institution of Chartered Surveyors passed; negotiations still continued, until on February 27 1975 the tenants took the point that the landlords were out of time, and therefore that the rent could not be increased. The landlords, in a letter dated March 5 1975, applied to the President of the Royal Institution of Chartered Surveyors to appoint an arbitrator, and the question is whether that application is valid or whether, as the tenants say, the landlords have lost their opportunity.

The authorities disclose that there are at least two kinds of rent revision clauses. The first kind has been analysed as "an option to the landlord to obtain a higher rent," and in that case, if the landlord does not comply with any time limits provided for the exercise of the option then he wholly fails. The time limits are said to be mandatory. The second kind of clause has been analysed as "creating an obligation on the landlords"—or sometimes the tenants as well—"to take the steps necessary to determine what the rent is going to be." If in the obligation cases the time-limits prescribed by the document are not complied with, then the court construes those time-limits as being purely directory, and provided that the tenant has not been prejudiced by any delay, then the rent is fixed after the time-limits have expired. The analysis of the option rent review clause is a triumph for theory over realism. In practice landlords insist on a rent review clause. They are only prepared to grant a 21-year lease if the rent is increased to keep pace with inflation every 7 years or so. The concept of the tenant granting the landlord an option and conferring benefits on the landlord does not accord with reality. The courts could have come to the conclusion that all time-limits imposed by rent revision clauses were directory and not mandatory, because an increase in rent is a condition imposed by the landlord for the favour of a long term. Certainly in the obligation clause cases the courts have had no difficulty in ignoring the express words of the clause and treating the time-limits as directory; but they have refused to do the same in the case of clauses which they have designated option clauses. The authorities begin in 1972. When the landlord before 1972 went to his draftsman to draft a rent revision clause the draftsman might, in all innocence, produce an option type of clause, or an obligation type of clause, or something which was neatly halfway between the two. Since 1972 the drafting cannot be done wholly in innocence, but can quite often be done in ignorance. The draftsman may lay a trap, and the landlord may fall into the trap by forgetting the time-limit. But that is the effect of the authorities.

In *C Richards & Son Ltd v Karenita Ltd* (1971) 221 EG 25, the rent revision clause provided that "if the landlords shall by giving notice in writing to the tenants at any time during the first three months of the seventh year of the term hereby created require a review of the rent payable hereunder such rent shall be revised with effect from the expiration of the seventh year." Goulding J rejected the argument that time was not of the essence, an argument adopted from the relationship of vendor and purchaser. In *Samuel Properties Developments Ltd v Hayek* [1972] 1 WLR 1064, if the landlords desired to have the rent reviewed by reference to the open market rental value, and served a notice in writing to that effect not later than two quarters before the expiry of the seventh and fourteenth years, then by agreement or arbitration the rent was to be increased. Whitford J at p 1070 said:

> Clauses of the character which I am considering run very closely with the type of option clauses . . . I think it right that a review to bring about an increase in rent at the end of seven and 14 years can reasonably be described as a benefit to one party alone. It might not be unreasonable for it to be described as a privilege, but I do not think that the particular words used are of much importance one way or the other. I do think that in a case of this character it is important that the party seeking to secure the benefit should comply strictly with the provisions

A stipulated just as in the case of option for renewal of leases, or for example options for the repurchase of shares.

After referring to *Hare v Nicoll* [1966] 2 QB 130, the classic case which reiterated that in the exercise of an option the terms must be strictly complied with if the option is properly to be exercised, the learned judge held that in the *Samuel Properties* case the rent revision clause was an option type of clause, and the terms not being complied with, the landlord had lost his opportunity. That decision was affirmed by the Court of Appeal in the same volume at p 1296, and at p 1302 Russell LJ referred to the arguments that a rent review clause should come within a general approach that B time provisions should not be construed as inflexible and mandatory unless there be special reason for so doing, and that there were no special reasons for importing inflexibility into provisions that were basically mere machinery for achieving what the parties must have regarded as an equitable alteration in rent to safeguard the lessor against the consequences of possible or indeed probable continuing depreciation of the value of money in relation to housing accommodation. Russell LJ said:

> I am not myself impressed by these arguments. The right or privilege of exacting an additional rent was conferred by the bargain between the parties as an express option which would be C effective if a condition precedent was complied with. It could be equated with an offer by the lessee to pay an increased rent only in certain circumstances which it lay in the power of the lessor unilaterally to bring about, which offer was not accepted in those terms. It was argued that there was a distinction (as to time-limits) between options to determine, or to renew, or to acquire the reversion, and a right such as the present. I do not see why this should be so. Accordingly in my judgment the time requirement . . . is to be treated as inflexible and mandatory.

That is binding on me. In *C H Bailey v Memorial Enterprises Ltd* [1974] 1 WLR 728 a rent review clause provided that "if on September 21 1969 the market rental value shall D be found to exceed the rent of £2,375 hereby reserved there shall be substituted from such date for the yearly rent hereby reserved an increased yearly rent equal to the market rental value so ascertained." There were provisions for the market rental value to be agreed or to be determined by arbitration. That rent review clause was to operate at the end of the first five years of the term, and unless it did so operate there would be no rent at all fixed by the lease or capable of being fixed. Eveleigh J held that the market rental value on September 21 1969 could be ascertained after that date, but that the increased rent only became payable on the quarter day following its ascertainment. The Court of Appeal held E that since the rent review clause provided for the increased rent when ascertained to be substituted from September 21 1969, it was payable retrospectively from that date. In coming to that decision, the Court of Appeal rejected the tenants' argument that the landlords had lost their right to increase the rent at all because it was not in fact ascertained on September 21 1969. This was not an option which the landlord had power to exercise but an obligation on both parties to determine what the rent should be as from September 21 1969.

In *Kenilworth Industrial Sites Ltd v E C Little & Co Ltd* [1974] 1 WLR 1069, before Megarry J, in a 21-year lease F the annual rent for the first five years was £2,900. A rent revision clause provided that "not more than twelve months nor less than six months before the expiration of the fifth, tenth and fifteenth years of the term the landlord shall serve upon the tenant a notice to agree the rent for the ensuing five years and thereupon the parties hereto shall agree a new rent." There was an express proviso that "any failure to give or receive such notice shall not render void the right of the landlord hereunder to require the agreement or determination of a new rent." Megarry J decided that the failure of the landlord to give the notice did not render void his right to an increased rent, and the Court of Appeal [1975] 1 WLR 143, agreed. Between the judgment of Megarry J in the *Kenilworth* case at first instance and the decision of the Court G of Appeal there occurred *United Scientific Holdings Ltd v Burnley Corporation* (1974) 231 EG 1543. A 99-year lease provided that during the year immediately preceding the first 10-year period of the term "the landlord and the tenant shall agree, or failing agreement shall determine by arbitration, the rack rent, and one-quarter of the sum total so ascertained, or £2,000, whichever is the greater, shall be the rate of rent reserved by this lease in respect of the next period." And then it provided for a reference to arbitration to be agreed by the parties or to be nominated by the President of the Royal Institution of Chartered Surveyors. H That was a rent review clause where at first blush the language appears to be language of obligation and not of option. In the event the parties failed to reach agreement during the year immediately preceding the end of the period and failed to go to arbitration. Pennycuick V-C assumed in favour of the landlords that the requirement that the arbitration should be during the relevant year was satisfied if a reference to arbitration were made during the specified period. It had been argued for the landlords that time was not of the essence, that "all that one was concerned to do was to quantify the amount of rent under the existing obligation to pay rent," and that there was no reason why that J should not be done outside the year. Pennycuick V-C continued, "Although the rent review provision was expressed merely as a provision for the quantification of additional rent, it was in substance a unilateral right or privilege vested in [the landlords] alone, the nature of the right being to increase the rent payable by the tenants. The tenants were entitled under the lease to possession of the property for the whole 99 years at [the original rent] unless the corporation elected to require a rent review." Similarly, says Mr Slade for the tenants in the present case, although there may be some indications in the instant lease which appear at first blush K to be language of obligation, on analysis it will be found that the lease created an option for the landlords to increase the rent above the original rent if they took the proper steps to do so.

The last authority is a decision of the Court of Appeal, as yet unreported decided on October 14 1974, the name of which is *Stylo Shoes Ltd v Wetherall Bond Street W1 Ltd.* A review clause provided that "the landlords shall be entitled to require the rent to be revised from the commencement of the eighth year, and if the landlords shall so require then the yearly rent payable during the residue of the said term shall be either the said sum of £2,750 or such amount L as may be agreed in writing between the landlords and the tenants before the commencement of the eighth year of the said term which represents the full rack rental value, or in the absence of agreement as aforesaid as shall be determined by arbitration, whichever amount shall be greater. The arbitrator shall in default of agreement between the parties be nominated by the President for the time being of the Royal Institution of Chartered Surveyors on the application of the landlords made not more than twelve months nor less than three months before September 29 1972 and the decision of the arbitrator shall be final and binding upon the parties." In the event there was no reference to arbitra- M tion before the time-limit required by the clause expired. The approach of the Court of Appeal was that the lease was in a form put forward by the landlords; it plainly provided that the landlords could not apply for the appointment of an arbitrator less than three months before September 29 1972. It showed, said Lord Salmon, that "if the landlord wants to ensure he will obtain a revised rent otherwise than by written agreement he has to initiate an arbitration not less than three months before the beginning of the eighth year." Mr Slade submits that the similarities between that case and the present are too plain to be ignored; in both cases the landlord and the tenant could agree a higher rent during a particular period; if there was no agreement the landlord had the right to set in motion arbitration machinery by applying to the President of the Royal Institution of Chartered Surveyors. This right was exercisable by the landlord; it was for the benefit of the landlord alone because the rent could not be decreased, it could only be increased, and the right to apply to the President was given a time-limit, and *mutatis mutandis* Mr Slade says that in the same way as the landlord was caught by his own defective drafting in *Stylo Shoes*, so in the present instance.

I come back to the present lease to determine whether it is in truth an option rent review clause or an obligation rent review clause. The first indication supports Mr Colyer, who appears for the landlords: there is not one rent of £2,750 for the whole term, there are the respective rents for three separate periods. During the first period there is the basic rent, and then the rent for the second period must be determined. I agree with him that this is some indication of an obligation. It is necesary to determine the rent for the second six years of the term. The rent is to be "whichever of the two following rents is the greater, that is to say," the basic rent, £2,750, or the amount which is agreed or determined by arbitration. Mr Colyer submits that in order to determine which of the rents is the greater there must be two rents which can be compared; that would seem the language of obligation. A rent must be calculated in order to make a comparison. But then the clause which deals with the ascertainment of the rack rent, the amount to be compared with the basic rent, provides that the rack rent is to be such an amount as *may* be agreed—and that is permissive—between the landlord and the tenant before September 25 1974, and if there is no agreement such amount as *may* be determined, not *shall* be determined. If there is a determination by an arbitrator he is to be nominated. The application of the landlords for his nomination is to be made within a certain time-limit, and *in case* of such arbitration the amount to be determined by the arbitrator shall be the rack rent as assessed by him at the ending of the eighth year. In my judgment, in the same way as in *Stylo* and in the *United Scientific* case, the indications of an obligation are overruled by contra-indications of option. The clause merely empowers the landlord, and the landlord alone, to go to arbitration if there is no agreement, and imposes on him the right to go to arbitration only if he applies to the President of the Royal Institution of Chartered Surveyors within the fourteen days, that is to say before October 9 1974. Dealing with the last period, the rent is said to be the rent "agreed or calculated on the same basis *mutatis mutandis* as set out above in respect of the eighth year." That reads in the whole of the language of the rent revision clause attributable to the second period, and the argument for the landlords is not advanced. Mr Colyer submitted that this was a case where, if there is no arbitration because the time-limit is exceeded, then there can be no rent for the second period, but I do not take that view. There is a basic rent and a provision whereby the landlord can, if he likes, obtain an arbitration with regard to a higher rent. If he fails to exercise his right then the basic rent remains; that basic rent compares with such amount as may be agreed or may be determined. In the events which have happened, nothing has been agreed, nothing has been determined, and therefore the basic rent is greater.

This is a clause in which there are some indications that could have led to the view that this was a rent revision clause of obligation rather than option, but construing the clause, and paying attention as I must to the decisions to which I have referred, it seems to me that I am bound to come to the conclusion that this is an option-type clause, and that it would only make the law more confused than it is if I were able on the wording of this clause to find some subtle distinction between it and the earlier authorities. In my judgment, in substance this is an option-type form of rent review clause, and the time-limit was mandatory. It is common ground that the landlords failed to comply with the time-limit, and it follows that the application for the appointment of an arbitrator was out of time, and out of order.

His Lordship made a declaration that on the true construction of the above-mentioned lease, and in the events which have happened, the plaintiffs were not entitled to apply and had not validly applied to the President of the Royal Institution of Chartered Surveyors for the appointment of an arbitrator. The defendants were awarded their costs.

COURT OF APPEAL

October 14 1974

(Before Lord SALMON, Lord Justice STEPHENSON and Mr Justice MACKENNA)

STYLO SHOES LTD v WETHERALL BOND STREET W1 LTD

Estates Gazette January 31 1976
(1974) 237 EG 343

Rent review clause—Complicated terms ending with requirement of nomination of arbitrator on application by landlords before a specified date—Time-limit not merely "machinery," but inserted for protection of tenants—Not to be ignored in a case where landlords failed to make application by due date

This was an appeal by Wetherall Bond Street W1 Ltd from a judgment of Blackett-Ord V-C on January 31 1974 holding that the respondents, Stylo Shoes Ltd, were entitled to revise the rent of premises at 9 Bond Street, Leeds, in accordance with a review clause contained in an underlease of December 31 1965.

Mr G Lightman (instructed by Rubens, Shapiro & Co) appeared for the appellants, and Mr L J Porter (instructed by Wurzal & Co, of Leeds) represented the respondents.

Giving judgment, LORD SALMON said: This appeal turns entirely upon the construction of a short clause in an underlease made on December 31 1965 between the present appellants (whom I will call "the tenants") and the respondents (whom I will call "the landlords"). The underlease granted to the tenants for a term of 14 years certain premises at 9 Bond Street, Leeds. The lease commenced on September 29 1965, and provided that for the first three years of the term the rent should be £2,600 a year and for the next four years £2,750 a year. The lease made no express provision as to what rent should be payable during the last seven years of the term. The clause to which I have referred is clause 1 (b), which reads as follows:

> The landlords shall be entitled to require the rent to be revised from the commencement of the eighth year of the said term being the 29th day of September 1972 and if the landlords shall so require then the yearly rent payable during the residue of the said term shall be either the said sum of £2,750 or such amount as may be agreed in writing between the landlords and the tenants before the commencement of the eighth year of the said term as represents the full rack rental value (as hereinafter defined) of the demised premises or (in the absence of agreement as aforesaid) as shall be determined by arbitration whichever amount shall be the greater. The arbitrator shall in default of agreement between the parties hereto be nominated by the President for the time being of the Royal Institution of Chartered Surveyors on the application of the landlords made not more than 12 months nor less than three months before September 29 1972 and the decision of the arbitrator (who shall be deemed to be acting as an expert) shall be final and binding upon the parties.

The clause is very ill-drafted and not at all easy to construe. It seems to me, however, that it provides for three things.

Firstly, by necessary implication, if the landlords do nothing to indicate that they require a revision of the rent before the beginning of the eighth year, the tenants shall go on paying rent for the residue of the term at £2,750 a year. Secondly, if the landlords require a revision of the rent, although the clause does not expressly oblige them to give any notice, it is to be implied that they can show that they require the rent to be revised only by giving the tenants notice of that fact; but the date when the notice is to be given is, so far as the express words in the clause are concerned, not stated. Thirdly, if the landlords require the rent to be revised, the clause clearly contemplates that the parties may come to some agreement as to the amount of the true rack-rent: if, however, they do not agree in writing on this point, then it may be settled by arbitration. The parties may either agree upon an arbitrator between themselves, or alternatively, if they do not agree upon an arbitrator, then an arbitrator shall be nominated by the President of the Royal Institution of Chartered Surveyors, but only on the application of the landlords made "not more than 12 months nor less than three months before September 29 1972."

It seems to me that that part of clause 1 (b) which provides what is to happen if deadlock is reached is of critical importance. The application of the landlords has to be "made not . . . less than three months before September 29 1972." I think those words are clearly put in for the protection of the tenants. They protect the tenants against the risk of having to face an increased rent at the very last moment before the last seven years start to run, or indeed at any time during the last seven years. Mr Porter (to whom we are indebted for a very able argument) was driven to concede that on his construction of the clause, which the learned Vice-Chancellor was persuaded to accept, the application could be made by the landlords at any time during the currency of the lease. This lease was in a form put forward by the landlords. The clause is shockingly badly drafted. The one clear thing about it is that the landlords cannot apply for the appointment of an arbitrator less than three months before September 29 1972. I think that that last sentence in the clause colours everything that goes before it. It shows that if the landlords want to ensure that they will obtain a revised rent otherwise than by a written agreement, they have to initiate an arbitration not less than three months before the beginning of the eighth year. It is implicit in this clause that if the landlords desire to establish a right to arbitration they must certainly notify the tenants, before those three months to which I have referred, that they require a revision of the rent. If they so notify the tenants prior to the beginning of the three months, and no written agreement has been reached with the tenants by that time, they can then, but not later, apply for arbitration.

We have been invited to read the clause in a reverse sense, as indeed the learned Vice-Chancellor read it. He said that the first part of the clause contained nothing to compel a request to revise, let alone an agreement for revision, to be made not less than three months before the beginning of the eighth year: without looking at the last sentence, there is nothing to prevent the landlords informing the tenants at the last moment of the seventh year, or even at any time during the succeeding seven years, that they require a revision of the rent: therefore, if the last sentence means what it says, and the landlords may be deprived of arbitration unless they apply within the three months to which I have referred, there is an inconsistency between the last sentence and the first part of the clause which ought to be resolved in favour of the landlords. Alternatively, the last sentence of the clause is only machinery, and accordingly the time-limit, clearly put in for the protection of the tenants, may safely be ignored. I am afraid that I cannot accept that argument. We have been referred to three authorities, from which I derive no help. Counsel were right to refer us to them, for they (or at any rate two of them) were relied on to some extent by the learned judge. In all of these cases the clause in question was, however, so strikingly different from the present clause that I think it is unnecessary for me to say any more about those authorities. I have come to the clear conclusion that, looked at broadly, this very slovenly-drawn clause has the meaning which I have already indicated. It has none of the inconsistency for which Mr Porter has contended and which appealed to the learned Vice-Chancellor. The construction which I favour makes sense, and preserves for the tenants the protection which the clause manifestly intend to afford them. It is the landlords' clause, and it would not be right to stretch its language in their favour so as to justify the time-limit for an application for arbitration being disregarded. I would accordingly allow the appeal.

STEPHEHSON LJ: I agree. Clause 1 (b) is badly drawn, but it is the landlords' own language which, in my judgment, defeats them and compels us to allow this appeal, for the reasons my Lord has given. As I read the clause, the landlords can take their time for requiring the rent to be revised, in the hope of getting an agreement in writing with the tenants on the revised rent, and if they get it at any time before the revised rent is payable in advance on September 29 1972 they can claim the revised rent. In the absence of agreement on the amount of the revised rent, they may be able to get the tenants to agree on an arbitrator, and that agreement they can, in my judgment as at present advised, get, although it is unnecessary to decide the point, at any time before or after September 29 1972. But in default of agreement on an arbitrator, the only way of getting a revised rent is by applying to the President of the Royal Institution of Chartered Surveyors, and that the landlords must do within the time-limit laid down by the clause which they put forward. If, therefore, the landlords are in any doubt whether the tenants will agree the amount of the revised rent or agree upon an arbitrator, they cannot take their time, but must require the rent to be revised in time for the application to be made to the President not less than three months before September 29 1972. I sympathise with the learned Vice-Chancellor's view that the time-limit was not of the essence and was inconsistent with the earlier part of this obscure clause; but I have come to the conclusion that he was wrong to make the declaration which he did, that it is too late for the landlords to require the rent to be revised, and that their action must be dismissed.

MACKENNA J: As I construe this unsatisfactory clause, if the landlords are to compel the tenants to pay a revised rent they must require them to do so before the three-months period begins. If they do, and if before that period begins the parties fail to agree to a revision, then, if they do not agree on an arbitrator, the landlords may apply to the President for the appointment of one. If they do apply to the President, they must make the application before the three-months period begins. It follows, if I am right, that unless the tenants agree to something different, the requirement to pay a revised rent, the failure to agree on a revision, and the application to the President for the appointment of an arbitrator, must all occur before the commencement of the period. In this case the landlords did not require the rent to be revised before the beginning of the period. As the tenants did not agree to a revision the landlords lost any right to a revision which they might otherwise have had. I agree that the appeal should be allowed.

The appeal was allowed, with costs in both courts.

CHANCERY DIVISION

May 23 1975

(Before Mr Justice GOFF)

ACCUBA LTD v ALLIED SHOE REPAIRS LTD

Estates Gazette February 14 1976

(1975) 237 EG 493

Rent review clause modelled on Form 8, 1 of the Encyclopaedia of Forms and Precedents held to contain provisions of the nature of "machinery" rather than an "option"—Landlords some 18 months late with their notice demanding a revised rent accordingly held entitled nevertheless to call the clause into operation—Court refers to C Richards & Son Ltd v Karenita Ltd (1971) 221 EG 25 and United Scientific Holdings Ltd v Burnley Corporation (1974) 231 EG 1543

This was a summons by Accuba Ltd, who sought a declaration that a notice served on them by the respondents, Allied Shoe Repairs Ltd, on September 24 1974 was invalid and ineffective to determine the rent payable for the second seven-year period of a 14-year underlease granted by the respondents to the applicants on October 16 1967. By a cross-summons the respondents sought a declaration that the notice in question was valid and effectual for that purpose, and that in any event the President of the Royal Institution of Chartered Surveyors was empowered under clause 5 (2) (c) of the underlease concerned to appoint an independent surveyor to determine the open market rental value of the demised premises.

Mr S J Sher (instructed by Rayner & Co) appeared for the applicants, and Mr J L Knox (instructed by Fitzhugh, Eggar & Port, of Brighton) represented the respondents.

Giving judgment, GOFF J said that the case concerned a rent review provision in an underlease dated October 16 1967 granted by Allied Shoe Repairs to Accuba Ltd. It was of importance to the parties because of the money involved, and also of considerable general importance, because the underlease was clearly modelled on Form 8, 1 in the Encyclopaedia of Forms and Precedents (1966) vol 12. The underlease was for 14 years beginning on March 25 1967. Clause 1 read: "yielding and paying therefor during the first seven years . . . the clear net yearly rent of £5,000 . . . and for the residue of the said term . . . a rent to be determined in accordance with the provisions of clause 5 hereof." Clause 5 provided:

> The reviewed rent . . . shall be determined in manner following that is to say it shall be whichever shall be the higher of the first reserved rent and the open market rental value of the demised premises for the review period provided that and it is hereby agreed as follows:
>
> (1) The expression the open market rental value as aforesaid means . . . the annual rental value of the demised premises in the open market on a lease for a term of seven years certain . . . with vacant possession at the commencement of the term. . . .
>
> (2) The said open market rental value shall be determined as follows:
>
> (a) It shall be such sum as shall be specified in a notice in writing by the landlord to the tenant at any time before the expiration of the period of six years after the commencement of the term hereby granted; or
>
> (b) As shall within three months after such notice be agreed between the parties in writing in substitution for the said sum; or
>
> (c) It shall be determined at the election of the tenant by counternotice in writing to the landlord not later than three months after the landlord's said notice (time to be of the essence hereof) by an independent surveyor appointed for that purpose by the parties jointly in writing or upon their failure to agree upon such appointment within one month after the date of the said counternotice then by an independent surveyor appointed for that purpose by the President for the time being of the Royal Institution of Chartered Surveyors. . . .
>
> (3) In the event of the determination by such independent surveyor not having been made and communicated to both parties hereto prior to the commencement of the review period for any reason whatever then in respect of the period of time (hereinafter called "the said interval") beginning with the said commencement and ending on the quarter day immediately following the date on which such determination shall have been made and communicated as aforesaid the rent payable hereunder shall continue to be paid at the rate of the first reserved rent provided that at the expiration of the said interval there shall be due as additional rent payable by the tenant to the landlord on demand a sum of money equal to the amount whereby the reviewed rent shall exceed the first reserved rent but duly apportioned in respect of the said interval.

Clause 6 was a suretyship clause which he (his Lordship) did not need to read in full, but it contained a provision under which, in the event of the underlease being disclaimed under any statutory or other power, the surety agreed to take from the landlords (if required by them) a grant of another underlease. The period of six years after the commencement of the term ended on March 24 1973, and the landlords should therefore have given notice some time before then, but due to an oversight for which they themselves were fully responsible, they failed to serve any notice until September 25 1974, when they wrote claiming a figure of £15,000 a year. The tenants said this was too late, though without prejudice to this contention they had in fact served a counter-notice under clause 5 (2) (c).

It was clear that the issue before the court was one purely of construction of the underlease: see the remarks of Lord Denning and Megaw LJ at p 732 in *C H Bailey Ltd v Memorial Enterprises Ltd* [1974] 1 WLR 728, and the passage at p 735 in which Sir Eric Sachs stated: "The objective of the courts in a case relating to office leases is naturally to determine the intended commercial effect of a particular agreement reached between the parties. In this respect a lease is no less a contract relating to the use of the premises than an agreement in relation to the supply of furniture for those premises is also a contract. It follows to my mind that the courts should in this class of case avoid resort, so far as practicable, to any of the highly technical points that stem from the intricacies of the ancient law of landlord and tenant." Further, although the construction of the underlease must depend on its own particular words, and he (his Lordship) was bound by decisions even of superior courts only in so far as they laid down any principle, still he did discern a dividing principle between the cases. If the words used imported an option or privilege for the landlord to increase the rent which would otherwise be payable, then any time conditions must be absolutely strictly observed. Examples of this type of case were *Samuel Properties (Developments) Ltd v Hayek* [1972] 1 WLR 1296; *C Richards & Son Ltd v Karenita Ltd* (1971) 221 EG 25; and *United Scientific Holdings Ltd v Burnley Corporation* (1974) 231 EG 1543. In those three cases there was a fixed rent throughout the whole term with a provision expressly granting an option or saying that the rent should be subject to increase. On the other hand, where the provision was mere machinery the time was not strict. Examples were *Kenilworth Industrial Sites Ltd v E C Little & Co Ltd* [1974] 1 WLR 1069, [1975] 1 WLR 143, and *In re Essoldo (Bingo) Ltd's Underlease* (1972) 23 P & CR 1. In the latter case the question of time being strict was not argued. The case turned on a point to do with reference to surveyors. It was of some significance, however, that the time question was not raised although the landlord's claim was more than a year out of time and more than 10 months into the new rent period. Also, very importantly, both in that case and in *Kenilworth* the original rent was reserved for an original period only, leaving the later rent to be agreed or ascertained.

On which side of the line did the language used in the present underlease fall? He (his Lordship) thought it was

A mechanics only. In *Kenilworth* there was of course an express provision that "any failure to give or receive such notice shall not render void the right of the landlord to require the agreement of a new rent," and although the provision in clause 5 (2) (c) of the present underlease afforded a supporting argument for the landlords' construction here, it was impossible to treat that provision as equivalent to the words quoted from *Kenilworth*. Nevertheless both Megarry J and the Court of Appeal in that case relied on those words merely as an additional or supporting reason, and arrived at their conclusion independently of it. Megaw LJ said at p 146: "The proviso, if it has any effect at all, is saying B that even though the time specified in the opening words of the clause may not have been observed by the landlord, and the landlord is thereby in breach of that provision of the contract, nevertheless the breach of the contract by him shall not render void his right to have the new rent for the succeeding five years agreed or, in the absence of agreement, determined by arbitration." The significant words there were, of course, "if it has any effect at all." Moreover there was in that case a gap if the rent review clause did not operate, and Megaw LJ said: "If the lease failed to make provision for any rent to be payable in the absence of notice being given within the proper time then plainly there would not C be the alternative which would be necessary to constitute an option." It was argued that there was no gap in the present case, because the agreement was to pay whichever should be the higher of the first reserved rent and the open market rental value for the review period determined in manner therein provided, so that if in the event there were no determination the first reserved rent remained payable. He (his Lordship) did not think that was right. It was not a covenant to pay that rent unless something else were determined, but one to pay the higher of something specific and something ex hypothesi unascertainable, and so there was a gap. In D this connection it was to be observed that in *Kenilworth* Megaw LJ refused at p 146 to imply a term to fill the gap, not only because of clause 5 (it happened to be clause 5 in that case also), but also because of clause 1, the reddendum, and he (Megaw LJ) added, "whether read separately or together." So far as clause 1 was concerned, that lease and the present underlease appeared to him (Goff J) wholly indistinguishable. Even however if there were no gap, and if in default of determination of the open market rental the first reserved rent continued, that would not be because of an accrued right to pay only that sum because there was an express covenant in clause 5 to pay that or something E higher. It would be because of a breakdown in the machinery, and even on that view of the matter it seemed to him (his Lordship) that he should regard this as mechanics only: see the point stressed by Megarry J in *Kenilworth* at p 1071, "In the *Samuel* case the Court of Appeal was concerned with an underlease for 21 years in which the rent had been agreed for the entire term. . . ." That appeared to him (Goff J) to be crucial.

In his judgment, therefore, time was not of the essence in clause 5 (2) (a), and the express stipulation to that effect in clause 5 (2) (c) was of course a supporting context of some strength. Counsel for the tenants had relied on clause F 6, where, he said, time must be of the essence, though not so stated. That might well be so, but it was not so strong as the express reference in clause 5 (2) (c) compared with the absence of it in the same clause in paragraph 2 (a). Counsel also relied on the provision in clause 5 (1) that the open market rental value was to be determined at the time of such determination, which, he said, was an impossible operation if one had to value *ex post facto*. The parties did, however, contemplate that that might have to be done, though only to cover a failure, likely to be short, to carry out all the necessary operations in the time allowed, not a failure to start at all which resulted in the determination having to be made much longer after the due date. This G was an important point which he (his Lordship) had carefully weighed, but he did not think it outweighed the basic view he had formed that clause 5 was machinery not in the nature of an option, and he accepted the argument of counsel for the landlords that the difficulty was a valuation problem, not an impossibility. Finally he had to consider whether even so the notice was too late. As Lord Denning MR had stated in *Bailey*'s case, delay might give rise to a defence by way of equitable estoppel. For that purpose, however, the tenant had to establish that he had altered his position: see the *Essoldo* case at p 6. The tenants here had filed no evidence on that point, and it had been agreed by counsel H that if he (Goff J) allowed the landlords' notice to take effect he should order that in applying it retrospectively the rent freeze should apply as if the rent had been fixed from the start, and subject thereto no case of equitable estoppel was put forward. He (his Lordship) accepted that agreement, and passed to the next question, which was how long a landlord might wait even when time was not of the essence. He thought the law was correctly stated in *Emmet on Title* 16th ed p 214:

> Even where the date for completion was not originally of the essence (either by express provision or in accordance with the J above-mentioned rules), if delay has been unreasonable, the other party can treat the contract as broken, without first giving notice as mentioned below (*Farrant v Olver* [1922] WN 47).

Was the delay here so long, then, as to be unreasonable? In the light of the *Bailey* and *Essoldo* cases, he (his Lordship) thought that though it was a borderline case the landlords were not too late in the context of a seven-year term. Accordingly he dismissed the tenants' summons, and granted the landlords relief on their cross-summons, subject to the agreed application of the rent freeze.

K

COURT OF APPEAL

March 1 1976

(Before Lord Justice BUCKLEY, Lord Justice ROSKILL and Lord Justice BROWNE)

UNITED SCIENTIFIC HOLDINGS LTD v BURNLEY CORPORATION

Estates Gazette May 15 1976 and May 22 1976
(1976) 238 EG 487 and 563

Rent review clause—Parties "shall agree or failing agreement shall determine by arbitration" new rent within specified L period—Rent neither agreed nor referred to arbitration within that period, landlord not entitled to an increase—Important observations on supposed distinction between "option" and "obligation" types of review clauses—Distinction impractical and illusory—To secure increases, landlords must in general comply with whatever time-limits have been agreed—Care necessary in drafting future revision clauses and administering those now current—References to four Estates Gazette reports—Stylo Shoes Ltd v Wetherall Bond Street W1 Ltd 237 EG 343 extensively quoted and in effect applied—Richards v Karenita 221 EG 25 cited for judge's opinion on issue of principle M

This was an appeal by Burnley Corporation from a judgment of Pennycuick V-C on May 13 1974 in the Chancery Division holding in favour of the respondent lessees, United Scientific Holdings Ltd, of Burnley, Lancashire, that the appellant lessors had no right to demand an increase in rent under a review clause in a lease dated August 31 1962 in respect of the period of 10 years beginning on September 1 1972. The judgment of Pennycuick V-C was reported at (1974) 231 EG 1543.

Mr H E Francis QC and Mr B C Maddocks (instructed by Turner, Peacock, agents for Smith & Smith, of Burnley)

A appeared for the appellants, and Mr A J Balcombe QC and Mr B Levy (instructed by Fremont & Co) represented the respondents.

Giving judgment, BUCKLEY LJ said: This is an appeal from a decision of Pennycuick V-C dated May 13 1974 on a rent review clause in a lease. The lease was dated August 31 1962. It was a building lease for 99 years of some land in Burnley in Lancashire. The lessors were the appellants, Burnley Corporation, and the lessee was the plaintiff company, under its then name of E Cooksen (Properties) Ltd. The habendum and reddendum of the lease are in the following terms:

B

> To hold the property (except and reserved as aforesaid) unto the lessee from the date hereof for the term of ninety nine years yielding and paying therefor the rents mentioned and referred to in the next succeeding clause in the manner and at the times therein stated. . . . The lessee shall pay to the corporation during the period from the commencement of this lease until the first day of January one thousand nine hundred and sixty three a peppercorn (if demanded) and thereafter until the expiration of a period of ten years from the commencement of this lease the yearly rent of nine hundred pounds and thereafter during the residue of the said term the yearly rent of one thousand pounds plus any additional rent payable under the provisions contained C in the schedule hereto without any deduction whatsoever other than landlord's property tax on the said rents each and all of such rents to be paid by equal half yearly instalments on the thirty first day of March and the thirtieth day of September in each year during the said term the first payment of the said rents or a proportionate part thereof as the case may be to be made on the rent day next succeeding the date whereon the said rents shall first commence to be payable.

The schedule there referred to is in the following terms:

> During the year immediately preceding the period of the second ten years of the said term and during the year immediately preceding each subsequent ten year period of the said term and during the year immediately preceding the last nine year period D of the said term (each of such periods being hereinafter referred to as a "relevant period") the corporation and the lessee shall agree or failing agreement shall determine by arbitration the sum total of the then current rack rent (which expression "rack rent" shall for the purposes of this schedule be deemed to mean the full annual value of the property and of all buildings and erections thereon and appurtenances thereto and including all improvements carried out to the same calculated on the basis of all rates taxes repairs and other outgoings being borne wholly by the occupier thereof) reasonably to be expected on the open market for leases of the property and all buildings and erections thereon and one quarter of the sum total so ascertained or one thousand pounds (whichever is the greater) shall be the rate of rent reserved E by this lease in respect of the then next succeeding relevant period. All arbitrations under or by virtue of this schedule shall be referred to the decision of a single arbitrator to be agreed by the parties hereto or failing their agreement thereon shall be referred to ,the decision of a person to be nominated by the President for the time being of the Royal Institution of Chartered Surveyors and such reference shall be deemed to be a submission to arbitration within the meaning of the Arbitration Act 1950 or any statutory modification or re-enactment thereof for the time being in force.

On September 29 1967 a supplemental lease was entered into between the parties relating to some adjoining land, F and the land comprised in the supplemental lease was demised for the same term as the land in the main lease. The rent was to be, from April 1 1966 until the buildings should be erected on the plot of land comprised in the supplemental lease, a peppercorn, and from the date upon which the peppercorn rent should cease the rent was to be £1,000 until the expiration of 10 years from the commencement of the lease, and thereafter and during the residue of the term the yearly rent of £1,000 plus any additional rent payable in accordance with the provisions for rent review contained in the schedule to the lease—that is the original lease—which provisions should equally apply to the rents G reserved by the supplemental lease. So the machinery of the lease of August 31 1962 was made to apply to the supplemental lease, and the basic rent, as one might say, in the case of the supplemental lease was £1,000 a year in the same way in which the £1,000 a year is referred to in the lease of August 31 1962. It will be convenient to deal with the matter with reference to the lease of August 31 1962, because no separate point arises on the supplemental lease, but it will be appreciated that the total rent payable under the two documents in respect of the period after the first 10 years of the term was £2,000, or £2,000 and such additional rent as might be ascertained by applying the H machinery provided in the schedule to the lease.

On May 10 1972 estate agents acting for the plaintiff company wrote to the borough council a letter saying that they had been instructed by the plaintiff company to enter into negotiations to agree the new lease rent for the land comprised in the two leases. Following that letter, a telephone conversation took place on July 31 between a member of the firm of estate agents and a representative of the borough council, in which it was arranged that the agents would supply particulars of the rents reserved by the underleases of the property. There were in fact nine underleases of this property subsisting at that time for a total rental, J I think, of £12,515; but particulars of those underleases were never in fact supplied to the borough council. On August 21 1972 the plaintiff company's solicitors wrote a letter to the town clerk of the borough council, in the course of which they said:

> We have carefully considered the provisions of the lease and supplemental lease and we are entirely satisfied that the provisions are not susceptible of any legally enforceable meaning and are void accordingly. In our view therefore the rent as from September 1 1972 is the sum of £2,000, being the yearly rent of £1,000 payable under the provisions of clause 3 of the lease and the yearly rent of £1,000 payable under the provisions of the supple- K mental lease.

The town clerk did not acknowledge that letter until after August 31 1972, which was the end of the first 10 years of the term limited by the two leases. He effectively answered the letter of August 21 1972 on October 12 1972 in a letter in which he said that the corporation had taken counsel's opinion on the question of the enforceability or otherwise of the rent review provisions, and that in the light of that opinion he was unable to agree that the position was as the plaintiff company had stated it in their letter and the corporation must insist upon the rent revision provisions L being implemented. That resulted in the plaintiff company saying that they proposed to take proceedings. The originating summons was issued on December 4 1972 by the company as plaintiffs, joining the corporation as defendants. There was some subsequent correspondence, but for the purposes of this appeal I do not think it is necessary for me to say anything about that. As I say, for the purposes of this judgment I propose to ignore the supplemental lease and deal with the matter in relation to the original lease.

The appellants (the corporation) assert that time is of the essence of a rent review clause only where it is in the nature of an option or confers a unilateral right on one party to M secure an alteration of the rent to that party's advantage. They say that this is not a lease for 99 years at £1,000 a year with a power for the landlord to call for an upward review. They contend that this is a lease at specified rents during the first 10 years and thereafter at a rent to be ascertained according to a formula, which cannot be less but may be more than £1,000 a year, with machinery provided for ascertaining the amount according to the formula set out in the schedule. To such a provision they say time is not essential. Mr Francis, who appears on their behalf, has contrasted what has been described in some of the authorities as an "option" clause with clauses of a kind which have also been described in the authorities as "obligation" clauses; that is to say, clauses in which the parties either are contractually obliged to carry out a review or where by the terms of the lease the review is an automatic operation. Mr Francis draws attention to the mandatory language in the schedule, which twice uses the verb "shall," and he contends that the provisions of the schedule are equally binding upon both parties, and that this is not a case in which it could be said that either party has any right in the nature of an option. It is true that an agreement to agree something cannot create a legally binding obligation to agree, but I think it is true to say in this case that there is a binding obligation. There is a binding obligation to submit the matter to arbitration in default of the parties agreeing. The respondents, on the other hand, say that a review clause in a lease is a clause of a commercial character and that time-limits contained in the clause should be strictly adhered to, in the way that time provisions in commercial clauses are normally regarded as being of the essence of the contract. They point out that uncertainty about the amount of the rent may seriously affect the saleability of the tenant's interest in the land and his decision whether he can afford to retain the property, or how he should deal with it, and they point out that a review clause is directed to ensuring that the landlord shall get a proper return for his investment throughout the life of the lease. These, they say, are commercial considerations which should attract the same sort of approach to time-limits as is adopted in other commercial cases.

This sort of question has arisen very frequently of late in the courts, no doubt because economic inflation has greatly increased the appetite of landlords for rent reviews, with the result that rent review clauses are nowadays probably much more common than they used to be. We have been referred to about a dozen cases decided in the High Court, some of which have come to this court, since October 1971 on clauses of this kind. A review of these decisions leads me to the conclusion that there is a danger of distinctions becoming established between one kind of clause and another on what appear to me to be narrow and somewhat artificial grounds, which may be capable of explanation on an intellectual level but are hard to justify on a practical level. A simple example of one class of cases might be a lease for 21 years at a yearly rent of £X provided that if within a period of (say) not more than nine months and not less than three months before the end of the first seven years of the term the landlord should give notice to the tenant requiring a rent review, the rent for the eighth to the fourteenth years should be such as before the end of the seventh year the parties should agree, or in default of agreement a valuer should determine to be the open market rent of the property, followed by a similar provision operative in the fourteenth year; or the clause may be in a more complex form of the kind to be found in *Mount Charlotte Investments Ltd v Leek, Westbourne & Eastern Counties Building Society,* which will be found reported in vol 237 of the ESTATES GAZETTE at p 339. Clauses of this kind, though not necessarily strictly option clauses, have been referred to in some of the cases as "option" clauses because time has been held to be of the essence of the clauses by analogy to a strict option clause for (say) a purchase of land or renewal of a lease.

A simple example of another class of cases might be a lease for 21 years at a rent of £X a year with an automatic or obligatory review at the end of the seventh year, providing that for the eighth to the fourteenth years the rent should be such a yearly sum as in default of agreement a valuer should determine to be the open market rent of the property, followed by a similar provision operative at the end of the fourteenth year in respect of the remainder of the term. A more elaborate example of this type of clause is to be found in *Accuba Ltd v Allied Shoe Repairs Ltd* [1975] 1 WLR 1559. Clauses of this kind have been referred to as "obligation" clauses or "machinery" clauses, and in some of these (of which the *Accuba* case was one) time has been held not to be of the essence on the ground that the time programme laid down relates merely to the machinery provided for carrying out the obligatory review and is not part of the essence of the obligation: see, for example, *Kenilworth Industrial Sites Ltd v EC Little & Co Ltd* [1974] 1 WLR 1069, per Megarry J at p 1071. This distinction will be found to be discussed and adopted in the judgment of Goff J in the *Accuba* case and of Templeman J in the *Mount Charlotte* case.

In the present case, the learned judge disposed of the matter upon the following grounds, which will be found at page 6 of the typescript of the note of his judgment:

> It is well established that in a contract for sale of land, time is not of the essence until made so. On the other hand, with regard to the exercise of an option, it is beyond question that time is of the essence. This appears from a recent case in which the principle was stated. I refer to the case of *Samuel Properties (Developments) Ltd v Hayek* [1972] 1 WLR 1296, more particularly to the judgment of Russell LJ at page 1301. The ground on which it is held to be of the essence, of course, in relation to the exercise of an option is that it represents a unilateral privilege residing in the party who procures it and who must exercise it, if at all, strictly within the time limited. In the present case the rent is not expressed as an option in favour of the council. The form in the lease is expressed to be £1,000 plus that additional rent under the schedule ascertained either by agreement or by arbitration. So Mr Maddocks contends that this provision as to rent review represents not a step to create an obligation but a method to quantify the amount. It seems to me that although expressed for the quantification of additional rent the rent review is in substance a unilateral right vested in the council alone, the nature of the right being to increase the rent payable by the tenant. The tenant is entitled under the lease to possession for 99 years at a rent of £1,000 unless the corporation elects to require a review, and his right to remain in the property is already assured. Therefore, it is for the benefit of the landlord solely. If the landlord does so require, the rent can only be increased and will not decrease, and that being the nature of the right it seems to me that the right is on the same footing as an option and the principle enunciated by Russell LJ applies to one no less than the other. Time is of the essence of the contract as it is a unilateral right and must be exercised strictly in the time set down by the lease. It would not be consistent with the substance of justice to allow the time to apply for a review to remain open for an indefinite period. The position here is wholly different from the simple case where goods were sold or services rendered for a sum to be determined. Here the plaintiffs already have the right to possession for 99 years. The rent review provides the unilateral right to increase.

The case of *Samuel Properties (Developments) Ltd v Hayek* there referred to was a case in which land was leased by an underlease for a term of 21 years from July 7 1964, with a proviso that the rent was subject to increase at the end of the seventh and fourteenth years of the term in accordance with the fourth schedule to the underlease. The first paragraph in the schedule referred to was in the terms, "The yearly rent reserved by this underlease shall be subject to review at the option of the lessors in the seventh and fourteenth years of the term hereby granted in the manner provided in the following paragraphs," and paragraph 2 read:

> If the lessors shall desire to have the said rent reviewed at the said time by reference to the open market rental value of the demised premises then prevailing and shall serve upon the lessee a notice in writing to that effect not later than two quarters before the expiry of the said seventh and fourteenth years respectively of the said term then within one month after the service

A of the said notice the parties thereto by themselves or by their respective valuers shall agree the amount of the said open market rental value of the demised premises and in default of such agreement within the period of one month from the service of the said notice the said amount shall be determined by a valuer (acting as an expert and not an arbitrator) to be appointed by the President of the Royal Institution of Chartered Surveyors.

Then paragraphs 3 and 4 of the schedule provided further detailed machinery. Russell LJ at p 1301 drew attention to the fact that clause 5 gave the lessee the power to determine the underlease on giving a time-limited notice, and he said that it was accepted that that clause was intended to and B did contain a strict time requirement. He then went on to draw attention to the fact that clause 1 of the schedule conferred an option on the lessor. He referred to the arguments which had been presented to the court, and he said at p 1302, just below letter C:

> I am not myself impressed by these arguments: the right or privilege of exacting an additional rent was conferred by the bargain between the parties as an express option which would be effective if a condition precedent was complied with: it could be equated with an offer by the lessee to pay an increased rent only in certain circumstances which it lay in the power of the lessor unilaterally to bring about.

C

In that case the rent review clause was expressly described in the underlease as being an option, and had the true characteristics of an option. Accordingly, the court held that the time requirements of the clause were not merely directory, but mandatory; that is to say, they were of the essence of the contract. In *C H Bailey Ltd v Memorial Enterprises Ltd* [1974] 1 WLR 728 a rent review clause provided that "if on September 21 1969 the market rental value shall be found to exceed the rent of £2,375 hereby reserved there shall be substituted an increased yearly rent equal to the market rental value so ascertained," and there were pro- D visions for the market rental value to be agreed or to be determined by arbitration. The question arose whether if, as occurred, the new rent had not been agreed or determined by September 21 1969 the tenant could be liable for the new and higher rent from that date or only, as Eveleigh J had held, from the time when it was either agreed or determined. The Court of Appeal held that since the rent review clause provided for the increased rent when ascertained to be substituted from September 21 1969, it was payable retrospectively from that date. In coming to that decision, the Court of Appeal rejected the tenant's argument that the landlords had lost their right to increase the rent at all because E it was not in fact ascertained on September 21 1969. But the decision in that case is not, I think, of much assistance to us in the present case, for it turned upon the construction of the words "shall be found to exceed." The review clause there did not really lay down any time programme for the review operation.

A review clause which permits of an alteration of the rent only in an upward direction seems to me to be very much the same in character whether by its terms it requires the landlord to initiate the review process by serving a notice or making a request, or whether it purports to make the review obligatory or automatic. The only person to F whose financial advantage the review can operate is the landlord. The tenant must, of course, have an interest in the outcome of any review. He may in some circumstances want to have his rent defined even if this may involve an increase; but the person who in any case in which an increase is probable will always have an incentive to obtain it is the landlord. So even where the rent review is obligatory or automatic, it is the landlord who is likely to take the initiative, and if he fails to do so, the tenant will probably let the sleeping dog lie. What bearing have these considerations upon the rules of equity relating to time clauses in contracts? It is of course notorious that the Court of Chancery adopted a different approach to time clauses from that of the com- G mon law courts. The common law held the parties to the strict terms of their bargain. Equity, on the other hand, did not regard stipulations as to time as being of the essence of the contract—that is, as bound to be strictly observed—unless this was expressly provided in the contract or the circumstances and nature of the contract were such that this intention was to be imputed to the parties. We are not presently concerned with time being subsequently made essential by notice. Since the fusion of equity and law in 1873, the equitable principles prevail: see now the Law of Property Act 1925, section 41. We have heard some discussion in this case as to whether the equitable principles ever H extended to contracts other than those in respect of which some form of equitable relief could have been obtained in the Court of Chancery. For the purposes of this judgment I am prepared to assume that the equitable rules about time clauses now apply to all kinds of contract. I think that the correct test of whether time is originally of the essence of a contract according to equitable principles is correctly enunciated in *Fry on Specific Performance* 6th ed p 502 para 1075:

> Time is originally of the essence of the contract, in the view of a court of equity, whenever it appears to have been part of the real intention of the parties that it should be so, and not to have J been inserted as a merely formal part of the contract. As this intention may either be separately expressed, or may be implied from the nature or structure of the contract, it follows that time may be originally of the essence of a contract, as to any one or more of its terms, either by virtue of an express condition in the contract itself making it so, or by reason of its being implied.

Implication is dealt with further in para 1079 and 1080, and in para 1081 the learned author says:

> And so, again, where the object of the contract is a commercial enterprise, the court is strongly inclined to hold time to be essential, whether the contract be for the purchase of land for such purposes, or more directly for the prosecution of trade. K

So where the subject-matter of a contract is the acquisition of a wasting asset or of a perishable commodity or of something which is likely to change rapidly in value, an intention that punctual observance of time provisions is intended to be an essential term of the bargain is to be imputed to the parties. In *C H Bailey Ltd v Memorial Enterprises Ltd*, all three members of the court emphasised the commercial character of rent review clauses: Lord Denning MR at p 732, Megaw LJ at the top of p 733, and Sir Eric Sachs at p 735 in a passage which I will read, starting at the beginning of his judgment:

> There are clearly now in current use in leases of business L premises a number of variants of a clause which provides for an increase in the rent of those premises as from a given date, if their market rental "shall be found" to be higher than that originally agreed upon the grant of the lease. The interpretation of each such clause depends, of course, on its own precise terms. Unfortunately, however, some of these clauses have been clumsily drawn in a way which unnecessarily opens a path to refined arguments which are bound to cause the lay parties themselves to throw up their hands in gloom. The objective of the courts in a case relating to office leases is naturally to determine the intended commercial effect of the particular agreement reached between the parties. In this respect a lease is no less a contract relating to the use of premises than an agreement in relation to M the supply of furniture for those premises is also a contract. It follows, to my mind, that the courts should in this class of case avoid resort, so far as practicable, to any of the highly technical points that stem from the intricacies of the ancient law of landlord and tenant.

The importance of discovering in good time what the rent of leasehold property is to be after a date from which the rate of the rent is liable to change is in my opinion self-evident. The tenant will want to know what his liability will be. He may want to know whether he can afford to retain his lease. He may want to know whether and at what price

A he is likely to be able to dispose of it; and he will in any event want to know what the effect of any change in the rent will be on his financial position generally. The landlord also will want to know what the effect of any change will be on his financial position; but since I am considering cases in which any change will be to his advantage, he may be more anxious than the tenant to have the position clarified as early as possible. These considerations, in my view, emphasise the commercial importance to the parties of the consequences of a rent review being known to them at latest by the first rent day in the period to which the new rent will apply.

B In *Stylo Shoes Ltd v Wetherall Bond Street W1 Ltd*, reported in vol 237 of the ESTATES GAZETTE at p 343, this court had to consider a rent revision clause in a lease for 14 years where the rent for the first three years was reserved at £2,600 and for the next four years at £2,750 a year. The lease made no express provision as to what rent should be payable during the last seven years of the term. Clause 1 (b) of the lease was in the following terms:

> The landlords shall be entitled to require the rent to be revised from the commencement of the eighth year of the said term being the 29th day of September 1972 and if the landlords shall C so require then the yearly rent payable during the residue of the said term shall be either the said sum of £2,750 or such amount as may be agreed in writing between the landlords and the tenants before the commencement of the eighth year of the said term as represents the full rack rental value (as hereinafter defined) of the demised premises or (in the absence of agreement as aforesaid) as shall be determined by arbitration whichever amount shall be the greater. The arbitrator shall in default of agreement between the parties hereto be nominated by the President for the time being of the Royal Institution of Chartered Surveyors on the application of the landlords made not more that 12 months nor less that three months before September 29 1972 and the decision of the arbitrator (who shall be deemed to be acting as an expert) shall be final and binding upon the parties.

D Lord Salmon, who delivered the leading judgment, said:

> The clause is very ill-drafted and not at all easy to construe. It seems to me, however, that it provides for three things. Firstly, by necessary implication, if the landlords do nothing to indicate that they require a revision of the rent before the beginning of the eighth year, the tenants shall go on paying rent for the residue of the term at £2,750 a year. Secondly, if the landlords require a revision of the rent, although the clause does not expressly oblige them to give any notice, it is to be implied that they can show that they require the rent to be revised only by giving the tenants notice of that fact; but the date when the notice is to be given is, so far as the express words in the clause are concerned, not stated. E Thirdly, if the landlords require the rent to be revised, the clause clearly contemplates that the parties may come to some agreement as to the amount of the true rack rent: if, however, they do not agree in writing on this point, then it may be settled by arbitration. The parties may either agree upon an arbitrator between themselves, or alternatively, if they do not agree upon an arbitrator, then an arbitrator shall be nominated by the President of the Royal Institution of Chartered Surveyors, but only on the application of the landlords made "not more than 12 months nor less than three months before September 29 1972." It seems to me that that part of clause 1 (b) which provides what is to happen if deadlock is reached is of critical importance. The application of the landlords has to be "made not . . . less than F three months before September 29 1972." I think those words are clearly put in for the protection of the tenants. They protect the tenants against the risk of having to face an increased rent at the very last moment before the last seven years start to run, or indeed at any time during the last seven years.

Then Lord Salmon goes on to deal with the arguments which were presented, and in the second column of p 345 he says:

> The construction which I favour makes sense, and preserves for the tenants the protection which the clause manifestly intended to afford them. It is the landlords' clause, and it would not be right to stretch its language in their favour so as to justify the time-limit for an application for arbitration being disregarded.

G That was, of course, a case in which the landlords were to initiate any revision of the rent, and their right to do so may therefore be said to have been in the nature of an option; but none of the members of the court in terms put his decision on that ground. They decided the case on the ground that the clause required in a mandatory way that if an arbitration became necessary, the application for the appointment of an arbitrator must be made before September 29 1972; and Lord Salmon clearly took that view because he thought that date was specified for the protection of the tenants. It seems to me that considerations of the kind which I have been discussing, concerning the business efficacy of a rent H review clause or at least its commercial convenience to the parties, afford just as cogent reasons for imputing to the parties an intention that a prescribed time-table shall be punctually observed in a case where the rent review is obligatory or automatic as in a case where the review process is dependent on one party taking the initiative. In my judgment, the court should not be slow to impute such an intention in either case.

This does not mean that time should be treated as of the essence in every rent review clause. The circumstances or the terms of the lease may negative any inference of such an intention in a particular case. This was so in *Kenilworth* J *Industrial Sites Ltd v E C Little & Co Ltd* [1975] 1 WLR 143 in this court. There, in a lease for 21 years, the rent reserved was limited to the first five years of the term. Clause 5 of the lease provided how the rent payable during the remainder of the term should be ascertained. The decision proceeded upon the basis that if for any reason clause 5 could not be operated, no rent would be payable. This would in itself clearly be a very strong argument against adopting a construction of the lease which might result in the rent becoming unascertainable in consequence of a failure to comply punctually with the time-table. Moreover, in that case clause K 5 contained a proviso which made it clear that a failure to comply with the time-table was not to deprive the landlord of his right to have an increased rent determined by arbitration. For these reasons, I would hold that the suggested dichotomy between "option" cases and "obligation" or "machinery" cases as producing different results in respect of time-limits is unsound. In my opinion, the court should be readily disposed to regard time as of the essence in both classes of case. In each class of case the circumstances and the nature of the contractual term should be considered in order to ascertain whether it is reasonable to impute to the parties an intention that time shall be of the essence, an important circumstance being the practical operation of the L clause and its impact upon the parties.

With deference to Graham J, I feel unable to agree with the view which he expressed in *Cheapside Land Development Co Ltd v Messels Service Co*, which is reported in *The Times* newspaper of February 4 this year, where the learned judge is reported as having said that the authorities "led too readily to the conclusion that time was *prima facie* to be regarded as of the essence in respect of all times in all rent review clauses, with only an occasional exception." I think that if parties to a lease agree a time-table for the purposes of a rent review, much less confusion is likely to ensue if the con- M sequence is that the parties must normally comply with that time-table. The time-table will probably be much more comprehensible to a layman, and possibly even to his legal adviser, than the rather esoteric rules of equity which sometimes regard time-limits as binding and sometimes do not. The landlord or the tenant is much less likely, I think, to be unwarily trapped if he knows he should be careful to comply with the time-table than if he thinks he may be able to ignore it. For these reasons, which are broader in their nature than the reasons relied on by Pennycuick V-C, I would dismiss this appeal. I do not, however, dissent from the narrower grounds on which he decided the case. In my

A judgment, the review clause in this case, although not framed as an option, is for the relevant purpose of determining whether time should be regarded as of its essence closely analogous to an option. It is, as the Vice-Chancellor pointed out, a unilateral right in the sense that it is capable of operating for the benefit of the landlord only, which he alone really has an incentive to invoke.

Pennycuick V-C dealt with another point at the end of his judgment, relating to a suggested waiver by the plaintiffs in correspondence of their right to insist upon the time-limit. That point has not been pressed in this court, and I see no B reason to differ from the Vice-Chancellor upon it. I would dismiss this appeal.

ROSKILL LJ: The huge capital investment in property over the last 20 years or more has naturally led to those making that investment seeking to ensure an adequate return on their capital. Where that return depended upon the grant of leases, especially of long leases, it was natural for investors to seek to secure that adequate return over the duration of those leases by ensuring that its value was not hopelessly eroded in real terms by the likely diminution in the value of money. That diminution in the value of money has unhappily accelerated at an alarming pace in recent years. It is not C therefore surprising that landlords have resorted to rent revision clauses in their leases to minimise the consequences of inflation upon the value of their investment and have insisted upon their inclusion in long-term leases. Equally it is not surprising that lessees and their advisers should have resorted to much ingenuity in an effort to defeat landlords' attempts to avoid the consequences of inflation at their lessees' expense. The result has been a large number of cases in the last five years, three of which before the present have reached this court but most of which have not gone beyond first instance. The rent revision clauses which have come D up for consideration are immensely varied in their terms. Their draftsmanship has been almost uniformly condemned judicially, not without justification. The draftsmanship of the relevant clause 3 and the schedule in the present lease might not unfairly be similarly criticised, though in fairness to some of the draftsmen it is to be observed that some clauses have emanated, with or without modification, from well-known precedent books. It is not, I hope, out of place to suggest that one result of the present appeal may be to lead to more careful drafting of these rent revision clauses and even to prompt, and in some cases highly necessary, revision of some of the precedents.

E All these rent revision clauses contain in one form or another stipulations as to time within which certain acts or matters are required to be done. Inevitably, human nature being what it is, some acts and some matters have not been timeously done. In consequence, lessees have strenuously resisted attempts by landlords belatedly to operate these clauses in their favour and against the interests of the lessees. Different cases have been decided different ways and, as will later emerge, not always consistently. But the courts have shown, rather surprisingly as I think in view of the current trend of statute law governing the relationship between landlord and tenant, a tendency—at least at first instance—to F construe these clauses as permitting such belated operation notwithstanding that, if looked at realistically and whatever the detail of their language, these are clauses for the benefit of the landlords and not of the lessees. Manfully grappling with the difficulties, judges of first instance have evolved an approach to the crucial question of construction which appears to require the court to ask itself whether a particular clause is (to quote one phrase used) " an option to the landlord to obtain a higher rent," and if so to construe the provisions as to time strictly, so that non-compliance with those time limits will defeat the landlord's claim; or, to quote another phrase used, " creating an obligation on the landlord to take the steps necessary to determine what the rent is G going to be," in which case non-compliance with the true limits will not defeat the landlord's claim. This reasoning in turn has led to a suggested dichotomy between two suggested classes of clause, the " option clause " and the " obligation clause." If a particular clause is given one label, it is to be construed one way; if it be given another label, it is to be construed another way. In two of the most recent cases at first instance in this field, *Mount Charlotte Investments Ltd v Leek, Westbourne & Eastern Counties Building Society,* decided by Templeman J on June 20 1975 (237 ESTATES GAZETTE 339) and *Cheapside Land Development Co Ltd v Messels Service Co Ltd,* decided by Graham J as recently as H January 29 this year and only so far reported in *The Times* on February 4 1976, both learned judges (rightly, as I think) deplored the state of the authorities but felt obliged to decide the cases respectively before them upon the footing that this dichotomy should be adopted. We are told that the second case is under appeal to this court, and I wish to make clear that what I later say about Graham J's judgment must not be taken as in any way directly affecting the actual question to be decided in that case, with which the present appeal is in no way concerned. I respectfully and entirely agree with both these criticisms of the authorities, though I think the evolution of the suggested dichotomy represents a valiant J judicial attempt to bring order out of confusion rather than an enthusiasm for the adoption of the supposed principle.

Mr Balcombe, for the respondent lessees, vigorously attacked the suggested dichotomy. He said that neither principle nor authority, properly understood, supported it, and he urged us to take the opportunity presented by this appeal to state or re-state the principles which we thought should be applied to the construction of rent revision clauses. It is legitimate criticism of the suggested dichotomy that it involves seeking to give each particular clause a label and then interpreting it according to the choice of label. This to my mind K is not a proper approach to a question of construction. The right question to ask in each case is, upon the true construction of the particular clause, did the parties intend that the particular stipulations as to time must be strictly adhered to or not; or if, as happens in so many cases, the parties have not expressly dealt with this question, must there be imputed to the parties an intention that the particular stipulations as to time must be strictly adhered to or not? It is, I think, impossible to state a rule which will prescribe that every such clause must be interpreted so as to produce the same result in every case, however desirable that might ideally be, in order to avoid a multiplicity of disputes. So to hold would L be to defy the principle that every document must be properly construed according to its own terms and not according to the terms of some other document. Questions of construction of rent revision clauses are not to be resolved by asking, " Is this case like that case?" They are to be resolved by asking the question I have just stated, and answering it by the application to the particular clause of the proper principles of construction.

I next turn to consider what those proper principles are. Mr Francis's argument for the appellant had the attraction of simplicity. Section 41 of the Law of Property Act 1925 provides: " Stipulations in a contract, as to time or otherwise, which according to rules of equity are not deemed to be or to have become of the essence of the contract, are also construed and have effect at law in accordance with the same rules." The present lease and underlease were contracts to which that section applied, and no ground was shown for departing from the rule to which the section gives statutory effect. As is well known, this section, which re-enacted section 25 (7) of the Judicature Act 1873, indeed gives statutory effect to the previous rules of equity which modified the rigidity of the old common law approach to the construction of particular classes of stipulation, especially as to

A time, in contracts. Mr Balcombe referred us to the history of these rules of equity, which now have had statutory force for over 100 years, as related in *Fry on Specific Performance* 6th ed (1921), and to some of the relevant passages to which my Lord has already referred. I would add references to paras 1071 to 1073 inclusive. In para 1075, which my Lord read, the learned author says:

> Time is originally of the essence of the contract, in the view of a court of equity, whenever it appears to have been part of the real intention of the parties that it should be so, and not to have been inserted as a merely formal part of the contract.

B Thus the rules of equity were a means of imputing to the parties a more restrictive intention regarding the interpretation of the language they had used in their stipulations as to time than the rigid and more literal interpretation of the common law courts would have accorded to those stipulations. But it was never a means of according to the language which the parties chose a different meaning from that which they were to be taken to have clearly intended in any particular case. The well-known exceptions to the rule of equity will be found listed in a chapter entitled "Specific Performance" in the present (27th: 1973) edition of *Snell on Equity* at pp 595 and 596. These include "mercantile contracts" and "contracts for the sale of leaseholds." This shows that the application of the rules of equity was by no means universal even where the subject-matter of the contract was land. The point is well made in *Halsbury's Laws of England* 4th ed vol 9 para 482 in a sentence which reads: "Broadly speaking, time will be considered of the essence in 'mercantile' contracts and in other cases where the nature of the contract or of the subject-matter or the circumstances of the case require precise compliance." In the well-known line of cases on mercantile contracts of which *Bowes v Shand* (1877) 2 App Cas 455 and *Reuter v Sala* (1879) 4 CPD 239 are among the most famous, parties to mercantile contracts have almost always been held strictly to their stipulations as to time, though even in mercantile contracts it is possible to conceive of stipulations as to time which would not necessarily be construed as of the essence.

C

D

In the end, therefore, one returns to the question what intention did these parties to these leases have, or what intention must be imputed to them, due regard being paid both to the language which they used and of course to section 41 of the Law of Property Act 1925. I would refer in this connection to a single passage in the speech of the first Lord Parker of Waddington in *Stickney v Keeble* [1915] AC 386 at 417, though it has to be remembered that this was in the context of a case involving the sale and purchase of land: "The section"—the section to which Lord Parker was referring was the former section 25 (7) of the Judicature Act 1873—"cannot in my opinion mean that the rules as to time laid down by courts of equity in certain cases, for certain purposes, and under certain circumstances only, shall be applied generally and without inquiry whether the particular case, purpose, or circumstances are such that equity would have applied the rules." A contract—and a lease is a contract which contains mutual obligations as well as creating an estate in land—does not automatically cease to be a commercial or mercantile contract because the subject-matter is land and not goods. Commerce today is not restricted to the buying, selling, financing and insuring of commodities, however much that may have been largely the case 100 years ago. This court, in *C H Bailey Ltd v Memorial Enterprises Ltd*, to which my Lord has already referred ([1974] 1 WLR 728), stressed the commercial nature of transactions involving long leases of land. At page 733 Megaw LJ said: "As I understood it, counsel for both the landlords and the tenants in this court accepted that this is a commercial document between commercial parties which ought to be construed as far as possible to give effect to commercial good sense"; and at page 735, in a passage which my Lord has already read but I venture to repeat, Sir Eric Sachs said: "The objective of the courts in a case relating to office leases is naturally to determine the intended commercial effect of the particular agreement reached between the parties. In this respect a lease is no less a contract relating to the use of premises than an agreement in relation to the supply of furniture for those premises is also a contract. It follows to my mind that the courts should in this class of case avoid resort, so far as practicable, to any of the highly technical points that stem from the intricacies of the ancient law of landlord and tenant." In his judgment, the Master of the Rolls had said at page 732: "The time and manner of the payment is to be ascertained according to the true construction of the contract, and not by reference to out-dated relics of medieval law."

E

F

G

H

Let me suppose the existence of two companies in a modern group of companies. One deals in land and the other in commodities. The former as lessor enters into a 21-year lease with a rent review clause providing for rent review at stated intervals and containing certain stipulations as to time. The latter as seller enters into a long-term commodity supply contract with a buyer, with a price review clause providing for price review at stated intervals in identical terms. When the time comes for such reviews of rent in the case of one company and of price in the case of the other, each company fails timeously to comply with those stipulations as to time. On the authorities, it can hardly be doubted that the second company would be held strictly to the terms of the bargain as to the time within which the price review clause must be invoked. Is the first company's contract to receive a different construction merely because it is a lease and not a purchase and sale contract? Such a result cannot, I think, be correct in principle.

J

There has been a good deal of discussion in the cases as to the nature of a rent review clause, whether it is for the landlord's benefit or the tenant's, whether it is truly an option or not, and the like. I respectfully agree with Templeman J's comment in the *Mount Charlotte* case that "the analysis of the option rent review clauses is a triumph for theory over realism." Questions of construction should be approached with realism and not resolved by over-analysis on a purely theoretical basis. The reality of such clauses is that the landlord insists on obtaining an upward review of rent at stated intervals at the lessee's expense and that unless the lessee agrees to that before the lease is executed, the lessee will not obtain the lease. I find it as difficult as did Templeman J to accept as realistic the concept that the lessee is giving the landlord an option and thus conferring a benefit on him. True in some cases the clause may be expressed as an option, as it was in *Samuel Properties (Developments) Ltd v Hayek* [1972] 1 WLR 1064 and 1296; and where it is so expressed, that is a very strong reason for giving the stipulations as to time the literal meaning which this court and indeed Whitford J did in that case. But that is not to say that unless the word "option" is used, or the language used in substance creates an option, the lessor is not to be held strictly to the strict performance of the stipulations as to time. It must depend in each case upon what, judging from the language used, the parties intended or must be presumed to have intended.

K

L

There are, as my Lord has already pointed out, strong commercial reasons in these cases for presuming that the parties intended that these stipulations should be strictly adhered to. A lessee faced with the possibility of a large rent increase may not be able to afford it and may in consequence be obliged to dispose of his leasehold interest. He cannot do so unless he knows the rent his assignee or underlessee is going to have to pay. This is especially true since the decision in *C H Bailey Ltd v Memorial Enterprises Ltd* that rent increases under rent review clauses are retrospective in their effect, overruling the earlier decision of Pennycuick V-C in *Re Essoldo (Bingo) Ltd's Underlease* (1971) 23 P &

M

A CR 1 in this respect. Similarly, the lessor may wish to dispose of his interest. He or his advisers cannot properly value that interest unless they know what return on capital their would-be purchaser would receive by way of rent. Further, if time-limits are not to be adhered to strictly, by when during the currency of the period in question must the rent review machinery be put into operation? It has been suggested—subject to questions of waiver or estoppel—that it must be within a reasonable time of the relevant date, and Goff J (as he then was) took this view in *Accuba Ltd v Allied Shoe Repairs Ltd* [1975] 1 WLR 1559. But questions of reasonableness, as that learned judge found in that case, are never easy B to resolve, and doubts must inevitably arise in any given case where latitude is permissible what is reasonable and what is not.

It seems to me, therefore, that there are very strong reasons first for approaching the construction of these rent review clauses as clauses arising under commercial or mercantile contracts in which the parties are likely to have intended the stipulations as to time to be strictly adhered to; secondly, unless authority otherwise requires, for avoiding attaching a particular label to such clauses and then construing them one way or the other by reference to the chosen label; and thirdly, for avoiding being too ready to construe C these clauses in such a way as to free a lessor who *prima facie* has failed to comply with their terms from the ordinary consequences of failure to comply with the condition precedent because of some supposed subjective consideration of hardship. If, as I think, the realistic view is that these clauses are in general clearly for the landlord's benefit, it follows that in general the stipulations as to time are for the lessee's benefit and protection, so that he may know where he is. This accords with Lord Salmon's view in *Stylo Shoes Ltd v Wetherall Bond Street W1 Ltd* (1974) 237 ESTATES GAZETTE 343 at 345, in the passage which my Lord, Buckley LJ, has D already read and which I will not repeat. In this respect I venture to differ from one passage in Graham J's judgment in *Cheapside Land Development Co Ltd v Messels Service Co Ltd,* to which I have already referred. As reported in *The Times,* the learned judge in the middle of the column said: "The authorities led too readily to the conclusion that time was *prima facie* to be regarded as of the essence in respect of all times in all rent review clauses, with only an occasional exception. That seemed to result from regularly placing them in the 'option' basket." If the authorities in future do so lead, I do not think it wrong that they should, and I respectfully agree with the criticism of that passage E which my Lord made a few moments ago in his judgment.

Thus far I have approached the question of construction on principle, with no more than a passing reference to some of the authorities. I now turn to consider whether authority compels a different approach from that which, as I venture to think, both principle and reason would suggest. I do not propose to refer to them all. I begin with *C Richards & Son Ltd v Karenita Ltd* (1971) 221 ESTATES GAZETTE 25, noting that in that case Goulding J disposed of the argument that time was not of the essence by saying that "he had never heard of the equitable doctrine, under which time in certain stipulations was not of the essence, being applied to F any clause of this nature or framed in this kind of way . . . and he could conceive of no stipulation more express to show that the observance of a time-limit was essential." In *Samuel Properties (Developments) Ltd v Hayek,* in a judgment affirmed in this court, Whitford J ([1972] I WLR 1064 at 1069) said:

> Accepting, as I do, that there is an equitable doctrine under which it may well be that in certain circumstances time should not be considered to be of the essence of the contract, the question in the end must still remain as to whether these principles have any application to a provision of this kind, and I have come to the conclusion that they have not.

In that same case, in this court ([1972] 1 WLR 1296) G Russell LJ, with whom Stamp and Edmund Davies LJJ both agreed, said at page 1301:

> One is familiar with the general approach to a time schedule laid down in a contract for sale of land, that the time schedule is taken, *prima facie,* as intended to establish guide lines rather than imperative requirements: and we were taken by counsel for the lessor, and I make no complaint of it, through a review of a number of authorities and he invited us to adopt language quite plainly limited to cases quite different from the present case as being general statements of principle applicable to the present case.

Then, after a reference to *Stickney v Keeble,* the Lord H Justice went on:

> Far too often, in my experience, attempt is made arguendo to apply to a situation words used in a previous judgment when the situation cannot have been in mind of the judge and is one to which the words used were never directed.

Later, at page 1302, after setting out the argument for the appellants in that case, Russell LJ said:

> I am not myself impressed by these arguments: the right or privilege of exacting an additional rent was conferred by the bargain between the parties as an express option which would be effective if a condition precedent was complied with: it could J be equated with an offer by the lessee to pay an increased rent only in certain circumstances which it lay in the power of the lessor unilaterally to bring about, which offer was not accepted in those terms.

I do not read the judgment of this court in that case as applicable only to a case where the lessors' rights were expressed as an option. The fact that that word was used was a compelling reason for treating the time limits (again to quote Russell LJ) as "mandatory and inflexible." The lessors lost; the lessees won. *Kenilworth Industrial Sites Ltd v E C Little & Co Ltd* [1974] 1 WLR 1069, affirmed in this court [1975] 1 WLR 143, was much relied upon by Mr K Francis, for that case was decided in the lessors' favour. Language supporting a distinction between "option" and "obligation" was first used by Megarry J at page 1071 of the report at first instance; but if I may say so, and as indeed my Lord pointed out during the argument, the phrase in the context in which it was used was entirely correctly used in connection with the clause which was then before the court. The lease there in question contained a proviso which, whatever the position might have been without it, clearly showed that the time-limits were not mandatory. Moreover, on the view taken in this court, this was what in argument was called a "gap" case; that is to say, there would have been L no rent payable for the period in question if the time-limits had been strictly adhered to, unless of course an implication that the earlier rent continued payable was made, an implication which this court was not prepared to make in that case. Whether any different view might have been expressed on the question of implication had this court been referred to the earlier but then unreported decision of another division of this court in *Stylo Shoes Ltd v Wetherall Bond Street W1 Ltd,* to which I have already referred, is not material. In the latter case this court felt no difficulty in filling the "gap" by implying such a term, but of course there was there no stipulation akin to the proviso in the *Kenilworth* case. M

I see nothing in the three decisions of this court, two decided in the lessees' favour and one (on very special wording) in the lessors', which compels a different result from that which both principle and reason suggests, or compels the adoption of the strict dichotomy between "option" and "obligation" in these cases and the adoption of one label or the other. It is to be observed that Lord Salmon in the *Stylo* case uses the phrase "option-type." I think he, like Russell LJ in the *Samuel Properties* case, was referring to option or option-type cases by way of analogy rather than for the purpose of exhaustive classification. It remains to consider the decision of Goff J in *Accuba Ltd v Allied Shoe Repairs Ltd* [1975] 1 WLR 1559, decided a few weeks before the *Mount Charlotte* case but not referred to by Templeman J, though it was referred to by Graham J in the *Cheapside* case. Mr Balcombe argued in the latter case, as before us, that the *Accuba v Allied Shoe Repairs* case was wrongly decided. Graham J expressed no view upon that question, but rightly said that a decision upon the construction of one document did not govern the construction of another. The main criticism of the decision in *Accuba v Allied Shoe Repairs* (and I leave aside certain criticisms of detail made by Mr Balcombe) was that the learned judge adopted a hard and fast distinction between "option cases" and "obligation cases" and elevated that phrase of Megarry J, entirely correct in its context, to a rule of law. With profound respect to the present Lord Justice, I think this criticism is justified, though one greatly sympathises with his attempt to rationalise the authorities. I am inclined to think that unless the decision can be supported on other grounds, it must be treated as wrongly decided.

Applying the principles I have endeavoured to state to clause 3 and the schedule in the present case, which my Lord has read in full and which I will not repeat, I have reached the clear conclusion that the stipulations as to time were here (to repeat Russell LJ's phrase) "mandatory and inflexible" and that section 41 of the Law of Property Act has no application to them. I would therefore dismiss this appeal. It will be seen that my reasons for so doing are wider than those of the learned Vice-Chancellor, who decided this case on the narrower ground that the appellants' rights were on the same footing as an option and that therefore Russell LJ's judgment in *Samuel Properties (Developments) Ltd v Hayek* could be directly and immediately applied. If I were wrong in my conclusion as to the true principles to be applied to the construction of the clauses in this agreement, I would in any event, like my Lord, have agreed with the reasoning of the Vice-Chancellor and would have decided this case against the appellants on the narrower ground upon which the case was decided by him. Our decision, if correct, involves that in the future far greater care must be taken over the drafting of these rent revision clauses than in the past, and that as regards those already agreed, landlords must ensure that stipulations as to time are timeously complied with. In future they cannot say that they have not been warned of the likely consequences in many of these cases if they fail so to do. Further, even if it is too optimistic to hope that our decision may avoid some of the litigation which has taken place on this subject in the past, one may at least hope that in future it will not be necessary for counsel laboriously to traverse an ever-increasing number of decisions at first instance in order to invite this court to consider whether "this case" is like "that case."

It only remains to thank counsel for their careful arguments, as well as for the unending supply of photostats of all these decisions, and the respondents' solicitor (if I may say so) for his splendid note of the Vice-Chancellor's judgment, which has amply filled the gap left by the failure of those concerned to see that a shorthand note was taken of the judgment appealed from. I would dismiss the appeal.

BROWNE LJ: I agree that this appeal should be dismissed, for the reasons given by my Lords. I hesitate to add anything, but I think that in deference to Mr Francis's most interesting argument I should state my own reasons, even though they are in substance the same as those of my Lords. If there were no authorities, and no equitable doctrine about time not being of the essence of contracts, I should think this a plain case. I need only refer to the main lease of August 31 1962, the effect of the supplemental lease of September 29 1967 being identical. By clause 2 of the main lease, the demise was for 99 years from the date thereof. Clause 3 provides that: "The lessee shall pay to the corporation during the period from the commencement of this lease"—that is, August 31 1962—"until January 1 1963 a peppercorn (if demanded) and thereafter until the expiration of a period of 10 years from the commencement of this lease"—that is, until August 31 1972—"the yearly rent of £900 and thereafter during the residue of the said term the yearly rent of £1,000 plus any additional rent payable under the provisions contained in the schedule hereto." Pausing there, it seems to me clear so far that if no additional rent becomes payable under the schedule, the rent is £1,000. I understood Mr Francis to accept that if nothing was done under the schedule, the rent would be £1,000; but anyhow, I think this is the position. This is therefore not what was called in the argument a "gap" case; that is, a case in which no rent at all would be payable in respect of the later part of the term unless the rent review provisions could be operated. I cannot accept Mr Francis's submissions that "the decisive clause is the schedule," nor that "the effect of the schedule is not correctly set out in clause 3." The two must of course be read together. Reading them together, this is in my view still not a "gap" case.

My Lord has read the schedule, and I need not read it in full again. Read literally, I think it clearly requires the "current rack rent" from which the actual rent is to be calculated to be ascertained during the year immediately preceding September 1 1972. It provides that: "During the year immediately preceding September 1 1972 . . . the corporation and the lessee shall agree or failing agreement shall determine by arbitration the . . . *then* current rack rent"—that is, during the year preceding September 1 1972—" . . . and one quarter of the sum total so ascertained or £1,000 (whichever is the greater) shall be the rate of rent reserved by this lease in respect of the *then* next succeeding relevant period"; that is, the period beginning September 1 1972. I agree, of course, that the rack rent during the year preceding September 1 1972 could in fact be determined after the end of that year, and that the phrase that the parties "shall determine by arbitration" during that year is not apt, both because the *parties* do not determine by arbitration and because the time when the arbitrator gives his decision is outside their control. But I am prepared to read "shall determine by arbitration" as meaning "shall refer to arbitration for determination." All the practical and business considerations seem to me to support the view that the words of the schedule are intended to have what I think is their natural meaning, namely, that the revised rent is intended to be settled *before* the beginning of the second 10-year period in respect of which it is to be in force. Obviously, the landlord will want to know what his rent is going to be. Equally, the tenant will want to know. There is no break clause in this particular lease, but if the tenant found that the rent was more than he could afford to pay, he might well want to sell the lease, and he obviously could not decide whether to sell, and neither he nor any prospective purchaser could decide at what price, until the rent for the second 10 years was known. Further, the result of the decision of this court in *C H Bailey Ltd v Memorial Enterprises Ltd* [1974] 1 WLR 728 might be that if the revised rent could be determined after September 1 1972, it would relate back to September 1 1972 and so might retrospectively impose a considerable additional liability on the tenant.

Mr Francis accepts that there is some essential time element in these provisions, because he accepts that the new rent must be determined "within a reasonable time" after the specified time. I think that such a construction would produce uncertainty and difficulty about what is a "reasonable time." But Mr Francis submits that there is a general rule of equity applicable to all contracts, and now applicable also at law because of section 41 of the Law of Property Act 1925, that time is not of the essence of any contract unless the case falls within certain recognised exceptions. I should want to hear further argument before I accepted this as a general proposition, but for the purposes of this case I am prepared to assume, without deciding, that it is right. Even so, in my view, rent revision clauses fall within one of the recognised exceptions. I agree on this point with what Goulding J said in *C Richards & Son Ltd v Karenita Ltd* (1971) 221 ESTATES GAZETTE 25, and with what Whitford J said in *Samuel Properties (Developments) Ltd v Hayek* [1972] 1 WLR 1064 at 1069, in the passage which my Lord, Roskill LJ, has already quoted. Mr Francis in argument referred us to *Halsbury's Laws of England* 4th ed vol 9, in the title "Contracts," first to para 481 at p 338, which says this:

> The modern law, in the case of contracts of all types, may be summarised as follows. Time will not be considered to be of the essence unless . . . (2) the nature of the subject-matter of the contract or the surrounding circumstances show that time should be considered to be of the essence.

Then para 482, part of which Roskill LJ has already quoted, goes on:

> Apart from express agreement or notice making time of the essence, the court will require precise compliance with stipulations as to time wherever the circumstances of the case indicate that this would fulfil the intention of the parties.

Then there follows a sentence which my Lord, Roskill LJ has already quoted. My Lord, Buckley LJ, has already quoted from *Fry on Specific Performance,* and indeed Roskill LJ has too. I will not repeat the quotations, but I refer particularly to para 1075, with its reference to "the real intention of the parties," and to para 1081, which my Lord, Buckley LJ, has quoted. For the reasons already given, I think that the practical and business circumstances surrounding a rent review clause will normally show that the intention of the parties is that time is to be of the essence of the contract, so that the court will approach such a clause at least with a leaning towards giving it such a construction. Of course, each case must depend on the construction of the particular clause in question, so that where (for example) there is a "gap," as the court construed the clauses in *Kenilworth* [1974] I WLR 1069, [1975] 1 WLR 143, and *Accuba* [1975] 1 WLR 1559, the court may construe the clause as not making time of the essence. It is not necessary in this case to consider or try to resolve the possible inconsistency between the decisions of this court in *Stylo Shoes* (1974) 237 ESTATES GAZETTE 343 and *Kenilworth* [1975] 1 WLR 143 as to the existence of a "gap" in those cases.

As to the suggested dichotomy between clauses of option and clauses of obligation, I agree with my Lords that in the present context this is a false dichotomy, for the reasons they have given. In other contexts it may of course be essential to decide whether an agreement is an option, giving a right to one party which he can choose whether or not to exercise, or whether it imposes an obligation on one or both parties, but I do not think this is an exhaustive test in relation to rent review clauses. For example, in *Imperial Life Assurance Company of Canada v Derwent Publications* (1972) 227 ESTATES GAZETTE 2241, the clause provided that the rent should be "automatically revised." In *Samuel Properties (Developments) Ltd v Hayek* [1972] 1 WLR 1064 and 1296, to which my Lord has referred, the clause was expressed as an option, with a condition precedent as to notice, and this court put its judgment on that basis, which was enough for the decision of that particular case. In my view, that case does not decide as a matter of general principle that time cannot be of the essence of a rent review clause unless the clause is expressly or in substance an option to the landlord. Nor do I think that the *Kenilworth* case lays down any such general principle. Megarry J and this court took the view that on the construction of the clause in that case there was a "gap," and it seems to me that the proviso to that clause in itself made it impossible to say that time was of the essence. I agree with my Lord, Buckley LJ, that *C H Bailey Ltd v Memorial Enterprises Ltd* does not affect the present case. To say that the test is whether the clause makes time of the essence or whether it is mere machinery is only to re-state the problem to be solved. I think that the decision in *Accuba* can be supported on the ground that it was a "gap" case, but so far as it goes further and relies on the option/obligation dichotomy, I do not agree with it.

It all comes back to the construction of the particular clause so as to ascertain "the real intention of the parties." In the present case, on the construction of clause 3 and the schedule in the light of the surrounding circumstances, time was in my judgment of the essence of this rent review clause and the decision of the Vice-Chancellor was right. If, contrary to my view, the option/obligation dichotomy is the right test, I should, like my Lord, be prepared to hold that the Vice-Chancellor was right in holding that this case fell by analogy on the "option" side of the line, for the reasons he gives. As I have said, I would dismiss this appeal.

The appeal was dismissed with costs. Leave to appeal to the House of Lords was granted.

QUEEN'S BENCH DIVISION

March 17 1976

(Before Mr DOUGLAS FRANK QC, sitting as a deputy judge of the division)

RANK CITY WALL LTD v GALLAHER & JOHNSTON ALLEN LTD

Estates Gazette May 29 1976

(1976) 238 EG 648

Review of rent of business premises—Operation of clause in lease affected by counter-inflation legislation—Letter from lessors' agents to RICS setting out position treated as "application" within meaning of clause—Further issue on date of letter—Delivery on September 25 1973 held "not less than six months before March 25 1974"

This was a summons by Rank City Wall Ltd, of London SW3, for a declaration that a letter of September 25 1973 operated as a valid application for appointment of a valuer to determine the rent payable from March 26 1974 under an underlease granted to predecessors in title of the respondents, Gallaher & Johnston Allen Ltd, of 3-9 Heddon Street, London W1, by predecessors in title of the applicants on December 29 1967.

Mr A D Dinkin (instructed by Richards, Butler & Co) appeared for the applicants, and Mr A G Steinfeld (instructed by Fortescue & Sons) represented the respondents.

Giving judgment, MR FRANK said that the lease in question was for a term of 14 years from March 25 1967, and provided that the rent to March 25 1974 should be £2,250 and from March 26 1974 for the residue of the term such amount (not being less than £2,250) as might be agreed in writing between the lessors and the lessees before September 29 1973, "or in the absence of such agreement as may be determined by an expert valuer to be nominated by the president for the time being of the Royal Institution of Chartered Surveyors on the application of the lessors made not more than 12 months nor less than six months before March 25 1974." No agreement for the second seven years' rent was reached before September 29 1973, nor had any negotiations taken place, but on September 25 1973 Goddard & Smith, the plaintiff's managing agents, delivered a letter addressed to the secretary of the Royal Institution of Chartered Surveyors in the terms that it was their opinion that the present-day open market rental value of the accommodation was £13,250 per annum exclusive and it was this figure which their clients would have wished to have substituted within the document with effect from the next March 25; nevertheless, the increase in rental would be affected by the provisions of the Counter-Inflation (Business Rents) Order, and their clients would not be seeking to collect any increase over the amount permitted by current legislation. On September 27 1973 an administrative secretary of the institution replied saying that the application had been put to one side until they heard further, but should it be wished to proceed with the case the letter stood as the original application.

The first question was whether the letter of September 25 was an application within the meaning of the rent review clause. The defendants took three points. First it was said that the letter by its terms did not apply for the appointment of a valuer, but rather summarised the effect of the rent review clause, and instead of asking for the appointment of a valuer contained an express request not to do so. That argument was unacceptable. It seemed that there could only be one reason for the letter, to preserve the lessors' position under the review clause; that was the purpose stated in the letter. That position could only be preserved by an application for an appointment of a valuer, and the fact that it was suggested that no appointment should be made for the time being seemed not to invalidate the application as such, but rather to make a commonsense suggestion. He (Mr Frank) had asked Mr Steinfeld what the position would have been if the application had been drafted in formal terms by counsel, but had had sent with it a letter along the lines of the last paragraph. Counsel had answered that in his submission the letter would cancel the application, but he (his Lordship) found that position untenable. Secondly, Mr Steinfeld argued that the application should have been brought to the attention of the right person, that is to say, it should have been addressed to the president of the RICS. This argument seemed to be technical in the extreme, and the answer to it was that technically the application was by the terms of the second paragraph brought to the attention of the president, on the assumption that the secretary, having read that paragraph, then carried out his assumed duty of passing the application on to the president. In any event, in the absence of evidence to the contrary, Goddard & Smith were in his (Mr Frank's) judgment entitled to assume that the secretary acted as agent for the president in such matters, and the partners, being members of the institution, would be well aware of the realities of the organisation. Thirdly, it was said that the application should have stated the name of the landlords and of the present tenants, but he (his Lordship) found nothing in the clause concerned which so required.

On the remaining issue, whether the notice was served not less than six months before March 25 1974, Mr Steinfeld argued that six clear months must have elapsed between the date and the date of service and that accordingly September 24 1973 was the last date for service. For the plaintiffs, Mr Dinkin pointed out that all the clause required was six months' notice, not more than six months', and that in computing that six months one should include the day of service but exclude the day by which the notice had to be served. In the absence of authority, he (Mr Frank) would have thought that according to ordinary understanding a period of six months beginning on September 25 ended on the following March 24, and that therefore the earlier date was not less than six months before the latter. However, there was authority, and it seemed to confirm the view just expressed. It was *Schnabel v Allard* [1967] 1 QB 627, where it was held that a notice to quit served on March 4 requiring the tenant to give up possession on April 1, the Friday four weeks from March 4, was valid both at common law and under the Rent Act 1957. Mr Steinfeld contended that that case did not go to the point, because the notice to quit would not have taken effect until midnight on April 1 and that therefore there were in fact 28 clear days. However, there was nothing in the judgments to support that proposition, and the Court of Appeal had held that *Thompson v Stimpson* [1961] 1 QB 195 had been wrongly decided. In that case the Divisional Court had decided that " not less than four weeks " necessitated that there should be four clear weeks exclusive both of the day of service and of the day on which the premises were to be vacated. The Court of Appeal held that the proper construction was that " not less than four weeks " meant inclusive of the day of service and exclusive of the day of expiry. Accordingly, he (Mr Frank) concluded that the application in the present case was served in time. Mr Dinkin argued, in the alternative, that time was not of the essence, but under the circumstances it was not necessary to decide that, and there would be judgment for the plaintiffs in the terms of the declaration sought.

NEGLIGENCE

A

QUEEN'S BENCH DIVISION

January 13 1975

(Before Mr Justice BRISTOW)

BOURNE v McEVOY TIMBER PRESERVATION LTD

Estates Gazette February 14 1976

(1975) 237 EG 496

Report on timber rot, including estimate for repairs, sought by vendor of property from timber preservation firm—" If estimate accepted, job done and paid for, otherwise firm got nothing "—Work ordered by purchaser, with vendor's agreement—Firm paid by purchaser, and purchaser's name written on 20-year guarantee—Firm held to owe purchaser a duty of care, but not, on the facts, to have been in breach of it—Observations on measure of damages in such cases

B

This was a claim by Mr Michael Alan Bourne and his wife, Mrs Sandra Joyce Bourne, of 11 Durham Road, Lower Edmonton, London N9, against McEvoy Timber Preservation Ltd, of Chadwin Road, London E13, for damages for negligence.

Mr G B W Nurse (instructed by Seifert, Sedley & Co) appeared for the plaintiffs, and Mr Patrick Phillips (instructed by L Bingham & Co) represented the defendants.

C

Giving judgment, BRISTOW J said: In May 1970, Mr Michael Bourne, the plaintiff in the action, who was then engaged to be married, was interested in buying no 11 Durham Road, London N9. The house was owned by a property company, London City & Westcliff Properties Ltd (" LC&W"); it was empty, and was in the process of being done up—or " tarted up," as a witness put it—with a view to sale. The price was £4,250, and Mr Bourne needed a 90 per cent mortgage, so he got in touch with his solicitors and they started negotiations with the Co-operative Permanent Building Society. On May 19, following a visit to the site by the building society's surveyor, the society wrote in these terms:

D

Dear Sir and Madam,

With reference to your recent application I have now received a report from the society's surveyor, and whilst the advance of £3,825 can be made, this will be subject to a sum of £150 being retained until the following work has been completed to the satisfaction of the society's surveyor:

1. Remedy dampness to back addition walls.
2. Obtain specialist report on timber rot and woodworm infestation and carry out their recommendations.
3. Provide 9 in x 6 in metal air gratings to back walls for underfloor ventilation.
4. Re-point open joints of brickwork and back wall parapet and chimney-back.
5. Overhaul and repair roof slating and felt valley gutters.
6. Discharge bathroom waste pipe directly over gulley.

E

No doubt you will let me know whether you wish to proceed on this basis.

On June 3 Mr Bourne's solicitors communicated the contents of that letter to LC & W's solicitors. They proposed that LC & W should share the cost of putting things right, which they estimated at £400-£500. Meanwhile LC & W instructed McEvoy Timber Preservation Ltd (the defendants) and K Fearnley Ltd, both of them companies specialising in the remedial treatment of timber affected by dry rot or beetles, to report and estimate for the eradication of woodworm, dry rot, etc. The defendants reported on June 11 1970, and their estimate was £128. Fearnleys reported on June 18: their estimate was £178. If your estimate was accepted, you got the job and were paid for the job; if not, you got nothing. On June 24 LC & W's solicitors wrote to Mr Bourne's solicitors in these terms:

F

G

Dear Sirs,

We write further to our letter of June 8 and our recent telephone conversation and advise that we have taken our clients' verbal instructions. We are informed that our clients have obtained an estimate for eradication of the woodworm with a 20-year guarantee in the sum of £128, and it is our clients' proposal that you should pay this sum and our clients will carry out the other necessary repair work.

Mr Bourne's solicitors communicated the effect of this letter to Mr Bourne. There is no evidence that he was shown the defendants' report. The defendants' evidence was that in the ordinary way LC & W themselves contracted with the defendants for the work to be done, paid the bill and were given the 20-year guarantee made out in blank in respect of the work carried out. The defendants regarded this guarantee as attaching to the house. In this case, however, Mr Bourne, with the agreement of LC & W, himself instructed the defendants to carry out the necessary work. On August 4 he completed and signed the acceptance of the estimate which had been enclosed with the report sent on June 11 to LC & W by the defendants. Mr Bourne paid the bill, and was named as the client on the guarantee. Until Mr Bourne instructed them to do the work, the defendants did not know of his existence, though they knew that their inspection and report must be in connection with a proposed sale of the house. They knew that a building society, before lending at any rate the full amount on mortgage, might want to be satisfied of the condition of the timbers, might want to know there was a 20-year guarantee and might well see the report. They did not know whether the prospective buyer would see and rely on the report or not, though, of course, if instructed by an occupier to inspect and report, they knew that that occupier would rely on them.

H

J

By August 19 the work was done and the guarantee was given. In early September the building society's surveyor, having had a sight of the guarantee, released the £150 retention money. The sale to Mr Bourne was completed and he went into occupation. In July 1971 the lavatory cistern in the bathroom in the back addition on the first floor fell from the wall. The cistern bracket screws pulled away from the wall because the wooden pad in which they had been fixed was in an advanced state of dry rot. The lead supply pipe to the cistern was bent, the ball valve stuck open and water under full mains pressure came into the house until Mr Somers, a local emergency service plumber who was called in, was able to retrieve the situation. Mr Somers at once

K

A recognised the dry rot in the pad and on the wall behind. On his advice Mr Bourne called in Mr Somers' old friend Mr Smith, of Timber Care Ltd, who are in the same line of business as the defendants. Mr Bourne also got in touch with the defendants and asked them if they regarded themselves as responsible and would put things right under the guarantee. Mr Smith did such opening up as he thought necessary, and reported on August 31 1971. He found dry rot in the flat roof above the bathroom, the ceiling joists, the lintel box frames, the skirtingboards and the flooring timbers. The brickwork of the flank and partition walls was infected, and so were the rear wall and attached timbers of B the first-floor centre room on which the back addition abuts.

The defendants also inspected after Mr Smith, taking advantage of the opening up which he had done. The only real difference between their report of September 3 and Mr Smith's report was that they did not find that the bathroom and middle room flooring had been attacked, though conditions were such that they thought an attack was likely. From the condition of the wall where the cistern had been, and the fact that a replacement cistern had by then been installed, they concluded that it was a leaky cistern which had been the cause of the outbreak. Neither they, Mr Smith nor Mr Somers discovered a further outbreak of dry rot at C the other end of the bathroom and over the passage into the bathroom where there is a trap door to the flat roof, or an extensive area of wet rot over the bedroom beyond the bathroom in the back addition. These did not come to light until Mr Somers was stripping the roof of the back addition in the process of dealing with the first outbreak. The cost of putting right the first outbreak is agreed to have been £455.67. The additional cost of making good attributable to the further dry rot outbreak was £248.81. No claim is made on Mr Bourne's behalf in respect of the cost involved in the wet rot trouble, approximately £150. In this action, Mr D Bourne claims that in making their inspection and report dated June 11 1970, on the instructions of LC & W, the defendants owed a duty of care to him, and that if they had performed it, they would have found the dry rot which was found in 1971. He claims the £704.48 cost of making good the dry rot trouble, plus unquantified and unparticularised general damage for the disturbance and inconvenience caused by the remedial work. The defendants say that in the circumstances of this case they owed no duty to Mr Bourne: that if they did owe a duty to him, they performed it, in that they made a careful examination of the house and did not find the dry rot, because there were no visible signs of E it then, or indeed until the cistern pulled away from the wall, and there were no suspicious circumstances which could have led a careful specialist inspector to do more than make a visual examination.

For Mr Bourne, Mr Nurse argued that he was the defendants' "neighbour" in the sense defined by Lord Atkin in *Donoghue v Stevenson* [1932] AC 562. He was a person so closely and directly affected by their act or omission, that they ought to have had him in contemplation as being so affected when they were directing their minds to their inspection or report. Mr Nurse relied primarily on *Dutton v Bognor Regis Urban District Council* [1972] 1 QB 373. F There the Court of Appeal unanimously decided that the council was liable to the buyer of a house built under their building by-laws made under the Public Health Act 1936 for the negligence of their surveyor in approving the foundations, which had in fact been built on the site of an old rubbish tip. As appears, for example, from the judgment of the Master of the Rolls at pp 391-2, and of Sachs LJ at p 406, the *ratio decidendi* was based upon the control over building work entrusted by Parliament to the local authority, which the court held carried with it a duty of care to subsequent purchasers and occupiers of the house, the very people whom Parliament intended to protect. Accordingly the decision in that case, and the review of the law geared to the G question in issue in that case, whether there was a duty situation in those circumstances, is of little help on the question whether there is a duty situation between the plaintiffs and defendants in the circumstances of this case. Nor do I get much help from *Clayton v Woodman (Builders) Ltd* [1962] 2 QB 533, also relied on by Mr Nurse. There an architect, who knew that the workmen would act on what he said, told a workman to cut a chase in a gable when the architect should have known it was unsafe for the man to do so without previously shoring it. The architect was held to owe a duty of care to the workman. The architect and workman were in direct, though not contractual, contact, H and the duty situation is easy to see.

On behalf of the defendants, Mr Phillips submits that the guidelines by which you can judge whether a "duty situation" exists for people in the position of the defendants were really set in the last century by two cases which were considered in *Hedley Byrne v Heller & Partners Ltd* [1964] AC 465. In *Le Lievre v Gould* [1893] 1 QB 491, mortgagees of a builder lent him money on the faith of certificates given by a surveyor employed by the builder and not in contractual relations with the mortgagees. The Court of Appeal held that the surveyor owed no duy to the mortgagees to exercise care in giving his certificates. In *Hedley* J *Byrne*'s case the House of Lords held that the basis of the decision in *Le Lievre*'s case, that there was no contract between the surveyor and the mortgagee, was wrong, but the decision itself might well be right, because the surveyor did not know anything about the mortgagees or the terms of the mortgage and his certificates were shown to the mortgagees without his authority (see Lord Reid at p 488, Lord Devlin at p 519 and Lord Pearce at p 535). In *Cann v Willson* (1888) 39 Ch D 39, valuers instructed by an intending mortgagor sent their valuation direct to the mortgagee's solicitors and the mortgagee acted on it, lent on mortgage K and suffered loss (when the mortgagor defaulted) because the valuation was careless. Chitty J held that the valuer owed a duty to the mortgagee, and the House of Lords, in *Hedley Byrne*'s case, decided that he had been right. In *Hedley Byrne*'s case itself, it is again easy to see the duty situation between the plaintiffs and the merchant bankers who were asked for the reference. They knew of the plaintiff, they knew what he wanted the reference for, and they knew that it was likely that the plaintiff would act upon it.

Applying the principle illustrated by the authorities to the facts of this case, in my judgment I have to test whether there was sufficient proximity between the plaintiff and the L defendants to give rise to a duty situation by asking myself this question: at the time the defendants made their inspection and reported to LC & W, their principals, did they know, or ought they to have known, that the purchaser of the house might well be affected in the decisions which he took by the contents of their report? In my judgment the answer to that question on the evidence in this case must be "yes." The defendants knew the house was being "tarted up" for sale. The defendants knew their report might go to the mortgagees. The fact that it might go to mortgagees meant that their findings must affect the value people would put on the house. What was the right value to put upon the M house must affect LC & W, the sellers, the mortgagees (if any), and most probably the third person concerned in the sale transaction, the buyer. The defendants regarded the buyer as the beneficiary of their work, if they got the job LC & W would fill in the buyer's name on the 20-year guarantee of their work, and the defendants would honour the guarantee in the hands of the occupiers from time to time of the house. In fact, before the sale contract and mortgage negotiations were concluded on July 30, Mr Bourne knew, though he had not seen the report, that as a result of the specialist examination to report on timber rot and woodworm, £128 was going to have to be paid for woodworm eradication to satisfy the building society conditions for the release of the last £150 of the loan.

Mr Bourne also claims that the defendants were obliged by the terms of their 20-year guarantee to make good the effect of any dry rot which ought as a result of a careful survey to have been disclosed in the survey report. Mr Nurse did not argue this head of the claim, but made it clear that he intends to do so on appeal, if so advised. It is enough to say that, in my judgment, on its true construction the guarantee extends only to subsequent trouble in timber treated as a result of the survey and not to omissions in the survey itself.

The duty of care being in my judgment established, was it carried out, or was it broken? The 1970 survey was carried out by Mr Peter McEvoy, who gave evidence about it. He had really no recollection of that survey, other than what appeared in his notes, transcribed in his report. He said that in houses of this age and type the trouble is usually with woodworm. Next comes wet rot, of which no complaint is made in this case. In only two per cent of the inspections he has carried out has he found dry rot. He did not note any damp places outside leading him to suspect trouble inside. He did not note that the blank wall of the back addition was made of place brick, which is not impervious to wet, and had no damp course at the top to stop water percolating down inside. He did not note that the flat roof of the back addition slopes in towards the main building, so involving a risk of damp getting into the main wall and the flat roof timbers if there was any defect in the waterproofing of the flat roof to the main wall joint, lying as it does beneath the pitched roof gutter level. He did not note the trap door in the roof of the back addition, always a danger point to the ingress of water. He made no check by probing the timbers which would have been at risk if damp had come in at the points which I have mentioned. He noted certain symptoms showing where conditions needed checking, and checked them. In no case did he find signs of dry rot or of a situation so conducive to dry rot as to be worth noting in his report. He says, and I accept, that there were no visible indications of the attack of dry rot which, in view of the situation in 1971, he accepts may well have already started by 1970. In short, Mr Peter McEvoy detected no signs of dry rot in 1970 and remains convinced that what was found in 1971 was not evident at his inspection in 1970.

Dr W P K Findlay, the author of several books and pamphlets on dry rot and an expert in this field, inspected the house in October 1974. Having regard to the extent of the outbreak found in 1971, he was of opinion that if Mr Peter McEvoy had prodded the exposed bathroom window-frame woodwork in 1970 it would have already been soft as a result of dry rot. He thought that the damp condition of the back addition wall ought to have made a surveyor look at the joists where they abutted on that wall, and that if he had, there would have been something there to see. He thought that had Mr Peter McEvoy prodded the pad holding the cistern, it would already have been soft. He himself would have tested the window-frames by prodding as a matter of routine. Mr Smith, of Timbercare Ltd, 18 years in the business and regarded by Mr Somers as a very keen surveyor and a reliable, trained observer, said that when he carries out a survey all he can do is to report on his suspicions if there is any evidence which causes him to form suspicions, for example the presence of damp in a wall, distorted timbers, blistered paintwork and certain other visible signs. When he was called in, in 1971, what he saw behind the site of the cistern pad, and the state of that pad itself, strongly suggested that the neighbouring timbers would be infected and led him to do the opening up that he did. But even in 1971, there was no visible indication of dry rot on what could be seen of the window-frame.

The problem for those in the position of Mr Peter McEvoy or Mr Smith is formulated succinctly in technical note no 44 of the Ministry of Technology Forest Products Research Laboratory, written by Dr Findlay. At p 4 he says:

> Since dry rot usually starts where timber is in contact with damp brickwork it often remains concealed from view until it has reached an advanced stage, and sometimes the collapse of a floor board is the first sign that there is anything wrong. When inspecting a house for dry rot, look out for any irregularity or waviness in the surface of panelling, skirting or window linings and test the suspected areas with a sharp tool.

There is no evidence of any such irregularity or waviness even in 1971. There is no evidence that when Mr Peter McEvoy and Fearnleys did their inspections in the summer of 1970 the place brick flank wall of the back addition or the joint between the back addition flat roof and the back wall of the house proper were damp. Even when he opened up as far as he thought necessary in 1971, Mr Smith did not spot the second outbreak of dry rot which was ultimately found, along with the wet rot in the back addition roof, by Mr Somers in the process of doing the remedial work. Neither Mr Peter McEvoy nor Fearnleys in 1970, nor Mr Terence McEvoy nor Mr Smith in 1971, spotted the trap door outbreak of dry rot. There was nothing to see which would have caused suspicion that it existed. It is for the plaintiffs to satisfy me on the balance of probabilities that Mr Peter McEvoy, when he made his inspection in the summer of 1970, did not exercise reasonable care. On the evidence called before me, I am not satisfied of this, and so the plaintiffs' claim in this action fails.

It is therefore not necessary to deal with the question of damages, but it may be helpful to the parties if I express my conclusion on what the true measure of damages would have been, if I had been satisfied that the defendants had been negligent. If Mr Bourne had known that the house was infected by dry rot to a substantial extent in the areas in which in 1971 the making good cost £704.48, he would have had the option of going on with the purchase of the house or of not buying it. He fairly said in evidence that he could not say what he would have done. If he had chosen to go on, it is reasonable to assume that something would have been knocked off the price by agreement. This is what happened in respect of the work required by the building society surveyor other than that called for in the defendants' report. How much would have been knocked off must be a matter of speculation, but the remedial work, with all its attendant inconvenience to the plaintiffs, would have had to be done anyway. In view of this, in my judgment, the plaintiffs have no claim in this case for general damages for the inconvenience arising out of the remedial work, and the whole cost of the dry rot remedial work cannot be the correct measure of damages. The remedial work arising from the wet rot, in respect of which no claim is made, must of course have contributed to the plaintiffs' inconvenience. In my judgment, what the plaintiffs would have been entitled to get by way of compensation is the difference in value in 1970 between 11 Durham Road without any dry rot, that is, in a condition indicated by the report calling for action over woodworm only following upon specialist examination for woodworm and rot, and 11 Durham Road with dry rot on a scale leading to what was found in 1971.

Authority for this approach is to be found in the decision of the Court of Appeal in *Philips v Ward* [1956] 1 All ER 874. Mr Nurse relied upon the judgment of Sachs LJ in *Dutton's* case at p 408 as authority for the proposition that the cost of making good the defects was the proper measure of damages, plus something for general inconvenience during repairs. *Philips v Ward* was not cited to the Court of Appeal in *Dutton's* case, where the argument was primarily con-

A cerned with the question of "duty situation" or not. It may be that in the peculiar circumstances of the duty situation in *Dutton's* case a different measure of damages was there appropriate. In my judgment, I am bound by the decision in *Philips v Ward,* and the fact that here the "duty situation" arises in the absence of contract does not affect the measure of damages. No evidence was led of the difference in value in 1970 of 11 Durham Road with and without dry rot, though Mr Phillips, for the defendants, said in argument that, if liable, his clients would be content with an award of £455.67, the cost of making good the dry rot found by Mr Smith when the cistern came away.

B There will be judgment for the defendants.

COURT OF APPEAL

November 27 1975

(Before Lord DENNING MR and Lord Justice Geoffrey LANE)

CAMPBELL v EDWARDS

Estates Gazette February 28 1976

(1975) 237 EG 647

C **Valuer instructed by landlord and tenant jointly to assess surrender value of lease—Law transformed by Arenson appeal in House of Lords—Parties bound by valuer's figure even though valuer be negligent—Aggrieved party's remedy is action against valuer—Landlord's proceedings against tenant to upset valuation held to disclose no cause of action**

This was an appeal by Mr John Colin Campbell, of Hertford Street, London W1, from an order of May J dated March 26 1975 confirming the decision of Master Warren of November 14 1974 and directing that the appellant's statement of claim in an action brought by him against the D respondent, Miss Irene Donalda Edwards, be struck out as disclosing no cause of action.

Mr A J Bateson QC and Mr S R Silber (instructed by Roney Vincent & Co) appeared for the appellant, and Mr J M Chadwick (instructed by Jaques & Co.) represented the respondent.

Giving judgment, LORD DENNING said: We are concerned here with a house in Trebeck Street in the very centre of Mayfair. It was built about 200 years ago as a private house, but is now converted to other uses. On the ground floor there is a restaurant. On the first floor there are offices. On E the second floor there is a residential flat. It has its own entrance from the street, and has three rooms, kitchen and bathroom. On June 28 1973 the landlords let this second-floor flat to Miss Edwards for seven years and two months free of rent. The tenant was liable to do repairs and there was a prohibition against underletting. The lease contained a special provision that if the tenant desired to assign the premises she had first to offer to surrender them to the landlord, and the price was to be fixed by a surveyor to be agreed upon by the parties. It is clause 11 (b), which says in the material part:

> F If the tenant desires to assign the whole of the premises as aforesaid she shall first by notice in writing to the landlord offer to surrender this lease . . . at a price fixed by a chartered surveyor to be agreed by the landlord and the tenant or failing agreement to be nominated by the President for the time being of the Royal Institution of Chartered Surveyors and the landlord may within 21 days of the service of such notice upon him accept such offer the acceptance to be in writing.

Six months after the lease was made Miss Edwards desired to assign the lease. Under that clause she had to offer to surrender it to the landlord. By a letter dated January 17 1974 her solicitors offered to surrender it "at a price to be fixed in the manner provided by the lease." By a letter of January 24 1974 the landlord accepted the offer. Some names G of surveyors were suggested but these were not agreed. Eventually, however, both sides did agree on a very eminent firm of surveyors, Chestertons, a firm who were recognised by both to be quite unimpeachable. On March 6 1974 the solicitors for each side signed jointly and sent a letter to Chestertons. It referred to the lease and said: "It has been agreed between the parties to instruct yourselves to assess the proper price for the surrender in accordance with the provisions of the lease." Chestertons duly inspected the premises. On March 21 they wrote this important letter, assessing the price at £10,000:

> H We confirm with thanks receipt of your letter of March 6 1974, giving your two firms' instructions to us to inspect the above premises in order to assess the proper price to be paid by the landlord to the tenant for the surrender of the residue of the term of the lease dated June 27 1973. . . . We have read the lease carefully and have taken into account, in assessing the price to be paid, those matters set out in clause 3 (ii) (c). . . . After carefully considering the matter, we are of the opinion that the proper price for the landlord to pay to the tenant for the surrender of the lease as at March 25 1974 is the sum of £10,000 (ten thousand pounds).

Following that assessment, the tenant went out of possession and the landlord went into possession of the flat. J There was some dispute about fixtures and fittings but undoubtedly there was a surrender by operation of law. The landlord then turned round and disputed the assessment of the price. He got new valuations for the value of the remainder of the lease from two other surveyors. Mann Winkworth made a valuation dated April 18 1974, putting the value as £3,500. Hinton & Co, on June 18 1974, put it at £1,250. I must say that, even to a layman, that seems a remarkably low assessment, especially when you remember that there were over six years to run of this lease and that no rent was payable under it. Having got those two new K valuations the landlord challenged Chestertons' valuation of £10,000. On June 17 1974 he issued a writ claiming that he was not bound by it. In the statement of claim he said: "The valuation was incorrect in that the true surrender value of the lease at the specified date was under £4,000. In the circumstances, Chestertons must be presumed to have assessed the price in an incorrect manner and their valuation is therefore vitiated." He claimed "(a) a declaration that he is not bound by the valuation," and (b) "an order that the defendant (Miss Edwards) concur with the plaintiff in obtaining a new valuation." Miss Edwards, by her solicitors, put in a defence and counterclaim, relying on Chestertons' price L and counterclaiming £10,000. She sought judgment for the £10,000, and she applied to strike out that statement of claim of the landlord, as disclosing no cause of action. The summons came before the master and before the judge. Both held that the claim by the landlord disclosed no reasonable cause of action. They struck it out and gave Miss Edwards judgment for £10,000. The landlord appeals to this court.

The appeal was stood over pending the decision of the House of Lords in *Arenson v Casson Beckman Rutley & Co* [1975] 3 WLR 815, which was decided only two weeks ago. The law on this subject has been transformed by two cases in the House of Lords, *Sutcliffe v Thackrah* [1974] AC 727 M about architects, and *Arenson v Casson Beckman Rutley & Co* about valuers. Previously, for over 100 years, it was thought that when vendor and purchaser agreed that the price was to be fixed by a valuer, then the valuer was in the position of a quasi-arbitrator and could not be sued for negligence. It is now clear that he owes a duty to both parties to act with reasonable care and skill in making his valuation. If he makes a mistake owing to want of care and skill he may be liable in damages. If he negligently gives a figure which is too high, he may be sued by the purchaser; if it is too low, he may be sued by the vendor. If he wants to avoid

A such a responsibility he must put in a special clause exempting him when he accepts the appointment. Unless he stipulates for exemption, he is liable for negligence. In former times (when it was thought that the valuer was not liable for negligence) the courts used to look for some way of upsetting a valuation which was shown to be wholly erroneous. They used to say that it could be upset, not only for fraud or collusion, but also on the ground of mistake: see for instance what I said in *Dean v Prince* [1954] Ch 409 at p 427. But those cases have to be reconsidered now. I did reconsider them in the *Arenson* case in [1973] Ch 346 at p 363. I stand by what I there said. It is simply the law of B contract. If two persons agree that the price of property should be fixed by a valuer on whom they agree, and he gives that valuation honestly and in good faith, they are bound by it. Even if he has made a mistake they are still bound by it. The reason is because they have agreed to be bound by it. If there were fraud or collusion, of course, it would be very different. Fraud or collusion unravels everything. It may be that if a valuer gives a speaking valuation—if he gives his reasons or his calculations—and you can show on the face of them that they are wrong, it might be upset. But this is not such a case. Chestertons simply gave the figure. Having given it honestly, it is binding on the parties. C It is no good for either party to say that it is incorrect. But even if the valuation could be upset for mistake, there is no room for it in this case. The premises have been surrendered to the landlord. He has entered into occupation of them. Months have passed. There cannot be *restitutio in integrum*.

I may add that Mr Chadwick put forward an additional argument. He said, "If this valuation is set aside, what is to take its place?" The answer is, nothing. The only surveyors on whom the parties have agreed are Chestertons. The parties are unlikely to agree on any other surveyors, and Chestertons cannot reasonably be asked to make another D valuation. So there would be nothing to take the place of this valuation. In my opinion, therefore, the landlord is bound by this valuation of £10,000. I would just like to add this. The position of a valuer is very different from an arbitrator. If a valuer is negligent in making a valuation, he may be sued by the party—vendor or purchaser—who is injured by his wrong valuation. But an arbitrator is different. In my opinion he cannot be sued by either party to the dispute, even if he is negligent. The only remedy of the party is to set aside the award, and then only if it comes within the accepted grounds for setting it aside. If an arbitrator is guilty of misconduct, his award can be set aside. E If he has gone wrong on a point of law, which appears on the face of it, it can be corrected by the court. But the arbitrator himself is not liable to be sued. I say this because I should be sorry if any doubt should be felt about it. This case is just a postscript to *Arenson*. The valuation is binding on the parties. The master and the judge were right, and we dismiss the appeal.

GEOFFREY LANE LJ: I agree. The parties here had agreed upon a valuer. They had agreed as to the terms on which the valuer was to value the property and on which he was to be instructed. The valuation took place, it was acted on, and the tenant surrendered the lease. It is a common law F situation in which there is no room for an equitable remedy. The most up-to-date and accurate statement of the law in the circumstances is in a passage of my Lord, the Master of the Rolls's judgment in *Arenson v Arenson* in this court [1973] Ch 346 at 363, which states as follows:

> "At common law—as distinct from equity—the parties are undoubtedly bound by the figure fixed by the valuer. Just as the parties to a building contract are bound by the architect's certificate, so the parties are bound by the valuer's valuation. Even if he makes a mistake in his calculations, or makes the valuation on what one or other considers to be a wrong basis, still they are bound by their agreement to accept it."

In this case Mr Bateson has argued that there is sufficient G discrepancy between the valuer's report and the subsequent valuations obtained by his clients to indicate that the valuers, Chestertons, must have acted on a wrong principle. He says, despite the fact that this is not a speaking report, that that wide discrepancy is sufficient to cast that doubt upon the valuation. I disagree. There is nothing to suggest that the valuer here did not take into consideration all the matters which he should have taken into consideration, and where the only basis of criticism is that another valuer has subsequently produced a valuation a third of the original one it does not afford, in my view, any ground for saying that Chestertons' valuation must have been or may have been H wrong.

The other matter which perhaps I should mention is this. Mr Bateson asked for leave to amend the statement of claim in order to join the valuer as a defendant as well as the tenant, alleging that the valuer had acted negligently in his work. This would have had the effect, desirable from Mr Bateson's point of view, of permitting him to obtain discovery, no doubt, of the valuers' (Chestertons') papers relating to their inspection of the premises, and so on, and then to use that information by way of attack against the tenant so that the valuation could be set aside *vis-à-vis* the tenant. J It seems to me that it is undesirable that valuers undertaking this type of duty should be subject to this additional peril. It is not only technically undesirable; it produces, as I see it, an anomalous situation, for this reason. If the discovery shows that there was no negligence on the part of the valuer, then the action against the valuer will of course fail. If on the other hand the discovery shows that there was negligence on the part of the valuer, then that means it would be possible, on Mr Bateson's argument, to set aside the valuation *vis-à-vis* the tenant, in which case no damage would have been suffered by the landlord, and likewise the valuers would not be liable in negligence. When that sort of anomalous K situation is produced, it cannot be right that this sort of amendment should be allowed. For the reasons I have already indicated and those adumbrated by my Lord, I agree that there is no fault in the reasoning or conclusion of the learned judge and that this appeal accordingly should be dismissed.

The appeal was dismissed with costs, and a stay of execution was refused.

L

QUEEN'S BENCH DIVISION

March 5 1975

(Before Mr Justice O'CONNOR)

HINGORANI v BLOWER AND OTHERS

Estates Gazette June 19 1976

(1975) 238 EG 883

Survey negligence—NHOA's surveyor should have realised an attempt had been made by vendor to disguise defects which required underpinning—Purchaser awarded agreed difference in value of £3,000 M

This was a claim by Mr Mohan Hardasmal Hingorani, of 2 Coleridge Walk, Golders Green, London NW11, against Mr Basil L Blower, honorary chairman of and honorary conveyancer to the National House Owners' Association; Mrs Jacquie Hickmott, secretary of the association, of White House, Ebford, Exeter; Mr Michael J King, treasurer of the association, of William Road, Queens Park, Bournemouth, Dorset; and Mr Jeffrey Reeves, surveyor to the association, of Long Lane, Hillingdon, Middlesex, for damages for breach of contract and/or negligence in or about a survey of 2 Coleridge Walk in December 1969.

A Mr M Myers (instructed by Brecher & Co) represented the plaintiff. The fourth defendant, against whom it was agreed the case should proceed in the first instance, appeared in person.

Giving judgment, O'CONNOR J said: In the autumn of 1969 the plaintiff, Mr Hingorani, decided to buy a house in the Hampstead Garden Suburb, 2 Coleridge Walk. He saw an advertisement in the newspaper by the National House Owners' Association offering to act on behalf of purchasers to carry out the necessary legal formalities which are a concomitant of the purchase of a house. He sent for their brochures, paid £1 to become a member of the association, B and engaged their services in the purchase of his house. The position at that stage was that he had agreed subject to contract to buy the house for £7,000. He had arranged tentatively for a mortgage with a building society who had put in their surveyor for that purpose, and he was strongly advised, and very properly advised, by the National House Owners' Association that he should have a structural survey of the property before he signed the contract. This was good sense. He was minded originally I think not to incur that expense, but it was not very great. The fee which was asked was £6, and after a conversation with a man named Feldman he agreed to a structural survey. He gave express instructions C orally, and confirmed in a letter, that if the survey revealed defects which would cost more than about £30 to put right the contract was not to be signed and the matter was to be referred back to him. That is found in letters which he wrote to the association on December 14 and again on December 20 1969. He then went away on holiday. On Christmas Eve, December 24, Mr Reeves, a qualified surveyor who was one of the surveyors acting for the National House Owners' Association, carried out the inspection of the property, and in due course he made his report. The relevant parts of the report deal with the external brickwork, and he had this to say about D it:

> At the time of the inspection this was thought to be structurally sound, the following points being noted: (1) Some joints of the brickwork require to be raked out and repointed in cement and sand mainly to the dampcourse joint and below. (2) It is thought that to prevent the possibility of dampness appearing on internal reveals if assisted by wind pressure the joint between the external side of window and door frames and the surrounding brickwork could be pointed with mastic putty compound. Estimated costs £12.

He completed his report with a general paragraph which reads:

> E Inspection of sample roof floor and structural timbers revealed no apparent signs of rot or pest infection, it being brought to the client's attention that most areas had fixed coverings that could not be lifted. Plaster surfaces that were papered covered up any cracks blemishes or blisters but in the normal inspection carried out no defects were apparent, it being thought the redecorations had been newly carried out obliterating any previous faults or defects.

Some other minor matters adding up to a little over £30 were noted in the report which are not relevant to this case. That report was submitted to the association, who were authorised by the plaintiff to sign a contract on his behalf, F because the structural survey showed that there were no major defects in the property. The association duly signed the contract on his behalf, and completion took place on January 28 1970.

The plaintiff moved into the house in April of 1970, and shortly afterwards cracks began to appear all over the place. They are illustrated in a bundle of photographs exhibited in the case, and by the summer of 1970 not only were there cracks all over the place in the newly-decorated walls of the interior of the house but cracks opened up in the external brickwork. The house is the end house of a terrace of houses, and when one looks at it facing the front door from the road G the ground slopes away to the right and the party wall of no 4 is to the left. Photograph no 3 shows a major crack running down to the first-floor window, the landing window, and down below it diagonally across the brickwork. There is a further substantial crack above the front door. Then one goes round to the back of the house, and now the ground is sloping away to the left. There is a major crack running down at the point of the party wall, which is in fact in a corner where there is a return wall, and it runs right down. It then moves diagonally inwards past a ventilator and then carries on down below. That crack is reflected internally in the kitchen and is illustrated in the photographs. The photograph H shows the crack running down at the level of the tiling which is there as a splash-board to the sink, but in fact it goes on down into the cupboard which is below the sink. The evidence is that the redecoration of the room which had been carried out by the vendor had not included the redecoration of the wall in the cupboard. These cracks all show that the house had moved and was structurally substantially defective, and in due course an estimate for underpinning was obtained from a specialist firm, Pynford Design Ltd, who ultimately carried out the work at the cost of about £1,444. The plaintiff obtained two estimates for repairing the superstructure. A great deal of work needed to be done to cut out the cracks J and so forth, apart from decoration, and the estimates in round figures from two different contractors were in the sums of £1,500 and £2,000. That of course was some time ago.

As far as the values are concerned, the evidence again is in a narrow compass. Two valuers have given evidence to the court with which the defendants do not really disagree, for the good reason that I do not think they can. Mr G W Mathews [senior partner of Daniel Watney Eiloart Inman & Nunn], a surveyor who has reported for the plaintiff and given evidence, said that in 1969 or about January of 1970, which is the relevant date for this purpose, in his view, if K the house had answered to the survey in its redecorated and modernised state it would have been worth approximately £6,750. In its true condition, with the major structural defects showing that it was subject to subsidence, he valued it at £3,500, making a difference of £3,250. Mr M A Roe, another surveyor and valuer, who knows a good deal about the house because his firm had acted in 1969 in circumstances to which I will refer in a few minutes, took the view that the difference in value at the relevant time was £3,000. When it comes to considering damage in a case such as this, the correct measure is the difference between the value of the house at the date of the contract or the date of the sale as it would have been L if it had been what it was represented to be and as it really was, and as I have said in round figures that is the sum of £3,000. It does not matter what it actually costs to put right in the end, and it is only a check to see that in fact £1,500 odd was spent on underpinning the house and there was another £1,500 odd which needed to be spent on putting it right. But as I have said, it is the difference of the values that is the correct way of approaching the issue of damage, and I am driven to the conclusion on the evidence before me that the correct measure of damage is the sum of £3,000 flat.

The plaintiff's case as originally pleaded was an action brought against the National House Owners' Association, and M it began as a representative action. That, for a variety of reasons which it is unnecessary for me to refer to, was changed, and the action continued against four named individuals, Mr Blower, the honorary chairman of the association, Mrs Hickmott, who was secretary at the time, Mr King, who was the treasurer at the time, and Mr Reeves, the surveyor who carried out the inspection on December 24 1969 and made the report on which the association and the plaintiff acted. The possible liability of the three members of the association raises difficult and complex questions of law, and there was an agreement between the parties about this. Here I may say that the defendants were unrepresented at the trial.

A Mr King and Mrs Hickmott did not come to court, and no criticism is to be made of them for their failure to appear, and I hope that their interests have been satisfactorily protected. Mr Blower did appear, and after the case had been opened it was agreed that the issue of fact should be tried out first and the liability of Mr Reeves, if any, established and all questions of the possible liability of the other defendants put over to see whether Mr Reeves could satisfy the judgment that might be given against him. It is in that fashion that the action proceeded, which had the effect of drastically shortening the cost of this litigation, and I hope that that saving may prove to be permanent. One other B general matter must be dealt with. The plaintiff's claim against Mr Reeves is effectively pleaded in negligence. Whether Mr Reeves was under a contractual duty through the association to Mr Hingorani does not in my judgment matter, because whether he was in a contractual relationship or not it is quite plain that he owed a duty in negligence to carry out his survey with reasonable care and skill as a competent surveyor and to report accordingly. If he failed so to do then he is liable in negligence. In my judgment there is no difference in the measure of damage which flows from the breach of duty whether it is claimed in contract or in tort.

C With those general observations I turn to the actual dispute in the case. Mr Reeve's defence in this matter is a straightforward one. He says that the terms of his employment were that he was to report on observable defects, and I accept that that was the correct approach to make in this case. He was to carry out a structural survey and report on observable defects in the structure. His case is that as at December 24 1969 there were no observable defects in the structure; that he carried out a proper, thorough, competent survey, and that he, along with anybody else who might have done work, was deceived, and so would any competent surveyor have been deceived, by the fact that the vendors had fixed the house D up and that they had successfully fixed it so that the defects which were undoubtedly there were not observable by a reasonably competent skilled surveyor making a competent skilled inspection. If he is right about that, that is an end of the case. The loss will fall on the plaintiff, because as so often happens in litigation of this kind the real culprit was the vendor of the property who had undoubtedly tried to fake it up. The plaintiff's case about this is that for whatever reason, on December 24 1969 Mr Reeves did not apply that degree of care which I have little doubt he normally does in carrying out his inspections of property, and that he missed signs which were there to be seen which would have put E him on his guard that the house was in a very damaged condition.

The evidence about this really starts with the inspection. which was carried out in January of 1969. At the time the firm of Gordon Hudson & Co, whose Mr Roe has given evidence in this case and presented a report, exhibit P7, was engaged in the sale and purchase of this house. There can be no doubt whatever that in January 1969 the house was in a very bad condition. Mr Roe instructed the firm of Pynford Design Ltd to inspect the house and give an estimate for underpinning it, because anybody looking at it in January 1969 would have seen large cracks showing that the house was F subsiding downhill, that its foundations were giving way, and that it needed underpinning and major structural repair. A Mr Pryke [of Pynfords] visited the house in January 1969 and he found internally and externally major cracks; in particular he found the cracks along the party walls to which I have already referred at the rear running down in the corner between the face wall of the house and the return wall which can be seen in photograph no 8. That crack ran right down and was plainly visible for all to see. On the front of the house he found a crack on the line of the party wall which can be seen in photograph no 3. The party wall crack is above the hedge which is just to the left of the front door, and the top of it can be seen where the gutters are painted G different colours. He found the crack which is visible on that photograph which runs as one faces the photograph above the lefthand corner of the first-floor window diagonally down across the face of the wall which is plainly visible for all to see. He also found a crack which is not quite so easy to see on the photograph (but it is in fact plainly visible) over the top of the righthand corner of the front door. Not only did he find these cracks, and they reflected on the inside of the house, but in order to give his estimate he had to draw a little plan of what he had found, and he had a contemporary note on which all these cracks were plainly shown. He gave evidence to say that when he inspected again in H March 1971 there were the cracks substantially the same as he had seen in January 1969. I accept Mr Pryke's evidence; it is really unchallenged on this topic. What had taken place in the meantime? Before I pass to that, there is one other piece of evidence which I also accept, and that comes from Mr Debere, the plaintiff's solicitor, who by chance happens to live opposite, at no 1 Coleridge Walk, and he used to visit an allotment which is on land to the rear of the plaintiff's house. He moved into his house in 1967 or 1968, and he told me that in 1968 and early 1969 this great big crack at the back of the house was plainly visible. He thought it was sufficiently large to allow one to stick one's fingers into it, J and I accept his evidence about that.

Now, what had taken place? Attempts to sell the house in early 1969 had failed. The evidence comes from Mr Roe's report that in one case the building society refused to mortgage because of the structural defects in the house and so forth, and in the end it appeared to have been sold to some people called Skerritt in about June of 1969. I believe that the purchase price was of the order of £4,000. Now these people set to work to fake up the house. They never lived in it. They carried out a complete redecoration internally. They modernised it by putting in a modern bathroom and a K kitchen, and completely at least repapered and repainted the interior walls, so any cracks that were there were either filled in or not—I neither know, nor does it matter; they successfully obliterated inside the house any signs of the structural defects which were apparent. As far as the cracks in the outside walls were concerned, the evidence is not terribly satisfactory. Mr Mathews, who inspected in September of 1970, did not at that stage apply his mind to the question as to whether the cracks in the outside walls had been filled up and had then reopened. He did in July 1971 make a further inspection and apply his mind to that issue, and his evidence was that the big long crack at the rear of L the house showed signs of having been filled but that he could find no evidence of new plaster work in the cracks on the front of the house, particularly that above the front door. Mr Debere, the solicitor, gave evidence that work was done to the outside of the house which he saw going on, and he is satisfied that the big crack at the back of the house had been filled up.

There can be no doubt whatever that in 1970 further movement took place, because the cracks inside all opened up again, and I have no doubt that the cracks outside were enlarged. What then was the position at the front end of the house? As I pointed out in argument, it seems to me M unlikely that people who were minded to fake up a house in order to sell it and conceal structural defects of which they must have been aware would have left a large crack on the front of the house plainly visible for all to see. If the answer is that at that stage the crack, although noticeable to a surveyor and noticeable to Mr Pryke in January 1969, was not anything like as big as it was or became in the autumn of 1970, it will be seen that it runs not only in the joins between bricks but actually across the bodies of a number of bricks. It is suggested that as this house was built about 20 years before, there is some evidence that it was

A built between the wars, but it was not a very new house and it needed modernising—the probability is that it was built between the wars, but it does not matter. It was built of a washed brick which is not all the same colour, and it was suggested by the defendants that what had happened was that in the course of faking it up it could be that the bricks had been cut out and new secondhand bricks had been inserted, so that as at December of 1969, when Mr Reeves carried out his inspection, there was no crack on the front of the house. It is quite plain that the crack at the back of the house had not been treated by having the defective brickwork cut out and new bricks put in. It is plain, and B I accept the evidence of Mr Mathews on this topic, that that showed signs of having been filled in with mortar. I am satisfied that the brickwork had not been cut out on the front of this house, and that if any attempt had been made to fill in the cracks on the front it had been made incompetently, because by 1971, and that was not all that long a time ahead, there was no trace left of any filling in those cracks.

I think the probability is, as I have said, that the cracks were not as noticeable as they have now become and that the vendors took a chance on them. I do not think in the end it matters, because what is quite plain is that there was a C large crack which had been filled in, and although I appreciate and paid due attention to the evidence that persons who are faking a property can colour the mortar so as to make it more difficult to see, the fact is that anybody looking at that corner of the wall, the whole length of it, even if the mortar had been coloured, would have seen a wide band of filling. And as Mr Reeves correctly pointed out in his evidence, when faced with an empty house such as this which had plainly been redecorated from top to tail internally, a careful surveyor is put on suspicion that the persons who have done it have been for good or bad reasons disguising D defects which are there, and he should be and is put on his guard. Mr Reeves says that he was on his guard, and I see no reason to doubt his evidence about that. He struck me as being both a competent and honest man, but I am satisfied that there was on the rear elevation this long crack which carried the straightforward tell-tale of having been refilled, a much wider line of mortar than ought to have been there, and he ought to have seen it. That he missed it I have no doubt. I think he missed it absolutely honestly. I do not know what sort of a day Christmas Eve 1969 was. It was in the middle of winter, at least. But he tells me that he spent two and a half hours there, and I accept his evidence about E that. The fact is that he missed the crack in the rear elevation, and the filling which was there was plain for all to see. I am satisfied that the crack must have run, including the little diagonal bit where it comes out from the corner, on to the face of the wall beside the ventilator and then run down to the ground. That must all have been there filled, and he ought to have seen it, and he did not.

That really is enough for the purposes of this case, because had he seen it he would have appreciated at once that there was a serious structural defect in the house, and he does not dispute that. I am satisfied that there were cracks which were either unfilled, or possibly an attempt had been made F to fill them, in the front of the house, and he missed those too for whatever reason. They cannot have been as big as they are now, because had they been anything like that I am quite satisfied that he would have seen them, and in addition I do not forget the evidence of Mr D V Weenen, who inspected the house on behalf of the building society for purposes of the mortgage loan, and he had gone there in November or early December, I think November, of 1969. He gave evidence and he said that he made a cursory inspection of the house; he was only there 20 minutes, and of course he was not carrying out a structural survey. He was only putting his eye over the house to value it for mortgage G purposes, and it is well known (and it is no disrespect to them) that surveyors are not expected to make a structural survey for that purpose. He tells me that if the crack had been anything like it is shown in photograph no 3 he must have seen it, and I accept his evidence about that. So I think it follows either that the crack as it was in 1969 was very much smaller than it is now and not plainly visible to somebody cursorily looking at it, or alternatively that some attempt had been made to disguise it by the vendors to make it less visible. But I am satisfied that it ought to have been visible on a careful inspection by the structural surveyor, and it is unfortunate that Mr Reeves missed it. It may possibly be that the degree of redecoration and modernisation inter- H nally, although it ought to have put people on guard, and Mr Reeves said it did put him on guard, had the effect of lulling him in his inspection of the exterior.

The fact is that tell-tales of major defects were there in December 1969 and ought to have been spotted. In fairness to Mr Reeves, I think it right to say in the judgment that I am satisfied that he really recognises that. The evidence for that comes from the note which he made when he inspected in September 1970. He told me in the witness-box that he went there and saw the condition of the house, and he was appalled, and for good reason. I think that the penny dropped at once that it must have been in a very defective J condition at the time that he looked at it, and that for one reason or another he had missed it. But despite the fact of asserting that the cracks were all new, which of course internally was partly true, he made a note for his own information really: "Since 24.12.69 subsidence has taken place causing serious cracking, these cracks definitely new, house movement and settlement towards right hand side requires underpinning." Then he raised the query: "Can this be referred back to local authority regarding by-law inspection?" that is the time that the house was built from the foundations. And secondly, "Can this refer back to vendor or original K builder?" I may say that it is common ground that at his inspection on September 26 1970, when Mr Hingorani was present, some conversation took place to the effect that in any event Mr Reeves was insured against any liability that might fall on him. I think it was quite plain to him that he was, as it was put in argument, in the firing-line in this case. Unfortunately, for reasons which I need not go into, there was at the crucial week no insurer on risk, for reasons which Mr Reeves has given in evidence, but it does not affect the issue in this case. It is a matter of grave misfortune to him, and he has my wholehearted sympathy. But that is a matter about which I can do nothing. The fact is that he ought to L have discovered the defective nature of the house in December 1969 and he did not.

There is one other topic which was dealt with in the evidence about which I should make a finding, namely the size of the crack internally which was present in 1969 below the level of the sink, that is in the cupboard in the kitchen underneath the sink. The evidence about it is that it was of sufficient size that it must have been present in January 1969, and the evidence is that it had not been redecorated in the summer of 1969 by the vendors who had faked up the house. I do not think Mr Reeves noticed it. He says that of course M he looked at the cupboard, and that if there was a crack it would have been only a very fine one, not a hairline crack but the sort of crack which appears in any plaster-work in a house such as this over a period of time and does not tell any tale of a serious defect. He says he is satisfied that he looked under the sink, because it is a place which any careful surveyor would certainly do in order to see any signs of dry rot or wet rot, whichever originates from defective bulkwork. I have no reason to suppose that he did not look in the cupboard for that purpose, but again whatever the nature of the crack that was present at that time it did not impinge itself on his mind. It is possible that some painting or other-

A wise had been carried out underneath the cupboard at that time. I do not think that it ultimately matters, because the real fault here was his failure to spot the signs which in my judgment were plainly visible at the rear of the house and his failure to spot the cracks which must have been visible on a careful inspection which ought to have been carried out on the front of the house. They were all there and he missed B them. In those circumstances there must be judgment for the plaintiff against Mr Reeves. I have already dealt with the issue of damage, and the answer is £3,000.

The action against the remaining defendants was adjourned with liberty to restore.

RATING

A

COURT OF APPEAL

October 14 1975

(Before Lord DENNING MR, Lord Justice JAMES and Lord Justice BRIDGE)

RAVENSEFT PROPERTIES LTD v NEWHAM LONDON BOROUGH COUNCIL

Estates Gazette January 3 1976

(1975) 237 EG 35

Rating—Unoccupied property—When is a newly-erected building "completed" for the purposes of the relevant provisions?—When it is ready for occupation, not when the structure alone has been erected—Rating authority's notice quashed, and court refuses, in the absence of appropriate evidence, to fix a date for notice to take effect

B

These were appeals by Newham London Borough Council from decisions of Judge Rankin at Bow County Court on December 18 1974 quashing notices served by the council, as rating authority, on the respondents, Ravenseft Properties Ltd, claiming that certain blocks of new office buildings constructed within the borough and collectively known as the Stratford Centre had been "completed" for the purposes of the First Schedule to the General Rate Act 1967.

C

Mr C S Fay (instructed by the council's chief executive) appeared for the appellants, and Hon D M Trustram Eve QC and Mr M F C Fitzgerald (instructed by Nabarro, Nathanson & Co) represented the respondents.

Giving judgment, LORD DENNING said: In the East End of London there is a fine new centre called the Stratford Centre which consists of two or three blocks of new air-conditioned offices. It has been developed by the London Borough of Newham in partnership with a property company called Ravenseft Properties Ltd. The office blocks are not occupied as yet, but the London Borough of Newham wish to charge Ravenseft Properties Ltd with rates upon them. At one time unoccupied premises did not bear rates at all.

D

"Voids," or "empties," as they were called, were not liable to pay rates during the time they were unoccupied. But since 1967 there has been a provision in the General Rate Act whereby, in the case of office buildings, they are free of rates for the first three months while they are unoccupied; but after three months, they are liable to pay rates even if they are unoccupied. That is why the London Borough of Newham wish to charge Ravenseft with rates. But—and here is the point in the case—the chargeability only arises when the buildings are "completed." The controversy is whether these buildings were completed or not. The London Borough

E

of Newham say that they are complete as soon as they are structurally complete, whereas Ravenseft Properties say that they are only complete when they are ready for occupation. That is the broad issue between the parties.

The London Borough of Newham on October 10 1974 served notices on Ravenseft Properties in which they gave notice that they regarded the 14 floors as completed: that the building was comprised in a "relevant hereditament," and that it was to be regarded as "completed" within the meaning of paragraph 8 of the First Schedule to the 1967 Act. Ravenseft Properties took a contrary view. They appealed to

F

the county court, saying that their offices were not completed and could not reasonably be expected to be completed by that date. The dispute depends on the state of the buildings. This has been agreed between the parties, and we have photographs to show it. The big block, block A, has 14 storeys of reinforced concrete construction. The whole of the structure, as a structure, is no doubt completed. There are large floors all the way up the 14 storeys. On each of these floors, at each end, toilets are installed. Central heating and air-conditioning systems are installed. These vast floors have not been divided into rooms or offices. There are no partitions in. Each of the 12 floors covers a rectangular

G

space of 8,385 sq ft, with nothing there except 10 structural columns in a line. There is a fuse-box on each floor with wiring for a power circuit, but no points have been installed for any outlets. There is no wiring for any lighting system. There is no telephone system. No GPO cables have been installed. No main distribution frame has been installed. Application has been made for a telephone line in different places, but it is nine months before the telephone can be installed. The main contest is as to these 12 floors. As I say, they are enormous floors with columns down the middle but no divisions whatsoever into rooms or into individual offices. It seems to me that they could not be occupied or

H

ready for occupation until they had been divided up by partitions and rooms and so forth. So there we have the contest. The London Borough of Newham say the buildings are complete because they are structurally complete. Ravenseft Properties say no, they are not complete until they are ready for occupation, and they are not ready. The county court judge held that they were not ready. He rejected the claim of the London Borough of Newham. Now there is an appeal to this court.

We have been taken through this Schedule 1 of the 1967 Act dealing with the rating of unoccupied property. We

J

have been taken to the distinction between a newly-erected building, on the one hand, and on the other hand a building which has already been used but has been structurally altered so as to become another hereditament. It does seem to me that the same test ought to be applied in each. I do not propose to analyse the schedule in detail. I am only going to take two cases which have come before the courts. The first is a case where there was a newly-erected building, *Watford Borough Council v Parcourt Property Investment Co Ltd* [1971] Rating Appeals 97 [(1971) 218 EG 1006]. There the Heinz company (the well-known company which

K

makes baked beans among its 57 varieties) were building a large office building in Watford. It was a newly-erected building. Much of the building was erected, but Heinz, when they took over the building, did a lot of work themselves. When they took it over it was devoid of any internal partitioning at all. There were big floors rather like the present case. It was held by Bridge J, as he then was, that this was not a completed building. Both sides there had agreed that the appropriate test was whether or not the building was ready for occupation. Bridge J held that it was an incomplete building and therefore the company were not liable to be rated in regard to it at all. If I may say so, that commends itself to me as being a proper application of the word "completed" in this context. The other case is on the other branch of the argument: where there is an existing building which has been structurally altered so as to become a different building. This was the case of *Easiwork Homes v Redbridge London Borough Council* [1970] 2 QB 406. The position there was that there was an old-fashioned block of 16 flats in Wanstead. The company decided to modernise the whole block of flats. The plumbing work had been removed; toilet fittings were being replaced; electric wirings were being renewed, and even in some cases flooring was being relaid. In that case there was an existing building, the old hereditament. The old valuation list, unless it was altered, continued to apply. It continued to apply to the building until there was a new hereditament constructed, and so constructed as to be complete in that sense, as ready for occupation. In that case the new modernised flats were not ready for occupation: they were incomplete, so the old valuation list and the old rating applied, as was held by the Divisional Court in that case. The court said that it would be very odd if a person could avoid paying rates simply by making a few removals of the sanitary fittings, part of the plumbing, and so forth.

Those two cases were, I think, correctly decided. They illustrate the problem which arises under this schedule in respect of the two aspects: first, the newly-erected building, and, secondly, the old building which has been structurally altered. Now to come to the wording of the schedule, the important paragraph here is paragraph 8 (1), which says:

> Where a rating authority are of opinion:
> (a) that the erection of a building within their area has been completed; or
> (b) that the work remaining to be done on a building within their area is such that the erection of the building can reasonably be expected to be completed within three months,
>
> and that the building is, or when completed will be, comprised in a relevant hereditament, the authority may serve on the owner of the building a notice (hereafter in this paragraph referred to as "a complete notice") stating that the erection of the building is to be treated for the purposes of this schedule as completed on the date of service of the notice or on such later date as may be specified by the notice.

It seems to me that the schedule uses the word "completed" there as something different from "structural alterations" in paragraphs 10 and 15, and from the words "substantially completed" in paragraph 9; and I am impressed by the reference to paragraph 6 which was made before us. The word "completed" in paragraph 8 (1) seems to me to mean completed in the sense of being ready for occupation. The test in this case is that the building should be ready for occupation. Once it is completely ready for occupation, there is a free period of three months during which rates are not payable, but after three months rates are payable. That applies not only to a new building, but also if there is an alteration to an old building.

The only remaining point is what is to be done. The judge allowed the appeal. That was all that he had to do, because the notice was not properly given: the building was not complete. But there is a provision in paragraph 8 (5) which says that if the appeal is in fact allowed "the erection of the building shall be treated for those purposes as completed on such date as the court shall determine." It seems to me plain the the court cannot apply that provision unless there is evidence on the matter and unless it is asked to do so; and I should have thought there ought to be evidence to show when it was completed so that the court could determine it. If it had the evidence it could determine the date on which it was completed But in the present case, where there was no evidence given us and no application until after judgment, it seems to me that that provision does not come into operation. We had considerable discussion as to the telephone cables and the electric wiring, and how far office buildings could be said to be complete if the electric wiring had not been completed and the telephones were not installed. I think that may give rise to difficult questions on other matters. I should like to have further evidence as to the condition of a particular building before giving any ruling on that matter. For the reasons I have given, I think this appeal should be dismissed.

JAMES LJ: I agree with the conclusion of my Lord that the appeal should be dismissed, and substantially for the reasons that he has given. I will avoid being repetitious by adding reasons of my own which merely conform to his. It seems to me that the appeal raises two short points, the first being the meaning to be given to the words "completed" and "completion" where they appear in relation to hereditaments in schedule 1 to the General Rate Act of 1967. That is a schedule that is not free from difficulty in construction, and if I may say so, not absolutely clear in places. But I see no warrant whatsoever for importing the word "structural" into paragraph 8 of that schedule merely because that word appears in paragraph 10, and there should be the same test for completion in respect of newly-erected buildings as for hereditaments which have already been on the valuation list and are subjected to structural alterations so that they become different or different parts of hereditaments. It seems to me that to adopt that construction, which is the basis of Mr Fay's argument, strains the meaning of the statutory words. On the other hand, if one looks at the following provisions, one finds there strong indications all pointing to the proposition for which the ratepayers contend in this case, namely, that the test of completion is capability of occupation and that that is the test to be applied to a newly-erected hereditament. Section 79 of the Act provides:

> (1) Subject to subsection (2) of this section and to the following provisions of this Act, namely, paragraph 6 (4) of schedule 1, paragraph 5 (4) of schedule 4, paragraph 8 (4) of schedule 5, paragraphs 10 (b) and 13 of schedule 6 and paragraphs 14 (b) and 15 of schedule 7, where an alteration is made in a valuation list by virtue of sections 71 to 78 of this Act, then, in relation to any rate current at the date when the proposal in pursuance of which the amendment so made was served on the valuation officer, or, where the proposal was made by the valuation officer, current at the date when notice of the proposal was served on the occupier of the hereditament in question, that alteration shall be deemed to have had effect as from the commencement of the period in respect of which the rate was made, and shall, subject to the provisions of this section, have effect for the purposes of any subsequent rate.
>
> (2) Notwithstanding anything in subsection (1) of this section, where an alteration in the valuation list:
>
> (a) consists of the inclusion in the valuation list of a newly-erected or newly-constructed hereditament or an altered hereditament which has been out of occupation on account of structural alterations; or
> (b) is made by reason of any of the events specified in section 68 (4) (b) to (k) of this Act,
>
> the alteration shall have effect only as from the date when the new or altered hereditament comes into occupation or as from the happening of the event by reason of which the alteration is made as the case may be.

Paragraph 5 (1) of schedule 1 is:

> Subject to the provisions of this schedule, the rateable value of a hereditament for the purposes of paragraph 1 thereof shall be the rateable value ascribed to it in the valuation list in force for the area in which the hereditament is situated or, if the hereditament is not included in that list, the first rateable value subsequently ascribed to the hereditament in a valuation list in force for that area.

Paragraph 6 reads:

(1) A rating authority may request the valuation officer to make a proposal for including in the valuation list in force for their area any unoccupied building in their area (together with any garden, yard, court or other land intended for use for the purposes of the building) which in their opinion is, or when completed will be, a newly-erected dwelling-house; and if the valuation officer thinks fit to comply with the request he may make a proposal for including the building (together with any such garden, yard, court or other land as aforesaid) as a dwelling-house in that list and for ascribing to it in the list such values as he considers are appropriate or will be appropriate when the building is completed.

(2) Where such a request is made by a rating authority and the valuation officer serves notice in writing by post or otherwise on the authority stating that he does not propose to comply with the request, the rating authority may, if they think fit, within the period of 28 days beginning with the date of service of the notice, make a proposal for including the building and any other land to which the request relates as a dwelling-house in the list aforesaid and for ascribing to it in the list such values as the authority consider are appropriate or will be appropriate when the building is completed.

(3) Where a new valuation list is prepared for any area, the valuation officer shall include in the list as transmitted to the rating authority:

(a) any dwelling-house included in the current list for that area in pursuance of a proposal under subparagraph (1) or (2) of this paragraph; and

(b) any building (with or without other land) in respect of which a proposal for its inclusion in the current list as a dwelling-house has been made by him under the said subparagraph (1) and has not been settled;

and if any such proposal is made by him after the new list has been so transmitted, shall cause that list to be altered so as to include the building (with or without other land) as a dwelling-house in the new list.

(4) Where a newly-erected dwelling-house is first occupied after its completion and a rateable value has, in pursuance of the foregoing provisions of this paragraph, previously been ascribed to it in the valuation list currently in force for the area in which it is situated, any different rateable value subsequently ascribed to it in that list and which, apart from this subparagraph, would have effect from the date when the dwelling-house is first occupied as aforesaid shall be deemed to have effect from the date on which the current list came into force or the date from which the previous rateable value had effect, whichever is the later.

One looks forward to the time, in considering a newly-erected hereditament, to the type of hereditament which is being required and sees whether at the date of the notice there is anything lacking which ought to be there in order to satisfy the nature of that hereditament. If there is something lacking and that which is lacking would, when done, fall to be part of the hereditament and taken into account for the purposes of the valuation, then there is no completion in the sense of capability of occupation.

On the other point, which arises under the wording of paragraph 8 (5) of the schedule, my view is that the trial judge rightly declined to determine a date upon which the hereditament should be regarded as completed. Paragraph 8 provides a number of methods whereby the date can be ascertained. Where the notice has been served under subparagraph (1) the date is that specified in the notice or the date of service of the notice. But there may be circumstances in which that has happened and then the rating authorities and the ratepayers agree on a different date. Paragraph 8 provides that the agreed date shall be the relevant one. Having served the notice, the rating authority may, by reason of ascertaining new facts, withdraw the notice and serve a new one. Then the date of service of the new one or the date specified in the new one is the relevant date. As I read subparagraph (5) of paragraph 8, that provides that, in the event of the notice not being withdrawn, an appeal being brought, and the appeal not being abandoned or dismissed (which is this case), then, although the word is "shall," the court can, if asked, short-circuit the means for serving a new notice and going through the procedure afresh. That I think is the proper construction to be applied to that subparagraph, and I think the county court judge rightly declined in this case to specify the date.

BRIDGE LJ: I also agree that this appeal should be dismissed, substantially for the reasons given in both the judgments which have been delivered, and I too shall seek to add the few observations that I wish to make in as short a compass as I may and without setting out *in extenso* the elaborate and complex provisions of the General Rate Act 1967 which have led me to the clear conclusion at which I have arrived. The essential question for decision is what is the appropriate test to be applied under paragraph 8 of the first schedule to the Act as to when a building is properly regarded as completed. The key phrase in the paragraph, in my judgment, is the phrase "and that the building is, or when completed will be, comprised in a relevant hereditament." "Relevant hereditament" is defined in paragraph 15 of the same schedule as meaning "any hereditament consisting of, or of part of, a house, shop, office, factory, mill or other building whatsoever, together with any garden, yard, court or other land used or intended for use for the purposes of the building or part." Bearing in mind that under the law as it stood for centuries before unoccupied property became capable of rating, occupation was always the test of liability, if construing this provision without having regard to its wider context, I should say without hesitation that what was contemplated was that the building should be completed so as to be capable of occupation for the appropriate purposes of the particular hereditament, ie as a house, shop, office, etc. If the building lacks features which before it can be occupied will have to be provided, and when provided will form part of the occupied hereditament and form the basis of the valuation of that hereditament, then I would take the view, unless constrained to the contrary, that that building was not within the meaning of the relevant provision of a completed building.

The two main arguments of Mr Fay to the contrary are based first upon the use of the phrase "the completion of the structural alteration" in paragraph 10 of the schedule, which I will not set out, and secondly upon the decision of the Divisional Court to which I was a party in the case of *Easiwork Homes Ltd v Redbridge London Borough Council* [1970] 2 QB 406. In assessing the strength of Mr Fay's submissions I take account of other aspects of the context in which the provisions to be construed are found, aspects which have been forcibly drawn to our attention by Mr Trustram Eve. The structural test of completion which is suggested to us by Mr Fay seems to me to be an extraordinarily difficult one to apply in the context of paragraph 9 of the schedule, which refers to "substantial completion" and contemplates additional works being done after substantial completion but before final completion. What on earth could they be, if Mr Fay's test of structural completion is the right test? Again, without looking at the detail of the provisions, it is clear that by the operation of paragraph 5 (1) of the schedule and sections 69 and 79 (2) of the Act, the first valuation which can reach the valuation list and on which the liability to the unoccupied rate pending occupation of a new building is retrospectively assessed is a valuation of the new building as it exists when it is occupied. That is in sharp contrast with the valuation possibilities which arise under the provisions of section 79 (2) (b) and section 68 (4) (b) in relation to an old hereditament undergoing structural alteration. It is clear that in a situation where an old existing hereditament has a valuation based on its occupiable value and is undergoing radical structural alterations, it can be the subject of a proposal for an alteration in the valuation list for, at all events, any substantial period when by reason of the alteration it is incapable of occupation. That seems to me to provide the answer to the problem of hardship to an owner which in the Divisional Court we felt could arise in the *Easiwork* case.

In the light of these considerations I come to the conclusion that capability of occupation is the test of completion which should be applicable both under paragraph 8 to a new building and under paragraph 10 to a new hereditament which comes into existence by structural alteration of an old building. To adopt the occupiability test advocated by Mr Trustram Eve, and not the structural completion test advocated by Mr Fay, to my mind does no violence to the language of paragraph 10, where the word "structural" is used to describe the nature of the operation to which the paragraph applies and not to indicate a test of completion of that operation. Similarly, the argument based upon the decision of the Divisional Court in the *Easiwork* case can be rejected for this reason. Mr Fay submits that that case establishes that a "relevant hereditament" can exist notwithstanding that it is not capable of occupation; and so it does. But in the special situation where an existing hereditament is under alteration, then either by direct application of paragraph 10 of the schedule or by an analogy with it, it is the clear result of express provision that the old hereditament remains in existence and remains (subject to the possibility of a proposal for any change) at the valuation shown in the valuation list until the moment when the new hereditament comes into existence. But it does not follow from that that it would be sensible or reasonable to apply as the test of completion of a new building under paragraph 8, or of a hereditament under paragraph 10, any other test than the completion of that building to the point where it is capable of occupation as such.

On the subsidiary point which has been raised as to the power of the county court judge where, as here, on abundant material entitling him to do so, he has concluded that the completion notice was incorrect in what it stated, namely, that these buildings were complete, it is odd at first blush to find that the only express power given to the court is the power to determine the date when the building should be deemed to be completed. Paragraph 8 (5) provides: "If the notice is not withdrawn and such an appeal is brought and is not abandoned or dismissed, the erection of the building shall be treated for those purposes as completed on such date as the court shall determine." In a case where the issue is canvassed and fully explored, and evidence is put before the court showing that the building has been completed, or if not, when it can reasonably be expected to be completed, and the court has reached a conclusion upon those matters and is asked by the local authority to substitute a new date for the date in the notice, I have no doubt that that power can and should be exercised. But all that happened here was that the appellants, the building owners, established their ground of appeal against the completion notices which had been served and showed that on the proper test the notices were incorrect; and at all events until after the county court judge had given judgment, the question of determining a fresh date of completion was never canvassed. In those circumstances I do not see how it was open to the county court judge to exercise in any way his powers under subparagraph (5). I think he was right not to do so.

The appeal was dismissed with costs. Leave to appeal to the House of Lords was refused.

HOUSE OF LORDS

April 30 1975

(Before Lord CROSS OF CHELSEA, Lord MORRIS OF BORTH-Y-GEST, Lord SIMON OF GLAISDALE, Lord EDMUND-DAVIES and Lord FRASER OF TULLYBELTON)

OXFAM v BIRMINGHAM CITY DISTRICT COUNCIL

Estates Gazette January 10 1976

(1975) 237 EG 107

Rating—Oxfam "leap-frogs" its test case claiming relief for its gift shops in the Birmingham area—Shops held not to qualify—Line must be drawn so as to exclude from relief premises used for fund-raising rather than for a charity's direct purposes

This was an appeal by Oxfam, a company limited by guarantee, directly from a judgment of Templeman J in the Chancery Division on June 10 1974 holding that a number of "Oxfam gift shops" operated by the appellants within the area of the respondent rating authority, City of Birmingham District Council, did not qualify for relief from rates under section 40 (1) of the General Rate Act 1967.

Mr D G Widdicombe QC and Mr C S Fay (instructed by Waterhouse & Co) appeared for the appellants, and Mr N C H Browne-Wilkinson QC and Miss E Appleby (instructed by Sharpe, Pritchard & Co) represented the respondents.

In his speech Lord Cross said that section 40 (1) of the General Rate Act 1967 provided:

> If notice in writing is given to the rating authority that (a) any hereditament occupied by, or by trustees for, a charity and wholly or mainly used for charitable purposes (whether of that charity or of that and other charities); or (b) any other hereditament, being a hereditament held upon trust for use as an almshouse, is one falling within this subsection, then, subject to the provisions of this section, the amount of any rates chargeable in respect of the hereditament for any period during which the hereditament is one falling within either paragraph (a) or paragraph (b) of this subsection, being a period beginning not earlier than the rate period in which the notice is given, shall not exceed one-half of the amount which would be chargeable apart from the provisions of this subsection.

Subsection (5) went on to give a rating authority power to reduce or remit the payment of rates chargeable in respect of (*inter alia*) any hereditament falling within paragraph (a) or (b) of subsection (1). Paragraph (a) was somewhat curiously worded. No body or trust could be a charity unless its objects were exclusively charitable, and if it was using premises of which it was in occupation for purposes for which it was entitled to use them, and not in breach of trust, it must be using them for some purpose or purposes of the charity. Yet the paragraph clearly contemplated that a charity might be properly using premises which it occupied for purposes which were not "charitable" purposes of the charity. A line had therefore to be drawn somehow or other between the user of premises for purposes which were charitable purposes of a charity within the meaning of the subsection, on the one hand, and their user for purposes which, though purposes of the charity, were not charitable purposes of the charity, on the other.

Oxfam's memorandum of association as a company limited by guarantee set out its objects in the following terms:

> To relieve poverty, distress and suffering in any part of the world (including starvation, sickness or any physical disability or affliction) and primarily when arising from any public calamity (including famine, earthquake, war or civil disturbance) or the immediate or continuing result of want of natural or artificial resources or the means to develop them and whether acting alone or in association with others; and in particular but without prejudice to the generality of the foregoing for that purpose: (i) to provide food, healing, clothing, shelter, training and education and to undertake or assist in work calculated directly to achieve that purpose; (ii) to promote research into medical, nutritional and agricultural matters related to the relief of such poverty, distress and suffering as aforesaid and publication of the results thereof.

Oxfam's activities both as regard fund-raising here and expenditure abroad had expanded and changed in character very much over the last 30 years. Mr Kirkley, the charity's director, had stated in an affidavit that there was now an increasing emphasis on development aid rather than emergencies. Dealing with fund-raising, he had said:

> Since 1960 the development of Oxfam in this country has been largely dependent upon the recruitment and organising of

voluntary helpers at the local level. As mentioned earlier, there are now 660 local groups of voluntary helpers, quite apart from over 8,500 voluntary collectors of monthly donations to the pledged gift scheme. Through this voluntary unpaid help Oxfam now derives about 50 per cent of its income. The activities of these groups in support of Oxfam are very varied, including the dissemination of information, the arranging of meetings and the usual traditional forms of fund-raising. Increasingly, however, our voluntary helpers have found that the most effective form of fund-raising is in the organisation and manning of gift shops. 508 of our groups now concentrate their efforts on the growing number of shops. The growth of this aspect of their activities is illustrated by the following figures. In December 1968 the groups were running 192 shops in the United Kingdom; in December 1969, 256 shops; in December 1970, 320 shops; in December 1971, 433 shops; and in December 1972, 508 shops. Of the last figure, 177 shops were based in rent-free premises loaned to the local groups for short and long terms by commercial firms. In the case of the remainder, the rent was usually the full commercial one and was generally paid by the local groups out of the shop takings. The gift shops are used for the following purposes:

(1) The reception and sorting of articles given to Oxfam.

(2) The sale of those donated articles which cannot be used in the work of Oxfam overseas. The purpose of the sales is to convert the donated articles into usable form, that is to say, money. Originally a substantial proportion of the donated goods, particularly clothing, was used in Oxfam's work overseas, but latterly most of the donated goods have been sold. The reason for this is that the cost of sorting and transporting clothing abroad is very high, and it has been found more effective to sell such items in the United Kingdom and use the proceeds of sale for Oxfam's purposes abroad. Sales of donated goods account for about 80 per cent of the sales at the gift shops taken as a whole.

(3) The sale of "village handicraft" articles. These are goods made in various parts of the developing world under Oxfam's "Helping by Selling" programme, the purposes of which are to encourage village industries and provide employment in poor countries. This category accounts for about 7 per cent of the sales of the gift shops taken as a whole.

(4) The sale of other articles, such as tea-towels and Biros advertising Oxfam's name, and, at Christmas time, Oxfam Christmas cards. These articles are produced by Oxfam Activities Ltd, a wholly-owned subsidiary of Oxfam formed for the purpose of trading. This company buys and resells, through the shops and by mail order, a limited variety of specially-made articles such as those mentioned above. The company employs 12 staff and its profits are totally covenanted to Oxfam. This category accounts for about 13 per cent of the sales of the gift shops taken as a whole.

The premises which were the subject of this appeal were nine shops in different parts of Birmingham. As an example, the first shop on the list, 372 Coventry Road, Small Heath, was opened in December 1971 and closed in January 1973. During that period it was open four days a week from 10 to four, and the rent payable was £8 a week. The sales area was 434 sq ft and the storage area 850 sq ft. The gross takings from the date of opening to September 30 1972 were £1,574 odd, and after deduction of rent, electricity charges and water rate amounting to £458 odd a sum of £1,116 odd was transferred to Oxfam. A breakdown of the sales showed that by far the greater part of the receipts came from the sales of donated goods, as opposed to the "village handicraft" articles and the articles provided by Oxfam Activities Ltd, and that by far the larger part of the donated goods consisted of clothing.

After the passing of the 1961 Rating and Valuation Act it was generally assumed by rating authorities that Oxfam shops were entitled as of right to 50 per cent relief and eligible for further discretionary relief under section 11 of the 1961 Act and later section 40 of the 1967 Act. In the year 1971-72, 201 shops were granted the mandatory 50 per cent relief, 29 relief between 50 per cent and 100 per cent and 135 total relief. But in the light of certain observations by Lord Denning and Winn LJ in *Aldous v Southwark Corporation* [1968] 1 WLR 1671 and an opinion of counsel obtained and circulated by the Association of Municipal Corporations, a number of authorities had recently taken the view that Oxfam shops did not in fact qualify for relief under section 40. By an originating summons issued on November 17 1972 by Oxfam against the Birmingham Corporation, which was in the nature of a test case, Oxfam claimed a declaration that the nine shops were entitled to the 50 per cent mandatory relief and eligible for the 50 per cent discretionary relief under section 40. Templeman J on June 10 1974 decided that Oxfam shops did not qualify for relief, and granted a certificate for a "leap-frog" appeal.

The wording of section 40 (1) showed that the legislature did not consider that the mere fact that the hereditament in question was occupied by a charity justified any relief from rates. That was justified only if the hereditament was being used for "charitable purposes" of the charity. So the first question that arose was what were "charitable purposes" of a charity, as distinct from its other purposes? The answer must be, so he (his Lordship) thought, those purposes or objects the pursuit of which made it a charity, ie in this case the relief of poverty, suffering and distress. Assuming that to be so, it might be argued that relief from rates could be granted only if the premises in question were being used for the actual giving of relief to those in need; if, for example, those in need came to them to receive food, clothing, money or shelter. But the decision of the House in *Glasgow Corporation v Johnstone and others* [1965] AC 609 showed that such a construction was too narrow. The charitable purpose there was the advancement of religion by the provision of services in a church owned by the trustees, and the question was whether a house belonging to the trustees and of which they were held to be in rateable occupation, but which was used as a residence by the church officer whose duty it was to look after the church, was being used by the trustees "wholly or mainly for charitable purposes" within the meaning of section 4 (2) of the Local Government (Financial Provisions etc) Scotland Act 1962. Lord Reid, with whom his colleagues (other than Lord Guest, who dissented) were in agreement, said that if the use which the charity made of the premises was directly to facilitate the carrying-out of its main charitable purposes, that was sufficient to satisfy the requirement that the premises were used for charitable purposes. Oxfam, therefore, was entitled to rating relief in respect of premises which it occupied and which were not being used for the actual relief of poverty or distress if—to quote Lord Reid—the use which it made of them was "wholly ancillary to" or "directly facilitated" the carrying-out of its charitable object, the relief of poverty or distress.

One example of such a use would be the head office of Oxfam. As Lord Donovan pointed out in *United Grand Lodge v Holborn Borough Council* [1957] 1 WLR 1080 at 1088, every organisation setting out to advance some cause must, if it was of any size, have an office where the necessary clerical and administrative work was done, and counsel for the corporation conceded that any office premises occupied by Oxfam, if they were wholly or mainly used for the organising and carrying-out of Oxfam's charitable activities, would be entitled to rating relief. On the other hand, a line had to be drawn somewhere, and a case on the other side which the appellants admitted to have been rightly decided was *Polish Historical Institution Ltd v Hove Corporation* (1963) 10 RRC 73. The institution was a charitable organisation whose object was to encourage the study of modern Polish history. It owned and occupied a house, 53 Brunswick Square, Hove, the rooms in which it let to lodgers who paid weekly sums for their use, and the net profits derived from this user of the house were used to promote the objects of the institution. Lord Wilberforce, before whom the case came when he was a judge of first instance, said then that the question was whether the activities carried on on the premises were activities for the carrying-on of which the organisation existed, and that they were clearly not. There was also *Belfast Association for Employment of Industrious Blind v Commissioner of Valuation for Northern Ireland* [1968] NI 21, where a shop for sale of goods manufactured by blind persons also sold goods bought in for resale, about 50 per cent of sales being of the latter type; and it was held that the premises did not qualify for exemption under an analogous statutory provision.

Counsel for the respondents in the present case had conceded that if the Oxfam shops in question had been mainly used for the sale of the "village handicraft" articles made under the "Helping by Selling" programme, they would have been entitled to relief, but of course only a very small percentage of the sales in the shops was attributable to such articles. Counsel for Oxfam, for his part, was prepared to concede that if the shops had been mainly used for the sale of the articles bought by Oxfam Activities Ltd they would not have been entitled to relief; but here again, only a small percentage of the sales was attributable to such articles. The premises were undoubtedly mainly used for the sale of clothing given to Oxfam, and the question was whether such user was a user "for" the charitable purposes of Oxfam. Each counsel had accepted some of the reasons given for the decision in *Aldous v Southwark Corporation,* already referred to. That was a case in which the Dulwich Estate secured relief from rates in respect of (1) the estate offices, (2) the estate workshop and stores used in connection with the maintenance and repair of the houses on the estate, (3) a house used as a residence by the estate bailiff, and (4) a cottage used as a residence by the wood-keeper. Lord Denning and Winn LJ had drawn a distinction in their judgments between getting in money for a charity by managing its existing property and raising money by appealing to the public for funds. Lord Denning said that a separate office set up by a charity from which it set out to launch an appeal for funds would not be wholly or mainly used for charitable purposes and would not be entitled to relief from rates, but that the management and administration of the Dulwich Estate was wholly ancillary to the carrying-on of the governors' main charitable purpose. The offices, workshops and cottages were essential to that management and administration, and were accordingly occupied wholly for charitable purposes. Counsel for Oxfam submitted that the decision in the Dulwich case was right, but that the distinction drawn by Lord Denning and Winn LJ was not the true one. He (counsel) maintained that the true distinction was between using premises for getting in or raising money for the charity, whether by management of property or appeals to the public on the one hand, and earning money for it by using the premises as the site of a business on the other. There were, he submitted, nine points of distinction between an Oxfam shop and an ordinary shop:

(1) The goods sold in an Oxfam shop had been given to the charity and had not been purchased like ordinary stock in trade.

(2) The staff was unpaid.

(3) The shops were not open for regular shop hours.

(4) Although the difference between the receipts from the sales and the outgoings could be called a profit derived by Oxfam from the shop, it was not an ordinary trading profit.

(5) The premises, as well as serving as a shop for the sale of the donated goods, also served to advertise the charity.

(6) In some shops at least, clothes were on occasion sold at less than the marked prices to persons who appeared to be in need.

(7) In some cases the shops were occupied rent-free or at reduced rents.

(8) In such cases the shops were occupied only for short periods, as for instance between two commercial lettings.

(9) It was reasonable to suppose that some purchasers of goods in the shops were actuated by a desire to help Oxfam as much as by a desire to possess the goods which they bought.

Counsel for the corporation agreed with counsel for Oxfam in submitting that the distinction between the getting in of money by managing charity property and its raising by way of appeals for funds could not be supported; but he, of course, argued that using the premises in the course of managing existing charity property was just as much on the wrong side of the line as using them to raise money by appeals or to earn money by carrying on a business. The true dividing line was, he said, that suggested by Lord MacDermott in the *Belfast Blind Association* case, namely, between the use of the premises for purposes which were directly related to the carrying-out of the charitable objects and using them to get in or raise money for the charity. A merely financial nexus was not enough. He (counsel) agreed that a strict application of this principle would mean that, in the not uncommon case of the office of a charity being used both for the purpose of carrying on the charitable work and for the purpose of managing its property and raising fresh funds, it would be necessary to see which was the main user; but he argued that this was a very slight difficulty as compared with the difficulty of distinguishing between user for purposes of management and user for the purpose of carrying on a business. He submitted that the Dulwich case must have been wrongly decided unless the various Acts of Parliament and schemes in question (which were not set out in detail in the reports) could be regarded as having set up a trust for the management and exploitation of the charity property which had to be regarded as itself a separate charitable trust. Finally he said that even if user for management or fund-raising justified relief, it was impossible to bring Oxfam shops under those headings and to distinguish them from ordinary shops on the grounds suggested.

He (his Lordship) was not prepared to put on section 40 (1) a construction which might involve the drawing of a wholly artificial distinction, and he agreed with counsel that the choice was between (a) drawing the line so as to exclude from relief user for the purpose of getting in, raising or earning money for the charity, as opposed to user for purposes directly related to the achievement of the objects of the charity, and (b) only excluding from relief user for the purpose of carrying on a business to earn money for the charity. If the second were the true view, the further question arose whether Oxfam shops could be distinguished, for the purposes of the section, from a shop run by a charity on ordinary commercial lines. He (Lord Cross) thought that the first alternative was to be preferred. He appreciated that its adoption might involve, in the case of an office used both for the supervising of the work of the charity and for managing its property and raising fresh money, an inquiry into what was the main user. But that objection seemed to be very slight as compared with the difficulties into which one got if one tried to draw a line between managing property and running a business. This difficulty was strikingly illustrated by the facts of the Dulwich case itself. The governors would have had to pay rates on their woodlands if woodlands had been rateable. How could the fact that the woods were not rateable justify relief in respect of the cottage occupied by the woodman whose duty it was to look after them? How could the fact that the houses let on leases were in the rateable occupation of tenants justify relief in respect of the workshops used to enable the governors to comply with their obligations of maintenance and repair, or in respect of the house in which the bailiff lived? The truth was that the running of the Dulwich Estate was analogous to the carrying-on of a business, and that all the hereditaments in question, including the estate office, were being used for activities which had as little to do with the furtherance of education as the running of a boarding-house in Hove had to do with research into modern Polish history. This was not to say that he (his Lordship) regarded the actual decision in the

A Dulwich case as wrong; it might be that it could be justified on the lines suggested by counsel for the respondents, though if so the case was a very special one which afforded no general guidance on the interpretation of the section. The difficulty of drawing any satisfactory line between managing property and earning money was further illustrated by the case of these Oxfam shops. Counsel for Oxfam said, truly, that the donated clothing was trust property, and argued that its sale could be regarded as management of trust property; but if the shop was being run on commercial lines, any stock which was purchased would equally be property held on trust for the charity. Yet it would be absurd to describe its sale in B the ordinary course of business as an example of the management of charity property.

For these reasons he (his Lordship) would dismiss the appeal. He did so with some regret. The question whether any, and if so how much, rating relief should be given to charities in respect of premises occupied by them was no doubt a vexed and difficult one; but it was unfortunate, and hard on the appellants, that Parliament had drawn the line so vaguely in the 1961 Act that rating authorities thought for a number of years that Oxfam was entitled to relief in respect of its gift shops to which, if their Lordships agreed, it was not in fact entitled. But his (Lord Cross's) construction of C the section in question had at least the merit of simplicity, and if Parliament considered that it bore too hardly on charities he hoped that any amending legislation would be framed with greater precision.

LORD MORRIS OF BORTH-Y-GEST said that in his opinion user "for charitable purposes" denoted user in the actual carrying-out of the charitable purposes: that might include doing something which was a necessary or essential or incidental part of, or which directly facilitated or which was ancillary to, what was being done in the actual carrying-out of the charitable purpose. There might, on the other hand, be D things done by a charity, or a use made of premises by a charity, which greatly helped the charity, and which must in one sense be connected with the charitable purposes of the charity and which were properly within the powers of the charity, but yet which could not be described as being the carrying-out, or part of the carrying-out, of the charitable purposes themselves. The nature of the user might not be sufficiently close to the execution of the charitable purpose of the charity. A charity might be entitled to occupy premises and to use them other than for its charitable purposes: only if to occupation by a charity there was added user "for charitable purposes" would the benefit given by the section E accrue. On this basis, he (his Lordship) concluded that Oxfam's shops were used for an activity which was not inherently charitable, and he would dismiss the appeal, though with a measure of regret.

LORD SIMON OF GLAISDALE, LORD EDMUND-DAVIES and LORD FRASER OF TULLYBELTON agreed with the speech of Lord Cross, and the appeal was dismissed. No order was made as to costs.

F

QUEEN'S BENCH DIVISION

February 14 1975

(Before Judge Edgar FAY, sitting as a deputy judge of the division)

BRITISH GAS CORPORATION v ANGLIAN WATER AUTHORITY

Estates Gazette March 27 1976

(1975) 237 EG 961

Rating—Gas board's methane terminal on Canvey Island not in the valuation list as such, river authority claims to value it and levy drainage rate on it—Notional "gas board hereditament" nevertheless entered in valuation list, and an element of general rate attributable to drainage rate G **paid upon it—Terminal not subject to assessment by river authority**

In this action the British Gas Corporation, successors to the North Thames Gas Board, sought a declaration that certain drainage rates which the defendants, the Anglian Water Authority, successors to the Essex River Authority, had purported to levy upon them were invalid. The defendants counterclaimed payment of the amount of the rates.

Mr G E Moriarty QC and Mr G R Bartlett (instructed by Linklaters & Paines) appeared for the plaintiffs, and Mr J H R Newey QC, Mr D G Knight and Mr G Hather H (instructed by Stanleys & Simpson North, agents for D S Akroyd) represented the defendants.

Giving judgment, JUDGE FAY said that there was no dispute about the facts put before the court. The salient matters were that on Canvey Island in Essex the plaintiff board occupied a methane terminal constructed for the reception of methane transported across the sea in liquid form from Algeria. The whole of the site was below high-water level at ordinary tides. The defendant authority exercised the functions of a drainage board for Canvey Island, among which were the provision and maintenance of sea defences which prevented land below high-water mark from being J submerged. The authority's revenues were derived from its rates, and on February 20 1968, under what it claimed to be its statutory authority, it determined the gross annual value of the methane terminal at £60,000. It followed this with a demand for drainage rates for the years commencing April 1 1966 and 1967, together amounting to £7,833 6s 8d. These rates were said by the board to have been levied without authority.

The defendants' case was that they had a general duty to impose a drainage rate on hereditaments within their area. They said that their rating powers constituted a code presenting many features quite different from the more K familiar local authority rating. For instance, drainage rates were levied on agricultural land and unoccupied land, and drainage rates were levied separately on owners and occupiers, whereas local authority rates were not levied on owners or on agricultural land nor, save when a half-rate system had been adopted, on unoccupied premises. Under this special drainage code agricultural hereditaments were valued at the schedule A value, land for which a rateable value was shown in the valuation list was valued at one-third of that rateable value, and other land was valued by the authority, with an appeal to the local valuation court. The core of the defendants' case was that the methane plant clearly being L land or a hereditament, they could find no rateable value for it in the valuation list, and therefore could value it themselves; under their code, they then had to levy rates on it. The plaintiffs, on the other hand, said that special provisions were made under the general rating system for dealing with property occupied by a gas board, and that this had in this instance resulted in a gas board hereditament appearing in the valuation list. This, they said, was the hereditament on which the authority should levy their rate. "No," said the defendants, "this gas board hereditament is a fictitious or notional hereditament introduced for the purposes of general rates and having no existence in the land drainage code, which M is concerned with actual land." The board added a claim that the defendants had disabled themselves from levying directly on the methane terminal because of an agreement they had made with the local rating authority, the Canvey Island Urban District Council.

Dealing with the latter point first, there was statutory power by which the defendants could make an agreement of the kind. Many of the hereditaments that were subject to drainage rate were also subject to general rates. Obviously there would be great saving of costs if the two rates were collected at the same time. Accordingly power was conferred on the two authorities to agree on two things. First, the drainage authority agreed not to levy drainage rate on rateable hereditaments within the rating authority's area, in other words, hereditaments included in the valuation list; and secondly, the rating authority agreed to pay the drainage authority a sum equal to the estimated aggregate of the sums which the latter had agreed not to levy. The statute was silent as to how the rating authority should recoup the payment, but the obvious course for an authority to take was to levy a sufficient general rate to cover this outgoing. And that was what happened at Canvey Island. Pursuant to an agreement with the urban district council, the defendants did not levy drainage rate on ordinary hereditaments appearing in the valuation list. Instead, they notified the Canvey Island council each year of the aggregate amount which they had thus forgone, and in due course they received payment of that amount. When the council sent out its rate demands it notified ratepayers that the total rate included an item which it described as "amount due Essex River Authority in lieu of drainage rate." In the year 1969-70, this amount was 8.28d out of a total general rate of 167d.

There was in the valuation list an entry described as the "gas board hereditament." Under its agreement with the river authority the council took the view, held and propounded in this case by the gas board, that this hereditament did not attract drainage rate. Each year the council notified the river authority of the total rateable value of the district, and each year it specifically deducted utility assessments from this figure (the other utility assessments were those of the EEB and CEGB). Each year the river authority omitted these hereditaments from its calculations, so that the aggregate sum notified to and paid by the council included nothing in respect of them. This was of course because the river authority expected to get their rate upon the land occupied by the gas board direct in the shape of the rates the subject of this action. But when the Canvey Island council came to levy its general rate, it could not omit the gas board hereditament, and the gas board had paid in their general rate the amount said to be in lieu of drainage rate. For example, in the year when rates were 8.28d in the £ they paid £11,862 11s 4d in rates, of which £588 was the drainage rate element. The result was that if the gas board were right the river authority did not cast a sufficient burden on the rating authority under the agreement, whereas, if the river authority were right, the gas board had had, by virtue of the mechanism of collection under the agreement, to pay the equivalent of a drainage rate on their gas board hereditament which did not attract a drainage rate, and to pay it in addition to the rate directly levied on the methane terminal. This extra sum had enured, in any event, not to the benefit of the river authority, but to lessen the burden on all the other general ratepayers. That, then, was the position under the local authority agreement. That agreement bulked large in the facts of the case, but did not appear to be essential to the controversy between the parties. If there had been no agreement, the river authority would still have purported to levy rates on the methane terminal, and the gas board would still be saying, "No, what you must levy on is the gas board hereditament appearing in the valuation list."

Section 24 (1) of the Land Drainage Act 1930 imposed a statutory duty on each drainage authority to levy rates upon the ownership and occupancy of all land within its area, subject to certain irrelevant exceptions and refinements. Section 22 (4) of the Land Drainage Act 1961 provided that in the case of land for which a rateable value was shown in the valuation list for the time being in force, the value on which any drainage rate was assessed should be one-third of the rateable value so shown. The position of the river authority here, therefore, was that they were bound to rate the board's methane terminal, which was undoubtedly land, but that if it was "land for which a rateable value is shown in the valuation list for the time being in force," they must take one-third of the rateable value there shown. Otherwise, by section 29 (2) of the 1930 Act they could assess the value themselves, and that was the course they had in fact taken: by a notice dated February 20 1968 they determined the gross annual value of property described as the "methane plant, Canvey Island" at £60,000, and followed this with a demand dated March 25 1968 for drainage rate at 2s 9d in the £ on the owner and 1s 1d in the £ on the occupier, which, upon a drainage rate value of £20,000, or one-third of the £60,000, produced a monetary demand for a total of £3,833 6s 8d. Then there was the General Rate Act 1967, which provided for gas boards who either supplied gas to consumers in a particular rating area or manufactured or produced it in that area to be rated as if they occupied in that area "a hereditament of a rateable value calculated in accordance with the provisions of Part I of Schedule VI to this Act." Part I of that schedule was headed "Calculation of rateable value of notional hereditament," and provided for the rateable value of a notional hereditament to be reached by an arithmetical calculation starting from a basic total of rateable values set out in paragraph 2 (1). Part II of the schedule, headed "Supplementary provisions," was also relevant. Paragraph 7 required a gas board to transmit the data necessary for the calculation to both the rating authority and the valuation officer. Paragraph 8 required the valuation officer to do the calculation and notify the rating authority of the rateable value, and paragraph 10 provided for inclusion of the necessary references in the valuation list.

It was important to note that these provisions first made their appearance on the statute book in 1955, gas board hereditaments having previously been valued on the profits basis. The new system of assessing rateable values was no doubt thought more appropriate to a nationalised industry. All gas board land in a generating area, other than the excepted showrooms, offices and dwellings, was lumped together as one hereditament described as "notional" and assigned a rateable value which seemed to be proportionate to the amount of gas manufactured and/or supplied in the relevant area. The notional hereditament could not exist, however, unless there was some occupancy of land as its foundation. What called it into being was either supply to consumers in the rating area or the manufacture of gas in the rating area. The latter could not in the nature of things be accomplished without land occupancy, and the pipes supplying the former had been held to be sufficient occupation of land to render a gas company rateable: see *R v Brighton Gas Light & Coke Co Ltd* (1826) 5 B & C 466. But although there could not be a notional hereditament without occupancy of land, the converse was not true. There could be occupancy of land without a notional hereditament being called into existence. This occurred if there was land occupied by the board in a rating area where they neither manufactured nor supplied consumers. Whether any such area existed in fact, he (Judge Fay) did not know, but the facts of the present case showed that its existence was possible, because the methane terminal itself was not devoted to manufacture, and if it had existed in a rating area where, unlike Canvey Island, no consumers were supplied and no gas was manufactured, there would have been a substantial hereditament existing in fact but no notional hereditament in the valuation list for the time being. As it was, the existence of the terminal was not a factor influencing the rateable value of the notional hereditament. That this might be a blemish in the system was recognised by the provision now forming subsection (5) of section 33 of the General Rate Act 1967, and the Minister had in fact made the Natural Gas Terminals (Rating) Order 1970 designating, *inter alia*, the Canvey Island terminal and providing for its valuation. There were now in the valuation list a series of notional hereditaments based on this terminal and deemed to be occupied by each of 11 gas boards, among whom were the plaintiffs.

Of all the statutory provisions cited, the most important seemed to be section 22 (4) of the Land Drainage Act 1961. Prior to this coming into force, the defendant river authority were clearly entitled to value the methane terminal and levy drainage rate upon it under section 24 (4) of the 1930 Act, but section 22 (4) brought that situation to an end. The crucial words of the subsection were, " land for which a rateable value is shown in the valuation list." The gas board's notional hereditament was shown in the valuation list, and shown with a rateable value. A hereditament was land: the terms in rating law had the same meaning, " land " being the generic name and " hereditament " the name of the unit, and this was no less so in drainage rating: see section 24 (4) of the Act of 1930, where all hereditaments were divided into agricultural land and " other " land. If the gas board hereditament was to be excluded from the ambit of section 22 of the 1961 Act, that must be because a notional hereditament was not land. But the whole object of creating the notional hereditament was to get it treated as land. This was made clear by the provision which first introduced it, section 6 of the Rating and Valuation (Miscellaneous Provisions) Act 1955. This was the forerunner of the present provision in section 33 of the General Rate Act 1967, which was a consolidating Act and could not be assumed to have changed the law as enacted in 1955. The 1955 provision was of course on the statute book when Parliament amended the Land Drainage Act and introduced section 22 of the 1961 Act. Accordingly, there appeared to be no justification for treating a notional hereditament as a non-hereditament. An argument to the contrary raised by counsel for the river authority was to the effect that the authority must apply the drainage code and were not necessarily bound by anything in the local authority rating code. but the fact was that the two codes were not wholly separate. Section 22 of the 1961 Act had engrafted some of what was now the general rate code on to the drainage code, as had also the agreement provision of section 25 of the 1961 Act. When in 1961 Parliament brought within drainage rating " land for which a rateable value is shown in the valuation list " it must have intended to include all hereditaments in that list, including the notional ones put there by section 6 of the 1955 Act. This view appeared to derive added force from the fact that Parliament in 1955 was not content with expressly requiring the general rating authority to levy on the notional hereditament: it also expressly required the rating officer to show the hereditament in the valuation list.

Mr Newey had called attention to the wording of section 33 (1) of the 1967 Act. That subsection provided that no premises occupied by a gas board should be liable to be rated or be included in any valuation list, and expressed this provision as being " subject to " subsection (2), which listed the excepted hereditaments, but as being " without prejudice to " subsection (3), which created the notional hereditament. It was suggested that the difference of wording reflected a difference in the character of the two subsections, and that " subject to " recognised a conflict, but that " without prejudice to " was consistent with there being no conflict between subsection (1) and subsection (3), and that there was no conflict if the notional hereditament was not land or premises. Here again it seemed that one must go back to the 1955 Act, where these phrases did not appear. They were words of caution introduced by the draftsman of the consolidating Act, and they were of no help in the construction, which must remain the same as that of the 1955 Act.

It followed that under section 22 (4) of the 1961 Act the river authority should have taken note of the gas board hereditament in the valuation list and should have rated it on a value equal to one-third that of the rateable value there shown. Having elected to make use of section 25, the authority should have included the appropriate amount in their claim upon the local council. This finding did not dispose of the action, however, because there remained the question whether the notional hereditament either included or was in substitution for the actual hereditament, namely the methane terminal. If it was neither of these things, its liability to drainage rate would not relieve the methane terminal of liability. Here it was desirable to describe the terminal more closely. The facilities included a jetty liquefaction plant, evaporation plant, compressor houses and storage facilities. Within the same curtilage was a naphtha reforming plant. It was to be noted that the defendants had treated this as a separate hereditament, and after this action was commenced had determined its annual value at £15,000 and demanded drainage rates for the years 1965/66 onward. The naphtha plant was not the subject of claim or counterclaim in this action, but its status could usefully be considered with that of the methane terminal. The plant had been used from 1965 to 1969 to manufacture town gas. This, together with the supply to consumers in Canvey Island, activated the machinery of section 6 (1) of the Rating and Valuation (Miscellaneous Provisions) Act and section 33 (3) of the General Rate Act, and led to the insertion of the notional hereditament in the valuation list. Since it had been held that the gas board must pay drainage rates on the notional hereditament, it would be odd if the board had also to go on paying on the hereditament which had called the notional hereditament into being and which, according to Part I of the schedule, was reflected in the rateable value assigned to the notional hereditament.

It was true that this consideration did not apply to the methane terminal, but the terminal's immunity from rating was recognised to be an anomaly by the taking of the power to rectify it now contained in subsection (5) of section 33 of the General Rate Act 1967, and since the making of the Natural Gas Terminals (Rating) Order 1970 the anomaly had ceased. As regarded the general rate, the position was clear: by section 6 (2) of the 1955 Act the new liability was in substitution for liability for any premises occupied (other than the excepted premises). This did not in strict terms apply to drainage rate, but it appeared that the effect of this subsection, and of its provisions as to the valuation list, and of the provision in Part II of the schedule as to the valuation list, was that the gas board were entitled to point to the gas board hereditament in the valuation list and say, " Apart from the excepted premises, this is the equivalent of all the property we occupy in this rating area." The matter became even clearer when one appreciated that since the Minister's order of 1970 the methane terminal as a hereditament was transformed into 11 notional hereditaments with the object of sharing the rates burden among the boards which took gas from the terminal. If the defendants were right, the drainage rate amount could not be so shared and must fall solely upon one of those boards, namely the plaintiffs.

Accordingly the plaintiffs' case succeeded. Little had so far been said about the effect of the agreement with the Canvey Island council. This figured large in the gas board's case as originally framed and in the argument, and part of the relief the board sought was a declaration based on the pragmatic argument that the section 25 agreement had led to the drainage rate being passed on to the gas board hereditament via the local authority's general rate, and that there was the injustice of double taxation if the methane terminal was still directly rateable. This argument was capable of affording support to the view he (Judge Fay) had formed, but he did not regard it as conclusive. True, the defendants' construction would produce the double taxation anomaly which the plaintiffs complained of, but agreements under section 25 produced other and perhaps greater anomalies, the most outstanding of which was the transference

A of the owner's drainage rate to the occupier, because the global sum payable by the local authority to the river authority included owner's rate, but the local authority could only recoup themselves, as they had done, by levying on the occupier. It was suggested that the occupier could pass the burden to the owner under s 26 (4) (c) of the Land Drainage Act 1930, as amended by the Land Drainage Act 1961. Some day a court might have to decide whether a tenant paying general rate might make a deduction from his rent under that provision; meanwhile, the situation stood as a serious anomaly depriving the plaintiff board's argument based on section 25 of much of its force. It was preferable for the B judgment to rest on the view formed of the effect of section 22 (4) of the Land Drainage Act of 1961. There would be judgment for the plaintiffs in the terms of the first declaration sought, and the counterclaim would be dismissed.

The gas board were awarded costs of both claim and counterclaim.

QUEEN'S BENCH DIVISIONAL COURT

December 15 1975

C (Before Lord WIDGERY CJ, Mr Justice PARK and Mr Justice MAY)

HARWICH DOCK CO LTD v COMMISSIONERS OF INLAND REVENUE

Estates Gazette April 17 1976

(1975) 238 EG 187

Rating—Permission obtained by dock company for construction of a jetty and ancillary buildings in Harwich harbour —Permission not shown to extend to use of the jetty etc— No evidence that dock company were carrying on their undertaking under any enactment—Jetty etc to be valued for rating purposes under the ordinary provisions to that D end, not under the Docks and Harbours (Valuation) Order 1971

This was an application by the Harwich Dock Co Ltd for an order of mandamus requiring the Commissioners of Inland Revenue and the valuation officer for the Colchester area to calculate the rateable value of a jetty and premises at the E Navyard Wharf, Harwich, in accordance with the provisions of the Docks and Harbours (Valuation) Order 1971.

Mr G Dobry QC and Mr F A Amies (instructed by Middleton, Lewis & Co, of Horley, Surrey) appeared for the applicants, and Mr B J Davenport (instructed by the Solicitor of Inland Revenue) represented the respondents.

Giving judgment, LORD WIDGERY said that the applicant company was incorporated in 1963 to carry on business at the Navyard Wharf. Pursuant to two Acts of Parliament, the Harwich Harbour Act 1863 and the Harwich Harbour Act 1865, the company applied to the Harwich Harbour Conservancy Board for permission to erect a new jetty and ancillary F buildings. In their reply the conservancy board offered no objection in principle, and the jetty and buildings were built. The respondent now contended that the jetty and buildings fell to be dealt with, for the purpose of calculating rateable value, in accordance with the General Rate Act 1967. The applicants contended that they were entitled to avail themselves of the provisions of the Docks and Harbours (Valuation) Order) 1971 as being a dock undertaking operating under the authority of two Public Acts.

By section 17 of the 1863 Act it was not lawful for any person within the jurisdiction of the board to erect any G building or jetty without consent, while section 26 of the 1865 Act required consent by licence for the construction of any jetty or building. It was conceded that the jetty and buildings comprised a hereditament and that the dock was a harbour undertaking. The dispute was whether the carrying-on of such an undertaking was carried on under any enactment. The applicants appeared to recognise that authority was required if they were to erect the jetty, but that did not mean that their undertaking was being carried on under any enactment. He (his Lordship) was not satisfied on the evidence before the court that any authority was needed to use the jetty and buildings. There was nothing to suggest that any H further authority was required, once permission had been given to erect the jetty, to use it. In his judgment the application should be dismissed.

PARK and MAY JJ agreed, and the application was dismissed with costs.

REAL PROPERTY AND CONVEYANCING

A

QUEEN'S BENCH DIVISIONAL COURT

January 23 1975

(Before Lord WIDGERY CJ, Mr Justice ASHWORTH and Mr Justice MICHAEL DAVIES)

PRICE v CROMACK

Estates Gazette January 3 1976

(1975) 237 EG 41

Owner who contracted to allow effluent to be dispersed on his land did not thereby "cause" pollution of river through cracks in walls of the dispersal system—Conviction quashed

B

—Distinction between "causing" and "knowingly permitting"

This was an appeal by Mr William Geoffrey Price, of Felton Grange, West Felton, Oswestry, Salop, against his conviction by Ellesmere magistrates on March 4 1974 on informations laid by the respondent, Mr Barry Norman Cromack, a district pollution prevention officer of the Severn River Authority, charging him with two offences of causing polluting matter to enter a stream.

Mr T I Payne (instructed by Underwood & Co) appeared for the appellant, and Mr A P Fletcher (instructed by Sharpe, Pritchard & Co, agents for the solicitor to the Severn-Trent

C

Water Authority) represented the respondent.

Giving judgment, LORD WIDGERY said that at the magistrates' court three informations were laid against the appellant. He was acquitted on one charge, but convicted on a charge that on September 27 1973, at Oak Farm, Bagley, Salop, he did cause to enter a tributary of the River Perry poisonous, noxious or polluting matter, contrary to section 2 (1) of the Rivers (Prevention of Pollution) Act 1951. He was also convicted of a similar offence under section 8 (1) of the Salmon and Freshwater Fisheries Act 1923, as amended in 1972, but it was necessary to deal only with the conviction under the Rivers (Prevention of Pollution) Act 1951, since

D

both convictions stood or fell together.

On September 27 1973 Mr Cromack inspected the River Perry, and found it foul-smelling and dirty, with dead fish floating in it. He traced the pollution to Oak Farm, Bagley, owned by Mr Price. The situation was that in 1971 an agreement had been reached between the appellant and two companies that the appellant would accept on his land certain effluents discharged by the companies on adjoining land. The contract contemplated the use of an existing sewer, and then the effluent spreading over the appellant's land and disappearing by saturation. In 1973 two lagoons were built on the appellant's farm and by that time the right to dis-

E

charge the effluent had become vested in a company called Ellesmere Animal Products Ltd. So far as it was relevant, the appellant was concerned with that company, he being the managing director up to 1972 and continuing as director thereafter. Evidently the existing method of disposal was causing trouble in 1973, and the lagoons were built to contain the effluent prior to its disposal by natural means. Mr Cromack traced the pollution to the lagoons. He found breaches in the walls of both, and the effluent was running from one lagoon to the other and then escaping into the

F

land and out into the river. The question for the court was whether, in the circumstances, it was proper for the justices to convict the appellant on a charge of "causing" the effluent to enter the river. It was to be noted that the 1951 Act contemplated two offences. One was causing poisonous, noxious or polluting matter to enter a stream, and the other was knowingly permitting it. The offence here had been framed in terms of "causing," and what the court had to consider was only whether the conviction of "causing" could legally stand.

The House of Lords decided a similar question in 1972 in *Alphacell Ltd v Woodward* [1972] AC 824, and it was import-

G

ant to observe that this same distinction between causing and knowingly permitting was very much in their Lordships' minds in that case. It was their overwhelming opinion that whatever else "causing" might involve, it did involve some active operation as opposed to mere passive standing by and looking on. One therefore had to begin by asking where was the positive act by the appellant which was said to be a cause, or perhaps the cause, of the effluent entering the river. Counsel for the respondent argued that the positive act by the appellant was his entering into an agreement with the next-door firm to receive the effluent on to his land. He (his Lordship) would have agreed that if the effluent had been

H

handed over at the boundary of the appellant's land and he had then been free to deal with it as he wished, it might very well have been that one would find in the appellant the essential feature of positive action which would justify his being said to have caused the ultimate entry into the river. However, that did not happen in this case. The effluent went on to the appellant's land by gravity, and found its way into the river by gravity, with no action on the appellant's part. He (Lord Widgery) sympathised with the respondent's approach. He could see that there should be no great difference between the man who generated effluent on his own land and one who voluntarily agreed to accept some-

J

body else's effluent and had it put on his land. However, he could not find that it amounted to "causing" effluent to enter a river when a landowner merely stood by and watched effluent pass over his land, even if he had committed himself by an agreement to allowing the adjoining owner so to act. Conviction on these charges was therefore not possible in law in the circumstances of the case.

ASHWORTH and MICHAEL DAVIES JJ agreed, and the appeal was allowed with costs.

K

CHANCERY DIVISION

December 11 1974

(Before Mr Justice GRAHAM)

BRACEWELL AND ANOTHER v APPLEBY

Estates Gazette March 6 1976

(1974) 237 EG 731

A right of way leading to a plot of land cannot be used for access to another plot, adjacent to the first, subsequently

A **acquired by the grantee of the right—Objectors who failed to take action until the grantee had built on the adjacent plot nevertheless refused an injunction and held entitled to damages only**

In this action Mr Colin Derek Bracewell, of 4 Hill Road, Heath End, Farnham Heights, Surrey, and Mr Guy Armand Charles Wright, of 1 Hill Road, sought a declaration that the defendant, Mr Edward Appleby, of 2a Hill Road, former owner of no 3, had no right of way over portions of the surface of Hill Road owned by the plaintiffs, and an injunction restraining trespass thereon, alternatively damages.

B Mr D Jackson (instructed by Charles Russell & Co, agents for W H Hadfield & Son, of Farnham) appeared for the plaintiffs, and Mr J Monckton (instructed by Slaughter & May, agents for Close & Son, of Camberley) represented the defendant.

Giving judgment, GRAHAM J said that this was a dispute between two neighbours about a right of way and was one which, if there had been better communication between the parties and no clash of personalities, might well not have come before the court. Nevertheless a question of law was raised which one would have expected to have found already clearly resolved. Did the law permit the creation of a right-C of-way, appurtenant to a particular close, for purposes of access not only to that close, but also to any adjoining land which was not part of such close at the date of the grant but which might at some future time come into the ownership of the owner of such close? Could a grant which, if that was its proper construction, purported to do such a thing have the effect of creating an easement in the true legal sense, so that it was appurtenant to the land, created a relationship of dominant and servient tenement, and passed with and on a subsequent assignment of the land?

The plaintiffs and the defendant all bought houses on part of what was known as the Farnham Heights estate in D Surrey, which was developed by Wrotham Estate Ltd in about 1962. This particular part of the estate was laid out as a cul-de-sac with six plots, fed by a private road known as Hill Road. The plaintiffs respectively bought plots referred to as no 43 or no 4, and no 40 or no 1. The defendant's predecessors in title, Mr and Mrs York, bought plot no 42, or no 3, and plots 38, 39 and 41 were bought by other parties. On the sale of each of the six plots there were grants and reservations of rights of way over Hill Road, and over portions of each plot. For example, the plan attached to the statement of claim showed a portion of Hill Road marked yellow, half of which belonged to Mr Wright E and half to Mr Bracewell, subject to the grants and reservations of rights of way (and obligations to repair) which were now in question. It was the construction and effect of the grants and reservations in the schedules to the conveyances, in the light of the further history of the matter, which was the determining factor.

The defendant was able to buy a piece of land adjoining his plot, no 3. There was no means of access to that piece of land except from no 3 itself or from the back gardens of houses in other roads. The site was land-locked, in very rough condition, and no use to anyone except the owner of no 3. The defendant incorporated it into his garden. He F then thought of building a new house, partly on the garden of no 3 and partly on the extra land, with access to the new house by means of a drive parallel to the north-west boundary of no 3 and leading to Hill Road and over part of the plaintiffs' yellow land to Alma Road. He then sold no 3 and moved into the new house, no 2a, and the whole dispute turned upon whether the original right of way he acquired when he bought no 3 also gave him a right of way to the new house built partly on land originally part of no 3 and partly on the new land. It was, of course, not quite so simple generally as might appear from his (Graham J's) short summary. The application the defendant had to make for planning permission was opposed by the plaintiffs and others. G There was no restrictive covenant imposed by Wrotham Estate Ltd when the original development took place which entitled the plaintiffs or any of the defendant's neighbours to object to the new house for that reason.

Objection to the defendant's proposal was formally raised in August 1972 on the basis that he did not have a legal right of way over Hill Road which would entitle him to access to the new house, such rights as did exist being limited to the six existing houses, and the plaintiffs gave him notice that an injunction would be sought if he proceeded. At first the defendant appeared ready to compromise, but later his H attitude hardened, and in effect he challenged the plaintiffs to take proceedings. The cold war hotted up when in January 1973 the defendant announced that he would shortly start operations. After a letter from the plaintiffs and a short exchange of fire about costs, the defendant started building, apparently early in April. The plaintiffs issued their writ and sought an interim injunction, which was refused, no doubt partly for the reason that the judge was not certain as to the proper construction of the grant in question and partly because the motion was brought only when the house was finished. He (his Lordship) would have reached the same conclusion on delay, the truth being that neither side felt J sure enough to test matters till driven to. The plaintiffs made threats, but stood by until the house was erected, and the defendant, insisting he had a right of way by virtue of the grant, never took any proceedings to obtain a declaration to that effect and pushed ahead until the house was a *fait accompli*. Furthermore, he (Graham J) would not grant an injunction now, even were he satisfied that the plaintiffs were entitled to relief, following the principles enunciated by Brightman J in *Wrotham Park Estate Co Ltd v Parkside Homes Ltd* [1974] 1 WLR 798 at 809-811. Although an injunction would be against trespassing upon the plaintiffs' land, and would not go to pulling the house down, it would K make no 2a uninhabitable and put the plaintiffs into an unassailable bargaining position. Mr Jackson, for the plaintiffs, no doubt realising the difficulty, offered that if such an injunction were granted he would not insist on enforcing it but would be prepared to grant a right of way on payment of £1,200 to each of the plaintiffs, that sum being 1/5 of the £6,000 which, he said, the evidence showed was the amount of the notional profit the defendant had made by his operations. It would be quite wrong of him (his Lordship) to accede to this suggestion, first in view of the unwillingness of the plaintiffs in fact to take effective action until the house was finished, so that they could not claim to be put into L the bargaining position they could have got into if they had taken action before the defendant had committed himself to his contract, and secondly, in view of the intolerable nature of a decision which would render uninhabitable a house which was a *fait accompli* and was now being lived in by the defendant.

That reduced to a considerable extent the matters which had to be decided. The grant in question created a "right of way of the fullest description" over Hill Road in common with the transferor and all other persons so entitled, and a "right of way with or without vehicles over" the yellow strip in common with the transferee and his successors in M title. He (Graham J) had reached the conclusion that the effect of the words used was to grant a private right of way of as extensive a nature as was legally possible, having regard to the fact that these were private houses on a building estate that was being developed. He thought that it could not have been intended that Hill Road should be treated as if there were a public right of way over it. The conveyance could quite simply have said "such rights as if Hill Road were a public highway" had that been the intention. *Harris v Flower* (1904) 74 LJ Ch 127 justified the assertion that in such circumstances the grant of access to no 3 did not enable the defendant to establish that he had a right to extend his right of way to the blue land, to which it was not appurtenant, thereby in practice doubling the burden on the servient tenements of the plaintiffs because of the existence of two houses and families using Hill Road from no 3 instead of one, as before. In the case cited, at p 132 Romer LJ remarked: "If a right of way be granted for the enjoyment of close A, the grantee, because he owns or acquires close B, cannot use the way in substance for passing over close A to close B." The circumstances in that case and in the present case were parallel, and the principle expressed by Romer LJ accordingly governed the present case. It followed that the defendant was not granted a right of way which entitled him to pass over Hill Road to and from his new house, no 2a. His right of way was limited and appurtenant to no 3 only.

That left the question of relief. As already stated, he (his Lordship) was unwilling in the circumstances to grant an injunction, but the plaintiffs had made out their legal right and so sought damages under Lord Cairns' Act. As he had been requested to do by both parties, he would assess a figure, and in doing so would follow the approach of Brightman J in the *Wrotham Park* case, beginning at p 812. The defendant must pay an amount of damages equivalent to a proper and fair price which would be payable for the acquisition of the right of way in question, not some proportion of the development value created. Here the plaintiffs, for amenity reasons, did not want an extra house built in the cul-de-sac, but for the purpose of estimating damages they and the other servient owners in Hill Road, albeit reluctant, must be treated as being willing to accept a fair price for the right of way in question and must not be treated as if they were in the extremely powerful bargaining position which an interlocutory injunction would have given them if it had been obtained before the defendant started operations and incurred expense. Such was the penalty of standing by until the house was built. On the evidence, the probable figure of notional profit the defendant had made, being the difference between the overall cost of the new house and its present-day value, seemed to be between £4,000 and £6,000, and it seemed fair to take £5,000 as about the accurate figure. The proper approach was to try to arrive at a fair figure which, on the assumption made, the parties would have arrived at as one which the plaintiffs would accept as compensating them for loss of amenity and increased user, and which at the same time, whilst making the blue land a viable building plot, would not be so high as to deter the defendant from building at all. The defendant was not a speculative builder, but wanted to live in (and in fact did now live in) 2a himself; he (his Lordship) thought he would have been prepared to pay what was relatively to his notional profit quite a large sum for the right of way in question and to achieve the building of his new home. This was a time of rising property values in which (as the court thought) he would have been prepared to pay £2,000 to get his right of way. The plaintiffs were entitled to their appropriate share of this figure, namely 1/5 each, so that each would be awarded £400 by way of damages for the exercise of a right of way over their respective pieces of land.

The plaintiffs were awarded costs.

QUEEN'S BENCH DIVISION

July 2 1975

(Before Mr Justice FORBES)

RAMAC CONSTRUCTION CO LTD v J E LESSER (PROPERTIES) LTD

Estates Gazette March 13 1976

(1975) 237 EG 807

Building contract—Builders begin action for balance of contract sums shortly before issue of architect's certificate fixing date for operation of clause providing for damages for late completion—Absence of a certificate at time of issue of writ no bar to an application for a stay under arbitration clause—Dispute in existence even before writ issued—In any case, builders themselves on shaky foundations in their action

This was an appeal by J E Lesser (Properties) Ltd, of Staines Road, Hounslow, from an order of Master Bickford-Smith granting Ramac Construction Co Ltd, of Stonecot Hill, Morden, summary judgment for £10,744 in an action brought by them against the appellants over a building contract.

Mr J A Tackaberry (instructed by McKenna & Co) appeared for the appellants, and Mr J F Uff (instructed by Masons) represented the respondents.

Giving judgment, FORBES J said: This case is concerned with a point affecting building contracts which both Mr Tackaberry and Mr Uff assure me not infrequently arises, and on which no ruling exists. They both, therefore, asked me to give judgment in open court, which I now do. The plaintiffs are the builders and the defendants the building owners or employers. The contract was one which incorporated the standard form of RIBA contract with certain modifications, the most important of which was that any reference in the standard form to "the architect" should be taken as a reference to "the employer's representative." The defendants explain this as a matter of convenience for them; it enables them to use their supervisory manpower in a more flexible manner because the change of "representative" can be done informally. It is, they say, a common practice adopted by many developers. The contract was one for the construction of an industrial workshop and office block in Hounslow. It provided for a completion date and for fixed sums for liquidated damages in the event of late completion. In fact there were delays, and completion was some time behind schedule. The employers' representative allowed certain extensions of time, but even with these extensions completion was late. The certificate of final payment, issued by the defendants on August 22 1974, showed a balance outstanding on the contract sum of £11,744 less a "contra charge" of £10,744 described as "Damages Total," leaving £1,000 as a balance due to the plaintiffs. This sum was paid. The plaintiffs sued for the sum of £10,744 and issued a summons for summary judgment under order 14. The defendants applied for a stay of proceedings so that the matter could be referred to arbitration under the contract. The master gave summary judgment for the plaintiffs under order 14 for the full sum, but stayed execution pending appeal on condition that the defendants brought £5,000 into court.

The argument turns on clauses 22 and 35 of the standard form of contract. Clause 22 is in these terms:

> "If the contractor fails to complete the works by the date for completion stated in the appendix to these conditions or within any extended time fixed under clause 23 or clause 33 (1) (c) of these conditions and the architect certifies in writing that in his opinion the same ought reasonably so to have been completed, then the contractor shall pay or allow to the employer a sum calculated at the rate stated in the said appendix as liquidated and ascertained damages for the period during which the works shall so remain or have remained incomplete, and the employer may deduct such sum from any moneys due or to become due to the contractor under this contract."

Clause 35 is a long clause, and is familiar to those concerned with building contracts, and I do not think I need to read the clause in full. The argument here turns really on the provisions of subclauses (1) and (3), since, as I shall show in a moment, subclause (2) is not applicable. At the time of the issue of the writ there was no certificate of the architect under clause 22, because one had not then been given, but before the order 14 summons was heard this omission was repaired by the issue of a certificate by a

A Mr Beatty, one of the defendants' directors, as "employer's representative."

In addition to the authorities which I shall mention in this judgment I was referred to *Russell v Pellegrini* (1856) 26 LJNS 75; *London & North Western Railway v Billington* [1899] AC 79; *Morgan v Johnson & Co* [1949] 1 KB 107; and *Hanak v Green* [1958] 2 QB 9. Mr Uff, for the plaintiffs, now concedes that a triable issue exists in the action and that he cannot support the order for summary judgment, but he does so because he concedes that under order 18, rule 9 the defendants are entitled to counterclaim liquidated damages equal to the full amount claimed, and to set off B this amount as a defence to the claim, although the cause of action on which the counterclaim was based arose after the issue of the writ. He says, however, that there can be no entitlement to a stay, because at the time of the issue of the writ there was no architect's certificate, and such a certificate is a condition precedent to a claim under clause 22. He relies on a passage from a judgment of the Master of the Rolls in *Dawnays Ltd v F G Minter and Trollope & Colls Ltd* [1971] 1 WLR 1205. The passage in question is at p 1210 between A and B:

> I ought to mention one further point. Under clause 8 (a) no sums are payable in respect of delay unless there has been a C proper certificate by the architect certifying that the works ought to have been completed within the specified time and any certified extension of it. The architect in this case did not finally determine what extension should be given. He originally gave three weeks, but afterwards was to reconsider it. So, in any case, no sums were payable for delay.

Mr Uff relies upon that passage as indicating that the architect's certificate is a condition precedent, but of course, in *Dawnays Ltd* the Master of the Rolls was dealing with a principle which the Court of Appeal was there laying down, a principle which is succinctly put on p 1209 at H:

> D An interim certificate is to be regarded virtually as cash, like a bill of exchange. It must be honoured. Payment must not be withheld on account of cross-claims, whether good or bad, except so far as the contract specifically provides.

Under this principle it was necessary for a defendant to be able to set against the architect's certificate showing payment due to the plaintiff another architect's certificate certifying the date on which work should have been completed, because it was only in those circumstances that the employer was entitled, under the contract, to "deduct such sum from any moneys due or to become due to the contractor under this contract." I ought to have added that the general principle enunciated by the Court of Appeal in *Dawnays'* case E appears to have been disapproved by the House of Lords in *Gilbert-Ash (Northern) Ltd v Modern Engineering (Bristol) Ltd* [1973] 3 WLR 421. As will be apparent in a moment, I would not think that the passage in which the Master of the Rolls referred to the certificate under what, for the purposes of this case, is clause 22, would have been covered by the disapproval of the House of Lords expressed on the general principle. Mr Uff's argument then proceeds in this way. He says that the counterclaim is here for a liquidated sum, namely the amount certified by the architect under clause 22. But at the time of the issue of the writ there was F no such certificate in existence; hence there could be no counterclaim, and thus, he says, no dispute to be referred to arbitration. He refers to section 4 (1) of the Arbitration Act 1950, which is in these terms:

> If any party to an arbitration agreement, or any person claiming through or under him, commences any legal proceedings in any court against any other party to the agreement, or any person claiming through or under him, in respect of any matter agreed to be referred, any party to those legal proceedings may at any time after appearance, and before delivering any pleadings or taking any other steps in the proceedings, apply to that court to stay the proceedings, and that court or a judge thereof, if satisfied that there is no sufficient reason why the matter should not be referred in accordance with the agreement, and that the applicant G was, at the time when the proceedings were commenced, and still remains, ready and willing to do all things necessary to the proper conduct of the arbitration, may make an order staying the proceedings.

Mr Tackaberry's point is that clause 35 is wide enough to allow the arbitrator to determine the substantive point at issue between the parties, namely the proper duration of the extension of time which ought to be allowed to the plaintiffs for completion of the work. Once this point is settled, the amount of liquidated damages is a mere matter of calculation, but, he says, the arbitrator is entitled to make an award in the sum so calculated. The power, continues Mr Tackaberry, to settle this difference between the parties is derived H from clause 35 and is irrespective of whether there has been an architect's certificate under clause 22. It should be noted in passing that the certificate of practical completion in this case was dated August 17 1973, so that the restrictions in clause 35 (2) no longer apply.

I find myself agreeing with Mr Tackaberry. The fallacy in Mr Uff's argument seems to me to be that he asserts that if no counterclaim had arisen before the issue of the writ there was no dispute to go to arbitration; but the general principle is, I think, plain. To succeed in an action on the contract the plaintiffs, or the defendants in their counter- J claim, will have to show that any necessary preliminaries to the claim have been fulfilled. An architect's certificate under clause 22 is such a necessary preliminary; in that sense it is a condition precedent to any action by the employer to recover the sums due. It is not, however, a condition precedent to the submission of a dispute to arbitration. Under the words of clause 35, once a dispute arises it is automatically referred to arbitration. I quote: "such dispute or difference shall be and is hereby referred." The words of clause 35 are very wide. In my view a dispute or difference arose between the parties the minute the plaintiffs refused K to accept that there was only £1,000 due to them on the final payment. All that is necessary under the contract is that the dispute should be, in the words of clause 35 (1),

> "as to the construction of this contract or as to any matter or thing of whatsoever nature arising thereunder or in connection therewith (including any matter or thing left by this contract to the discretion of the architect or the withholding by the architect of any certificate to which the contractor may claim to be entitled. . . ."

Under clause 22 the architect has to certify in writing that "in his opinion the same ought reasonably so to have been completed." In other words the work ought reasonably to L have been completed within the extended time fixed. The fact that he has not done so does not mean that no dispute has arisen: a dispute has arisen: it is as to the proper extension of time. The arbitrator has the power to decide this question irrespective of whether there has been an architect's certificate under clause 22. He also has power to award any sum due as a result of his determination of this question.

In fact the plaintiffs in this case have proceeded on an entirely false basis throughout. Mr Uff has argued that the defendants should not have a stay because the counterclaim was not maintainable at the date of the writ, as no certificate M under clause 22 was then in existence. I have already dealt with the confusion which this argument involves between what are the necessary conditions precedent to the bringing of an action and a reference to arbitration respectively. But the confusion goes much deeper. In fact it is the plaintiffs whose tackle is not in order. The necessary condition precedent to a claim under the contract in the circumstances which here prevail is that there should have been an architect's final certificate under clause 30 (6) (b). This alone creates the debt due from the employer to the contractor. But there is here only one document which I have been

A shown which could do duty as a final certificate. This is the document of August 22 1974, which shows, not £10,744 as a balance due to the plaintiffs, but £1,000, and the latter sum has already been paid. Indeed, in the statement of claim endorsed on the writ the plaintiffs asserted that that document was the final certificate on which they relied. In my view they cannot accept and deny the document; they cannot accept the parts of the document which are in their favour and deny those which are to their disadvantage. It has to be taken as a whole. It is either a final certificate or it is not, and as far as I can see the only sum that certificate shows to be due is the sum of £1,000, and that has been B paid. The true answer to the plaintiffs' claim in this action is thus not that there is an available set-off, but that the necessary condition precedent to a claim under the contract has not been fulfilled.

However that may be, I am not concerned with what might be the position in the action. I am concerned with whether or not a dispute or difference has arisen between the parties and whether, if it has, this is an appropriate case for a stay. As I have indicated, a dispute has arisen and has already been referred to arbitration, although, of course, the appointment of an arbitrator has not yet been agreed between the parties. The only remaining duty is to satisfy myself of the C matters referred to in section 4 of the Arbitration Act 1950, namely " that there is no sufficient reason why the matter should not be referred in accordance with the agreement, and that the applicant was, at the time when the proceedings were commenced, and still remains, ready and willing to do all things necessary to the proper conduct of the arbitration." There is no dispute about these matters, and in the circumstances I think this is a proper case for a stay. Accordingly I allow the appeal and discharge the master's order, substituting therefor an order that there should be a stay of proceedings so that the arbitration may proceed under section D 4 of the Arbitration Act 1950. The plaintiffs' summons for summary judgment is dismissed, and there should be an order for payment out to the defendants' solicitors of the £5,000 in court and accrued interest.

The defendants were awarded costs above and below.

CHANCERY DIVISION

July 11 1975

(Before Mr Justice FOSTER)

E BASILDON DISTRICT COUNCIL v MANNING

Estates Gazette March 20 1976

(1975) 237 EG 879

Poultry-farmer claims to have fenced off scrubland and dumped manure there for dispersal—No evidence of discontinuance of possession by owners who had on two occasions applied for permission to develop the land—Acts of fencing etc too trivial to found a claim to adverse possession—Farmer held not to have acquired a possessory title

This was a claim by Basildon District Council, as successors F to Basildon Urban District Council, against Mr Arthur Frederick Manning, for an injunction to restrain him from entering certain land at Queen's Park, Billericay, Essex, alternatively for an order for possession of the land. The defendant counterclaimed a declaration that he had a possessory title to the land.

Mr J M Chadwick (instructed by Sharpe, Pritchard & Co, agents for J C Rosser, manager, administrative and legal services, Basildon) appeared for the plaintiffs, and Mr R E Denman (instructed by Oswald Hickson, Collier & Co, agents for Wortley, Byers & Co, of Brentwood) represented the defendant.

Giving judgment, FOSTER J said that the land in question, G two plots, each having a frontage of 25 ft to a then non-existent road known as Victoria Road and a depth of 150 ft, was conveyed to one H A Cavalier in 1917 for £12. Mr Cavalier died in 1954, and the land was later conveyed to his eldest grandson, Mr Henry Samuel Podd, by a conveyance of December 21 1960. In fact in 1956 an application had been made to develop the land, but that application was withdrawn for some reason which never became apparent. However, Mr Podd on April 17 1969 made a similar application which was refused, and the result was that the council purchased the disputed land on September 4 1970. On November 6 1937 the defendant purchased the property known as "Novoli," H Victoria Road, for £150. As early as 1928, he had started keeping poultry on land to the south west which subsequently, in 1946, he purchased from his father, and he had since continued that activity under the name " Novoli Poultry Farm." He was an accredited breeder of poultry under the Ministry of Agriculture, and in the early 1940s, on one of the regular visits of an inspector from the Ministry, the inspector told him that after 12 years' occupation he could claim a possessory title to land. Since then he had acquired a considerable further area of land in the vicinity of " Novoli," some of it by purchase and some by acquiring and registering posses- J sory titles which had in some instances now become absolute. He claimed to have been in adverse possession of the disputed land since 1944, but the only use he claimed before 1954 was the grazing of a pony, and this claim he had dropped, since it was clear from his evidence that he had had Victoria Road made up in 1948 and that no ponies had grazed on the disputed land since then.

The writ in the action was issued on September 15 1972, and under the Limitation Act 1939 the defendant had to show that he had been in adverse possession for 12 years, so that September 15 1960 was the vital date in this case. One of the problems of poultry-farming was the disposal of K poultry manure. The disputed land in 1954 was just scrubland, described by more than one witness as " a jungle," consisting of high hawthorn bushes and brambles and completely overgrown. The evidence for the defendant was that in 1954 he fenced off the disputed land and had since used it for agricultural purposes. From 1954 to 1967 poultry manure was dumped on the disputed land, burnt and then dispersed. It was suggested that the disputed land had been cleared of scrub by 1960, but he (his Lordship) had no doubt that the land was not really cleared until 1967. In so far as the poultry manure heap became too large, further parts of the scrub may have been cleared, but the evidence against any L substantial clearing having been done before 1967 was overwhelming. There could be no doubt, however, that 11 contiguous plots including the two plots of the disputed land were cleared in or about 1967 and were enclosed by a chicken-wire fence, and grass and clover were then sown. From that time the defendant had placed on the land movable chicken huts and had kept free-range hens on it.

The relevant section of the Limitation Act 1939 was section 5, which contemplated two possible cases, discontinuance and dispossession. For the defendant it was submitted first that Mr Cavalier and his successors in title had discontinued possession, or—as he (his Lordship) thought it must amount M to—abandoned any interest in the land. In *Wallis's Cayton Bay Holiday Camp Ltd v Shell-Mex & BP Ltd* [1975] QB 94 at p 103 the Master of the Rolls said:

Possession by itself is not enough to give a title. It must be *adverse* possession. The true owner must have discontinued possession or have been dispossessed and another must have taken it adversely to him. There must be something in the nature of an ouster of the true owner by the wrongful possessor. . . . When the true owner of land intends to use it for a particular purpose in the future, but meanwhile has no immediate use for it, and so leaves it unoccupied, he does not lose his title to it simply because some other person enters on it and uses it for some temporary purpose, like stacking materials; or for some seasonal purpose, like growing vegetables.

Ormrod LJ in his judgment said at p 114 that the same point had been made by Bramwell LJ in *Leigh v Jack* (1879) 5 ExD 264, at p 272. In his (Foster J's) judgment, Mr Cavalier until his death in 1954 intended one day to build upon the disputed land, and the planning applications made in 1956 and in 1969 negatived any intention on the part of his successors in title to abandon their title or discontinue possession. As to whether there had been a dispossession, he (his Lordship) thought that the acts relied on by the defendant to constitute adverse possession from 1954 until 1967 were much more trivial that the acts which were done by the plaintiffs in *Wallis's Cayton Bay* case, and even the acts done by the defendant on the land since its clearance in 1967 were not sufficient to constitute adverse possession when compared with those in the *Wallis's Cayton Bay* case. It followed that the defendant was a trespasser on the land.

His Lordship granted an injunction to restrain further acts of trespass, and awarded nominal damages of £2. He dismissed the defendant's counterclaim, and awarded the plaintiffs the costs of the action and counterclaim.

CHANCERY DIVISION

June 16 1975

(Before Mr Justice WALTON)

ESSO PETROLEUM CO LTD v ALSTONBRIDGE PROPERTIES LTD AND OTHERS

Estates Gazette March 20 1976

(1975) 237 EG 881

Solus agreement coupled with mortgage—Supplemental deed of release and covenant on sale of garage business—Construction of the two deeds on default by a subpurchaser

This was a procedure summons by Esso Petroleum Co Ltd, plaintiffs in an action against Alstonbridge Properties Ltd, Gordon Henry Golding Thompson, Michael Charles Borne and Acemoor Ltd, seeking (1) payment of all moneys due under the joint effect of the covenants (a) in a mortgage dated January 28 1972 made between Maintenance Garage Ltd, Lily Marie Draper as owner, Lily Marie Draper and Frank Arthur Draper as sureties, and the plaintiffs, and relating to Maintenance Garage, Willow Grove, Chislehurst, Bromley, Kent, and (b) in a deed of release and covenant dated January 1 1973 made between Maintenance Garage Ltd, Lily Marie Draper as owner, Lily Marie Draper and Frank Arthur Draper as sureties, the first, second and third defendants, and the plaintiffs; and (2) possession of Maintenance Garage.

Mr C A Brodie (instructed by Durrant Piesse) appeared for the plaintiffs; Mr C Aldous (instructed by D J Freeman & Co) for the first defendants; and Mr G Lightman (instructed by J D Langton & Passmore and Lambert & Co) for the second, third and fourth defendants.

Giving judgment, WALTON J said that the summons was concerned with a mortgage of garage premises on the south side of Willow Grove, Chislehurst, Bromley, Kent, known as Maintenance Garage and referred to as the "south premises." Opposite these premises was another garage, the "north premises." In 1972, the south premises were owned by Mrs Lily Marie Draper and the north premises by her husband, Frank Arthur Draper. A garage business was carried on on both premises by a company known as Maintenance Garage Ltd, all the shares in which were held by Mr and Mrs Draper. Maintenance Garage Ltd ("the company") had no estate or interest in either of the premises, being merely a licensee of the owners. On December 22 1971 the company entered into two solus agreements in the usual form with the plaintiff company commencing on January 28 1972, one in respect of each of the south and north premises. At one stage in the argument Mr Brodie, for the plaintiffs, submitted that the company were in breach of the solus agreement affecting the south premises, because it was provided by clause 4 (3):

> Before completing any sale or transfer of the service station premises or business or making any other arrangement under which any person commences to carry on business there in succession to the dealer to notify Esso in writing and procure such person to enter into an agreement with Esso and the dealer whereby such person is substituted for the dealer for all future purposes of this agreement in relation to the interest transferred including this subclause.

The premises and business had changed hands, said counsel, yet the company had not notified the plaintiffs in writing and had not made any such arrangements for their successors to enter into a written agreement with the plaintiffs as therein provided. However, as far as the evidence went, all that had happened was that in consequence of the events hereinafter noticed, the licence of the company to trade on the south premises had been withdrawn, in consequence whereof it had been unable to continue to do so. In the circumstances there could not have been a breach of this clause by the company.

On the same January 28 1972 a mortgage of the south premises was effected to secure a loan by the plaintiffs to the company of £38,000, Mr and Mrs Draper also acting as guarantors. The precise terms of this document were important, and it was necessary for him (Walton J) to read a considerable portion of it in full:

> THIS MORTGAGE is made on January 28 1972 between [the company] of the first part [Mrs Draper] ("the owner," which expression except where the context otherwise requires shall include the person or persons from time to time deriving title under the owner) of the second part [Mr and Mrs Draper] ("the sureties") of the third part, and [the plaintiffs] ("the lenders") of the fourth part.
>
> WHEREAS:
>
> (1) The owner is seised in fee simple free from incumbrances of [the south premises] and the company carries on thereat the business of petrol filling and service station proprietors with the licence of the owner as she hereby testifies but no tenancy has been granted in respect thereof;
>
> (2) [Mr Draper] is seised in fee simple of [the north premises] and the company as licensees carry on thereon the business of garage proprietors and service station (which premises with the property hereby mortgaged are hereinafter together referred to as "the garage premises");
>
> (3) The lenders have agreed to lend to the company the sum of £38,000 bearing interest at the rate of 6 per cent per annum on the principal moneys from time to time outstanding on the terms of a memorandum of agreement dated December 22 1971. . . .
>
> NOW THIS DEED WITNESSETH as follows:
>
> 1 (i) In consideration of the said sum of £38,000 so advanced as aforesaid (the receipt whereof the company hereby acknowledges) and subject to the provisions of clause 6 of the third schedule hereto the company and the sureties hereby jointly and severally covenant with the lenders to repay to the lenders the said sum of £38,000 with interest thereon from the date hereof at the rate of £1,314.72 per annum by equal monthly instalments [and then those instalments are broken down in respect of principal and interest] on the 28th day of every month in every year until the whole of the said sum of £38,000 with interest as aforesaid shall have been repaid . . . ;
>
> (ii) Provided that if any of the said monthly instalments shall be unpaid for 21 days after the time hereinbefore appointed for payment thereof or if the company and/or the sureties shall fail to perform any of their obligations under this mortgage other than their obligation in regard to the repayment of principal moneys and interest the company and/or the sureties will pay to the lenders on demand so much of the said sum of £38,000 as shall then be unpaid together with interest thereon calculated in accordance with clause 2 hereof.

A

2 In the event of the said sum of £38,000 (or such part thereof as shall for the time being remain unpaid) becoming repayable or being repaid under the terms of this deed there shall be paid by the company and/or the sureties to the lenders such additional sum by way of interest as shall together with interest already paid be equivalent to interest at [a certain rate].

3 For the consideration aforesaid and as security for the covenants on the part of the company and the sureties herein contained the owner as beneficial owner hereby charges by way of legal mortgage [the south premises]. . . .

4 . . .

B

5 It is hereby agreed and declared that although as between the company and the sureties the sureties are only sureties for the company yet as between the sureties and the lenders the sureties shall be considered as principal debtors for all the principal and other moneys and/or interest hereby secured [and consequent matters which need not be gone into].

6 The company the owner and the sureties hereby jointly and severally covenant with the lenders that during the continuance of this security the company and where appropriate the owner or the sureties will observe and perform the stipulations set forth in the second schedule hereto.

7 It is hereby mutually agreed and declared between the parties hereto in the terms set forth in the third schedule hereto.

C

It was necessary to refer to two of the clauses in the second schedule:

2 To occupy the garage premises and to conduct and keep the same open for business as retailers of motor fuels;

4 To purchase exclusively from the lenders all motor fuels which the company and/or the sureties may require for consumption or sale on the garage premises (and on any premises now or hereafter owned occupied or controlled by the company and/or the sureties or either of them which may be adjoining or otherwise physically connected with the premises hereby mortgaged) so long as the lenders shall be ready to supply the same. . . .

D

In the third schedule, the first clause dealt with the power of sale conferred by the Law of Property Act 1925, which was applicable with certain variations and extensions; that was to say, the power was to become immediately exercisable without notice or other restriction if:

(ii) Any instalment of principal moneys and/or interest shall not be paid within 21 days after the day hereby appointed for the payment thereof or

(iii) the company and/or the sureties shall fail to comply with a notice given by the lenders under the terms of clause 6 (b) of the schedule or

E

(iv) the company the owner or the sureties shall fail to observe any of the covenants or provisions on their part herein contained. . . .

Clause 6 of the schedule provided:

Notwithstanding the covenants as to repayment contained in clause 1 (i) of this deed:

(a) the company the owner and/or the sureties shall be entitled to redeem this security at any time after five years from the date hereof. . . .

(b) The lenders shall be entitled at any time after five years from the date hereof upon giving to the company and/or the sureties 12 months' prior notice in writing to require the company and/or the sureties to repay to the lenders the whole of the principal moneys. . . .

F

Later in 1972 Alstonbridge contracted to purchase (i) the south premises from Mrs Draper, (ii) the north premises from Mr Draper, and (iii) the whole of the issued share capital of the company from Mr and Mrs Draper. These sales and purchases were apparently completed on January 1 1973, and a deed of release and covenant supplemental to the mortgage was entered into between the interested parties. Once again, he (Walton J) must read the relevant parts of this document:

THIS DEED is made on January 1 1973 between [the company] of the first part [Mrs Draper] ("the owner") of the second part [Mr and Mrs Draper] ("the sureties") of the third part [the second and third defendants] ("the new sureties") of the fourth part [Alstonbridge] of the fifth part, and [the plaintiffs] of the sixth part.

G

WHEREAS:

(1) [Here the effect of the mortgage (the "principal deed") was recited];

(2) The garage premises as described in the principal deed have been transferred to Alstonbridge by transfers dated of even date herewith and also by transfers dated with today's date Alstonbridge are the beneficial owners of the whole of the share capital of the company;

H

(3) The transfer of the garage premises is subject to the principal deed and the principal moneys and interest thereby secured and the company are to continue to occupy the garage premises as licensee of Alstonbridge but no tenancy is to be granted in respect thereof;

(4) The lenders have been requested to release the sureties and the owner from their obligations under the terms of the principal deed which they have agreed to do upon the new sureties and Alstonbridge entering into the covenants hereinafter appearing;

NOW THIS DEED WITNESSETH as follows:

J

(1) In consideration of the covenants on the part of the new sureties and Alstonbridge hereinafter appearing the lenders hereby release the owner and the sureties from all the covenants and obligations on their part contained in the principal deed in respect of any breaches thereof occurring from and after the date hereof.

(2) For the consideration aforesaid the new sureties and Alstonbridge hereby jointly and severally covenant with the lenders to henceforth observe and perform all and singular the covenants stipulations and agreements on the part of the sureties and the owner contained in the principal deed as if they were successors thereof respectively.

(3) All the definitions contained in the principal deed shall apply to this deed save that Alstonbridge shall be deemed to be a successor of the owner and the sureties shall be deemed to include the new sureties.

K

Alstonbridge were registered as proprietors of the south premises on March 8 1973 and remained registered at all material times, but they appeared to have sold the south premises to the fourth defendants, Acemoor, some time in March 1973, according at any rate to the terms of a letter of March 27 1973 that Acemoor's solicitors wrote to the plaintiffs. After saying that their clients had purchased the freehold interest in the south premises, the solicitors added that before completion they had obtained from the solicitors for Alstonbridge a notice of redemption of the mortgage on the property, and they enclosed that notice, which read:

L

We, Langton & Passmore, solicitors on behalf of [Alstonbridge], give you notice that [Alstonbridge] intend at the expiration of three months from the date of service hereof to pay off all the moneys then owing on the security of [the mortgage].

This letter and notice might give rise to interesting questions, but he (his Lordship) did not think they affected anything he had to decide. Whatever notice was given to the plaintiffs, no tender was at any time made to them of the outstanding principal, interest and costs, so that the question of the validity of the notices did not in law arise. However, from shortly after the date of that letter, namely April 4 1973, no mortgage instalments were received by the plaintiffs from anybody until January 31 1974, when Acemoor tendered a cheque for £3,000, which was refused by the plaintiffs. On May 16 1974 the plaintiffs issued an originating summons naming as defendants Alstonbridge, the sureties under the deed of release and covenant and Acemoor, asking as against Alstonbridge and the sureties payment of the principal sum, interest and costs and as against Alstonbridge and Acemoor possession of the south premises.

M

The first question was whether the plaintiffs had any claims against Alstonbridge at all. The claims made were for payment and possession because Alstonbridge, as registered proprietors of the south premises, were in the situation of owners of the equity of redemption in the premises subject to the

A mortgage, and therefore *prima facie* the persons entitled to possession. They were not actually in possession, and had not been at any rate since March 1973. It was not their fault that Acemoor had not registered the transfer to that company which Alstonbridge had executed about that time. He (his Lordship) was of opinion that if the claim to possession had been the only one advanced, Alstonbridge would never have been a proper party to the summons. It was established by *R v Judge Dutton Briant* [1957] 2 QB 497 and *West Penwith Rural District Council v Gunnell* [1968] 2 All ER 1005 that a mortgagee's application for an order for possession was simply an order for the recovery of land and not proceedings B for enforcing the mortgage. Accordingly it had been decided in *Alliance Building Society v Shave* [1952] 1 All ER 1033 that an order for possession could be made against a complete stranger to the title who happened to be in actual possession, and could be so made in the total absence of the mortgagor or mortgagors. Since the plaintiffs knew that Alstonbridge were not actually in possession and that no relief in this regard was or could have been sought against them directly, in his (Walton J's) judgment they were never necessary or proper parties to the claim for possession. This was a point which would only go to costs, but in the event nothing could turn upon this in the present case, since the C real contest between the plaintiffs and Alstonbridge was to the liability of the latter for payment of the moneys due under the mortgage.

As to this, Mr Brodie pointed to clause (2) of the deed of release and covenant and said, truly, that that clause contained a joint and several covenant by the new sureties and Alstonbridge to perform all the covenants on the part of the sureties and the owner contained in the mortgage, and that that was but one covenant; from which it followed, he said, that Alstonbridge had undertaken the covenants on the part of the sureties to the mortgage, ie Alstonbridge were liable as D sureties for the mortgage. But Mr Aldous, for Alstonbridge, said that this would undoubtedly have been the position had that clause stood alone in that form, but it did not. First, the clause itself continued with the vital words "as if they were successors thereof respectively," and, secondly, there was clause (3), which showed quite clearly that the whole purport and intent of the deed was to substitute Alstonbridge for "the owner" and "the new sureties" for "the sureties." And picking up that last point, Mr Aldous said that as for the fact that clause (3) stated, "the sureties shall be deemed to include the new sureties," that was obviously a drafting infelicity. Mr Aldous said that, looking at the deed as a E whole, there could really be no doubt but that the proper construction of clause (2) was that it really contained two covenants: one by Alstonbridge substituting it completely for "the owner" in the mortgage, and one by the new sureties, jointly and severally, substituting themselves for the sureties in the mortgage. He (his Lordship) had come to the conclusion that Mr Aldous's arguments were to be preferred. Accordingly it appeared to him that the only covenants in respect of which Alstonbridge were liable to the plaintiffs were the covenants on the part of the owner contained in the mortgage; and that as those never extended to liability for F payment of the mortgage debt or any part thereof either as principal or surety, the plaintiffs had no claim against Alstonbridge in this regard, and so far as concerned Alstonbridge, therefore, the originating summons fell to be dismissed with costs.

The next claim was that of the plaintiffs against the new sureties—the second and third defendants—for payment of all moneys due under the mortgage. Mr Brodie's case here, because he was unable to show any demand in the terms of clause 1 (ii) of the mortgage for payment of the unpaid principal, being only able to establish such a demand as having been made against Alstonbridge, was that the service of the originating summons herein was the demand, which the plain- G tiffs were fully entitled to make, because the monthly instalments had been unpaid for upwards of 21 days, and the plaintiffs had not, either by acceptance of such instalments or by any other action on their part, waived their right to make such a demand. Mr Lightman, for the second, third and fourth defendants, disputed this, and also argued that the kind of demand referred to in clause 1 (ii) must be a demand on the principal debtor (the company in the present case), upon whom no such demand had ever been made. It would, he maintained, be absurd that the obligation on the principal debtor should remain that of paying by instalments while the obligation of the sureties should have been trans- H muted into that of having to pay a single lump sum. He (Walton J) felt the force of this submission, but with the best will in the world was unable to give to the phrase "the company and/or the sureties will pay to the lenders on demand" etc any other meaning than that the mortgagee was entitled to demand repayment in full from either the company or the sureties, as it thought fit. The mortgagee was entitled to treat all the sureties as principal debtors (see clause 5), and he (his Lordship) thought that if in these circumstances a demand on the sureties and not the principal debtor created problems for the principal debtor, this had to be accepted, though he did not think that in practice J any real problems were likely to be created.

With regard to the demand point, however, he (Walton J) was of the opinion that in the present circumstances the service of the originating summons was not a sufficient demand for the purposes of clause 1 (ii) of the mortgage. He fully accepted that where there was a pre-existing debt which was payable "on demand," such a demand (other than the service of proceedings) was not a prerequisite to the bringing of an action to recover that debt. The case of *Re Brown's Estate, Brown v Brown* [1893] 2 Ch 300 showed that where the character in which payment was required was that of surety, K a demand was, in general, necessary; but he assumed for present purposes (without finding it necessary to decide) that the provisions of clause 5 of the mortgage, equating the liability of the sureties to that of a principal debtor, was effective to obviate the necessity for a demand merely on this ground. His difficulty, nevertheless, was that the demand in the present case was a demand which, of its own intrinsic nature, changed the nature of the liability. It turned a liability to pay by instalments into a liability to pay the whole at once. Under these circumstances he thought that even as against a principal debtor a demand antecedent to the issue of proceedings was a necessary prerequisite of the whole L cause of action. He put it this way: that where the pre-existing obligation was to pay the debt by instalments, the demand that it be paid in one lump sum was an act which radically changed the nature of the debtor's obligation, and so was an essential ingredient of any cause of action to recover the lump sum. He thought that the case was precisely equivalent to the type of case envisaged by Scrutton LJ in *Bradford Old Bank v Sutcliffe* [1918] 2 KB 833 at p 848. Accordingly, for those reasons, he concluded that a demand on the sureties was an essential ingredient of any cause of action against them for the whole mortgage debt (as distinct from any unpaid instalments), and that therefore the claim M for payment against them fell to the ground.

The final claim was that of the plaintiffs against Acemoor for possession. Mr Lightman, for Acemoor, resisted that, arguing (1) that a term ought to be implied in the present mortgage that the morgagee was not entitled to possession unless the mortgagor defaulted, and (2) that if such an implication were made, the mortgagee would only become entitled to possession so long as the mortgagor was in default, and that if the mortgagor corrected any default (as it did here by tendering all arrears) the mortgagee thereafter lost all rights to possession once again. He (his Lordship) accepted

A that the court would be ready to find an implied term in an instalment mortgage that the mortgagor was to be entitled to remain in possession against the mortgagee until he made default in payment of one of the instalments. But there must be something upon which to hang such a conclusion in the mortgage other than the mere fact that it was an instalment mortgage, and he (Walton J) could find nothing in the present mortgage which would enable the court to find any such implication here. Even had he been able to discover any such implication, it could only have been an implication until some default by the borrower, and this default quite clearly took place when the instalments were unpaid. He did not B think that there was any such general doctrine as that for which Mr Lightman contended, that if the mortgagor after default then tendered the arrears the implication that he was entitled to possession revived. He thought that if such an implication were made, it would be an implication that lasted until default; once default had been made, he thought the implication went completely, and the mortgagee then became entitled to pursue its ordinary fundamental right to claim possession at any moment it thought fit. He would add that it appeared to him that there had been a tactical battle here between the parties, with Acemoor wishing to redeem the mortgage when the financial situation was such that it would C have been possible for them to borrow money elsewhere more favourably, the plaintiffs correspondingly resisting redemption, and the position then dramatically changing when the financial climate itself dramatically changed. He did not think there was anything in the plaintiffs' conduct upon which Acemoor was properly entitled to rely in order to remain in possession, and accordingly, in his judgment, the claim for possession against Acemoor succeeded.

The summons against Alstonbridge and the second and third defendants was dismissed with costs. An order for possession was made against Acemoor with a stay of execution pending possible notice of appeal.

D

COURT OF APPEAL

July 10 1975

(Before Lord Justice BUCKLEY, Lord Justice STEPHENSON and Lord Justice LAWTON)

BACAL CONSTRUCTION (MIDLANDS) LTD v NORTHAMPTON DEVELOPMENT CORPORATION

Estates Gazette March 27 1976

(1975) 237 EG 955

E

Building work on large site reprogrammed at request of development corporation on the basis that the contractors would be secured closure of a road crossing the area—Contractors begin work informally, and subsequently insist on recognition of their claim for loss caused by delay in closure of the road as a condition of execution of formal contract—Construction of letter from corporation recognising validity of contractors' claim—Held to apply to prospective as well as existing loss—Further point on design of foundations also resolved in favour of contractors

This was an appeal by the Northampton Development F Corporation from a judgment of O'Connor J finding in favour of the respondents, Bacal Construction (Midlands) Ltd, building contractors, of Northampton, on issues as to liability arising out of a contract for the development of a site at Weston Favell, Northampton, known as "Lumbertubs I."

Mr K F Goodfellow QC and Mr A J Butcher (instructed by Sharpe, Pritchard & Co, agents for the solicitor to the corporation) appeared for the appellants, and Mr R S Alexander QC and Mr S D Brown (instructed by Ray & Vials, of Northampton) represented the respondents.

Giving judgment, BUCKLEY LJ said that in April 1970 the appellant corporation invited tenders for the development of one part of the general area of development with which they G were concerned, a part known as Lumbertubs I. The development was to comprise 518 dwellings, 131 garages, and drainage, footpaths and site works. The area upon which the works were to be carried out was at Weston Favell, approximately three miles north-east of Northampton. It comprised some 34½ acres and was rectangular in shape, having its longer axis from north to south and being divided roughly in half by a road running from north to south known as Lumbertubs Lane. The corporation supplied those invited to tender with a layout plan indicating the sites and respective characters of the several blocks to be erected. The tenderers were invited to submit their own designs. The contractor was H required to provide a programme of starts, completions and hand-over dates. It was stipulated that initially access to the site would be by way of Lumbertubs Lane, and it was stated that it was expected that a section of a new road, called the new transverse road, which was to run down the western side of the western boundary of the Lumbertubs I site, would be completed during the early part of the contract and so allow another means of access to the site. The plans for the development involved the closure of Lumbertubs Lane. The new transverse road was not expected to be available until about September 1970. The intention was that Lumbertubs Lane should not be closed until the new transverse road was J open, and that the programme for the works would be arranged accordingly.

Bacal (formerly called Adkins & Shaw (Builders & Contractors) Ltd) were the successful tenderers. They offered to complete the works in 65 weeks from commencement and to start handing over completed houses in week 43, which meant that houses would thereafter come forward at about 24 per week. The corporation were anxious to take delivery of fewer houses per week without extension of the 65-week period, and they asked if first hand-overs could be brought forward to week 34, which would reduce the average weekly K hand-over to 16 or 17 houses each week. Bacal said that this would only be possible if they reprogrammed the work to start building the blocks on the line of Lumbertubs Lane early in the contract, which involved the closure of Lumbertubs Lane before the new transverse road was ready. They accordingly prepared a new programme, the "34-week programme," for carrying out the works upon these lines. Bacal entered on the site on April 27 1970 and in agreement with the corporation began to operate the 34-week programme. Lumbertubs Lane, however, was not closed until September 14 1970. This was too late to enable Bacal to adhere to the 34-week programme. Accordingly Bacal's plan for the L development of the site was disrupted and much delayed, and Bacal asserted that they had suffered a substantial loss in consequence. At this time, although Bacal had been at work on the site for a considerable period, no formal contract had yet been signed. Bacal refused to sign the contract unless the corporation accepted liability for the loss which Bacal claimed to have suffered in consequence of the late closure of Lumbertubs Lane. This was eventually dealt with by means of a letter dated March 11 1971 from the corporation to Bacal which was bound up with the contract documents, although not expressly referred to in them. The contract as signed stipulated that the first houses should be M handed over in week 43. The letter in question was in the following terms:

> The Northampton Development Corporation hereby acknowledge the existence of a valid claim (the amount of which remains to be negotiated) by Adkins & Shaw (Builders & Contractors) Ltd in respect of alleged abortive expenditure incurred through the delayed closure of Lumbertubs Lane, which did not occur until September 14 1970, which said abortive expenditure is claimed to have arisen as a result of a request by the said Northampton Development Corporation made prior to March 5 1970 for the said Adkins & Shaw (Builders & Contractors) Ltd to commence handing over houses in week no 34 on the basis of

A Lumbertubs Lane being closed to enable this to be implemented in lieu of week 43 as specified in house buildings and road sewers contracts as made between the said parties and dated today's date which said request nevertheless retained the overall period of 65 weeks for the completion of the said contract and the said Northampton Development Corporation hereby undertakes to instruct its advisers to enter into negotiations with a view to amicably settling the said claim.

The contract in fact bore the date May 7 1971, but nothing turned upon this fact. The first matter of dispute in the action was the true construction and effect of the letter. B There was also another wholly distinct dispute between the parties relating to the foundations of certain of the buildings.

When Bacal had prepared the 34-week programme, a meeting, described as a programme meeting, was held on March 5 1970, at which the corporation was represented by (among others) Mr Redfern, its chief architect and planning officer. Bacal's two representatives included Mr A F Stockwell. The architects and the quantity surveyors were also represented, as were certain statutory undertakers concerned with the Lumbertubs I development. The trial judge said that even without explanation it must have been apparent to all present that if the 34-week programme was to be C agreed and adopted Lumbertubs Lane would have to be closed before the work started, for it called for an immediate start on three particular blocks, one of which actually encroached upon the carriageway of Lumbertubs Lane. The judge accepted that Mr Redfern did say something the effect of which was to indicate that the road would be closed before the work started. The minute of the meeting, prepared in the corporation's office, recorded that Mr Redfern reported that closure of Lumbertubs Lane would start on March 16 1970. The judge said that he was satisfied that Mr Redfern did not say this at the meeting, but that nobody objected to the minute, and that in the light of what had taken place at D the meeting it was not surprising that Mr Stockwell understood the minute to assert that physical closure of Lumbertubs Lane would start on March 16. Throughout the period until the trial, the corporation maintained that Mr Redfern had given no such assurance at the meeting, and this question was at all material times before the trial an issue between the parties. The corporation had contended in the Court of Appeal, as below, that the letter of March 11 1971 was a means of putting an end to this dispute and was in the nature of a compromise. They asserted that under it the corporation assumed liability only for abortive expenditure already E incurred by the date of the letter.

The trial judge cited a long passage from the speech of Lord Wilberforce in *Prenn v Simmonds* [1971] 1 WLR 1381, commencing at 1383H and going down to 1385H, and in his (Buckley LJ's) opinion rightly directed himself that he must construe the letter in the light of the factual background known to the parties at its date, including evidence of the "genesis," and objectively the "aims," of the transaction. He (the trial judge) reviewed the relevant evidence and reached the conclusion that the genesis of the letter of March 11 1971 was the fact that Bacal's contract was disrupted by the failure to close Lumbertubs Lane, and that the objective F of achieving first hand-overs in week 34 was consequently rendered impossible, with financial loss to Bacal. He went on to hold that the aim of the letter was to produce a formula binding on both parties by which the loss sustained by Bacal could be calculated by the quantity surveyors and agreed between them. He declined to accept the corporation's contention that the expenditure for which liability was accepted must have been incurred before the date of the letter. He construed the letter as an agreement by the corporation to meet any extra expenditure which flowed directly from the attempt by Bacal to implement the 34-week programme whenever such expenditure was incurred, so long as it could be shown that it was caused by the non-closure of Lumbertubs Lane. G

He (Buckley LJ) thought that the use in the letter of the past tense was of no particular significance. Any particular item of expenditure claimed by Bacal must be shown to have flowed from the cause indicated in the letter and to have been rendered abortive through the delayed closure of Lumbertubs Lane, but provided it fulfilled both these requirements, in his opinion it was of no significance at what date that particular item of expenditure had been incurred. Bacal might well have incurred or become irrevocably committed to expenditure exclusively referable to their embarking on H the 34-week programme, or might have become unavoidably involved in expense arising out of the disruption of that programme, which could not accurately be said to have been contractually incurred before March 11 1971. Nevertheless, if any such expenditure was shown to have been rendered abortive by the late closure of Lumbertubs Lane, he (his Lordship) thought that it was covered by the terms of the letter. In other words, he agreed with the view of the trial judge that the liability accepted by the letter was not confined to expenditure incurred before the date of the letter. This interpretation of the letter appeared to him to be a natural one in the light of the surrounding circumstances. Bacal were J claiming that they had been and would be put to unnecessary expense by reason of the breakdown of the 34-week programme. The letter appeared on its face to be an attempt to define that claim and to provide machinery for its amicable quantification. There was nothing on the face of the letter to indicate that it was intended to be in the nature of a compromise of a dispute. There seemed to him to be nothing inconsistent in this with the objective of settling the claim amicably. A foreseeable future expense was as capable of quantification as one which had already been incurred. Accordingly, in his judgment, the appellants failed on this part of the appeal. For his part, he should regard any K expenditure as "abortive" within the meaning of the letter which resulted from Bacal's embarking on the 34-week programme, including expenditure resulting from the disruption of that programme due to the delayed closure of Lumbertubs Lane, which would not have been incurred at some stage of the works if Bacal had throughout followed their original 43-week programme.

The other issue in the appeal related to the foundations of the various buildings erected on the site. It was common ground that the designs of foundations prepared by Bacal were adequate having regard to the hypotheses upon which L they were required to design them. None of the information given by the corporation to Bacal gave any warning of the presence of tufa. The tender documents contained an express statement that the general information was that the site was a mixture of Northamptonshire sand and upper lias clay. In the course of the works, however, it was discovered that in various areas of the site there were subsurface strata or patches of tufa. This was described by the trial judge as being a soft spongy material. Its presence necessitated the redesigning of the foundations wherever it was found. Additional excavation had to be carried out, and additional building work had to be done. Bacal contended that it was M an implied term of the contract, or an implied warranty on the part of the corporation, that the ground conditions would accord with the hypotheses upon which they were instructed to design the foundations. The corporation, on the other hand, contended that there was no room here for any implied term or warranty, because the terms of the contract sufficiently covered the facts. To understand the argument it was necessary to set out some of the clauses of the contract. Clause 12 read as follows:

(1) The quality and quantity of the work included in the contract sum shall be deemed to be that which is set out in the

contract bills which unless otherwise expressly stated in respect of any specified item or items shall be deemed to have been prepared in accordance with the principles of Standard Methods of Measurement of Building Works, Fifth Edition, Imperial, revised March 1964 by the Royal Institution of Chartered Surveyors and the National Federation of Building Trades Employers, but save as aforesaid nothing contained in the contract bills shall override, modify or affect in any way whatsoever the application or interpretation of that which is contained in these conditions, provided that the additional clauses set out in the contract bills shall be deemed to be contained in the conditions and not in the contract bills.

(2) Any error in description or in quantity in or omission of items from the contract bills shall not vitiate this contract, but shall be corrected and be deemed to be a variation required by the architect or supervising officer. Clause 12 (2) shall not have effect so as to enable the correction of an error or omission arising from an error, omission or inconsistency in the contractors' documents to be treated as a variation required by the architect.

In connection with clause 12, it was necessary to read clause 1 (3) and (4), which were as follows:

(3) Subject to the prior written approval of the architect in that behalf, the contractor shall have power to amend, modify or alter that part of the works designed by the contractor insofar as the same shall be necessary for the satisfactory completion of the works in the manner aforesaid.

(4) No such amendment modification or alteration shall entitle the contractor to any additional payment whatsoever in respect thereof, or be construed as a "variation" (whether or not the subject of an architect's instruction) as hereinafter defined.

The words "in the manner aforesaid" in subclause (3) referred back to clause 1 (1), which read:

The contractor shall upon and subject to these conditions carry out and complete the works shown upon the contract drawings and described by or referred to in the contract bills and in these conditions in every respect to the reasonable satisfaction of the architect/supervising officer.

The corporation said that in consequence of the discovery of tufa, the contract bills in relation to the foundations were in some cases found to be in error or to contain omissions, and that they consequently fell to be corrected in accordance with clause 12 (2). Consequently, they said, such corrections were to be deemed to be variations required by the architect. The provision which excluded the application of clause 12 (2) in the case of an error in any contractor's document did not, they said, apply to the present case, because Bacal made no error in designing the foundations in accordance with the hypotheses which they were required to assume. Since the additional works occasioned by the discovery of tufa constituted "deemed variations" within the provisions of clause 12 (2), Bacal was entitled to additional payment only to the extent allowed under clause 11 (6). Bacal relied upon clause 1 (4). They necessarily agreed that the foundations were part of the works defined by the contractor. They said that in consequence of the discovery of tufa they were in many instances compelled to amend modify or alter their designs pursuant to clause 1 (3). They said that clause 1 (4) in wide, clear and unqualified terms provided that no such "amendment modification or alteration" should be treated as a variation for the purposes of the contract. Since these amendments were not to be treated as variations, Bacal said that clause 11 could not apply to them. They (Bacal) questioned whether clause 12 (2) could apply in these circumstances at all, for they said that error in description or in quantity in, or an omission of items from, the contract bills was not the same as inadequacy of a design resulting from unforeseen physical conditions. The contract bills in respect of the foundations related only to specimen foundation designs prepared to meet hypotheses laid down by the corporation. It was common ground that there was no error or inadequacy in the designs in relation to the hypotheses, nor was there any error in the bills of quantity in relation to those designs.

Clause 1 (3) and (4) were not, or so he (his Lordship) imagined, inserted in the contract in order to meet such conditions as had arisen in respect of the foundations: they were primarily intended, he suspected, to ensure that the contractor should not get extra remuneration in respect of alterations in design necessitated by design defects for which the contractor was responsible. The language of these two subclauses, however, seemed to him to be clear and unequivocal and to fit the circumstances which had arisen as the result of the discovery of tufa. He saw no ground for construing them so as to relate them exclusively to defects in designs due to some fault on the part of the contractor. It would be a strange result if some other clause in the contract were to render something a deemed variation which was expressly excluded from ranking as a variation by clause 1 (4). The corporation had contended that Bacal's construction of clause 1 (4), which was itself a typed amendment inserted in the printed form of contract, would result in inconsistency with other parts of the contract which were also inserted as typed amendments of the printed form. There was a very large number of such typed amendments, and the contract was in consequence both confusing to read and in some instances confusing to construe. For himself, he (Buckley LJ) could not find elsewhere in the document any contrary indication which was so clear as to induce him to interpret clause 1 (4) otherwise than in accordance with what seemed to him to be its plain meaning. In his judgment, Bacal's analysis of the contract was correct, so that the redesigning and additional work occasioned by the discovery of tufa did not rank as variations for the purposes of the contract, and in particular for the purposes of clause 11. Accordingly, no provision was to be found in the contract entitling Bacal to additional remuneration for their additional work. In these circumstances, Bacal submitted that there were strong commercial reasons for implying such a term or warranty in the contract as they had suggested. First, before designing the foundations for any building it was essential to know the nature of the soil conditions. Secondly, where the contract was for a comprehensive development of the kind here in question, the contractor must know the soil conditions at the site of each projected block in order to be able to plan his timetable and to estimate his requirements for materials. These were matters which related directly to the contract price. Thirdly, if the work was interrupted or delayed by unforeseen complications the contractor was unlikely to be able to complete his contract in time. Clause 22 of the contract required the contractor to pay or allow liquidated damages in such an event. The corporation had in fact retained a very substantial sum by way of liquidated damages to which they claimed to be entitled because Bacal did not complete the contract within 65 weeks. The trial judge accepted Bacal's argument and held that the corporation was liable on an implied term or warranty. He (his Lordship) agreed, and accordingly, in his judgment, the corporation failed on this part of the appeal also.

STEPHENSON and LAWTON LJJ agreed, and the appeal was dismissed with costs.

COURT OF APPEAL

December 19 1975

(Before Lord Justice BUCKLEY, Lord Justice SCARMAN and Lord Justice GOFF)

GROSS v FRENCH AND ANOTHER

Estates Gazette April 3 1976
(1975) 238 EG 39

House bought by mother but conveyed to mother, daughter and son-in-law as beneficial tenants in common—Judge's finding that this was done purely for convenience left

A **undisturbed—Mother entitled to whole property on a resulting trust, subject only to charges in favour of son-in-law for moneys expended by way of improvements, etc**

This was an appeal by Mrs Joan Gertrude French and Mr Frank Edgar French, both of 62 Warrington Road, Harrow, Middlesex, from a judgment of Walton J dated October 7 1974 awarding Mrs Rosetta Gross, also of 62 Warrington Road, a declaration that the beneficial interest in that property vested in her alone, and dismissing the appellants' counterclaim for a declaration that the property was vested in all parties as joint tenants in common.

B Mr D Rice (instructed by Michael Kramer & Co) appeared for the appellants, and Mr R A Payne (instructed by Eric Cheek & Co) represented the respondent.

Giving the judgment of the court, SCARMAN LJ said that Mrs French was Mrs Gross's daughter and married Mr French some time before 1949, when Mrs Gross lost her husband and decided to buy a house in Paddington. She divided that house into three parts. She herself occupied the bottom part, and her daughter and son-in-law occupied the top, for which they paid her a rent of 25s a week; the middle part she let furnished. By 1963 the neighbourhood was deteriorating, and Mr and Mrs French, who by now had a daughter aged nine, wished C to move. Mrs Gross was willing to move, and offered to sell the house and, with the proceeds of sale, to buy another one. They found 62 Warrington Road, Harrow. Mrs Gross provided the purchase-money. The house was leasehold. A solicitor, a Mr Harris, was instructed, and in December 1963 the leasehold title was acquired and registered in the names of Mr French, Mrs French and Mrs Gross. By the deed of transfer, "in consideration of £5,100" the transferor transferred the house to all three as "beneficial tenants in common." The family moved into the house in 1964, Mr French paying Mrs Gross a weekly sum of £3. He also discharged D the rates and spent considerable time and money improving, renovating, and repairing the property. A time came when it proved possible to acquire the freehold reversion, and Mr French raised the £700 needed for its purchase. When acquired, the reversion was put into the three names, just as the leasehold title had been.

All went well until in 1967 a quarrel arose between Mrs Gross and Mr French, in the course of which he told her, in effect, to get out or keep out of his house. This apparent claim to ownership of an interest in the house alarmed her. The case for Mrs Gross was that she provided the purchase-price of the house and never intended to make a gift, in her E lifetime, of the house, or any part of it, to her daughter or son-in-law; the property was put into their names as well as hers "simply for convenience." There was therefore, she contended, nothing to rebut the ordinary presumption of a resulting trust in favour of herself as the person who provided the purchase-price: *Dyer v Dyer* (1788) 2 Cox Eq Cas 92. She recognised, however, that Mr French was entitled to a charge in respect of the £700 he advanced for the freehold, and her counsel conceded in this court that Mr French was also entitled to a lien in respect of his expenditure on the house, though there was an issue as to the subject-matter and conditions of the lien. Mr and Mrs French contended that F Mrs Gross had made a gift to them of two-thirds of the leasehold interest; that if the evidence was unclear, there was a presumption of a gift by her to her daughter and son-in-law by way of advancement; and that by acquiescing in Mr French's expenditure of time and money upon the house, including the provision of the £700 for the freehold, she had estopped herself from denying that he had a beneficial interest in the property. Walton J had found for Mrs Gross, though subject to orders in favour of Mr French granting him (1) a lien over the house for a sum equalling the value of the work done by him to repair and improve the property, to be payable upon sale or the death of Mrs Gross, whichever should be the earlier; and (2) a charge enforceable forth- G with over the house to secure to him the repayment of £700 expended by him in the acquisition of the freehold reversion, with interest thereon. Although it found no place in the order of the court, Walton J had in fact concluded that the defendants were tenants protected by the Rent Act 1968 of that part of the house which, with the consent of Mrs Gross, they occupied. It was no part of the case either of Mrs Gross or of the defendants that any tenancy existed, and Mrs Gross cross-appealed against that finding.

It was not in doubt that unless the appellants could establish a gift, a binding contract, or an estoppel, the respondent must succeed. The crux of the case was the course of events H and the discussions between the three parties and their solicitor which preceded the registration of the leasehold title in January 1964. Mr Harris clearly thought that Mrs Gross initially intended to convey the whole beneficial interest in the house to her daughter and son-in-law in consideration of the payment of an annuity of £3 per week for life and of their looking after her for the rest of her life. Indeed, in December 1963, at the crucial meeting in his offices, he felt he must advise Mrs Gross against parting with all her interest; and it was not disputed that he suggested the course they eventually took, putting the house into the names of all three of them as tenants in common. There were two documents, J one contemporaneous with the events of 1963-64 and the other of much later date, which appeared to lend support to Mr Harris's belief as to the nature of the transaction. These were his bill of costs (rendered, however, not to all three parties but only to Mrs Gross) dated January 2 1964, and a letter of May 24 1967 which he wrote her after the quarrel had arisen between her and Mr French. In the bill he referred to "the property to be conveyed into the names of yourself, your Mr French and his wife," and against this item Mrs Gross had noted in her own handwriting, "So I am the owner of one-third instead of the whole." And in the 1967 letter K Mr Harris wrote, "I am quite sure that I did explain to you at the time that the property would be vested in the three of you as beneficial tenants in common."

Their Lordships had been troubled by the judge's finding that Mrs Gross never intended that either her daughter or her son-in-law should have a beneficial interest in the house, and by his rejection of Mr Harris's view of the nature of the transaction in which he was advising Mrs Gross. But he (Walton J) had seen and heard the witnesses, all of whom were speaking of events that occurred long ago. In the circumstances, the court was not prepared to disturb the finding that Mrs Gross never did intend to convey any beneficial L interest in the house, and that she allowed it to be put into the three names for the sake of convenience, so that Mr French could handle the business side and she be left in peace. That finding, once accepted, was basic and negatived any possibility of a presumption of advancement, even if one assumed that which must at least be open to doubt, namely, that in these days of sex equality the presumption operated against a mother in favour of her child.

Associated with the basic finding was the judge's further finding that Mr and Mrs French became the tenants of Mrs Gross when they moved into the house. Neither Mrs Gross M nor Mr Harris nor Mr French would have it that they were tenants, but the judge had been impressed by the fact that Mrs Gross did speak of the £3 per week as "rent." This one fact was not enough, in their Lordships' judgment, to displace the weight of the evidence as a whole that the £3 was agreed as an annuity pursuant to a family arrangement under which Mrs Gross would allow Mr and Mrs French to live in the house provided they paid the annuity and the outgoings (eg the rates) and looked after her during her life. There remained the plea of an estoppel. In their Lordships' judgment, it failed *in limine*. Knowledge by Mrs Gross of the mistake had to be proved, but according to the judge's finding as to Mrs

A Gross's state of mind and intentions (a finding which this court was not prepared to disturb), at no material time did Mrs Gross have any knowledge of the existence of Mr French's claim to a share in the house.

Counsel for Mrs Gross conceded that Mr French was entitled to a lien over the property for a sum limited to the amount by which the value of the house had been increased by the work he had done to it. He (counsel) contended, however, that this sum was not to be payable until sale or the death of Mrs Gross, whichever should be the earlier, and he sought to reject the case for Mr French that the lien should not be limited to enhancement of value but should B be for the total of his expenditure of time and money on the house, and be immediately enforceable. Walton J thought such a lien could arise, limited to the enhanced value of the house as a result of Mr French's expenditure and enforceable only on sale or the death of Mrs Gross. Of course, a trustee who expended money upon the trust property in the proper discharge of his duties was entitled to a lien for the immediate recoupment of the whole cost. Mr French, however, was not quite in that position. He was a bare trustee, and certainly one with no express power to improve the property. Nevertheless, the judge reached the conclusion that there was such a lien, limited as he said, on the footing that Mr French was, C on the special facts of the case, to be treated as a trustee discharging the duties he was instructed to perform.

In their Lordships' view, having regard to the concession, they ought to uphold the judge's conclusion. But they also thought that the concession and the judge's decision alike might be supported by analogy with the case of a constructive trustee, or with that of joint owners on partition or sale: see Snell, 27th ed at p 449, and *Rowley v Ginnever* [1897] 2 Ch 503, where Kekewich J had to consider a case in which a constructive trustee of property had expended money on it in permanent improvements, and held that he was entitled to recoup his expenditure to the extent of the improved value. D There being no dispute, if the judge were right as to the intentions of Mrs Gross, that Mr French had a charge for the £700 he advanced for the acquisition of the freehold, their Lordships' conclusion was that they must uphold the judgment of Walton J and dismiss the appeal. The judge, in their view, had made only one error, his finding that Mr and Mrs French were tenants; and since this finding was not reflected in the order of the court, no variation of the order was required.

E

QUEEN'S BENCH DIVISION

January 19 1976

(Before Mr Justice OLIVER)

CHARLTON v LESTER

Estates Gazette April 10 1976

(1976) 238 EG 115

Protected tenant buys freehold with help of her daughter and son-in-law, property conveyed into all three names—Daughter and son-in-law buy their own new home and F **seek order for sale of the old one—Mother persuaded to buy in the first place only on the basis that she would always keep her home—" Did not expect to have to move willy-nilly to wherever daughter and son-in-law chose to live "—Hardship on mother with son in hospital nearby—Order for sale refused and mother awarded costs**

This was a claim by Mr Norman Charlton and his wife, Mrs Patricia Charlton, of Davidson Road, East Croydon, against Mrs Anne Lester, of 13 Hafer Road, Battersea, London, for a declaration that 13 Hafer Road should be sold and the proceeds of sale divided.

Mr J Cherryman (instructed by Hyde, Mahon & Pascall) G appeared for the plaintiffs, and Mr P Langan (instructed by Kingsbury & Turner) represented the defendant.

Giving judgment, OLIVER J said that Mrs Charlton was Mrs Lester's daughter. She and her husband had gone to live with Mrs Lester in Battersea in 1968. Mrs Lester was a protected tenant paying £5 a week rent. In 1970 the freeholder offered to sell the house to Mrs Lester at a favourable price of £2,000. She was persuaded to buy by Mr and Mrs Charlton, who wanted their own home. They got a mortgage from the Greater London Council with Mrs Lester named as joint owner. Contracts were exchanged in September 1970. Mr and Mrs Charlton paid the mortgage and paid for repairs H and modernisation of the house. It was accepted that although Mrs Lester made no cash contribution, her protected tenancy meant the difference between £2,000 and the value of the house on the open market; there was however a dispute between the parties about the open-market value of the property at the time and the amount by which the Charlton's share should be increased because of the money they spent on repairs. Doing his best on the evidence as to value, he (his Lordship) thought that the open-market value in 1970 was £5,200, and Mrs Lester's contribution was therefore worth £3,200. He thought that the Charltons should recover J only the cost of repairs they had to carry out as a condition of the GLC mortgage being granted, plus the solicitors' fees at the time of the sale. These items amounted to £385, making the total value of the house £5,585. The Charltons were entitled to 149/349 of that value, and Mrs Lester to 200/349.

The Charltons wanted the house sold and the proceeds divided. They were still paying the mortgage, and were making additional mortgage repayments on their new house. It was estimated that the house was worth £15,000 or £16,000 today. A sale would mean, however, that Mrs Lester would lose her home. She did not want to leave, and said that a sale would be a breach of a clear undertaking she was given K by her daughter at the time of purchase that she would keep her home. Mr and Mrs Charlton had a great burden of debt, but the circumstances had to be looked at before the usual order for sale was made. Mrs Lester had a seriously ill son, aged 28, who was receiving treatment at Tooting Hospital. He came home for long weekends, and was able to walk to Hafer Road to see his mother for a few hours every afternoon. Mrs Lester did not want to move away from the hospital. She was very lukewarm about buying the house in the first place, because she knew she was safe with a protected tenancy. She had very slender financial means. Mrs Charlton felt that her mother would be happy as long as they always gave her a L home, and when they (the Charltons) first intended to move they had plans for an extension to be built to Mrs Lester's specifications. However, he (his Lordship) did not think that Mrs Lester expected when they bought the Hafer Road house that she would have to move willy-nilly to wherever her daughter and son-in-law chose to live.

It was a case in which misunderstanding arose when Mr and Mrs Charlton wanted to move, and suspicions bred suspicions; relations had deteriorated, and it was now unrealistic to expect the parties to live in the same house. Nevertheless he (Oliver J) felt he had to dismiss the plaintiffs' claim, because he felt that M Mrs Lester was only persuaded to buy on the understanding that she would keep her home. He did not think she should now be turned out, and he must accordingly give judgment for her with costs. He thought it tragic that the family had been split assunder by a series of misunderstandings, and he was worried about Mr and Mrs Charlton's debts being increased by their having to pay the legal costs of the case, since they were not legally aided.

QUEEN'S BENCH DIVISION

February 9 1976

(Before Mr Justice EVELEIGH)

KING AND ANOTHER v TAYLOR AND OTHERS

Estates Gazette April 24 1976

(1976) 238 EG 265

Nuisance by encroachment of roots—"No damage, no nuisance" a misconception—Damages and an injunction awarded against neighbouring owner despite evidence as to clay soil and substandard construction of buildings affected

This was a claim by Mr Basil King, a chartered surveyor, of "Tregeagle," Green Lane, Northwood, Middlesex, and his wife, Mrs Mabel King, against their neighbours, Mr Raymond Taylor and his wife, Mrs Elsie Taylor, of 17 Dene Road, Northwood, and against A J A Taylor & Co Ltd, of Chancery Lane, London WC2, for damages for encroachment by roots of trees and for an injunction to restrain further nuisance.

Mr B W T Leech (instructed by William Charles Crocker) appeared for the plaintiffs, and Mr D G Rice (instructed by Muriss, Saywell & Co, of Northwood) represented the defendants.

Giving judgment, EVELEIGH J said that there had been a question as to the ownership of 17 Dene Road, but it was now agreed that Mrs Elsie Taylor was owner, and she was therefore the sole defendant in the case. The plaintiffs' case was that they bought their land in 1952 and their bungalow was completed in 1954. The site was first cleared of a number of trees, and the soil was the typical London clay. Mr King himself built the boundary wall with 17 Dene Road in 1954, on the north side of his land. In 1958 a crack appeared in the northern boundary wall, and Mr King wrote to Mr Taylor, drawing his attention to the fact that tree roots were encroaching upon the plaintiffs' property. Mr Taylor replied: "I don't think any of our trees would do any damage to your property. The roots of my trees will not extend far down the slope towards your property." On another occasion in 1958 Mr Taylor wrote: "I don't think the roots of my lime trees will ever do any damage to your house. I think the trees contribute very much to the scene. It would be an act of vandalism to cut them down, and I don't propose to do so." In 1971 the north-east wall of the plaintiffs' garage cracked and then the floor, followed by the other walls. At the same time the path to the north of the bungalow cracked and became uneven. A complete renewal of the garage and path would be necessary today; repair would not be sufficient. The drains on the north-east side of the bungalow cracked, and roots were found growing in a manhole behind the bungalow.

For the defence, it was contended first that London clay soil was known to shift with the changes in the seasons and with the action of the elements, and that irrespective of any action of tree roots, therefore, the damage to the drains and path would not be surprising. However, expert evidence had satisfied him (his Lordship) that both elm and lime tree roots had played a substantial part in causing the shifting in the soil at the back of the plaintiffs' bungalow. That conclusion emerged clearly from the evidence of both the defence and the plaintiffs. There had been substantial root growth, and experts from both sides found roots below the foundations of the bungalow. The bungalow itself was not damaged, but the fact that the roots had reached the foundations would make it not surprising if the path and drains nearer the trees had been adversely affected. He (Eveleigh J) bore in mind the possibility of damage being caused by the natural course of events and the action of the clay soil, but expert witnesses had said that the soil would be affected by trees taking water from the soil. He had also to bear in mind that the path constructed in the same way on the other side of the bungalow was unaffected. He came to the conclusion that the action of the tree roots, both lime and elm, had been substantially responsible for shifting the soil and damaging the drains and path. With regard to the boundary wall, it was clear that this had cracked as a result of tree root action, but the defendants contended that the clay soil would be expected to cause some damage unless the foundations went down at least four feet. That observation did not detract from the fact that the roots had played a substantial part in the collapse of the wall. One could view the defendant's position with some sympathy. The plaintiffs having built the wall where it was, an undue burden might be said to have been placed on the neighbours by expecting them to pay for a wall that was likely to crack. Nevertheless he (his Lordship) found it impossible to come to any other conclusion than that the tree roots were a substantial cause of the wall cracking.

Although no damage had been done to the bungalow, the plaintiffs sought an injunction to restrain further encroachment from the roots. The defence argued that if there was no damage, there was no nuisance, but this was a misconception. The word nuisance had many different shades of meaning. It could accurately be said that there could be no claim for damages unless it was established that damage had been caused, but he (his Lordship) thought it was sufficient to establish damage that the tree roots were shown to be abstracting water from the soil and making the soil less suitable than it was before they encroached. Then the defence contended that the bungalow had been built on insufficient foundations to cope with the climatic conditions of the clay soil. This bungalow, however, had stood since 1954 without damage in spite of the fact that the roots had been operating in the vicinity. There was nothing to justify the defence in its claim that the plaintiffs had forfeited their protection through the law of nuisance. There was nothing unreasonable in granting an injunction if the threat to the stability of the bungalow was a real one. The evidence showed that there were elm roots underneath the bungalow. A defence expert witness had said that it was clear that roots from the defendant's elms and limes had reached the foundations of the bungalow. That witness could see little damage to the foundations, and did not think that felling the trees would help because of the disturbance that would cause when the soil settled and heaved. It was clear, in his (Eveleigh J's) view, that even the defence witnesses accepted that the soil under the foundations of the bungalow had been affected. The presence of the roots did present a serious threat to the structure, and the roots presented all the ingredients necessary for damage to foundations. Therefore, as the roots presented a threat, the plaintiffs were entitled to their injunction.

There remained the question of the garage. He (his Lordship) had come to the conclusion that the elms on Mrs Taylor's property had substantially interfered with the soil conditions under the garage. The trees had been saplings when the garage was built, but they had since grown to a substantial size. He accepted the defendant's evidence that the construction of the garage was substandard; however, the building had stood until 1971. Whatever its structural faults, it had withstood the climatic changes in the London clay from 1954 to 1971. He thought the plaintiffs were entitled not to have the support of their garage weakened by the action of roots. He found the roots were a substantial cause of the damage to the garage. Although there was some dispute about which neighbour's trees had caused cracking in the garage, he thought it was similar to a case of a passenger being injured when two cars collided at a crossroads. The passenger was entitled to claim from both drivers for injuries, and the drivers had to argue over their respective contributions. Therefore the plaintiffs were entitled to claim damages for the garage, and he (Eveleigh J) assessed the various sums due at £450 for the drains, £240 for the wall, £151 for the path and £2,000 for replacing the garage. There would be an injunction restraining the defendant or her servants or agents from causing or permitting roots from any tree on the defendant's

A land to encroach on to the plaintiffs' land so as to cause a nuisance.

Costs were awarded against the defendant.

QUEEN'S BENCH DIVISIONAL COURT

February 26 1976

(Before Lord WIDGERY CJ, Mr Justice KILNER BROWN and Mr Justice WATKINS)

R v FENNY STRATFORD JUSTICES, EX PARTE WATNEY MANN (MIDLANDS) LTD

B Estates Gazette May 8 1976

(1976) 238 EG 417

Noise abatement order may impose a decibel level, but must require a suitable standpoint from which the meter reading is to be taken, or it will be void for uncertainty

This was a motion by Watney Mann (Midlands) Ltd, of Northampton, for an order of certiorari to bring up and quash that part of an abatement order made by justices sitting at Fenny Stratford, Buckinghamshire, on May 15 1975 which required that the noise level generated within premises known as the Fingals public house, Bletchley, Buckingham- C shire, should not exceed 70 decibels. The respondents were a number of local residents aggrieved by noise emanating from the public house.

Mr A F B Scrivener QC and Mr R J Rundell (instructed by Bower, Cotton & Bower, agents for Becke, Phipps & Co, of Northampton) appeared for the applicants, and Mr A B Hidden (instructed by Sharpe, Pritchard & Co, agents for S A J Levene, of Milton Keynes) represented the respondents.

Giving the first judgment, WATKINS J said that the background of the case lay in the fact that the premises in question had adjoining flats and dwelling-houses. At one D time the owners of the public house had employed a live band to entertain customers, but following complaints from nearby residents the services of the band were dispensed with and a juke box installed. On May 15 1975 justices sitting at Fenny Stratford made an order under section 99 of the Public Health Act 1936 on an application by the respondents, three aggrieved persons. The order required the applicants to cease causing a nuisance by generating noise in the public house which exceeded 70 decibels. The applicants, who did not challenge the finding by the justices that there had been a nuisance by noise, now contended that the justices acted in excess of jurisdiction in requiring that the noise should not E exceed 70 decibels. They also claimed that that part of the order was void for uncertainty. They contended that justices F making an abatement order had three alternatives. They could make an order *simpliciter,* an order in general terms, or an order indicating works to be undertaken.

It was plain from the authorities that there was a complete answer to these contentions. In the case of *Millard v Wastall* [1898] 1 QB 342 it was held that a notice dealing with black smoke from a chimney was good and that the justices had been fully entitled to add words relating to the extent of the smoke level which would be permitted. That decision had not been overruled, and in *Nottingham City District Council v Newton* [1974] 1 WLR 923 Lord Widgery CJ had said that justices were fully entitled, when making a nuisance G order, to use their discretion when requiring a respondent to do certain work. That had been approved in *Salford City Council v McNally* [1975] 2 All ER 860. He (his Lordship) thought, indeed, that it was often very helpful for justices to attempt to provide not only applicants, but also the aggrieved with some guidance upon a question such as noise and how the problem could be solved. Justices making an abatement order were fully entitled to add to their simple "Thou shalt not," provided that the additional terms or conditions were practical and easily understood.

That being said, it was clear that that part of the order made by the justices in the present case which related to the H 70 decibels permitted noise level was void for uncertainty. According to the terms of the order a reading of the noise level could be taken at any place in the public house itself. The true test for ascertaining whether a nuisance existed, however, was not by standing inside the premises from which the nuisance was said to be emanating. It was to be ascertained by standing outside the premises or inside adjoining premises. Before making such an order, he (his Lordship) would have visited the premises and then stated from where the sound reading should be taken. In his opinion the application should be allowed.

Agreeing, LORD WIDGERY said that the case threw up for J the first time the interesting question whether justices, making a noise abatement order could introduce into their order the concept of decibels. He thought that they could and should hear evidence relating to decibel readings, and that if they took the trouble to do so they could go on to refer to decibels when indicating in their order what would be an acceptable level of noise. Justices should take advantage of the advances of science, and he (his Lordship) thought that for the reasons given by Watkins J the application should be allowed.

KILNER BROWN J also agreed, and an order was made K quashing that part of the order in question which related to decibels. The applicants were awarded costs.

TAXATION

A

QUEEN'S BENCH DIVISIONAL COURT

October 20 1975

(Before Lord WIDGERY CJ, Mr Justice O'CONNOR and Mr Justice LAWSON)

BRITISH AIRPORTS AUTHORITY v COMMISSIONERS OF CUSTOMS AND EXCISE

Estates Gazette June 5 1976

(1975) 238 EG 723

VAT and occupation of land—British Airports Authority cannot charge VAT on supply of shop premises at Heathrow, even though licensee company bound to sell certain goods at certain times

B

This was an appeal by the British Airports Authority against a decision of the London Value Added Tax Tribunal dated March 13 1975 holding that a service provided by the authority under the terms of an agreement made by it with Hills London Shops Ltd for the provision of duty-free and duty-charged shops at London Airport did not attract VAT. The respondents were the Commissioners of Customs and Excise.

Mr P Whiteman (instructed by M W J Nott) appeared for the appellants, and Mr J Rogers QC and Mr H K Woolf (instructed by G Krikorian) represented the respondents.

C

Giving judgment, LORD WIDGERY said that the appellants' case was that they provided shops at London's Heathrow Airport, both on what was called the "land side" and on the "air side" of the no 1 terminal, and that under the terms of an agreement, Hills London Shops Ltd traded from the shops. Under the agreement Hills were to sell certain goods and to keep the shops open for certain hours. The repair and maintenance of the shops was the responsibility of the appellants. The appellants did not provide any fixtures or equipment for the shops, and by clause 5 of the agreement it was stated that nothing therein should be deemed to create

D

a tenancy. The appellants, who had to pay VAT on services provided for them in preparing the shops, had sought to include a VAT element in their concession agreement with Hills. The question was whether the supply of the services contained in the concession agreement was a taxable supply within the meaning of the Finance Act 1972. It was not in issue that the supply of the privilege was a supply made by a taxable person, and was therefore on its face a taxable supply. But the tribunal had rejected the appellants' claim, holding that Hills obtained a right to occupy premises and sell from them, and that that was a right to occupy land, which was not chargeable. Counsel for the appellants now

E

urged that the agreement gave two rights, one of which attracted VAT, and that therefore there should be an apportionment. It was claimed that the chargeable element was the right to sell. Counsel cited *Commissioners of Customs and Excise v Automobile Association* [1974] 1 WLR 1447 in support of his claim.

In his (Lord Widgery's) judgment, the true nature of the agreement was to give Hills a right to occupy land. The fact that Hills, when occupying the land, had to sell certain goods did not prevent that which had been supplied from being a right to occupy land. In his opinion the tribunal had reached the correct decision and the appeal should be dismissed.

F

Agreeing, O'CONNOR J said that there were no grounds in the present case for holding that the tribunal was wrong, but it did not necessarily follow that there would be no case where there was an agreement to occupy land and an apportionment was possible.

LAWSON J also agreed, and the appeal was dismissed with costs. The appellants were given leave to appeal.

G

CHANCERY DIVISION

November 24 1975

(Before Mr Justice FOX)

TEMPEST ESTATES LTD v WALMSLEY (INSPECTOR OF TAXES)

Estates Gazette June 5 1976

(1975) 238 EG 723

Land-holding company formed to avoid estate duty and pay portions to children—Substantial sales of developable land over the years to some extent forced on company—Commissioners entitled to reach conclusion that company was engaged in a trading venture—" Arguments both ways "

H

This was an appeal by taxpayers, Tempest Estates Ltd, of Mawdsley Street, Bolton, Lancashire, from a decision of special commissioners for income tax determining that the company's acquisition of part of the Tempest Estate, near Bolton, Lancashire, constituted a trading venture. The respondent was Mr G Walmsley, an inspector of taxes.

Mr C N Beattie QC and Mr H H Lomas (instructed by Broadbent, Heelis & Liptrott, of Bolton) appeared for the appellants, and Mr B J Davenport (instructed by the Solicitor of Inland Revenue) represented the respondent.

J

Giving judgment, Fox J said that the appellant company was incorporated on October 4 1946. Its objects included power to deal in land. It had a nominal capital of 40,000 shares of £1 each. At the date of the company's incorporation the eleventh Baroness Beaumont was tenant for life of the Tempest estate, comprising about 1,300 acres near Bolton, which had been in her family for a very long period. The commissioners found that the prime purpose of the formation of the company was to avoid estate duty on Lady Beaumont's death by transferring to the company part of the lands of the Tempest estate which were on the outskirts of Bolton, and

K

therefore likely to increase in value for development, and consequently likely to incur estate duty at the full rate and not the agricultural rate. The secondary purpose, the commissioners found, was to supplement the portions of Lady Beaumont's younger children (of whom there were seven) by allocating to them shares in the company which would automatically increase in value with the value of the lands. The directors were Lord Howard, who was Lady Beaumont's husband; Lord Beaumont, her son; and Mr Liptrott, a solicitor. Lord Beaumont and Mr Liptrott gave evidence

A before the commissioners; Lord Howard was by then dead.

By the arrangement that was made, the trustees of the family estates of Lord Howard appointed £21,050 out of these estates to Lord Beaumont (Lady Beaumont's eldest child), and that sum was used by him in subscribing for shares in the company. Those shares were then allotted to trustees for the children of Lady Beaumont. Secondly, the company itself, some time in 1946, purchased from the trustees of the Tempest estate certain farms, cottages and undeveloped land on the outskirts of Bolton comprising about 493 acres for £43,165, at which price the land was valued by a valuation in August 1946. The gross rents were about £1,200 B per annum. Thirdly, of the purchase price of £43,165, a sum of about £28,000 was raised on mortgage from the trustees of the Beaumont settled estate. After the purchase the company's land agents (who were also agents of the Tempest estate) were instructed that the company's policy in relation to the land was to be the same as when the land was part of the Tempest estate; in other words, all agricultural land was to be retained unless required by the local authority under the threat of a compulsory purchase order, and the policy of reletting to relatives of outgoing tenants was to be continued. The commissioners found that when the company was formed there was then the prospect of eventual develop- C ment of the lands because of land scarcity in Bolton. Until 1957, however, disposals were trifling. There were many inquiries by would-be developers, but they were turned down. Land was never offered for sale by the company. In 1959 the company received an offer to sell certain lands, referred to in the case as the Deane land. This was refused. Lord Howard, in a letter on behalf of himself and Lord Beaumont to Mr Liptrott, said, "There is no pressure for us to sell now, and we feel that a much better price will be obtained in a few years' time."

In 1961 there were sales to the Bolton Corporation under D threat of compulsory purchase orders, the prices realised amounting to about £61,000. In 1963 the mortgage was repaid. Lord Howard, who was chairman of the directors until 1965, viewed with displeasure the selling of any family land. In 1963, however, he was persuaded by Lord Beaumont and others to agree to a more active development policy. Lord Beaumont and the other persons involved were concerned about possible nationalisation of development land. They were particularly concerned about the Deane land, which was designated in the Bolton development plan for residential development. At a meeting of the directors on October 8 1963 it was decided that an immediate application should be E made for outline planning permission for the Deane land with a view to its development. It was also resolved (a) that if planning permission was granted the company should either develop the land itself or enter into an arrangement with a builder for its development, and (b) that in principle the company should consider the development of land capable of development as and when occasion permitted, with a view to realising the potential of the land to the best advantage. As appeared from the company's accounts dated March 31 1966, the Deane land and two farms, Clough farm and Eatocks farm, were valued at £155,000, £47,500 and £9,250 respectively and were expressed to be transferred from F investment land to trading stock. In November 1964 the Deane land was sold for £160,500 to the company's subsidiary, Ladybridge Developments Ltd. The company admitted liability to tax on its sale by reference to the difference between the sale price of £160,500 and the figure of £155,000 at which it had been valued as trading stock. In 1971 the company sold further land for £31,000. Thus the company, having purchased its lands for about £43,000 in 1946, had by 1971 made sales amounting to about £270,000 and still retained the Clough and Eatocks lands and three farms of about 120 acres, of which one, Dob Hill, about 37 acres, was never development land.

The commissioners said that eventually the shares of Lord G Beaumont's younger brothers and sisters (there were seven of them) were purchased for cash by the trustees of the Beaumont settled estates, each brother or sister receiving £35,000. The company never purchased any land save for the purchase in 1946 from the Tempest estate. In these circumstances the commissioners found that the acquisition of the land in 1946 was a trading venture. The company, on the other hand, said that on the facts the commissioners found, the only reasonable conclusion contradicted their determination. The question, it was submitted, was whether the property was acquired for the purpose of trade or not. This property, it was asserted, was manifestly not acquired for H trading purposes. Admittedly what the company acquired in 1946 was land likely to increase in value because of its development potential. But the purpose for which it was acquired, said the company, was saving estate duty and supplementing portions. As to the statement by the commissioners in their decision that the directors must have known that "it was only a matter of time before the lands would have to be disposed of, voluntarily or compulsorily, or developed," the company said that valuable building land did not have to be disposed of or developed. It could simply be retained. Given a family of extensive landowners like this one, retention J would be a perfectly normal course. In support of this, the company pointed out that the commissioners found that the agents were instructed that the company's policy in relation to this land was to be the same as when the land formed part of the Tempest estate. The policy of reletting to relatives of outgoing tenants was to be continued. There were in fact many relettings. In particular the Deane golf course, part of the Deane land, was relet in 1951 for 31 years from 1949. The commissioners, the company said, had no justification for their inference that the company expected to exploit the land to its best advantage. Lord Howard was not anxious to sell any land. As regards the letters written by Lord K Howard in 1959, these, said the company, merely showed a landowner aware of the value of his land and not prepared to sell at an unfavourable price. No doubt, it was said, the company hoped ultimately to sell at a splendid profit, but that did not make it a dealer.

These were arguments of substance. It might well be that the commissioners could properly have come to the conclusion that the company was not trading. As in many of these cases, there were arguments both ways. However, the question before the court was not whether the company was trading or not. The question was whether the only reasonable conclusion from the facts contradicted the commissioners' conclusion. It L was established, first, that the company had power to deal in land. Secondly, it must have been perfectly plain to the directors when the company acquired the land in 1946 that it was almost certain to go up in value. The commissioners found that when the company was formed to take over the lands there was then the prospect of their eventual development because of their proximity to Bolton and because of land scarcity in Bolton. The company bought the land in 1946 for full value, over half of which was raised by mortgage. Having regard to the company's exceedingly low income (only a little over £1,000 a year on properties costing £43,000), then, if that mortgage was to be paid off, the most likely way of M doing it was by sales of land. That was what in fact happened: by 1963 the company had sufficient money in hand from sales to pay off the mortgage. As early as 1958, the company made a sale amounting to about £14,000 without pressure by threat of compulsory acquisition or otherwise. The fact that the company was formed for the purpose of estate duty saving and supplementing portions was not necessarily inconsistent with acquisition of the land for trading purposes. Both purposes could be achieved even if the company was trading. A policy of reletting vacant land to the family of the deceased tenant was not necessarily inconsistent with a long-term policy of exploiting the land by sale. Making full allowance for the facts that the land had been in Lady Beaumont's family for generations, that it was transferred to the company as part of an estate duty saving scheme, that the company was in no hurry to sell and until sale dealt with the land as it had been dealt with when settled, he (his Lordship) could not in the circumstances say that the only reasonable conclusion to which the commissioners could have come was that the land was not acquired for trading. It seemed to him that there was evidence upon which the commissioners could come to the conclusion they did, and he accordingly dismissed the appeal.

The company were ordered to pay the costs.

CHANCERY DIVISION

June 26 1975

(Before Mr Justice BRIGHTMAN)

ECC QUARRIES LTD v WATKINS (INSPECTOR OF TAXES)

Estates Gazette June 12 1976

(1975) 238 EG 797

Cost of unsuccessful applications for permission to win gravel held capital expenditure and not deductible in computing profits for tax purposes

This was an appeal by ECC Quarries Ltd, a wholly-owned subsidiary of English China Clays Ltd, from a decision of special commissioners for income tax holding that the cost of certain unsuccessful applications for planning permission for extraction of sand and gravel was a capital expenditure which could not be deducted when the company's profits were computed for tax purposes.

Mr C N Beattie QC and Mrs H Watson (instructed by Robbins, Olivey & Lake, agents for Stephens & Scown, of St Austell) appeared for the appellants, and Mr M Nolan QC and Mr Brian Davenport (instructed by the Solicitor of Inland Revenue) represented the Crown.

Giving judgment, BRIGHTMAN J said that the company owned leasehold land at Blackhill and Colaton Raleigh, Devon, a freehold site at Rockbeare, a few miles away from Blackhill, and a number of other sites from which extraction was at present proceeding. It was estimated that Rockbeare and the permitted area of Blackhill would be exhausted by the mid-1970s, and in 1965 the company bought land at Straitgate Farm and Lowlands Farm, Ottery St Mary, Devon, as a medium-term replacement for the Rockbeare deposits. In February 1967 the company lodged applications for planning permission for the winning and working of sand and gravel from 151 acres of the Straitgate and Lowland farms, from a further 626 acres at Blackhill, and alternatively from 570 acres at Colaton Raleigh. If granted, these applications would have enabled the company's activities to be pursued at the three sites well after the year 2000. In the summer of 1969 the Minister, who had called in the applications for decision, refused all three, though he indicated that subject to certain qualifications they might be renewed at some future time. The applications had cost the company about £24,000, and it was now sought to charge such expenses, which had become more substantial than formerly, to the profit and loss account for tax purposes. Counsel for the company did not contend that the expenses of a planning application would be allowable against revenue in all circumstances. He said that here the deduction was permissible because of the combination of the following four circumstances: first, the company's object was not to create a new asset or addition to any existing asset; secondly, planning permission would not have created a new asset or right, but merely removed a fetter from an existing asset; thirdly, such expenditure was chargeable against revenue according to ordinary principles of commercial accounting; and fourthly, the money was spent on an attempt to free replacement assets and preserve the continuity of the company's trading activities.

It seemed to him (his Lordship) that the company had spent money for the purpose of securing a permanent alteration to the nature of the land it owned or occupied; in other words, a change from land confined to its existing use and of little or no value to the company for the purposes of its trade to land capable of being turned to account in pursuance of the company's trading activities. It was, he thought, unbusinesslike to say that if planning permission had been granted no new asset would have belonged to the company. The company's assets would have been radically and enduringly changed, and could have been written up in value in the balance sheet. On commonsense principles, and with the benefit of the guidance of the authorities to which he had been referred, including *Pendleton v Mitchells & Butlers Ltd* [1969] 2 All ER 928 and *Pitt v Castle Hill Warehousing Co Ltd* [1974] 1 WLR 1624, he thought the expenditure was of a capital and not an income nature. The planning permission, if obtained, would in some sense have been an intangible asset of a capital nature, and if that was right, money spent in seeking to acquire such an asset must equally be expenditure of a capital nature. The only remaining question was whether the unchallenged evidence as to accountancy practice ought to persuade him (Brightman J) to a different view. There were unqualified statements of high authority in the cases which established that this question, capital or income, was ultimately one of law, which meant that he need not be persuaded by the accountancy evidence if he did not want to be. He had formed his own view, based on the decided cases, that the expenditure here was of a capital nature, and the accountancy evidence was not sufficient to persuade him to alter that view. In reaching this conclusion, he did not intend to say that it would be erroneous, in drawing up a balance sheet for the purposes of the Companies Act 1948, to debit the cost of unsuccessful planning applications against revenue or to appropriate the balance of profit towards that expenditure. Accounts drawn up to comply with the Act and show shareholders what was the divisible profit involved a somewhat different form of exercise from the drawing-up of accounts on strict accountancy principles, and he was not concerned with the statutory form of accounts.

The appeal was dismissed with costs.

TOWN AND COUNTRY PLANNING

A

QUEEN'S BENCH DIVISIONAL COURT

October 28 1975

(Before Lord WIDGERY CJ, Mr Justice O'CONNOR and Mr Justice LAWSON)

TESSIER v SECRETARY OF STATE FOR THE ENVIRONMENT AND ANOTHER

Estates Gazette January 10 1976

(1975) 237 EG 117

Barn used by sculptor as studio does not acquire the character for planning purposes of a "general industrial building"—This is so notwithstanding that the barn had been equipped by the sculptor with benches, an anvil etc, that lead, brass and other material had been worked on, and that some of the finished work had been sold—Secretary of State entitled to hold that studio use came within a use class of its own

B

This was an application by Mr Frederick Arthur Leslie Tessier for an order of certiorari to bring up and quash a decision of the first respondent, the Secretary of State for the Environment, upholding with minor modifications an enforcement notice served on the applicant by the second respondents, the New Forest District Council, formerly the New Forest Rural District Council, requiring him to cease using a barn at Pear Tree Farm, Emery Down, Lyndhurst, Hampshire, for the purpose of vehicle maintenance and servicing.

C

Mr M S Rich (instructed by Lovell, Son & Pitfield, agents for Williams, Thompson & Co, of Christchurch) appeared for the applicant, and Mr H K Woolf (instructed by the Treasury Solicitor) represented the first respondent. The second respondents took no part in the proceedings.

Giving judgment, LORD WIDGERY said that for some years prior to the appearance of the applicant on the scene the barn had been used by a Mr Sven Berlin as a studio for his work as a sculptor and painter. The applicant had eventually acquired the premises and had begun to use the barn for the service and maintenance of motor vehicles. The local planning authority's enforcement notice required him to discontinue such user on the ground that it constituted a material change of use for which planning consent had not been granted and would not be granted if an application were made. Against that notice the applicant appealed, there was an inquiry, and the applicant now sought an order to quash the decision of the Secretary of State confirming the notice. The point taken was that there had been no material change of use, in that the previous use as a studio was a use as a general industrial building under class IV of the use classes order of 1972. It was common ground that the present use of the building was as a general industrial building, so that the issue was simply whether the previous use was within class IV.

D

E

The inspector who conducted the inquiry came to the conclusion on the evidence that the earlier activities in the barn were such as to cause it to be used as a general industrial building. The Secretary of State took the other view. He accepted that the barn had been equipped with benches, an anvil and other equipment, and that lead, brass and other material had been worked on, and that some of the finished sculptures had been sold. But in his decision letter, he nevertheless described the use of the barn as a studio as not being within class IV. He decided that the use as a studio was a use with its own class, and that the primary object was the carrying-on of artistic work. He (his Lordship) thought that that decision should be upheld. The use of the barn as a studio could be fairly described as the Secretary of State had described it, and the Secretary of State had a wide discretion in deciding what was a proper use having regard to all the facts. The application should accordingly be rejected.

F

G

O'CONNOR and LAWSON JJ agreed, and the application was dismissed, no order being made as to costs.

QUEEN'S BENCH DIVISIONAL COURT

October 3 1975

(Before Lord WIDGERY CJ, Mr Justice O'CONNOR and Mr Justice LAWSON)

WHOLESALE MAIL ORDER SUPPLIES LTD AND OTHERS v SECRETARY OF STATE FOR THE ENVIRONMENT AND ANOTHER

H

Estates Gazette January 17 1976

(1975) 237 EG 185

Permission for Luton cash-and-carry warehouse refused on appeal because of the possible impact on council's project for central shopping area—Secretary of State did not need evidence on which to base this conclusion—Entitled to rely on opinion and experience—Further point that in another such appeal in the same district permission was given—Court cannot investigate facts of another application, even if inconsistent decisions do disclose an error of law—Possibly a further application of some kind would bring the two matters before the Secretary of State together—Owners' appeal dismissed and leave to appeal refused

J

This was an appeal by Wholesale Mail Order Supplies Ltd and John Evans (Furnishings) Ltd, of Bilton Way, Luton, and by two directors, Mr Lawrence Alfred Kostick and Mr Alan Francis Kostick, from decisions of the first respondent, the Secretary of State for the Environment, dismissing their appeals against enforcement notices served on them by the second respondents, Luton Borough Council, requiring them to cease using premises at Bilton Way as a "cash-and-carry" warehouse.

K

Mr L Lait (instructed by Anthony & Co) appeared for the appellants, and Mr A Fletcher (instructed by Sharpe, Pritchard & Co, agents for the solicitor to the council) represented the second respondents. The first respondent took no part in the proceedings.

Giving judgment, LORD WIDGERY said: This is an appeal, purporting to come to this court under section 246 of the Town and Country Planning Act 1971, by Wholesale Mail

Order Supplies Ltd, another company, and officers of those companies, against the Secretary of State for the Environment and the Luton Borough Council, arising out of enforcement notices served by the respondent planning authority upon the appellants and relating to premises at nos 70, 72 and 74 Bilton Way, Luton, in Bedfordshire. The appellants are the owners of the premises, which were initially used as a warehouse and distribution depot. The proposal was to use them for retail trading. The change having been factually made, and the planning authority having concluded that it was a material change of use and that no planning permission had been granted for it, they proceeded to serve the enforcement notices to which I have referred. The appellants, as they were perfectly entitled to do, then appealed to the Secretary of State under the Act, and they appealed on a number of grounds. But in the end, as the matter is presented to us, there are only two grounds which require consideration at all. One is a ground which was mentioned at the hearing before the inspector and featured in the inquiry below, and appears in the appellants' statement of grounds of appeal in this court. The other is an entirely novel ground to which I shall have to refer more fully in a moment.

To take the first ground first, if one goes to the decision letter, which is dated February 28 1975, the essential paragraph comes in the reasons for the decision under the figure 10. At this stage the Secretary of State has left the technical aspects of the case and come to consider the planning merits. These paragraphs are relevant because amongst other heads of appeal relied upon by the appellants is paragraph (a) of section 88 (1) of the Act, which is a ground upon which the appellants ask that planning permission should be granted. The Secretary of State discusses the disadvantages of the new use to which the appellants have put the site in a planning sense, and considers the argument in favour of their being allowed nevertheless to continue with it. He narrows his decision down to one ground, which is expressed in this way, that "it would be both contrary to and detrimental to the council's policy of establishing a central shopping area which is already in the course of construction." It is in the end on that ground that the Secretary of State decided to refuse planning permission under ground (a) of the appeal. The views of the inspector, as adopted by the Secretary of State, are as follows:

> It is my opinion that it is the last which must be the deciding factor. It is not a planning function to protect individual traders, but it is a planning responsibility to ensure that there is a balanced layout between shopping, residential, industrial, etc development. In the case of Luton this is achieved by the production of the town plan which is the result of considerable study by the authorities concerned. If there is to be a substantial departure from this plan, there must be good reason for it, or the whole plan would become meaningless. I consider that the grant of planning permission for a large retail store in this otherwise industrial estate would amount to a substantial departure. Furthermore, if the proposal were permitted, it would be very difficult for the council to refuse other similar applications for cash-and-carry establishments dealing with a variety of commodities including foodstuffs, and there appears to be space available, bearing in mind the new development under construction to the west. A proliferation of this form of trading here could well jeopardise the viability of the town redevelopment scheme.

That is what the inspector said, and the Secretary of State adopted it, and therefore it becomes the basis for refusal of planning permission in the first instance. Mr Lait attacks this conclusion on the ground that there is no evidence, or there was no evidence before the Secretary of State, to enable him to come to that conclusion. The actual words of objection are:

> The Secretary of State erred in law in deciding that it would be contrary to and detrimental to the council's policy of establishing a central shopping area to grant planning permission. There was no evidence on which the Secretary of State could so decide.

The short answer to that objection is that in the planning sphere a great many decisions are not based on facts and evidence but are based on opinion and experience; and when the inspector first, and Secretary of State secondly, expressed the views which I have just read out they did not require to call witnesses to support those views. They were giving the opinions of people experienced in this sort of business whose opinions were worthwhile and upon whose opinions actions could take place. It is a complete misconception to take the view that matters of professional opinion, in planning in particular, require the sort of factual support in evidence which is required in proving the existence of a criminal case. The objection taken that there was here an absence of evidence, with the implication that there should have been further evidence, is in my judgment wholly misconceived.

The other ground to which Mr Lait seeks to draw our attention is one which has been put in writing before us, and I think I can summarise it without doing injustice to it. He seeks to contend before this court that the Secretary of State has committed a further error of law in his approach to these problems in that he has recently considered a similar application in another part of Luton and has taken a view there favourable to the developers, and has taken the view, further, that such development is not inconsistent with the proper development of the centre of Luton. So, says Mr Lait, we have here two conflicting decisions based on substantially the same facts and showing that something has gone wrong. He invites us to say, having gone through that tortuous process of reasoning, that thus is disclosed an error of law of the kind which entitles this court to interfere under section 246. The principal difficulty here, of course, is that this court cannot possibly investigate facts of another and wholly different planning application which was never referred to in the proceedings below and upon which the Secretary of State has never expressed an opinion. How, first of all, we can be apprised of the facts is hard enough, because there is already authority which indicates the general tendency in these proceedings not to embark on a retrial of facts on the basis of fresh evidence admitted: I am referring, of course, to *Green v Minister of Housing and Local Government* [1967] 2 QB 606. So not only are we faced with the problem of how, with propriety, this court is to be informed of the facts in the other case, but we are also faced with the far more serious problem that the Secretary of State has never been asked to consider that case in relation to the present, and it would be quite wrong, in my judgment, to consider any arguments of this kind in the absence of the Secretary of State and without his having been joined in the proceedings. Whether or not other procedures can be adopted, whether or not a further application of some kind may not bring these two matters before the Secretary of State together, I do not know, but certainly this court would not be a promising forum in which to achieve that, because it must be remembered first and last that our authority is only in regard to errors of law, and I am very far from accepting, for myself, that an error of law is disclosed merely because one finds the Secretary of State or the planning authority reaching inconsistent conclusions in two cases. It may be an error of law; I do not express a final view upon it, but I am by no means certain it is. In my judgment, what is clear in this case, however, is that the appeal must be dismissed, and I would dismiss it.

O'CONNOR J: I agree.

LAWSON J: I agree.

The appeal was dismissed with costs, and leave to appeal to the House of Lords refused.

A

QUEEN'S BENCH DIVISION

March 11 1975

(Before Mr Justice PHILLIPS)

AI & P (STRATFORD) LTD v LONDON BOROUGH OF TOWER HAMLETS

Estates Gazette February 7 1976

(1975) 237 EG 416

Planning permission made subject to a condition requiring existing office accommodation, at present used for extraneous purposes, to be ancillary to the buildings on the development site—Condition valid—Intended to promote planning authority's office policies both outside and inside the area covered by the permission—Held fairly and reasonably to relate to the permitted development

B

This was a summons by AI & P (Stratford) Ltd, of 9 Grafton Street, London W1, seeking a declaration that a condition attached to a planning permission granted to the company by the defendants, the London Borough of Tower Hamlets, was void and of no effect.

Mr M H Spence (instructed by Wigram & Co) appeared for the plaintiffs, and Mr A D Dinkin (instructed by the solicitor to the council) represented the defendants.

C

Giving judgment, PHILLIPS J said: The plaintiffs, AI & P (Stratford) Ltd, claim by originating summons a declaration that condition 5 of a planning permission granted by the defendants and dated June 28 1974 is void. By their defence the defendants, the council of the London Borough of Tower Hamlets, who are the relevant planning authority, deny that the condition is void, and in the alternative claim that if it is void the planning permission itself is null and void by reason of the invalidity of the condition.

The planning permission relates to premises at East Cross Centre, Waterden Road, Tower Hamlets. The development which is permitted is the erection of two warehouse units and one industrial unit, being the first phase of development proposals for the site. Condition 4 provides that the welfare accommodation in the existing welfare office block shall be retained as welfare accommodation and not used for any other purpose. Condition 5, which is the disputed condition, provides that the office accommodation in the existing welfare office block shall only be used as ancillary to non-office uses on the site. It is common ground that the word "site" in that condition refers to the whole area, including what I may call the new units and the old units. Condition 6 provides that the portion of the proposed building set apart as offices shall not be used otherwise than as ancillary to the main use of the premises.

D

E

There is a plan, which is exhibit 2 to the affidavit of Mr T J Davies [the development manager to the plaintiff company], and is a copy, I think, of the plan annexed to the permission. What that shows is that before the permission there was an existing industrial building subdivided into parts and that there were existing offices. On the plan the existing offices are outlined in red, and there is additionally shown in dark solid grey the newly-permitted development, which consists of two blocks of warehouses and some blocks of industrial units. Only the first one of these was permitted by this permission, being the first phase of the proposed development. The whole of that is outlined by a black line, and the whole of the area within that black line, as I understand it, is the "site" referred to in condition 5. Therefore, it is to be seen that there is within the site both old development—for which there were presumably permissions which had been in operation for some time—and the newly-permitted development. It is also the case that included in the newly-permitted development was a small element of office accommodation to be used as ancillary to the newly-permitted warehouse and industrial units, but that is not separately shown on the plan.

F

G

The effect, then, of condition 5 is that what I have called the existing offices (which are so marked in the plan) are required thenceforth only to be used as ancillary to non-office uses on the site. That is to say, they are to be used as offices ancillary to existing industrial buildings, or to the newly-permitted warehouse and industrial units. What is complained of is this: it is said that the condition attached to this permission cuts down the plaintiffs' rights as owners in respect of their own existing offices, which are not part of the newly-permitted development. The relevant statutory provisions are sections 29 and 30 of the Town and Country Planning Act 1971. That is a consolidation Act, and the provisions to which I am about to refer date from the beginning of the post-war extension of the town and country planning legislation in the Act of 1947. Section 29 provides:

H

> Subject to the provisions of sections 26 to 28 of this Act and to the following provisions of this Act, where an application is made to a local planning authority for planning permission, that authority in dealing with the application shall have regard to the provisions of the development plan, so far as material to the application, and to any other material considerations, and (a) subject to sections 41, 42, 70 and 77 to 80 of this Act, may grant planning permission either unconditionally or subject to such conditions as they think fit, or (b) may refuse planning permission.

J

Section 30 provides:

> Without prejudice to the generality of section 29 subsection (1) of this Act, conditions may be imposed on the grant of planning permission thereunder (a) for regulating the development or use of any land under the control of the applicant (whether or not it is land in respect of which the application was made) or requiring the carrying out of works on any such land, so far as appears to the local planning authority to be expedient for the purposes of or in connection with the development authorised by the permission.

It is common ground that the wide words of section 29 and section 30 have to be interpreted in the light of the decided cases. I have been taken by Mr Spence and Mr Dinkin most helpfully through the various cases. There is, I think, no need for me to make any citation from them, because there is no real controversy as to what the true position is. But they, and in particular *Pyx Granite Co v Ministry of Housing and Local Government* [1958] 1 QB 554, *Kingston upon Thames Royal London Borough Council v Secretary of State for the Environment* [1973] 1 WLR 1549, and the cases therein cited, with the others to which I have been referred, establish that the wide words of sections 29 and 30 have to be read subject to these limitations, that a condition cannot be justified, and will if necessary be declared invalid by the courts, if it is arbitrary or whimsical, if it is inserted for some ulterior purpose, or if it is unreasonable; provided always that in the latter case it is so unreasonable that it can be said that no reasonable planning authority would attach it to the permission. Putting the same thing positively, in order to be valid a condition must be reasonably related to the development in the planning permission which has been granted; or, to use other similar language from the cases, the condition must fairly and reasonably relate to the permitted development. For example, in most cases, to require the applicant for planning permission to abandon the use of different and distant premises owned by him as a condition of the grant of permission for other premises would usually not be a proper condition.

K

L

M

Mr Spence, on behalf of the plaintiffs, puts his argument under two heads. First, he says condition 5 does not reasonably relate to the development to be carried out, because (a) it is an attempt to require the plaintiffs to restrict the use of the office building (that is, the existing office building) as a condition of obtaining permission to carry out the development applied for. I think that that contention is too widely framed, because the mere fact that such a requirement—that is to say, the restriction of the use of existing

A premises—was attached as a condition of the obtaining of permission would not necessarily, by itself, render the condition void, unless it can further be said that the condition so attached did not relate to the permitted development. And I think that in the end Mr Spence accepted that comment on this part of his submissions, and he broadened submission (a) by saying that what the defendants were here trying to do was to obtain an incidental advantage of a purely negative character by preventing in the future the existing offices being used for general office purposes without any benefit to or connection with the permitted development; and that in effect what they were doing was trying to obtain B by other and improper means what could only properly be obtained by a section 52 agreement. The submission continues in (b) of its first head: "because the condition is not expedient for the purposes of, or in connection with, the development authorised by the permission." Secondly, Mr Spence submits that condition 5 is unreasonable in that it seeks to deprive the owners of their existing use rights in another building on the site. I think he may want to add, or anyhow make the rider, "and to do that without compensation." That contention, apart from being stated, has not been further advanced in argument in this court, for the reason that there is authority to the contrary in the Divi- C sional Court in the case of the *Kingston Royal London Borough Council*. But Mr Spence puts it forward because if this case should go further he may wish to argue the point hereafter.

So really the matter comes down to (b) of counsel's first head, which I will repeat because it is the heart of the case: "the condition is not expedient for the purposes of, or in connection with the development authorised by the permission." Before going any further, it is necessary to note that there is here an unresolved question of fact, that is to say, the extent to which the existing offices, or the existing welfare D office building, have over the last few years been used for purposes ancillary to the existing industrial building, and how far they have been used for what I would call extraneous office purposes. That is a matter in dispute between the same parties. Not only is it in dispute between them, but it is the subject of different litigation proceeding elsewhere. It is obviously a matter of great importance, because (without going into details) the extent to which it has been used for extraneous office purposes may determine the extent to which the plaintiffs will be able to use the existing office building for such purposes in the future. That question remains to be determined by the pending litigation between E these parties, and during the course of this case I have been somewhat concerned whether I can properly decide the question before me without knowing the answer—the factual answer to that matter. In the end I have come to the conclusion that I can, and that it is possible to determine the validity of this condition without knowing the answer to this question. From a practical point of view, I think it makes the problem more difficult, because one's mind the whole time is slightly wondering precisely what the actual user in the last few years has been. And to some extent, of course, that question lies at the back of Mr Spence's submissions, because in effect what he is saying here is that F it is wrong for the planning authority to be able to tell the owner of a building, who has used it for extraneous office purposes, in the future to cease such user and to use it for office purposes ancillary either to the old existing industrial buildings, or to the new ones. But certainly I think no injustice can be done if I test the validity of the plaintiffs' contentions by assuming for present purposes what really is the most favourable position from their point of view, namely that the existing offices have for some time past not been used (anyhow exclusively) as ancillary to the existing industrial buildings. On that assumption, and against the background where I have indicated, I come to the question whether condition 5 is fairly and reasonably related to the G permitted development, reminding myself that the permitted development is the two blocks of warehouse units and the first phase of the industrial units, and bearing in mind, as Mr Spence points out, that the existing offices are part of the old development and are at the opposite end of the site from the newly-permitted development.

The plaintiffs' contentions, in elaboration of the submissions which I have read, are that this condition has been inserted as a means of supporting and advancing the defendants' policy as the planning authority of the restriction of office development and of confining it as far as possible to certain particular selected areas. To some extent that is H obviously factually true, but the real question is this: whether that is the only purpose of the condition, or whether as well as having that purpose it relates also to the permitted development. There are available to the court in the determination of this question as well as the matters which I have already indicated, certain additional matters in the planning permission, and certain matters in the affidavit of Mr John David Hume, who is the director of development of the defendant planning authority. I have already read conditions 4, 5 and 6, but the planning permission also sets out the reasons for the insertion of those conditions. Condition 4 is the con- J dition requiring the welfare accommodation to be retained, and the reason given is to ensure that the welfare accommodation is retained for the use of employees of future occupiers of the site. The site includes, as well as the old existing industrial building, the newly-permitted development, and the reasons given for conditions 5 and 6 are to ensure that the development complies with this authority's policy of restricting further office development, and to ensure that these elements of the scheme conform to the provisions of the initial development plans of Greater London in which the site is allocated for industrial purposes.

Those reasons, particularly 5 and 6, are perhaps not as K clear as they might have been, and Mr Spence draws attention to the different treatment of the welfare accommodation and the office accommodation, because condition 4 and reason 4 make it clear that the welfare accommodation is to continue to be available for the purposes of the new development whereas condition 5 and reasons 5 and 6 (so counsel contends) make it clear that the purpose of condition 5 is to restrict the use of the existing office accommodation in pursuance of the planning authority's policy for offices, and that this has nothing to do with the newly-permitted development. It seems to me, however, that it is at least possible from a reading of conditions 4, 5 and 6 L and reasons 4, 5 and 6 that the planning authority had *two* purposes in mind: a general purpose, namely to reduce office accommodation except in the appropriate areas, and a particular purpose, namely to ensure that the existing office accommodation remained available to be used for the purposes *inter alia* of the new development. It seems to me that support for that interpretation is to be found in the affidavit of Mr Hume, to which I have referred. In paragraph 9 Mr Hume says:

> In my view it would not have been a good time for planning practice to consider any part of the proposals in the centre in M isolation. In considering the section 72 application regard was had, *inter alia*, to (1) the need to retain existing amenities and services on the site for the use of future occupiers, and also (2) the need to ensure that development proposals would not result in increase in office space off the site which would have been contrary to the office policy of this council and the Greater London Council—

and then I omit the rest of that paragraph. And in paragraph 11 Mr Hume says:

> In the light of these two factors, when the section 72 application was reported to this council's development committee, I

A recommended that no notice be served but that the applicants be informed that in the event of a varied planning application being submitted, a number of conditions should be imposed, including a condition limiting the use of the existing ancillary office welfare building to office and welfare units ancillary to the main non-office uses of this site.

Then, and I interpolate, he goes on to explain why:

My concern was that if the offices were let independently [and that is what I call extraneous offices] future industrial occupiers of the centre would need to provide their own ancillary office space. This need to provide additional office space could lead to the existing and proposed industrial and warehouse space B being used for office purposes thereby resulting in a net increase in office floor space and a reduction in the industrial and warehouse floor space. This would be contrary to the council's planning policy of retaining and encouraging industrial employment in the borough.

Mr Spence, of course, comments that despite all that, what the planning authority are really trying to do is to compel the plaintiffs to use their own property for purposes for which they do not wish to use it. That is being done as a condition of the grant of planning permission. In paragraph 16 Mr Hume continues:

C This application [that refers to the planning application which followed the section 72 application] in fact was made before the result of the section 72 application was known. This application [and that of course is the one to which this permission relates] is considered as the first phase development for the whole centre, and I therefore recommended to the development committee the conditions which previously had been approved by the committee and of which the applicants had been informed as being suitable conditions to be imposed in the event of a varied planning application being submitted.

The chronology then was this: first of all the section 72 application, followed by the planning application, and before D that was granted, the result of the section 72 application was made known to the plaintiffs (whether or not they had known it before, I do not know), followed on June 28 by the planning permission. The net result of all that is, it is plain from this affidavit, that the same considerations led to the imposition of this condition as led to it in the section 72 application.

That evidence by Mr Hume has not been challenged, and there has been no application to cross-examine him. As I have already indicated, independently of that evidence (so it seems to me) a possible interpretation of reasons 5 and 6 is along the lines to which he deposes in the paragraphs of his E affidavit which I have read. Thus, it seems to me, one can see that the reason for the attachment of this condition was that it was first of all intended to serve and promote the policy of the planning authority and the Greater London Council for the control and the restriction of office accommodation, but that it was intended to do this not only in the existing offices themselves (and therefore outside the particular area of the planning permission), but that it was intended to implement those policies within that part of the site which is within the area of the planning permission. It was hoped thereby to prevent what was intended to be used in the future as warehouse and industrial accommodation from F having to be used, in part at least, for ancillary office purposes. Now, if that is the right interpretation of the facts—and I find that it is—it seems to me that inevitably condition 5 is a condition which in fact does relate to the permitted development. That being so, the ground upon which it has been attacked must fail. It is not said that this condition is so unreasonable that no reasonable planning authority could have attached it; and, if it had been, I do not think that that could have been made good. So I come to the conclusion that this is a valid condition. In those circumstances I do not think that it is necessary for me to say much about the contention of the defendants that if it had been void, the whole planning permission itself would also be void. The G reason I do not think it would be useful to say anything about that is that one can only deal with this contention when one knows precisely why, if it were the case, the condition was void. Accordingly the conclusion is that in my judgment the plaintiffs are not entitled to the declaration which they seek.

The summons was dismissed with costs.

QUEEN'S BENCH DIVISIONAL COURT

October 22 1975 H

(Before Lord WIDGERY CJ, Mr Justice O'CONNOR and Mr Justice LAWSON)

PERCY BILTON INDUSTRIAL PROPERTIES LTD v SECRETARY OF STATE FOR THE ENVIRONMENT AND ANOTHER

Estates Gazette February 14 1976

(1975) 237 EG 491

Outline planning permission for development of 22-acre industrial site granted in 1952—Construction of various buildings contemplated, and at various stages—Separate buildings in fact constructed by a number of independent owners between 1952 and about 1970—Development held J to have been "begun" for the purpose of preserving the outline permission after 1967—Secretary of State bound to hear appeals from local planning authority's deemed refusals of applications made in 1973 for detailed permission to develop further portions of site.

In these proceedings Percy Bilton Industrial Properties Ltd, of Park Street, London W1, moved for orders of mandamus directed to the first respondent, the Secretary of State for the Environment, requiring him to hear and determine appeals against the deemed refusal in August and September 1973 by the second respondents, St Albans District Council, of detailed planning permission for the erection of industrial premises K on parts of a 22-acre site at Colney Street, Radlett, Hertfordshire, outline planning consent for the development of which had been granted by St Albans Rural District Council in 1952.

Mr W J Glover QC and Mr M B Horton (instructed by C M Crichton) appeared for the applicants, and Mr H K Woolf (instructed by the Treasury Solicitor) represented the first respondent. The second respondents took no part in the proceedings.

Giving judgment, LORD WIDGERY said: In these proceedings Mr Glover moves for orders of mandamus directed to the Secretary of State for the Environment requiring him to L hear and determine certain appeals brought by the present applicants, Percy Bilton Industrial Properties Ltd under the town and country planning legislation. The history of the matter begins on October 8 1952, when some 22 acres of land at St Albans, then belonging to Handley Page Ltd, were the subject of an outline planning permission under the Town and Country Planning Act 1947. The permission is in what are now familiar terms, but it is necessary to look a little closely at one or two of the provisions, and I emphasise that within this single permission are embraced the whole of the 22 acres with which this case is concerned, even though a railway separated the 22 acres into two unequal parts. The M description of the development which is sanctioned by the permission of October 8 1952 is this. The form says: "Development of land for industrial purposes at Colney Street, near St Albans, Herts." The permission goes on to make it clear that it is what we now call an outline planning permission, the details of which have to be filled in later. It is made an outline permission by a condition in these terms: "The approval of the local planning authority is required before any development is commenced to its (a) siting; (b) design; (c) external appearance; (d) means of access." So the approval of the planning authority is required before any development is commenced and the approval is concerned with siting, design, external appearance and access. After the grant of that planning permission, a good deal of activity took place on these 22 acres. There were a number of independent and separate buildings erected, and a number of independent and separate transactions entered into. From time to time what would happen was that the owner of the land for the time being, not always Handley Page Ltd, as will be appreciated, would apply to the planning authority for approval of the details which had been left undetermined when the outline permission was granted. That of course is the obvious and logical way of dealing with such a situation when, following an outline permission, the landowner is in a position to put forward his proposals for details. In a number of instances what happened was that the details were put forward for approval and approved, and then a building was erected on the appropriate part of the site. A good example of that happening is in connection with a building which we have referred to as B1, where we have before us the actual formal approval of the details which had been submitted in respect of the proposed building, and it is perfectly clear that what is being done is obtaining approval under the existing outline permission. In other cases, however, for reasons which may have varied considerably from case to case, fresh planning applications were made, and sometimes a fresh planning permission was granted which embraced part of the original 22 acres and so on, but the planning authority and the several owners were obviously content that this substantially should be developed piecemeal in this way, and between 1952 and 1970 or thereabouts this sort of activity was going on.

In 1973 two applications were made to the planning authority which gave rise to the present motions. At that time the building B1 had long since been built. It was completed on August 14 1967, and other buildings were either complete or in the process of construction. On August 31 1973 the present applicants submitted to the planning authority an application for approval of details on a site called the "Dee" site. It is not necessary to identify the location of the Dee site. It was an area within the 22 acres which as yet had not been built upon. It enjoyed, so the applicants thought, the advantage of the outline permission which originally attached to all the 22 acres, and so they put forward a perfectly straightforward application for approval of details of the buildings proposed to be put on the site. The following month, on September 24 1973, a similar application was made in respect of what we are to call the "Stinchar" site, again part of the 22 acres and enjoying, on the applicants' submission, the benefit of the original outline permission. Application is made, logically enough as it seems, for approval of the details of the development. The planning authority made no decision in regard to either of those applications, and under the general machinery of the Town and Country Planning Act the effect of the planning authority taking no action was that they were deemed to have refused the application and deemed to have rejected, therefore, the proposed details in respect of these two sites, the Dee site and the Stinchar site. In the ordinary way, when a local authority refuses, either expressly or by an operation of the Act, to grant a right which is sought there is an appeal to the Secretary of State for the Environment, and it was an appeal to the Secretary of State which the applicants now sought to pursue in respect of both these applications. But when the matter was put before the Secretary of State he did not deal with it, and the reason why he did not deal with it is clearly set out in a letter which he writes to the solicitor for the applicants on October 4 1974 and in which he contends that the submission of the details was out of time and in breach of the conditions attaching to the appropriate planning permission, and that accordingly he would not deal with the application.

Therefore the matter comes to rest, as it were, when the Secretary of State declines to play the part which the applicants expect him to play, and it then becomes a matter for us to consider whether as a matter of law the Secretary of State is wrong and should be instructed to proceed to hear the appeal.

The reason any sort of difficulty arises is that in the Town and Country Planning Act 1971 a good deal of important provision is made which is designed, I think I can fairly say, to clear up outstanding planning permissions which seem not to be going to be implemented. One finds one such provision in paragraph 20 of the 24th Schedule to the Act, I must read the first subparagraph of that paragraph carefully:

> Subject to subparagraph (2) of this paragraph, where before April 1 1969 outline planning permission (as defined by section 42 of this Act) has been granted for development consisting in or including the carrying out of building or other operations, and the development has not been begun before the beginning of 1968, that planning permission shall be deemed to have been granted subject to conditions to the following effect: (a) that in the case of any reserved matter (as defined in that section) application for approval must be made not later than the expiration of three years beginning with April 1 1969. . . .

If one approaches that subparagraph from the end and works backwards, it becomes apparent that the effect of the rule in situations to which it applies is to impose a limit of three years from April 1 1969, and thus to limit to a period expiring on April 1 1972 the submission of details in respect of applications to which the subparagraph applies. To what applications does the subparagraph apply? It is perfectly easy to say. It applies to those cases where outline planning permission was granted before April 1 1969 for building and the development referred to in the section had not been begun before the beginning of 1968. If those conditions are satisfied, the case falls within the subparagraph, the obligation to obtain the approval of details within three years becomes operative, and the failure to obtain or apply for those details within three years would give the Secretary of State, in my judgment, the right to reach the conclusion which he did. Thus the whole case turns on whether the facts here before us bring the matter within paragraph 20 or not.

Looking again at the words of the paragraph (and everything in the end depends upon them), the first thing which is necessary to be shown in order to bring the matter within paragraph 20 is that outline planning permission had been granted for building development before April 1 1969. Well, it had. The outline planning permission was the one which I have already read, the one on October 8 1952, and it dealt with outline permission for building development. We move on to the next requirement, "and the development has not been begun before the beginning of 1968." Factually what had happened, as I have already indicated, is that before the beginning of 1968 some development had taken place. In particular, the building B1 had been completed in 1967, and other buildings were in various stages of progress. If you look at the 22 acres as a whole, and ask yourself whether development has begun pursuant to the outline planning permission of October 1952, there can, as it seems to me, be no argument but that the building has begun. On the other hand, if it is possible in some way to break up the 22 acres into separate parcels, it may well then be possible to argue that on certain parcels the building has begun and on other parcels it has not. So again as one narrows the issue, that is really what lies between the two parties in this case. Mr Woolf's argument, it seems to me, must start with the proposition that when the outline planning permission was granted in October 1952 it was there and then a collection of separate permissions and not a single and dissoluble permission. I know he does not accept that that is what his argument amounts to, and I accept what he says. But it seems to me that that is really the basis of it all, and that is a basis which

A would seem in my judgment to be a very sensible one. I fully recognise that when a planning permission has been granted it can be varied in a number of ways subsequently. It may be varied by a new application. It may be varied by the course of dealing between the parties, and lots of things like that, but I can see absolutely nothing in the present instance to suggest that there had been any kind of variation since the initial permission was granted. I do not think that the mere fact that some work has been done of the kind which would be covered by the outline permission on part of the land in any sense segregates that part of the land and requires it to be separately treated. If that which is done is B work which would normally be done pursuant to the outline application, I can see no reason why merely because the work is done one is justified in severing the area into individual and separate parcels.

That really is an end of the case in my judgment, but I have not overlooked, and I must mention it in order to show I have not overlooked it, the argument put forward in the correspondence on behalf of the Secretary of State that the building work which had taken place on these 22 acres cannot have been lawful because it was in breach of the condition of the outline permission to which I have already referred. C The words in question were, "The approval of the local planning authority is required before any development is commenced to its (a) siting. . . ." It is argued that since no development of any kind can take place until agreement in regard to the siting of the buildings has been obtained, the entire development in this case was unlawful. It seems to me that that cannot possibly be sustained. There is no obstacle to the parties agreeing to the approval of details by stages. I do not express any view as to whether every party can insist on this, but unquestionably if they want to do it that way, the approval of details can be done by stages. If one stage is approved, and it is within the consent of the local D authority that the building relevant to that one stage should take place, then it can take place, and to suggest after what has happened that there is anything unlawful in the planning sense about the buildings seems to me to be entirely incorrect. For all those reasons, it seems to me that in so far as the Dee site was concerned the Secretary of State should have accepted the appeal and dealt with it. Similar considerations apply to the Stinchar site, save only for the fact, as I understand it, that there is an extension of time of one month required there before the appeal can be submitted at all. It is clear that the Secretary of State, for reasons which seemed good to him, has not considered any of these matters E as yet, and I think the order of the court should require him to do so. That would mean in regard to the Dee site that mandamus should go to hear the appeal, and in regard to the Stinchar site that the Secretary of State should give consideration to the question of whether an extension of time should be granted. If he decides to grant an extension of time, he should then hear the appeal which in consequence would become alive again.

O'Connor J: I agree.

Lawson J: I also agree.

Orders were made accordingly, and the Secretary of State was directed to pay the applicants' costs.

F

QUEEN'S BENCH DIVISIONAL COURT

July 7 1975

(Before Lord WIDGERY CJ, Mr Justice MILMO and Mr Justice WIEN)

WEST CHESHIRE CARAVAN CO LTD v ELLESMERE PORT BOROUGH COUNCIL AND ANOTHER

Estates Gazette February 21 1976

(1975) 237 EG 573

Three enforcement notices prohibiting use of adjacent pieces of land for sale of caravans—Use for that purpose found established on two pieces of land, but third notice upheld— G **Owner claims that the three pieces should have been treated as a single planning unit despite his withdrawal of the relevant ground of argument at the inquiry—Secretary of State not bound to "work the ground out for himself" despite its abandonment and the want of evidence with regard to it—Cheshunt Urban District Council v Minister of Housing and Local Government [1965] EGD 229 cited and distinguished**

This was an appeal by West Cheshire Caravan Co Ltd, of Telegraph Road, Heswall, Wirral, Merseyside, against a decision of the second respondent, the Secretary of State for the Environment, communicated on November 15 1974, H upholding an enforcement notice served by the first respondents, Ellesmere Port Borough Council, requiring the appellants to cease using land at the rear of Ledsham Filling Station, Welsh Road, Ledsham, for display and sale of caravans.

Mr A B Dawson (instructed by Thompson, Quarrell & Megaw, agents for Oliver & Co, of Ellesmere Port) appeared for the appellants, and Mr A P Fletcher (instructed by Sharpe, Pritchard & Co, agents for the solicitor to the council) appeared for the first respondents. The second respondent took no part in the proceedings.

J Giving judgment, Lord Widgery said: This is an appeal under section 246 of the Town and Country Planning Act 1971 brought by the West Cheshire Caravan Co Ltd against the Ellesmere Port Borough Council and the Secretary of State for the Environment in respect of a decision by the Secretary of State communicated in his decision letter of November 15 1974 wherein, having ordered that certain other enforcement notices should be quashed, he upheld what in the proceedings had been described as notice C relative to a rectangle of brown-coloured land on the plans submitted to this court and delineating an area near Ledsham railway station on the London Midland & Scottish & Great K Western Joint Railway line at Ellesmere Port. There is at this point what for many purposes can reasonably be described as a single piece of land which in its entirety is roughly triangular in shape and has a frontage to the railway to which I have already referred. For the purposes of the various planning applications relating to the land it has been treated as divided into four portions and given four different colours. Fronting on the public road, Welsh Road, is an area which is coloured white and in respect of which the planning authority made no complaint about the user by the appellants. Immediately to the west and south of the land L coloured white is a triangular area coloured green. There follows, moving towards the south and west, a rectangular area coloured blue, and then finally the brown rectangular area to which I have already referred. On the white land and the frontage to Welsh Road there has at all material times been a petrol filling station, and on the white land adjoining the petrol filling station caravans have been displayed and sold at all material times. The local authority moved against the caravan use on the other parts of the site on March 16 1971, when they served three enforcement notices each alleging a material change of use from use to caravan sales after 1964—the first, notice A, relating to the M green triangle, the second, notice B, relating to the blue rectangle and the third, notice C, relating to the brown rectangle. The Secretary of State held the usual inquiry, and in regard to the white land there was, as I have already said, no move by the planning authority to restrict the use of the premises for the sale of caravans, so no decision was required from the Secretary of State. On the green and blue land the Secretary of State, accepting the report of his inspector, took the view that the use of those two portions for the purposes of caravan display and sale had occurred before January 1 1964. Accordingly there was an established right to continue that use, and those two enforcement notices, A and B, were

A quashed on that ground. All that remained, therefore, was the notice C relating to the brown land, and in this respect the Secretary of State upheld the notice with modifications that do not call for any comment in this court at this stage.

How did it come about that the Secretary of State regarded the third use on the brown land as being varied when he had rejected the other two? The reason was that it was not possible for the appellants to contend that they had begun to use the brown land for the sale of caravans before January 1964. We have not been required to go in detail into what their difficulties were, but it is quite evident that they could not sustain that as part of their argument. They B had originally based their objection to enforcement notice C (the one relating to the brown land) on a number of grounds. But in the course of the argument before the inspector grounds (b) and (d) were withdrawn. I remind myself what grounds (b) and (d) were. They are to be found in section 88 of the Act of 1971. Ground (a) provides "that planning permission ought to be granted for the development to which the notice relates or, as the case may be, that a condition or limitation alleged in the enforcement notice not to have been complied with ought to be discharged." Ground (b): "that the matters alleged in the notice do not constitute a breach of planning control." Further down, in (c), we have: "in C the case of a notice which, by virtue of section 87 (3) of this Act, may be served only within the period of four years from the date of the breach of planning control to which the notice relates, that that period has elapsed at the date of service." Finally, ground (d) provides: "in the case of a notice not falling within paragraph (c) of this subsection, that the breach of planning control alleged by the notice occurred before the beginning of 1964." It is perfectly understandable that the appellants should have abandoned ground (d), because they admitted at an early stage, if not from the outset, that they could not prove user since before 1964; but D what is very relevant in the present instance is that they also abandoned ground (b).

The way the argument is put before us today is this: It is said that by 1969, let alone by the date when the enforcement notice was served, all four of these portions of the land were in the single occupation either of the appellants or of their predecessors in title, and that at that time they constituted a single planning unit, which is the contention before us in this court today. It is said, therefore, that when the brown land went over to use for caravan display and sales, whenever that was, it was the change of use in part of a planning unit, and not in its entirety, the same unit embracing the E white, blue and green land. The next stage in the argument is that since the change of use affecting the brown land affected only part of the planning unit, then it was for consideration, as a matter of fact and degree, whether it had created a material change of use in the whole of the planning unit; and that is a factor which was never gone into before the inspector, or in the Secretary of State's reasons, for the very good reason, I think, that this matter had been abandoned when ground (b) was withdrawn. When one goes back to ground (b), it is, as I have already said, that the matters alleged in the notice do not constitute a breach of planning control. That was the basis upon which the single F planning unit argument would have been based. The argument would have been that having regard to the fact that the four parts were all constituents of a single planning unit. the change in the brown land alleged in the third notice did not constitute a breach of planning control. That very ground was given away, and that, no doubt, is the reason why the Secretary of State pays no attention to it in his decision letter.

What is said now by Mr Dawson on behalf of the appellants is that notwithstanding that this ground was thrown away at the inquiry, and notwithstanding that no mention of it appears in the Secretary of State's decision letter, yet the Secretary of State's failure to work out this ground for himself (notwithstanding that it had been abandoned) and G to apply it is an error of law which entitles the appellants to have the decision sent back for reconsideration by the Secretary of State. Counsel supports that contention by referring us to the somewhat abbreviated report of a case called *Cheshunt Urban District Council v Minister of Housing and Local Government* [1965] EGD 229. It was a decision of this court in which, on an issue of material change of use or no, it became relevant to consider whether the change from selling cars to selling caravans was a material change of use. Lord Parker, in giving the leading judgment, referred to this issue and said at p 230:

> H The Minister decided that storage and sale of caravans in this case was not a substantially different use from the storage and sale of cars, and did not amount to a material change in the use of the land. It was contended for the appellants that the Minister decided the appeal on a ground that had been withdrawn. He held that the change from the sale of cars to the sale of caravans did not constitute or involve development, and that ground was specifically withdrawn before the inspector. But that argument failed, because although that ground was withdrawn—for whatever reason—it was plain that it remained a live issue to the extent to which it was dealt with by the Minister.

Therein is contained the whole difference between that case and this one. I think that the *Cheshunt* case must be J regarded as exceptional in any event, but it wholly differs from the present instance, because there the Minister himself kept alive that issue because he thought it necessary in order to do justice and notwithstanding that it had been abandoned by the parties. It is a very long step from there to say that the Minister must find out for himself a possible ground for supporting the appellants' argument, even though there has been a deliberate withdrawal of that argument before the inspector, and even though the relevant facts have not been found upon which a consideration of that argument can depend. It seems to me abundantly clear that the argument now sought to be put before this court is one which K should not be accepted in view of the history of the matter as I have recounted it, and for my part I would dismiss the appeal.

MILMO J: I agree.

WIEN J: I agree also.

The council was awarded costs.

QUEEN'S BENCH DIVISIONAL COURT L

July 14 1975

(Before Lord WIDGERY CJ, Mr Justice MILMO and Mr Justice WIEN)

JOYCE SHOPFITTERS LTD v SECRETARY OF STATE FOR THE ENVIRONMENT AND ANOTHER

Estates Gazette February 21 1976

(1975) 237 EG 576

Enforcement notice—Use of major part of site for industrial purposes established by 1963—Secretary of State can require owners to pull down extensions built since then, but cannot prohibit industrial use of the portions of the site on which they stood—Further point on site of two M cottages demolished by owners since 1963—Upholding the notice with regard to this area would involve treating that portion of the site as a distinct planning unit—Matter remitted for reconsideration by Secretary of State in light of court's observations.

This was an appeal by Joyce Shopfitters Ltd, of Starts Hill, Farnborough, Kent, against a decision of the first respondent, the Secretary of State for the Environment, on January 24 1975 upholding an amended enforcement notice served by the second respondents, Bromley Borough Council, in October 1973 requiring the appellants to cease using for the purposes

A of their shopfitting business certain land between Starts Hill Road and Willow Walk, Farnborough.

Mr G E Moriarty QC and Mr M B Horton (instructed by Kidd, Rapinet, Badge & Co, agents for William J Wade & Co, of Sidcup) appeared for the appellants, and Mr M H Spence (instructed by J H Stevens, of Bromley) represented the second respondents. The first respondent took no part in the proceedings.

Giving judgment, LORD WIDGERY said: This is an appeal under section 246 of the Town and Country Planning Act 1971 brought by the occupier of land, a company called Joyce Shopfitters Ltd, against a decision of the Secretary of State B for the Environment contained in a formal decision letter of January 24 1975 whereby he upheld, to a limited extent, an enforcement notice which had been served on the appellants by the London Borough of Bromley as the local planning authority.

The historical background to the matter is this. The appeal site lies between Willow Walk and Starts Hill Road in Farnborough, Kent. It has had a somewhat chequered history. In its entirety it comprises about 8,000 sq ft, and within that total area there is a group of buildings stretching over two storeys which was described at the inquiry as a two-storey building. The present appellants acquired the land by stages C in the latter part of the 1960s and early part of the 1970s, and in their hands significant changes in the use of the land have occurred. There was, as I have said, from the beginning this two-storey building situate on the northern part of the appeal site, and when the land came into the hands of the present appellants they proceeded to build extensions to the two-storey building on the north, west and south sides, and these extensions went up in the 1970 to 1972 period. There also stood from a much earlier date in about the middle of the appeal site a pair of cottages called 8 and 9 Willow Walk, which the appellants had demolished in or about the year D 1972. Having made those physical changes to the building which I have recited (the building of the extensions and the demolition of the cottages), the appellants proceeded to use the entirety of the site for the purposes of their business as shopfitters.

The land had had a history of industry before. The inspector who held the inquiry in the course of normal practice and the Secretary of State both accepted this change, and what seems to have happened is that at the end of 1963, the significance of which of course is that this is a date at which a development can become protected by lapse of time, the two-storey building was being used as a workshop, partly E in connection with a fencing contractor's business and the manufacture of garden furniture, and partly as a workshop for car repairs and car breaking. The remainder of the appeal site (which means the unbuilt-upon land) was being used in conjunction with and ancillary to these two separate uses of the building. Thus in the early stages we have this situation: two uses on the land, the building being used for both of them and the unbuilt-upon land being used as an ancillary to those two uses. Then, as I have already described, the present appellants, when they had the power to do it, built these extensions and knocked down the cottages, and thereafter used the entirety of the property for their business of F shopfitters.

In paragraph 8 of the decision letter is the Secretary of State's consideration and opinion of the consequences of those events which I have described so far. He says:

"It is considered that the change of use which took place in 1968 and 1970 from a fencing contractor's business, the manufacture of garden furniture and rustic work and for car repairs and car breaking to a use for a joinery and shopfitting business did not constitute development by virtue of the provisions of section 22 (2) (f) of the 1971 Act."

That is a very important decision, and speaking for myself I have no doubt it was a correct one. The effect of those provisions and the Town and Country Planning (Use Classes) G Order 1972 is that there can be a change of use from one form of industrial activity to another, and that is what the Secretary of State is saying was the fact in 1968 and 1970. Accordingly the use of the premises for shopfitting has become a secure use and one which cannot be enforced against so far as it extended before the intervention of the appellants to the appeal site which I have described.

Now one comes to the notices themselves to see what it is which has given rise to complaint by the planning authority, and there are two notices in question. The first one is called notice A and is dated October 12 1973. It is directed to the additions to the two-storey building to which I have referred. H The complaint in notice A is that these additions have been put up without permission, that the erection of the additions is itself development, and this development having been carried out without permission, the planning authority is calling upon the appellants to pull down the unauthorised additions. This is no longer in dispute, because in the course of the inquiry the appellants stated that they had no answer to the first notice, and so that will stand, and in due course the extensions will come down if they have not come down already. It is notice B upon which the remaining contentions centre, and notice B is concerned with the whole of the land J —the entire site—surrounded by the red line upon the plan attached to the notice, and it complains that the land is being developed by a change of use, namely, a use for the purposes of a joinery and shopfitting business; the notice calls upon the occupiers (the present appellants) to cease that use and restore the land for the purposes for which it was formerly used.

Once it is accepted, as I have already said it has been accepted, that the shopfitting business was in the same use classes order as the previous industrial use, it follows that notice B is doomed to failure in so far as the land to which K it is sought to be attached enjoyed the previous industrial use, because to that extent the change from the previous industrial use to shopfitting would not be a material change of use at all. So we come back to see what the Secretary of State has done in regard to the various parts of this site in the light of their previous planning history. So far as the two-storey building plus the land which has never been built on is concerned, the Secretary of State has accepted that that is protected from enforcement because it had an industrial use before the change was introduced by the appellants, and for the reasons I have already given that industrial use covers them in their shopfitting activity. But so far as the L extensions are concerned, the Secretary of State takes the view that the industrial use never applied to them, because they have only been in existence for some few years; they have not been up long enough, as it were, to aspire to any established or secure planning use. The Secretary of State has accepted the view of the planning authority that so far as they are concerned they are to be regarded as having started with a nil use, and the adoption of the shopfitting use is a change of use which is the subject of enforcement. Similarly, the Secretary of State has taken the view that the sites of the cottages were not used industrially before the cottages were pulled down, and here again, as I understand M his decision, he takes the view that the sites of the cottages started with a nil use in about 1970 when the cottages were pulled down and that there has not been time for them to aspire to and acquire a secure use on their own account. Thus he has amended enforcement notice B, and the effect of his amendments is to support the notice and thus require the cessation of the shopfitting use on the areas previously covered by the extensions to the two-storey building and by the two cottages.

That is the only area in which contention remains between the parties, and Mr Moriarty says the Secretary of State was not entitled to uphold the notice, even to that degree, on the facts found by the inspector. Mr Spence supports the opposite. I read paragraph 9, which is the crucial paragraph in the Secretary of State's decision. He says:

> "With regard to ground (d) it is clear from the evidence"— I remind myself that that is the ground that the use in question has gone on since before 1964—"that the site of the two former cottages was not used for the joinery and shopfitting business prior to 1970 at the earliest, as the cottages were used residentially until 1970 and were not demolished until 1972. The use of this part of the site has not therefore become established. Furthermore, the extensions to the brick buildings were erected between 1970 and 1972, and as their planning history starts from the date of their completion and a material change of use occurred on their first use in the business, the use of the extensions has not become established. It is concluded that there has been no change of use constituting development since the end of 1963 on the remainder of the site and the use of this part of the site has become established within the meaning of section 94 of the 1971 Act. The appeal on ground (d) succeeds to this extent."

Taking first of all the extensions to the two-storey building, I think, with deference to the Secretary of State, that he was wrong to allow the enforcement notice to stand in regard to them. Those extensions had to go by virtue of enforcement notice A. They were to be pulled down, and once they were pulled down the site—the land thereby disclosed—in my view would have the same industrial use which the rest of the land surrounding it had had. Neither the land nor the extensions had been used for anything other than this industrial purpose, and when the extensions have gone by virtue of enforcement notice A, then the appellants will be entitled to use the site of the extensions for the purposes of their industrial business. In those circumstances I think it was wrong, and indeed oppressive, to have an enforcement notice relating to the use of this land, because the enforcement notice could only take effect when the buildings had gone, and once the buildings had gone the industrial use was a use which could properly be practised upon the extension sites. So I would send this matter back to the Secretary of State with certain advice. The first point which I would venture to advise him upon is that it was inappropriate to apply the second enforcement notice B in regard to the sites of the extensions because when the extensions came down, as they had to come down, the land thereby disclosed would enjoy the right of industrial user which enforcement notice B seeks to stop.

What about the site of the cottages? The Secretary of State in my opinion has unquestionably treated the site of the cottages and the rest of the land as though they were two different planning units. If they could properly be described as two different planning units, his conclusions, if I may say so with respect, would have been impeccable, because it would have been perfectly correct that the sites of the cottages had not acquired a separate use of their own, and perfectly correct that the remainder of the land would have had a protected industrial use. But no one questions the proposition followed by the inspector that the planning unit in this case is the entirety of the land, and once that is established then the question material change of use or no cannot be determined finally by looking at the individual parcels of land comprised in the unit. It can only be determined finally by looking at the unit as a whole. True, as Mr Spence says, in considering the position over the unit as a whole, the Secretary of State would go and look at the component parts. He would know that the houses had not had an industrial use before. He would know that the sites of the extensions would have an industrial use for the reasons I have already given. All this material would go into his mind, but in the end he would have to say in regard to the unit as a whole whether there had been a material change of use or not. I do not think he has ever reached that stage in his consideration of the matter. So, in sending it back to him, I would advise him that he must consider this final point. If he concludes that there has been no material change of use in the area as a whole, then of course the enforcement notice must go. It cannot stand in regard to part if there is no material change of use in regard to the whole. If, on the other hand, he comes to the conclusion that there has been a material change of use in the area as a whole, quite different considerations would arise, and indeed I would not seek to suggest that that is a conclusion which I would expect him to find, because it seems to me it is contrary to the inspector's decision and all the basic facts of this case.

In summary, therefore, I would allow this appeal. I would send the matter back to the Secretary of State for reconsideration, inviting him to reconsider his conclusions in regard to enforcement notice B, having particular regard (a) to the fact that the sites of the extensions would enjoy an industrial use in any event, and (b) to the fact that the question of material change of use or no must be determined ultimately by reference to the circumstances affecting the planning unit as a whole. It is not the obligation of this court to go through all the points in the notice of motion. We can only deal with the matter which has been raised before us, and I think that has been done.

MILMO J: I agree with the judgment that has just been delivered.

WIEN J: I agree also.

The appellants were awarded costs.

QUEEN'S BENCH DIVISIONAL COURT

October 6 1975

(Before Lord WIDGERY CJ, Mr Justice O'CONNOR and Mr Justice LAWSON)

MORRIS v SECRETARY OF STATE FOR THE ENVIRONMENT AND ANOTHER

Estates Gazette February 28 1976

(1975) 237 EG 649

Enforcement notice—Repair of vehicles on northern half of a site whose southern half is used for growing fruit—Notice may specify whole site as planning unit—Further point on defect in notice—Sale of vehicles referred to, as well as repair, but discontinuance of sale not required—Inspector's action in amending notice to require sales to cease endorsed by court

This was an appeal by Mr Edward Morris, of Portland Lodge, Brentwood Road, Bulphan, Essex, against a decision of the first respondent, the Secretary of State for the Environment, upholding with an amendment an enforcement notice served on the appellant by the second respondents, the Thurrock District Council, requiring him to cease using land at Portland Lodge for storage and maintenance of vehicles.

Mr J M Sullivan (instructed by Kenneth Elliott & Rowe, of Romford) appeared for the appellant, and Mr K H T Schiemann (instructed by G W Plater) represented the second respondents. The first respondent was not represented and took no part in the appeal.

Giving judgment, LORD WIDGERY said: this is an appeal under section 246 of the Town and Country Planning Act 1971 brought by Mr Edward Morris, who is the owner of the affected land, against a decision of the Secretary of State made through the hand of one of his inspectors in regard to an enforcement notice served by the local planning authority upon the appellant and relating to land and premises known as Portland Lodge, Brentwood Road, Bulphan, Essex. The history of the matter can be put quite simply. There exists in the urban district of Thurrock the house and premises to which I have already referred, the total area involved being something of the order of five acres. The house and buildings

A are situate in the northern half of the area, and the southern half consists of agricultural land which is used at present for growing fruit. Following the arrival of the present appellant as the owner and occupier of Portland Lodge, a motor sales and service business has been started by him on the northern half of the premises. The vehicles for sale are stored in the open on the land there, and repairs and respraying have taken place as required. The business has been an extremely thriving one. In a very short time it has built up to such an extent that when the inspector called at the premises there were 400 vehicles there, which gives one some idea of the amount of effort which has been put into this B affair by Mr Morris. But no one could doubt that the use of this land for the sale, repair and respraying of motor vehicles was a material change of use for the purposes of the Town and Country Planning Act, and since the appellant had not sought, or at least had not obtained, permission to make that material change of use, it is not altogether surprising that an enforcement notice was in due course served. The notice is dated March 7 1974. It recites that the Thurrock District Council act on behalf of the local planning authority in respect of these premises in Essex. It recites further:

> It appears to the council that there has been a breach of planning control after the end of 1963 in that development consisting C of use of the premises for the purpose of (a) the sale of motor vehicles, and (b) the repair and respraying of motor vehicles, has been carried out without the grant of permission required in that behalf under Part III of the Town and Country Planning Act 1971.

The notice then goes on to require the landowner upon whom it is served within the period specified to remove from the premises all motor vehicles other than those incidental to the enjoyment of the dwelling-house as such, and to discontinue the use of the premises for the purpose of the repair and respraying of motor vehicles other than the repair or respraying of motor vehicles incidental to the enjoyment of the dwelling-house as such. An inspector was appointed D to consider the appellant's appeal against that notice, and in due course he made a report dealing, if I may say so, very fully with the merits of the matter, because one feels that the principal argument before the inspector was whether planning permission ought to be granted for this activity. However, he came down against the appellant in that respect, and the outstanding points which are then open for argument in this court are matters of law.

First of all, it is contended that the enforcement notice should have been restricted in its effect to that part of the appellant's premises upon which the motor-car selling and E repair business was actually carried on. In other words, it is contended that the notice should not have been directed to the whole five acres, but only to the northern 2½ acres or thereabouts upon which at the time of service of the notice the business was being carried on. Secondly, it will be observed from what I have already said that the enforcement notice did not require action to remedy the breach of planning control which was directly consistent with the breach complained of. In particular, although the breach complained of the sale of motor vehicles as well as the repair and respraying of vehicles, the requirement of action did not mention sale at all, although it seems to have been accepted F at the hearing before the inspector that that was an error on the part of the local authority, and one can hardly doubt that it was anything else. But as the notice stands, it does not prohibit the sale of motor vehicles, and theoretically after an interval of time vehicles could be brought back and put up for sale, although that would merely attract another enforcement notice. The inspector, appreciating the futility, I think, of allowing the notice to stand as it was, acceded to an argument from the local authority in that he treated the notice as containing an error and proceeded to cure that error by inserting appropriate words to prohibit the sale of motor vehicles upon the site. It is said that he erred in law in both those respects, and I will try to deal with them G individually.

First of all, on the question whether the notice could properly apply to the whole five acres, it is necessary to remember that the first step in deciding whether any breach of planning control has taken place is to ascertain the planning unit concerned and ask in relation to that unit whether the alleged change of use is a material one for present purposes. Once the planning unit has been ascertained, and once it is clear that a material change of use in regard to that unit has occurred, then in my judgment it is open to the planning authority to bring enforcement proceedings, either in respect of the whole planning unit or in respect of some H smaller portion upon which the offending change of use has occurred. There is some authority on this point in the case of *Hawkey v Secretary of State for the Environment and Another* (1971) 22 P & CR 610. That indeed was a case where the planning authority sought to enforce against an area smaller than the planning unit itself, and the decision of this court was to the effect that such activity was permissible on the part of the planning authority. I think that that situation exactly covers the one with which we are faced today. I know of no principle which prevents the local authority from enforcing against the whole planning unit; J and indeed a ridiculous situation might otherwise exist, because on any other basis enforcement would apply to an area of the land steadily increasing in size as the activity itself increases, and it seems to be eminently sensible and entirely just that once enforcement action has become possible because there has been a material change in the use of the whole unit the planning authority should be entitled to enforce against the whole unit. In my judgment, therefore, there is nothing in the first point.

The difficulty about the second point is simply this. Section 88 of the Act of 1971 contains provisions which have been in this legislation now for some years and which make pro- K vision on an appeal against an enforcement notice for the correction of errors in the notice and variation of the notice. Subsection (4), paragraph (a) of section 88 is in these terms: "On an appeal under this section . . . the Secretary of State may correct any informality, defect or error in the enforcement notice if he is satisfied that the informality, defect or error is not material." That is a power to correct defects. In subsection (5) of the same section there follows the provision: "On the determination of an appeal under this section, the Secretary of State shall give directions for giving effect to his determination, including, where appropriate, directions for quashing the enforcement notice or for varying L the terms of the notice in favour of the appellant." I repeat the words "in favour of the appellant." What is argued here is that the inspector has himself described his addition to the enforcement notice referring to the sale of vehicles as being a "variation." He uses that word in his report, and so Mr Sullivan argues that what has happened is a variation under section 88 (5), but that the variation is unlawful, because section 88 (5) only allows a variation in favour of the appellant, and this is in the opposite direction. Mr Schiemann, for the local authority, says, "No, this is not a matter upon which subsection (5) becomes material; it is all to be dealt with under subsection (4)." He says, citing Lord M Denning, that the correction made by the inspector in the enforcement notice is legitimate as long as it is one which could be made without injustice. No one can doubt that it can be made without injustice, because it was a pure error, and obviously the words now inserted should have been in the notice from the start.

The words of Lord Denning, which are always cited in relation to this matter, are to be found in *Miller-Mead v Minister of Housing and Local Government* [1963] 2 QB 196 at 221. Dealing with the form in which the power to correct errors then existed in 1962, Lord Denning says

A of the Minister's powers: "I think that it gives the Minister a power to amend, which is similar to the power of the court to amend an indictment. He can correct errors so long as, having regard to the merits of the case, the correction can be made without injustice." In my knowledge and understanding, that test has been applied ever since 1963, and I think it should continue to be applied, because I think it specifies the real point of subsection (4) and distinguishes it from subsection (5). The correction of the error which took place in this instance was one which was made without any sort of injustice to the appellant, and I think that it was properly remedied or rectified by the inspector under subsection (4) B and that the enforcement notice can stand with that error rectified in that way. It seems to me, therefore, that there is no substance in this appeal, and I would dismiss it.

O'CONNOR J: I agree.

LAWSON J: I agree also.

An order for costs was made against the appellant.

QUEEN'S BENCH DIVISIONAL COURT

March 4 1975

C (Before Lord WIDGERY CJ, Mr Justice BRIDGE and Mr Justice EVELEIGH)

SNOOK v SECRETARY OF STATE FOR THE ENVIRONMENT AND OTHERS

Estates Gazette March 6 1976
(1975) 237 EG 723

Enforcement notice—Builder's yard used mainly for storage of building materials with some sales from premises—Yard bought by demolition contractor and used essentially for the same purpose, storage of building materials, sale now being the primary object—No material change of use— D "Means of disposal of building materials when they leave the site of no importance in land-use terms"—No evidence before Secretary of State on which he could differ from his inspector and support the notice

This was an appeal by Mr Frederick William Ivor Snook, demolition contractor, of Vallis Road, Frome, Somerset, from a decision of the first respondent, the Secretary of State for the Environment, dated December 11 1973, dismissing his appeal from an enforcement notice served upon him by the second respondents, West Wiltshire District Council, successors to Bradford & Melksham Rural District Council, E as agents for the third respondents, Wiltshire County Council, requiring him to cease using a site in Trowbridge Road, Hilperton, Wiltshire, for the purposes of his work.

Mr G Eyre QC and Mr C S Rawlins (instructed by Hiscox & Co, of Trowbridge) appeared for the appellant, and Mr H K Woolf (instructed by the Treasury Solicitor) represented the first respondent. The second and third respondents took no part in the appeal.

Giving the first judgment, BRIDGE J said: This is an appeal under section 246 of the Town and Country Planning Act 1971 from a decision of the Secretary of State for the Environment given by letter dated October 16 of last year, F dismissing an appeal by the present appellant against an enforcement notice dated December 11 1973 which had been served on him by the second respondents, the local planning authority, alleging a material change in the use of land. The history of the matter can be quite shortly related. The land the subject of the enforcement notice is a small site with a frontage of 60 ft to the main road between Trowbridge and Devizes, and a depth of some 80 ft. It is situated in a village called Hilperton. This site was used from 1951 or 1952 to 1969 by a firm of builders called Mattock & Wells as a builder's yard. They stored on the site building materials which they used in their own business, mostly new building materials, but a proportion of them, said to have been about G 25 per cent, reclaimed building materials, arising no doubt from the demolition of old buildings. They used the bulk of the materials stored on the site in their own building operations elsewhere, but something like 25 per cent of the turnover of material in Mattock & Wells's time was eventually sold to third parties, both to other builders and to members of the public to whom the yard in question was open. In 1969 the present appellant came upon the scene and acquired the site from Mattock & Wells. His business is not that of builder but that of demolition contractor. He evidently did quite a lot to clear and tidy the site, but nothing turns upon that. He has used the site without a break since he acquired H it for the storage of materials accruing from his business as a demolition contractor. As he demolishes buildings elsewhere he burns what is combustible of the rubbish, disposes to scrap merchants what is scrap, and transports to the site the subject of this appeal what is of value for re-use in the building trade. Approximately 95 per cent of the materials which, under the appellant's control, find their way to the site in question, when they leave the site are to be used by third parties, having been sold on by the appellant. The appellant himself conducts his business, no doubt both in relation to his demolition operations and in relation to these sales of reclaimed goods to the trade, not from the site at J all, but from his home at Frome, and his business with other builders is carried on over a wide area, having a radius of some 50 miles in three counties, Somerset, Wiltshire and Dorset. At the site itself, no one is present except on Mondays, Wednesdays and Fridays and for two hours on Saturdays. On those days, however, a gentleman, who we are told is an old-age pensioner, is employed on the site, sorting and stacking materials, and on those days the site in the appellant's occupation, as it was in the occupation of Mattock & Wells, is open to casual members of the public who may be interested in purchasing any of the reclaimed K building materials stored there. There are in fact two small notices on the road frontage to the site, one which indicates that salvaged building materials are sold there, and the other which gives the appellant's name and address in Frome.

Against that background of fact, the inspector who held the inquiry at which these facts emerged came to the conclusion that there had been no material change of use. In expressing his conclusions, the inspector said: "It seems to me that the essential element of the present use of the land, the storage there of building materials, is the same as that of its use during the period 1951-52 to 1969 as a builder's L yard. I am not persuaded by the argument that there is a material change of use because previously sale of building materials was merely an incidental facet of the builder's yard use, whereas it is a primary object of the present use. The use of the land for 'storage, sorting, treating and burning' of building materials is similar to its pre-1969 use, and the means of subsequent disposal of materials when they leave the site seem to be of little importance in land-use terms." When the Secretary of State came to deal with the matter, in dissenting from that conclusion, he expressed himself in paragraph 6 as follows: "The inspector found as facts which are accepted that from 1951-52 until 1969 M the appeal site was occupied together with the adjacent land to the west as a builder's yard by Mattock & Wells, who stored building materials for their own use and for sale to the public. About 25 per cent of the materials stored were reclaimed secondhand goods. The evidence shows that your client, who is a demolition contractor, has occupied the appeal site since about 1970. Large quantities of building materials from the buildings he demolishes are sorted and stacked on the site and sold from there, mainly to builders for re-use, but some sales are made to the general public. A builder's yard is considered to be a composite use, containing many elements which may comprise amongst other

A elements storage, industrial elements, ie the manufacture of joinery items, and possibly some ancillary sales, but the primary use of the yard is the storage and preparation of materials for use in various building projects. The view is taken that your client's use of the premises is primarily a sales use to which end salvaged building materials are sorted and stored on the appeal site." Then after reference to his disagreement with the inspector, the Secretary of State concludes: "It is considered that this use is different in character from the former builder's yard use and that, as a matter of fact and degree, its introduction amounted to a material change of use constituting development for which planning B permission was required."

Now in this court Mr Eyre submits that that conclusion by the Secretary of State is vitiated by an error of law on the footing that the Secretary of State, as the tribunal of fact, if properly directing himself as to the relevant legal considerations to be applied, could not upon the factual material which was before him properly or reasonably reach such a conclusion. We tend to think that we know the old authorities in this field very well, but we have, if I may say so, been helpfully and usefully referred to one of the most familiar and to two short passages from that very familiar C authority, *East Barnet Urban District Council* v *British Transport Commission and another* [1962] 2 QB 484, where Lord Parker CJ, adverting to the considerations of law which are applicable to the determination of the question whether a material change of use amounting to development has taken place, said this at p 491: "It seems clear to me that . . . what is really to be considered is the character of the use of the land, not the particular purpose of a particular occupier." On the previous page, at page 490, referring to the use of the word "material" in the relevant part of the definition of development, he said: "The word 'material' came in for the first time in the definition in the Act of D 1947, but that must be referring to material as material for planning purposes." We have also been referred to a decision of this court in *Lewis* v *Secretary of State for the Environment and another* (1971) 23 P & CR 125 in which my Lord, Lord Widgery CJ, said: "The actual activity on the land so far as the Secretary of State refers to it appears to be identical both before and after and it is not my understanding of the law that, if the activity is exactly the same throughout the relevant period, a material change of use can occur merely because of a change in the identity of the person carrying out that activity. Similarly, I am not prepared to accept that, if the use throughout the relevant period is the repair of E motor vehicles, a material change of use can occur merely because the ownership or source of supply of those motor vehicles has changed."

Mr Eyre submits that *mutatis mutandis* those passages and principles are applicable in the present case. In *Jones* v *Secretary of State for the Environment* (1974) 28 P & CR 362, to which Mr Woolf has drawn our attention, my Lord, the Lord Chief Justice, at p 367, indicated the limitations of the passage cited from *Lewis's* case and pointed out that if an activity formerly carried out as a use ancillary to another and primary use itself reaches the level of a primary use, that may justify an allegation that a material change of use has F occurred. The question here is whether, upon a proper analysis of the factual material, there was evidence of such a change in the proportion of the relevant activities on the site one to another as to justify the Secretary of State's conclusion that there had been a material change of use. I say "the relevant activities on the site," because in the circumstances of this case I would draw a sharp distinction between sales being effected on the site, either to members of the public visiting the site casually on the days when it was open, or indeed to members of the trade going to the site to look at, and if they liked to buy, some of the goods they found there, on the one hand, and on the other hand sales being effected to his customers by the appellant from his G home and office at Frome. Once the appellant had sold a lorry-load of building materials to a buyer and that buyer went to the appeal site to collect it, I can see no planning significance whatever in the circumstance that that load when it left the site was destined for use by a purchaser and not for use in the business of a builder who occupied the site as his own yard. It is that consideration, it seems to me, which underlies and confirms the good sense of the inspector's conclusion in paragraph 55 of his report, to which reference has already been made, when he said, "The means of subsequent disposal of materials when they leave the site seem to be of little importance in land-use terms." I would H would have gone further, and said "of no importance in land-use terms."

Now if the element of what I will call off-site sales as opposed to on-site sales which appeared in evidence here had been excluded from consideration, and if the Secretary of of State had concentrated his attention upon on-site sales and had come to the conclusion that on-site sales as a relevant and constituent activity on the site itself had so far increased in proportion in relation to their previous proportion since the arrival of the appellant, it might very well be that he could have come, or if the matter should develop in future J could yet come, to the conclusion that there was a material change of use. But I can see no material whatever anywhere in his inspector's report, which was of course the factual report before the Secretary of State, which leads to the conclusion that the element of on-site sales had increased in the appellant's time as compared with Mattock & Wells's time at all. For those reasons, I have reached the conclusion that this was one of those very exceptional cases—and they are exceptional—when this court is driven to say that there was no material before the Secretary of State on which, directing himself properly in law, he could reach the conclusion which he did. I would allow the appeal and send K the matter back to the Secretary of State for reconsideration in the light of the judgment of this court.

EVELEIGH J: I agree.

LORD WIDGERY: I agree also.

The appellant was awarded costs.

QUEEN'S BENCH DIVISIONAL COURT

February 10 1976

(Before Lord WIDGERY CJ, Mr Justice KILNER BROWN and Mr Justice WATKINS)

SCURLOCK v SECRETARY OF STATE FOR WALES AND ANOTHER L

Estates Gazette April 3 1976

(1976) 238 EG 47

Enforcement notice—Most of three-storey Georgian house used as residence, but two ground-floor rooms used by occupant for purposes of her estate agency business—Premises not a "dwelling-house" within terms of general development order, notice requiring reinstatement of sash windows accordingly effective

M This was an appeal by Mrs Grace Scurlock, of 17 Hamilton Terrace, Milford Haven, Dyfed, from a decision of the first respondent, the Secretary of State for Wales, upholding an enforcement notice served by the second respondents, Preseli District Council, requiring the appellant to remove a single window installed on the ground-floor storey of her property and to replace it with two sash windows.

Mr C Fay (instructed by Rutland & Cranford, agents for Price & Kelway, of Milford Haven) appeared for the appellant, and Mr H K Woolf (instructed by the Treasury Solicitor) represented the first respondent. The second respondents took no part in the proceedings.

Giving the first judgment, KILNER BROWN J said that the premises concerned were Georgian in style and three storeys high, and it was obvious that the local planning authority and the Secretary of State were of opinion that they required careful treatment and a degree of planning control. The appellant lived in the upper part of the premises and used the front two downstairs rooms for her work as an estate agent. Her contention was that as she occupied the upper part of the premises as her home the premises were, for all purposes, to be regarded as a dwelling-house. There came a time when she was minded to have repairs and decorations done to the property, and in the course of that work it was discovered that there had at one time been one long window at the front instead of two small sash windows. That large window must have been removed and replaced by the two sash windows to make the whole front of the house look elegant and Georgian. The appellant, who was minded to replace the two sash windows with one large window, had some discussion with the local authority and got the impression she had permission to go ahead with the proposed development. This she did, only to find that the authority and the Secretary of State took the view that there was not, and should not be, permission for the change.

The sole question raised on appeal was the definition of "dwelling-house." The appellant said that the premises were a dwelling-house within the meaning of the Town and Country Planning General Development Order 1973 and that the development was permitted development under paragraph 3 of that order. No specific definition of "dwelling-house" could be found in the order, and the proper test was the factual approach. That was the approach adopted by the inspector and the Secretary of State. The appellant submitted that the Secretary of State had erred in law in holding that the premises were not a dwelling-house, but he (his Lordship) thought the Secretary of State had assessed the situation correctly. He had said that development as defined in section 22 (1) of the 1971 Act was involved, but that it was necessary to look at the use of the appeal property as a whole before it could be determined whether the provisions of the general development order were applicable. He had gone on to decide that the premises had a dual use, residential and business, and that there had accordingly been development for which planning consent was required. That was an entirely correct view, and there was no error of law. The appeal should be dismissed.

LORD WIDGERY and WATKINS J agreed, and the appeal was dismissed with costs.

QUEEN'S BENCH DIVISIONAL COURT

February 3 1976

(Before Lord WIDGERY CJ, Mr Justice KILNER BROWN and Mr Justice WATKINS)

CATTON v SECRETARY OF STATE FOR THE ENVIRONMENT AND ANOTHER

Estates Gazette May 1 1976

(1976) 238 EG 335

Enforcement notice—Appellant failed to make out pre-1964 development—Inspector's conclusions on factual issues supported—Appeal against confirmation of notice dismissed

This was an appeal by Mr Arnold Catton against a decision of the first respondent, the Secretary of State for the Environment, upholding, with certain modifications, an enforcement notice served upon the appellant by the Neston Urban District Council, the predecessors of the second respondents, the Ellesmere Port Borough Council, requiring him to discontinue the use of land owned by him at Heath Farm, Dunstan Lane, Burton, Wirral, for parking or servicing vehicles or storing vehicle parts.

Mr J Hugill (instructed by Gregory Rowcliffe & Co, agents for Steggles & Mather, of Chester) appeared for the appellant, and Mr H K Woolf (instructed by the Treasury Solicitor) represented the first respondent. The second respondents took no part in the proceedings.

Giving judgment, LORD WIDGERY said that by an enforcement notice dated April 23 1973 the Neston Urban District Council, the predecessors of the Ellesmere Port Borough Council, required the appellant to discontinue the use of 90 acres of land at Heath Farm for the purpose of parking or servicing of heavy goods vehicles and the storage of vehicle parts. The ground for the notice was that the use of which complaint was made represented a material change of use since January 1964 for which planning consent had not been given. The notice, with certain amendments, had been upheld by the Secretary of State after an inquiry held by one of his inspectors. The sole issue at the inquiry was whether the material changes had taken place after the material date, January 1964. If they had taken place before that date there had been no breach. The inspector had had no easy task, and he had taken great care in reaching his conclusion that there had been some development in breach of planning control. Questions of fact and degree were for the inspector. It had been a difficult case, but the inspector was entitled to reach the conclusion that the appellants had not discharged the onus on him of showing that in the relevant respects there had been no breach. There had been no error of law shown, and the appeal should be dismissed.

KILNER BROWN and WATKINS JJ agreed, and the appeal was dismissed with costs.

COURT OF APPEAL

May 22 1975

(Before Lord DENNING MR, Lord Justice BROWNE and Lord Justice GEOFFREY LANE)

MOLTON BUILDERS LTD v WESTMINSTER CITY COUNCIL AND ANOTHER

Estates Gazette May 8 1976

(1975) 238 EG 411

Use of Whitehall Court suite as offices contemplated by lease from Crown Estate Commissioners but subject to planning control—Occupier converts to office use without permission from planning authority, and commissioners give the necessary consent to service of an enforcement notice —Commissioners not liable for derogation from grant—Judgment of Willis J confirmed

This was an appeal by Molton Builders Ltd, of 32 Old Burlington Street, London W1, from a judgment of Willis J in the Queen's Bench Division on July 31 1974 dismissing their claim against the first respondents, Westminster City Council, and the second respondents, the Crown Estate Commissioners, for a declaration that an enforcement notice served by the council on August 9 1971 with the consent of the commissioners was void and of no effect, alternatively (against the commissioners) for damages for derogation from grant. The judgment of Willis J was reported at (1975) 234 EG 115.

Mr J Harman QC and Mr L Hoffmann (instructed by Beer, Dunnett & Co) appeared for the appellants; Mr I D L Glidewell QC and Mr A B Dawson (instructed by E Woolf) for the first respondents; and Mr G Slynn QC and Mr P L Gibson (instructed by the Treasury Solicitor) for the second respondents.

Giving judgment, LORD DENNING said: Whitehall Court occupies one of the finest sites in London. It is on the

A Embankment and near the Houses of Parliament. It is a big block of buildings, containing 110 flats or suites of rooms. Some are occupied by Members of Parliament, others by private persons, and yet others by well-known public companies. It is owned by the Crown Estate Commissioners. As long ago as 1931 the commissioners let it to Whitehall Court Ltd for 99 years at a rent of £10,760 a year. There was a covenant which restricted its user. It was to be used and kept " for private dwelling houses, first-class clubs, residential or professional chambers and offices only." There would be a restaurant where the occupiers could entertain their friends. Nearly 40 years later, on June 5 1968, Whitehall Court Ltd B let the whole building on an underlease to Clabon Developments Ltd for the period of the head lease less 10 days. The rent, was £11,200 a year, but Clabon paid a premium of £1,370,000. The underlease contained the same covenant as the head lease restricting the user of the flats or suites. Two years later, in September 1970, Clabon Developments Ltd sublet one of the suites to a company called Molton Builders Ltd for nearly the rest of the term. It was suite no 139, and the rent was £290 a year. In this sublease the covenant as to user was a little different. The sublessee covenanted " to use the suite as a suite of offices or for the purpose of private residence in one occupation only." There was further a C covenant not to carry out any development within the meaning of the Town and Country Planning Act without the previous licence in writing of Whitehall Court Ltd.

Previously this suite, no 139, had been used as a private residence. But when it was sublet in 1970 to Molton Builders Ltd they allowed Mr Jack Dunnett, a member of Parliament, to occupy it as a licensee. It had eight rooms and a large hall. Mr Dunnett decided to alter the use of it. Instead of a private residence, Mr Dunnett altered it to a suite of offices. He happened also to have three adjoining flats numbered 137a, 138 and 138a. He converted those three flats into one D unit as a residence containing ten rooms, four bathrooms, three toilets and a kitchen for himself and his family. When he turned no 139 from a residence into offices, he did not apply for planning permission. The planning authority, the Westminster City Council, thought that the change was a breach of planning control, but as Whitehall Court was Crown land, they could not serve an enforcement notice without the consent of the Crown Estate Commissioners. On July 23 1971 the commissioners wrote a letter to the Westminster City Solicitor in which they said: " The commissioners consent to the service by your council of an enforcement notice under section 15 of the Town and Country E Planning Act 1968 requiring the discontinuance of the unauthorised office use of flat 139." On August 9 1971 the Westminster City Council served an enforcement notice on the owner and occupier of flat 139. It said that the use of flat no 139 for office purposes was a breach of planning control, and it required the discontinuance of the use. On September 10 1971 Mr Dunnett appealed to the Minister. His ground of appeal was " that planning permission ought to be granted " for the change of use. He pointed out that other flats in the same block had been devoted to office use; that he had done a lot of work in converting no 139 into a suite of offices; and that he had given notice to the district F surveyor and the health department. That appeal has never come on for hearing. It has been adjourned pending the legal proceedings now before us, which I must now relate.

The legal proceedings have been started by the underlessees, Molton Builders Ltd, and not by Mr Dunnett, as he is only a licensee. On October 31 1971 Molton Builders Ltd issued a writ against the city of Westminster. Afterwards they joined the Crown Estate Commissioners as defendants. They seek a declaration that the enforcement notice is void and of no effect; alternatively, they claim damages against the Crown Estate Commissioners. They allege that when the Crown Estate Commissioners gave their consent to the enforcement notice, that act was a derogation from their G grant. It is put in this way, that when the Crown Estate Commissioners in 1931 granted the lease for 99 years, it contained permission for this suite to be used not only for residential purposes but also for professional chambers and offices. Molton Builders say that when the Crown Estate Commissioners consented to the service of an enforcement notice requiring the discontinuance of the use for office purposes their act was a derogation from that grant, with the result that the consent was invalid in itself, so that the enforcement notice is invalid: or alternatively, that they ought to be able to claim damages from the Crown Estate Commissioners for giving their consent. So the case is put H entirely on derogation from grant. At one time there was a suggestion of a breach of a covenant for quiet enjoyment, but that has been abandoned, because Molton Builders Ltd were only underlessees, and there was no privity of contract between them and the commissioners, so they cannot sue the commissioners in covenant. But a derogation from grant is said to be a property right of which underlessees can take advantage.

In the first place, it is necessary to consider the position of Crown land under the planning permission. At the relevant time the current Act was the 1962 Act, but the material sections are re-enacted in the 1971 Act, so for convenience J I will quote them from the 1971 Act. The Act deals with Crown land in sections 266 to 268. The result is this, that when the Crown itself (by itself or by a Government department) is in occupation of Crown land, it is exempt from planning control, but that when Crown land is leased to others, then those others, who are in occupation of it, are subject to planning control. They must not make any material change of use unless they get planning permission. That appears from section 266 (1) (b), which says that Part III of the Act applies to Crown land " to the extent of any interest therein for the time being held otherwise than by K or on behalf of the Crown." So Part III applies to the leasehold interest in Whitehall Court. It is that part that contains the provisions for planning control. It includes section 23 (1), which says, " Planning permission is required for the carrying out of any development of land." And " development " includes any " material change of use ": see section 22 (1). So neither Molton Builders Ltd nor Mr Dunnett could make any material change of use of no 139 unless they got planning permission. So when Mr Dunnett and Molton Builders Ltd changed the use from residential use to office use, they required planning permission. Molton Builders Ltd did not apply for it, nor did Mr Dunnett. They L did not get planning permission and have not got it. That was plainly a breach of planning control. Another provision about Crown land is section 266 (2) (a), which says: " Except with the consent of the appropriate authority "—and that is in this case the Crown Estate Commissioners—" no order or notice shall be made or served under any of the provisions of [*inter alia*] section 87 of this Act in relation to land which for the time being is Crown land." Now section 87 gives power to serve enforcement notices, so the effect of section 266 (2) (a) is that the planning authority, the Westminster City Council, were not allowed to serve an enforcement notice except with the consent of the Crown Estate Commissioners. M In this case, as I have said, there was a breach of planning control by changing from residential use to office use, but as it was Crown land, the Westminster City Council could not serve an enforcement notice unless they got the consent of the Crown Estate Commissioners.

On July 23 1971 the Crown Estate Commissioners gave their consent in the letter which I have read. Now Molton Builders Ltd say that the Crown Estate Commissioners, by giving their consent, acted in derogation of their grant. This makes it necessary to consider the scope of the doctrine of derogation from grant. The doctrine is usually applied to

A sales or leases of land, but it is of wider application. It is a general principle of law that if one man agrees to confer a particular benefit on another, he must not do anything which substantially deprives the other of the enjoyment of that benefit, because that would be to take away with one hand what is given with the other. It is said to be "a principle which merely embodies in a legal maxim a rule of common honesty": see *Harmer v Jumbil (Nigeria) Tin Areas Ltd* [1921] 1 Ch 200 at p 225 per Younger LJ. Sometimes it is rested on an implied term in the contract, but this is not correct; it is a principle evolved by the law itself. Applied to sales or leases of land, it means that when a man B has sold land, or granted a lease of it, and expressly or impliedly agrees that the other party shall be at liberty to use it for a particular purpose, then he must do nothing actively to render the premises unfit, or materially less fit, for the particular purpose for which they were sold or let: see *Browne v Flower* [1911] 1 Ch 219 at pp 225-7. The obligation can be enforced against him and his successors, not only by the original buyer or lessee, but also by those claiming under them. It is in this respect akin to a right of property. The principle is well stated in *Megarry & Wade's Law of Real Property*, 3rd ed, at p 683:

> C It is a principle of general application that a grantor must not derogate from his grant, he must not seek to take away with one hand what he has given with the other. This obligation binds not only the grantor himself but persons claiming under him; and the right to enforce it passes to those who claim under the grantee.

And at page 817:

> It is sometimes said to rest upon an implied promise, but it is in truth an independent rule of law, and has nothing to do with restrictive covenants or the equitable doctrine of notice.

Mr Harman relied on the principle there stated. He says that in 1931, when the Crown Estate Commissioners let D Whitehall Court to the head lessees, it was contemplated that the flats or suites could be used for the stated purposes, namely, for residential or professional chambers or offices. He says that, by consenting to the enforcement notice, they are derogating from their grant because they are preventing no 139 being used for the purpose of professional offices.

There is I think a short answer to this contention. When a man sells or lets land for a particular purpose, it is always subject to the proviso that it is lawful to carry on that purpose. Neither the buyer nor the tenant can pray in aid the doctrine so as to enable him to do something that is unlawful. That is shown by *Pwllbach Colliery Co Ltd v* E *Woodman* [1915] AC 634, where a landowner let a piece of land to a butcher on which to build a sausage factory. The same landowner had also let adjacent land to a colliery for the purpose of mining coal. The colliery company put up an apparatus which deposited coal dust on to the sausage factory. It was held that the butcher could sue the colliery company for nuisance causing injury to his business. The grant to the colliery company did not authorise the making of a nuisance. So here the grant in 1931 to the lessees for "residential or professional chambers or offices" was subject to it being lawful to carry on any particular one of those purposes. Until the planning Acts were passed, it was per- F fectly lawful for the lessees to use the suites or flats for any of those purposes; but when the planning Acts were passed in 1947 it became unlawful to make any material change of use without getting planning permission. It was not lawful in 1970 for Mr Dunnett or Molton Builders to change the use of no 139 from residential to office use without planning permission. They never got planning permission for that change of use. It was therefore unlawful. Long before the enforcement notice was served, it was already unlawful. Later on, when the Crown Estate Commissioners gave their consent to the service of the enforcement notice, that was simply a consent to the law taking its course: to the law being enforced against a person who had changed the use without G planning permission. It cannot be a "derogation from grant" simply to allow the law to take its course in stopping something which was unlawful.

In any case, Mr Dunnett and Molton Builders Ltd have appealed to the Minister against the enforcement notice. The appeal operates as an application for planning permission. So they are in fact in the same position as if they had done what they ought to have done in the beginning: they are applicants for planning permission. The action by the Crown Estate Commissioners in giving consent to the enforcement notice has not prejudiced them in the least. It has only made them seek planning permission, as they ought to have done. H It cannot therefore be said to be a derogation from grant. Even if this be wrong, however, there is a further point. The Crown cannot contract itself out of its public duty. In *Commissioners of Crown Lands v Page* [1960] 2 QB 274 on p 291 Devlin J said:

> When the Crown, or any other person, is entrusted, whether by virtue of the prerogative or by statute, with discretionary powers to be exercised for the public good, it does not, when making a private contract in general terms, undertake (and it may be that it could not even with the use of specific language validly undertake) to fetter itself in the use of those powers, and in the exercise of its discretion. J

So it can be said that the doctrine of derogation from grant cannot be so applied as to fetter the Crown Estate Commissioners in the use of the powers which it has to exercise for the public good. But I prefer not to go into this point. It might involve a discussion as to the distinction, if any, between Crown lands and the Duchy lands. It is sufficient that here there was no derogation from the grant, because no grant can be said to give permission to the grantee to do an act which turns out to be unlawful. The turning of this flat from residential to offices was unlawful because permission had to be, and should have been, obtained, and it was not. I find K myself in agreement with the judge, and I would dismiss the appeal.

BROWNE LJ: I agree that this appeal should be dismissed. Once one escapes from the seductive labyrinth of Mr Harman's argument, it seems to me that the appeal fails on a fairly simple ground which has already been stated by my Lord. The plaintiffs' case is based on derogation by the Crown Estate Commissioners from their grant made to Whitehall Court Ltd by the lease of May 9 1931, and made in particular by that clause which provides that the premises shall be used only for purposes specified, including professional chambers or offices. Assuming that what the Crown L Estate Commissioners did in this case was a derogation from that grant, I find great difficulty in seeing how Mr Harman's clients, Molton Builders Ltd, are entitled to sue the commissioners in respect of such a derogation, but in the view I have formed it is unnecessary for me to express any opinion on that point, and I do not do so. It is common ground that there was no implied warranty in the 1931 lease that the uses referred to in the relevant clause, clause 9 (c), could lawfully be carried on in 1931, still less that they would remain lawful throughout the 99 years of the lease; if any authority is needed for this view, see *Hill v Harris* [1965] 2 QB 601, per Diplock LJ at pp 614 and 615 and Russell M LJ at p 617. But Mr Harman says it is enough if both parties intended and contemplated (as no doubt they did in this case) that Whitehall Court, including the suite now in question, no 139, could and would be used for any of those purposes. He says that if the landlord acts in such a way as to prevent or damage one of these contemplated uses, he is derogating from his grant. In my judgment, Willis J was plainly right in saying, "I think that [counsel for the commissioners] is correct when he says that the right to use the premises for the purposes set out in the 1931 lease must be construed as subject to the qualification that such uses must

A not be otherwise than in accordance with the general law" [see (1975) 234 EG 119]. Counsel's argument on this point, which I also accept, is set out at (1975) 234 EG 117. I do not see how the parties could have intended or contemplated anything else.

In 1931 there was no effective planning legislation, but since the Town and Country Planning Act of 1947 planning permission has been required for any material change in the use of land. On the hearing of this appeal everyone has for convenience referred to the provisions of the Town and Country Planning Act of 1971, which reproduces all the provisions of the earlier Acts which were in force in 1970 B and are relevant for present purposes. Leaving out the position of the Crown and its tenants for the moment, the relevant provisions of the 1971 Act are as follows. Section 22 (1) defines "development" as including the making of any material change in the use of any buildings or other land. Section 23 (1) provides, "Subject to the provisions of this section, planning permission is required for the carrying out of any development of land." Sections 25 to 35 deal with applications to the local planning authority for planning permission, and section 36 provides for an appeal to the Secretary of State against the refusal by a local planning C authority to grant permission. Section 87 (1) provides that "where it appears to the local planning authority that there has been a breach of planning control after the end of 1963 . . . they may serve a notice under this section (in this Act referred to as an 'enforcement notice') requiring the breach to be remedied." Subsection (2) provides, "There is a breach of planning control if development has been carried out, whether before or after the commencement of this Act, without the grant of planning permission required in that behalf in accordance with Part III of the Act of 1962 or Part III of this Act" (which contains section 23). In my judgment, those words in subsection (2) defining "breach of planning con- D trol" are simply a definition of that phrase for the purposes of subsection (1) of that section. I do not think it affects the situation that a person who carries out development without permission is in breach of his duty to obtain permission under section 23 (1), even before an enforcement notice is served. Section 88 provides for appeals to the Secretary of State against enforcement notices. Subsection (1) provides various grounds for appeal, and as my Lord has pointed out, Mr Dunnett's notice of appeal in this case relates only to grounds (a) and (g), (a) being "that planning permission ought to be granted for the development to which the notice relates." Subsection (3) of section 88 provides, "Where an E appeal is brought under this section, the enforcement notice shall be of no effect pending the final determination of the appeal." Subsection (5) provides for the powers of the Secretary of State on an appeal, including among other things, power to "grant planning permission for the development to which the enforcement notice relates." And finally, by subsection (7), "Where an appeal against an enforcement notice is brought under this section, the appellant shall be deemed to have made an application for planning permission for the development to which the notice relates. . . . "

When in 1970 the plaintiffs (or Mr Dunnett) wished to make a change in the use of this suite from residential to F office use, they were required by section 23 (1) to make an application for planning permission, but they did not do so. Apart from the special provisions as to Crown land in section 266 of the Act, an enforcement notice could have been served on them and they would then have had a right of appeal to the Secretary of State. The general effect of section 266 (1) (b) and (2) (a) seems to be clear. The Crown is not bound by the Town and Country Planning Acts, and does not have to get planning permission in respect of its own interest in Crown lands, but its tenants do have to get planning permission in respect to their interests: see *Ministry of Agriculture, Fisheries and Food v Jenkins* [1963] 2 QB 317, especially what my Lord said at p 325. Section 266 (1) (b) G refers to Part III of the Act, which contains section 23, and to Part V, which contains section 87. The plaintiffs (or Mr Dunnett) were therefore required by section 23 (1) to make an application for planning permission for their change of use, but did not do so. The only relevant limitation on the obligation or liability of a tenant or subtenant holding under the Crown is section 266 (2) (a), which prevents an enforcement notice under section 87 being served without the consent of "the appropriate authority," which is in this case the Crown Estate Commissioners. The effect of granting such consent will no doubt probably be that an enforcement notice will be served, as it has been in this case. But when it is, the H person on whom it is served has a right of appeal to the Secretary of State, just as he would have had if he had originally applied for planning permission and it had been refused by the local planning authority. This right of appeal has been exercised in this case, and by virtue of section 88 (7) this is deemed to be an application for planning permission.

As Willis J said, and I have said, the original grant of 1931 was in my judgment subject to the qualification that the use of the premises must be in accordance with the general law. I think that the "general law" in this context clearly includes J not only the common law (as in *Pwllbach Colliery Co v Woodman* [1915] AC 634) but also statute law. I think it is also clear that it includes not only the law as it was in 1931 but also any changes in the law during the currency of the term. When the plaintiffs (or Mr Dunnett) made their change of use in 1970, they were required by section 23 (1) to apply for planning permission; if they had applied and been refused, they would have had a right of appeal to the Secretary of State. Now that an enforcement notice has been served and they have appealed to the Secretary of State, they are deemed to have made an application for planning permission. As Willis J said, they are only being put in the position of hav- K ing to obtain now permission which they should have applied for in the first instance: see judgment (1975) 234 EG 119. The grant made by the Crown Estate Commissioners in 1931 was subject to the qualification that the uses referred to in clause 9 (c) should not contravene the general law. Before the change of use was made in 1970, the plaintiffs (or Mr Dunnett) were required by section 23 (1) to get planning permission. They did not do so, and the change contravened that section. The effect of the giving of consent by the Crown Estate Commissioners was to give effect to the implied qualification in the grant and to allow the general law to take its course, thereby putting the plaintiffs in the same position L as if they had done what the law required them to do and applied for planning permission before they made the change of use. If on the appeal against the enforcement notice planning permission is granted, the office use can continue. If it is refused, the plaintiffs are in no worse position, so far as use is concerned, than if they had complied with the requirement of section 23 (1) in the first instance and their application had been refused by the local planning authority and by the Secretary of State on appeal. In my judgment the granting of consent by the Crown Estate Commissioners was in these circumstances not a derogation from their grant. M On that ground I would dismiss the appeal. This conclusion, as my Lord has said, makes it unnecessary to consider the other points argued on this appeal, and I express no opinion about them.

GEOFFREY LANE LJ: The primary question to be decided in this case is whether the Crown Estate Commissioners derogated from their grant of 1931 by giving consent to the service of an enforcement notice when they had an option as to whether that notice was to be served or not. There is no dispute that at the time of the lease of 1931 both parties envisaged that the premises might be used either as offices or as a private dwelling; and there is no dispute that the

A effect of an enforcement notice may be, if the appeal is unsuccessful, to prevent the present occupier, Mr Dunnett, from using the premises as an office if he wishes to. A landlord by permitting a certain type of user under the terms of his lease does not warrant that such user will always be legal or possible. He does not even warrant that such user is legal at the time when the lease is granted: *Hill v Harris* [1965] 2 QB 601. The effect of the various Town and Country Planning Acts has been to curtail many of the rights which tenants enjoyed under their leases; but the tenants' rights are, of course, subject to the overriding consideration that they must be exercised in accordance with the law in existence at the time, whether it be common law or whether it be the law enacted by statute: *Pwllbach Colliery Co Ltd v Woodman* [1915] AC 634. That was a case of nuisance under the common law. The same principle applies in respect of statute law. So if the law changes, and the tenant's rights, if he exercises them, are in breach of the law, then the landlord incurs no liability to the tenant under normal circumstances under the terms of the lease.

So much would appear to be self-evident. There is nothing in the 1962 Act—which is the operative Act we have to consider—that exonerates the lessee from the terms of section 13 (1) of the 1962 Act, which is now section 23 (1) of the 1971 Act, namely, " Subject to the provisions of this section, planning permission is required for the carrying out of any development of land." Indeed, by virtue of section 199 (1) of the 1962 Act, now section 266 of the 1971 Act, it is provided, " Notwithstanding any interest of the Crown in Crown land, but subject to the following provisions of this section, any restriction by Part II of this Act [which includes section 13] shall apply to Crown land to the extent of any interest therein for the time being held otherwise than by or on behalf of the Crown." In short, despite the exemptions of land occupied by the Crown itself, tenants of the commissioners and those claiming under them require planning permission before development. The change of user in the present case amounted to development, and it could not be carried out without planning permission under section 13 (5). No permission was obtained, so that was a breach of section 13 (1). It is true that the consequences of the breach, namely, the serving of an enforcement notice and all that may follow thereafter, could not by virtue of section 199 take place unless consent was obtained from the Crown Estate Commissioners; but that does not mean that the breach was cured or ceased to be a breach. Thus what the commissioners were here doing was to give their consent to the law taking its course in respect of a breach of the planning regulations in the 1962 statute. It would be a very strange thing if such an act, the giving of consent in those circumstances, were to amount to a derogation by the landlord from his grant, entitling the tenant or his successors in title to a declaration against the local authority that it was of no effect and entitling him to damages against the landlord and the Crown Estate Commissioners. In my judgment it does not. There must be read into any covenant of this sort—a covenant as to user—a proviso that the user is in accordance with the law. This user was not. There was accordingly no derogation by the landlord; and in so far as the landlord did take away anything in this case, it was something which he had not given. I agree with the reasoning and the conclusions of the learned judge, and I agree that the appeal should be dismissed.

The appeal was dismissed, with an order for costs in favour of each respondent. Leave to appeal to the House of Lords was refused.

QUEEN'S BENCH DIVISIONAL COURT

April 17 1975

(Before Lord WIDGERY CJ, Mr Justice ASHWORTH and Mr Justice MAY)

LEIGHTON & NEWMAN CAR SALES LTD v SECRETARY OF STATE FOR THE ENVIRONMENT AND ANOTHER

Estates Gazette May 22 1976

(1975) 238 EG 571

Enforcement notice prohibits display or sale of cars on " forecourt " of garage premises—Definition of " forecourt "—Not limited to area immediately surrounding pumps—Further point on agreement between advocates at planning inquiry—Additional evidence as to established use would have made no difference—Decision of Secretary of State upheld

This was an appeal by Leighton & Newman Car Sales Ltd, of 271 Mare Street, Hackney, London E8, from a decision of the first respondent, the Secretary of State for the Environment, upholding with modifications an enforcement notice served by the second respondents, the London Borough of Hackney, requiring the appellants to cease selling cars, or displaying them for sale, on the forecourt of their premises.

Mr B Payton (instructed by Clinton Davis & Co) appeared for the appellants, and Mr H K Woolf (instructed by the Treasury Solicitor) represented the respondents.

Giving judgment, LORD WIDGERY said that on the corner site of Mare Street and Richmond Road, Hackney, there had been for a number of years a garage and petrol filling station. The occupiers of the site desired to modernise and rebuild, and they put in an application for planning permission. This was duly granted in terms which, so far as material, permitted the construction of a petrol and filling service station comprising a single-storey building with lubrication bays, office, stores, lavatory, a petrol sales area with pumps and canopy, underground petrol storage, a paraffin vending machine and the provision of 13 car parking spaces. It would be seen from those words authorising development that this was to be a petrol filling station in a conventional form with the natural fittings that went with it, and that there was to be car parking for 13 vehicles. The requirement with regard to car parking was emphasised by the conditions to which this development was subject. Condition 8 required that the whole of the car parking accommodation should be provided and retained solely for accommodation of the vehicles of the occupiers and users of the premises. In other words, in order to avoid congestion of the premises by an absence of proper parking facilities, these 13 spaces had to be provided for people whose vehicles came on business to the premises. Condition 9 was that no vehicles should be sold or displayed for sale on the forecourt. The associated plan showed the main road of Mare Street. It also showed the return frontage to the site going down Richmond Road. It showed that petrol pumps were to be installed in the right-hand half of the site as seen from Mare Street, and it specified 13 parking spaces running round the left-hand edge of the site, namely on its Richmond Road frontage and for some distance into the Mare Street frontage.

Having received planning permission, the occupiers of the premises carried out the development but proceeded to use the parking places for cars displayed for sale. Assuming that the relevant area was properly described as " forecourt," that seemed to be a use in direct contravention of condition 9. An enforcement notice was duly issued by the planning authority which drew attention to the condition and required the discontinuance of the unauthorised use. The matter was taken by way of appeal to the Secretary of State, who on July 25 1974 issued a decision letter upholding the notice with certain modifications which did not need to be considered in any detail. The case now came to the Divisional Court under section 246 of the Act of 1971, which gave the court authority to deal with matters of the kind if it could be shown that the decision below erred in a point of law. The court had had the advantage of considerable argument from Mr Payton covering a very wide area. In particular, he had dealt with one aspect of planning law, which tended to determine that where a new planning permission was granted in respect of a site and that permission was carried out, the issue of the new permission and its being carried out cancelled, as it were, the planning record of the site and started it off afresh as a new hereditament with a nil use. The court had heard a great deal from Mr Payton as to how far the authorities had reached that stage in a simpler case than the present one, where there was no express condition in the planning permission requiring that a previous use should be discontinued. He (his Lordship) was not going to follow Mr Payton through those somewhat intricate paths, because in this case there was in the planning permission a specific condition requiring that the display of cars for sale should not take place even though everyone knew it was a use to which the land had previously been put. In his judgment, the law was that where there was an express condition of that kind which was clear and unambiguous, then effect must be given to it, even though the net result was that a former established use of the land was no longer to be followed. He (the Lord Chief Justice) found nothing outrageous or unreasonable about that, because the recipient of planning permission need not exercise it. When permission was given subject to conditions, the landowner had the choice of either carrying out the development and complying with the conditions, or not carrying out the development. In this instance there was an unpopular condition, the development was carried out and the obligation to comply with the conditions was perfectly clear. But of course, in order to say that that justified the present enforcement notice, one had to be clear on the allegation that the condition was breached, and this brought him (his Lordship) to the terms of the condition in which reference was made to the fact that there should be no display of cars for sale on the forecourt.

Mr Payton's principal argument was that in this case the place where the cars were displayed was not on the forecourt, and if it was not on the forecourt, then of course the display of the cars was not in breach of the condition. He (counsel) contended that this was not a forecourt, first of all by saying that it was the common understanding of men that the forecourt of a garage was the area round the pumps and not an area which could extend to the edge of the site. This was a proposition that he (Lord Widgery) was just not prepared to accept. What was a forecourt in a given case must depend on the circumstances, and he was not prepared to accept that the forecourt was, as a matter of law, confined to some ill-defined area in the neighbourhood of the pumps themselves. Secondly, Mr Payton argued that this could not be a forecourt because the terms of the condition implied that there must be a power to display cars for sale somewhere. The prohibition, so the argument went, was only on the forecourt, and the implication was that somewhere else on the site there was an area where cars could be displayed for sale. The result of that argument, it was contended, was that one must be able to find somewhere on the site for selling cars, and that meant that the area marked off for parking spaces was not part of the forecourt. Again he (the Lord Chief Justice) found that argument too intricate. The Secretary of State had taken the sensible view that in this particular case, the area of the forecourt did extend to the parking places, and he (his Lordship) could see no reason at all why in law that conclusion should be said to be false.

Accordingly the appeal against the enforcement notice, so far as it was based on section 88 (1) (b) and (d) of the Act of 1971, must fail, because there was no error of law in the Secretary of State's action on this notice. There was, however, another and rather an unusual aspect of this particular appeal, and that was that the court was asked to consider the position under section 88 (1) (a) as well. This was of course a subparagraph whereby the Secretary of State might give planning permission for the carrying out of the disputed development even though the enforcement notice was a perfectly valid one, as in the present case. Normally the court was not troubled with matters arising under subparagraph (a) because they were very rarely points of law. The subparagraph was concerned with planning opinion and discretion and things of that kind, with which the court was not concerned. Mr Payton in this case, however, said that there was a point of law in the Secretary of State's disposition of the matter under paragraph (a). Apparently there was some discussion between the advocates before the inquiry began as to whether there was any dispute as to the previous use of the site for car sales. Something was agreed, as to which the court had no evidence but merely an explanation from counsel, which resulted in some of the witnesses going home, witnesses who were there to speak to the prior use for car sales. In the result, so it was submitted, the Secretary of State never had it explained to him that there was a prior use of this land for car sales which went back to the appointed day in 1948. It was said that that was a fundamental matter because it would have affected the Secretary of State's discretion had he known that there was such an ancient and respectable use for the car sales as that. Mr Payton went on to say that the inspector's failure to make this point clear meant that his report was no report at all in law, and that that meant that the Secretary of State acted without any report from his inspector, and that that meant that the Secretary of State erred in law.

While he admired Mr Payton's ingenuity in building up this complicated structure, he (Lord Widgery) thought that it crumbled because it was built on foundations which failed. It was true that the Secretary of State was not informed that the prior use for car sales was as extensive as the appellants maintained. He was clearly informed that there had been a prior use for car sales, but his information did not go to the length of saying that it went back to 1948 or any such date. Nevertheless this was really a matter of no consequence at all. The Secretary of State was considering whether as a matter of discretion there should be an authorisation of car sales on the new hereditament as it appeared at the date of the inquiry, and although he might have been somewhat influenced by the knowledge that there had been a previous use for car sales, he (the Lord Chief Justice) could not believe that the inspector's report was rendered defective to the point of nullity merely because it did not elaborate this particular point. There was no substance in this argument, and the appeal should be dismissed on all the grounds on which it was promoted.

ASHWORTH and MAY JJ agreed, and the appeal was dismissed with costs. Leave to appeal was refused.

COURT OF APPEAL (CRIMINAL DIVISION)

December 1 1975

(Before Lord Justice LAWTON, Lord Justice BRIDGE and Mr Justice CROOM-JOHNSON)

R v ENDERSBY PROPERTIES LTD

Estates Gazette June 12 1976

(1975) 238 EG 795

Demolition of house in conservation area—Town Clerk's letter prohibiting demolition expressed to take effect that day—Clear implication that that was because it was expedient for the direction to take immediate effect—Absence of an explicit reference to expediency immaterial—Dictum

A

of Lord Simonds in East Riding County Council v Park Estate (Bridlington) Ltd not "holy writ"—Fine of £10,000 confirmed

This was an appeal by Endersby Properties Ltd against their conviction at Snaresbrook Crown Court on May 5 1975 of causing the execution of works for the demolition of the White House, 39 Snaresbrook Road, Wanstead, London, without authorisation, contrary to the provisions of section 55 (1) of the Town and Country Planning Act 1971. The court imposed a fine of £10,000, with costs not exceeding £1,000.

B

Mr G N Eyre QC and Mr D W Keene (instructed by Malcolm Ellicott & Co) appeared for the appellants, and Mr C S Fay (instructed by the solicitor to the council) represented the respondents, the London Borough of Redbridge.

Giving the judgment of the court, LAWTON LJ said that the White House, which was situated in the Snaresbrook conservation area, could formerly have been demolished without the consent of anybody. But following the 1972 amendment of the 1971 Town and Country Planning Act, buildings in conservation areas acquired the same kind of protection that was afforded to listed buildings. The provisions of the later Act allowing local planning authorities

C

to bring conservation areas within the terms of section 55 of the 1971 Act were enshrined in section 8, subsection (3) of which called upon an authority making a direction under the section to submit it to the Secretary of State for the Environment for confirmation. It was appreciated, however, that local authorities might suddenly become aware of demolition works and that there would then be no time to await the confirmation of the Secretary of State. Section 8 (4) thus provided for a direction to take immediate effect if it contained a declaration by the local planning authority that that was expedient. In effect, therefore, Parliament

D

envisaged two distinct categories of directions, those which were immediately operative and those which awaited the Secretary of State's confirmation.

On March 13 1973 the Town Clerk of the London Borough of Redbridge wrote to the secretary of the appellant company, pointing out that the demolition of buildings in a conservation area was subject to control in order to preserve the appearance of the area, and directing the cessation of demolition works in respect of the White House. The letter pointed out that the building was to be treated as if it were a listed building, and the direction was to take effect that day, March 13. There was no dispute but that the direction was served

E

upon the company and demolition work ceased. But the company sought legal advice, and later advised the local authority through a surveyor named Hawkins that demolition would proceed. Demolition of the building was completed by April 3 1973, and the White House ceased to exist in any

F

shape or form on that day. The basis for that advice, and for the argument now before the court, was that the direction contained no reference to its being expedient that it should take immediate effect. It was submitted that the absence of any such reference amounted to a fatal flaw, that the direction thus became a direction under section 8 (2) requiring the confirmation of the Secretary of State, and that no confirmation as such having been obtained, no offence had been committed.

It could not have been clearer from the form of the direction that demolition work was to cease after March 13. It was also conceded that any reasonable person reading

G

it would appreciate that the local authority thought it expedient to act at once. But it was contended that that was not enough, and that the notice did not comply with the express words of the statute. Accordingly it was said that however useless the words might be, if Parliament saw fit to make it a provision of some Act that "magic words" had to be used, then magic words had to be used. The case of *East Riding County Council v Park Estate (Bridlington) Ltd* [1957] AC 223 was cited to the court in support of that proposition. In that case, the House of Lords held that an enforcement notice issued by the local authority under the provisions of the 1947 Town and Country Planning Act was

H

invalid because it did not specify that the complaint was one of development before the coming into operation of the 1947 Act. Lord Simonds said at p 233, "The court must insist on a strict and rigid adherence to formalities. This, as a general proposition, commands assent, and not the least because disregard of an enforcement notice is an offence involving sufficiently serious penal consequences." But it was important to bear in mind the context of Lord Simonds' remarks. In the *East Riding County Council* case, procedural consequences were of the greatest importance, and the comments of Lord Simonds with regard to such a subject-matter should not be regarded as holy writ applicable to all cases.

J

Later cases had shown that notices under this legislation had to indicate to recipients what it was they were alleged to have done, and specifically, what they had to do to put themselves in the right. The problem in the present case was whether the magic words really added anything to the enforcement of the notice. Since it was conceded that any reasonable person would think that the direction was to take effect immediately, the fact obviously was that he would appreciate that that was because it was expedient for it to do so. In other words, the declaration of expediency was implicit in the form of the direction. In those circumstances,

K

as a matter of the construction of this particular direction, the court was of the opinion that the ruling of the judge was right and that the appellants had no defence to the prosecution.

The appeal was dismissed.

VENDOR AND PURCHASER

A

COURT OF APPEAL

February 24 1975

(Before Lord DENNING MR, Lord Justice ORR and Lord Justice SCARMAN)

YATES BUILDING COMPANY LTD v
R J PULLEYN & SONS (YORK) LTD

Estates Gazette January 17 1976

(1975) 237 EG 183

Notice exercising option to buy building plots to be sent "by registered or recorded delivery post" to vendors or their solicitors—Requirement not mandatory—Ordinary post good enough in a case in which the notice is actually received

B

This was an appeal by Yates Building Co Ltd, of Enfield, from the dismissal by Templeman J in the Chancery Division on December 12 1973 of their claim for a declaration that a contract for sale by the defendants, R J Pulleyn & Sons (York) Ltd, of York, of plots of land at Haxby, Yorkshire, had been validly created by exercise of an option contained in an agreement of September 21 1971 and ought to be specifically performed.

Mr J Harman QC and Mr R Reid (instructed by Beer & Partners) appeared for the appellants and Mr A L Price QC and Mr D A Lowe (instructed by Kenneth Brown, Baker, Baker, agents for Harrowells, of York) represented the respondents.

C

Giving judgment, LORD DENNING said: At Haxby in Yorkshire there is an area of land belonging to R J Pulleyn & Sons Ltd which in 1971 was laid out as a building site with 153 building plots. The Yates Building Co Ltd were interested in acquiring building plots. Negotiations took place by which they were to take up the plots in four portions spread out over four years. In the result four agreements were signed on September 21 1971. One agreement provided for the first portion to be taken up for £18,900 paid down then and there. In addition, on the same day three separate agreements were signed by which the buyers were given options to take up the remaining three portions: one portion for £18,900 between April 6 1972 and May 6 1972; another portion for £18,900 between April 6 1973 and May 6 1973; and the final portion for £18,900 between April 6 1974 and May 6 1974. The date "April 6" in each agreement was significant. It suited the sellers for tax reasons to receive the money as an annual profit against which they could set off their expenses. It also suited the buyers, because they would only have to find the money in instalments. The agreements did not, however, bind the buyers to purchase. Each of them gave the buyers an option. The buyers exercised this option quite validly in 1972 and 1974. But the question is whether they exercised it validly in 1973. I will read the option clause:

D

E

> The option hereby granted shall be exercisable by notice in writing given by or on behalf of Yates to Pulleyns or to Pulleyns' solicitors at any time between April 6 1973 and May 6 1973 such notice to be sent by registered or recorded delivery post to the registered office of Pulleyns or the offices of their said solicitors.

F

Now let me say what happened. The buyers had until May 6 1973 to exercise the option. At a time when there was still one week in hand, namely on Monday April 30 1973, the London solicitors for the buyers wrote this letter to the York solicitors for the sellers:

> On behalf of our clients, Yates Building Co Ltd, we are writing to formally exercise the option under the agreement of September 21 1971 for the purchase of plots 15-18, 38-50, 52-83 and 55-63. A cheque for £1,890 for the deposit is enclosed.

That was posted on Monday April 30 1973. But here is the point. It was sent by ordinary post and not by registered or recorded delivery post. Nevertheless it got there all right. It was well in time. It was opened by the York solicitors for the sellers at some time on or before Friday May 4 1973. Then on that Friday they wrote back saying:

G

> We write to acknowledge receipt today of your letter of April 30 1973 with its enclosure. You will recall that clause 2 of the option agreement provides for notice to be sent by a registered or recorded delivery post. Your letter was not so sent.

They returned the cheque and said:

> We also enclose for your information a copy of your envelope enclosing the letter.

H

They enclosed a photograph of the envelope. It showed that it was sent on April 30 1973 by ordinary post and not by registered or recorded post. The sellers' solicitors posted that reply on Friday May 4 1973, so that it did not reach the buyers' solicitors until Monday, May 7 1973. It was then too late for the matter to be rectified, because the latest date for exercising the option was May 6 1973. The buyers were very upset by this turn of events. This portion was an integrated part of the whole housing estate. They had made roads and services for the whole estate, including this portion. Yet the sellers claimed to withhold this portion. The buyers brought proceedings for specific performance, but the judge refused it. He held that this requirement that the letter had to be sent by registered or recorded delivery post was a requirement which must be complied with, and as it had not been complied with, there was no contract.

J

It seems to me that this depends on the construction of the option clause. The option is an offer: an irrevocable offer. When a person makes an offer, he does sometimes prescribe the method by which it is to be accepted. If he prescribes it in terms which are mandatory or obligatory, the acceptance is only good if it complies with the stated requirements. Thus in the present case the notice of acceptance *must* be in writing, and *must* be given to Pulleyns or to Pulleyns' solicitors, and *must* be given between April 6 1973 and May 6 1973. But the question is whether the words "such notice to be sent by registered or recorded delivery post" are mandatory or directory. That test is used by lawyers in the construction of statutory instruments, but it can also be used in the construction of other documents. The distinction is this: a mandatory provision must be fulfilled exactly according to the letter, whereas a directory provision is satisfied if it is in substance according to the general intent (see *Howard v Bodington* (1877) 2 PD at 210-211). In

K

applying this rule of construction, you must look to the subject-matter, consider the object to be fulfilled, and then see whether the provision must be fulfilled strictly to the letter or whether the substance of it is enough. So in the present case the question is whether the letter of acceptance *must* be sent by registered or recorded delivery post, else it is bad; or whether it is sufficient if it gets there in time, as, for instance, by ordinary post or by special messenger. Orr LJ gave this instance in the course of the argument. Suppose there were a postal strike during the last week, and the buyer, to make sure it was in time, sent the letter by special messenger, would this not be sufficient? Looking at the object of this provision, it seems to be this. It is inserted for the benefit of the buyer so that he can be sure of his position. So long as he sends the letter by registered or recorded delivery post, he has clear proof of postage and of the time of posting. But if the buyer sends it by ordinary post, he will have no sufficient proof of posting, or of the time of posting. In that case, if the seller proves that he never received it, or received it too late, the buyer fails. None of those reasons apply, however, when the seller does receive it in time. So long as he gets the letter in time, he should be bound. So I would hold, simply as a matter of interpretation, that if the letter did reach the sellers in time, it was a valid exercise of the option.

There are only a few cases on the point, and they support what I have just said. There is *Tinn v Hoffmann & Co* (1873) 29 LT 271. There was an offer which contained the words "waiting your reply by return of post." The court held that that did not mean a reply had to be sent by *letter* by return of post. A reply by telegram, or by verbal message, or by any means which arrived not later than a letter sent by post would reach its destination would equally satisfy the requisition. The next case is an old one from the United States, *Eliason v Henshaw* (1819) 1 Wheaton 225, in which the offeror at Harper's Ferry wrote to the offeree at Mill Creek: "Please write by return of wagon whether you accept our offer." The wagon was due to return to Harper's Ferry. The letter of acceptance was not sent by return of wagon to Harper's Ferry. It was sent by ordinary mail to Georgetown and took longer to get there. The Supreme Court of the United States said:

> The meaning of the writers was obvious. They could easily calculate by the usual length of time which was employed by this wagon, in travelling from Harper's Ferry to Mill Creek, and back again with a load of flour, about what time they should receive the desired answer, and, therefore, it was entirely unimportant, whether it was sent by that, or another wagon, or in any other manner, provided it was sent to Harper's Ferry, and was not delayed beyond the time which was ordinarily employed by wagons engaged in hauling flour from the defendant's mill to Harper's Ferry. . . . The place, therefore, to which the answer was to be sent, constituted an essential part of the plaintiffs' offer.

The Supreme Court there looked to see what was the essential part of the offer, what was important or not important. It said that the manner of sending was "entirely unimportant," so long as it got to the proper place at the proper time. The only remaining case is *Manchester Diocesan Council of Education v Commercial & General Investments* [1970] 1 WLR 241, where this very point was considered by Buckley J. He gave this guide to construction on p 246, letter c:

> Where, however, the offeror has prescribed a particular method of acceptance, but not in terms insisting that only acceptance in that mode shall be binding, I am of opinion that acceptance communicated to the offeror by any other mode which is no less advantageous to him will conclude the contract.

It seems to me that Buckley J was there adopting the same test as I have stated. If the offeror uses terms insisting that only acceptance in a particular mode is binding, it is mandatory. If he does not insist, and it is sufficient if he adopts a mode which is no less advantageous, it is directory. At any rate, adopting Buckley J's test in this case, there were no words insisting that only registered or recorded delivery post would do, and the sending by ordinary post was no less advantageous to the sellers than sending by registered post, so long as it got there in time. In my opinion this option was perfectly well exercised and there was a binding contract accordingly. I would allow the appeal.

ORR LJ: I agree.

SCARMAN LJ: I agree, subject, however, to one minor point. I am not convinced that the term "directory" has any application to the field of contract. Contractual provisions seem to me to be either obligatory or permissive, and the term "directory"—which is, of course, borrowed from the statute law—does not seem to me to be helpful in this context. I agree with the Master of the Rolls that the one question before the court is the interpretation of clause 2 of the option agreement. I read that agreement as requiring the option to be exercised by a notice in writing which is to be actually received by Pulleyns or Pulleyns' solicitors. When later in the clause one comes to the words which have to be construed in this case "such notice to be sent by registered or recorded delivery post," I think they are a clear indication, and are intended as such to the offeree, that if there is to be any issue as to whether or not the notice has in fact been received, he had better use registered or recorded delivery post if he wishes to put it beyond doubt. Of course, if there was any such issue, the burden would be upon the party seeking to exercise the option to prove that his notice had been received. The clause is a clear indication that one would most easily and most efficaciously discharge that burden by using registered or recorded delivery post. I wholly agree with the rest of the judgment of the Master of the Rolls.

The appeal was allowed with costs above and below. An order was made in the usual terms for specific performance. The court refused leave to appeal to the House of Lords, and refused a stay of execution pending an application to the House. Lord Denning observing, "We never give that."

CHANCERY DIVISION

February 7 1975

(Before Mr Justice GRAHAM)

WATTS v SPENCE AND ANOTHER

Estates Gazette January 17 1976

(1975) 237 EG 187

Contract of sale signed by one joint tenant without the authority of the other—Signatory held to have represented to purchaser that he owned the house and was in a position to sell it—Substantial damages may be awarded for such a misrepresentation—Damages not limited by the rule in Bain v Fothergill—Gosling v Anderson [1972] Estates Gazette Digest 709 cited and passages from Lord Denning's judgment quoted

This was a claim by Mr Nigel Francis Kingsford Watts against Mr John Lloyd Spence and his wife, Mrs Phyllis May Spence, both of 121 Clapham Manor Street, London SW4, for specific performance of a contract dated February 8 1972 for sale of 121 Clapham Manor Street by them for £7,000. By an amendment allowed at the hearing, the plaintiff claimed in the alternative damages for misrepresentation.

Mr J R Cherryman (instructed by H B Wedlake, Saint & Co) appeared for the plaintiff, and Mr J F Parker (instructed by Clintons) represented the defendants.

Giving judgment, GRAHAM J said: In this case the plaintiff, Mr Watts, who is a fine-art dealer and is also interested in a property company, Watts Bowden Property Holdings Ltd, is suing the defendants, Mr and Mrs Spence, who are coloured people and came to this country from Jamaica some 18 years ago. The defendants jointly own a terraced house at 121 Clapham Manor Street, London SW4, and the action is about that house. The plaintiff claims that he is entitled to specific performance of a contract dated February 8 1972 and signed by Mr Spence to sell the house to him, and failing a decision in his favour on that head, to a number of alternative heads of relief against both or one of the defendants. The defendants' main answer is that the house is jointly owned by them and that Mrs Spence never signed the contract or authorised Mr Spence to do so on her behalf.

The decision on the issue of specific performance by both the defendants to convey the whole of the title in the house to the plaintiff depends on a resolution of the conflicting evidence in the case, a task which I have not found easy, since there is little in the way of confirmatory written evidence and the balance of probabilities does not point clearly in any direction. Certain matters are, however, agreed or not disputed, as follows: Mr Spence signed the contract on February 8 1972, at the house, and after discussion with the plaintiff. The contract is exhibit P3, and consists of two sheets of paper each containing the essential terms and each signed, and apart from Mr Spence's signatures, the document is in the handwriting of the plaintiff. The total price mentioned in the contract is the figure of £7,000, and at that interview a sum of £500 was paid to the defendant, Mr Spence, by the plaintiff. He gave Mr Spence a cheque for that amount and wrote out the contract there and then in longhand. It also now appears that at the same interview the plaintiff wrote out nearly identical copies of the two sheets of the contract and gave these copies to Mr Spence for retention. These copies, exhibit P4, were not disclosed by the defendant until half-way through the trial, and I will deal with the evidence given by the plaintiff in respect of them later. It is agreed that no suggestion can be made, and none was made, that £7,000 consideration for the house was anything other than a fair price at the time for the house in its existing condition. The plaintiff was anxious to get it at that price so that in due course his property company could renovate it and make a profit thereby. The plaintiff's evidence was that the £500 was paid as a deposit, but Mr Spence says it was not a deposit but a loan, as mentioned hereafter.

Although paragraph 4 of the defence alleges that Mrs Spence was ill in hospital on February 8 1972, her answers to interrogatories Nos 1 and 3, ordered on July 3 1974, show that this allegation is incorrect and that she was in fact ill in bed on February 8 1972 at 121 Clapham Manor Street. This is now not disputed. The evidence of what happened at the interview is however disputed. The interview took place in the front room of No 121, and only the plaintiff and Mr Spence were present. The plaintiff says that Mr Spence made it clear that he would not agree to sell the house without his wife's concurrence, and in particular her agreement as to the date of giving up possession. The original proposal which the plaintiff made, as he says, was for possession within two months, and (the plaintiff goes on) Mr Spence went and consulted his wife, who was, it turns out, in the front bedroom above that where the interview took place. He returned, saying that his wife objected to so short a period as two months for giving up possession, and eventually the date of two years mentioned in P3 was agreed upon. The plaintiff says Mr Spence left the room three times for periods of "about eight minutes," and the plaintiff says he assumed that he had been upstairs to consult his wife and to get her consent. When he gave his evidence for the first time, before P4 was discovered, the plaintiff was very precise in his story as to his actions, particularly in relation to the drafting of P3 and his insistence that Mr Spence should have a copy of the second sheet of P3, which was signed by both of them for record purposes. There was at that time no suggestion from him that he could not remember the precise details of the interview. On the second occasion, after P4 had been discovered, he gave evidence again and in relation to it was much less precise, being uncertain whether he had drafted P3 and P4 at the same time and whether the four pages were all signed together, or whether P3 had been drafted and signed first and had been followed by the drafting and signing of P4. Mr Spence, on the other hand, said that he never left the room except on one occasion to put the kettle on because he had to go to work shortly thereafter. He denied that he had ever consulted his wife and indeed insisted that he had never told her anything about the sale.

These stories conflict and cannot be reconciled, unless it be that Mr Spence in fact left the room on more than one occasion and, though he may have given the impression of having visited his wife to consult her, in fact did not do so. I am, however, at this point bound to say that Mr Spence was an unsatisfactory witness and I find it difficult to place any real reliance on anything he said unless I can find something which corroborates it. He was vague, incoherent, and often inconsistent. His main theme of what took place at the interview was that the plaintiff was, he thought, a very nice man and was anxious to lend him £500 (the amount of the deposit) to make improvements to his house. When asked why he thought the plaintiff should do this, he replied that the plaintiff owned a house next door but one to No 121 and wanted him to do up No 121 so that there would be less likelihood of the council taking over any of the houses in the terrace, including that already owned by the plaintiff. I cannot give any credence to most of Mr Spence's utterances, which I found unconvincing. Turning now to Mrs Spence, she was, it is agreed, in the front bedroom ill in bed at the time of the interview. She gave her evidence in a careful, restrained and dignified way and I believe what she said. She stated positively that she had never given her husband authority to sell her share of the house. She said she was happy in the house, which suited her and her children, and she certainly did not want to move them on now. She said her husband "never told her anything," by which she clearly was referring to business matters, and that he certainly did not come and consult her on the day of the interview as presumed by the plaintiff. It is also clear from the evidence that both Mr and Mrs Spence were very hard-working and did not have much time to discuss business matters. He was holding down two jobs and working shifts, and she was cooking in an hotel till late in the evening or the early hours of the morning. She positively denied that this was a case where she had originally agreed to the sale of her share in the house but had afterwards had second thoughts, and said that the first she knew of the proposed sale was when she saw the transfer form (correspondence p 10) which was sent with the plaintiff's solicitors' letter of February 17 1972 (p 8) to the defendants' solicitors and sent on by them to Mr Spence, as stated in their letter of February 22 1972 (p 12). She says she protested to her husband, and it is certainly true that the defendants' solicitors wrote their letter of March 6 1972 to the plaintiff's solicitors (p 14) which is consistent with her protest having been made and transmitted to them.

The plaintiff sought to corroborate his story by the evidence that he, in company with his brother-in-law, Mr Bowden, who is a shareholder in the property company of which Mrs Bowden, his wife, is also a director, visited No 121 to have a look around it one evening, about 7 to 8 pm, in November 1971. They both said that Mr and, more important, Mrs Spence showed them round. Mr Bowden said there was no discussion about prices. When this was put to Mrs Spence she firmly denied that she was present at any such visit and said she had never seen Mr Bowden before he came into the witness-box. At that time she was working in the evenings as a cook in an hotel, and did not get home till very late. She

A then said she had a sister, Monica Lewis, living in the house at that time, and possibly the plaintiff and Mr Bowden might have seen her. Monica Lewis got married about three years ago, and she was not called, so it is impossible to check the story further. On the whole, I am not prepared to disbelieve Mrs Spence on this point, and it may well be that the plaintiff and Mr Bowden were mistaken in thinking that it was Mrs Spence who had shown them round on that occasion. They had no reason to take particular notice of Mrs Spence, since neither of them had any idea that she was part-owner of the property. The circumstantial evidence given about the smell of cooking in the house at the time of the visit does not B unfortunately give any clue to the identity of the cook. The truth of the matter, as appears from the evidence, may well be that Mr Spence took it upon himself to sell the house without, at first at any rate, bringing his wife into the matter, hoping no doubt that, if successful, he would be able to get her to agree, willingly or unwillingly. Whether this is so or not, and whatever the precise truth of what happened at the interview of February 8 1972, I am satisfied that Mrs Spence never agreed to or gave her authority to Mr Spence to sell her share of the house on her behalf.

The other ground on which it is sought to show that Mrs C Spence authorised the sale is that of estoppel. The allegation is to be found in paragraph 3 of the reply to defence of the second defendant, and is based on her conduct, and that of her solicitors, in not promptly informing the plaintiff that her husband had no authority to sell on her behalf when she knew that the plaintiff was incurring expense in investigating title, believing that such authority had been given. That conduct, it is said, amounted to a representation of the existence of such authority which induced the plaintiff to act to his detriment by continuing his investigation of title. I am, however, satisfied, and it was not really disputed by Mr Cherryman for the plaintiff, that there is no real evidence to show that D Mrs Spence ever made any such representation. She never saw Cyril Ralton & Co, the solicitors, until later, and there is no evidence that they made any such representation themselves. In the circumstances I hold there is no basis for the estoppel pleaded. The above position being as set out above, there is an end of the claim for specific performance and for damages for loss of the bargain against both the defendants jointly.

There could not properly be a decree of specific performance against Mr Spence alone for his part or share of the house in the present case. First, he could not convey the legal estate, as he is only one of the two trustees for sale. The E only beneficial interest which he could convey is in the proceeds of sale under the trust for sale. The conveyance of such a beneficial interest could not be said to be a conveyance of a part of the fee simple in the land. As was said by Farwell J in *Rudd v Lascelles* [1900] 1 Ch 815 at 819: "The court should confine this relief (part specific performance) to cases where the actual subject-matter is substantially the same as that stated in the contract, and should not extend it to cases where the subject-matter is substantially different." It is not strictly necessary, therefore, to say any more about specific performance or damages in lieu of it. However, apart altogether from the reasons just given, it would be unreason- F able to decree specific performance here, since quite clearly Mrs Spence, a third party interested in the property, would be seriously prejudiced: see the reasoning of Lord Langdale MR in *Thomas v Dering* (1837) 1 Keen 729 at 747-8. The court would not therefore exercise its discretion to grant specific performance in the case in any event.

Since Mrs Spence goes out of the case from the point of view of liability, and since specific performance is out of the question, it remains to consider whether the plaintiff can recover against Mr Spence and on what basis. As the pleadings originally stood, the plaintiff was suing for breach of contract, and it was argued by Mr Parker for the defendants that on the authority of *Bain v Fothergill* (1874) LR 7 HL 158 G the plaintiff must be limited to recovering damages measured only by the cost of the expenses incurred, that is of investigating title, and it was said that he could not recover the loss of his bargain. Fraud, it was said, was not alleged, and fraud was the only exception to *Bain v Fothergill* which entitled a plaintiff to recover also for his loss of bargain. An examination of *Bain's* case shows that the House approved the principles of *Flureau v Thornhill* (1776) 2 WBl 1078, decided almost exactly 100 years earlier, and confirmed that in the absence of fraud, the mere failure to make out a good title to real estate only gives rise to damages measured by the cost of expenses incurred, that is, to the cost of investigating H such title. At page 207, Lord Chelmsford stated that "if a person enters into a contract for the sale of a real estate, knowing that he has no title to it, nor any means of acquiring it, the purchaser cannot recover damages beyond the expenses he has incurred by an action for the breach of the contract; he can only obtain other damages by an action for deceit." These words at first sight might seem to go further than those of Pollock B at page 170, where he says: "Where the vendor, *without his default,* is unable to make a good title, the purchaser is not by law entitled to recover damages for the loss of his bargain." The same words, "without his default," were also used by Denman J at page 176 and by Pigott B at page J 193, and of course came in fact from the first question proposed for consideration of the judges (see page 170). They do not, if taken out of context, seem to me necessarily to connote fraud sufficient to found an action for deceit. On the other hand, when one looks at *Flureau v Thornhill* it is clear that the court held that the presence of fraud was necessary before the purchaser could get damages for the loss of his bargain: see the words of Grey CJ at page 1078:

> Upon a contract for a purchase, if the title proves bad, and the vendor is (without fraud) incapable of making a good one, I do not think that the purchaser can be entitled to any damages for K the fancied goodness of the bargain which he supposes he had lost.

It follows, I think, that the words "without default" in *Bain's* case must be taken to mean "without fraud," and that on the authority of that case, the purchaser must be able to go as far as proving fraud before he can recover for the loss of his bargain. In a recent case, *Wroth v Tyler* [1974] Ch 30, Megarry J was able to distinguish the case before him from *Bain's* case on the basis of the facts and in particular on the nature of the charge there in question. He was dealing with a case where the wife of the defendant, against whom specific performance was being sought, had entered on the L land charges register a notice of her rights of occupation under section 1 of the Matrimonial Homes Act 1967. That Act, he held, gave her a personal and non-assignable statutory right not to be evicted from the matrimonial home during marriage, and such right constituted a charge on the estate of the owning husband which required registration to obtain protection against third parties (see p 417D-E). At p 426 *Bain v Fothergill* was discussed, and at p 426 Megarry J concluded that the wife's rights under the statute were not dependent on the vicissitudes of a particular title to property. He said: M

> The charge is *sui generis.* . . . If her rights are rights of property at all they are at least highly idiosyncratic. They do not seem to me to fall within the spirit or intendment of the rule in *Bain v Fothergill.*

In the case before me, unfortunately, I find it impossible to say that there is present some defect in title not contemplated in and covered by the principles of *Flureau v Thornhill* and *Bain v Fothergill.* As already stated, fraud was not alleged in the statement of claim here as originally delivered, and in the absence of any amendment adding a plea of misrepresentation, I think the plaintiff here would have been limited to

A recovery of damages on the restricted basis of *Flureau v Thornhill* and *Bain v Fothergill.* Having regard to the nature of the evidence, however, it appeared to me at its conclusion that I ought to give Mr Cherryman, for the plaintiff, an opportunity of considering whether or not he should apply for amendment of his pleading by adding an allegation of misrepresentation so as to enable, if thought fit, reliance to be placed on the Misrepresentation Act 1967 and in particular on the possible effect of section 2 (1) of that Act upon *Bain v Fothergill.* If this were not done, it seemed to me that the case might well have to be decided without the real issue between the parties being dealt with. Amendment at this B late stage was objected to by Mr Parker, for the defendant, *inter alia* on the basis that if it was made he ought to be permitted to reopen his defence again and cross-examine further all the witnesses called for the plaintiff. Although of course amendment at this late stage is unusual and will only be allowed in exceptional cases, I was satisfied that justice here could only be done by permitting it, and I gave leave accordingly. My reasons were that the evidence given satisfied me that Mr Spence, by his conduct, clearly made a false representation to the plaintiff that he was the owner of the house in question and therefore able to sell to the plaintiff. C The plaintiff relied on this representation and was induced to enter into the contract by it. Mr Spence, as I find, in the words of the section, " had no reasonable ground to believe, nor did he believe up to the time the contract was made, that the facts represented were true." I do not think that any further evidence by or cross-examination of the witnesses would change my conclusions on this matter, nor do I think that there it could properly be said that in the circumstances permitting the amendment was unfair to the defendants. By it they were faced, albeit for the first time, with the real point in the case, and it would I think have been unfair to the plaintiff not to have allowed him to bring out the legal consequences of the evidence which had been given. I therefore D adjourned the case to enable the matter to be considered. In due course, the amendments were put forward, I allowed them, and they are set out in the amended statement of claim and in the amended defence.

It is now necessary to consider and come to a conclusion upon the effect, if any, of the Misrepresentation Act 1967 on *Bain v Fothergill.* The material facts I have already found above mean that Mr Spence represented to the plaintiff that he owned the house in question and was in a position to sell it. That representation was false at the date it was made; that is, the date of the signing of the contract. It was false E to the knowledge of Mr Spence, who knew perfectly well that he and his wife were joint owners of the property and that he could not sell without her agreement. It may well be, and I would have been quite prepared to believe, that Mr Spence, though he did not in fact say so, was at the time of the contract confident that he would be able to persuade his wife to sell her share of the house by the time of completion, and that he had no intention to cheat the plaintiff, but that does not make the representation any less false at the time it was made. Although, on the assumption made above as to the state of mind of Mr Spence, it would be right to say that he was not intending to be fraudulent, nevertheless, F legally, it seems to me the true position is that he was guilty of fraud within the definition of deceit by Lord Herschell in *Derry v Peek* (1889) 14 App Cas 337 at 374, where he said:

> To prevent a false statement being fraudulent there must I think always be an honest belief in its truth. And this probably covers the whole ground, for one who knowingly alleges that which is false has obviously no such honest belief. Thirdly, if fraud be proved the motive of the person guilty is immaterial. It matters not that there was no intention to cheat or injure the person to whom the statement was made.

Here, as already stated, the plaintiff has never suggested that Mr Spence made a fraudulent representation in the sense of wishing to cheat him, and fraud is not pleaded. Although, G therefore, if it had been, it would be necessary to deal with the case on that basis, it is fortunately not necessary to do so. Can the plaintiff, then, on the facts found, rely on the Misrepresentation Act 1967, and if so, what is the result of his being able to do so? Section 2 (1) of the Act reads as follows:

> Where a person has entered into a contract after a misrepresentation has been made to him by another party thereto and as a result thereof he has suffered loss, then, if the person making the misrepresentation would be liable to damages in respect thereof had the misrepresentation been made fraudulently, that person shall be so liable notwithstanding that the misrepresentation was not made fraudulently, unless he proves that he had reasonable ground to believe and did believe up to the time the contract was made that the facts represented were true. H

In *Gosling v Anderson, The Times,* February 8 1972, also now reported in [1972] ESTATES GAZETTE DIGEST 709, at p 713 Lord Denning said:

> This is the first case we have had under the Misrepresentation Act of 1967. It gives a cause of action for innocent misrepresentation just as if the misrepresentation had been made fraudulently.

He then quotes section 2 (1), and says:

> That is a very long sentence, but it means that if there is an innocent misrepresentation which leads another person to enter into a contract, then he can recover damages for it, unless the person making the misrepresentation can show that he had reasonable ground for believing it to be true. This Act has a considerable impact on the law, and especially in a case like the present. J

Gosling's case was one where, on the sale of a flat and plot, there was a representation that planning permission for the erection of a garage on the plot had been obtained. The representation was made by an agent, Mr Tidbury, who honestly believed in its truth, but where his principal, Mrs Anderson, knew quite well it was false. Lord Denning continued: K

> Before this Act Miss Gosling (the plaintiff) would have failed unless she proved that one or other of them was guilty of fraud: see *Armstrong v Strain* [1952] 1 KB 232. Now there need be no question of fraud. Sufficient that the agent, Mr Tidbury, made a statement which was in fact untrue, although he believed it to be true.

An inquiry was ordered on the basis of the difference in value of the land with and without planning permission. Now here it seems to me that the words of section 2 (1) cover the present case on the facts as I have found them. Mr Spence made a false statement which induced the plaintiff to enter into the contract. Legally that statement was, I think, fraudulent and would entitle the plaintiff to secure damages on that basis. If, however, as here, the representation is not treated by the plaintiff as fraudulent, it is none the less false, and the defendant has no defence under the last part of the section, because he did not believe it to be true nor had he reasonable ground for any such belief. L

The so-called " exceptional rule " in *Bain v Fothergill,* and its rationale, are discussed at length in chapter 21 of *McGregor on Damages* 13th ed, starting at p 465. In para 667 on p 471 the learned author discusses the effect of *Bain's* case in finally closing the loophole opened by *Hopkins v Grazebrook* (1826) 6 B & C 31, and mentions that subsequent decisions M have set up other limitations upon the application of the restrictive rules established by Lord Chelmsford in *Bain's* case. Megarry J's decision in *Wroth v Tyler* is a case in point. The Misrepresentation Act is, however, not discussed in chapter 21 of *McGregor,* and so far as I know the present case is the first in which the bearing of the Act upon *Bain v Fothergill* has been considered. Its effect, in my judgment, is considerable, and the legislation has altered the law as stated by Lord Chelmsford in that case. If the Misrepresentation Act had been in force at the time of *Bain v Fothergill,* it seems at least probable that the vendor's immunity against damages

A for loss of bargain ought to have been confined to cases where there was no misrepresentation, innocent or fraudulent, which induced the contract; for example, where the defect in title was something which was unknown at the time of entering into the contract and which was only found out on investigation. In that event Lord Chelmsford's words at 7 App Cas 207 quoted above, "If a person enters into a contract for the sale of real estate knowing that he has no title to it nor any means of acquiring it," and so on, down to the words, "other damages by an action for deceit," might well have read:

B In those cases where a person enters into a contract for the sale of real estate knowing that he has no title to it and has made no representation, innocent or fraudulent, that he will be able to acquire it by the time for completion, thereby inducing the purchaser to enter into the contract, the purchaser cannot recover damages beyond the expenses he has incurred by an action for the breach of the contract; he can only obtain other damages by an action based on innocent misrepresentation or deceit where such exists.

The law should I think be so stated now the Misrepresentation Act is in force, and I so hold. The truth of the matter is that *Bain v Fothergill* limits the damages for breach of contract; it does not limit damages for fraudulent misrepre- C sentation. The Misrepresentation Act 1967 for the first time enables a plaintiff to sue for innocent misrepresentation, a cause of action now made akin to an action for damages for fraud. The Act of 1967 has thus created a new cause of action, one with which *Bain v Fothergill* never had anything to do. The practical effect is, however, that some purchasers who would have been caught by *Bain v Fothergill* if the 1967 Act had not been passed can now, by suing on the new statutory right, get damages for loss of their bargains which they could not have recovered before. It follows that in my judgment the present case falls outside the restrictive rule of D *Bain v Fothergill* as that rule should now be limited in the light of the Misrepresentation Act 1967. If the representation here had been treated as fraudulent, the plaintiff would, on the authority of *Bain's* case itself, have been entitled to recover for loss of his bargain, and in my judgment he is equally outside the case and equally entitled to recover for the loss of his bargain on the ground that there was a representation which was in fact false and which the defendant, Mr Spence, had no ground for believing to be true and which he did not in fact believe to be true.

No figures are agreed in this case, so it will be necessary to order an inquiry. On the basis on which I think the E plaintiff is entitled to recover, such inquiry should arrive at the value of the house at the dates of the contract and of completion. The plaintiff will then, in my judgment, be entitled to damages equal to the rise in value, if any, of the house between those dates. The costs of such inquiry will be reserved.

His Lordship ordered the plaintiff to pay the costs of both defendants up to the date of his amendment, subject to a set-off for costs due from Mrs Spence in respect of proceedings to strike out her defence. The plaintiff was awarded the costs of the restored hearing. An order was also made for repayment of the plaintiff's deposit, with interest at 4 per cent F from February 8 1972, the damages, deposit, interest and plaintiff's costs to be secured by a lien on the property.

G

CHANCERY DIVISION

February 17 1975

(Before Mr Justice MEGARRY)

WOODS AND OTHERS v MACKENZIE HILL LTD

Estates Gazette January 24 1976

(1975) 237 EG 267

Contract of sale of land incorporating Law Society's conditions, including condition permitting service of notice to complete—Presence of this condition does not exclude normal contractual obligation to complete—Vendor entitled to specific performance merely on neglect by purchaser to H **complete on day fixed for completion or within a reasonable time thereafter—Four months "at least prima facie substantially more than a reasonable time" in a case where a 28-day notice could have been served**

This was an order 86 summons by Francis Jack Woods, Esther Helena Tatford Woods and Mildred Oliver Woods, owners of freehold properties in West Street, Osborn Road and Malt House Lane, Fareham, Hampshire, against Mackenzie Hill Ltd, for specific performance of a contract dated July 6 1973 by which the defendants agreed to purchase those properties.

J

Miss H Williamson (instructed by Kingsford, Dorman & Co, agents for Blake, Lapthorn, Rea & Williams, of Fareham) appeared for the plaintiffs, and Mr W Blum (instructed by Slowes) represented the defendants.

Giving judgment, MEGARRY J said: This is a vendor's summons under order 86 for specific performance of a contract dated July 6 1973 for the sale of land and buildings at Fareham, Hampshire. There is no dispute on the facts, most of which are admitted on the pleadings, but a far-reaching submission has been made on the law. The contract incorporated the Law Society's General Conditions of Sale (1973 Revision) and provided for completion on September 30 K 1974. A special condition in the contract provided for completion on the last days of May, June, July or August 1974 if the vendors gave the purchasers a 28-day prior notice, but this condition never took effect, and September 30 remained the contractual completion date. On that date the purchasers, a limited company, did not complete, and the vendors on the same day served on the purchasers a 28-day completion notice under condition 19 of the Law Society's conditions. It is common ground that this notice was invalid, if only because it purported to be given by only two out of the three vendors. The purchasers continued not to complete, and on November 8 1974 the vendors issued a specially-endorsed L writ against the purchasers claiming specific performance and alternatively damages. On December 13 the summons under order 86 was issued, whereby the vendors claimed specific performance or alternatively directions as to pleadings. It is that summons that is now before me, and on behalf of the vendors Miss Williamson submitted that it was a clear case in which specific performance should be decreed.

On behalf of the purchasers, Mr Blum conceded that the case was one in which the court could make the declaratory part of an order for specific performance, although he said that the case was one in which the court ought not even to M do that, since it was unusual to do so on an application for summary judgment, and the court would never make a declaration on a matter which was not an issue between the parties. But his main submission was that where, as in this case, a contract provides for the service of a notice making time of the essence of the contract, there is no obligation to complete on the date fixed for completion or within a reasonable time thereafter, since the provision for a completion notice overrides any such obligation. Even if years go by without completion, the purchaser will never be in breach of contract unless and until a completion notice has been served and expired. If, say, 10 years went by in inactivity,

A possibly the contract would be treated as having been rescinded by mutual agreement: but subject to that, and to any possible effect of the Limitation Act 1939, nothing save a completion notice would suffice. Mr Blum accepted that such a notice could have been served at any time, and could be served now, and that if it were served the purchasers would be bound to complete in accordance with it: but no valid notice had in fact been served. In any case, he said, there was no evidence before the court as to a reasonable time having elapsed since the contractual completion date. On the authority of *Hasham v Zenab* [1960] AC 316, Mr Blum accepted that no breach of contract need B be established in order to found a decree of specific performance, but he said that it would be wrong to include in any order any of the directions as to computing the sum to be paid on completion, or delivering a conveyance and the title deeds against payment of the proper sum on completion. Nothing more than, at most, the declaratory words stating that the contract should be specifically performed and carried into execution should be included in the order. Mr Blum expressly disclaimed, I may say, any contention that the writ had been issued prematurely.

That was the argument; and I have no hesitation in rejecting it. I do not for one moment think that the inclusion of C express provisions for completion notices, as now contained in both the Law Society's conditions and the National Conditions of Sale, has the effect of excluding the contractual obligation to complete on the date fixed for completion, or within a reasonable time thereafter. In my judgment, such provisions add to the remedies available against a defaulting party without driving out the existing remedies, or altering the existing structure. I can see nothing in condition 19 of the Law Society's conditions which is in any way inconsistent with the contractual obligation to complete on the day fixed by the contract for completion, or within a reasonable D time thereafter. Condition 19 avoids the uncertainty as to what is a reasonable time, and confers and spells out specific rights against the defaulting party; but there is no trace of any intention to exclude the rights and remedies otherwise existing at law and in equity if no such notice is relied upon. I wholly reject any notion that the contractual completion date has lost its potency and that the service of a completion notice is now a prerequisite to the enforcement of any contract which contains provisions enabling such notice to be served. In this case, as I have mentioned, the contractual date for completion was September 30, nearly 15 months after the date of the contract, with provisions (which the E vendors did not operate) which would allow the vendors to substitute completion dates up to four months earlier. Condition 19 of the Law Society's conditions provides for a 28-day completion notice, a provision which two of the vendors tried without success to operate. The case now comes before me well over four months after the contractual completion date has passed, and Mr Blum contends not only that a reasonable time has not in fact elapsed but also that there is no evidence before me to show that it has. He concedes, as he must, that at any time during the four months the vendors could have served a completion notice which, by the terms of the contract by which the purchaser has bound F itself, would have bound the purchaser to complete within 28 days. Having regard to the time scale that the parties have chosen, the period of over four months seems to me at least *prima facie* to be substantially more than a reasonable time for completion, and I should require cogent evidence before I would reach a contrary conclusion. The purchasers have adduced no evidence whatever on the point, and indeed there is nothing before me to indicate that the purchasers were even going to raise the point. I think it plain that a reasonable time has elapsed.

In the result, therefore, this seems to me to be a case in which there is nothing that can be called an arguable defence to the plaintiffs' claim to a full decree of specific performance. Furthermore, I can see no issue or question in dispute that ought to be tried, or any other reason why there ought to be a trial of the action: I speak, of course, in the language or order 86, rule 4. In my judgment, the contract ought to be performed, and the appropriate consequential directions as to performance ought to be given; and I so order. G

CHANCERY DIVISION H

May 2 1975

(Before Mr Justice GOFF)

MICHAEL RICHARDS PROPERTIES LTD v CORPORATION OF WARDENS OF ST SAVIOUR'S PARISH, SOUTHWARK

Estates Gazette March 13 1976

(1975) 237 EG 803

Sale by tender already effectively in contract form, with no terms remaining to be negotiated—Words "subject to contract" mistakenly typed by a secretary on vendor's acceptance of tender—Parties behaved as though a contract existed until purchaser was informed of prospective compulsory acquisition of property—Typed words held meaningless in their context—Goff J disclaims any intention of causing "alarm bells" to ring in solicitors' offices—Contract held to have come into existence and purchasers' deposit forfeited J

This was a claim by Michael Richards Properties Ltd against the Corporation of Wardens of St Saviour's Parish, Southwark, for the return of a deposit paid on an agreement made in October 1972 for the purchase of property owned by the defendants in Sydenham, London SE26. The defendants counterclaimed a declaration that the deposit was forfeit. K

Mr A J Balcombe QC and Mr B Levy (instructed by Nabarro, Nathanson & Co) appeared for the plaintiffs, and Mr H E Francis QC and Mr T R F Jennings (instructed by Simpson, Palmer & Winder and W R Millar & Sons) represented the defendants.

Giving judgment, GOFF J said that in October 1972 the defendants advertised certain property at Sydenham for sale by tender. The tender documents contained full particulars and special conditions of sale, and the special conditions incorporated the National Conditions of Sale 18th ed with amendments including the deletion from condition 22 (3) of the words "(unless the court otherwise directs)." Special conditions 5 and 7 read: L

5. Every person desiring to purchase the property described in the foregoing particulars shall fill in and sign with his name and address the form of tender printed at the foot of these conditions, and shall send a copy of the foregoing particulars and these conditions with the said form of tender so filled in and signed still atached thereto to reach the Clerk, the Corporation of Wardens of St Saviour . . . in an envelope marked "tender" not later than 12 noon on the 27th day of October 1972 when the tenders will be opened. No tender may be withdrawn before the date specified in paragraph 7 hereof. M

7. The person whose tender is accepted on the 27th day of October 1972 shall be the purchaser, subject to the approval of the Charity Commissioners, and shall be informed of the acceptance of his tender by letter sent by registered post or recorded delivery post addressed to the address given in his tender and any letter of acceptance so sent shall be deemed to have been received in the due course of post.

On October 26 1972 the plaintiffs sent in a completed form of tender offering £110,000 for the property in question, and on October 27 the defendants' agents, Richard Ellis & Son, sent a letter of acceptance of that tender. The letter, which required payment of a 10 per cent deposit, was signed with the firm's name, immediately underneath which appeared, typed in capitals, the words "SUBJECT TO CONTRACT." The evidence was that an assistant surveyor, Mr Ryder, had either dictated that letter to his secretary or given her general instructions what to say, and that he did not intend the words "subject to contract" to be used: she however had added the words because of a general (and understandable) practice in the office by which all offers and acceptances in private treaty cases were to be so safeguarded. On or about October 31 the plaintiffs paid the deposit of £11,000 to the defendants' solicitors. Mr Ryder did not realise what had happened, and he (his Lordship) was satisfied that the words "subject to contract" were used by mistake. The matter proceeded in the normal way. The vendors did not submit a draft contract, nor did the purchasers ask for one, and it was difficult to see what a contract could have contained.

That same October 27 1972 the local authority sent the vendors a letter saying that its officers had it in mind to recommend the compulsory purchase of the property, and suggesting that the vendors might think it ethically right to inform the successful tenderer. In fact this information was not passed on to the plaintiffs till November 29, but they made no complaint of that, accepting that it was merely the result of the wardens not having met in the meantime. Both parties meanwhile proceeded on the basis that there was a contract. On November 1 the defendants' solicitors sent the plaintiffs' solicitors various formal documents. On November 8 the plaintiffs' solicitors acknowledged these, and on November 21 they submitted requisitions. That letter crossed with one sending a draft of the order the Charity Commissioners proposed to make. On November 29 the plaintiffs' solicitors approved the draft order, and subject to replies to requisitions enclosed a draft conveyance. However, that letter, too, crossed with one from the defendants referring to the letter from the local authority, and this caused consternation and a prompt reply to the effect that the local authority's letter must have been received by the vendors on October 30 or 31, and if the contents had been made known to the plaintiffs on either of those days, the plaintiffs would still have had time to consider the matter further before "lodging with you the ten per cent deposit and concluding the contract." That last statement was not accurate on any showing, but was entirely inconsistent with the matter having remained in negotiation. On December 4 the vendors approved the draft conveyance, and on December 20 they enclosed replies to requisitions and asked for the conveyance for execution. On January 1 1973 they advised that the Charity Commissioners' sealed consent was available. On January 15 they served notice to complete, but on the same date the plaintiffs' solicitors delivered by hand a letter requesting return of the deposit, pointing out that the vendors' agents' letter of acceptance was stated to be "subject to contract," and that, as no unconditional offer and acceptance had taken place, no contract existed. The plaintiffs did not comply with the notice to complete.

He (his Lordship) would dispose of one point at once. In his opinion a contract expressed, as here, as conditional on the consent of the Charity Commissioners being forthcoming was effective upon that consent issuing. It was therefore not made without consent, and did not offend against section 29 (1) of the Charities Act 1960. On the main point, if the words "subject to contract" stood and had to be regarded, the plaintiffs were clearly right. It was not possible to say that the words referred only to the special condition relating to the Charity Commissioners' consent. A further argument, that the effect of the words was capable of being waived, and was waived, he (Goff J) thought could be advanced only in the House of Lords, having regard to *Tiverton Estates Ltd v Wearwell Ltd* [1975] Ch 146. It was true that the question there was whether the qualification "subject to contract" could be disregarded so as to enable a writing so limited to be treated as a note or memorandum of a contract, but the question whether it could be waived so as to make a contract appeared to be *a fortiori*. The point on which the case finally turned, therefore, was whether the words "subject to contract" in the letter of acceptance ought to be rejected. He (his Lordship) could not infer, as he had been invited to do by counsel for the defendants, that the plaintiffs knew the words had been inserted by mistake. After all, the surveyor concerned signed the letter, and did not himself notice it. Even if he (Goff J) did make this inference, he could not see how it would avail the defendants. He could not make a contract for the parties if by mistake they had failed to make one for themselves. The approach on this basis must be a claim for rectification, but then where was the prior agreement to which by mutual mistake the letter of acceptance failed to give effect?

Counsel for the defendants further contended, however, that the words "subject to contract" should be rejected as meaningless in the context, relying on *Nicolene v Simmonds* [1953] 1 QB 543, and in particular on words of Denning LJ at p 552 and Hodson LJ at p 553. That case was different on its facts, but he (his Lordship) thought the principle applied. He hoped that this judgment would not ring warning bells in solicitors' offices, as Lord Denning had put it in *Tiverton Estates v Wearwell Ltd* (*supra*) when referring to *Law v Jones* [1974] Ch 112. He (Goff J) was not casting any doubt on the meaning, effect and protection of the words "subject to contract" in the cases in which they were used in normal conveyancing practice and everyday life, viz when estate agents were negotiating a sale, solicitors negotiating a contract or otherwise acting for a vendor or purchaser on a proposed sale by private treaty, and individuals meeting, making a written or oral offer to buy or sell, or an agreement subject to contract. He would not wish to do so, even if he could, and in any case it was clear on the authorities in the Court of Appeal that it would be quite wrong. His decision was on the particular facts of this case. Here was a sale by tender: nothing remained to be negotiated, there was no need or scope for a further formal contract, and it was difficult to see how one could be drawn. Nobody ever thought there was. The vendors did not submit a draft contract, nor were they asked to do so, and the matter proceeded with the necesary steps, not to negotiate or finalise a contract, or even to put it into further form or shape, but with the steps required for completion. In the context of a tender document setting out all the terms of the contract, and requiring to be annexed to the tender offer, it seemed to him (his Lordship) that the words "subject to contract" in the acceptance were meaningless, and could be rejected in accordance with the decision in *Nicolene v Simmonds (supra)*.

As to whether he ought to exercise his discretion to order a return of the deposit under section 49 of the Law of Property Act 1925, the plaintiffs advanced two reasons for his doing so: that the defendants would otherwise have made a profit, and that the warning concerning the local authority's views could have been communicated before the deposit was paid. The first was inherent in cases of forfeiture, the second was not really significant because the contract was concluded before then by the letter of acceptance. The parties had, by deleting the words "unless the court otherwise directs" from condition 22 (3) of the conditions of sale, agreed between themselves to the forfeiture, and the court should not lightly go behind their agreement, though it was common ground that there was jurisdiction so to do. Moreover, the plaintiffs had deliberately refused to perform their contract. Some sympathy might be felt for them, because they had discovered a matter which might seriously affect the value of the property, and if they had known it before they might not have tendered at all, or might have tendered

A less, but the fact remained that they had contracted and had deliberately repudiated their contract, thinking they had an escape route which, on the view he (his Lordship) had reached, they had not. The action therefore failed, and on the counterclaim, the defendants would take a declaration that the deposit was forfeited.

B

CHANCERY DIVISION

December 18 1974

(Before Mr Justice GOULDING)

NEW HART BUILDERS LTD v BRINDLEY

Estates Gazette March 27 1976

(1974) 237 EG 959

Option to buy land for development signed by landowner, subsequently altered by addition of an extra term above his signature—Altered agreement held not to meet requirements of section 40, Law of Property Act 1925—Signature not affixed to, or recognised by words or gestures as applying to, the altered terms—Applications for planning C permission not acts of part performance, option agreement accordingly unenforceable.

This was a claim by New Hart Builders Ltd against Mr John Brindley, sued as administrator of the estate of his father, the late John Brindley, for specific performance of an agreement for sale of Allen's Rough Farm, Willenhall, Staffordshire, for £5,000 an acre.

Mr A C Sparrow QC and Mr G W Seward (instructed by Bartlett & Gluckstein) appeared for the plaintiffs, and Mr M J Albery QC and Mr T L G Cullen (instructed by Sharpe, Pritchard & Co, agents for Haden & Stretton, of Walsall) D represented the defendant.

Giving judgment, GOULDING J said that the late John Brindley, a farmer, lived at and owned Allen's Rough Farm between Wolverhampton and Walsall. By an agreement in writing (the "option agreement") of November 25 1966 he gave the plaintiff company an option to buy the farm for £5,000 an acre. The option agreement distinguished different parts of the farm by reference to a plan, the land with which the action was concerned, something above 20 acres in extent, being edged blue on the plan and conveniently described as "the blue land." The option agreement provided that the option in respect of the blue land must be renewed every E calendar year. The plaintiff company claimed to have renewed the option in or about December in each year from 1967 to 1970 inclusive, and to have finally exercised the option by letter of October 26 1971. John Brindley died on February 6 1970, and the defendant became his personal representative on November 16 1970. The plaintiff company now sued the defendant for specific performance of a contract of sale of the blue land constituted by exercise of the option contained in the option agreement. In answer, the defendant had ultimately relied on three specific points. He said first that the contract on which the plaintiffs sued was unenforceable, because there was no sufficient writing to F satisfy section 40 of the Law of Property Act 1925. Secondly he alleged that the plaintiffs had omitted to renew the option on the first occasion when renewal was required, namely in 1967, and thirdly he said that they had failed to renew on the third occasion, in 1969.

Section 40 of the Law of Property Act 1925 required a contract of sale of land either to be in writing or to be evidenced bv a memorandum in writing. In either case the writing had to be signed by or on behalf of the party to be charged. The only signature which had been relied on for this purpose by the plaintiffs was the late John Brindley's signature on the option agreement, and to understand the

G defendant's first contention one must know the history of that document in some detail. In 1966 Mr Dean, of the plaintiffs, had a number of discussions with the late Mr Brindley and the defendant, and he introduced them to Mr Hartley, also of the plaintiff company. No planning permission for development of Allen's Rough Farm having yet been obtained, Mr Dean and Mr Hartley had an option in mind rather than an immediate purchase. A document was typed out in the plaintiff company's office and taken on November 25 1966 to Allen's Rough Farm, and after some discussion which took place mainly between Mr Hartley and the defendant the blanks in the document were filled H in so as to constitute an option agreement for the purchase of "yellow" and "red" portions of the land at £5,000 an acre, the consideration for the option to be £100. At some stage this document, so completed, was signed by the late Mr Brindley, and after signature it was handed to Mr Dean, who gave the late Mr Brindley a cheque for £100 in return. A period of conversation followed, after which the defendant asked for the option agreement and plans, saying that he wanted to check them. When he got them, he wrote some further provisions into the agreement in the space between the end of the typed text and his father's signature. Using a blue pencil, he drew the blue edging which now appeared J on the plan referred to in the document, and this area was mentioned in the added provisions, which read as follows:

The area edged blue extending to an area of approximately 23 acres the option on this land to be renewed every 12 months and to run from January 1 to December 31 1967. The option on the remainder of the land will terminate December 31 1968 on any land for which planning permission has not been granted. thereafter a new agreement will be needed.

Mr Dean was not at all pleased with what he felt was this high-handed proceeding, but in the end he and Mr Hartley left, taking the amended option agreement and the plans K with them. He (his Lordship) found that before they left, Mr Dean and Mr Hartley on behalf of the plaintiff company, and the late Mr Brindley on his side, had agreed to the document as amended. The late Mr Brindley did not sign or initial the alterations made in the defendant's handwriting, nor was anything said about his signature after it was first affixed.

On these facts, the defendant now argued that although the contract on which the plaintiff company sued was in writing, it was not signed by the late Mr Brindley. The signature was affixed to the option agreement as it stood before altera- L tion by the defendant, and the terms of the option agreement in that form were finally agreed, and became binding, when the cheque for £100 was handed to the late Mr Brindley. The offer contained in the option agreement as it then stood was not accepted, because the agreement was shortly afterwards discharged when the parties accepted the amendments made in the defendant's handwriting. This was an unattractive argument in the mouth of the defendant, but as a matter of positive law to be decided on principle and not on general merits his objection had much intrinsic force. On the evidence, the late Mr Brindley's signature was intended by him, at the instant when it was written on the paper, to authenti- M cate a different option agreement from that which the plaintiff company claimed to have exercised. If he (his Lordship) said that in agreeing orally or by conduct to be bound by the subsequent alteration of the document the late Mr Brindley in effect re-executed his signature, that would be to place a construction of law on the words of section 40. Counsel for the defendant had conceded that in certain cases an alteration made to a document after signature and approved by the party to be charged was sufficiently authenticated for the purposes of section 40 by the original signature, as where the signed document was altered to correct a mistake in the written statement of an existing contract.

A Where however the document was altered before the parties were contractually bound at all, merely to accept the altered writing was not specific enough: appropriate words or gestures must be directed to the signature as a signature. There were statements in the authorities which clearly supported this argument, and he (Goulding J) thought it his duty in a matter of the kind to follow a strong current of judicial opinion, even though it might not strictly bind him and even though it drew what he for his part conceived to to be an illogical distinction. He accordingly accepted the defendant's submission that there was here no sufficient B writing to satisfy section 40 of the Law of Property Act 1925.

That raised the plaintiff company's alternative argument based on the well-known equitable doctrine of part performance. The acts relied on as constituting part performance were the planning applications made and prosecuted by the company from December 1966 until permission was obtained in June 1970. On any view, these applications could not rank as acts of part performance unless by their own character they were referable to the existence of an agreement between the plaintiff company and the late Mr Brindley. Counsel for the company had laid a great deal of stress on the fact that the Brindleys gave assistance to the company in C pursuing the applications. However, there appeared to be nothing improbable or unusual in the co-operation of a would-be developer and a landowner in obtaining planning permission in the hope that both would eventually profit, without necessarily having bound themselves by contractual terms. Thus the planning activities did not prove the existence of a contract, and the plaintiffs had failed to make out their allegation of part performance. The defence under section 40 was therefore a good answer to the plaintiff company's claim. In addition, as a matter of construction of the option agreement he (his Lordship) had concluded that written notice to the landowner was required of renewal D of the option, and it was common ground that no such notice was served in 1967. Oral renewal had been relied on as the basis of an alternative claim of promissory estoppel or quasi-estoppel, but he (Goulding J) was of opinion on the facts that no oral renewal had been proved. On the balance of probabilities, the plaintiffs had established that the option was renewed in 1969, but that was of course of no avail to them in the circumstances, and their action must be dismissed.

The defendant was awarded costs.

LANDS TRIBUNAL
COMPENSATION

A

LANDS TRIBUNAL FOR SCOTLAND

MENZIES MOTORS LTD v STIRLING DISTRICT COUNCIL

(LTS/APP/3/44)

(H M BRAINE WS and WM HALL FRICS)

September 2 1975

Estates Gazette January 10 1976

(1975) 237 EG 121

Acquisition of motor workshop and garage at Stirling—Claim on alternative bases of development value and existing use value plus disturbance—Planning permission for supermarket held unlikely because of traffic problems—Disturbance claim for move to temporary and permanent premises—Allowed items include (a) heating of temporary premises, despite council's "value for money" argument; (b) adaptation of temporary premises on jobbing basis without detailed costs; (c) difference between statutory interest on unpaid compensation and overdraft interest due prior to payments to account by acquiring authority—Disallowed items include (i) overdraft interest on disturbance claim expenditure incurred after payments to account; (ii) architect's fee for new building and costs of site investigation; (iii) surveyor's fees for inspecting alternative premises before notice to treat—Existing use value plus disturbance "as high if not higher than development value"

B

C

Mr W L K Cowie QC and Mr P Fraser (instructed by J & R A Robertson, of Edinburgh) appeared for the claimants, and Mr D B Weir QC and Mr T C Dawson (instructed by Robson McLean & Paterson, of Edinburgh) represented the acquiring authority.

Giving their decision, THE TRIBUNAL said: This reference arises from a dispute as to the compensation payable for a workshop and garage with land attached at Orchard Place, Stirling, known as Menzies Motors Ltd and extending to 0.76 of an acre (3,667 sq yds).

D

The claimants are Menzies Motors Ltd, whose interest as owner-occupier was compulsorily acquired by the Royal Burgh of Stirling under the Burgh of Stirling (Comprehensive Development Area—No 1) Compulsory Purchase Order 1967 confirmed by the Secretary of State for Scotland on July 26 1971. Notice to treat was served on August 19 1971. It was agreed that July 9 1973 was the date of valuation.

Buildings Demolished

E

At the date of the hearing the buildings had been demolished. The buildings had been reconstructed in 1942 to accommodate a car showroom, petrol filling station and a workshop for repairs, service and maintenance. The buildings occupied the greater part of the site with a frontage to Orchard Place of 80 ft and to Thistle Street/Goosecroft Road of 365 ft. Part of the Goosecroft Road frontage was to an area of open land extending to 751 sq yds. This land was at a lower level than Orchard Place (about 12 ft) and was not normally used by the garage for parking of cars.

The claim submitted was supported by evidence from Mr Henry J Crone FRICS FRTPI and was stated in alternative forms.

F

Firstly, it was claimed that as the site was suitable for the erection of a supermarket the market value with that potential was £220,020, being 3,667 sq yds at £60 per sq yd, and, in addition, that the costs involved in moving to temporary premises should be paid as compensation. If it were held that a supermarket could not be established, it was claimed that the site was suitable for other developments, eg warehousing, in which case Mr Crone maintained that a rate of £60 per sq yd was still appropriate.

Secondly and alternatively, the existing use value of £168,145 plus the disturbance claim should be determined.

G

It was submitted that the higher figure of these alternatives should be determined as the compensation.

The acquiring authority's contention, supported by valuation evidence from Mr Alexander T S Cargill ARICS, senior valuer in the Inland Revenue Valuation Office in Stirling, was that the value of the site for development purposes was £73,340, being 3,667 sq yds at £20 per sq yd. The existing use value was £100,000 and in addition he was prepared to accept a disturbance claim of £10,000. As the £110,000 was higher than the value as a development site the £110,000 should be the compensation.

H

In determining the development value of the site, the question of whether planning permission for a supermarket would be obtained was a fundamental issue. Planning permission, in turn, depended on whether vehicular traffic could enter or leave the site without creating a traffic hazard.

Evidence from Mr Colin A J Beckett FRICS was led to establish that the site was a suitable commercial one for a supermarket. He visualised the design being such that there would be a basement area with a frontage to Goosecroft Road, which would be used jointly for customer car parking and service facilities. He had not, however, considered in any detail the solution to the traffic problems which Goosecroft Road entry and exit would cause.

J

Mr Crone recognised that there would be traffic problems but considered that a solution to these would be a matter for the purchasers. He maintained that traffic engineers could, by the design of slip roads, solve any problem which might arise. Servicing of the site could take place from Orchard Place as an alternative.

A contrary view was expressed by two witnesses for the acquiring authority.

Traffic Problem

K

Mr A F Goodwin MSc MICE, burgh surveyor at the time, said that his advice was sought as to the traffic problem if a supermarket were to replace the garage relative to the traffic conditions at July 1973. If there had been no comprehensive development of the area, he said, Thistle Street would have remained. The cost of altering the public services would be a strong argument for that decision. He said that if a plan demonstrating a method of entry and exit to and from the site had been produced with a planning application for a supermarket he would have had to consider it, but in regard to the standards laid down, he did not consider that a visibility attainment sufficient to allow approval could have been devised. This view was derived from the curve of Thistle Street, the habitual lane-changing by owners and the danger to pedestrian use which would result from the removal of a sufficient length of pavement to allow the formation of a slip lane in and out of the site. It was difficult, in his view, to visualise a scheme which would allow sight lines sufficient to achieve the requisite stopping distances. He was against Orchard Place being used as a service access.

Mr G A McLennan, who is now depute director of planning, Clackmannan District Council, but was formerly with the Burgh of Stirling, stated that, in his view, planning permission for a supermarket would not have been approved of a part from the comprehensive development proposals. The main objection would be the question of access. His advice to the planning authority would be to refuse the application on traffic grounds. He agreed that permission had been given for a Littlewoods store and a Fine Fare supermarket with only a temporary solution to servicing access but that these permissions were in the context of the central area being comprehensively developed.

Having regard to the apparent traffic difficulty in connection with use of the site and the conflicting views of the claimants and the acquiring authority on such a vital matter affecting the value, we consider that it is for the claimants to persuade us that the traffic problem can be solved. The claimants provided no plan to demonstrate that a solution could be reached, and in absence of such plan it is our view that it would be difficult, if not impossible, to provide a satisfactory access and exit for servicing a supermarket development to and from Goosecroft Road. We also consider that servicing from Orchard Place would not be satisfactory or acceptable because of the congested nature of this cul-de-sac.

We cannot find, therefore, that it was likely that planning permission would have been obtained for the erection of a supermarket on this site.

Mr Crone considered that even if a supermarket was not allowable, the site could be developed for offices or warehousing. Before dealing with the value of the site for such a purpose, we prefer to deal with the existing use value and disturbance and thereafter to test that figure against the development value of the site.

Existing Use Value

We firstly set out the conflicting values of Mr Crone and Mr Cargill adjusted to the agreed areas and related to capital rate 5 per sq ft.

	sq ft	@	*Mr Crone*	@	*Mr Cargill*
(i) Showroom and offices	2,272	£10.50	£23,856	£5.50	£12,500
(ii) Upper floor	2,352	£2.60	£6,115	£1.50	£3,528
(iii) Garage	23,391	£5.25	£122,802	£2.75	£64,325
(iv) Forecourt (v) Pump-tanks	2,831	£2.60	£7,360	£2.58	£7,300
(vi) Land	751 sq yds	£12.00	£9,012	£15.00	£11,265
			£169,145		£98,918
Deduct ground burdens of £30 x 10 YP					£300
					£98,618
			£169,145		say £100,000

Mr Crone did not try to support his valuation by any comparable sales or lets and thought, in any event, that any comparisons would not be of much help in valuing the special property of Menzies Motors.

Mr Cargill relied on the comparison of a sale of a central site garage in Stirling in April 1974. The garage at Allan Park, Stirling, together with four houses and an area of land was sold for £120,000 in a transaction between the owners and Stirling Town Council. It was said to be a market transaction and without compulsory powers. Mr Cargill's analysis of this transaction is as follows:

Showroom and offices	2,170 sq ft	@	£10.00	£21,700
Garage—workshop	21,304 sq ft	@	£3.00	£63,912
Forecourt (covered) including lock-ups	5,602 sq ft	@	£2.25	£12,604
Upper-floor office	400 sq ft	@	£2.00	£800
				£99,016
Add				
Four 3-apartment flats with vacant possession £4,000 each			£16,000	
Ground 381 sq yds @ £15			£5,715	£21,715
				£120,731
				say £120,000

In arriving at the value of Allan Park, Mr Cargill had regard to his analysis of two sales in Bishopriggs and Airdrie in 1969 which, so far as the buildings were concerned, brought out rates, adjusted for the time factor, at £3 and £3.20 per sq ft overall. His analysis of an overall rate for the Allan Park buildings was £3.90 per sq ft and for Menzies Motors £2.90 per sq ft. We think the £3.90 is wrongly calculated and should be £3.67.

Mr Crone said that his firm had valued Allan Park when it was on the market at £100,000. There appeared to be some doubt in his mind as to whether the houses had been rent restricted. In any event, he considered the ultimate sale price of £120,000 " was probably just about right."

Neither valuer considered the throughput of petrol sufficiently high to value the forecourt on a throughput basis nor did they consider it necessary for a town centre garage to have land attached to it. This was in contrast to the valuation approach of the valuers in the recent tribunal case of *Park Automobile Co Ltd v City of Glasgow District Council* (1975) 235 EG 307, 385.

Allan Park is the only comparable we have before us and we consider it as close a comparison as one may expect to have. The sale is about the same date. The property is in the town centre of Stirling and is about the same size. We therefore accept the sale as a comparison to be used in valuing Menzies Motors.

Mr Cargill considered that the showroom and offices of Allan Park were superior in both standard and location to Menzies Motors. Mr Crone did not give an opinion on the differences between the two properties. Having inspected Allan Park we can see that there is a distinction in standard, but we are not convinced that so far as a car showroom is concerned the location would influence a purchaser, provided that it is reasonably central. The assessor has apparently placed a much higher gross annual value on Allan Park, but this may be due to his placing a shopping rate on the showroom consistent with the rates he has applied to adjacent shops. This analysis was not revealed to us. We consider that the differential determined by Mr Cargill is too large and determine a rate for the showroom of £7 per sq ft. He had made a slight distinction in the garage parts. We have inspected Allan Park and do not consider the distinction justified. As Menzies Motors has been demolished we cannot make a comparison by visual inspection. From the information available, however, we do not consider that the rate for the workshop should be less than £3 per sq ft. The upper-floor office at Allan Park appears superior and we determine the upper floor of Orchard Place at £1.50 per sq ft.

The two valuations of the forecourt of Menzies Motors were about the same. Mr Cargill had placed a higher value on the vacant land than Mr Crone.

A Our determination for the existing use value is therefore as follows:

Showroom—offices	2,272 sq ft	@	£7.00	£15,904
Garage	23,391 sq ft	@	£3.00	£70,173
Upper-floor stores	2,352 sq ft	@	£1.50	£3,528
Forecourt	2,831 sq ft	@	£2.58	£7,300
Land	751 sq yds	@	£15.00	£11,265
				£108,170
Deduct ground burdens of £30 x 10 YP				£300
				£107,870
				say £108,000

B

Disturbance Claim

We now deal with the disturbance claim. Mr Arthur Willcox-Jones, the managing director of Menzies Motors Ltd, explained that steps had been taken over a number of years since the compulsory purchase order was made to find suitable alternative premises. A permanent site in Goosecroft Road has been offered by the acquiring authority, and provided that suitable terms can be reached it is proposed, if financially viable, to erect new garage premises on it. He C considered it essential to stay near the centre of Stirling to maintain his business. Pressure was put on him to vacate Orchard Place in July 1973 and he moved to temporary premises in Goosecroft Road made available by the town council. No financial terms were or are agreed so far for the use of these premises. The disturbance items relate to the temporary move and the permanent move as and when it takes place.

A number of items have been agreed and amount to £3,174.63. The remaining items are in dispute and we shall refer to them by item numbers as shown in the claimant's schedule.

D *Items 11 and 12.* These were for heating the temporary premises and consist, firstly, of the hire of heaters at £145.75 and, secondly, the installation of a more permanent heating installation by the gas board at a cost of £3,751, being £1,493 for the installation and £2,258 for the supply of the units.

These items are not admitted by the acquiring authority, first as being extravagant and second as representing "value for money" in that heating was included in the value of Orchard Place and that the rent for the temporary premises would reflect that they were unheated and the rental saving would cover the heating costs.

The "value for money" argument applies not only to this E item but also to other items, so we shall deal with it first. We accept that heating has been included in the value of Orchard Place, and in the normal way if heating had to be provided in purchased premises or premises taken on lease, such expenditure, if it increased the value of the interest, would not be allowed. If a reduction in rent were given for a temporary let and that reduction equated with the costs involved, then these costs might not be allowed. In this case, however, we were not informed what the rent terms are and therefore cannot determine whether the costs will equate with any reduction in rent. It is clear, however, and accepted by Mr Cargill, that the expenditure has not increased the value of F Menzies Motors' interest in the temporary premises and that on removal the money spent will be lost. There was no indication that the owners will reimburse Menzies Motors for any expenditure on the premises. We do not consider the "value for money" argument valid in this case.

With regard to the charge of extravagance it appears to us that the hire of heaters was not unreasonable and that the sum of £145.75 should be accepted.

The gas board account is more difficult. Mr Willcox-Jones himself expressed some dismay at the amount of the account. Owing to a misunderstanding it had been thought that a sum of £1,642.30 was involved compared with a tender for oil-fired units at £1,139. The units can probably be reused in G the final premises and therefore a credit against the total cost should be made for their value. We therefore allow the actual cost of the temporary installation amounting to £1,493 together with a sum of £1,250 for the units, a total of £2,743.

Adaptation of Temporary Premises

Item 13. The structural alterations, repairs and renovations were carried out on a jobbing basis costing £8,915.68. Details of the work proposed to be carried out were provided to us. No details of the work done or the detailed cost was supplied to us at the hearing. We accept that as the work was done on a jobbing basis it was difficult to provide the cost of each H item.

Again there is criticism of this item by the acquiring authority on the basis of "value for money" and that more was done than was necessary for a temporary period. We have dealt with the "value for money" argument in the previous item and consider that what we have said applies to this item as well. Mr Cargill made his own estimates against the various items which he was prepared to concede and proposed that £637 be paid under this heading.

We have inspected the work carried out on the temporary premises. Our inspection revealed that the total cost included items which were not on the list given to us. In particular, J a substantial amount of electrical work had been carried out. We did not find any signs of extravagant expenditure on the premises and it all appeared necessary to allow the business to continue and to meet the requirements of the appropriate authorities and in particular the fire officer. Some of the items such as lighting fittings may be able to be used in the permanent premises, but the value of these items could be offset by the cost of dismantling and removing.

Counsel for the acquiring authority conceded that as no detailed costs had been provided by his witness, he was not challenging the claimants' figure if we rejected his arguments K on extravagance and value for money. We allow the cost of £8,915.68.

Item 14. Architect's fees for temporary accommodation at £1,950. This is accepted in principle by the acquiring authority, but the amount was rejected on the basis that the work on which the fee is based included items which were not admitted. As we have accepted that the work carried out on the temporary premises is allowable we shall allow the claim. We accept that the claimants were entitled to employ an architect to arrange and supervise the work.

Item 15. Supplying and fitting compressor, steam cleaner and two hoists at £1,890.81. This item was criticised by the L acquiring authority on the grounds that a new compressor was not necessary and that the hoists should have been removed from the old premises. The steam cleaner was accepted.

Mr Willcox-Jones explained that the compressor was necessary for all air lines and that this together with the two hoists had to operate until removal. It was essential to have the hoists and compressor functioning at the temporary premises to continue the work. The compressor in Orchard Place was six/seven years old and worth £50/£60. He has this in his possession. The lifts in Orchard Place were fairly old and worth about £50; they had been left. M

We consider that the new compressor can be used in the new permanent premises and the reserve one used during the changeover. So far as the new lifts are concerned, we consider that these will be utilised in the new premises. We are prepared to allow the installation of the lifts at the temporary premises and the removal to the new premises. We shall allow the removal of the compressors. Taking into account the inconvenience during these changeovers we shall allow a sum of £900 to cover this item.

Item 16. This is for resiting the lubricating equipment and is agreed at £362.76.

Overdraft Interest

Item 17. This is a claim for overdraft interest on the amounts expended on the disturbance items. At the date of the hearing it had amounted to £3,132.71.

This claim is rejected by the acquiring authority in view of the fact that two payments to account had been made, one of £50,000 on November 27 1973 and one of £53,500 on May 3 1974, which payments exceed the expenses incurred in removal.

Interest will accrue on the total compensation figure from the date of entry at the statutory rate. The payments to account will be reflected in the calculation of interest. It was suggested by the claimants that the payments to account were for the property, but, as counsel for the acquiring authority pointed out, section 48 of the Land Compensation (Scotland) Act 1973, which authorises such payment to account, refers to " any compensation " without specifying property or disturbance. We consider, therefore, that these payments to account cover the outlays which have been made and that no claim for overdraft interest for the period following the first payment to account can be allowed. Interest is, of course, still accruing on any balance of the final compensation as determined.

There appears to be an overdraft interest amounting to just under £60 due prior to the first payment to account, but any claim can only be for the difference between the statutory rate of interest and the overdraft interest. We have no information as to what this amounts to, but assuming a 2 per cent difference and an overdraft interest rate of 14 per cent the amount involved is one-seventh of £60, say £9. We accordingly allow this £9. In effect, the cost of bridging finance is offset by the accrual of statutory interest.

Items 18-19. Architect's fees for new building and costs of site investigation amounting to £13,503.65.

These are rejected by the acquiring authority as being a part of the cost of the new building. If the basis of compensation was the reasonable cost of equivalent reinstatement under section 12 (5) of the Land Compensation (Scotland) Act 1963, then these could be allowed. In the present case, however, where compensation is being assessed under section 12 (2), this is not so.

Counsel for the acquiring authority further argued that if the claimants chose to spend their compensation on building new premises that was the claimant's own decision. Accordingly the acquiring authority were not liable for any costs involved in a new building. We agree with this view and cannot allow the item.

If the claimants were unable to re-establish themselves with the compensation for the property, it would have been open to them to pursue a claim for extinguishment of business as an alternative. No such claim was made.

Surveyors' Fees

Item 20. This is Henry J Crone & Partners' account for £4,693.33. Leaving out the scale 5 fee, which is calculated on the amount of the compensation, the net amount is £2,456.33. A substantial part of this amount is in connection with inspecting alternative premises, some of which inspections were done prior to the date of the notice to treat. These latter items were conceded by counsel for the claimants as being irrecoverable in law because they were incurred before August 19 1971, although a plea was made to us to include them in the special circumstances of this case. No legal basis, however, was argued before us to justify this and we are left having to exclude them. The acquiring authority admitted four items (it should be five) amounting to £890 and we accept that.

A sum of £615 is charged for meetings and correspondence in connection with alternative sites. The acquiring authority consider this excessive and, in any event, state that some of this account is relative to work done prior to the notice to treat. They propose £300. We allow a sum of £400.

A sum of £455 is claimed as the fee for negotiating the Goosecroft Road site. No details have been given to us about this negotiation. The acquiring authority say it is incompetent because no agreement has been reached. We cannot accept that, as it is an expense which the claimants will have to meet as a consequence of the compulsory purchase. If the claimants had purchased alternative premises, professional fees would have been reimbursed. No alternative figure has been put forward, so we accept the £455. Expenses amounting to £100.08 are also claimed. No reasons were given for challenging that amount and we allow it. The total under this item is accordingly £1,845.08.

Expenses of Second Removal

Items 21-30. These are the expenses involved in the final move and are a repeat of the earlier items. The same amounts have been put forward with an increase of 25 per cent, being the anticipated cost at the estimated time of removal. The acquiring authority have allowed most of the items in principle, although Mr Cargill has taken the same figure as the cost of the first removal on the basis that the projected cost would have to be deferred to take into account the accruing interest. Interest will accrue on the sums awarded from the date of entry and if we took the projected costs we would require to defer these back to July 9 1973 to take into account the interest accruing from that date. The position is complicated in this case by the payments to account, but on the whole we consider that the way Mr Cargill has dealt with it should be accepted. If we took the projected cost and did not defer, the claimant would get more than he is entitled to.

Items 21-27 amount to £2,910.41 and on the basis stated above we allow that.

Item 28 is rejected by the acquiring authority. This is a sum of £46.90 for fire appliances being removed and upgraded. It was, however, allowed in the first move. Mr Cargill said he had been generous in the first place and that servicing was a continuing process. Having accepted the principle for the first move, it is difficult to see how we can reject it in the second move. We accordingly allow it.

Items 29-30. These are items involving the installation by the gas board of the units from the temporary premises and the architect's and surveyor's fees if the revised scheme is proceeded with. The acquiring authority reject these as incompetent. We consider the installation of heating into the new premises as part of the building costs and, therefore, as explained before, not allowable. We also accept that the architect's fees are disallowable for the reasons set out previously in relation to the cost of a new building.

The determination for the existing use value and disturbance is therefore as shown at the top of the next column.

A calculation was made by Mr Cargill purporting to show the value as a going concern by analysing sales relative to turnover and there was then added disturbance items arising from extinguishment of the business. This produced a figure of £88,200, but it was not applied as his estimate of the compensation for the loss was £110,000. We do not therefore feel it is necessary to consider this analysis. We should say, however, that we consider it would be necessary to investigate the profit and loss accounts of the various transactions before accepting an analysis on turnover.

We still have to consider whether the value of the site available for commercial development which creates no traffic problem is greater than the existing value plus disturbance stated above. Our determination of compensation of £130,857 relative to the site of 3,667 sq yds represents a rate of just over £35 per sq yd.

Mr Crone's £60 per sq yd was derived from two sales:

A

(a) Property		£108.000
(b) Disturbance		
Agreed items	£3,174.63	
Heating (temporary)	2,743.00	
Adaptation of temporary property	8,915.68	
Fees (architect)	1,950.00	
Compressor etc	900.00	
Lubrication	362.76	
Overdraft interest	9.00	
Fees (valuer)	1,845.08	
Second removal	2,957.31	22,857.46
		130,857.46
Add surveyor's fee—scale 5 of the RICS scale of charges		1,401.75
		£132,259.21

B

(1) A sale in 1970 to Fine Fare in Murray Place at a price of £85,050. This related to an area of 1,494 sq yds, representing about £57 per sq yd. This was for a supermarket, planning permission having been received and satisfactory arrangements made about servicing and parking.

C

(2) A sale of the Plaza Cinema in May 1973 to the Post Office at a price of £55,000 for a site of 571 sq yds: the rate per sq yd is about £96. Mr Crone had been of the opinion that the rate was £61.90 per sq yd, but he did not appear to dispute Mr Cargill's information about the price or area.

Mr Cargill's £20 per sq yd was based on the following:

(1) The Picture House, Orchard Place (adjoining Menzies Motors) in August 1972 purchased by the acquiring authority for £27,500, representing a rate of about £25.50 for 1,077 sq yds.

D

(2) The TA Centre, Goosecroft Road purchased by Samuel Properties in July 1972 for £75,000—a rate of £16 per sq yd.

(3) The Gas Works site east of Goosecroft Road purchased by the acquiring authority in September 1973 for £40,000, representing a rate of £14 per sq yd.

(4) May Day Yard, Thistle Street, purchased by the acquiring authority from British Rail for £40,000 in May 1971, representing a rate of £6.57 per sq yd overall or £20 per sq yd for first 40-ft zone. There was an area of 1.258 acres.

(5) Area adjacent to Menzies Motors with various owners. The total cost was £54,950 and the price represented an overall rate of £10 per sq yd.

E

(6) Littlewoods purchased a site in September 1970 for £169,000 at the corner of Port Street and Dumbarton Road. The area was 2,253 sq yds and the price represents £75 per sq yd. Mr Cargill considered that, relative to 1973 values the price could be equivalent to £85/£90 per sq yd.

(7) If the sale of Allan Park garage is analysed as a development site it represents £25.30 per sq yd.

We have considered all these transactions. The two highest sales have planning permission for intensive retail developments, and as we have determined that an intensive retail development would not be allowed on the Menzies Motors site we do not accept them as direct comparisons. The high rate brought out for the Plaza Cinema, which is not a prominent retail site, confirms the view expressed by Mr Cargill, and accepted in the record by the claimants, that the price paid was for its value as a bingo hall. The fact that the Post Office may use it for redevelopment purposes does not establish that the price is for a development site. Statutory authorities, for technical reasons, have often to purchase at values for existing use, including disturbance, which are more than the value as a cleared site available for development.

F

The rates produced by Mr Cargill tend to support his £20 per sq yd and in any event do not justify a rate of more than £35 per sq yd.

G

We therefore consider that the compensation of £130,857.46 is as high if not higher than the development value of the site and determine it at that figure. Including surveyor's fees the sum is £132,259.21. We meantime reserve the question of expenses.

W CLIBBETT LTD v AVON COUNTY COUNCIL

(REF/233/1974)

(DOUGLAS FRANK QC, President)

May 21 1975

H

Estates Gazette January 24 1976

(1975) 237 EG 271

Leasehold interest in factory premises of precision engineering firm at Bristol—Total extinction of business—Loss of goodwill claim—No concern with market value—President decides not to determine each of several items of disagreement between valuers—Awards sum which "in all the circumstances is reasonable"—Reiteration of statement that tribunal's previous decisions relevant only to arguments on law or procedure—"Assessment of compensation must be decided on, and only on, the evidence"

J

Mr W M Huntley (instructed by Hewetson & Co, of Yate, Bristol) appeared for the claimants; Mr W Evans (assistant solicitor of Avon County Council) for the acquiring authority.

Giving his decision, THE PRESIDENT said: This is a reference to determine the compensation payable for the compulsory purchase of the leasehold interest in factory premises at Park Road/Gilbert Road, Kingswood, Bristol, and in particular the amount of compensation for disturbance. The material facts are as follows:

K

A business was founded by a Mr Clibbett at Keynsham, near Bristol, in 1935 for manufacturing or processing by precision engineering. In 1947 that business was formed into a limited company. In 1959 the business was moved to the premises which are the subject of this reference. Generally, the nature of the work carried on is to convert steel bars to nuts, bolts, screws, rivets and so on, and to machine parts and equipment supplied by other manufacturers. In 1967 Mr Clibbett died but his two employees carried on the business with the help of his widow, who kept the books and did other administrative work, which took about 15 hours a week. The company had regular customers and did not have to seek work and in 1973 had a turnover of about £12,000.

L

No Alternative Accommodation Found

In 1971 a new lease for seven years was granted at a rent of £240 per annum. In 1972 the compulsory purchase order was made and was confirmed on September 1. Thereupon Mrs Clibbett, with professional help, did everything practicable in order to secure alternative accommodation, but their efforts proved fruitless. It is agreed that the claim should be treated on a total extinction basis. Notice to treat was served in November 1973, and on January 21 1974 possession was taken pursuant to a notice of entry, and the machines and stock were auctioned. It is agreed that the value of the lease is £275.

M

At all material times since the death of her husband Mrs Clibbett has owned 2,999 out of the 3,000 shares issued in the company. She drew as director's fees in effect the whole of the net profit of the company, in 1970 £3,724, in 1971 £2,629, in 1972 £3,000 and in 1973 £4,742.

Most of the items of claim, being for reimbursement of expenditure, but including the value of the lease, have been agreed. There are, however, two items out of 18 which are

A the subject of disagreement and for my determination. They are:

1. "Loss made in respect of rivets delivered to Parnell found to be over-sized and returned on day of sale therefore unable to be remachined and the amount claimed is £121."

As to that, the claimants were bound to, and did, give credit to Parnell and there is no doubt that a loss was incurred It was, however, conceded that the amount of £121 included some element of profit, and I find that the proper amount allowable is £105.

B 2. The second item of disagreement, and this is more difficult, is the loss of goodwill, which is another term for loss of expected profits.

The valuers for both parties adopted the traditional method of assessment, that is to arrive at an adjusted net profit averaged over a period of years and, after allowing for the proprietor's remuneration and interest on capital, to apply a multiplier.

Five Questions in Issue

The valuers, however, are in issue on five questions; they are:

C (1) Whether to average over three years, as Mr T P Maggs FRICS, the claimants' valuer, says, or four years, as proposed by Mr S R Hoggett ARICS, the district valuer;

(2) Whether certain professional charges should be taken into account, although having heard the evidence Mr Hoggett virtually conceded that they should not;

(3) Whether plant should be assumed to have depreciated or, on the other hand, as appears to have been the case, appreciated.

D (4) Whether allowance for proprietor's remuneration should be £800, Mr Maggs' assessment of the value of Mrs Clibbett's work, or £1,500, the amount paid to each of the other employees, as Mr Hoggett contends; and

(5) Whether the appropriate years' purchase is 4.5 (Mr Maggs), or 2.5 (Mr Hoggett).

Mr Maggs' calculation resulted in the goodwill valuation of £11,601, whereas Mr Hoggett's is £4,318. It is common ground that the amount of compensation should equal the monetary loss to the claimants and that I am not concerned with market value, that is to say what amount the goodwill E if offered in the market would have realised; indeed that would have been an impossible exercise.

Having regard to the well-known dictum of Lord Justice Scott in *Horn v Sunderland Corporation* [1941] 2 KB 26, namely that compensation should be such amount as, so far as money can do it, would put the claimant in the same position as if his land had not been taken from him, I find it difficult to understand how a capital sum of £4,318 can compensate for income averaging about £3,000 a year.

That difficulty was conceded by Mr Hoggett and by Mr Evans, the solicitor who appeared for the county council, there being no evidence of any similar suitable business F being available in the market in the vicinity. On the other hand it would not be right to award a sum to produce on investment or by annuity a similar annual amount as the past profits, for that would disregard, for example, the risks attached to the business and taxation and, therefore, would exceed the claimants' true loss.

"Robust Approach"

I was asked to look at certain previous decisions of the tribunal, but, as I have said and emphasised before, decisions are relevant only to arguments on law or procedure. The assessment of compensation must be decided on, and only G on the evidence. Useful though the respective valuers' calculations are in assisting me to reach a decision in this matter, I do not propose to determine each item in dispute. How can I, for example, say what is the appropriate years' purchase when on the evidence I can choose any number between 1.5 and 10 without any evidence which I should choose other than the *ipse dixit* of each of the valuers? I propose, therefore, to adopt a robust approach similar to that used by the courts in assessing general damages and to award a sum which, in my judgment, in all the circumstances is reasonable. It seems to me that my function is comparable to that of the courts in assessing damages for loss of future H earnings.

Doing the best I can, I think that the right award for loss of goodwill or profits here is £9,000. It follows, after having made the necessary adjustments to the claim, that the total amount of my award is £12,605 plus surveyor's fees based on scale 5a of the scale of fees of the Royal Institution of Chartered Surveyors.

It was conceded that the amount of my award is in excess of the amount in the sealed offer, and accordingly the acquiring authority will pay to the claimants their costs, which if not agreed shall be taxed by the Registrar of the J Lands Tribunal on the High Court Scale.

ROY v WESTMINSTER CITY COUNCIL

(Ref/163/1974)
V G WELLINGS QC)
October 17 1975

Estates Gazette January 31 1976
(1975) 237 EG 349 K

Compensation for goodwill of medical practice established for 19 years in freehold house/surgery in Paddington, London—Practice continued after acquisition for short period in subject premises and thereafter under bare licence in "unhygienic and unsuitable" temporary accommodation —Council contends goodwill worth nil—Held that bar on sale of goodwill under National Health Service Act 1946, s 35, does not necessarily have effect that goodwill of no value to claimant—Claimant "fully justified" in rejecting suitable alternative accommodation offered on "onerous" leasing terms with one condition seemingly "oppressive"— L No reasonable prospect of obtaining alternative freehold premises in neighbourhood—Extinction basis of compensation correct—Tribunal prefers "robust approach" adopted in "Clibbetts" case to application of YP multiplier "where there is no market"—Award of £11,000 against £28,770 claimed

The claimant appeared in person; Mr G R G Roots (instructed by the city solicitor, Westminster City Council) for the acquiring authority.

Giving his decision, MR WELLINGS said: In this reference the dispute relates to the amount of compensation to which M the claimant is entitled for the goodwill of his medical practice, which he formerly carried on at 92 St Stephen's Gardens, London W2, a freehold dwelling-house/surgery owned by the claimant. That property was one of 175 residential properties in the vicinity purchased by the acquiring authority under the City of Westminster (St Stephen's Gardens, Paddington) Compulsory Purchase Order 1967, which was confirmed without modification by the Minister of Housing and Local Government on September 2 1969. The claimant has been paid the sum of £14,500 for the property. The acquiring authority say that under rule (6) of section 5 of the Land Compensation Act 1961 the claimant is entitled to £14.85,

A the amount of his removal expenses, but that he is entitled to nothing for the goodwill of his medical practice, the balance of compensation to which he is entitled, therefore, being £14.85. The claimant claims that sum for his removal expenses but in addition claims £28,770.60 as compensation for goodwill of his practice. The total compensation which the claimant claims is therefore £28,785.45.

The primary facts are not in dispute. The claimant, Dr P Roy, carried on his medical practice in the subject property from the year 1954 until some date in July 1973. At all material times he carried on his practice alone and was registered as a practitioner under the National Health Service, his B remuneration being paid by the local family practitioner committee. Notice to treat was served on him on January 28 1972 and the acquiring authority formally took possession of the property on December 19 1972 but permitted Dr Roy to remain in occupation as a licensee continuing to carry on his practice until July 1973. The acquiring authority then found temporary accommodation for his practice at 102 Great Western Road, W2, about 200 yds, as the crow flies, from the subject property.

Ten Years to Establish Practice

C It took Dr Roy about 10 years before his practice became established. The numbers of patients registered with him on the dates mentioned below were respectively as follows:

October 14	1954	48	April 1	1969	2,233
June 9	1955	133	April 1	1970	2,137
October 11	1956	342	April 1	1971	2,133
February 11	1960	936	April 1	1972	2,179
July 28	1960	999	April 1	1973	2,016
December 9	1961	1,314	April 1	1974	1,918
April 1	1965	1,658	July 1	1974	1,957
April 1	1966	1,894	October 1	1974	1,968
April 1	1967	2,023	January 1	1975	1,902
D April 1	1968	2,167	April 1	1975	1,914

The net profit of his practice in each of the years mentioned below (each year ended on April 5) was:

1966	:	£1,590	1970	:	£2,718
1967	:	£1,793	1971	:	£3,732
1968	:	£2,236	1972	:	£3,203
1969	:	£2,798	1973	:	£3,183

Mr Guy Roots, on behalf of the acquiring authority, in contending that Dr Roy's goodwill was worth nil to him, said that a disturbance claim or other claim under rule (6) E is an element in the value of the land: see *Horn v Sunderland Corporation* [1941] 2 KB 26. That element represented the value to the owner of the goodwill above the market price of the land. Loss of income was not a head of claim. Dr Roy could not realise that element in the value of his land which reflected goodwill: see section 35 of the National Health Service Act 1946, which provides:

"(1) Where the name of any medical practitioner is . . . entered on any list of medical practitioners undertaking to provide general medical services, it shall be unlawful subsequently to sell the goodwill or any part of the goodwill of the medical practice of that medical practitioner . . .

F (2) any person who sells or buys the goodwill or any part of the goodwill of a medical practice which it is unlawful to sell by virtue of the last foregoing subsection shall be guilty of an offence. . . ."

The effect of these provisions, according to Mr Roots, was that the claimant cannot realise his loss and that therefore the goodwill of the practice has no value to him. The decision by the learned President in *W Clibbett Ltd v Avon County Council* (1975) 237 EG 271 (that in valuing goodwill one is not concerned with market value) was distinguishable for that reason. He, Mr Roots, would, having regard to section 35 (2), *supra*, of the Act of 1946, contend that the acquiring authority would commit an offence under that subsection if it paid compensation to Dr Roy for his goodwill. G

Decision on Point of Law

If Mr Roots is right on his point of law that is, of course, an end of the matter. It is convenient that I should straightaway proceed to decide that point of law.

In one sense the fact that it is an offence to sell or buy a National Health doctor's goodwill presupposes that the goodwill has some value in the market. On the assumption which, however, no doubt ought to be made, namely that Dr Roy and any prospective purchaser would not disobey the law, that value is nil. Mr Roots did not argue before me H that goodwill is incapable of existing in relation to a medical or other professional practice. His approach was to assume that Dr Roy in his practice possessed goodwill but that it had no value in the market or to him. I do not accept that merely because Dr Roy's goodwill has no market value it has no value to him. In *Remnant v London County Council* (1952) 160 EG 209 at 211 Mr J P Done FRICS, the member of the tribunal, expressed the point thus:

"It is not the market value of the claimant's local connection that is to be ascertained but the quantum of loss suffered by him in having to sacrifice potential business arising from an established practice. . . ." J

See also *Nielsen v Camden Borough Council* (1968) 206 EG 483 (a decision of Mr R C G Fennell FRICS). See further the decision of the learned President in *W Clibbett Ltd v Avon County Council, supra*, where he said:

"It is common ground that the amount of compensation should equal the monetary loss to the claimants and that I am not concerned with market value, that is to say what amount the goodwill if offered in the market would have realised. . . ."

In my opinion the fact that Dr Roy cannot sell his goodwill in the market does not necessarily have the effect that it is of no value to him. That resolves the first point, namely K the point of law put forward by Mr Roots.

The next question is: on what basis should compensation for Dr Roy's goodwill be assessed? Dr Roy and his accountant, Mr Baprasad Dé ACA, both of whom gave evidence before me, said that the basis should be one of total extinction of the practice. Mr Roots said that that basis ought not to be applied in the present case because, if Dr Roy had accepted the acquiring authority's offer of alternative accommodation, Dr Roy would have retained his patients and his goodwill. He had acted unreasonably in not accepting that offer and had failed properly to mitigate his damage. The offer concerned a ground-floor flat in a large block of flats belonging L to the acquiring authority. The flat in question is situate on Westbourne Park Road and is about 260 yds, as the crow flies, from the subject property. The acquiring authority has taken note of the alterations to the flat which would be necessary in order to enable Dr Roy to carry on his practice there and, at Dr Roy's behest, has carried out the following alterations:

(i) a wc has been provided in the bathroom, thus permitting the existing wc to be used exclusively by patients;

(ii) a hand wash basin in each of two rooms has been provided so as to enable them to be used respectively M as a consulting room and an examination room;

(iii) a ramp and a new doorway at the rear of the flat has been provided in order to give a separate access for patients.

I have seen the flat and there can be no doubt that it is suitable for a doctor's practice and that if Dr Roy's practice were carried on from there he could expect to retain his patients. He does not contest this at all. He objects to the terms upon which the flat is offered. The terms were set forth in a letter from the acquiring authority to Dr Roy dated July 5 1974.

They are as follows:

(a) Dr Roy to take a lease of the flat for a fixed term of 20 years; the rent for the first five years of the term should be £1,250 exclusive of all outgoings; there should be a rent review at the end of the fifth, 10th and 15th years; the rent to be payable quarterly in advance on the usual quarter days.

(b) The flat not to be used otherwise than for the purposes of a general medical practitioner's surgery.

(c) Change of use would not be permitted.

(d) The lessee would be required to obtain all necessary planning and other consents.

(e) The lessee to be responsible for interior repairs and decoration, repair and renewal of fixtures and fittings, maintenance of the garden area and of the walls and fences bounding that area, and the payment of all outgoings.

(f) The acquiring authority to be responsible for the maintenance of the structure and for the external decorations excluding the walls and fences bounding the garden area.

(g) Subject to the provisions of section 38 of the Landlord and Tenant Act 1954 the lessee shall not be entitled to compensation under sections 37 or 59 of that Act.

(h) The lessee to comply with the requirements of the Offices, Shops and Railway Premises Act 1963 or any statutory modification or re-enactment thereof.

(i) The lessee not to be permitted to assign the premises except to another general medical practitioner.

(j) Such other conditions as the acquiring authority's solicitor may require.

I agree with Dr Roy that the above terms offered to him were onerous in that:

(a) the acquiring authority was unwilling to seek approval of the district valuer to the rent demanded, notwithstanding that Dr Roy could not expect to be reimbursed under the National Health Service in respect of rent paid by him which had not been so approved;

(b) the acquiring authority, although demanding three rent reviews during the term of 20 years, would not agree to give to Dr Roy the right to break the term. He is aged 50 and has a wife and two sons. It is unreasonable to expect him (as the acquiring authority expect him) as original covenantee to be liable to pay a substantial rising rent and bear other substantial obligations until he is aged 70 even in circumstances where during the term he has retired from practice and/or lawfully assigned the premises.

(c) the terms as to assignment and user were unduly restricted; they would not permit him even to assign or sublet the flat for the purposes of residential user or user by a person carrying on some other profession.

(d) the last condition ("such other conditions as the council's solicitor may require") seems to go far beyond "the usual covenants" and to be oppressive.

For these reasons, it appears to me that Dr Roy was fully justified in refusing to accept the terms put forward with respect to the flat.

Mitigation of Loss

Mr Roots' next point was that Dr Roy had failed to mitigate his damage in that he had taken insufficient steps to ascertain whether suitable freehold premises in the neighbourhood were available. No evidence on that issue was offered by the acquiring authority. The only evidence thereon was by Dr Roy himself. Mr Roots said that the onus of proof on mitigation was on Dr Roy. In my opinion that submission is wrong: the onus is on the party alleging failure to mitigate: see *McGregor on Damages* (13th edition) p 149. Moreover, I am satisfied from Dr Roy's evidence that there is no reasonable prospect of his obtaining alternative freehold premises in the neighbourhood. A very large area has been demolished under the compulsory purchase order and the chance that such alternative freehold accommodation does exist is remote.

Mr S J Pook BSc (Lond) FRICS FRVA, deputy city valuer in the acquiring authority, in evidence said that if Mr Roots were wrong on his point of law and if Dr Roy was justified in refusing the acquiring authority's terms for the flat and if I were satisfied as to the non-availability of other accommodation, the proper basis of assessing compensation would be an extinction basis with no allowance being made on account of profits received at the premises in Great Western Road. In the light of the opinions which I have expressed with regard to Mr Roots' point of law and the question of mitigation of loss, that in my judgment is the correct basis of assessing compensation.

Notwithstanding that the claimant has been permitted by the acquiring authority to continue carrying on his practice, first for a short period in the subject premises and thereafter under a bare licence in Great Western Road (premises which are unhygienic and unsuitable as a doctor's surgery), the proper inference, in my opinion, is that looking at the matter as at any relevant date the practice would be extinguished within a short period. Having regard to his age, it is rather late in the day to expect him to set up in practice in a different area. I think that his contention that if he did so it would take him some 10 years in which to become established again is not an unreasonable estimate.

It was not argued before me that the claimant must give credit for profits received after the acquiring authority took possession of the subject premises. Moreover it appears to me that he need not give such credit, because the loss which is being compensated for is the loss of a capital asset and not a loss of income. The loss of income may be the measure of the loss of the capital asset but it is not the subject of the claim. If the extinction basis is not correct, then I have been given no assistance by either side in calculating the amount of compensation to which Dr Roy is entitled for loss of goodwill, at all events, if it be assumed that the amount of that compensation is not nil.

Evidence on Years' Purchase

Mr Dé said that in assessing compensation I should apply to the average net profit (£3,373.333 for the years 1971 to 1973) eight years' purchase. This was because the profits were tending to rise. He agreed that the normal multiplier at all events in the market was three years' purchase.

Mr Pook said that he had never known more than three years' purchase applied in the case of goodwill. That is the multiplier which he would apply in the event of the claim being assessed on the basis of extinction.

I have some doubt as to whether it is correct or useful to apply a multiplier of a number of years' purchase in a claim for compensation for goodwill in a case where it is clear that there is no market. I prefer the robust approach adopted by the learned President in *W Clibbett Ltd v Avon County Council, supra.* Adopting that approach I award to the claimant, Dr Roy, the sum of £11,000 by way of compensation for goodwill.

I do not accept that there was, prior to the compulsory purchase the potential for growth upon which Mr Dé relied. Dr Roy's accounts and the figures as to the numbers of his patients at particular times (to which I have referred) do not justify Mr Dé's views. No evidence was given showing that potential existed in the neighbourhood for an influx of new patients. Indeed the demolition which had taken place might suggest the contrary.

Finally, I do not accept Dr Roy's contention that he is entitled to put back into his net profit the amount deducted from fees paid to him on account of superannuation. Those deductions are common to all doctors. More particularly, however, they are of benefit to Dr Roy even if he should not continue in practice until pensionable age: see National Health Service (Superannuation) Regulations 1961, regulation 35.

The total amount to which Dr Roy is entitled after the addition of £14.85 for his removal expenses is £11,014.85. I award that sum to him.

The acquiring authority will pay to the claimant his costs of this reference in the sum of £70.20 in addition to the amount of the hearing fee as laid down in Schedule 2 to the Lands Tribunal Rules 1975.

TRAGETT v SURREY HEATH BOROUGH COUNCIL

(REF/114/1975)

(V G WELLINGS QC)

November 12 1975

Estates Gazette February 7 1976

(1975) 237 EG 423

Sports shop at Camberley held on lease with some four years unexpired—Inference that tenancy renewable under Landlord and Tenant Act 1954—Claim for losses on forced sale of stock and loss of goodwill—Mark-up of 7½ per cent applied to stock—Monopoly value of suppliers' agencies—Calculation of growth potential—Tribunal again adopts "robust approach" of Clibbetts case in award of compensation for goodwill

Mr R T Tragett appeared by leave of the tribunal on behalf of himself and his wife, Mrs A F Tragett, joint claimants; Mr J M Sullivan and Mr D R P Mole (instructed by the solicitor to the Surrey Heath Borough Council) for the acquiring authority.

Giving his decision, MR WELLINGS said: This is a reference to determine the amount of compensation to which the claimants are entitled by reason of a confirmed blight notice relating to their shop and premises known as Camberley Sports House, 1 Princess Way, Camberley, Surrey. In 1966 the claimants took an assignment of a lease dated April 10 1963 of the shop and premises. The lease granted a term of 14 years from March 25 1963 at the yearly rents of £600 for the first seven years and £700 for the remainder of the term. After taking the assignment the claimants fitted out the shop as a sports shop. They provided it with good-quality fittings and furnishings. It had ample storage space and changing rooms. The business carried on by them in the shop, though owned by them jointly, was in practice conducted by Mrs Tragett, her husband having another employment elsewhere.

Blight Notice Confirmed

By November 16 1972 the shop and the surrounding area had become subject to planning blight attributable to the published proposals of the Frimley and Camberley Urban District Council and the Surrey County Council to make that area a comprehensive development area, their purpose being the redevelopment of the central area of Camberley. On that date the claimants served on the urban district council a blight notice under the provisions of section 193 of the Town and Country Planning Act 1971. On November 28 1972 the clerk of the council wrote confirming that the blight notice had been accepted by the council.

The amount of compensation claimed before me by the claimants was, in addition to professional fees, a total of £29,700.06, made up as follows:

Stock, stock losses, fixtures and fittings (as at Jan 1 1973)	£8,531.06
Hardship and loss of earnings (up to the same date)	1,500.00
Hardship and loss of earnings since that date and for "goodwill and agencies"	16,069.00
Leasehold, 4½ years plus continuation of at least 7 years	3,600.00
	£29,700.06

According to the acquiring authority, the total compensation to which the claimants are entitled is £11,011, comprising £1,520 for the value of the leasehold interest, £6,041 for losses on a forced sale and £3,450 as compensation for goodwill.

The claimants gave and called no evidence in support of their head of claim for hardship and loss of earnings in the sum of £1,500. Even if it can be said that that head of claim is admissible at all (for example as an alternative claim to the claimants' claim for compensation for goodwill), the claimants have failed to show that they are entitled to that sum or any part of it. Accordingly I reject this part of their claim.

Before I consider the claims which remain (for the value of the leasehold interest, for losses on a forced sale and for loss of goodwill) I must decide two other issues which arose at the hearing.

The material date. The parties are in agreement that the acquiring authority has entered the subject premises and indeed has demolished them. They are, however, in dispute as to the date on which entry was effected. The claimants' case is that entry was effected on January 1 1973. They believe this (and it cannot be characterised as greater than a belief) because their agents, Chancellor & Sons, in their letter to the urban district council accompanying the blight notice, suggested that date for completion and because:

(1) On December 13 1972 Mr G A Scott of Chancellor & Sons told Mrs Tragett that he was going to hand over the keys of the shop to the council on January 1 1973;

(2) She handed the keys of the shop to him on that date.

After that date she did not return to the shop, but considerable amounts of stock remained in it for some months until they were sold. Neither Mr Scott nor any other representative of Chancellor & Sons was called as a witness before me and I am satisfied by the evidence of Mr Ronald Churchouse, formerly senior legal assistant to the clerk of the urban district council, that the keys of the shop were delivered to the council for the first time by a representative of Chancellor & Sons on July 27 1973 and that entry into the premises was effected by the council on that date. That date, accordingly, in my judgment, is the material date for the purposes of assessment of compensation.

Security of Tenure. It is the contention of the claimants that the urban district council had given a firm undertaking to traders displaced by the redevelopment scheme to relocate them in alternative accommodation in the new shopping centre when built but that that undertaking had not been fulfilled notwithstanding the erection of that new centre. The only proof of the terms of the undertaking offered by the claimants consisted in the fact that the urban district council had unequivocally accepted the blight notice and that the blight notice, in a schedule, described the claimants' lease as "4½ years unexpired at £700 pa exclusive with unfulfilled undertaking to rehouse by Frimley and Camberley UDC." I doubt whether the council's acceptance of the blight notice can fairly be said to be an acceptance of every fact stated in it and more particularly facts which are unrelated to the entitlement of the server to serve a blight notice. The undertaking as described in the blight notice is, however, unenforceable and not capable of valuation: it does not specify the alternative premises nor the rent nor the term of the lease or other interest to be enjoyed under the undertaking. It was proved to my satisfaction on behalf of the acquiring authority that the actual undertaking given by the urban district council was that in the event of displacement of traders becoming necessary " the council would use their best endeavours to assist in the satisfactory relocation of businesses in so far as the proposals for the comprehensive development permit." I can think of other possible reasons for rejecting this part of the claimants' case (such as the *Pointe Gourde* principle), but I will not go into them. Suffice it to say that I reject this part of the claim.

Application of 1954 Act, s 47

The claimants' position under Part II of the Landlord and Tenant Act 1954 is rather more promising. It is common ground that section 47 of the Land Compensation Act 1973 applies in the present case. Subsection (1) of that section provides that where a tenancy to which Part II of the Act of 1954 applies is compulsorily acquired, the right of the tenant to apply under Part II of that Act for the grant of a new tenancy is to be taken into account in assessing the compensation payable by the acquiring authority; " and in assessing that compensation it shall be assumed that neither the acquiring authority nor any other authority possessing compulsory purchase powers have acquired or propose to acquire any interest in the land."

The tenancy created by the lease dated April 10 1963 was undoubtedly a tenancy to which Part II of the Act of 1954 applied. By virtue of section 24 of the Act it could not have been brought to an end except in the manner permitted by the Act. Upon an application to the court for the grant of a new tenancy duly made, the claimants would *automatically* have been entitled to the grant of a new tenancy for a term not exceeding 14 years from the coming to an end of the original tenancy at the market rent and otherwise on the terms of the lease dated April 10 1963 or other terms approved by the court, unless the landlord succeeded in establishing to the satisfaction of the court any of the grounds mentioned in section 30 (1) of the Act. Of those grounds there is in the circumstances of the present case only one which could possibly be relevant: that contained in paragraph (f), which provides:

> "That on the termination of the current tenancy the landlord intends to demolish or reconstruct the premises comprised in the holding or a substantial part of those premises or to carry out substantial work of construction on the holding or part thereof and that he could not reasonably do so without obtaining possession of the holding."

New Tenancy of " At Least Three Years "

In his evidence before me, the district valuer, Mr S J Rance ARICS, said that in the absence of the redevelopment scheme the subject premises and neighbouring properties would have been ripe for redevelopment. He made that statement notwithstanding that he had not seen the subject premises before they were demolished. That statement represents the only evidence which I received of this particular matter. I will assume, without deciding, that it was correct. Nevertheless I heard no evidence as to what the intentions, immediate or otherwise, of the landlords were and in the circumstances and having regard to the assumption required to be made by section 47 of the Act of 1973, I think that I am entitled to infer that after the expiration on March 25 1977 of the tenancy created by the 1963 lease, a new tenancy would have been granted for a term of years which could reasonably have been expected to be at least three years with the undoubted possibility that it might be considerably longer. I draw that inference.

The value of the interest acquired. The district valuer's evidence under this head of claim was unchallenged. He said that as at the date of entry, July 27 1973, the open market rent of the subject premises was £1,500 pa That represented a profit rent of £800. At 10 per cent and 3 per cent (tax adjusted) for 3¾ years the appropriate multiplier was 1.90 years' purchase. £800 × 1.90 = £1,520. There was no additional value in the fact that there was security of tenure under the Act of 1954 or in the undertaking given by the council because the rent payable under any new tenancy, whether under the Act of 1954 or otherwise, would be at the open market rent. The profit rent would therefore come to an end on March 25 1977. The value of the leasehold interest was in the profit rent. No purchaser would pay anything extra for the possibility that a profit rent might arise during a new tenancy granted under the Act of 1954. The proper place to take security of tenure into account was under the head of compensation for goodwill. That was where he had taken it into account. I accept his evidence on this part of the case and accordingly, in my judgment, the compensation to which the claimants are entitled for their interest acquired is £1,520.

Compensation for losses on forced sale. A sale of some of the claimants' stock took place at reduced prices prior to their closing down the business on December 31 1972. It has throughout been agreed that £2,667 represents the compensation to which the claimants are entitled in respect of their loss on that sale. On January 11 1973, Hollingsworth, a firm of valuers experienced in valuing stock in the sports trade, in a written valuation, valued the fixtures and fittings and the remainder of the stock in the shop. They were instructed by the claimants' agents, Chancellor & Sons. Hollingsworth's valuation was accepted by the district valuer, who had no direct experience of the valuation of sports equipment. The valuation was as between a willing buyer and a willing seller and was £640.60 in respect of the fixtures and fittings and £3,602.64 in respect of the stock = £4,243.24 in total. Subsequently the district valuer and Mr Scott of Chancellor & Sons agreed that the sum of £3,602.64 should be increased by 7½ per cent in order to ascertain the retail value of the stock, thus producing a total valuation of £4,243.24 + £270.19 = £4,513.43. When the stock, the subject of the valuation, was sold at auction it realised £1,351.06 only, representing a loss, as compared with £4,513, of £3,162. As I have stated, the claimants' claim for losses on a forced sale amounts to £8,531.06. It is plain from their letter dated June 7 1975 to the acquiring authority that they arrived at that sum simply by adding together the sums of £2,667, £4,513 and £1,351.06. In effect they were claiming title to the sum of £1,351.06 twice, once from the purchaser and once from the acquiring authority. On the figures their true loss was £2,667 + £4,513 − £1,351.06 = £5,828.94. However, at the hearing the claimants sought to justify their claim for £8,531.06 in a different way; they alleged that the mark-up in the sports equipment trade was 50 per cent and not 7½ per cent. If that were so the higher mark-up would justify not £8,531.06 but £7,360: that is to say: £2,667 + £4,243 + £1,801 (50 per cent of £3,602) − £1,351.

It has not been proved to my satisfaction that the mark-up in the sports equipment trade is 50 per cent. I see no reason why I should not apply the rate of 7½ per cent agreed between the district valuer and the claimants' agents. I propose to award the sums offered by the district valuer: that is to say, £2,667 for losses incurred before December 31 1972, £3,162 (£4,513 − £1,351) for losses incurred thereafter, plus £212 (this item not originally claimed by the claimants) for expenditure on rates and rent for a three-month period to clear stock after closing, making £6,041 in all.

Compensation for loss of goodwill. The claimants' accounts for the years 1969, 1970 and 1971 disclose *inter alia* the following figures:

Year ended December 31	*Gross turnover*	*Net profit*
1969	£20,949	£ 421
1970	£22,307	£1,542
1971	£28,490	£1,647

Until they closed the business at the end of 1972, the claimants had the benefit of sole agencies for the sale in Camberley of sports equipment manufactured by 27 suppliers, including Dunlop, Slazengers and other well-known names. Shortly before the claimants took over the business in 1966, Mrs Tragett, at the suggestion of some of the suppliers whose agents the claimants later became, was apprenticed to the previous owner of the business for a period of three months in order to familiarise herself with the business. It would seem that in 1971 and 1972 the business suffered some damage (I do not know how much) to its turnover from two or three competing businesses which had set up in shops in Camberley. If the claimants' business had continued after the end of 1972, it is, I think, doubtful whether this competition would have continued to damage the claimants' business, because the shops in question did not have the benefit of any of the sole agencies to which I have referred and which accordingly gave the claimants' business a monopoly value. Indeed, in the event, notwithstanding the closure of the claimants' business, only one of the competing businesses now remains extant.

Accountant's "Novel" Assessment

The claimants, in support of their claim for £16,069 under this head, called their accountant, Mr I W Fotheringham TD MA FCA. His method of assessment was novel. It was as follows:

(1) The actual gross turnover for the years 1969, 1970 and 1971 represented a growth rate of 16.6 per cent compound.

(2) He took the actual gross turnover of £20,949 for 1969, applied to it the rate of 16.6 per cent compound, thus producing an adjusted figure for 1970 of £24,427; he made a similar adjustment to that figure and to the figures obtained for each of the years following until, by this process, a figure of £45,150 for the year 1974 was obtained.

(3) He adjusted the actual net profit for each of the years 1969, 1970 and 1971 in order to reflect what he thought was the true net profit, and the adjusted figures for those years became £328, £544 and £1,370 respectively.

(4) By the process of extrapolation or doubling up, the expected net profit for the years 1972, 1973 and 1974 would be £2,624, £5,248 and £10,496 respectively. That represented a total loss of net profit of £18,368.

(5) In order to find the value of the goodwill as at January 1 1973, the net profit envisaged for the years 1973 and 1974 ought to be discounted by 10 per cent, producing a final total of £16,069 as being the loss of net profit attributable to the deemed compulsory purchase order.

(6) However, that was a computed figure and the true compensation for loss of goodwill lay somewhere between the figure of £3,450 supported by the district valuer and that figure of £16,069 but rather nearer to the latter figure than to the former. If he were pressed, he would give it as his opinion that the compensation ought to be £12,000.

Mr Fotheringham further expressed the opinion that if the tribunal were to apply what he called the normal basis of assessment, that is to say, were to apply a number of years' purchase to the average net profits over three years, he would take the average (£1,203) of the actual net profits for the years 1969, 1970 and 1971, and multiply that average by eight years' purchase. That would of course result in compensation of £9,624. So high a multiplier would be necessary in order to take account of the exceptional growth in this business. He would not, however, resort to this normal method unless forced to do so.

The district valuer's approach was a variant of the method described by Mr Fotheringham as normal. The district valuer first adjusted the net profit as shown by the accounts for the years 1969-1971 by putting back into the net profit amounts shown in the accounts as expenditure in respect of proprietor's salary and amortisation and insurance (in part) which the district valuer thought had been wrongly deducted. He also adjusted the accounts against the claimants in order to take account of the profit rent and interest on capital not shown in the actual accounts. In this way he adjusted the net profit for the three years to £903, £961 and £1,244 respectively. He, however, did not take the average (£1,036) of these figures. In order to allow for the upward trend in the business he took a figure, £1,150, nearer to the net profit (£1,244) for the year 1971. He applied three years' purchase to £1,150 = £3,450.

Mr Rance said that the figure of three years' purchase was sufficient to take account both of the monopoly value of the agencies from the suppliers of sports equipment and of security of tenure enjoyed by the claimants under the Landlord and Tenant Act 1954. He said that there was no relationship between the number of years' purchase applied in relation to the value of goodwill and the return on his investment which a purchaser would expect: that is to say, that where a purchaser offers three years' purchase he is not saying that he is in fact expecting a return on his investment of 33⅓ per cent. Mr Rance agreed that in some cases the value in the open market of goodwill represents the minimum value to the claimant. Thus it would not be right to say that merely because in a particular case the goodwill had no market value, a claimant whose property had been compulsorily acquired should receive no compensation for goodwill. In the present case he thought that four years' purchase would be too much.

Mr Jeremy Sullivan of counsel, on behalf of the acquiring authority, expressly acknowledged that the tribunal had power if it thought it right so to do, as it were, to pluck a figure out of the air; see the decision of the learned President in *W Clibbett Ltd v Avon County Council* (1975) 237 EG 271.

Tribunal's Finding on Goodwill

I reject the approach of Mr Fotheringham as being artificial and having no relevance which I can detect to any question which I have to decide. I accept the adjustments to net profit made by the district valuer and his starting figure of £1,150. I do not think, however, that, from the point of view of the value of the goodwill to the claimants, the district valuer's valuation adequately reflects the growth potential of the business, the monopoly value of the agencies, or the security of tenure to which the claimants were entitled under Part II of the Act of 1954. It may be that I take a more optimistic view of the extent of that security of tenure than did the district valuer. I have received no evidence of the multiplier which a purchaser of this type of business would expect to pay in the open market. The district valuer's evidence in respect of the multiplier is based on his general experience rather than particular experience of this trade. In all the circumstances, I have decided to apply the approach, the robust approach, of the learned President in *Clibbett's* case. See also *Roy v Westminster City Council* (1975) 237 EG 349 (my decision). I am nevertheless grateful to the district valuer for his evidence, because it assisted me to direct my mind to the bracket in which my finding should lie. The compensation to which the claimants are entitled for loss of goodwill is in my judgment £4,500.

I award to the claimants £1,520 + £6,041 + £4,500 = £12,061 in the aggregate, together with their surveyors' fees in accordance with scale 5 of the scales of the Royal Institution of Chartered Surveyors and their proper legal costs (if any) incurred before the date of the reference herein.

Having read the decision in this matter and having then opened the sealed offer of compensation lodged by the acquiring authority, I find that the amount thereof is less than the amount of my award; accordingly the acquiring

A authority will pay to the claimants their costs of this reference, such costs, if not agreed, to be taxed by the Registrar of the Lands Tribunal on scale 4 of the county court scales of costs.

BUCKINGHAM STREET INVESTMENTS LTD v GREATER LONDON COUNCIL

(REF/24/1975)
(V G WELLINGS QC)
November 12 1975

B

Estates Gazette February 14 1976
(1975) 237 EG 503

Freehold houses in Camberwell, London, subject to controlled statutory tenancies—Tenants rehoused by acquiring authority after notice of entry—Houses padlocked and secured a week after tenants vacated, but no overt act by authority of entry or taking possession meantime—Right to possession held to vest in claimant owners until date of padlocking—Compensation to be assessed on vacant-possession basis.

C Mr J M Sullivan and Mr D R P Mole (instructed by Arnold & Co, of Watford) appeared for the claimants; Mr J C Harper and Mr E R E Caws (instructed by the director of legal services of the Greater London Council) for the acquiring authority.

Giving his decision, MR WELLINGS said: This is a reference to determine the amount of compensation to which the claimants are entitled for the compulsory acquisition of their dwelling-houses 23 and 25 Westhall Road, London SE5. Valuations are agreed, the only question in the case being whether compensation ought to be assessed on the basis of a sale of the claimants' freehold interests with vacant D possession or on that of a sale thereof subject to sitting tenants. The compulsory purchase order was made under section 90 of the Education Act 1944, the purpose of the acquisition being the provision of a new secondary school.

Certain facts were agreed in writing between the parties as follows:

(1) 23 and 25 Westhall Road, London SE5, were included in the Greater London (Archbishop Michael Ramsey C of E Secondary School, Southwark) Compulsory Purchase Order 1970, made by the acquiring authority on April 28 1970 under the provisions of the Education E Act 1944 and confirmed by the Secretary of State for Education and Science on August 20 1970.

(2) A notice to treat in respect of both properties was served upon the claimants on October 22 1970.

(3) The claimants' interest in both properties at the date of the notice to treat was freehold, subject in each case to a controlled statutory tenancy of the whole of the property.

(4) At the date of the notice to treat No 23 was let on a controlled statutory tenancy to Mr F W Foot at a weekly rent of £2.61 and No 25 similarly to Mr B H F Flain at a weekly rent of the same amount.

(5) A formal notice of claim dated April 13 1971 was sent on behalf of the claimants to the acquiring authority.

(6) Notice of entry was served upon the claimants and occupiers of both properties on January 14 1971 authorising the acquiring authority to enter on and take possession of the properties at any time after the expiration of a period of 14 days from the date of service of the notice.

(7) The acquiring authority granted tenancies of housing accommodation elsewhere to Mr F W Foot on June 28 1971 and to Mr B H Flain on May 31 1971 and G they vacated their respective properties, Nos 23 and 25, on those dates. This is not a case to which section 50 (2) of the Land Compensation Act 1973 applies.

(8) Demolition of both properties commenced on November 23 1971 and was completed on December 8 1971.

(9) The full compulsory purchase value of the freehold interest on a sitting-tenant basis is £1,500 for each property.

(10) The full compulsory purchase value of the freehold interest with vacant possession is £3,000 in respect of each property.

H It will have been observed that the statement of agreed facts does not expressly purport to specify the date on which possession of the subject properties was taken by the acquiring authority.

Mr Sullivan, counsel for the claimants, called no evidence. He was content to rely on the inference which he sought to draw from the statement of agreed facts, namely, that the acquiring authority took possession of the subject properties on the date (November 23 1971) when demolition of them commenced. If it be right to draw that inference, there was in each case an interval of over four months after the occupying tenants vacated and before the acquiring authority J took possession. Mr Harper, counsel for the acquiring authority, called Miss Jean Kathleen Harding, administrative officer in the acquiring authority's housing department. She produced a record, in longhand, which on its face shows that 23 Westhall Road was padlocked and secured by an agent of the acquiring authority on July 5 1971 and that No 25 was similarly padlocked and secured on June 7 1971. I know of no reason why those entries should not be accepted as being correct and I accordingly find the facts to be in accordance with them.

Miss Harding's evidence also showed that in making offers of alternative accommodation to the occupying tenants of the K subject properties the acquiring authority considered their wishes; that had they objected to the alternative accommodation first offered to them, other alternative accommodation would have been offered; and that the process would have continued until they voluntarily accepted alternative accommodation which they regarded as suitable to them. There was no evidence whatsoever that the acquiring authority's attitude in respect of alternative accommodation for the occupying tenants was ever other than persuasive. More particularly there was no evidence that the acquiring authority in its dealings with the occupying tenants so much as mentioned to them its statutory powers of eviction under section L 13 of the Compulsory Purchase Act 1965.

Earlier Decisions Considered

In these circumstances Mr Sullivan contended that the case was indistinguishable from the decisions of the tribunal in *Banham v London Borough of Hackney* (1971) 217 EG 202; *Bradford Property Trust v Hertfordshire County Council* (1973) 229 EG 1226; *Metcalfe v Basildon Development Corporation* (1974) 230 EG 1140; *Midland Bank Trust Co v* M *London Borough of Lewisham* (1975) 235 EG 59. On any basis the occupying tenants, he said, had vacated their houses before the acquiring authority took possession. The date on which that authority took possession was at the earliest the date on which the houses were padlocked and secured and at the latest the date when demolition commenced.

Mr Harper submitted that the acquiring authority took possession of the subject houses on the date when the occupying tenants were rehoused. Although no steps were in fact taken under section 13 of the Act of 1965, the power given by that section lay in the background. That distinguished

A the present case from the cases cited by Mr Sullivan. There were other distinguishing factors:

(1) In the *Banham* case the owner himself managed to deliver vacant possession to the acquiring authority without the intervention of any notice of entry;

(2) In the *Bradford* case the acquiring authority again took vacant possession of the premises, the occupying tenant having been rehoused by another authority before service of the notice of entry;

(3) In *Metcalfe's* case the occupying tenants were rehoused by the acquiring authority before expiration of their B notice of entry; the acquiring authority therefore took vacant possession;

(4) In the *Midland Bank* case the facts were very similar to those in *Metcalfe's* case;

(5) In the present case the acquiring authority took possession following the expiration of the notice of entry;

(6) Their acts were factually sufficient because they were done in the exercise of statutory powers and in a situation which they had created themselves rather than one which had been inherited and could not be controlled.

C Mr Harper said that in the present case the process of obtaining possession had been a continuous process, beginning with rehousing of the occupying tenants and ending with padlocking and securing the premises. He also put the matter in another way: when the occupying tenants were rehoused, they were in effect dispossessed by the acquiring authority; that dispossession was confirmed by the act of padlocking and securing the premises. He relied upon the decision of Kilner Brown J in *Harris v Birkenhead Corporation* [1975] 1 All ER 1001. In that case after expiration of the acquiring authority's notice of entry, the occupying tenant quitted the premises, preferring to make her own arrangements for D alternative accommodation rather than accept an offer of alternative accommodation from the acquiring authority. After she had left, the property was severely damaged by vandals and became dangerous, at all events to small children. One small child entered the property, fell from a window and received grievous injuries. I need not go into all the facts of that case nor into all the matters there considered. It is sufficient to say that the learned judge held that the acquiring authority, in that case, although it had not taken physical possession of the property prior to the accident, was an occupier within the meaning of the Occupiers' Liability Act 1957 and was liable in damages to the child. Mr Harper E said that the learned judge, without expressly saying so, appeared to regard the acquiring authority as having taken possession from the moment when the occupying tenant left and before they took any physical step to secure the premises.

I agree with Mr Sullivan that Kilner Brown J in *Harris's* case was considering matters very different from those which I have to decide. He was concerned with the meaning of the word "occupier" in a statute in which it is regarded as "simply a convenient word to denote a person who had a sufficient degree of control over premises to put him under a duty of care towards those who came lawfully on to the premises": see per Lord Denning in *Wheat v E Lacon & Co* F [1966] 1 All ER 582 at 593. See also 594 where he said: "If a person has any degree of control over the state of the premises it is enough." Much the same conclusion is to be derived, I think, from the other speeches in the House of Lords.

No Evidence of Keys Handed Over

I do not accept Mr Harper's submissions, able and well sustained though they were. There is no evidence that when the tenants in the present cases were rehoused the keys were handed to the acquiring authority. In my opinion the proper inference is that upon rehousing and vacation by the tenants of the properties, the legal possession (or right to possession) G of the properties vested in the claimants as owners: the reason for that is that where no one else is in possession, possession follows title: see per Lord Denning in *Newcastle City Council v Royal Newcastle Hospital* [1959] 1 All ER 734 at 736; see also *Burson v Wantage Rural District Council* (1974) 229 EG 1600 (my decision). Moreover, some overt act on the part of the acquiring authority, however small, was required in order to "enter on and take possession of" the subject properties (the process authorised by section 11 of the Act of 1965 and the notice of entry served under that section). No servant or agent of the acquiring authority set foot, according to the material before me, in the subject properties prior H to the dates on which they were padlocked and secured. The proper finding which I should make is, in my judgment, that those were the dates on which they took possession. On those dates and for one week previously each property had been vacant. In the circumstances the correct basis on which to assess compensation is to value the freehold interest as being with vacant possession.

For these reasons I award to the claimants the sum of £6,000 together with their surveyor's fees in accordance with scale 5 (a) of the scales of the Royal Institution of Chartered Surveyors together with their proper legal costs incurred prior J to the date of the reference herein.

I ought to add that even if I had accepted Mr Harper's argument that possession was taken on the dates when the occupying tenants were rehoused, the question whether the vacant-possession basis ought to be applied would not necessarily have been decided in Mr Harper's favour but for a concession made for the purposes of the present case by Mr Sullivan.

The acquiring authority will pay the claimants their costs of this reference, such costs, if not agreed, to be taxed by the Registrar of the Lands Tribunal on the High Court scale.

K

BRADD v LONDON TRANSPORT EXECUTIVE

(Ref/129/1975)

(R C WALMSLEY FRICS)

December 8 1975

Estates Gazette February 21 1976

(1975) 237 EG 583

Tunnel for Victoria Underground line constructed beneath small, inner-terrace, owner-occupied house in Walthamstow—Claim for injurious affection arising from noise and vibration—Relevant date September 1968, when line opened—Both parties cite comparables of sales several years later of properties without tunnel beneath—Structure of subject house not affected and no injurious affection in today's climate, but "real local anxieties at the time"—Reduction of GV by 12 per cent on relevant date owing to presence and use of line "persuasive evidence"—Award of £350—Tribunal attaches no weight to exercise carried out by estate agents to "test the market"—Uncertainty in such cases whether "low" instead of "full" offer being sought in reality L M

Mr R W Belben (instructed by Romain Coleman & Co) appeared at the hearing for the claimant; Mr G Seward (instructed by Mr V J Moorfoot) for the compensating authority.

Giving his decision, Mr Walmsley said: During the late 1960s the London Transport Executive constructed their Victoria Line out to Walthamstow, a distance of some 10 miles. Service began on September 1 1968. The Victoria Line runs under some 1,300 properties, of which about 900 are residential. Arising out of all this vast construction work

A only four claims in respect of injurious affection by reason of noise and vibration were received, and one of these was later withdrawn. Two of the other three claims have already been referred to the Lands Tribunal for determination, and have been disposed of. In respect of one of these, which concerned a house in Gibson Square, N1, the tribunal, in June 1973, made a nil determination (*Pepys v London Transport Executive,* Ref/216/1971). And a nil determination was also made, in July 1975, in respect of the other referred claim, which concerned the Westbury Hotel in W1 (*Knott Hotels Co of London Ltd v London Transport Executive* (1975) 236 EG 64). The present reference relates to the fourth B and last of the claims received in respect of injurious affection. There can be no further such claims because time has expired for their lodgement.

The subject house is 7 Ickworth Park Road, Walthamstow, E17, and the claimant is the owner-occupier, Mr Frederick George Bradd. On April 14 1965, in consideration for a payment of £17, Mr Bradd entered into a grant of easement with the then London Transport Board enabling part of the subsoil beneath his house to be used for the purpose of constructing and using tunnels and other works. Two tunnels were duly constructed, one of which lies wholly underneath the house; the other tunnel lies partly beneath. There is C 58 ft of cover over the top of the two tunnels, and the actual rails are 68 ft below ground level.

"A Standard House with Standard Faults"

No 7 Ickworth Park Road is within easy walking distance of the new Blackhorse Road station on the Victoria Line. It is a small inner-terrace, two-storey house typical of a great number of houses built in Walthamstow at or around the turn of the century. Upstairs there are two bedrooms; downstairs there are two living-rooms, a kitchen, bathroom and wc. At the front is a small forecourt, and at the back a D small walled garden. The valuation witness for the claimant, Mr McHardy, described the property very fairly as "a pretty standard house, very common in Walthamstow . . . in a road in the poorer part . . . its structural condition in keeping with its age . . . a standard house with standard faults."

The claimant gave evidence himself, and evidence on his behalf was called from two witnesses: Mr R A McHardy FSVA, sole principal in the firm of McHardy & Sons of E17; and Mr M F A Tolley BSc CEng MIStructE ARICS, a member of the firm of Gooch & Wagstaff.

For the authority evidence was called from five witnesses: E Mr H E Levi FSVA FRVA, senior partner in the firm of Harold Levi & Co, of E17; Mr W E A Bull CVO FRICS, until 1974 senior partner in the firm of Vigers; Mr John Richards BSc CEng, assistant scientific adviser (physics and engineering) on the staff of the authority; Mr G A Sullivan FRICS, estate manager of the authority; and Mr D F L Fox, train services superintendent. Mr D J Goodwin FRICS, district valuer and valuation officer for Waltham Forest, gave evidence on subpoena.

The measure of injurious affection is agreed to be the difference, if any, between (i) the price that the subject house would have commanded in the open market on September 1 F 1968 with the tunnels of the Victoria Line underneath (*viz* its value in "the with-scheme world") and (ii) the price it would have attracted on the same date if the Victoria Line had never been constructed (*viz* its value in "the no-scheme world").

Comparables in 1970-71 and 1975

Both Mr McHardy and Mr Levi produced comparables, being sale prices of other terraced houses in Walthamstow in 1975 and in 1970-71 (but not in 1968). None of the cited comparables has a tunnel beneath; but it goes without saying that they all enjoy the benefit of the improved travel facilities G afforded by the Victoria Line. Mr McHardy gave it as his opinion (a) that the prices of houses such as 7 Ickworth Park Road had increased by, say, 20 per cent from September 1968 to 1970-71; (b) that from 1970-71 to 1975 there had been a further increase of a little more than 100 per cent; and (c) that 7 Ickworth Park Road, without any tunnels underneath, would have fetched £4,200 in 1970-71 and £9,500 in 1975. Mr Levi's opinion was (a) that between September 1968 and 1970-71 the general increase in house prices had indeed been about 20 per cent as stated by Mr McHardy, but for houses in Walthamstow within easy walking distance of a station on the Victoria Line there had been a greater H increase, say of 33 per cent; (b) that between 1970-71 and 1975 the general price increase had indeed been about 100 per cent as stated by Mr McHardy, but for houses in Walthamstow near one of the new stations there had been a greater increase, say of 125 per cent; and (c) that 7 Ickworth Park Road, without any tunnels underneath, would have fetched £4,000 in 1970-71 and £9,000 in 1975.

I have inspected the subject house and I have looked at the houses cited as comparables. A great deal of evidence had been given about all these during the course of the three-day hearing. In the light of the evidence and of my inspection I find myself in agreement with Mr Levi on his J figures of £4,000 and £9,000, and these two figures corroborate his opinion of the percentage increase in prices between 1970-71 and 1975. I adopt also Mr Levi's opinion as to the percentage increase in prices between September 1968 and 1970-71. The result is a finding that the value of 7 Ickworth Park Road has progressed as follows:

Value at September 1968	£3,000
Add 33%	1,000
Value at 1970-71	£4,000
Add 125%	5,000
Value at 1975	£9,000

K

The first figure above, of £3,000, having been derived from prices for houses without a tunnel beneath, represents the value of the subject house in September 1968 with the Victoria Line completed and in operation but leaving out of account the fact that the tunnels of the Underground had been constructed underneath the house. In as much as Mr Levi gave further evidence (which I also accept) that the beneficial effect of the Victoria Line on house prices in Walthamstow only became noticeable *after* 1968, I am able L to adopt the same figure as representing also the second of the two values which fall to be ascertained. I find that the price that 7 Ickworth Park Road would have commanded in the open market on September 1 1968 if the Victoria Line had never been constructed to be £3,000.

There remains to be ascertained the price that would have been obtained for 7 Ickworth Park Road if it had been sold on September 1 1968 just as it was. The claimant and his valuer Mr McHardy consider that a much reduced price would have had to be accepted: in 1975 terms Mr McHardy put the injurious affection at £1,000. Mr Levi, on the other hand, thought the presence of the tunnels underneath the M house made no difference at all to its value: there were, he said, not many such houses to sell; the demand for them, mainly from first-time buyers, was "enormous," and he would anticipate no difficulty in obtaining his figure of £9,000, being the price corresponding with the market for similar houses without any tunnel underneath. Mr Bull, also, was confident that the house would sell for as much as if there were no tunnels below.

Evidence of Noise and Vibration

The claimed reduction in market value is attributed principally to noise but, in less measure, also to vibration.

A The noise of the trains, as heard in the house, was described to me by various witnesses, in lay terms and in technical terms. I have also listened myself. I find the noise level in the house to be low. It is not such as would arrest conversation; it is less than that of passing aircraft, or of the self-starter of a car in the street outside; and it is less also than the normal noise from a television or radio set.

But, on the other hand, it is a noise which cannot be switched off, and it is heard for 19 hours out of the 24 (from 5.30 am to 12.40 am), at intervals of 4 to 4½ minutes during peak hours and at longer intervals off peak.

B When the Victoria Line was being constructed there was vibration, and various small defects occurred in the structure of the house; a payment of £25 was made by the authority to Mr Bradd in reimbursement of the cost of making good these defects. Since 1968 various further defects have appeared, but I am satisfied, on the evidence of Mr Richards (which was in fact corroborated by the evidence of Mr Tolley), that the structure has not at all been affected, nor is it likely to be affected, by vibration from the Victoria Line. I did not myself " feel " the trains.

C If I were deciding this case in the climate of 1975 I should have little difficulty in agreeing with the compensating authority that the claimant had failed to discharge the onus of establishing an injury measurable in money terms. However, the circumstances with which I am concerned are those of 1968, not those of 1975.

Exercise to " Test the Market "

D Before going back to 1968 I propose looking back only to 1970-71, because at that period Mr Bradd went to considerable trouble to set up his house for sale. According to a lodged bundle of correspondence, the claimant, in September 1970, engaged a local firm of estate agents, Beken & Stokes, to " test the market." That firm advertised the house each week over a period of about six months in their normal card advertisement in the local press; particulars were circularised to some 30 people on their waiting list, quoting a price of £4,500; and the property was subsequently offered to other applicants, making a total of about 50 persons. Arising from all this activity there resulted one offer of £3,625 from a Mr Keown who wrote to Mr Bradd on February 21 1971 as follows:

> " Thank you for showing us over your house. The place is suitable in size and very convenient for getting to and from work. However I do feel that the price asked is not realistic for two reasons. Firstly the rumbling of the trains every few minutes E from early morning till late night is not a very pleasant sound. Secondly the continued vibration must, in the long term, have a detrimental effect on the structure.
>
> I would for these reasons be willing to pay £3,625 for the property and would be pleased to hear your views on my offer."

The compensating authority, who had been kept fully informed of these happenings, declined to acknowledge that injurious affection had been thereby established.

F Mr Seward submitted that the above exercise was wholly unsatisfactory; any exercise to " test the market " would, in order to be valid, have to be undertaken strictly under control and with strict propriety. Counsel invited me to note *inter alia*:

1. That (on evidence given by Mr Bradd) the asking price of £4,500 had been a figure fixed by the claimant himself.
2. That the agents, after advertising and circularising, had left the claimant to continue the exercise unaided.
3. That no representative of Beken & Stokes had been called to give evidence.
4. That Mr Keown had not been called to prove his letter of February 21 1971.
5. G That although this letter had clearly invited a continuance of negotiations, Mr Bradd (on his own evidence) had not even replied to it.

H I accept Mr Seward's submission that no weight can be attached to the exercise described. With an exercise carried out as this one was, there must always remain some uncertainty as to whether, instead of seeking a *full* offer which might result in a sale, what was in reality being sought was a *low* offer in order to demonstrate a greater degree of injurious affection. In this connection it may be noted that the asking price fixed by Mr Bradd was £500 higher than I have found it should have been on the evidence of the 1970-71 comparables.

J I can now return to September 1968. It by no means follows that because no injurious affection has been established in today's climate, there was no injurious affection at the time when the Victoria Line had just been completed and was being opened for service. At that time there were real local anxieties. Mr Sullivan's evidence was that, when trains began to run, concern was expressed by a number of residents whose houses were over the tunnels or contiguous, and they consulted their MP. Mr Goodwin, valuation officer, produced extracts from his records, and these show that various residents, including Mr Bradd, lodged proposals for a reduction in their rating assessments. And where the house or flat was above, or was close to, a tunnel the local valuation court on the advice of the then valuation officer had made reductions. The assessment on 7 Ickworth Park Road was reduced, as from September 1 1968, from gross value £95 to gross value £83. (Mr Goodwin was called to attend and give evidence on the shortest of notice, and I expressed my appreciation to him at the hearing.)

Reduction in GV " Persuasive Evidence "

K I readily accept the point made by Mr Seward regarding this reduction in assessment, that the London Transport Executive had no say in the local valuation court decision, which could not be regarded as binding either on the authority or on the Lands Tribunal. Nevertheless I find this persuasive evidence. Here is the valuation officer, an independent and impartial valuer, looking at conditions on exactly the date which is relevant to my purpose, and taking the view that in rental terms the value of the subject house had been reduced some 12 per cent by the presence and use of the Victoria Line. It may be that the anxieties of 1968 have now proved groundless: Mr Sullivan said that the residents' L concern " died down fairly rapidly. People got used to it." And it may be that those anxieties always were irrational: but the market for two-bedroom terraced houses in Walthamstow is apt to be made—I apprehend—essentially by buyers and sellers who may accord less than proper respect to the informed opinions of experts. Thus, when it was put to Mr Bradd that the compensating authority's expert, Mr Richards, using a B&K piezoelectric accelerometer type 4334 magnetically clamped to an 8 lb steel plate, had established that the vibration in 7 Ickworth Park Road arising from the operation of trains on the Victoria Line was imperceptible, the claimant's only observation, regrettably, was:

M " You must be joking! I have a stone-deaf sister-in-law who feels the trains through her feet."

I take the view that the valuation officer, carrying out his duties of assessment in the actual climate of September 1968, was more likely to have correctly interpreted the feel of the market, albeit in rental terms, than is even the most able and practical surveyor valuing with hindsight in 1975. It is on this basis that I find the price that the subject house would have commanded in the open market on September 1 1968 with the Victoria Line tunnels underneath to be £350 less than the price it would have attracted on the same date if the Victoria Line had never been constructed.

A The compensating authority will pay the claimant compensation in the sum of £350 plus a surveyor's fee based on scale 5A of the scales of professional charges of the Royal Institution of Chartered Surveyors.

The compensating authority will also pay the claimant his costs of this reference, such costs, unless agreed, to be taxed by the Registrar of the Lands Tribunal on scale 3 of the County Court Scales of Costs, with discretion in relation to solicitors' charges for "preparing for trial" and as to counsel's fees.

The costs of the said Frederick George Bradd are to be taxed in accordance with the provisions of Schedule 2 to the B Legal Aid Act 1974 and shall be so taxed by the Registrar of the Lands Tribunal.

PERKINS v WEST WILTSHIRE DISTRICT COUNCIL (REF/113/1975)

(R C WALMSLEY FRICS)

December 11 1975

Estates Gazette February 28 1976

C **(1975) 237 EG 661**

Pasturage land at Trowbridge—Allocated as "white" land on approved town map and as public open space in draft review—Residential development refused—Two grounds of objection in counternotice: (1) no proposal to acquire, (ii) conditions of s 193 (1) (c) and (d) not complied with—Whether exercise of statutory powers of acquisition necessary for achievement of public-open-space allocation—Claimant professionally advised against making endeavours to sell for development—Acceptance of that advice precludes him from invoking s 193—Objection upheld on D second ground

Mr W D R Spens (instructed by Middleton & Upsall, of Warminster) appeared for the claimants; Mr C Clayton (instructed by solicitors to West Wiltshire District Council) for the local authority.

Giving his decision, MR WALMSLEY said: The question for determination in the present reference is whether an objection by the West Wiltshire District Council, under section 194 (1) of the Town and Country Planning Act 1971, to a blight notice served on behalf of the claimant, Mr David Ernest Perkins, under section 193, in respect of land comprising an E agricultural unit, part OS406 at Upper Studley, Trowbridge, in Wiltshire, is well founded.

No Statutory Status for Draft Review

The subject land is a level field of about 5.4 acres, owned and occupied by the claimant, who uses it for pasturage. On the Trowbridge town map, formally approved by the then Minister of Housing and Local Government in 1966, the land is shown as unallocated (or "white" land), indicating that its existing use was intended to remain for the most part undisturbed. The draft review of the Trowbridge town map, approved by the Wiltshire County Council also in 1966 (but F not submitted to the Minister and with no statutory status) and accepted by the then Trowbridge Urban District Council, shows the land as proposed public open space. The town map and draft review are policy documents reaffirmed and notified formally by the Wiltshire County Council to the West Wiltshire District Council in 1974 on local government reorganisation, in accordance with the development control scheme. The district council thus acts on the provisions of the draft review in considering development proposals submitted to it as district planning authority.

On December 2 1971 a firm of architects on behalf of the claimant submitted to the urban district council an application for residential development of the land in question. On May G 17 1972 this application was refused by the county council for the following four reasons:

1. The proposed development would be contrary to the provisions of the approved Trowbridge town map on which the site is unallocated and to the draft review of that plan on which it is allocated for public open space.

2. It is considered essential that the land be reserved for the latter purpose being the only suitable and adequate area of undeveloped land in the area of compact residential development to the north of the A361 Frome Road.

3. The development of the land would in itself result in further need for public open space and in a serious deficiency of land H for that purpose, creating an unsatisfactory overall development in this sector of the town.

4. There is no deficiency of land for residential development in Trowbridge which would warrant the development of this site for residential purposes.

An appeal against this refusal, by the claimant to the Secretary of State for the Environment, was later dismissed. The department's decision letter, dated May 14 1974, includes the following sentences:

"Though the draft review of the Trowbridge town map, on which the site has been allocated for public open space, has not J been approved, it is considered that there is clearly a need for more public open space, particularly in the southern half of the town. Despite the fact that the former Trowbridge Urban District Council declined to purchase the appeal site when it was offered to them for public open space, it does not follow that it should be released for another purpose. It appears suitable for public open space and was considered to be so by the local planning authority."

On January 22 1975 the claimant served on the district council his blight notice requiring the authority to purchase his interest. The authority's counter-notice is dated March 14 1975 and it sets out two grounds on which objection is taken. One ground, under section 194 (2) (b), is "that the council K do not propose to acquire any part of the affected area, in the exercise of any relevant powers." The other ground, under section 194 (2) (g), is "that the conditions specified in paragraphs (c) and (d) of section 193 (1) of the Town and Country Planning Act 1971 are not fulfilled." I will take these in turn.

Section 194 (2) (*b*). In respect of this ground, section 195 (3) requires that the authority's objection is not to be upheld unless it is shown to the satisfaction of the tribunal that the objection is well founded.

Mr Cedric Clayton, appearing for the authority, sought to discharge this onus of proof by calling evidence from Mr L Frank Archer MICE MAPHI, Director of Environmental Services, West Wiltshire District Council, and from Mr Stephen Blades BA DipTP MRTPI, Senior Development Control Officer, West Wiltshire District Council.

Mr Archer was concise. What he had to say was:

"The acquisition of land by the council for leisure and recreation purposes lies in the first instance with its environmental committee. I am responsible for presenting to that committee details of estimated capital costs and revenue budget implications of such acquisitions. I confirm that no provision for the purchase of this land has been made in current estimates and I know of no proposals to make such provision in any future estimates." M

Mr Blades spoke to planning policy and to the planning history of the subject land, and he concluded his evidence-in-chief by saying:

"In the planning context the council as district planning authority undoubtedly has the objective in view of utilisation of the site for public open space purposes. When and how this is to be achieved remains to be determined. In the meantime the long-existing agricultural use of the land can continue."

It was put to Mr Blades in cross-examination that the planning history made clear the authority's policy that the subject land should become publicly owned, and that this

A policy in itself indicated an intention to acquire—and he agreed that such an intention was "implicit."

On the above evidence of his two witnesses, Mr Clayton submitted that Mr Blades had been speaking in the capacity of a planner; the position, however, was that any proposal to acquire would have to come not from "the planning people" but from "the environmental people." It was not necessary, for achievement of a public-open-space allocation, that the authority should itself exercise statutory powers of acquisition; other bodies and other persons, including local residents, were able, if they wished, to obtain grants for the purchase and provision of a public open space. In support B of this argument Mr Clayton relied on *Bolton Corporation v Owen* (1961) 181 EG 269.

"Bolton" Case Distinguished

With respect, I do not think the *Bolton* case helps the authority in the present reference; unlike in the *Bolton* case, the onus of proof here is not on the claimant but on the authority—a distinction which Lord Evershed MR referred to (in a passage at p 271) as being "important." If section 194 (2) (b) had been the only ground to be considered, the authority's objection could not be sustained—I have not been C satisfied that, on this ground, it is well founded.

There remains the second ground of objection, to which I now turn.

Section 194 (2) (*g*). This ground puts in question whether the following conditions for the service of a blight notice are fulfilled:

"193 (1). Where the whole part of . . . (an) agricultural unit is comprised in land of any of the specified descriptions, and a person claims that:

(a) he is entitled to an interest in that . . . unit; and . . .

D (c) he has made reasonable endeavours to sell that interest; and

(d) in consequence of the fact that the . . . unit . . . was . . . comprised in land of any of the specified descriptions, he has been unable to sell that interest except at a price substantially lower than that for which it might reasonably have been expected to sell if no part of the . . . unit were . . . comprised in such land. . . ."

Mr W D R Spens, appearing for the claimant, sought to discharge the onus of proof (that the above conditions had been fulfilled) by calling evidence from Mr Perkins himself and from Mr G J Cook FRICS and Mr D A Brown FRICS.

Mr Perkins said:

E "On November 12 1974 I telephoned Mr Godfrey Cook of Hartnell Taylor & Cook of Bristol and instructed him in the disposal of the land. He came to see me on December 19 1974 and we discussed the whole planning position. He advised me that it would be quite impossible to sell the land at a figure approaching residential development prices in view of the proposals of the local authority to use the land as a public open space. Relying on this advice and on the previous history of the land I made no further attempts to sell it as this would obviously have been a complete waste of time."

"Pointless Exercise"

F Mr Cook, who is a senior partner in the firm of Hartnell Taylor & Cook, with nine offices in the Bristol/Bath area, confirmed the above dates; he said he had arranged a meeting with Mr Perkins' architect and the local planning authority to try to find a compromise between residential development and public open space use; in this he had failed, and he had advised the claimant that it would be a pointless exercise to try to sell; advertising would be a waste of time and money because any developer would lose interest on being informed that the land was allocated as a public open space.

Mr Brown, who is a partner in the firm of Tilley & Culverwell, of Trowbridge, told me he would have given similar

G advice; if Mr Perkins had tried to sell the subject land in 1974-75 he could not have expected to obtain a price much above its agricultural value.

Having regard to the nature of the above evidence I accept the submission made for the authority that the conditions specified in paragraphs (c) and (d) of section 193 (1) are not fulfilled. What the claimant has done is to take professional advice as to the likely outcome if he were to try to sell his land for residential development. But this is not what the statute requires; the requirement is that the claimant "has made reasonable endeavours to sell" his interest *simpliciter*. He must have put his interest in the H market and he must have tried to sell it just as it is, with all its advantages and disadvantages. In the present reference the claimant's interest has a number of disadvantages including:

(i) that the subject land is, as it happens, landlocked. A prerequisite of any residential development would be the coming to terms with the local authority for the provision of access over other land which is at present in public ownership.

(ii) that there are, apparently, certain drainage difficulties.

J (iii) that in the draft review of the town map the land is shown as proposed public open space.

The simple fact is that the claimant was advised against making endeavours to sell his interest; he accepted that advice, and his acceptance precludes him, in the present reference, from invoking the provisions of section 193.

On this second ground it has not been shown to my satisfaction that the authority's objection is not well founded, and accordingly I uphold the objection.

One further matter should be mentioned. On July 31 1975 the authority served a further notice, stated to be under section 194 (1), objecting to the blight notice. This fresh K counter-notice was similar to the earlier objection dated March 14 1975, save that it also included the ground, under section 194 (1) (a), that "no part of the agricultural unit to which the notice relates is comprised in land of any of the specified descriptions." At the hearing, counsel for the authority sought to develop this additional ground, but I ruled that the counter-notice dated July 31 1975 could not be the subject of the present reference because the document had not been served within the two-month period specified in section 194 (1).

The claimant will pay the costs of the authority, such costs failing agreement to be taxed by the Registrar of the Lands L Tribunal on scale 4 of the County Court Scales of Costs.

LEY v KENT COUNTY COUNCIL

(REF/221/1975)

(R C WALMSLEY FRICS)

January 19 1976

Estates Gazette March 6 1976

M **(1976) 237 EG 735**

Freehold detached house at Pembury, near Tunbridge Wells —Line of proposed road cuts across freehold right of way providing access to house and grounds—Objection that no part of hereditament qualifies for protection—Whether rights appurtenant "part of hereditament"—Meaning of "aggregate of the land"—Objection upheld

Mr R G Greenslade, solicitor, of Buss, Stone & Co, of Tunbridge Wells, appeared for the claimants; Mr C W V Hopkins, solicitor, of the Kent county solicitor's office, for the authority.

A Giving his decision, MR WALMSLEY said: This is a reference to determine whether the objection stated in a counternotice served by the Kent County Council ("the authority") on William Claude Ley and Barbara Mary Ley ("the claimants") under section 194 of the Town and Country Planning Act 1971, objecting to a blight notice served by the claimants under section 192 of the Act of 1971 requiring the authority to purchase the claimants' interest, is well founded.

The subject hereditament is described in the notice of reference as: "Freehold land and house (together with right-of-way) situate and being Little Crofters, 6 Tonbridge B Road, Pembury, Tunbridge Wells, Kent." No 6 Tonbridge Road is a substantial five-bedroom detached house lying just outside the village of Pembury, about five miles south-east of Tonbridge on the main Hastings-Tonbridge road (the A21). The authority are proposing to construct a road by-passing Pembury village. This new road (the B2015 by-pass) is intended to pass within about 50 yds of the claimants' house and, at the point where the proposed by-pass joins the A21, a new traffic roundabout is planned (also within about 50 yds of the house). The completion date for this road scheme is 1978.

C The claimants' blight notice, dated July 2 1975, stated *inter alia*:

(a) that the hereditament had been included in land falling within paragraph (e) of section 192 (1), being "land shown on plans approved by a resolution of a local highway authority as land comprised in the site of a highway as proposed to be constructed, improved or altered" by the authority.

(b) that since the date of the above resolution the claimants had made reasonable endeavours to sell their interest but had been unable to sell it except at a price substantially lower than that for which it might reasonably have been expected to sell if no part of the hereditament had been so included.

(c) that their interest qualified for protection because the D annual value of the hereditament did not exceed the prescribed limit of annual value and their interest was that of owner-occupiers.

The authority's counternotice, dated August 4 1975, objected to the blight notice on the ground that no part of the hereditament was comprised in any of the specified descriptions. No evidence was called by either party.

One Issue Outstanding

I was told that the authority had accepted (i) that because of the authority's highway proposals the claimants had been unable to sell except at a substantially reduced price; (ii) E that the hereditament fell within the prescribed limit of annual value; and (iii) that the claimants were owner-occupiers. There remained in fact only one issue outstanding between the parties.

It is necessary, in order to appreciate the nature of the issue which remains for determination by the tribunal, to describe the claimants' property in a little more detail. No 6 Tonbridge Road stands in grounds of about three-quarters of an acre, and this is the area edged red on their title plan as being the extent of their freehold ownership. The access to the house is by way of a further strip of land over which the claimants have a freehold right-of-way and which is F coloured green on their title plan. I shall refer to the two areas so coloured as "the freehold land" and "the freehold access" respectively. No part of the freehold land is within the land to be acquired for the proposed road, but the line of that road cuts right across the freehold access.

The question is whether, in these circumstances, the claimants' interest qualifies for protection. The relevant statutory provisions are to be found in sections 192, 207 and 290 of the Act.

Section 192 (3) provides that interests qualifying for protection are "interests in hereditaments or parts of hereditaments."

Section 207 (1) defines "hereditament" as meaning "the G aggregate of the land which forms the subject of a single entry in the valuation list for the time being in force for a rating area."

Section 290 (1) defines "land," in the context of Part IX of the Act, as meaning "any corporeal hereditament, including a building."

"Shell-Mex & BP v Langley" Cited

In the present case the "single entry in the valuation list" merely states "6 Tonbridge Road." The contention for the claimants is that the freehold access stands part of this H "hereditament." Mr Greenslade's first line of argument was that the claimants' circumstances fell squarely within the general ambit of the Act, which contemplated that where a person had property which was directly affected by a proposed highway and was unable to sell at a fair price, he could enforce acquisition forthwith instead of having to wait on the convenience of the authority. His second argument was related to the definition of "hereditament": the definition above does not speak of "*the land* which forms the subject of a single entry," it speaks of "*the aggregate of the land* which forms the subject of a single entry"; the inclusion of the words "the aggregate of" must, said Mr J Greenslade, enlarge the meaning of "the land," and a fair and proper interpretation of the phrase would be "the land, including any rights over other land which are appurtenant to and inseparable from the land." Such a meaning would accord with the decision of the Court of Appeal in *Shell-Mex & BP Ltd v Langley (VO)* (1962) 9 RRC 249, where it was held that in valuing a hereditament for rating purposes there must be taken into account any rights appurtenant to the letting and inseparable from it. Mr Greenslade developed a third argument by drawing an analogy with the legislation relating to agricultural tenancies: in the Agricultural Holdings Act 1948 "agricultural holding" is defined as meaning K "the aggregate of the agricultural land comprised in a contract of tenancy . . . ," whereas in the Agricultural Holdings Act 1908 "holding" had been defined as meaning "any parcel of land held by a tenant, which is either wholly agricultural or wholly pastoral, or in part agricultural and as to the residue pastoral . . ."; the effect of this change of wording had been considered by the courts in the case of *Dunn v Fidoe* [1950] 2 All ER 685 and *Howkins v Jardine* [1951] 1 All ER 320, reviewing the earlier case of *Re Lancaster and Macnamara* [1918] 2 KB 472.

Mr Hopkins, for the authority, accepted none of the above L arguments. The Act of 1971 was, he said, not drawn in general terms; those interests which qualified for protection were defined with particularity. The decision in the *Shell-Mex* case made it clear that rights "appurtenant to" a hereditament were not "part of" that hereditament, and the decision therefore afforded no support to the present claimants. As to the phrase "the aggregate of," said Mr Hopkins, whatever may have been the intention of Parliament it would not have been an intention to enhance or enlarge the meaning of the controlling noun "land" to include something *other* than land.

M For my part, and for the following reasons, I find myself unconvinced by the arguments for the claimants.

(1) In section 290 (1) the full definition of the word "land" is:

"any corporeal hereditament, including a building, and, in relation to the acquisition of land under Part VI of this Act, includes any interest in or right over land."

The claimants, it seems to me, are asking that the phrase "aggregate of the land" as used in Part IX of the Act, should be treated as being analogous to the word "land" as used in Part VI. I see no justification for this.

(2) One dictionary meaning of "aggregate" is "sum

A total," from which it seems at least possible that the phrase "the aggregate of the land" may mean simply "the sum total of the land."

(3) To the extent that the decision in the *Shell-Mex* case assists at all, I regard it as tending to support the authority rather than the claimants.

(4) As regards Mr Greenslade's drawn analogy with the wording of the Agricultural Holdings Acts, I have examined the legislation referred to and read the cited cases but have derived no assistance therefrom.

Having considered the matters set out in the notice served B by the claimants and the grounds of objection specified in the counternotice, it has not been shown to my satisfaction that the objection is not well founded. Accordingly I uphold the objection.

The claimants will pay the authority their costs of the reference, such costs unless agreed to be taxed by the Registrar of the Lands Tribunal on scale 4 of the County Court Scales of Costs.

C

HOVERINGHAM GRAVELS LTD v CHILTERN DISTRICT COUNCIL

SAME v BUCKINGHAMSHIRE COUNTY COUNCIL

(REF/7/1974; REF/149/1974)

(DOUGLAS FRANK QC, President, and J R LAIRD FRICS)

December 18 1975 and January 30 1976

Estates Gazette March 13 1976

(1976) 237 EG 811

Purchase notice—Long, narrow enclosure at Chesham, Bucks, zoned as green belt—Notice confirmed in two parts: county council to purchase front land affected by road improve- D **ment line and district council to purchase back land—Three section 17 certificates for residential development issued: (1) by Minister for whole site three years before purchase notice; (2) by county council for front land after purchase notice; and (3) by county council for back land later but out of time—Certificate for whole site held extant but irrelevant—Front and back land to be valued separately—Improbability of marriage—Strong expectation by prospective purchaser of back land of consent for residential development, but would have to await road-widening or buy access over front land—Tribunal awards £21,915** E **agreed valuation of front land (0.652-acre) on basis of apportionment of residential value of whole site but considers it wrong in law—£13,750 awarded for back land (1.810 acres), valuation being discounted by half to allow for risk of non-grant of planning permission and access difficulties**

Mr I D L Glidewell QC and Mr D M W Barnes (instructed by Rollit, Farrell & Bladon, of Hull) appeared at the hearing for the claimants; Mr W J Glover QC and Mr M B Horton (instructed by Blaser, Mills & Lewis, of Chesham) for Chiltern District Council; and Mr D W Keene (instructed by the solicitor to Buckinghamshire County Council) for the county F council.

Giving their decision, THE TRIBUNAL said: The land the subject of these references (ordered to be heard together) is a long, narrow enclosure of some 2.462 acres (hereafter called the "said enclosure") situate about one mile south-east of the centre of Chesham, Bucks, bounded on the north by Latimer Road (the B485) and on the south by a narrow stream known as the River Chess; south of the river is low-lying land zoned in the Chesham Town Map (approved in 1961) as "Min gravel" and situated within the "Metropolitan and local green belt"; to the south of that is some housing development. The said enclosure has a frontage of about 950 ft to Latimer Road, an average depth of about G 120 ft and forms part of a larger area of disused land, in all some 17 acres, of which about 15 acres carry planning permission for the extraction of sand and gravel.

Résumé of Agreed Facts

When the claimants purchased the 17 acres in early 1964 it was shown in the Chesham Town Map as "Metropolitan and local green belt"; that zoning is still in force today. The claimants took the land with the burden of a tenancy granted by their predecessors in title to W Priest & Son (Chesham) Ltd (hereafter called "Priests"), the proprietors of a garage, showrooms and workshop situated on the north H side of Latimer Road, opposite the south-eastern end of the Latimer Road frontage of the said enclosure. The land comprised in that tenancy was about a quarter of an acre. The tenancy was granted for the purpose of permitting Priests to park staff cars, but they used it for purposes other than that for which the tenancy and planning permission had been granted. In consequence, about 1966, the Bucks County Council sought to bring the unauthorised use to an end by enforcement procedure.

Following Priests' failure to secure planning permission for the quarter-acre site of which they were tenants, the claimants applied to the Buckinghamshire County Council J for planning permission to develop the said enclosure of 2.462 acres with six pairs of semi-detached houses and garages; the application was refused on the grounds that:

"The site is included within the green belt on the approved Chesham Town Map and that the proposal if permitted would result in an undesirable form of sporadic development along this class II road which would be detrimental to the character of the Chess Valley at this point."

The claimants then appealed to the Minister of Housing and Local Government who, on February 28 1968, dismissed the appeal on green belt grounds. K

The claimants then sought to open negotiations with the Chesham Urban District Council for the sale to them of the said enclosure for development by the district council for its declared intention of "a public open space not later than five years hence." The district council expressed a willingness to negotiate but declined to accept that the proper basis for compensation was the value of the said enclosure for residential development. The claimants then applied to the county council for a certificate of appropriate alternative development pursuant to section 17 of the Land Compensation Act 1961 (hereafter referred to as the "section 17 certificate"), specifying residential development at a density L of six houses to the acre.

Certificate for Six Dwellings per Acre

On July 22 1969 the county council issued a negative certificate, so the claimants appealed to the Minister of Housing and Local Government under section 18 of the Act. On January 19 1970 the Minister allowed the claimants' appeal, and on June 15 1970 he issued a new certificate (the "June 1970 certificate") certifying that planning permission might reasonably have been expected to be granted for residential development of the said enclosure at a density not exceeding six dwellings per acre subject to access being from M a service road constructed within the site with access from Latimer Road at one point only and subject to the usual conditions as to approval of detailed siting, design, etc, of buildings and means of access.

There then followed two further applications for planning permission (one by Priests and the other by the claimants), refusals by the county council, and appeals to the Minister (which were dismissed). In July 1971 the claimants submitted their second application to the county council for planning permission for the erection of six pairs of semi-detached houses on the said enclosure; as a decision was not given by the county council within the statutory period of two months the claimants appealed to the Secretary of State for the Environment. That appeal was dismissed on June 29 1972, principally for the reason that a road improvement line proposed by the county council would affect the frontage of the appeal site and that the remaining or back area was inadequate for six pairs of semi-detached houses. On the following day the claimants served on the district council a notice under section 180 of the Town and Country Planning Act 1971 requiring the district council to purchase their interest in the said enclosure.

By notice dated September 18 1972, the district council informed the claimants that the council was unwilling to comply with the purchase notice and had not found any other authority to take its place and stated:

"4. The reason why the council are not willing to comply with the purchase notice is that in their view another local authority (viz the Bucks County Council) who have not expressed willingness to comply with the notice should be substituted as acquiring authority for all the land in accordance with the provisions of section 183 (4) of the Town and Country Planning Act 1971 on the following grounds:

(i) The purchase notice arises as a direct result of exercise of the functions of the Bucks County Council.

(ii) The paramount reason for the decision of the Secretary of State to refuse planning permission for the development proposed in Planning Application C/293/70 as indicated in paragraph 3 of the decision letter dated June 29 1972 is the intention of the Bucks County Council as highway authority for Latimer Road (B485) to widen and improve the road for which purpose part of the land will be required. The extent of the land required has not yet been finally determined and the work to improve the road is not in a programme of work.

(iii) A large part of the land will be needed for a road improvement scheme, the cost of which will be a charge initially against the Bucks County Council. In the circumstances it would be manifestly wrong for the burden of purchasing the land to be imposed now upon the ratepayers of the Urban District of Chesham and carried by them indefinitely until decisions are taken by the county council upon the design and timing of the improvement scheme.

(iv) On the premise that the land is shown on the Chesham Town Map as metropolitan and local green belt and it is the policy of the county council as local planning authority to preserve and enhance the green belt, the financial burden of implementing that policy should lie with the county council.

(v) Were it not for the possibility of obtaining financial assistance the urban district council would not contemplate the purchase of the land for public open space because of the existence of a conditional certificate of appropriate alternative development for residential development at a density not exceeding six houses to the acre. Such certificate would make the cost of acquisition of the land an intolerable burden on the urban district council unless a substantial grant or grants were forthcoming to assist them. Whilst the possibility of obtaining such a grant or grants in respect of the use of the land as public open space is being pursued, no such grant is yet assured.

(vi) Having regard to the probable ultimate use of the land it is contended that the Secretary of State should consider it expedient if he confirms the notice to modify it by substituting the Bucks County Council (being another local authority) for the urban district council as purchasing authority in accordance with subsection (4) of section 183 of the Act."

When the Secretary of State intimated his proposed decision on the purchase notice on its being referred to him, he afforded an opportunity to the parties concerned to make representations to him at a hearing to be arranged, if necessary, for that purpose. The claimants by their solicitors replied that they did not wish to be heard further with regard to the matter. The county council also decided that they did not wish to be heard and would raise no objection to the Secretary of State confirming the purchase notice so far as they were concerned. But their view was that the purchase notice should not be confirmed so far as the district council was concerned but that the Secretary of State should issue a direction, pursuant to subsection (3) of section 183 of the Town and Country Planning Act 1971, that planning permission for residential development of that part of the site should be granted in the event of an application being made in that behalf. The district council expressed a desire to be heard.

The hearing took place on February 7 1973, and the Secretary of State's decision following that hearing was given on March 13 1973. The Secretary of State confirmed the purchase notice with the modification that the county council as local highway authority should be substituted as purchasing authority in respect of the land hatched green on the plan accompanying the letter, and confirmed the purchase notice without modification in respect of the remainder of the land shown hatched red on the plan. The effect of that decision was to require the county council to purchase 0.652 of an acre of road frontage land and the district council to purchase 1.810 acres of back land, the two parts making up the 2.462 acres being the said enclosure. We shall refer to those areas as the "front land" and the "back land" respectively.

Deemed Service of Separate Notices to Treat

The Secretary of State further directed that separate notices to treat should have been deemed to have been served respectively by the county council (front land) and district council (back land) on the date of his decision. Negotiations for the assessment of compensation commenced, but the district council asserted that the "June 1970 certificate" did not apply to the back land with which they were concerned, and that in the absence of a valid certificate, the basis of compensation must be a purpose appropriate to the green belt. On December 18 1973 the claimants then made two applications for section 17 certificates to the county council, for the front land and the back land respectively, but without prejudice to the claimants' contention that the "June 1970 certificate" still applied. On December 20 1973 the claimants gave notice of reference to the Lands Tribunal in respect of the back land, but notice of reference in respect of the front land was not given by them until August 5 1974. On February 21 1974 the county council issued a section 17 certificate (the "February 1974 certificate") that if it were not proposed to acquire the front land, planning permission might reasonably have been expected to be granted for residential development of it at a density not exceeding six houses to the acre subject to the following conditions:

"(1) the approval of the county council shall be obtained to the siting, design and external appearance of the buildings and the means of access thereto before the development is commenced.

(2) application for approval of all the matters referred to in the last preceding condition must be made not later than the expiration of three years from today.

(3) the development to which this permission relates must be begun not later than whichever is the later of the following dates:
(i) the expiration of five years from today; or
(ii) the expiration of two years from the final approval of the matters referred to in condition (1) above or, in the case of approval on different dates, the final approval of the last such matter to be approved.

(4) the development hereby permitted shall be carried out comprehensively with the land at the rear of the site shown hatched black on the plan forming part of the application."

On April 1 1974 the functions of the Chesham Urban District Council were taken over by the Chiltern District Council; we shall continue to refer to them as the "district council."

On November 8 1974 the county council purported to issue a section 17 certificate (the "November 1974 certificate") in respect of the back land that planning permission might reasonably have been expected to be granted for residential

A development comprising not more than five dwelling-houses and ancillary domestic garage accommodation subject to the following conditions:

"(1) the approval of the county council shall be obtained to the siting, design and external appearances of the buildings and the means of access thereto before the development is commenced.

(2) application for approval of all the matters referred to in the last preceding condition must be made not later than the expiration of three years from today.

(3) the development to which this permission relates must be begun not later than whichever is the later of the following dates:

(i) the expiration of five years from today; or

B (ii) the expiration of two years from the final approval of the matters referred to in condition (1) above, or, in the case of approval on different dates, the final approval of the last such matter to be approved.

(4) the site shall be served by a private drive laid out to the requirements of the highway authority with one point of access from the classified road towards the eastern end of the site."

The district council appealed to the Secretary of State against the issue of that certificate but later withdrew the appeal. It is common ground that the certificate was given out of time and has no effect under section 17 of the Act.

That completes our résumé of the agreed facts and we now C turn to consider the issues between the parties.

Date of Valuation

We can deal with this dispute very shortly. Mr Glidewell argued in favour of the date of valuation being the date of the reference to the Lands Tribunal; Mr Glover for the district council and Mr Keene for the county council argued that the correct date is the date of the hearing (ie July 1975). We deferred the giving of our decisions in these cases pending the decision of the Court of Appeal in *W & S (Long Eaton) Ltd v Derbyshire County Council* (1975) 236 EG 726, knowing that it was pending. Judgment in that case has now D been given and that requires us to hold that the correct date is the date of the award which, as a practical matter, means the last day of hearing before the Lands Tribunal.

The "June 1970 Certificate"

Mr Glidewell argued that the interest in the land which is the basis for a certificate under section 17 of the 1961 Act is the interest of the claimants in the whole of the said enclosure. The fact that the interest is now to be acquired in two parts by two local authorities does not lessen the validity of the "June 1970 certificate." It is analogous to E a planning permission for a larger area in relation to part of that area. Hence, if planning permission for the whole 2.462 acres at six houses per acre had been granted and the county council had taken the front land (0.652-acre), planning permission would remain for the back land (1.81 acres) at the density of six houses per acre: *F Lucas & Sons Ltd v Dorking and Horley RDC* (1964) 17 P & CR 111 and *Inglewood Investment Co Ltd v Minister of Housing and Local Government* (1961) 179 EG 497. There is nothing in the section which prevents a certificate for a larger area applying to a smaller area. The proper approach is to look at sections 17 and 22 for the circumstances for which a certifi- F cate can be granted and section 14 for its effect. Alternatively he submitted, referring to section 14 (3), that the tribunal must consider what planning permission would have been granted for the back land; he then argued that a potential purchaser would regard residential planning permission as a high probability because he would know of the "June 1970 certificate," the district council's representation that planning permission should be granted, the "February 1974 certificate," and the "November 1974 certificate" (albeit out of time), which, although not binding, indicated the county council's attitude. Mr J C Rackham BSc (Est Man) FRICS, the valuer called by Mr Glidewell, supported that contention and said that a purchaser would pay £42,500 less £10,625 G (being 25 per cent discount), to allow for the fact that there is no consent but strong expectation that permission would be granted.

Mr Glover submitted that a certificate is only good for the acquisition for which it was granted and one given for the whole cannot relate to part. There are two acquisitions, so we have to determine the amount the claimants are entitled to receive from each local authority. He argued that under section 17 there must be a proposal to acquire, and one has to know the land to be acquired, and regard must be had to the circumstances existing at the date of the proposal H to acquire; *Jelson v Ministry of Housing and Local Government* [1970] 1 QB 243. There must be a proposal to acquire a particular interest in particular land, ie the land in question mentioned in subsections (3) and (4) of section 17. The "June 1970 certificate" applied to the relevant land (that is the 2.462 acres) and relevant interest but did not have regard to circumstances at the date of the deemed notice to treat of March 13 1973. It related to a different proposal and in part to a different purchaser, whereas section 17 applies to the totality. He referred to *Pilkington v Secretary of State for the Environment* [1974] 1 All ER 283 at 288. Arguing against Mr Glidewell's alternative submission, he J reviewed the facts and submitted that no purchaser would assume that planning permission would be likely to be granted and would be so uncertain that he would assume that no planning permission would be granted.

Mr Keene argued that the "June 1970 certificate" is extant and relevant. Nothing in the Act says certificates cease to operate when a further proposal to acquire is made. A section 17 certificate assumes that planning permission would have been granted at the time when it was issued and deals with circumstances at that date. He also referred to the *Inglewood Investments* case (*supra*). He said that the certificates are valid *pro tanto* and that that principle applies K whether a certificate is positive or negative. As to Mr Glidewell's alternative submission, the out-of-time certificate of November 1974 implies that the county council, as local planning authority, favoured development of the back land together with the front land.

Section 14 of Land Compensation Act 1961

We approach this matter by looking first at the statutory provisions. Section 14 (1) of the Land Compensation Act 1961 provides:

"For the purpose of assessing compensation in respect of any L compulsory acquisition, such one or more of the assumptions mentioned in sections 15 and 16 of this Act as are applicable to the relevant land or any part thereof shall be made in ascertaining the value of the relevant interest."

and then at section 15 (5):

"Where a certificate is issued under the provisions of Part III of this Act, it shall be assumed that any planning permission which, according to the certificate, might reasonably have been expected to be granted in respect of the relevant land or part thereof would be so granted, but, where any conditions are, in accordance with those provisions, specified in the certificate, only subject to those conditions and, if any future time is so specified, M only at that time."

Certificate Relates to Whole or Part

We find nothing in the statutes nor in the "June 1970 certificate" to support the proposition that the certificate is only good for the acquisition for which it was granted. Moreover, we do not consider that the acquiring authority's intention can be relevant. It may be that conditions could be imposed, or even be deemed by statute to be imposed, for the purpose of limiting the life of the deemed permission. However, that question does not arise here, nor was it canvassed. We accept that applications under section 17 must relate to a particular area of land, but we do not understand how the certificate can relate only to a particular interest in that land. Further, if the application includes land not to be acquired, then under subsection (8) of section 17, the certificate is to be limited to the land acquired. But, if notwithstanding that subsection, additional land is included in the certificate, it is to be disregarded, per Megaw J in the *Inglewood Investments* case (*supra*). We conclude, therefore, that the "June 1970 certificate" is extant.

The next question is whether the "June 1970 certificate" is effective as to part of the said enclosure. We say "effective" because in the *Inglewood Investments* case it was said that a certificate can relate to part. In our view, we should cease to look at the certificate as such, but rather to examine the planning permission deemed to have been granted by it as if it had been an actual planning permission, and to ask whether at the date of the deemed notices to treat to what extent, if at all, it could have been implemented. In that respect only is *Jelson's* case relevant. Having looked at each site separately, we find that each has the benefit of the same planning permission and we also find that the permission clearly contemplates one scheme of development. Then the only remaining question is whether at the date of the notice to treat anything had happened on the land to prevent the carrying-out of that scheme to which the answer must be "no." The fact that implementation would require single ownership and agreement between different owners is *nihil ad rem* and cannot invalidate a planning permission actual or deemed. However, as will transpire, this does not help the claimants. We now turn to the next issue as named by counsel.

Method of Valuation

Mr Glidewell submitted that if in truth one interest is being acquired, the correct method of assessing compensation is by one valuation followed by apportionment. On that basis he claimed on valuations as at the date of the notices of reference £124,000, apportioned as to the front land £32,860 and the back land £91,140. If treated as two acquisitions, he said, the fair thing to do is to value as a whole, otherwise it would be necessary to make provision for acquiring access and for awarding compensation for severance and injurious affection.

Mr Glover submitted that one must value what has happened, ie acquisitions by two different local authorities of two adjacent but different pieces of land. The sum of the values of the two areas is not necessarily the value of the whole, and therefore the sale price of the whole is an invalid approach. The correct method is to value the back land knowing that the front land would not be available as it was to be used for road widening and could not be married with it. Further, when valuing the front land there must be taken into account the impossibility of marriage of sites because of the proposed acquisition by the district council of the back land. He said that the right course is to value the back land in isolation on the basis of a use appropriate to a green belt at the date of the hearing and that the figure is £1,810, ie £1,000 per acre.

Mr Keene, supporting the county council's method, said that it assumes both parcels being sold separately to a single purchaser, possibly at simultaneous sales, for development as one entity. Even if there were two purchasers they would merge their interests. He referred to the decision of the tribunal in *Barstow and Others v Rothwell Urban District Council* (1970) 215 EG 879. He contended that where one owner serves a single purchase notice his compensation should not be reduced because part was to be acquired by another local authority. Adopting the claimants' basis of assessment, but substituting the date of hearing, the figure is £82,700, apportioned as to front land £21,915 and the back land £60,785.

We find this a very difficult case. On the one hand, one of the principles commonly derived from *Horn v Sunderland Corporation* [1941] 2 KB 26 implies that compensation should not be reduced by the fact of there being two local authorities. On the other hand, we think we are compelled to assume separate sales, in which case the hypothetical purchaser of each would know of the compulsory acquisition of the other and therefore the improbability of marriage. Further, we do not think that the principle in *Pointe Gourde Quarrying & Transport Co v Sub-Intendent of Crown Lands* [1947] AC 565 can be applied to an acquisition other than one in pursuance of the scheme concerned.

The planning permission deemed by the "June 1970 certificate" clearly permits and envisages a comprehensive scheme for the whole of the said enclosure of 2.462 acres, subject to the approval of the reserved matters. The local planning authority, confronted with an application showing only development on the back land, would be bound to say that that was not the development contemplated by the permission and further that it had become impossible to carry out the authorised development (see *Pilkington v Secretary of State for the Environment* (*supra*) and *Percy Bilton Industrial Properties Ltd v Secretary of State for the Environment* (1975) 237 EG 491). Further an application relating only to the front land would evoke a similar response. Our conclusion, therefore, is that having taken into account the "June 1970 certificate" it does not add to the amount of compensation, because in the hypothetical circumstances we have to assume that development could not be carried out.

We should perhaps mention again Mr Glidewell's reference to the case of *F Lucas & Sons Ltd v Dorking and Horley RDC* (*supra*). The facts in that case were rather special; it has not been followed since and we think it must be of limited authority. In any event, the *ratio decidendi* was that the planning permission granted was for the development of sites of land delineated upon a plan respectively and separately by the erection upon them of dwelling-houses to be occupied as such with those areas of land, and so that case clearly is distinguishable and is not inconsistent with the conclusion to which we have come, namely, that a permission for the comprehensive development of an area does not permit piecemeal development not forming part of a scheme for the whole area. Thus in the hypothetical circumstances we have to consider the "June 1970 certificate," although extant, is irrelevant, and the "front" land and the "back" land must be valued separately.

The Valuations

We must now deal with the alternative submission made by the claimants, namely, that but for the proposed acquisition a prospective purchaser would regard a residential planning permission as a high probability because he would know of the "June 1970 certificate," of the district council's representation that planning permission should be granted, and of the two certificates of 1974, albeit that one of them was out of time. We are persuaded by those arguments, and in view of the evidence we think that a prospective purchaser of the back land would have a strong expectation that permission would be granted. However, he would have to make some allowance for risks and, more important, he would also have to take into account that either he would have to wait until Latimer Road were widened in order to gain access to the back land or, alternatively, he would have to buy a means of access over the front land. The out-of-time "November 1974 certificate" was for residential development by not more than five detached dwelling-houses and we start our valuation on that assumption.

Mr Rackham, the valuer called by the claimants, produced what he described as four comparable sale transactions in support of his valuation which he had based upon a price of £8,500 per plot at the time of the references (in fact

A August 1974). We have examined that schedule and the sites concerned and are of the opinion that three of them do not assist us. That is because two of the sales were of single plots for single houses situated in much better residential areas two years before the relevant date here, that is July 1975. Another was of a much larger area (11.5 acres) near Amersham, with planning permission for 69 plots, some 2½ years prior to our relevant date. His fourth transaction, at Alma Road, Chesham, was also cited by the county valuer; it was a site of 1.29 acres with planning permission for 21 town houses, sold in April 1974 for £84,000, equivalent to only £4,000 per plot.

B The county valuer produced a schedule of what he called five comparable sales of which three were for much larger developments (21 and 36 houses), while the fourth was for land zoned industrial; two (at Tylers Green near High Wycombe and at Wendover) were some distance away from the said enclosure. We find that his fifth comparable provides by far the best evidence. It lies close to the claimants' land, is not too large in area nor in scale of development, and is not too distant in time. The transaction concerned 0.757 of an acre—the site of the old Black Horse public house, now demolished—in Latimer Road, Chesham, only about 300 yds west (nearer Chesham) from the claimants'

C land; it was purchased by the district council in September 1974 for £26,000, equivalent to £34,346 per acre or £3,250 per plot. We have seen this site, which is on rising ground and does not present the same difficulties of development as the claimants' land.

As to the said enclosure, we find that it is disused and overgrown with weeds, young trees and scrub. It is a low-lying damp site bounded on its south side by the River Chess, which is at about the same level; the site is well below Latimer Road at its western end where a 6-ft-high brick retaining wall some 100 yds long is necessary in order to support the highway, which is narrow, busy and noisy. On

D considerably higher ground on the opposite side of Latimer Road there are poor, old, terraced dwellings with slated roofs, looking directly down upon the said enclosure. Undoubtedly these drawbacks of situation and physical difficulties would not assist development and our inspection confirms that Mr Rackham overvalued the claimants' land. We therefore put his valuations on one side and now turn to our awards.

Tribunal's Awards

Ref/149/1974. 0.652-acre Front Land—Buckinghamshire

E *County Council*

This land is so narrow in depth that it could not be developed residentially save for purposes ancillary to the development of the back land, in particular to provide a means of access. The only valuation before us is that of Mr A L Bennett BA FRICS, the county valuer, and that amounts to £21,915 at the date of the hearing. It was arrived at by the apportionment by area of the residential value of the whole of the said enclosure at £35,000 per acre less an agreed allowance for pumping sewage. We think that method of valuation is wrong in law, because, in our opinion, separate sales must be assumed, and further, in the sale of

F the front land it must be assumed that as the back land is being acquired for green belt purposes under a different scheme, the "marriage" of the two areas could not be contemplated.

However, the only valuation left to us is the county valuer's £21,915, accepted and agreed to by the claimants, and so we cannot do other than award it. To that will be added a surveyor's fee based upon scale 5 (a) of the Scales of Professional Charges of the Royal Institution of Chartered Surveyors, and legal costs properly incurred from the date of the deemed notice to treat up to the date of the notice of reference. In so determining we have had regard to the

G "February 1974 certificate" under section 17 but having regard to the shallow depth of the front land coupled with the need for a building line so that the planning permission could only be exercised in the residential sense for ancillary purposes such as providing access and gardens to houses, but that it would not be practicable to erect houses on that front.

Having read the decision in this matter and having then opened the sealed offer of compensation lodged by the acquiring authority, we find that the amount thereof exceeds the amount of our award; accordingly the acquiring authority will pay the claimants their costs of this reference up to the

H date of the sealed offer and the claimants will pay the costs of the acquiring authority from that date such costs, unless agreed, to be taxed by the Registrar of the Lands Tribunal on the High Court Scale.

Ref/7/1974. 1.810 acres Back Land—Chiltern District Council

In our judgment the correct assumptions are (1) that there is a strong possibility that planning permission would be granted for five houses, (2) that there is no question of joint development with the front land, and (3) that the purchaser would either have to buy a means of access through the front land or await the widening of the highway. Hence we

J adopt the county valuer's figure of £5,500 per plot, which we derive from his lodged valuation, thus arriving at £27,500. We discount that amount by 50 per cent to allow for the risk of not being granted planning permission and, more particularly, for the difficulties of access. Thus our award is £13,750, to which will be added a surveyor's fee based upon scale 5 (a) of the Scales of Professional Charges of the Royal Institution of Chartered Surveyors and legal costs properly incurred from the date of the deemed notice to treat up to the date of the notice of reference.

The acquiring authority will pay to the claimants their costs of this reference such costs, unless agreed, to be taxed

K by the Registrar of the Lands Tribunal on the High Court scale.

ST PIER LTD v LAMBETH LONDON BOROUGH
(REF/82-84/1975)

(DOUGLAS FRANK QC, President, and J H EMLYN JONES FRICS)

January 22 1976

Estates Gazette March 20 1976

L **(1976) 237 EG 887**

Land in several separate parcels containing shops and dwelling-houses at Brixton—Claim on basis of development with offices, shops and flats at 3:1 plot ratio on an enlarged site including adjoining properties—Council's scheme carried out on the enlarged site and on additional land formerly owned by public authorities which private developer could not have acquired—Council planners' contention for 2:1 plot ratio influenced by knowledge that most satisfactory development incorporated public authority sites—Reasonable expectation in "no-scheme" world of development

M **with 3:1 plot ratio—"Pointe Gourde" doctrine—Diminution in value caused by council's scheme to be disregarded —Residual valuations "only valuation approach" before tribunal**

The Hon D M Trustram Eve QC and Mr J C Taylor (instructed by Knapp-Fishers) appeared for the claimants; Mr M H Spence (instructed by the chief solicitor of Lambeth London Borough) for the acquiring authority.

Giving their decision, THE TRIBUNAL said: These three references, which were heard together, are concerned with the compensation to be paid to the claimants, St Pier Ltd, arising from the acquisition by the acquiring authority, the London Borough of Lambeth, of their freehold interest in certain parcels of land in Brixton within the borough. The three references stem from three separate compulsory purchase orders, two made under Part III of the Housing Act 1957 and one under section 28 of the Town and Country Planning Act 1968 (now section 112 of the Town and Country Planning Act 1971). Notices to treat were served on August 10 and 11 1972 and entry was made by the acquiring authority on to the land covered by the two orders made under the Housing Act on November 15 1972. Entry on to the rest of the land was made as to part on July 2 1973 and as to the remainder on February 21 1974.

The parties are agreed that the land should be considered as a whole for the purpose of assessing compensation and that November 15 1972 is to be taken as the effective date of valuation of that whole. They are also agreed as to the value to be attributed—as part of that whole—to the two parcels of land acquired under the "planning" compulsory purchase order at the relevant dates in 1973 and 1974. It is further agreed that although the site value provisions contained in section 59 (2) of the 1957 Act would normally apply, they are overtaken in the present case by the ceiling provisions contained in the Second Schedule to the Land Compensation Act 1961. Both parties therefore put forward valuations on the assumption that the existing buildings are standing.

The acquiring authority's case is that the claimants' interest in the subject land and buildings is worth £69,500, being the standing value of the properties as they existed at the relevant dates.

The claimants accept that figure as correctly reflecting the existing use value of the property but contend that the land would be worth substantially more to a developer, who would anticipate a development of the subject land together with other land, and that such a development would be likely to obtain planning permission and would be economically feasible. Such a developer seeking to buy the claimants' interest in the land in the open market would be prepared to pay the sum of £200,000, and that, say the claimants, is the compensation to which they are entitled together with surveyor's and legal costs.

The subject properties lie behind and to the east of the Brixton Road and immediately to the north of Brixton Station Road, which itself runs along the north side of the Southern Electric Railway line leading to and from Victoria. Brixton Station on the Victoria Underground line is also nearby. The land comprised the following properties:

1. *Under the "planning" compulsory purchase order*

(i) 15, 17 and 19 Brixton Station Road: Three interconnecting shops occupied by a lessee trading as Brixton Pet Supply Co Ltd.

(ii) 35 and 45 Brixton Station Road: Two single-storey shop units occupied by the Station Road Furniture Stores Ltd.

(iii) 16 Popes Road: Also occupied by Station Road Furniture Stores Ltd and forming an integral part of 51 Brixton Station Road, which together with other adjoining properties was in the ownership of Station Road Furniture Stores Ltd.

2. *Under the Housing Act Order (A)*

(i) 16, 17, 18, 19, 20 Industry Terrace: Five terraced dwellinghouses, two of them semi-derelict.

(ii) 1, 2, 3, 4, 5 Alders Cottages: Five terraced dwellinghouses, one of them semi-derelict.

3. *Under the Housing Act Order (B)*

12, 13, 14, 15 Popes Road: Four terraced dwellinghouses (one end-of-terrace).

These are the subject properties, which in their existing state at the relevant dates had an agreed value of £69,500. In order to carry out any worthwhile development on the subject land it would, in the view of the claimants, be necessary for a developer to acquire other land in order to fill in the gaps and to present a site of reasonable size and shape. For this purpose the claimants' witnesses put forward a scheme involving the acquisition of the following properties:

(i) 2 and 4 Beehive Place (2 Beehive Place immediately adjoined 15, 17 and 19 Brixton Station Road). This property was owner-occupied and was used as a shop-restaurant known as Weston's Fish Restaurant.

(ii) 6 and 8 Beehive Place, the Volunteer Public House (immediately adjacent to 20 Industry Terrace).

(iii) 21 and 23 Brixton Station Road. Two single-storey shops used as a Continental grocer and delicatessen. The proprietor of the business was the freeholder of 21 and held 23 under a lease from the London Borough of Lambeth.

(iv) 25 to 33 (odd numbers), 37, 39, 41, 49 and 51 Brixton Station Road and 15a Popes Road, occupied by Station Road Furniture Stores Ltd. This company was also the freeholder of a passageway between 15 and 15a Popes Road and a stretch of land at the rear of 1, 2 and 3 Alders Cottages and a passageway at the rear of 35 to 45 Brixton Station Road.

(v) 43 Brixton Station Road. Pat's Café, held on lease from the London Borough of Lambeth.

(vi) 47 Brixton Station Road. Shop, owner-occupied, used for the sale of ladies' clothing.

(vii) 10 and 11 Popes Road. Two dwellinghouses forming part of the terrace known as 10-15 Popes Road. The London Borough of Lambeth is the freeholder and both properties were vacant at November 15 1972 in view of the authority's scheme of redevelopment.

The claimants' land enlarged by the addition of these further properties—the whole being referred to as "the green site"—provides the starting point for the valuation put forward by the claimants. The development envisaged assumes a mixed user of offices, shops and flats. The valuer called for the acquiring authority also put in a valuation assuming a similar mixed development on the same enlarged site but arrived at the conclusion that there was no possibility of producing a feasible scheme of development, mainly because he assumed much less floor space for office development.

The whole area has now been cleared and the acquiring authority is carrying out its own scheme of development, involving substantial office development and a recreation facilities centre with a swimming pool and sports hall. This development is being carried out on the enlarged area of land to which we have already referred—ie, the green site—together with a further area of land to the north, part of which was formerly in use as a depot by the Greater London Council and the remainder being in the occupation of London Electricity Board as a transformer station together with a number of dwellinghouses which the board had been holding for future expansion.

Compensation is to be assessed in accordance with rule (2) of section 5 of the Land Compensation Act 1961, that is to say "the amount which the land if sold in the open market by a willing seller might be expected to realise." Neither party suggested that any assumptions regarding planning permission were to be made under either section 15 or section 16 of the Act, nor does any question arise of a certificate of appropriate alternative development under section 17. The subject land is zoned under the approved development plan for the County of London for commerce,

A that is to say, broadly speaking, for warehousing and uses of a similar character. There was at the relevant date no planning permission in force in respect of the relevant land and there had been refusals of a number of applications made both by the claimants and by the owners of the Station Road Furniture Stores. Both parties proceeded on the assumption, however, that planning permission for some form of mixed development could reasonably have been expected, the issue between the parties being related to the scale of any such proposed development.

We are therefore faced with the problem of determining what the subject land might have been expected to realise B on a sale in the open market by a willing seller, taking into account its location, its physical characteristics, and the sort of development which a prospective purchaser might reasonably expect to be able to carry out having regard to the climate of opinion in the property world in November 1972. If, however, there should be any increase or diminution in the value of the relevant interest attributable to the development of the adjoining land compulsorily taken by the acquiring authority, such increase or diminution is to be ignored in accordance with the provisions of section 6 of and the First Schedule to the Land Compensation Act 1961. Furthermore, no account is to be taken of any increase in the value C of the subject land which is entirely due to the scheme underlying the acquisition (see *Pointe Gourde Quarrying & Transport Company v Sub-Intendent of Crown Lands* [1947] AC 565). In our opinion not only the increase in value as found in the *Pointe Gourde* case but also any diminution in value is to be ignored—see the words of Lord Denning MR in *Salop County Council v Craddock* [1970] RVR 63 at p 64, where he said:

> "The root principle of all compensation is that you disregard the scheme under which the acquisition is being made."

D

Evidence on planning matters was given for the claimants by Mr John Trustram Eve FRICS and for the acquiring authority by Mr L C McConnell ARICS MRTPI, a principal planner in the planning department of the Greater London Council and by Mr S A J Lear MRTPI, the chief planning officer for the Lambeth Borough Council.

The evidence on the two sides revealed a clear-cut divergence. In the opinion of the authority's planners, a potential developer of the green site (ie the enlarged area but excluding the property of the London Electricity Board and the Greater London Council) could reasonably have expected permission E for a mixed development of offices, shops and flats with an overall plot ratio of 2:1. Mr McConnell took the view that any development on the green site, if taken by itself without relating it to the larger comprehensive scheme for the Brixton Town Centre, would have constituted piecemeal development and any greater density than 2:1 would give rise to buildings of unsuitable bulk.

Mr Lear expressed a similar view and listed in particular, as disadvantages of developing the green site, (a) congested layout, (b) poor access in relation to surrounding highways, (c) difficulties to be encountered in incorporating walkways which the planning authority would consider to be most F important, and (d) difficulties arising from the application of daylighting and sunlighting standards. He also referred to the planning refusals, as indicating quite clearly what the attitude of the planning authority and of the Minister on appeal would have been; and he did not recognise that there were sufficient planning advantages in the suggested scheme put forward by the claimants to justify any departure from a plot ratio of 2:1.

Mr Eve, on the other hand, thought that a plot ratio of 3:1 would be a reasonable expectation. He referred to the demand for office accommodation in locations removed from the centre of London. He derived support from the amended revision statement published by the Greater London Council G in May 1972 and, in particular, paragraph 4 of that statement, which set out the sort of planning benefits which would influence the granting of permission for office development in various locations in London among which Brixton was specifically referred to.

We are impressed by the evidence of demand for office development at centres outside central London at the relevant time in November 1972 and on balance we are inclined to the view that Mr Eve's estimate of a 3:1 plot ratio is more acceptable. But in our opinion the matter is concluded for us by an examination of the law and the authorities. The evidence of Mr McConnell and Mr Lear, which was clearly H honest and sincere, was that they would be unwilling to recommend a development involving a plot ratio of more than 2:1. The considerations which carried them to this conclusion, however, were influenced by the knowledge that the most satisfactory development of the subject land was that which was in fact contemplated and which involved the incorporation within the land for development of the sites owned and occupied by the London Electricity Board and the Greater London Council. That these additional areas of land came to be developed together with the green site was entirely due to the special status of the acquiring authority. J

We are satisfied, on the evidence, that no private developer would have been successful in acquiring this additional land from these other public authorities. The assumption therefore of a 2:1 plot ratio for the green site, which by general consent leads to a scheme which is not economically feasible, flows directly in our judgment from the prospect of development of this adjoining land by the acquiring authority. At the same time, the entire development now being carried out by the authority is the scheme underlying the acquisition. If, therefore, one ignores that scheme the planning authority are faced with two alternatives—either to relax their attitude regarding plot ratio or to allow some other more modest K development, possibly in accordance with the existing zoning. We should say that we heard no evidence from either side as to any other alternative for the development; and in our opinion the most likely outcome in these circumstances would have been the approval of a development something like the scheme put forward by the claimants based on a plot ratio of 3:1.

Quite clearly, therefore, the existence of the local authority's scheme, by inhibiting development for offices, shops and flats at a density of more than 2:1, diminishes the value of the subject land. Such diminution is to be disregarded in arriving at the market value either under section L 6 of and the First Schedule to the 1961 Act or by virtue of the *Pointe Gourde* doctrine, or indeed on both counts. It seems to us to follow that, in either event, the land is properly to be valued on the assumption that a prospective purchaser operating in the no-scheme world at the relevant date could reasonably look forward to development of the green site with a plot ratio of 3:1.

We turn, therefore, to a consideration of the valuation evidence and set out here the evidence of Mr T J Taylor FRICS, the expert valuation witness called on behalf of the claimants. His valuation was made on the residual method, that is to say he envisaged a probable scheme of development M of the green site, estimated its value when completed and let, and then deducted the costs of development, leaving a value for the whole site. He then apportioned this value between the claimants, who owned the largest single area of the whole, and the owners and occupiers of the remaining parcels of land within the green site, and, finally, from the total relating to the claimants' land he made a further deduction of £60,000, which is agreed with the acquiring authority as being the compensation necessary to buy out the existing statutory and business tenants on the subject land. His valuation was as follows:

Value of shops and offices completed and let		
Shopping 6,030 sq ft at £3.00	£18,090	
Offices 42,755 sq ft at £2.75	117,576	
	135,666	
Y.P. perpetuity at 5½%	18.18	
	2,466,408	
Less selling costs (solicitor and agent) at 2½%	61,660	
Value completed and let	2,404,748	
say	2,400,000	
Deduct developer's profit	400,000	
Leaves for total development cost		£2,000,000
Development cost		
Building: Shopping 6,700 sq ft at £8	£53,600	
Offices 53,445 sq ft at £12	641,340	
	694,940	
Site works, etc	40,000	
	734,940	
Architects', quantity surveyors' and engineers' fees at 12½%	91,867	
	826,807	
Finance for 18 months at 10% on half costs	62,010	
	888,817	
Letting and promotion costs	18,000	
	906,817	
	say	910,000
Leaves for site plus interest and costs		£1,090,000
x = value of site		
0.33x = interest at 10% for 3 years		
0.025x = acquisition fees at 2½%		
1.355 x = £1,090,000		
x = ...		804,428
Plus value of site for 9 flats and a public house, say 10 x £4,000 =	£40,000	
Deduct costs and interest as before		29,520
		£833,948
	Site value, say	£835,000

In order to apportion the site value between the various interests, Mr Taylor looked first at the agreed standing value of the various properties, which showed a figure for the claimants' land of £69,500 and a total for the remainder of £280,000. These were actual or agreed figures excluding disturbance. He also apportioned the total on the basis of the relative site areas. In Mr Taylor's opinion the first method would leave a margin for the claimants' land which would not have been big enough to persuade them to proceed. Instead, he suggested, they would have proceeded to acquire other properties with a view, presumably, to carrying out the development themselves. The second method left a margin which Mr Taylor considered was inadequate to tempt the other owners to sell. Mr Taylor's evidence continued:

"A settlement would have been reached somewhere between these figures and, while this is difficult to quantify, a prospective purchaser would have wished to buy out the major freeholder quickly in order to establish himself as the developer. The St Pier interest was a property company interest and the others were mainly retailers who could be re-established in the scheme after completion if required, and this also would affect the purchaser's consideration.

A reasonable settlement for the St Pier ownership would have been £260,000, to include the cost of obtaining vacant possession. This would have left a sum of £575,000 for the other properties—over twice their existing value.

From the figure of £260,000 must be deducted the cost of obtaining vacant possession (£60,000), giving the St Pier ownership in November 1972 a value of £200,000."

The evidence on value for the acquiring authority was given by the borough valuer, Mr David Bookman FRICS. His primary valuation, as we have already indicated, was on the basis of the standing property value and gave a figure of £69,500. His alternative valuation, on the residual method, arrived at a negative value, mainly because the scheme of development which he envisaged was related to a plot ratio of 2:1. Since we have rejected that basic assumption, we do not derive any direct help from Mr Bookman's valuation, but we have looked at it again in the light of our findings as to the probable development which a prospective purchaser might have anticipated. We have also taken into account Mr Bookman's evidence regarding the climate of the market at the relevant date and, in particular, his criticism of Mr Taylor's basic yield, which he took at 5½ per cent. In Mr Bookman's view, the rate of interest which would have been taken in the market at the relevant date would have been at least 6½ per cent.

Both valuers drew our attention to indications from current material, from which we are satisfied that in November 1972 very low yields were acceptable on "prime rate" developments. We have finally come to the conclusion, in the light of the evidence, that Mr Taylor's estimate of 5½ per cent does not adequately reflect the potential difficulties likely to arise in carrying out the proposed development, having regard to its mixed character, the provision of substantial residential accommodation with the probable introduction of a further party to deal with that side of the matter, complications which might be expected in relation to such matters as daylighting and walkways and the difficulties of site assembly. At the same time we think Mr Bookman has overestimated the difficulties and our conclusion is that a prospective purchaser of the claimants' interest would proceed on a 6 per cent basis.

We have looked again at Mr Taylor's valuation, making the necessary consequential amendments. We have also looked at a valuation put in by Mr Bookman which adopted the form of development assumed by the claimants but taking 6½ per cent as the yield rate. For apportionment purposes, Mr Bookman applied the site area basis, which gave greater value to the claimants' portion of the total site. The result was that the amount remaining for the rest of the site was clearly insufficient. We prefer the approach of Mr Taylor at this stage of the valuation and we also think that Mr Bookman has overprovided for the cost of finance for funding the development and the acquisition.

Both valuers, as will be seen, put forward valuations carried out on the "residual" method, a method which the Lands Tribunal has always regarded with some reserve. In the present case it is, in effect, the only valuation approach before us; and we must therefore base our findings on it. We presume that if either party could have found evidence of actual transactions of comparable land in the market they would have produced it.

After studying the valuations of the two valuers, and after making the adjustments necessary to give effect to our findings on plot ratio and on the basic rate per cent, and one or two minor consequential amendments, we have come to the conclusion that the compensation payable to the claimants in respect of the acquisition of their freehold interest in the subject land is the sum of £152,500 to which will be added fees for services of a valuer in accordance with scale 5A and of a second expert in accordance with scale 5B of the scales of professional charges of the Royal Institution of Chartered Surveyors, together with appropriate legal costs at the (agreed) rate of 0.75 per cent of the award. For the calculation of these costs the total of the award is to be taken as a single sum. For purposes of the order to be made that sum is to be apportioned on the agreed basis as follows:

15, 17 and 19 Brixton Station Road	£16,000
35 and 45 Brixton Station Road	£18,000
16 Popes Road	£118,500
12-15 (incl.) Popes Road	
16-20 (incl.) Industry Terrace	
1-5 (incl. Alders Cottages	

Having read the decision in this matter and having then opened the sealed offer of compensation lodged by the acquiring authority, we find that the amount thereof is less

A than the amount of our award; accordingly the acquiring authority will pay to the claimants their costs of these references, such costs failing agreement to be taxed by the Registrar of the Lands Tribunal on the High Court Scale.

SIRELING & BRODER LTD v TOWER HAMLETS LONDON BOROUGH

(REF/67/1975)

(DOUGLAS FRANK QC, President, and J R LAIRD FRICS)

B February 18 1976

Estates Gazette March 27 1976

(1976) 237 EG 967

Leasehold interest in light-industrial premises at Spitalfields, E London, used for manufacture and sale of ladies' outerwear garments—Value of interest with allowance for effect of inflation on profit rent and for security of 15½ years' unexpired term—Disturbance claim on basis of total extinguishment of business—Loss on forced sale of made-up garments—Claim for surveyor's fees based on global amount of award, which included accountant's and trade C valuer's fees, not accepted—" Wrong that fees should be paid on fees "

Mr B Marder (instructed by Lewis Cutner & Co) appeared for the claimants; Mr A Dinkin (instructed by the solicitor to the London Borough of Tower Hamlets) for the acquiring authority.

Giving their decision, THE TRIBUNAL said: This is a reference to determine the compensation payable for the compulsory acquisition of the leasehold interest in factory premises known as 27 Brick Lane and 57 Flower & Dean Street, London E1. The claimants, Sireling & Broder Ltd, held a full repairing and insuring lease for a term of 21 years D from June 24 1969, at the rent of £2,200 pa for the first seven years with rent reviews at the end of the seventh and 14th years. The lessees covenanted to decorate the interior of the premises in every seventh year of the term and to paint the exterior in every third year. The property was acquired by the acquiring authority, the London Borough of Tower Hamlets (hereafter called the " council "), under Part III of the Housing Act 1957 and the London Borough of Tower Hamlets, Spitalfields Area, Compulsory Purchase Order 1967 confirmed with modification on January 7 1970 by the Minister of Housing and Local Government. The reference properties E were included in the order as lands outside the clearance area, the acquisition of which was deemed to be necessary in order to achieve a cleared area of convenient shape and size for redevelopment. Notice to treat and notice of entry were served by the council on March 6 1970. The council accepted the keys of 27 Brick Lane from the claimants on October 7 1974, which is the agreed date for the assessment of compensation, which falls to be assessed under the Land Compensation Act 1961 and section 59 (4) and Part III of the Third Schedule to the Housing Act 1957.

The reference properties, in light industrial user, were situated in an area zoned for residential use with shopping F frontage on the initial development plan. They stood on the corner of Brick Lane and Flower & Dean Street with a frontage of 20 ft to the former and a return frontage to the latter of about 70 ft. The accommodation was on three floors, namely, a basement (6 ft headroom, used for storage and occasional cutting room—815 sq ft), a ground floor (showroom and office—753 sq ft), and a first floor (workrooms and staff room—1,270 sq ft). All main services were available.

A " Stock House "

The business carried on by the claimants involved the manufacture and sale of ladies' outerwear garments for the wholesale trade on a seasonal basis—known in the trade as G a " stock house." It involved the making-up of sample dresses, an initial order normally being for a number less than 50, but the minimum number actually to manufacture was 150; the balance above the initial order was held in stock. A follow-up order might come at short notice, and a stock house has to have available stock at 24-hours' notice. The claimant company was established in 1946 by its managing director, Mr S Feldman, who died in February 1970, leaving Mr Henry Caplin, formerly the sales director, as the sole director of the company.

It is agreed that compensation for disturbance is to be H assessed on the basis of total extinguishment of the business. Apart from the claim in respect of estimated loss suffered on the sale of finished garments below normal prices, during the period of January 1974 to the end of August 1974, losses on the forced sale have been agreed as follows:

Stock in trade, including plastic covers and hangers (per Beecroft Sons & Nicholson's valuations) ...	£6,407.88
Plant, fixtures and fittings (per Beecroft's valuations)	£5,200.69
Specialist's valuation fees of Beecroft incurred by the claimants between the date of notice to treat and the date of the reference (excluding VAT) ...	£366.21
In addition, the following other items of compensation have been agreed:	
Solicitors' fees incurred by claimants between notice to treat and reference dates (excluding VAT) ...	£107.00
Redundancy payment to an employee	£105.00
Maintenance and rental payment for burglar alarm system (AFA Minerva)	£53.21

J

It is also agreed that in estimating average adjusted net annual profit, interest on capital invested in the business shall be calculated at the rate of 8 per cent.

As we have said, the council served a notice of entry on the claimants dated March 6 1970 which stated:

K " Now therefore the council . . . hereby gives you notice that it may at any time after the expiration of 14 days from the date of the service of this notice enter upon and take possession of the said land."

Industrial Action by NALGO

By letter dated November 3 1971, the council informed the surveyors at that time acting for the claimants that it was anticipated that possession of the reference premises would be required in approximately two years' time. On January 3 1974, the council's solicitor wrote to the claimants confirming a telephone conversation to the effect that the premises would be required for redevelopment at the beginning of March L 1974. Subsequently, at the request of the claimants, the council extended the date for possession until the end of April 1974, and this was acknowledged by the claimants' solicitors on January 24 1974. On March 27 1974, Mr G W Mason FRICS, senior partner in Bunch & Duke, by then acting as surveyors to the claimants, wrote to the council stating that he had been given to understand that possession would be required on April 30 1974, and asked for formal confirmation to such effect. On April 24 1974 Mr Mason wrote to the council requesting a reply to his letter of March 27, and also to the district valuer; on April 29 Mr M Mason received an undated circular letter from the council's solicitor stating that, because of an industrial dispute (industrial action by NALGO), no work could be undertaken on the matter for the time being. That dispute commenced on April 8 1974 and continued until August 12 1974. Eventually the council was asked by the claimants to take possession of the reference properties on September 2 1974, but due to the necessary removal by the claimants of abandoned fixtures and fittings from the premises, and subsequent bereavement in Mr Caplin's family, the council did not receive the keys of the reference properties from the claimants until October 7 1974.

A There are three main matters in dispute, namely:

(1) the value of the interest acquired,

(2) the amount of compensation for disturbance, and

(3) the amount of professional fees properly to be included in a claim for disturbance.

We shall deal with each of these in turn.

(1) *Value of interest acquired*

In October 1974 the claimants' lease had about 15½ years to run. Mr Mason said that he estimated the market rental value of the reference properties in October 1974 at £4,500 pa, so that until the first rent review (due 18 months later), B there was a profit rental of £2,300 pa. To that he applied five years' purchase, on the ground that the term had 15 years and 8 months unexpired in respect of a substantially-constructed light-industrial factory and showroom premises, with ample basement storage, situated in an area where such accommodation was in short supply just north of Whitechapel High Street and within easy walking distance of Aldgate East Underground Station. Hence he valued the interest at £11,500.

On the other hand, Mr N R J Coleman ARICS, a senior valuer in the district valuer's office, assessed the market rent at £3,375 pa and thus a profit rent of £1,175 pa for one year C and nine months (up to the first rent review), which he valued at 14 per cent and 2½ per cent (adjusted for tax at 35p in the £1), producing a multiplier of 0.98, arriving at a value for the claimants' leasehold interest of £1,150.

Having heard the evidence, including matters relating to comparable properties, we accept Mr Mason's assessment of the annual market rental of the reference properties, but we do not accept that he is right to disregard the fact that an increased rent would have become payable following the first rent review due one year and eight months ahead. On the other hand, so long as an inflationary period lasted, that rent D could be expected increasingly to become less than the market rent, and we think that some allowance should be made for that and for the fact that an assignee of the lease would have the security of a 15-year tenancy in an area where premises were in short supply. After giving the matter due thought we think that the hypothetical purchaser would pay £7,000 for the profit rent and the security of a 15½ years' unexpired term.

(2) *Disturbance*

We now turn to the assessment of compensation for disturbance which, as we have said, is agreed should be on the basis of the total extinguishment of the business. Both valuers approached this problem by ascertaining an average adjusted E net annual profit. Further, although they vary considerably in their calculation thereof, there is less than 10 per cent difference in their final figures; as it happens, Mr Coleman's estimate is £5,731, whereas Mr Mason's is £5,221. Thus the former is more favourable to the claimants and we adopt it. However, Mr Mason, in his lodged valuation, applied 5 years' purchase to his net profit figure, producing £26,105 (say £26,000) compensation for total extinguishment of the business, whereas Mr Coleman applied 2 years' purchase to his £5,731, which produced £11,500 for what he described as a "valuation of goodwill on the basis of total extinguishment."

F

Established Nearly 30 Years

We are satisfied by Mr Caplin's evidence that the business concerned contains a high measure of speculation, and to a considerable extent its success depended on the skill of the managing director, as has been shown by the record of profits; on the other hand it has been established nearly 30 years. We have come to the conclusion that the fair amount of compensation for the total extinguishment of the business payable to the claimants is £15,000.

The next disputed item of claim is for £10,020.73, being the loss on the forced sale of made-up garments sold below normal prices. This item is particularised in Mr Mason's claim as:

"Garments sold below normal prices, between January 1974 G due to intimation that the council would require possession at the end of April 1974, and the eventual need to close down the business at the end of August 1974, it not being practicable to purchase stock for the autumn trade."

This item was based upon the normal selling price of the made-up garments estimated at £45,323.40, whereas the amount realised as shown by invoices totalled £35,302.67. For the council, it was conceded that compensation could be payable for losses incurred before entry, but it was contended that the losses must follow directly from the compulsory purchase and must not arise from failure of the claimants to mitigate the loss. The council further contended that the H claimants made no effort to reduce the scale of their business or to run down stocks and that from January 1974 they should have mitigated.

We for our part see the matter in this way: the council, having served notice of entry on March 6 1970, were at liberty to stop the claimants carrying on the business at any time by entering upon the premises and, so long as they failed to do so, the claimants were fully entitled to carry on the business in such a way as to earn the maximum profit for themselves. There was no obligation upon them to run down the business and thereby reduce its profit in the expectation that the council might at any time enter on the reference J properties. In the event, however, the council, on January 3 1974, had said that the premises would be required at the beginning of March 1974; that date was later postponed at the request of the claimants to April 30 1974; then, due to an industrial dispute within the council's office, that date was subsequently postponed until August, and then for other reasons until October 1974. For some time the claimants were actively seeking suitable alternative premises, but without success. For the whole of that period the claimants were aware that from the end of April 1974 onwards, the council might be expected to take possession and therefore they sold the stock in hand at the best possible price and they argue K that they did so in mitigation of their loss.

Bearing in mind that the claimants' business catered, in advance, for the four seasons of the year and that had they succeeded in finding alternative premises they would have mitigated further, we hold that this claim is good in principle. It is further said, for the council, that the correct measure of loss is the difference between the cost to the claimants and the actual sale price of the made-up garments. We hold that is not so, for what the claimants lost was not only that difference but also, as long as they continued to trade, the proper profit on their efforts. Accordingly, we hold that this item is L payable in full as claimed, that is to say, £10,020.73.

The next item of disturbance in dispute is the loss on the forced sale of motor vehicles, which the claimants put at £1,380 in respect of four motorcars and one trade van for gowns; Mr Coleman's figure was £405 for one motorcar and the gown van. We find that the claimants have failed to satisfy us that all five vehicles were or were necessarily employed on the claimants' business and we think that Mr Coleman's £405 is sufficient compensation for this item.

(3) *Professional fees*

We turn now to the last disputed matters which concern professional fees and fall under two heads. First, the quantum M of accountant's fees up to the date of the notice of reference; the claim was for £972 exclusive of VAT, supported with a fully itemised bill. The council say that this amount is excessive and should be limited to £200 exclusive of VAT, but they were unable to show that the claim was unreasonable and so it must be paid in full.

The second dispute on professional fees is a matter of principle. The claimants contend that the surveyor's fees payable to their valuer, Mr Mason, should be based upon the global amount of our award which, in turn, includes the items agreed between the parties, calculated in accordance with

A scale 5 (a) of the scales of professional charges of the Royal Institution of Chartered Surveyors. That global sum, they say, should include not only the accountant's fees of £972 as awarded but also the agreed fees of £366.21 (excluding VAT) which the claimants had paid to Beecroft, Sons & Nicholson, trade valuers, in respect of the following itemised accounts rendered by Beecrofts to the claimants:

"*29th April 1974*

To services in attending at 27 Brick Lane, London E1, scheduling the plant, fixtures, fittings, etc, and stock-in-trade, and giving a valuation on the basis of a going concern in the sum of £8,297.53 (as at 18th April 1974) ... £228.45 (excluding VAT)

B *27th September 1974*

To scheduling the stock-in-trade and valuing same at 27 Brick Lane in the sum of £5,511.04 ... ... £148.80"

"Masterminding" Role

Mr Mason did not accept that he was seeking to charge fees for himself on someone else's work, contending that his role had been to "mastermind" the whole matter on behalf of the claimants.

The council disagree; Mr Coleman told us that he had accepted Beecroft's two valuations *in toto,* save for a minimal C reduction of £200 which he had agreed direct with them; Mr Mason did not value, or negotiate, these particular items. The council had agreed to pay Beecroft's valuation fees amounting to £366.21 and will be paying the accountant's fee that we have determined; Mr Coleman thought that both sets of fees should be excluded from the global sum on which Mr Mason's fees are to be calculated, since he thought it wrong that fees should be paid on fees.

We think that Mr Coleman is right.

In the result our award of compensation is as follows:

1. Value of leasehold interest	£7,000.00
D 2. Compensation for total extinguishment of business	15,000.00
3. Loss on forced sale of made-up garments ...	10,020.73
4. Loss on forced sale of motorcars	405.00
5. Maintenance and rental payment for burglar alarm system (agreed)	53.21
6. A surveyor's fee based on scale 5(a) of the scales of professional charges of the Royal Institution of Chartered Surveyors (excluding VAT) on the above total of, say, £32,479 ...	368.55
7. Loss on forced sale of stock (£6,407.88), plant, fixtures and fittings (£5,200.69) valued and E agreed by Beecroft	11,608.57
8. Beecroft's valuation fees (excluding VAT) (agreed)	366.21
9. Redundancy payment to employee (employer's 50% contribution, agreed with accountant) ...	105.00
10. Accountant's fees up to date of reference (excluding VAT)	972.00
11. Legal costs up to date of reference (excluding VAT) (agreed)	107.00
Total amount of award	£46,006.27

F Having read the decision in this matter and having then opened the sealed offer of compensation lodged by the acquiring authority, we find that the amount thereof is less than the amount of our award; accordingly the acquiring authority will pay to the claimants their costs of this reference, such costs unless agreed to be taxed by the Registrar of the Lands Tribunal on the High Court Scale of Costs.

WILLIAM IVENS & SONS (TIMBER MERCHANTS) LTD G v DAVENTRY DISTRICT COUNCIL

(Ref/85/1975)

(DOUGLAS FRANK QC, President)

February 6 1976

Estates Gazette April 10 1976

(1976) 238 EG 127

Purchase notice—Land at Watford, Northants, formerly occupied by detached house—Used for dumping spoil from M1 construction and factory waste, raising level by 20 ft—Tribunal holds notional right to rebuild house incapable H of being exercised and valueless

Mr A B Dawson (instructed by Frere Cholmeley & Co) appeared for the claimants; Mr M B Horton (instructed by Shoosmiths & Harrison, of Northampton) for the acquiring authority.

Giving his decision THE PRESIDENT said: This is a reference to determine the compensation payable for the deemed compulsory purchase of land at Welton Station, Watford, Northamptonshire. The claimants had served a purchase notice under section 129 of the Town and Country Planning Act 1962 on the ground that the land was incapable of bene- J ficial use. The council had objected to the confirmation of the purchase notice on the ground that the land could be used for agriculture. The claimants said, however, that use would not be practicable because they were not engaged in agriculture and because access on three sides was barred to them because they were not owners or lessees of the adjoining land, and on the fourth side planning consent for an access was not forthcoming. Those contentions by the claimants were accepted by the Secretary of State and accordingly he confirmed the purchase notice and directed that notice to treat was to be deemed to have been served on June 26 1973. The following is a statement of the facts as K agreed by the parties:

Agreed Facts

"1. The reference relates to about 0.93-acre of land on the north-western side of Station Road (B4036).

2. The claimants' interest is the freehold.

3. From about 1850 until 1964 there stood on the reference land a detached house with an access to the highway B4036.

4. In about 1966 the highway authority dumped spoil from the construction of the M1 on the reference land, and the claimants' parent company were using the land for dumping L factory waste.

The extent of the deposit raised the land by 20 ft all over.

5. In 1967 the highway authority for the purpose of widening the B4036 acquired the strip of land lying between the reference land and the then north-western edge of B4036, shown on the agreed plan.

6. The said strip has now been incorporated into the B4036.

7. At the request of and at the expense of the highway authority after the widening the claimants, who are timber M merchants, erected in the highway on its north-west side a continuous fence: the said fence remains in position and there is in fact no access from the highway to the reference land.

8. On June 26 1973 the Secretary of State for the Environment confirmed a purchase notice in respect of the reference land served by the claimants on the acquiring authority.

9. The agreed value of the reference land with the benefit of an assumed planning permission to rebuild the house referred to in paragraph 3 and with a right of access from the reference land directly to the B4036 is £9,000.

A 10. The agreed value of the reference land without any assumed planning permission and with no access directly to the B4036 is £300."

The parties are unable to agree the value of the land with an assumed planning permission to rebuild the house referred to in paragraph 3 but without any access to the highway, and if it is decided that is the correct method of valuation a further hearing will be required.

The claimants' case, put shortly, is that in valuing the land for compensation purposes it must be assumed that planning permission would be granted to rebuild the house which stood on the land until 1964. That submission arises firstly B from subsection (3) of section 15 of the Land Compensation Act 1961, which provides:

"It shall be assumed that planning permission would be granted, in respect of the relevant land or any part thereof, for development of any class specified in the Third Schedule to the Town and Country Planning Act 1947 (which relates to development included in the existing use of land)."

The Third Schedule to the 1947 Act is now the Eighth Schedule to the Town and Country Planning Act 1971, and that specifies as a class of development for which planning permission is to be assumed the following [in para 1]:

C "The carrying out of any of the following works, that is to say: (a) the rebuilding, as often as occasion may require, of any building which was in existence on the appointed day, or of any building which was in existence before that day but was destroyed or demolished after January 7 1937, including the making good of war damage sustained by any such building."

Mr Horton, for the council, disputed the relevance of those rights, partly on the ground that there is no access to the land and planning permission would be required for the formation of an access and, alternatively, on the ground that the relevant land no longer exists. Further, that the rights under the Eighth Schedule have been abandoned by the filling of D the land.

I propose first to deal with the submission that the relevant land, as Mr Horton put it, no longer exists, but I think more accurately the contention should be that it is not possible to build a house on the relevant land. He says that the relevant land is that part of the ground upon which the former house stood and it is not possible to build on that because it is now covered with spoil to the depth of 20 ft.

Mr Dawson for the claimant argues, however, that the assumption in the Eighth Schedule must be made and that it is not inconsistent with the tipping on the land before the carrying out of the building operation. He says by the defini- E tion of land in section 39 of the 1961 Act, that land there means the whole of the area "from the gates of Hell to the gates of Heaven" and that the raising of levels has no effect.

War Damage Commission's Test

I find myself unable to accept the second part of that submission. It seems to me that in any given case it is a question of fact and degree whether a building is a rebuilding of a pre-existing building or whether it is a new building; see, for example, *In re Walker's Settled Estate* [1894] 1 Ch 189. I gain some help from the practice notes issued by the War F Damage Commission in which they set out the test they applied in deciding what is meant by making good war damage, and they said that it is: "looking at the works executed, can the property be fairly described as still the same property as before the war damage although altered or added to?" That test was expressly approved in the case of *City of London Real Property Co Ltd v War Damage Commission* [1956] Ch 607. Adapting that test to the present case I asked myself this question: "Looking at a notional new building of the same design as the demolished house on the site as it exists today and in the same vertical plane as the demolished house, could it reasonably be said that the new house is the same property as before although rebuilt, G or is it an entirely new house?"

Another test one could apply is to assume that there had been an express planning permission to build a house on the site as it existed before being filled and to ask whether that permission would authorise the building of a similar house on the site as it is now. In my view the answer is clearly "no," and if authority is required for that it is the case of *Shemara v Luton Corporation* (1967) 18 P & CR 520. Therefore I conclude that the notional right to rebuild is incapable of being exercised and therefore valueless.

I follow the Divisional Court in the recent case of *Pilkington v Secretary of State for the Environment* [1974] 1 All H ER 283, and base my decision solely on the impossibility of exercising the notional right to rebuild.

It follows from what I have said that in my judgment the notional right to rebuild is valueless and accordingly I hold that the proper amount of compensation payable is £300 plus surveyor's fees on scale 5a of the RICS scale of fees. It is unnecessary under the circumstances for me to deal with the questions of abandonment or access.

The claimant will pay to the acquiring authority its costs of this reference, such costs, unless agreed, to be taxed by the Registrar of the Lands Tribunal on scale 4 of the County J Court Scales of Costs.

VAYNOR DEVELOPMENTS LTD v WALES GAS BOARD

(REF/10/1975)

(J H EMLYN JONES FRICS)

January 28 1976

Estates Gazette May 1 1976
(1976) 238 EG 347 K

Reference by consent—Land near Merthyr Tydfil sterilised by gas main construction—Purchased by claimants as agricultural land with six-year-old planning consent for 15 houses on part and subject to Gas Board's rights over 40-ft-wide strip—Ten weeks later board proposes to extend rights to 320-ft-wide strip—Claimants abandon negotiations for agreement with building contractors for development with 69 houses for which planning permission "virtually assured"—Land ripe for development but half area below level of sewers—Claimants' valuation of £51,000 on basis of near-agreement for development discounted for uncertainty, risk and deferment—Board's valuation of £11,500 on sale-price basis held not adequately reflecting true value as ripe land—Tribunal awards £27,000 L

Sir Derek Walker-Smith QC and David Woolley (instructed by Myer Cohen & Co, of Cardiff) appeared for the claimants; the Hon David Trustram Eve QC and Patrick Webster (instructed by the solicitor to the Wales Gas Board) for the authority.

Giving his decision, MR EMLYN JONES said: This is a reference by consent of the parties for the determination of M compensation to be paid to Vaynor Developments Ltd, the claimants, owing to the sterilisation of certain land at Trefechan near Merthyr Tydfil in Mid-Glamorgan following the construction of a gas main by the Wales Gas Board, the compensating authority. There is no compulsory purchase order; but it is agreed that compensation is payable "against the background of statutory powers" and according to the rules governing compulsory purchase.

The subject land [at High Trees Estate] extends to some 7¼ acres or thereabouts and was formerly owned by a farmer, Mr D G Thomas, and occupied by him for many years for agricultural purposes. In 1965 Mr Thomas obtained outline

A planning permission for residential development over the whole of the land. In the following year detailed planning permission was obtained for the erection of 15 dwelling-houses. The land continued to be used for agricultural purposes. On June 30 1969 Mr Thomas entered into an agreement by deed with the Gas Council, who were the predecessors of the Gas Board, whereby the council obtained an easement for a gas main running under a 20-ft strip of land through the middle of the 7¼ acres with a 10-ft strip on either side which was to be kept free for access purposes. There was thus a 40-ft wide strip of land over which it became effectively impossible to build.

B

Every Prospect of Feasible Scheme

On August 4 1971 the present claimants entered into a contract to buy the whole of the subject land subject to the easement for £10,500; and during the following two months entered into negotiations through their agents with local authorities and statutory undertakers with the object of developing the whole site for housing. By the end of September there seemed every prospect that a feasible scheme of development could shortly be put in hand.

While this was going on the claimants were in negotiation with a firm of building contractors with whom they had C co-operated on developments in other parts of South Wales, and were on the point of entering into an agreement with them. On October 15 1971, however, the claimants' agents were informed by the Gas Board that the board were now interested in acquiring rights over a larger area of land extending in all to a total width of 320 ft in place of the original 40-ft strip which had been covered by the earlier deed of grant. This proposal put an end to any immediate negotiations for the development of the land, and the plans being discussed at that time were abandoned.

As a result of the Gas Board's latest proposals it is common ground that some 5.0385 acres out of the total of the D land in the claimants' ownership have now been sterilised and are incapable of being developed. The remaining 2¼ acres or thereabouts have subsequently been developed by the erection of some 22 bungalows. The dispute between the parties concerns the amount of compensation payable for the loss sustained by the claimants by virtue of the sterilisation of the rest of their land. The date of valuation is agreed as October 15 1971. The claimants say that the compensation owing to them is the sum of £45,855 plus £5,250 in respect of architect's fees, making a total sum of £51,105. The expert evidence called on their behalf sought to justify these figures E by reference to the agreement which was about to be entered into in October 1971 when the Gas Board's proposals first brought those negotiations to an end. The authority, for their part, suggest that the compensation properly due is £11,500, and this sum is founded on the price which the claimants contracted to pay for the purchase of the land in August 1971. In addition, they accept that the further amount of £5,250 is due in respect of abortive architect's fees.

The subject land is situated at Trefechan, a small village on the outskirts of Merthyr Tydfil—some three or four miles by road from the town centre. It is undulating land, which slopes generally downwards from south-west to north-east, F the difference in elevation between the highest and lowest points being about 75 ft. It is bounded on the west by a country road—the main road through the village of Trefechan; on the east by a disused railway line running along the valley Cwm Taf Fechan; on the southern side is a small estate of local authority houses; and on the north a large limestone quarry and its ancillary buildings.

Drainage Difficulties

For the claimants Sir Derek Walker-Smith called Mr P R G Mathias, the architect who had advised the claimants in connection with the acquisition of the subject land. He gave evidence regarding the negotiations on planning and G by-law requirements on drainage and sewage treatment and on discussions with the Electricity Board, Water Board and Gas Board regarding the supply of services to the subject land. The only one of these matters which presented any difficulty was that of drainage. Because of the slope of the land about half the area of the land lay below the level of the existing sewers. Mr Mathias had investigated the possibility of installing a pumping station to deal with this low-lying part of the land, and the local authority had indicated that whereas it would be possible to connect to the council's existing sewers they would not be prepared to take over any pumping installation nor be responsible for its operation and H maintenance. From the evidence of Mr Mathias it seems clear that at the end of October planning permission for the development of the whole of the claimants' land by the erection of 69 dwelling-houses could reasonably have been anticipated, subject only to resolving the difficulties of drainage. In view of the intervention by the Gas Board this application was subsequently withdrawn, and, as I have already indicated, the development of the unimpeded part of the land was subsequently carried out in pursuance of planning permission granted on a later application.

Sir Derek called evidence on valuation from Mr R C Cotsen FRICS, who practises in Cardiff and in Blackwood. J Mr Cotsen had been brought into the case immediately following the indication given by the Gas Board that they were seeking to acquire rights over the 320-ft wide strip. Mr Cotsen gave details of the negotiations for the development of the land immediately prior to that date and produced correspondence and other documents showing the nature of the agreement which the claimants were seeking to reach with a Birmingham development company known as Gilbert & West Bros Ltd. The terms of this agreement were not set out in detail in the correspondence but were said to be "on the usual premium basis the contract of which will be precisely the same as Blackwood." The Blackwood agreement K was one reached between Hillandale (Wales) Ltd, an associated company of the claimants, and Gilbert & West Bros (Wales) Ltd.

Terms of Agreement

The exact status of these companies was not explained to me, but it was accepted that the parties to the later negotiations on the subject land were aware of the terms negotiated at Blackwood. Under that agreement Gilbert & West, who were described as the builders, were authorised to enter upon certain land in order to build a number of dwelling-houses for which detailed planning permission was to be granted. L They were to complete the erection of the buildings within a period of 30 months from the date on which planning permission was received and to provide roads, footways, parking areas and street lighting in accordance with approved plans to a standard suitable for adoption under the Highways Act 1959. The builders were also to pay to Hillandale (Wales) Ltd a premium of £58,000 "by payments being proportionate with the number of units to the premium on the grant of each lease. The interest shall be payable on any outstanding premium due after the said period of 30 months at the rate of 12½ per cent." Hillandale also undertook to grant to the builders or to any person nominated by them leases of the M said land and buildings for a term of 99 years at a yearly ground rent per plot calculated at the rate of "1 per cent of the selling price of the dwelling-house or such lesser sum as the vendors may agree but the minimum ground rent receivable in respect of the development shall not be less than £1,800 pa." The builders were to be responsible for preparing the lease plans, but there was no reference in the agreement to any payments for the preparation of architect's plans or fees in connection with such matters as planning applications, by-law applications or negotiations with gas, water, electricity and drainage authorities.

A In valuing the land sterilised in the present case Mr Cotsen looked first at two transactions of building land on the outskirts of Merthyr Tydfil—one sale of what he regarded as inferior land on the Old Swansea Road of approximately 13½ acres sold in February 1973 for £75,000 and the second sale of 3.296 acres at Pontsticill in July 1972 for £25,000. The first-mentioned transaction worked out at about £5,555 per acre and the other at £7,600 per acre. In the light of these transactions Mr Cotsen, who considered that the subject land was considerably better than the Old Swansea Road land and slightly more valuable than the land at Pontsticill, considered that the value of the land sterilised by the Gas Board in the B present case, should be £8,000 per acre, giving a total of £40,000.

Claimants' Valuation

On the basis, however, of what he regarded as the virtually completed agreement with Gilbert & West in respect of the subject land, Mr Cotsen put in a valuation in the following form:

(a) *Value of building premium*		
Original layout, as approved, of 69 units at an average of £900 per unit	£63,200	
C Reduced to 23 units by the Gas Council's intervention. 23 x £900	20,700	
Loss of 46 plots		£42,500
(b) *Value of lost ground rents*		
46 plots at an average of £20 pa ...	£920	
YP in perpetuity at 10%	10	
		9,200
		£51,700
In addition compensation should be paid on the abortive architects' fees		
D incurred amounting to		£5,574
		£57,274

During the course of the hearing Mr Cotsen made certain alterations to these figures. He first took £62,100, which is the arithmetical result of 69 units at £900 per unit, as his starting point instead of £63,200, which represents the total premium asked for by the claimants in their negotiations with Gilbert & West. He also deferred for 12 months at 8 per cent the total of premiums and ground rents; from that E figure he deducted £1,000 as being the residual value of the land retained and then added the agreed figure for architect's fees of £5,250. These calculations produced an amended total valuation of £51,105, which, in Mr Cotsen's opinion, was the compensation to which the claimants were entitled.

For the authority Mr Trustram Eve called Mr J A W Protheroe BSc (Est Man) FRICS, who has been a consultant valuer to the board and their predecessors for over 20 years. Mr Protheroe accepted that planning permission for the erection of 69 houses on the subject land would have been forthcoming had it not been for the intervention of the Gas Board, but he considered that Mr Cotsen had not paid due F account to the difficulties of drainage for the lower parts of the land. He did not disagree with a sum of £3,500 to £4,000 as an estimate of the cost of providing a pumping station and pump but he thought that running and maintenance costs would have been double Mr Cotsen's estimate of £300 to £400 pa. He did not attach any weight to the negotiations with Gilbert & West because they were not completed and, in his opinion, were too imprecise in their terms. In his opinion the transaction was highly speculative and some considerable time would elapse before any moneys at all would have been received—hence the need for deferment which Mr Cotsen subsequently accepted.

Mr Protheroe also referred to an abortive agreement for G the subsequent sale of that part of the land which was not affected by the Gas Board's requirements where the company agreeing to buy that land and put up the houses, although signing the conveyance and starting the work, had been unable to complete their operations because of financial difficulties. In Mr Protheroe's opinion the best evidence of the sale of the subject land was the price paid for it by the claimants when they bought it from Mr Thomas, the contract being dated August 4 1971, only two months before the relevant date. He also relied on the value of £2,000 per acre put on the subject land by a valuer acting on behalf of Mr Thomas in 1969, when agreement was reached on the H compensation payable on the grant of the earlier easement.

"Before and After" Valuation for Gas Board

He approached his valuation by taking a "before and after" value as follows:

"Before" value			
2¼ acres at £3,000 per acre	...	£6,750	
5 acres at £2,500 per acre ...	...	12,500	
			£19,250
"After" value			
2¼ acres at £3,000 per acre ...	...	£6,750	J
5 acres at £200 per acre ...	...	1,000	
			£7,750
	Depreciation		£11,500

On to this figure Mr Protheroe accepted that there should be added a further amount of £5,250 in respect of architect's fees.

It will be seen that the two valuers relate their values to two widely differing pieces of evidence regarding the subject land itself. Mr Protheroe says, in effect, that the best indication of the value of the five acres lost at the relevant date K on October 15 1971 is to be obtained from the price agreed to be paid for the 7¼ acres just over two months earlier on August 4 1971. Mr Cotsen, for his part, looks to the negotiations current at the relevant date between the claimants and Gilbert & West which, as he analysed them, justified a valuation for the whole 7¼ acres about five times as great as that which had been paid two months previously.

In my opinion neither of these two sources is particularly reliable. The sale by the farmer to the claimants was a sale of agricultural land on which outline planning permission for residential development had been given some six years previously and permission for 15 houses on part of the site L some five or six years previously, although neither of these planning permissions had been acted upon. The land at the relevant date, however, could properly be described as building land ripe for development, with the granting of detailed planning permission for 69 houses virtually assured and having the benefit of the negotiations carried out through the expertise of the claimants and their professional advisers dealing with such matters as by-law permission and satisfactory arrangements for the supply of public services.

Agreement "Far too Nebulous" M

Nevertheless, the so-called agreement with Gilbert & West is far too nebulous to be taken as it stands as evidence of the value of the land at the relevant date. Although the claimants were to be described in the Blackwood agreement as vendors, they were not, strictly speaking, agreeing to sell the land as a whole to Gilbert & West. The agreement had more of the character of a joint enterprise in which the landowners and the builders combined to produce a development of some 69 houses which were then to be sold individually on long leasehold. The builders were to recover the purchase price of the houses as their reward, the landowners were to receive premiums from the builders as their price for entering the scheme, and in addition they were to be in receipt of the ground rents reserved under the leases. Such a scheme carried with it many of the commercial risks of development, and it is clear that such risks were a reality in the present case by a glance at the later history of that part of the land which has subsequently been developed. In any event the return to the landowners would not be made immediately available, and taking the Blackwood agreement as the model, up to 30 months might elapse before all the premiums became payable.

Furthermore, the only detail of the proposed agreement comes from the Blackwood agreement. That agreement, however, contained its own provisions regarding ground rent which quite clearly could not apply on the subject land. There are certain other areas of doubt—for example, the cost of providing roads and sewers, which could be of some importance in view of the drainage difficulties to be anticipated in developing the lower parts of the subject land. The correspondence itself was far from clear in other ways. For example, on October 6 the claimants wrote to Gilbert & West with suggested premiums payable in respect of the various plots of land and using the expression " the usual premium basis, the contract of which will be precisely the same as Blackwood." In a letter dated October 19 1971 Gilbert & West wrote: " We have investigated this estate and we feel that this would contain the viability as mentioned by you and agreed in our telephone conversations." They then went on to suggest a reduction in premium of £50 on 12 of the plots, a suggestion accepted by the claimants in a letter dated October 21. Nothing was said at the hearing in amplification of the telephone conversations referred to. Finally, the Gilbert & West negotiations were dealing with the whole site, and the part which has been lost included most of the low-lying land, which would on the evidence have been more difficult and more costly to develop. I think Mr Cotsen is wrong in averaging the value of the land over the whole, since the value of the land lost is below that average for two reasons. Firstly, as I have indicated, it would have been more expensive to develop and, secondly, by involving a much larger development it would involve a longer period of deferment.

In the light of these uncertainties, I have to look further for guidance. Mr Cotsen produced evidence of two other transactions in the Merthyr Tydfil area and I have looked at these pieces of land. The Old Swansea Road site is not yet developed. It is less attractively situated than the subject land, but both valuation witnesses agreed that it was in any case a " bad buy " and that the price paid was to some extent influenced by the extravagant climate which existed in the early months of 1973. I agree with Mr Protheroe that the land at Pontsticill is in a much more pleasant setting than the subject land. It is significant that here also, as in the case of the subject land, there were two levels of value, one in March 1971, when the land was sold at a price equivalent to £3,340 per acre, and again in November 1972, when it was sold at a price equivalent to about £7,600 per acre. In my opinion neither of these two transactions substantiates the level of value adopted by Mr Cotsen.

Turning to Mr Protheroe's valuation, I have already indicated that the value at the end of October 1971 could well have been appreciably higher than the price which the claimants agreed to pay in August. I do not consider that the price agreed for the purchase of the land from Mr Thomas properly reflected the full development potential of the land as it existed in the middle of October. The doubts which I have expressed as to the relevance of the August 1971 value apply with greater force to the values which were taken two years earlier, when compensation was agreed following the grant of the earlier easement.

So, on the one hand, I have the evidence of the Gilbert & West negotiations—the figures for which must, in my judgment, be heavily discounted for uncertainty, for risk and for deferment. On the other hand, the two earlier values taken for the subject land and the earlier sale of the land at Pontsticill do not appear to me adequately to reflect the true value of the land in question, ripe for development as the subject land was at the relevant date. In between these two extremes are the actual transactions in the sale of land at Old Swansea Road and at Pontsticill which took place some time after the relevant dates, at a time when, as both valuers agreed, land values were rapidly increasing.

Award

In the light of all the evidence, I have finally come to the conclusion that the compensation payable to the claimants is the sum of £27,000, which I reach by adopting a value of £4,500 per acre, adding thereto the agreed sum of £5,250 for abortive architect's fees, and deducting the agreed figure of £1,000 which represents the residual value of the affected land retained in the ownership of the claimants.

The relevant date of valuation, as I have already stated, is agreed as October 15 1971, and the compensation which I award represents the loss to the claimants calculated as at that date. The claimants accordingly ask for interest to run from that date, and as I understand it, the authority do not resist that claim. Since this is a reference by consent under section 1 (5) of the Lands Tribunal Act 1949, the tribunal is acting as an arbitrator, and in that capacity I award interest to run from the relevant date—as representing part of the loss suffered by the claimants and therefore forming part of the disputed claim referred to the tribunal for arbitration. As to the calculation of interest, since the parties are agreed that compensation is to be assessed in accordance with the rules governing compulsory purchase, I take October 15 1971 to be a deemed date of entry and determine that interest shall run from that date at the rates laid down from time to time under the Acquisition of Land (Rate of Interest after Entry) Regulations, as provided in section 32 of the Land Compensation Act 1961.

Having read the decision in this matter and having then opened the sealed offer of compensation lodged by the authority, I find that the amount thereof is less than the amount of my award; accordingly the authority will pay to the claimants their costs of this reference, such costs failing agreement to be taxed by the Registrar of the Lands Tribunal on the High Court Scale.

J BRESGALL & SONS LTD v HACKNEY LONDON BOROUGH

(Ref/182/1975)

(J D RUSSELL-DAVIS FRICS)

April 2 1976

Estates Gazette May 22 1976

(1976) 238 EG 577

Disturbance claim for removal of manufacturing business from Hackney workshop premises to Hendon railway arch —Adaptation of alternative premises under terms of seven-year lease held to be value for money—Expenditure on additional electric-light points and erection of partitioning partly allowed—Installation of wc an improvement and disallowed—Tribunal not impressed by " long-stop " valuations

J Sullivan (instructed by Rose & Birn) appeared for the claimants; C J Lockhart-Mummery (instructed by the solicitor to the London Borough of Hackney) for the compensating authority.

Giving his decision Mr. RUSSELL-DAVIS said: This reference was initiated by the claimants for the determination of the amount of compensation to be paid to them in respect of the compulsory acquisition of their leasehold interest in the workshop, premises and yard at 28a Smalley Road, Hackney, London N16 ("no 28a"), which comprised 1,270 sq ft on two floors and has now been demolished. The claimants, J Bresgall & Sons Ltd, are manufacturers of catering hire equipment and no 28a was used for their business, including the repair and maintenance of catering furniture. No 28a was included in a compulsory purchase order made in 1969, which was confirmed on June 2 1971. Notice to treat was dated January 21 1972, to which particulars of claim were effectively returned on June 11 1972. Entry was finally made on August 6 1973, which is the relevant date of valuation. The claimants held a lease of no 28a for seven years from December 31 1969, and the value of that interest has been agreed at the sum of £1,350.

It remains to determine compensation for disturbance under rule (6) of section 5 of the Land Compensation Act 1961 consequent on the removal of the claimants to other premises, which they failed to find until July 1973. They entered into a lease from August 6 1973 until September 28 1980 from British Waterways Board at a rent of £1,500 pa, of a railway arch, no 9, at Adrian Avenue, North Circular Road, Hendon, NW2 ("no 9"). These premises had an effective area of 2,320 sq ft covered space, with an open yard of 580 sq ft. Under this lease the claimants are under covenant to carry out the following work at their expense:

1. To provide means of carrying away moisture seepage from internal face of the arch.
2. To take down the existing north-east closing end of the arch and to erect in permanent materials a wall surmounted with glazed panels and to have a 10 ft x 14 ft doorway.
3. To repair and make good or renew the glazing and glazing bars over the south-west brick closing end of the arch.

There is a right of re-entry in favour of the lessors on six months' notice at any time for the purpose of the undertaking. The user is restricted to the claimants' business, except with consent.

The following items of compensation, all in connection with the move to no 9, have been agreed:

	£
(a) Removal expenses	475.00
(b) Positioning and bolting down of machinery	155.00
(c) Legal and lessors' costs for new lease of no 9	120.35
(d) Rent and rates for no 9 during alterations from 1.8.73 to 1.11.73	417.38
	£1,167.73
(e) Cost of electric power wiring and installation, agreed shortly before the hearing ...	564.00
	£1,731.73

The matters remaining at issue, as claimed, are as follows:

	£
(A) Cost of adaptation of the arch at lower estimate: this includes substantially that work which the claimants are under covenant to carry out	2,127
(B) Electric light installation	594
(C) Installation of one wc	850
(D) Erection of brick and chipboard partition walls internally for the purpose of carrying on business	838
	£4,409

The authority admit that the work (except for item (C)) has been carried out and that the cost is reasonable. They do not accept liability, shortly, on the ground that it amounts to an improvement and that the new premises are better than the old.

"In a Deplorable State"

Mr Lewis Briscoll, a director of the claimant company, described no 28a. He said that he had been responsible for taking the lease of no 9, which, although in a deplorable state, was the only alternative accommodation he could find. It was full of debris; all the work under head (A) above was essential to make no 9 usable. At no 28a there was a total of nine light points on each floor; in order to achieve the same standard of lighting for a lofty arch, 36 points were necessary. There was a wc at no 28a, but none at the arch, and until one was installed it was impossible fully to use it; owing to shortage of funds this has not yet been done. The only partitioning at no 28a was for an office, but there were two floors which constituted some subdivision. At no 9 they had built some 10 or 11 partitions in order to keep machinery and storage separate and had enclosed an office, and there was a central van avenue for ease of loading to take the place of the hoist at no 28a into which vans backed. The partitioning was roofed over to provide storage platforms. Mr Briscoll agreed that working conditions at no 28a were cramped, and that at no 9, with a clear height of 18 ft, storage was much easier.

Mr Sullivan for the claimants called expert evidence from Mr G W Mason FRICS, who was acquainted with both no 28a and no 9. He maintained that all the work carried out at no 9 was essential to make it usable for the claimants' business, but no better than at no 28a; in a sense it amounted to what might be termed "equivalent reinstatement." But he agreed that the expenditure in items (A), (B) and (C) did enhance the value of the lease, and that the need for it would have been reflected in the lease rent of £1,500—in other words the claimants were getting value for their money in respect of the lease; if the landlords had done the work at their expense the rent could have been expected to be higher. He accepted that if and when the landlords resumed occupation, the tenants could claim for the work in (C) as an improvement under the Landlord and Tenant Act 1927.

Mr Mason did not consider that the work under (D) appreciably enhanced the value of the lease, since it was only of special value to the claimants.

If these arguments were not successful, he put in a valuation, as a hypothetical exercise, showing that the capital value of the lease, the work having all been done, was less than the cost of the work. This valuation was amended during the hearing and gave a value of the lease of £3,473 as against an expenditure of £4,409, a net shortfall of £936.

Mr Lockhart-Mummery, for the authority, called evidence from Mr G Sinclair FRICS, the borough valuer for Hackney, who dealt separately with each item of claim.

The work under (A), he said, was an express condition of the lease: the work had to be carried out by the tenants and was reflected in the rent of £1,500.

Under (B) he had conceded £564 for the electric power installation, since this was in the nature of special adaptation. That existing was single phase, and the claimants' machinery required 3-phase. This was not the case with the light wiring; that existing was defective and unsafe through damp and had to be renewed under the full repairing terms of the lease, since to keep in repair implied first putting in repair. This differed but little from the position under item (A).

Under (C) the installation of a wc was an improvement to no 9, and notwithstanding that it might be required under the Factories Acts, the lessees were not precluded from claiming it as an improvement under the Landlord and Tenant Act.

He recognised the partitioning erected under (D) to be substantial, and that the sum of £101, the cost for timbers for the office, amounted to an improvement to the premises.

The balance of £737, however, did not constitute an improvement but was for the benefit of the tenants, the claimants, in that it provided convenient subdivision and made use of the extra height for stacking purposes. On reflection, he did offer to contribute two-elevenths of the cost of £737 as that necessary to separate machinery from storage; this amounted to £135.

He had carried out a similar "long-stop" valuation exercise as Mr Mason, and he showed a deficiency of £223, using rather different figures, because, he said, he had taken account of the value of the yard, which Mr Mason had not. But he did not rely on this valuation.

In his concluding address Mr Lockhart-Mummery submitted that the claim had to satisfy five tests. It had to be a direct and natural consequence of the displacement from no 28a; it had not to be too remote; it had to be necessary for the continuance of the particular function of the business; it had not to be reflected in any other interest—in other words no double compensation; and it had to be assumed that the price paid for the alternative accommodation represented market value—in other words that the claimants were getting money's worth.

"M & B" and "Smith" Cases

Money's worth was reflected by item (A), and he relied on the decisions of *M & B Precision Engineers v London Borough of Ealing* (1973) 225 EG 1186 and *Smith v Birmingham Corporation* (1974) 232 EG 593, which disallowed claims where the claimants were getting value in the alternative premises. The same consideration applied to item (B), and the *Smith* case dealt directly with the cost of power and light installations, which claim was disallowed. Item (C) also failed the test of money's worth, and wcs and sanitary fittings were dealt with in both *Smith* and *M & B* and disallowed. Item (D) did not satisfy any of the tests, apart from the £135 conceded. The claimants had not enjoyed partitioning of this extent at no 28a.

The two long-stop valuations displaced the value-for-money argument; neither valuer supported his unit prices by comparable transactions and offered only his experience.

Mr Sullivan invited me to disregard the *M & B* case; money spent on leasehold property can seldom be recovered unless the lease were offered for sale, and this was a hypothetical consideration because the claimants had acquired no 9 simply to carry on their trade and not to speculate with. The expenditure incurred was solely to allow them to continue their business. He reminded me that the partitioning did provide for van-loading, which was available at no 28a. He referred me to *Wilrow Engineering Ltd and Another v Letchworth UDC* (1974) 231 EG 503 in which it was argued that the erection of partitions in the new premises was a direct consequence of the displacement, and the whole cost of £340 was allowed.

I have viewed the premises at Arch no 9 and I recognise many of its disadvantages for the claimants' business. On the other hand it is much more spacious than the old premises, which I can judge, from the photographs put in by Mr Sinclair, to be typical outworn workshops as are frequently to be found in back streets of East London.

On the four elements of claim:

Item (A) £2,127. I accept the value-for-money argument, and that this work was carried out under the express covenant in the lease; and I follow the *M & B* and *Smith* cases in disallowing this item.

Item (B) (now) £594. From an examination of the electrical contractor's account this figure is the cost of wiring for 36 points at £13 per point, £468, and supplying 28 fluorescent fittings, £126, or £4.50 each. The same consideration of value for money could be said to apply in so far as it covers renewal of a defective or unsafe installation. Mr Briscoll said that he achieved the same standard of illumination as at no 28a by the 36 points; but no 9 (2,320 sq ft) is almost twice as large as no 28a (1,270 sq ft). On this basis, had no 28a been the same size as no 9, on a crude calculation it would have required 33 points. The remaining three, therefore, might be said to be in the nature of special adaptation. I think it is fair, therefore, to allow three points at £18 each: total £54.

Item (C) £850. I do not think it can be denied that the installation of a wc would be an improvement for the enhancement of the value of the lease of no 9. Had it existed when the lease was entered into, I do not believe there is any doubt that the rent would have been higher. At the termination the improvement can be claimed under the Landlord and Tenant Act 1927. This is enough then to dispose of this item of claim, but it may be said that a similar claim was disallowed in both the *M & B* and *Smith* cases.

Extensive and Permanent Partitioning

Item (D) £838. Mr Briscoll said that at no 28a, with floors of low height, there was no partitioning other than for the office. He may have implied there was no need for it, and no need in the circumstances to divide machinery from storage, because the floors were small, storage mainly upstairs and machinery downstairs, so that in a sense the floor acted as a partition, albeit at a sacrifice of convenience, having to work on two levels, compared with no 9.

No 9 has been quite extensively partitioned, the work being permanent and becoming a landlord's fixture, and its arrangement does afford stacking and storage space; this may be considered to give an improvement over no 28a, but on the other hand no 9 is inconveniently lofty for good working conditions, and without at least some partitioning I do not believe that these conditions would be as good even as at no 28a. On this account I am inclined to award something towards the cost. I find the *Wilrow* decision to be persuasive, but it can be distinguished because, although the partitioning at no 9 does not appreciably enhance the value of the lease, there was virtually none (other than for the office) at no 28a which would have been included in the compensation for that lease. I have come to the conclusion that I should allow one-third of the balance of £737, say £250, in addition to the £135 already conceded; this makes £385 for the item.

Finally, I should say that I am not very impressed by the "long-stop" valuations put in by Mr Mason and countered by Mr Sinclair. I think they are unreliable in that they differ widely, they do not analyse what are and what are not improvements and neither valuer appears to place much confidence in them.

The compensation due to the claimants thus becomes:

Lease	£1,350.00
Agreed items	1,731.73
Electric light installation (B)	54.00
Partitioning (D)	385.00
	£3,520.73

Accordingly the respondent authority will pay to the claimants the sum of £3,520.73 by way of compensation under all heads of claim, together with a surveyor's fee in accordance with no 5 of the scale of charges of the Royal Institution of Chartered Surveyors, and any legal costs properly incurred from the date of notice to treat up to the date of reference.

Having read this decision and having opened and read the sealed offer lodged on behalf of the authority, dated February 20 1976, I find that the amount thereof is less than my award; accordingly the authority will pay to the claimant costs of this reference, such costs if not agreed to be taxed by the Registrar of the Lands Tribunal on the High Court Scale.

WALTERS AND OTHERS v SOUTH GLAMORGAN COUNTY COUNCIL

(REF/144/1975)
(DOUGLAS FRANK QC, President)
April 8 1976

Estates Gazette June 5 1976
(1976) 238 EG 733

Preliminary point of law—Date of assessment of compensation for disturbance—Blight notice served and accepted after CPO confirmed but before vesting declaration made or notice to treat served—Premises vacated by claimants before date of deemed notice to treat—Claimants' contention of entitlement to disturbance compensation under section 38 (1) (a) of 1973 Act fails—CPO irrelevant—By serving blight notice claimants fixed date of notice to treat

John Prosser (instructed by Phoenix Walters & Co, of Cardiff) appeared for the claimants, Joseph Lewis Walters, Cyril Ernest Brett and John Park; Vernon Pugh (instructed by the solicitor to South Glamorgan County Council) appeared for the acquiring authority.

Giving his decision, THE PRESIDENT said: This is a hearing on a preliminary point of law to determine the date on which compensation for disturbance is to be assessed.

The facts relied upon by the claimants can be summarised as follows:

The subject property of the reference, 8 and 9 Dumfries Place, Cardiff, was occupied as offices by the claimants, a firm of solicitors. I shall refer to this property as "the offices." The respondent acquiring authority is the successor to the Cardiff City Council, and I shall refer to them both as "the council." At all material times the offices have been included in the council's central area redevelopment proposals, for the implementation of which their acquisition was necessary. A compulsory purchase order was made in June 1970, and the area of the order included the offices.

On August 24 1971, the council wrote to the claimants referring to the compulsory purchase order and stated that the Secretary of State for Wales' decision on it was expected in the early part of 1972. They said that their proposals provided for the construction of a new office building and inquired whether the claimants would be interested in taking accommodation in it. They also enclosed a questionnaire on relocation and mentioned that they had a vacant site to let, stating that priority would be given to occupiers affected by the proposed redevelopment.

On September 14 1971, the claimants returned the questionnaire.

On November 3 1972, the council wrote to the effect that the Secretary of State's approval was imminent and stated: "You may consider it advisable in the interest of the continuity of your business to make arrangements for obtaining alternative accommodation as soon as possible if you have not already done so."

On February 2 1973, the council repeated what they had previously stated, namely, that the possession of the offices would be required at the latest by the end of March 1974 and again referred to alternative accommodation.

On January 2 1974, the compulsory purchase order was confirmed, and on April 5 the council stated that it was their intention to use the vesting declaration procedure as opposed to the usual service of a notice to treat. On October 30 1974, the claimant served on the council a blight notice pursuant to section 193 of the Town and Country Planning Act 1971, which on December 12 1974 was accepted as a valid blight notice.

On November 1 1974, the claimants vacated the offices and moved to other premises which they had obtained.

It is common ground that no vesting declaration was made, nor any notice to treat served pursuant to the compulsory purchase order made by the council. It is further agreed that pursuant to section 196 of the Act of 1971 the council were deemed, by accepting the blight notice, to have been authorised to acquire the claimant's interest in the property compulsorily and that notice to treat is deemed to have been served at the expiration of two months from the service of the blight notice, that is at about the end of December 1974.

The claimants contend that notwithstanding that the offices were vacant on the date of the deemed notice to treat they are entitled to full compensation for disturbance assessed as at one of the dates I have mentioned, namely August 1971, November 1972, January 1974 or April 1974. Mr Prosser, for the claimants, argued that the claim falls under section 38 (1) (a) of the Land Compensation Act 1973, that is to say that the amount of a disturbance payment shall be equal to the reasonable expenses of the person entitled to the payment in removing from the land from which he is displaced. He says that means what it says, namely that if the compulsory purchase order is confirmed, all damage that directly flows from it, if not too remote, is compensated for. He drew a distinction between the words in paragraph (a) and those in paragraph (b); the latter provided for compensation for a person having to quit the land, and in that respect the case was distinguishable from *Bostock, Chater & Sons v Chelmsford Borough Council* (1973) 226 EG 2163, 2359; 227 EG 141. He further submitted that the council had encouraged the claimants to seek alternative accommodation, although he later said that if there had been no such encouragement his argument would have been the same.

It seems to me that as there has only been one notice to treat, namely that deemed pursuant to the blight notice, I have to have regard to that alone. It necessarily follows that the compulsory purchase order made by the council, and all the correspondence which went with it, is irrelevant and inadmissible. It further seems to me that the law is quite clear, namely that in assessing compensation, one must have regard to the property and the circumstances as they were on the date of the notice to treat *rebus sic stantibus*. Thus, as Lord Denning, Master of the Rolls, said in *Bailey v Derby Corporation* [1965] 1 All ER 443 at p 445:

> ". . . you must ascertain the compensation for disturbance, as it is called. That must be ascertained by looking at what has in fact happened since the notice to treat."

I can find nothing in section 38 of the 1973 Act which calls for a departure from that principle. In my judgment, the claimants by serving the blight notice took themselves outside the compulsory purchase order proposed by the council and thereby themselves fixed the date of the notice to treat and the consequences which flow from that.

Mr Prosser, for the claimants, hinted that this was a case which might be decided on some grounds of equity. Quite apart from the fact that it may well be that there were advantages to the claimants in moving when they did, it seems to me that the law is quite clear, and it would be wrong for me to admit obscurity where there is now certainty. In the circumstances, I determine that compensation shall be determined following the date of the notice to treat deemed to have been served pursuant to the service of the blight notice, namely some time in December 1974.

Since this is a preliminary point of law, I think the right course is for costs to be reserved.

HAROLD W WALKER & SONS (WALTHAM ABBEY) LTD AND ANOTHER v ESSEX COUNTY COUNCIL

(Ref/176/1975)

(J D RUSSELL-DAVIS FRICS)

April 2 1976

Estates Gazette June 12 1976

(1976) 238 EG 807

Forecourts of insurance brokers' premises acquired for road widening—Value of lost private car-parking spaces—Injurious affection claim for sound-proofing windows allowed at half cost—Disturbance claim—Renewal and cleaning of carpets partly allowed—Surveyors' valuation fees for preparing details for tribunal disallowed as "in effect qualifying to give evidence"—permanent loss of profits due to loss of passing and casual trade during works also disallowed

P J Boyce, solicitor, of Budd, Martin, Burrett, of Chelmsford, appeared for the claimants; J R M Brown, a senior assistant solicitor with Essex County Council, for the compensating authority.

Giving his decision, MR RUSSELL-DAVIS said: This is a reference initiated by the two claimants, Harold W Walker & Sons (Waltham Abbey) Ltd and Sewardstone (E4) Investments Ltd, for the determination of compensation payable in respect of the compulsory acquisition by the respondent authority, the Essex County Council (hereinafter referred to as "the authority"), of 62 sq yds of land forming a forecourt used for car parking at 2, 3 and 4 Sewardstone Road, Waltham Abbey, Essex.

The acquisition took place to allow for the widening and improvement of Sewardstone Road (A112) at the junctions of Sun Street and Farmhill Road under section 214 of the Highways Act 1959. The compulsory purchase order was confirmed by the Secretary of State for the Environment on June 14 1972. Notice to treat was dated July 5 1972, in reply to which a statement of claim was returned jointly by the two claimants on August 15 1972, the day after possession had been taken by the authority; the latter date is agreed to be the effective date of valuation for the purpose of this reference. The road works were finally completed by March 28 1974.

Nos 2 and 3 Sewardstone Road comprise offices on the ground floor in the occupation of the first claimants ("Walkers"), who are insurance brokers, offices and a flat upstairs and are a pair of semi-detached houses standing at the junction of Monkswood Avenue, and no 4 is in residential use, the forecourts of all three properties being used by Walkers for the parking of private cars in connection with their business. There is a passageway, agreed to be 5 ft 11½ in in width at its narrowest point, between the flank walls of nos 3 and 4 leading to the rear of the premises available for parking as far as it is accessible. This rear land is otherwise landlocked, access via Monkswood Avenue being denied.

Increased Traffic-flow

The effect of the road scheme is to make the former 15-ft, one-way, southbound carriageway into a 40-ft, four-lane, two-way carriageway. Traffic emerging from Sun Street is, as before, one-way, east-bound and most of it turns either north up Crooked Mile or south down Sewardstone Road, the east arm of the crossroads, Monkswood Avenue, being a cul-de-sac. Traffic-flow past the property was found to have increased by 54.8 per cent from 15,100 vehicles in 1967 to 23,381 vehicles in 1972 for an adjusted 16-hour day, and heavy goods vehicles have also increased considerably. In place of a triangular reservation there are now three traffic islands and a full set of traffic lights, one standard being immediately outside no 3.

The shortest distance between the carriageway and the front wall (no 2) has been reduced from 18 ft to 9 ft.

Accommodation works carried out include a new 2-ft dwarf boundary wall at the back of the pavement. There was and is a 16-ft access and pavement cross-over.

Before the scheme, according to the claimants, five visitors' cars could be accommodated on the forecourt, three end-to-end parallel to the road in front of nos 2 and 3, one similarly in front of no 4 and one with its nose into the narrow passageway between nos 3 and 4. It is apparent that parked cars would have been immobilised until each in turn was moved. Since the scheme only two small cars can be accommodated, one in front of no 4, albeit in a narrower width, and one at right angles nose into the passage. Since the entrance is on top of the traffic lights, driving in and out and manoeuvring is very much more difficult for visitors coming from any direction. Staff cars negotiated the narrow width of this passageway and parked in the former back garden and still do, part having been asphalted. Evidence was given that at the date of the hearing the second claimants ("Sewardstone") are the freeholders of the whole. Walkers are the lessees but without formal tenancy agreement. At the date of notice to treat both companies had the same directors. Both parties in their valuations appear to have treated the two claimants as one entity. Although no submissions were made to me, it seems clear that Sewardstone are entitled to be compensated for the land-take and injurious affection, and Walkers for disturbance, and I have accordingly apportioned the respective valuations, as amended, and spoken to by the parties as follows:

Head of Claim	*Claimants*	*Authority*
Sewardstone		
1. Land taken	£3,000	£600
2. Injurious affection—sound-proofing and insulation costs	970.80	356
Total	£3,970.80	£956
	Say	£960
Walkers		
3 *Disturbance*		
Agreed items		
(a) Replacement of damaged trade sign	£32.56	£32.56
(b) Fees for structural survey	50.00	50.00
(c) Photographs taken during progress of works	5.00	5.00
(d) Cost of alternative parking during progress of works (not formally claimed but agreed during hearing)	50.00	50.00
(e) *Harvey v Crawley* costs Survey and valuation fees	54.00	54.00
Total agreed items	191.56	191.56
Disputed items		
(f) Renewal of carpets damaged during progress of works	371.90	—
(g) Cleaning of ditto	—	50.00
(h) Replacement of box sign removed during sound-proofing	9.00	—
(i) Loss of rent from letting	283.80	—
(j) Valuation fees	27.00	—
(k) Permanent loss of profits—cost of parking three cars	—	288.00
(l) Permanent loss of profits due to loss of clients during progress of works	3,257.00	—
	£4,140.26	£529.56
	Say	£540

Note. The authority's total valuation was rounded up to £1,500.

The total of the agreed items comes to £191.56. There are seven issues between the parties unresolved at the end of the hearing, since item 3 (k) can in effect be said to arise from a different valuation approach adopted by the authority. These issues will be considered in turn against the background of the road scheme.

Item 1. Value of land taken. Mr P J Boyce, for the claimants, called evidence from Mr Laurence Wilkinson FRICS, a partner in R Cheke & Co, chartered surveyors of East London and suburbs. By arithmetical calculation, Mr Wilkinson reckoned that, pre-scheme, 4½ cars could have been accommodated; he had frequently seen three to four cars parked and thought that with care five could be fitted in. He therefore valued the land for five spaces at a rental of £1 per week each, or £260 pa in all, which he capitalised at 12 years' purchase, making £3,120, which he rounded down to £3,000. He supported his rental by the parking charge made by the Abbey Filling Station on the corner of Farmhill Road 50 yds away of £1 per week; and by another letting of spaces behind offices close by, which, however, was subject to conditions which make it difficult to apply as a comparison.

In view of the danger and difficulty of entry and manoeuvre, he considered that, post-scheme, the value of the forecourt was merely nominal.

Mr J R M Brown, for the authority, called Mr John R Robinson FRICS, the deputy county estates officer and valuer, who considered that, pre-scheme, five cars were possible, albeit with the danger of backing out into the highway; post-scheme two spaces were the practical maximum, with the same backing danger.

He submitted two valuations. In the first he put 558 sq ft at £1 per sq ft, making £560. He supported this by a settlement in the same scheme of a forecourt on the corner of Sewardstone Road and Rue de Ste Laurence nearly opposite, for which the authority paid £738 in August 1972 for 738 sq ft of land taken. I find this to be a valid comparison because the land was used for car parking for a small commercial building but, post-scheme, vehicles can be parked in the roadway of the cul-de-sac outside the property, and in this way it differs from the subject land.

Alternatively, he valued three spaces lost at £200 each, £600, and he adopted this, the higher valuation. It was supported by another settlement for the loss of two car spaces in front of a public house at North Weald, eight miles away, which had accommodated some 30-40 cars. I find that this comparison has no validity.

To his second valuation he added, as a head of claim under disturbance (3(k)), a permanent loss of profits arising from the need for parking three cars in the Abbey Filling Station at £1 a week, £156 pa, from which he deducted the annual equivalent of the value of the land taken, £600 at 10 per cent, £60, capitalising the difference of £96 at three years' purchase, making £288; and he offered £888 in this respect.

Having viewed the land and spent some time observing traffic conditions, and having studied photographs of conditions, pre-scheme, put in by both parties, I judge that parking of five cars on the forecourt was just possible. Even making allowance for busy traffic-flow and morning and evening rush hour, I find it is possible now to park two small cars (up to 1,300 cc) and to this extent the forecourt, as it is now, has a remaining value. To value a loss of three cars is a liberal approach and up to this point I go along with Mr Robinson.

Subject Land Valued in Isolation

But both valuers have valued the subject land in isolation, Mr Wilkinson using 12YP, which is the multiplier he said he would have applied to the whole property of which the forecourt is an element. Mr Robinson (under 3(k)) in effect used 10YP for the car-parking element and he agreed that with a mixed office and residential use he would have used the same multiplier for the whole property.

I disregard the use of the rear car park as having no relevance to the loss of the forecourt, since it is too difficult of access for visitors. I think that measureable loss has occurred because visitors with large cars can neither use the rear car park nor be expected to go to the filling station; but I differ from both valuers, since in my view the more realistic valuation approach is the before-and-after method for the whole property. However, this approach has not been adopted, and I prefer Mr Wilkinson's method; I find that the value of the three lost spaces, which are all related to the office element of the property, amounts to £156 pa at 12YP, making £1,872. It is thus inappropriate to take account of the loss of profits as applied by Mr Robinson under 3(k).

Item 2. Injurious affection

The claimants incurred expense in sound-proofing the ground-floor office windows to the front of nos 2 and 3 at a cost of £712. The authority admits that the work was done and the expense reasonable but contend that betterment has taken place in that noise and vibration conditions are better than they were before the works; they do, however, concede half the cost of £356.

The cost of £258.80 was incurred in providing a new pair of glazed double doors to the outside of the entrance porch of no 2. There were already a solid front door and glazed inner vestibule doors giving on to a hall not used as an office, there being an inner door to the reception office in the front room of no 2. The entrance door was habitually left open or ajar to indicate to clients and visitors that the offices were open for business. It was claimed that the new doors are necessary as a sound barrier.

Having viewed the premises I find that present conditions are better than tolerable for sound, and I am not persuaded that traffic vibration is necessarily more discernible at 9 ft than 18 ft. I am satisfied that half-cost as offered by the authority is fair to the claimants. I am, however, not satisfied as to the necessity for the double doors.

I find that £356 is due to the claimants under this head.

Disturbance. 3 (f) and (g) carpets

The claimants contended that the carpets in the hall and front reception office, which were in August 1972 only two years old, were so fouled by mud and dirt during progress of works that they had been renewed at a cost of £371.90. The authority maintained that £50 for cleaning was sufficient.

There is a clear conflict of opinion here, but having heard the evidence, and considering that much of the works were undertaken during winter months, I have come to the conclusion that the claimants should be given the benefit of the doubt as regards £209.90 for the renewal of the hall carpet and I disallow £162 for that in the reception office, for which I award £25 for cleaning.

Accordingly I find for £234.90 under this head of claim.

3 (h). Box sign

I have not been satisfied that this claim for £9, although it arose from the sound-proofing work, is not too remote. I disallow this item.

Item 3 (i). Loss of rent £283.80

Evidence was given that Mr Edwards, the tenant of the top-floor offices of no 2 on monthly tenancy, left in March 1974 after completion of the works because post-scheme working conditions were intolerable. The offices were not relet immediately until the first-floor tenant vacated, so that the two floors could be let as a whole.

Having read the correspondence in which the following words appear "Since the recent redevelopment of the roads I have found it difficult to continue my work with the offices rented by ourselves and regret having to give notice having found quieter premises." I am not wholly convinced that the real reason was in fact that conditions were intolerable; the tenant had endured the turmoil of the works themselves; moreover, I do not think that a real effort had been made to relet until the first floor became available and advantage was taken of the requirements of good practical estate management to find one tenant to occupy the whole upper part. I cannot find, therefore, that this loss of rent is an item which was the direct result of the road scheme.

3 (j). Valuation fees £27

This forms part of R Cheke & Co's account and reads "inspection, preparing of details and submissions for the Lands Tribunal during January and February 1976."

I find that this work, in so far as it is not included in that

A which is covered by scale 5, is in effect qualifying to give evidence at this hearing, and costs follow the awards of compensation. These fees cannot be allowed as an item of claim. The claimants' *Harvey v Crawley* costs due to Mr Wilkinson's firm have already been agreed at £54.

3 (l). Permanent loss of profits £3,257

Detailed and fairly involved evidence of the nature of Walkers' business and their clients was given by Mr T M Ball FCA, their chartered accountant. A greater proportion of commissions earned by the firm were derived from the major or established clients who relied on the telephone or personal calls from the members of Walkers; and it is not B claimed that income from this source was affected by the road works in progress. There was, however, a growing business from "smaller" clients, who included passers-by or casual trade. The income from this business increased between 1970-71 and 1971-72 by 197 per cent; in the next year the increase was smaller at 16 per cent; and in the third year to August 1974 actually decreased by 9 per cent: the works period fitted into these last two years; the effect of small business is always more apparent in the second year due to renewals of insurance, etc, when the casual becomes an established client.

C Walkers attributed the cause of this progressive decline to the actual progress of the works during which time on a few occasions it was impossible to reach the offices, and for a long period, at best, conditions deterred potential clients from approaching. It was noticeable that after completion of the works between 1974 and 1975, the upward trend was resumed by an increase of 30 per cent over the previous year.

For the two affected years 1972-73 and 1973-74, Mr Ball calculated that there was a deficiency of income from this source of £332 and £2,925 respectively, making £3,257, the amount claimed; such figures were not offset by any increase in cost of overheads for servicing such trade. He admitted D that the basis of his calculations was "somewhat contentious."

Mr P A Wright CIPFA, chief technical assistant in the County Treasurer's Department, considered that the company's accounts did not distinguish between commissions receivable and premiums payable on behalf of clients; it was the former out of which the expenses of the business were met and profits derived. Nor did Mr Ball's evidence distinguish between the classes of small clients, that is to say those who came through personal introduction and those who were passers-by or casuals; the former, he thought, do in fact search out the company and would not be deterred from E an intended visit by the state of the works. It was only the latter who might be deterred, but they had not been identified. It was therefore impossible reliably to detect a change in the amount of business transacted.

Monthly bankings, which did not distinguish between premiums and commissions, were plotted on a graph on a three-month moving average which levelled out cyclical or seasonal fluctuations, and this did in fact show a statistical surplus between mid-July 1972 and end-March 1974 over the general upward trend. But he thought that no meaningful conclusion could be drawn from the bankings.

I find myself in agreement with Mr Wright that Mr Ball's F evidence failed to identify the passing or casual trade which it is claimed was lost, although in nature of things, I believe it is possible that some potential clients may well have been deterred. On the other hand the graph shows a continuous uptrend of cash flow, with even a statistical surplus over the critical months based on the moving average of bankings.

I find that Mr Ball's approach to the problem, through no fault of his own, because the information was not available, is much too crude for me to find proved that there was a measurable loss of business due to the execution of the works. In the absence of such proof I fear I cannot find for the claimants under this head of claim.

G

Tribunal's Award

Accordingly, I find that the following sums are due to the respective claimants under all heads of claim:

Sewardstone	
Land taken	£1,872.00
Injurious affection	356.00
	£2,228.00
Walkers	
Agreed items	£ 191.56
Renewal and cleaning of carpets ...	234.90
	£ 426.46

H

The authority will therefore pay these sums to the respective claimants, together with surveyors' fees in accordance with no 5 of the scale of charges of the Royal Institution of Chartered Surveyors and any proper legal costs incurred from the date of notice to treat up to the date of notice of reference.

Having read the decision and having opened and read the sealed offer of compensation lodged by the authority dated February 25 1976, I find that the amount thereof is less than the amount of my award; accordingly the authority will pay J the costs of the claimants of this reference, such costs if not agreed to be taxed by the Registrar of the Lands Tribunal on Scale 4 of the County Court Scales of Costs, but with no allowance for the expert witness, Mr Ball, on the issue of permanent loss of profits (head of claim 3 (1)).

BURLIN v MANCHESTER CITY COUNCIL

(Ref/68/1975)

(J STUART DANIEL QC and E C STRATHON FRICS) K

February 9 1976

Estates Gazette June 19 1976 and June 26 1976

(1976) 238 EG 891 and 974

Revocation of planning permission for development of land at Rusholme, Manchester, with restaurant, residential club, flats and garages—After consent site declared within conservation area and large building on it "listed"—Council's contention of implied condition in permission that pre-existing buildings be demolished before any development not upheld—Value after revocation held to be Schedule 8 value—Parcel reserved for restaurant and club valued at equivalent residential value—Abortive costs allowed without interest—Claim for interest on total compensation sum "reluctantly" disallowed—Tribunal raises question why in suitable case it should not have power to award interest antecedent to award—Present power "seems to verge on the absurd"—Award of £157,281 L

Iain D L Glidewell QC and A B Dawson (instructed by Alexander, Tatham & Co, of Manchester) appeared for the claimants; J Fitzhugh QC and P B Keenan (instructed by the director of administration of Manchester City Council) for the compensating authority. M

Giving their decision, THE TRIBUNAL said: In this case the claimants ask for compensation for the revocation of a planning permission. They ask for £270,825. The compensating authority, the City of Manchester, deny that they are liable to pay any compensation at all. Alternatively, they say that the compensation should be assessed at some much lower figure.

The claimants were represented by Mr Glidewell and Mr Dawson. They called two witnesses, Mr L D Clegg FRICS, of the firm of Leslie D Clegg, Morgan & Co, and Mr Alexander Schulz, a developer. The city corporation were represented by Mr Fitzhugh and Mr Keenan. They called four witnesses, Mr R C McCormack, principal planning officer with the corporation, Mr F W Marshall, an architect specialising in civic design, Mr R J Warburton RIBA, an architectural and planning consultant, and Mr G R Jackson ARICS, senior assistant city valuer with the corporation.

In the course of the hearing, issues of causation, inference and valuation were raised. To see how these arose it is necessary to summarise the facts. We will leave over for the moment the detailed questions of valuation, but it will be convenient now to recite the basic facts and events which led up to the revocation of the planning permission and the claim for compensation which followed it. This summary comprises the principal events set out in the agreed statement of facts which was read to us and certain other facts as found by us from the evidence.

The claimant, Mr A H Burlin, and the trustees of a settlement created by him (also claimants) owned an area of some 4.5 acres at Anson Road, Victoria Park, Rusholme, Manchester. Anson Road is one of the main radial streets leading from the centre of the city to the south east. The land, which is irregular in shape, has a frontage of about 500 ft (as measured by the tribunal) to Anson Road. On the east is Hanover Crescent. On the north is Daisy Bank Road and Victoria Park itself. On the west is the Christian Science Church. In the centre of the Anson Road frontage there debouches Hope Road, a short road which curves away northwards and serves two comparatively large buildings, Milverton Lodge and The Gables. At one time there existed another large house at the north end of Hope Road. This was Wyncote, which has now been demolished. The Burlin family live in The Gables. Milverton Lodge became a residential club in the 1950s and it so remains. The several different ownerships of the claimants need not be distinguished, for in 1967 they agreed, as between themselves, that the whole of the area should be redeveloped as a single unit.

We will now set out a diary of events in date order:

May 26 1965. Outline planning permission (no 47440 TP) was granted for the development of a restaurant and residential club containing 32 service flats with garage and parking spaces, landscaping and access road, and 20/24 dwelling-houses and garages (this permission was revoked in 1968 in order to reserve a frontal strip of land for future widening of Anson Road; compensation of £11,000 and £925 for abortive expenditure was paid by the corporation).

March 1967. Meanwhile (knowing that the above revocation was pending) the claimants obtained, on March 1 1967, an outline permission (no 51910 TP) and on April 5 1967 a detailed planning permission (no 52066 TP) for the development of the whole of their land except for the frontal strip. This land ("the subject land") extended to 3.19 acres.

The terms of the outline permission (no 51910 TP) were as follows:

"Notice is hereby given that pursuant to the provisions of the above-mentioned Act and Order the Council have had under consideration an application for permission to develop land as follows, viz:

Erection of restaurant and residential club, 46 service flats with garages and garage spaces, 64 flat units and garages, after demolition of existing property, site bounded by Anson Road, Daisy Bank Road and Hanover Crescent, Rusholme.

The council hereby grant the permission applied for in accordance with the plans and particulars submitted, subject to the following conditions:

(1) Before the development is commenced the applicant shall submit to the Council, and obtain their approval under the Town and Country Planning Act of, detailed plans and elevations of the proposal and detailed particulars of the materials to be used in external elevations.

(2) The site to be landscaped and subsequently maintained to the satisfaction of the Corporation; details of the landscaping to be first approved before the development is commenced.

(3) The consent of the Highways Committee to be obtained to the size and design and siting of the proposed crossings over the footpath in Hanover Crescent before the development is commenced.

(4) No trade or business to be carried on in the garages.

(5) The requirements of the Medical Officer of Health to be complied with.

(6) The requirements of the Corporation to be complied with regarding means of escape in case of fire.

(7) No trees to be felled or lopped without the prior consent of the Corporation.

The reasons for the imposition of the foregoing conditions are set out below:

(1) To ensure that the proposed development is carried out in such a manner as will secure amenity and convenience.

Dated this 1st day of March 1967."

The terms of the detailed planning permission (no 52066 TP) were as follows:

"Notice is hereby given that pursuant to the provisions of the above-mentioned Act and Order the Council have had under consideration an application for permission to develop land, as follows, viz:

Erection of restaurant and residential club, 46 service flats with garages and garage spaces, 64 flat units and garages after demolition of existing property, Anson Road, Hanover Crescent, Daisy Bank Road, Victoria Park.

The Council hereby grant the permission applied for in accordance with the plans and particulars submitted, subject to the following conditions:

(1) The consent of the Highways Committee to be obtained to the location, size and design of the proposed crossings over the footpath.

(2) The site to be landscaped and subsequently maintained to the satisfaction of the Corporation; details of the landscaping to be approved by the Corporation.

(3) No trade or business to be carried on in the garages.

(4) The requirements of the Corporation to be complied with regarding means of escape in case of fire.

(5) The requirements of the Medical Officer of Health to be complied with.

The reasons for the imposition of the foregoing conditions are set out below:

(1) To ensure that the proposed development is carried out in such a manner as will secure amenity and convenience.

Dated this 5th day of April 1967."

In the development mentioned in these permissions the service flats (including a penthouse) would have provided a total of 88 habitable rooms and the flat units would have provided 160 habitable rooms. Among the details approved in this permission were drawings which showed four blocks of flats and a block of service flats partly superimposing a building showing the restaurant and a banqueting suite. Garaging and parking spaces were shown in detail as were also many mature trees.

April 5 1967. Certain trenches were dug on the site with the object of precluding the possibility of the development sanctioned by permission no 52066 TP attracting a liability for betterment levy.

March 1 1972. The corporation designated Victoria Park (including the subject land) as a conservation area for the purpose of section 277 of the Town and Country Planning Act 1971.

May 2 1973. Milverton Lodge (which is an attractive mid-19th-century villa with interesting Gothic fenestration) was placed on the statutory list of buildings of architectural or historic interest.

May 1973. There was a period of uncertainty. It is clear from Mr Clegg's evidence that the claimants decided not to press on with development until things were clearer and the corporation's policy was known; they had intended, as part of their scheme, to demolish Milverton Lodge; but they then

A heard that Milverton Lodge was listed and decided to test the matter by applying, as they did on July 2 1973, for "listed building consent" to demolish the Lodge. During the same period another doubt arose, namely whether the permission no 52066 TP of April 5 1967 would under the planning legislation survive beyond March 31 1974 if no further development occurred meanwhile; in November 1973 the claimants applied for a renewal of that permission and (*ex majori cautela*) also carried out some site clearance and excavation for one of the blocks of flats, block "A" of the proposed development (see entry under March 20 1974).

October 8 1973. Mr Clegg wrote to the city planning B officer as follows:

"In granting to your Council the enclosed extension of the statutory time limit during which a decision on our Application for Listed Buildings Consent should be communicated to us, we are asked to explain that the Listing of Milverton Lodge as 'a building of special architectural or historic interest' in the spring of this year has frustrated our clients from implementing as they had intended in the past six months the Planning Permission 52066 T.P. which was granted by the Council for 'the erection of a restaurant and residential club, 46 service flats with garages and garage spaces, 64 flat units and garages after demolition of the existing property at Anson Road, Victoria Park.'

C The existing property of which demolition was envisaged included Milverton Lodge.

In view of the inability of your Council promptly to decide the application for Listed Buildings Consent for the demolition of Milverton Lodge, we are to request that the current Planning Permission 52066 T.P. which is due to expire on the 1st April 1974 be renewed forthwith for a period which would expire on the 31st March 1975.

Would you please tell us whether this Application for renewal of permission can be determined at an early date on the terms of this letter, or indeed whether you wish us to prepare and submit to you copies of the drawings already approved, along with a formal planning application.

D Your early reply would be much appreciated."

February 6 1974. The corporation decided:

(i) to refuse permission to extend the validity of permission no 52066 TP beyond March 31 1974;

(ii) to refuse permission to demolish the listed building Milverton Lodge;

(iii) to make orders under section 45 of the Town and Country Planning Act 1971 to revoke the two permissions of March 1 and April 5 1967 and to submit orders to that end for confirmation by the Secretary of State.

E *February 19 1974.* The corporation made an order revoking the permission of March 1 and April 5 1967 and submitted it to the Secretary of State. The reasons for making the order were expressed as follows:

"1. Since the planning permissions for the development of the above-mentioned land were granted on 1st March 1967 and 5th April 1967 respectively circumstances have changed as a result of the taking of the following action:

(i) On the 1st March 1972 the City Council designated an area in Victoria Park including the site to which the planning permissions relate as a Conservation Area for the purposes of Section 277 of the Town and Country Planning Act 1971.

F (ii) On the 2nd May 1973 the Secretary of State for the Environment compiled the fifteenth List of Buildings of Special Architectural or Historic Interest in the City of Manchester relating to Milverton Lodge which is within the site covered by the planning permission.

2. On the 2nd July 1973 and 22nd November 1973 applications were submitted on behalf of Mr A. H. Burlin for Listed Building Consent to demolish Milverton Lodge and for the renewal of the planning permission granted on 5th April 1967 respectively.

3. The Manchester Conservation Areas and Historic Buildings Panel, who have been appointed by the City Council to advise the Council on the designation of Conservation Areas and matters affecting Conservation Areas and Listed Buildings, were consulted about both applications. The Panel considered that the scheme G which was the subject of the application for renewal of permission was unacceptable in the Victoria Park Conservation area in terms of its density, height, scale and massing and that the loss of many mature trees was undesirable. So far as Milverton Lodge is concerned the Panel felt the building was a good example of the type of large scale family residence originally erected in Victoria Park and that the loss of the building would radically alter the character of the Conservation Area. Furthermore they felt that the development which was intended to replace the listed building was totally inappropriate in the Conservation Area. The Victorian Society, the Ancient Monuments Society and the Rusholme and Followfield Civic Society also opposed the granting of Listed Building Consent for the demolition of Milverton Lodge. H

4. The City Council were of the opinion that the development proposals would be out of scale and character with the Conservation area which comprises an area of low density development with a mature landscaped setting and they therefore decided to refuse to renew the permission granted in April 1967. The Council also felt that Milverton Lodge formed an integral part of the Conservation Area and accordingly refused permission on the application for Listed Building Consent for its demolition.

5. The Council also decided that in view of their decision not to renew the permission and to refuse Listed Building Consent they should revoke the planning permissions granted for the development of the site in March and April 1967." J

March 11 1974. The corporation approved under the Building Regulations the drawings for the flats contained in block A of the approved development.

March 20 1974. (This was the second episode of operations on the site.) The claimants served a notice on the corporation saying that trenches were ready for inspection.

March 22 1974. The Secretary of State confirmed the revocation order.

April 16 1974. The solicitors acting for the claimants gave notice to the corporation claiming compensation under section 164 of the Town and Country Planning Act 1971.

It was agreed during the hearing that there was a gap or K clearance of 8 ft between the north-west corner of the curtilage of Milverton Lodge and the nearest of the buildings proposed to be developed under the revoked permission (52066 TP), that is to say the proposed club and restaurant building. It was further agreed that ordinary vehicles could have circulated between this building and the curtilage of Milverton Lodge but that large vehicles such as fire engines would not have been able to get through the gap if they had come from Hanover Crescent and had wished to pass westwards along the south of the club building. It seems to us, however (as the claimants suggested at the hearing), that there would have been no difficulty in such vehicles reaching L that part of the new building from Hope Road.

During 1974 and 1975 correspondence passed and meetings took place between the claimants and their representatives and representatives of the corporation. The corporation were, as Mr Fitzhugh described it, "benevolent" in the sense that they were, albeit with some delay, perfectly willing to discuss a scheme of development replacing that which had been revoked. Between November 1974 and September 1975 no fewer than three advisory schemes were discussed. In respect of one of these the claimants applied for another planning permission. We heard on the fourth day of the five-day hearing before us that this permission had been M granted. However, having regard to the fact that it was agreed between the parties that valuation should be assessed as at March 22 1974, and having regard also to certain conclusions we have come to (as will appear later) as to the method of valuation, we do not think it necessary to set out an historical account or description of these subsequent events.

In order to be entitled to compensation for revocation of a planning permission, a claimant must establish, in accordance with section 164 (1) of the Town and Country Planning Act 1971, that he:

A (a) has incurred expenditure in carrying out work which is rendered abortive by the revocation or modification; or

(b) has otherwise sustained loss or damage which is directly attributable to the revocation or modification.

Then in section 164 (4) it is provided:

"in calculating, for the purposes of this section, the amount of any loss or damage consisting of depreciation of the value of an interest in land, it shall be assumed that planning permission would be granted for development of the land of any class specified in Schedule 8 to this Act."

B Schedule 8 broadly corresponds to the old Third Schedule to the Act of 1947; it among other things assumes permission for the rebuilding of certain buildings in certain circumstances. It is not necessary to refer to its provisions *in extenso*.

As will have appeared at the outset of this decision the differences between the parties are so great that it is not surprising to find that there was a major and fundamental difference as to the correct method of quantifying the depreciation, if any, of the value of the claimants' land directly caused by the revocation. The effect of section 164, said Mr Glidewell for the claimants, was to require two valuations; first, the value of the land with the permission; secondly C (after the revocation), the value of the land on the assumption that planning permission would be granted for Schedule 8 development (in this case agreed to be worth £78,000). The difference between these two figures was the measure of the depreciation of the value of the land. The effect of section 164 (4) was in the clearest and most express terms that, as regards the possibilities of planning permission, the assumption to be made post revocation was the Schedule 8 assumption; when a permission was revoked it was necessary to see whether any permission survived; of course, if there were two permissions and one was revoked, then the live permission would be valued, but if the sole actual permission had D been revoked it was not to be assumed (as apparently the corporation would argue) that some other assumption or hope of planning permission would be brought in, for that would be contrary to the statute, which limited the assumption to that in section 164 (4). There was nothing surprising or difficult about this, for if the slate had been wiped clean (apart from Schedule 8) and then some other new or different development was permitted, then any compensation paid on account of revocation could (or the appropriate amount thereof could) be recovered under the procedure provided for subsequent recovery of compensation under sections 158 to 160; this procedure was expressly applied in the case of E revocation by section 166 (5).

Mr Fitzhugh submitted in the first place that the revoked permission should be held to have been a permission for a single indivisible scheme of development which was subject to a condition that all the pre-existing buildings on the site should be demolished before any new building could be started; and it was the listing of Milverton Lodge, not the revocation, which had prevented the permission (so constructed) from fructifying. For this contention he relied in particular on three matters:

1. The claimants had agreed between themselves to F redevelop the site as a single unit.
2. The words of the planning permissions. In both the outline and the detailed permission the permission sought had been specified as being certain named buildings "after demolition of existing property," and that had been the permission granted.
3. Mr Clegg, by his letter of October 8 1973, had stated in words that the listing of Milverton Lodge had frustrated the claimants' scheme of development.

Mr Fitzhugh referred to three cases at this stage of his argument: *Francis v Yiewsley & West Drayton UDC* [1958] 1 QB 478; *Ellis v Worcestershire CC* (1961) 178 EG 103, LT; and *Lucas (F) & Sons v Dorking & Horley RDC* (1964) 17 G P & CR 111.

In *Francis v Yiewsley* the landowner had, on appeal to the Minister, obtained a permission to use some land as a caravan site for six months. The local authority served an enforcement notice seeking to prevent the continued use of the land as a caravan site. The enforcement notice was held to be invalid for reasons which need not be set out. In the course of his judgment Parker LJ (as he then was) said:

"The Minister by his decision . . . has permitted the user of this land as a caravan site for a period of six months. It is said (and said with some force) that no express condition is therein H set out that at the end of the six months' period the caravans are to be removed. But as I read it, such a condition must be implied."

Lord Parker then proceeded to explain why, in his view, having regard to the provisions of sections 14, 15 and 16 of the Act of 1947, the Minister really had no power to achieve the granting of a temporary permission except by the grant of a permission subject to a condition.

In *Ellis v Worcestershire CC* two planning permissions had been given on the same site. The earlier of the two was revoked, and in the course of proceedings claiming compensation for the revocation a question arose as to the scope of J the permission. It was contended for the landowner that in the absence of a limiting condition as to user an ordinary dwelling-house could have been built. At p 105 of the report Mr Simes said:

"The application sets out that the proposed development is a 'farmhouse' and that the proposed use of the land and/or buildings is 'farming.'

The permission which was granted permits 'development comprising erection of a farmhouse . . . in accordance with the application and plans.'

No conditions are imposed other than in regard to the design of the building, but, in my view, the permission granted was for K the erection of a farmhouse for use for farming the area of land shown on the plan as amended. In my opinion the use of the word 'farmhouse' and the reference to the application limit the occupation of the building as effectively as if a specific condition had been imposed, while the plan seems to me to indicate the area of land with which the building is to be occupied as effectively as a plan annexed to a permission for an ordinary dwelling-house showing the land to be occupied with it as garden land."

Thus, said Mr Fitzhugh, it was quite clear that in Mr Simes' opinion a condition could be implied into a permission (or into a permission eked out by the relevant application).

In *Lucas v Dorking* the developers had obtained permission L in 1952 to build on a plot of land 28 houses in a cul-de-sac, 14 to the north and 14 to the south. Later they obtained another permission—to build six detached houses fronting the main road and two of these had been built; then they partially reverted to the 1952 scheme, proposing to build the houses on the south of the cul-de-sac. The authority contended that the 1952 permission was no longer valid or effective. The developers sought a declaration that it was effective and entitled them to carry out all or any of the building or other operations to which it related. Winn J granted the declaration sought, finding that the 1952 permission was not to be regarded in law as a permission to develop the plot as a M whole provided that the approved layout was completed but on the contrary as permission for *any* of the development therein comprised. At p 119 the judge pointed out that the proviso suggested in the authority's contention could have been embodied in a condition but in fact had not been. Mr Fitzhugh said that there was here an important distinction to be made. In *Lucas'* case the land had been vacant land and the 1952 permission in effect gave 28 separate permissions to build 28 houses each in its own curtilage. In *Ellis's* case on the other hand the land on which the farmhouse was to be built was a single area and not severable. This was also

A true in the instant case, which was very much closer to *Ellis's* case than to *Lucas v Dorking*. Thus the position was reached that on the authorities it was permissible to imply a condition into a planning permission, and in the present case it was clearly necessary in all the circumstances to imply the condition that all the pre-existing buildings should be demolished before any new building could be started. But the impossibility of doing this had been caused by the listing of Milverton Lodge; the subsequent revocation merely confirmed this. Mr Fitzhugh then referred to the detailed provisions of sections 54 and 55 of the 1971 Act (dealing with buildings of special architectural or historic interest). He B mentioned in particular section 54 (9), which provides that anything within the curtilage of a building shall be treated as part of the building. There were two separate points here: first, the listing of Milverton Lodge made the implied condition impossible, secondly, the retention of Milverton Lodge would have caused serious problems of access and car parking (if it was suggested that the scheme could have proceeded while keeping the Lodge); there was a serious difficulty as regards large vehicles, and it was equally serious that, according to Mr McCormack's evidence, some 25 car spaces would have been lost.

C If this were right, then there was no depreciation in value flowing from the revocation. But in case the tribunal did not agree, Mr Fitzhugh addressed us as to the construction of section 164 of the 1971 Act. The claimants' contention was that after revocation the land must be valued as having its existing use value (or Schedule 8 value) and nothing else; this was wrong; after revocation the land should be assessed at whatever was its true value in the market, having regard to the hopes and possibilities of obtaining another planning permission; in isolation section 164 (4) might suggest that no assumption should be made about planning permission beyond development specified in Schedule 8; but then it was necessary D to look at section 178. Section 178 (1) and (2) provide as follows:

> "(1) For the purpose of assessing any compensation to which this section applies, the rules set out in section 5 of the Land Compensation Act 1961 shall, so far as applicable and subject to any necessary modification, have effect as they have effect for the purpose of assessing compensation for the compulsory acquisition of an interest in land.
>
> (2) This section applies to any compensation which, under the preceding provisions of this Part of this Act . . . is payable in respect of depreciation of the value of an interest in land."

E Thus section 5 (2) of the Land Compensation Act 1961 came into play and the value of the land was to be assessed on the open market; of course if a permission was revoked because, for instance, a motorway was to be constructed on the land, then the land might (post revocation) be worth very little in the market; then the compensation might be heavy, for there was no potentiality left so far as the landowner was concerned. The provisions dealing with repayment of compensation were concerned with cases in which in truth all potentialities seemed to have gone (and therefore compensation was paid) and then because of some unexpected turn of events some chance of development value reappeared. But F here there had always been potentialities, for the corporation were benevolent, as was shown by their three advisory schemes. The potentialities should therefore be assessed as at March 22 1974 in accordance with section 5 (2) of the Act of 1961.

Mr Glidewell in reply said that in the claimants' scheme there had been four blocks of buildings together with road-works. In his submission it was open to the claimants to build one or two or three of the blocks as per *Lucas v Dorking;* a planning permission was not to be construed like a contract; no doubt there was an intention to demolish Milverton Lodge, but blocks A and B were on vacant land and block C only impinged on the Gables, which could be G demolished anyway; there was no reason why some of the development could not have been carried out, deferring other parts; if it had really been the intention to prohibit any new work till all demolition had been done, surely there would have been an express condition with reasons (as required by law). As to *Francis v Yiewsley,* it was true that Lord Parker spoke of implying a condition, but this was said *obiter* and it was not said by the other two judges. Mr Glidewell then referred to the case of *Wilson v West Sussex CC* [1963] 2 QB 764, where conditions and limitations were discussed. He submitted that there could be an express condition or an H express limitation, ie a permission limited in scope, eg as to time or user, but not, *pace* Lord Parker, an implied condition; if that were right, then the ordinary express meaning of the permission allowed different parts of the development to be done at different times; if the suggested condition were implied, then once a listed building consent had been refused there was no need to revoke, yet a revocation order was made.

Returning to the question of section 164 (4), Mr Glidewell said that the reference to section 178 had caused a misunderstanding. It was important to remember that under the Lands Clauses Act 1845 and also under the Acquisition of Land (Assessment of Compensation) Act 1919 compensation had J been assessed with all the potentialities of the land. Then by the 1947 Act development value was taken away and that remained the case till the 1959 Act, when market value was reinstated. For the purpose of assessing compensation for compulsory acquisition under the rules comprised in section 5 of the Act of 1961, provisions are made in sections 14 to 16 of that Act telling the hypothetical purchaser what assumptions are to be made as to the likelihood of planning permission being granted. But these provisions are not applied in the case of revocation; there is instead the clear provision in section 164 (4) of the 1971 Act that the assumption as to planning permission would simply be that Schedule 8 develop- K ment would be permitted.

On the questions so far canvassed we can state our conclusions fairly shortly. Whether or not it can in principle be possible for a condition to be implied in a planning permission we cannot see any reason for such an implication in the present case. No doubt at the time when the permission was sought and obtained the claimants intended that Milverton Lodge and the other existing property should be demolished. This is common ground. And it is true that the permission for the new buildings was expressed to be as "after demolition of existing building." But the condition sought L to be implied is that all existing buildings shall be demolished before any of the new development is started. We can see no reason why such a condition should have been required. We think the words in the permission indicate no more than an indication of a probable sequence of events. We cannot believe for a moment that some alteration in this intended time-table would have been met by an enforcement notice. Indeed, ironically enough, the early works intended to avoid betterment levy were quite clearly new works and not works of demolition, so that on the corporation's reading of the planning permission the claimants were in breach of it on the very day it was granted. If such a bizarre arrangement had M really been intended we think that it would have been expressed and reasons given for it. As was suggested during the hearing, large schemes of this kind are often implemented in phases. We cannot think that it was intended to stipulate that one phase was on no account to be started unless all demolition had been completed.

On this view of the matter it is not necessary to analyse again the various reported cases to which we were referred, interesting though that might be.

We are satisfied, therefore, that the listing of Milverton Lodge did not cause the claimants' scheme to founder because of any impossibility of compliance with an implied condition, nor do we think that the listing of the Lodge prevented the scheme being carried out as a practical possibility. The tribunal was not particularly impressed with the beauty of the Lodge's gateway in Hope Road, but even assuming that it and the wall and everything in the curtilage were sacrosanct, all the new buildings could still have been constructed. It seems to us that problems of access and car parking could easily have been dealt with had not the corporation made up their minds that the scheme which they had approved in 1967 was now, in 1974, unsuitable in terms of " density, height, scale and massing." It seems to us that the conservation area was really the key to the matter. It was to safeguard this that the corporation refused to renew the permission (though this became irrelevant because of the operations on the site) and refused permission for listed building consent, and finally decided to revoke the permissions of March and April 1967. We think that if they wished to safeguard the conservation area the corporation, properly and of necessity, had to make a revocation order, and that it was this order which caused the claimants' scheme to founder. We do not think it is right to say that the revocation merely confirmed the listing of Milverton Lodge.

We do not overlook Mr Clegg's letter of October 8 1973 in which he wrote that the listing of Milverton Lodge had " frustrated " the claimants from implementing the planning permission. In evidence Mr Clegg said that in fact he had not thought that the permission had been frustrated by the listing. It was Mr Barnett and Mr Parnell, officials in the city planning department, who had first used this word in conversation with him and he had merely followed suit without meaning the word to have particular significance. We do not think that Mr Clegg's use of the word at that date should be regarded as of any great materiality.

Then the question arises whether the value of the land after revocation should be taken, as the claimants say, at its Schedule 8 value, or whether it should be taken with all its potentialities in the market. On this question we agree with Mr Glidewell. It was of course quite right for Mr Fitzhugh to refer us to section 178, and it is quite right also that in accordance with that section the compensation is to be in accordance with the rules set out in section 5 of the Act of 1961, that is to say the value of the land shall be taken to be the amount which it would fetch if sold in the open market by a willing seller. So, it is said, an open market is to be assumed and the land will be assessed in that market, first with the planning permission and then without the planning permission, and if the second value is less than the first, then that is the compensation. Now Mr Glidewell counters with the proposition that in a true compulsory acquisition the assessment is clothed with the various assumptions as to planning permission which appear in sections 14 to 16 of the Act of 1961: but in section 178 of the 1971 Act these assumptions are not mentioned. These assumptions are of various kinds and are tailored to suit different sets of circumstances. Some of them, as in section 15 of the 1961 Act, provide for an uncluttered assumption that a permission would be granted, others provide for grants of permission for certain classes of land coupled with the rider that permission for the development in question must be shown to be such as could reasonably have been expected to be granted.

Now section 178 of the Act of 1971 expressly states that the rules set out in section 5 of the 1961 Act are " so far as applicable and subject to any necessary modifications to have effect as they have effect for the purpose of assessing compensation for the compulsory acquisition of any interest in land." We have wondered whether these words could reasonably be read as importing by reference the assumptions which have to be made under sections 14 to 16 of the Act of 1961. We do not think this will do. In the first place the application and necessary modification, if one seeks to translate a situation of true acquisition to a situation of revocation, are full of difficulty. But stronger than that is the point which Mr Glidewell makes, that if the chances of obtaining planning permission are to be assessed, then having regard to the express provision in section 164 (4) that Schedule 8 development shall be assumed, it is inconceivable that it was also intended to import various other assumptions. It is perhaps worth pointing out that on a true compulsory acquisition one of the assumptions which has to be made under section 15 of the Act of 1961 is that " it shall be assumed that planning permission would be granted . . . for development of any class specified in the Third Schedule to the Town and Country Planning Act 1947 (which relates to development included in the existing use of land)." The Third Schedule to the 1947 Act, after amendments in the Acts of 1951 and 1954, became in effect the Third Schedule to the 1962 Act, which with some further amendment and rearrangement became the Eighth Schedule to the Act of 1971. If, therefore, section 15 of the Act of 1961 was to be regarded as engrafted into section 178 of the Act of 1971 there would have been overlapping, perhaps duplicating, matter which should surely have needed express reconciliation. We think the better view is that the reference to Schedule 8 in section 164 (4) is intended to be an exclusive definition of what assumption is to be made as to the planning permission which would be granted.

Having travelled so far we can now turn to questions of valuation. The particulars of claim and Mr Clegg's valuation were presented by him as a single document, which is reproduced as follows:

1. *Diminution in the value as at March 22 1974 of claimants' land.*	
Value of land with the benefit of planning permission 52066 TP as detailed in Appendix A	£285,000
Less value of land when deprived of the benefit of planning permissions 51910 TP and 52066 TP as detailed in Appendix B	78,000
Diminution in value	£207,000
2. *Expenditure incurred by claimants for the benefit of their land and rendered abortive* by the revocation of planning permissions 51910 TP and 52066 TP as detailed in Appendix C	4,582
3. *Loss of profit suffered by claimants after March 22 1974* when their development had to be suspended and no other permission for development was available, thereby precluding them from earning interest on the compensation sum. Say two years at 14% per annum on £211,582	59,243*
Total compensation	£270,825

Plus the claimants' professional costs in the formulation and assessment of their claim for compensation.

Note: The quantum of the item marked * would be amended in the event of the compensation being received by the claimants other than on March 22 1976.

Following our findings as to the correct basis of approach, we come to the valuations relating to Item 1 of the claim. We set these out opposite in a form which embraces the calculations as submitted and as expanded in evidence.

It can be seen that Mr Clegg values the land with the benefit of planning permission 52066 TP at £285,000 and that Mr Jackson's comparable figure is £173,800. The value after revocation was agreed at £78,000. Thus, on the basis that we have found to be the correct approach, the amount of compensation for this part of the claim was estimated by Mr Clegg to be £207,000 and by Mr Jackson at £95,800.

Offer to Purchase

Mr Clegg adopted, as primary evidence of value, an offer to purchase by Alexander Dawson Developments Ltd; it was made by its managing director, Mr Alexander Schultz, in a letter to Mr Arnold Burlin on January 18 1974. It read as follows:

> "This is to confirm the offer which we made to you last Tuesday when in Manchester. We make a firm offer of £295,000 for the entire buildings and land at Milverton Lodge and the Gables and all other lands therein. This is of course subject to the current planning consent. We look forward to seeing you once again on Tuesday the 22nd January 1974, when final negotiations and a programme for the vacation of the properties and the commencement of building can be agreed."

In evidence Mr Schultz said that he had first become interested in the subject land about six years ago and over this period the price talked about had varied between £200,000 and £300,000. The parties agreed that the offer had been made "subject to contract" and that it had not been accepted. Mr Clegg's approach had been (a) to accept the offer as evidence of value close to the date of valuation, (b) to devalue the offer and (c) to rely upon comparable transactions in support of the devaluation details. The details contained a deduction of £10,000 on account of an estimated payment by the claimants to provide for relocation of the tenancies of "Milverton Lodge" and "The Gables"; thus Mr Clegg's valuation of £285,000 accords with the offer as so adjusted. Mr Jackson relied for support of his valuation upon comparable transactions.

We will deal first with the value of the part of the subject land allocated for residential development. A composite schedule of the comparables of both witnesses is set out opposite.

It will be seen that Mr Clegg's comparable transactions occurred during the period from June 1972 to January 1974, before the date of revocation and valuation, March 22 1974. Mr Jackson identified himself with two of these and in addition cited two transactions and an offer, all three of which related to times which were post the date of valuation. We were told that by comparison with the subject land some comparables were to be rated better in quality and/or location and some inferior, and although there was some disagreement about this there was agreement that the sale prices of land had increased from 1972, to have reached a peak in August 1973, and thereafter had declined. It was Mr Clegg's view that, by the date of valuation, prices had reverted to about the 1972 level, but Mr Jackson thought the rate of descent had been more dramatic.

The tribunal inspected both the subject land and all the comparables and, having reviewed the details of the transactions together with the evidence as to quality and location of the several comparables, we assess the value of the residential part of the subject land at £800 per habitable room. This value compares with Mr Clegg's valuation of £900 and Mr Jackson's valuation of £600. We therefore ascribe to the residential area of 1.29 acres a value of £128,000, being 160 habitable rooms multiplied by £800, which is equivalent to £67,365 per acre.

Restaurant and Club Site

The next parcel to consider is the site reserved for the construction of a restaurant and residential club, valued by Mr Clegg and Mr Jackson at £135,000 and £77,800 respectively. Mr Clegg derives his basis of valuation from two purchases of contiguous areas of land, amounting in all to 1.75 acres, by Trust Houses Forte Ltd in November 1970 and February 1971 for a total consideration of £171,314. The site has been developed and there now stands upon it an hotel called the Northenden Post House. Mr Clegg relied on a devaluation of this transaction to support his valuation of the 1.2 acres of the subject land, which he valued in two parts: (a) by applying to the planned floor area of the restaurant and banqueting space a value of £10 per sq ft and (b) to the service flats a value of £1,000 per habitable room.

Mr Jackson found this transaction to be of no assistance in making his valuation of the relevant parcel of the subject land. The Northenden Post House was, he said, situated just off the M56 motorway; it was an international type of hotel similar to those to be found on many of the motorway networks; and the land was purchased at a time when the Government was encouraging the building of hotels by contributing grants of £1,000 per bedroom towards the cost of development. It was his view that for the development of the kind proposed on the subject land an adequate value was not in excess of £50,000 per acre, which was the equivalent of the value of land for residential development. He had so valued an apportioned area of 0.5 acre at £25,000, to which he had added the residential element at £52,800, being 88 habitable rooms multiplied by £600.

We accept Mr Jackson's view as to how to go about making this valuation. We assess the value of the part attributable to the restaurant and banqueting suite at £33,500 (0.5 acre at an equivalent residential value of £67,000 per acre) and the residential part at £70,400 (88 habitable rooms at £800). The total of £103,900 compares with Mr Clegg's valuation of £135,000 and Mr Jackson's £77,800.

We were told by Mr Clegg that the third item in his valuation "garden strip in front of improvement line" had been included because it had formed part of the area covered by Mr Schultz's offer. It comprises the area in front of the improvement line and it contained 0.74 acre, which Mr Clegg valued at £8,000 per acre. We do not consider that, as part of the valuation of the whole of the subject site, any additional value should be ascribed to this particular area.

Summarising these findings, therefore, we assess the value of the claimants' interests with the benefit of planning permission 52066 TP at £231,900 being:

The residential parcel	£128,000
The restaurant and banqueting suite parcel	£103,000
	£231,900

We recognise that this amount falls short of Mr Schultz's offer of £295,000 (subject to Mr Clegg's estimated £10,000 payable by the claimants), but on all the evidence we are satisfied that the difference is justified.

The value after revocation, restricted to the provisions of the Eighth Schedule to the 1971 Act has been agreed at £78,000.

The amount of compensation payable under this head of claim is therefore £153,900 (£231,900 less £78,000).

Abortive Costs

We will now deal with head of claim no 2, described by the claimants as "Schedule of costs incurred by the claimants and rendered abortive by reason of the revocation of planning permissions 51910 TP and 52066 TP."

The schedule is here reproduced:

Schedule of costs incurred by claimants and rendered abortive by reason of the revocation of planning permissions 51910 TP and 52066 TP

1. Costs incurred in 1967 on the preparation of drawings, the supply of information to the City Planning Officer and the obtaining of Outline Permission 51910 TP and of Detailed Permission 52066 TP—as attached copy of fee note dated 3 May 1967 of Leslie D Clegg. Morgan & Co	£1,785.00	
Add interest from 1 June 1967 to 22 March 1974 ..	* 1,062.91	£2,847.91
2. Costs incurred in 1967 in cutting out of the Claimants' land for parts of the foundations of the proposed buildings—as attached copy account dated 5 April 1967 of J F Goodwin & Co Ltd	112.00	
Add interest from 17 April 1967 to 22 March 1974 ..	67.87	179.87
3. Costs incurred in 1967/69 by Quantity Surveyors in providing estimates of the likely cost of construction of the building proposed by Planning Permission 52066 TP—as attached copy of fee note dated 26 August 1969 of Perry & Perry	170.00	
Add interest from 1 October 1969 to 22 March 1974 ..	* 69.27	239.27

VALUATIONS
(Item 1 of claim)

A. *Valuations of the claimants' interest at March 22 1974 with the benefit of planning permission 52066 TP (the subject of revocation)*

Valuation Parcel	*Description*	*Number or Measurement*	L D Clegg FRICS (for the claimants): *Value or (Equivalent Value) per unit*	*Total values*		G R Jackson ARICS (for the compensating authority): *Value or (Equivalent Value) per unit*	*Total values*	
			£	£	£	£	£	£
Residential	Area	1.9 acres	(75,789)			(50,526)		
	Flats (and garages)	64	2,250		144,000			
	Habitable rooms	160	(900)			600		96,000
Restaurant and residential club	Area	1.29 acres	—			0.5 acres at £50,000	25,000	
	Public rooms	4,700 s.f.	10	47,000				
	Service flats	46	—					
	Garages	115	—					
	Habitable rooms	88	1,000	88,000		600	52,800	
					135,000			77,800
Garden strip in front of improvement line	Amenity land sterilised from development	0.74 acres	8,000		5,920	—		—
					284,920			
			Total	(say)	£285,000			£173,800

B. *Valuations of the claimants' interest after revocation of planning permission T.P.52066 but with the benefit of rights conferred by Town & Country Planning Act, Schedule 8*

	L D Clegg	G R Jackson
Existing use values of two buildings, "Milverton Lodge" and "The Gables," and Schedule 8 rights in respect of three (demolished) houses. Total value (agreed)	£78,000	£78,000
Difference between valuations A and B	£207,000	£95,800

COMPARABLES
(Residential Land)

Claimants: "C"
Compensating Authority: "A"

Reference on plan	*Address*	*Date of transaction*	*Purchase price*: *Auction*	*Private treaty*	*Purchase by compensating authority*	*Purchase price devalued as:* *Per sq. yd.*		*Per acre*	*Per flat unit*	*Per habitable room*
			£	£	£	Sq. Yd.	£	£	£	£
2C	232 Booklands Road Brooklands	June 1 1972	31,500			2,675	11.78		2,625 (12)	955 (33)
3C	1002 Burnage Lane, East Didsbury	Oct. 4 1972	33,500			2,565	13.06		1,861 (18)	798 (42)
4C	36 Denison Road and Sherwood House Upper Park Road, Victoria Park	Nov. 9 1972	61,625			5,900	10.45		1,868 (33)	880 (70)
5C	80 Palatine Road, Didsbury	Sept 26 1972	45,355			2,522	17.98		3,240 (14)	1,031 (44)
6C	38 Cavendish Road, Urmston	Feb 27 1973	25,000			1,380	18.12		2,778 (9)	926 (27)
7C	72/74 Waterpark Rd, Broughton Park	July 24 1973	70,000			3,770	18.57		3,500 (20)	1,167 (60)
8C and 6A	7 Stanton Avenue, Didsbury	Aug 8 1973	51,000			2,150	23.72		5,100 (10)	1,645 (31)
9C and 5A	1 Derby Road, Fallowfield	Jan 30 1974	30,216			2,750	11.00		2,014 (15)	673 (45)
(Reference site. Date of valuation March 22 1974)										
7A	Moor End, Northenden	June 1974			90,000			45,000 (2.0)		750 (120)
8A	Edge Lane, St. Clements Road and rear of Kingsmill Road, Chorlton	Jan 1975			60,000			36,675 (1.636)		612 (98)
10A	Land in Victoria Park: Conyngham Road—Oxford Place Upper Park Road	Sept 1975		On offer 75,750				46,759 (1.620)		

A

4. Costs incurred in 1973/74 upon the preparation of drawings, the supply of information to the City Architect, the obtaining of approval under the Building Regulations for Block A of the flats incorporated in Planning Permission 52066 TP and the staking out on site of the foundations for Block A—as attached copy of fee note dated 1 October 1974 of Leslie D Clegg, Morgan & Co	692.25	
Add 8% VAT	55.38	
		747.63
5. Costs incurred in 1974 on site clearance and excavations for the foundations of Block A of the intended flat development—as attached copy account dated January 1975 of W Davis & Hughes	525.00	
Add 8% VAT	42.00	
		567.00
Total abortive costs		£4,581.68

B Note: Interest on items marked * has been computed in accordance with the rates prescribed by the Acquisition of Land (Rate of Interest after Entry) Regulations.

Mr Jackson at first accepted that each of the items 1 to 5, but excluding the sums of interest, were proper items of claim at the amount shown; but towards the end of the hearing he expressed doubt about item 2, which, he said, had been incurred for the purpose of avoiding payment of betterment levy. We find that all the items are proper matters for claim and that, excluding interest, the five sums amount to £3,381.

As to interest, Mr Clegg said that the claims were made because the claimants had been deprived of the right to earn C interest on the several items of expenditure. Mr Jackson, however, considered that interest was not properly claimable because it had been the choice of the owner to wait for development to proceed. We shall later come to the issue and arguments relating to a claim for interest on the total award of compensation, but on the separate issue with which we are now dealing we prefer and accept the evidence of Mr Jackson and we accordingly limit our assessment under this head to £3,381.

We now summarise our findings on the first two heads of claim, namely:

D

1. Diminution in the value of the claimants' interest	£153,900
2. Expenditure incurred and rendered abortive	3,381
	£157,281

Interest on Compensation Sum

There remains item 3, described by the claimants as "Loss of profit suffered by the claimants after March 22 1974 when their development had to be suspended and no other per- E mission for development was available, thereby precluding them from earning interest on the compensation sum." This major claim for interest was quantified at the sum of £59,243. This sum was reached by taking 14 per cent for two years on the aggregate of the sums claimed in items 1 and 2. Item 3 is therefore adjustable, depending on (a) the amounts of the sums finally awarded under items 1 and 2 and (b) the question of the rate of interest and the period of time over which it should be taken to have run. But before we look again at the figures it will be convenient to consider whether there can be, in principle, any liability to pay interest in the relevant circumstances.

F The claimants say that overnight they lost a large sum of money and thereafter were without the use of it until the compensation was paid. It was suggested that this fell squarely within the meaning of "loss and damage" in terms of section 164.

Mr Fitzhugh disputed this proposition and relied on the recent Lands Tribunal decision of *Hobbs (Quarries) Ltd v Somerset County Council* (1975) 234 EG 829 (the President and Mr Russell-Davis). In that case (which was a case of revocation) the tribunal decided that the compensation should be assessed by reference to loss of profits. Then there was a question about interest. There was no dispute that interest could be ordered running from the date of the award. But G there was a dispute whether interest could be made payable (as the claimants contended) as from the date of the revocation order. Mr Widdicombe, for the claimants, referred to various statutory provisions and referred to the case of *Jefford v Gee* [1970] 2 QB 130 in which the matter of interest was considered; this, he said, established that if the present case was a common law case the claimants would get loss of profits plus interest and it would be wrong if the compensation law was different. Mr Alan Fletcher, for the county council, on the other hand said that there was a well-established principle that until compensation was quantified no interest was payable; he referred to *London, Chatham & Dover Railway Co v South-Eastern Railway Co* [1893] AC H 429, *The Caledonian Railway Co v Sir William Carmichael Bart* [1870] LR 2 HL (SC) 56, *Re Richard & Great Western Railway* [1905] 1 KB 68, and *Swift v Board of Trade* [1926] AC 620. In their decision the tribunal said:

"In our view we have to assess the amount necessary fully to compensate the Claimants as at the date of the revocation and discontinuance orders, and having done that we are *functus officio*. Such compensation is limited to the loss arising from the service of the notice and order. It follows from those considerations that we are entitled neither to compensate for a loss which arose not from the making of the order and notice but from the J fact that the compensation moneys were not paid on or immediately following the date of service, nor to compensate for a loss which in truth is not part of the damage to the interest in the land but arises from the fact that the claimants had not had the use of the money. We think that sufficient authority for our conclusion is to be found in *Richards v Great Western Railway (supra)*, and we also get some support from the recent case of *Wallerstiner v Moir, The Times*, January 29 1975. In the latter case it was said by the Master of the Rolls that interest was awarded not under the statute but under the equitable jurisdiction of the court. The Lands Tribunal, however, is a creature of statute and has no equitable jurisdiction. Accordingly we find that no sum is payable in respect of interest." K

Mr Glidewell said he accepted that the *Hobbs* case was contrary to his contention, but he invited us to reconsider the question and to follow the submissions made by Mr Widdicombe in that case.

The tribunal is not bound by its own decisions but finds them highly persuasive, more particularly when, as in the *Hobbs* case, the President was party to the earlier decision. We have carefully looked through the reports of the cases cited by Mr Fletcher and Mr Fitzhugh (except *Swift v Board of Trade*, which we have been unable to trace). We have reluctantly come to the conclusion that they do indeed establish the proposition propounded by Mr Fletcher. We L say "reluctantly" because we can see no reason in principle why the sterilisation (but for Schedule 8 purposes) of land ripe for development should not earn in a suitable case an award of interest as well as a capital sum for the depreciation of the value. The same kind of reluctance was felt, we think, by Lord Herschell in the *London, Chatham & Dover* case. The ambit of that case was very wide, for it decided not only that no interest could be recovered under 3 and 4 Wm 4, c 42 s 28, since there was no "debt" or sum certain payable by virtue of a written instrument at a certain time, but also that damages could not be given by way of damages for detention of a debt, "the law upon that subject, unsatisfac- M tory as it is, having been too long settled to be now departed from."

Statutory Provision Relating to Interest

Before we leave this topic there is something we should like to add. In the case of compulsory acquisition there are special statutory provisions relating to interest, as was noted in the *Hobbs* case. Thus where entry has been taken special provision is made in section 32 of the Act of 1961. As regards awards (and this applied to "awards" generally) rule

A 50 of the Lands Tribunal Rules 1963 purported to give power to the tribunal, if it thought fit, to direct that interest should run from the date of the award. Some doubt was expressed whether this power was *intra vires* and the matter was dealt with in another way in the 1975 rules, where, in the new rule 38, section 20 of the Arbitration Act 1950 is incorporated by reference. This has the effect that a sum directed to be paid by an award shall, unless the award otherwise directs, carry interest as from the date of the award at the same rate as a judgment debt. Thus under the new rule interest will run automatically unless the contrary is expressly directed by the tribunal. But clearly it can only run from the date of the B award. There is another provision we would like to mention; by virtue of section 63 of the Land Compensation Act 1973 compensation under section 68 of the Lands Clauses Act 1845 or section 10 of the Compulsory Purchase Act of 1965 (compensation for injurious affection where no land taken) carries interest from the date of the claim until payment. These various provisions which direct or enable that interest shall run from certain dates in certain circumstances do seem to indicate a pattern under which express provision is made when it is so intended, and therefore, by inference, when it is not provided there is neither duty nor power to award interest. For, as the President has said, the tribunal is a C creature of statute and needs a statutory power before it can act.

Arbitrators' Powers to Award Interest

Whether this pattern is a logical or a reasonable one is another matter. In the case of *Chandris v Isbrandtsen-Moller Co Inc* [1951] 1 KB 240 a question arose whether an arbitrator had power to award interest (that is to say interest for damages, and therefore antecedent to the award). The headnote reads:

"... the power of an arbitrator to award interest was derived D from the submission to him, which impliedly gave him power to decide 'all matter in difference' according to the existing law of contract, exercising every right and discretionary remedy given to a court of law; that the Law Reform (Miscellaneous Provisions) Act 1934, which repealed s 28 of the Civil Procedure Act 1833, was not concerned with the powers of arbitrators; and that E the plaintiff was entitled to the interest provisionally awarded by the arbitrator."

But with respect it is not quite as simple as that. In his judgment at p 264 Cohen LJ said:

"In my opinion, the right of arbitrators to award interest was not derived from ss 28 and 29 of the Civil Procedure Act 1833, but from the rule that arbitrators had the powers of the appropriate court in the matter of awarding interest. In my opinion, therefore, the effect of the Act of 1934 is that, after it came into force, an arbitrator had no longer the powers of awarding interest on damages conferred on juries by ss 28 and 29 of the . . . Act of 1833, but he had the power conferred on the appropriate court F in the Act of 1934 described as a 'court of record. . . .' "

This power, namely to award interest antecedent to the award on money representing either moneys due under a contract or damages for breach of a contract, is commonly exercised by arbitrators in building and engineering disputes, and for all we know in commercial arbitrations. We mention this not as suggesting that the Lands Tribunal has a similar power but in order rather to raise the question why, in a suitable case, it should not have such a power. The question is the more poignant when it is remembered that under section 1 (5) of the Lands Tribunal Act 1949 the tribunal may act as an arbitrator under a reference by consent. The G tribunal may therefore either have or not have the power to award interest antecedent to the award, depending on whether or not it is sitting in open court or behind the locked doors of the arbitration room. This seems to us to verge on the absurd.

We therefore determine that the amount of compensation payable by the compensating authority to the claimants will be £157,281 to which will be added any legal costs properly incurred from the date of revocation up to the date of the notice of reference and a surveyor's fee based on scale 5a of the scales of charges of the Royal Institution of Chartered Surveyors.

Submissions were heard as to costs. The compensating H authority will pay the claimants their costs of this reference such costs unless agreed to be taxed by the Registrar of the Lands Tribunal on the High Court Scale.

LANDS TRIBUNAL
LEASEHOLD REFORM

A

NORFOLK v TRINITY COLLEGE, CAMBRIDGE
(LR/106/1975)
(W H REES FRICS)
April 1 1976

Estates Gazette May 8 1976
(1976) 238 EG 421

First reference under s 118 of Housing Act 1974—Price of freehold of detached, five-bedroomed house at Cambridge with RV over £500 and 59 years of lease unexpired—Tenant's bid held not to be excluded—Marriage value equally divided—Effect of improvements on value—Price determined at £3,750

B

The tenant, Major J D Norfolk, appeared in person; Nigel Hague (instructed by Boodle, Hatfield & Co) for the landlord, the Master, Fellows and Scholars of Trinity College, Cambridge.

Giving his decision, MR REES said: This is a reference to determine the price which the tenant should pay to acquire the freehold of his house, 33 Barrow Road, Cambridge, under section 9 of the Leasehold Reform Act 1967, as amended by section 118 of the Housing Act 1974, and is the first reference under that amendment.

C

The essential facts and other matters are not in dispute and are as follows. The house is detached and is one on an estate of similar properties developed in the 1930s. The subject house stands on the corner of Barrow Close and Barrow Road, which is off Trumpington Road (A10), about a mile and a half south of the centre of Cambridge. It is constructed of brick and tiles and the accommodation originally consisted of: porch; lavatory; cloakroom; hall; two reception rooms and a study; kitchen with scullery recess, larder, wc and fuel stores; on the first floor: five bedrooms, bathroom and wc. Alterations were made by the tenant and are referred to below. The agreed floor area is 2,050

D

sq ft plus outbuildings and a garage and the site area is agreed at 0.375 acre. The lease is for 99 years from December 25 1934 at a ground rent after the first two years of £17.75 per annum, and the value of the freehold in possession (as the house stood at the date of the hearing, ie with the accommodation altered) is agreed to be £37,000. A fair rent is also agreed at £1,065 per annum, tenant paying rates, landlord doing all repairs. The rateable value with effect from April 1 1973 is £563. Facts regarding some 11 comparable properties and plans and photographs of the subject house were also agreed. The relevant date is agreed as February 7 1975, the date of the tenant's claim.

E

Works of Improvement

I find that the tenant had carried out works to the subject house at his expense between 1959 and 1975, the main items being the installation of central heating in two stages, the installation of an electric ring main, enlargement of windows on both floors, the provision of fitted wardrobes and vanitory units and enlargement of the study to convert it into a dining room (with appropriate alterations to windows and doors), and rearrangement of the kitchen, scullery recess, larder and wc to provide a larger modern kitchen and boiler room. The house is in a very good state of repair. At the rear of the subject property is an area of open land of about 2.6 acres. This area is allocated for residential development in the approved town map. Just before the relevant date an application for planning permission for residential development had been refused, but by the date of the hearing a second application had been granted for the erection of nine houses on it, a density of 3.25 houses per acre compared with the density in Barrow Road—2.3 houses per acre.

F

The basis of the price to be paid (by the tenant to purchase the freehold) in this instance is laid down in section 118 of the Housing Act 1974 and is as follows:

G

"(4) In section 9 of the Leasehold Reform Act 1967 (purchase price of enfranchisement) there shall be inserted after subsection (1):

'(1A) Notwithstanding, the foregoing subsection, *the price payable for a house and premises,* the rateable value of which is above £1,000 in Greater London and £500 elsewhere, *on a conveyance under section 8 above, shall be the amount which at the relevant time the house and premises, if sold in the open market by a willing seller might be expected to realise on the following assumptions:*

H

(a) *On the assumption that the vendor was selling for an estate in fee simple, subject to the tenancy, but on the assumption that this Part of this Act conferred no right to acquire the freehold;*

(b) on the assumption that at the end of the tenancy the tenant has the right to remain in possession of the house and premises under the provisions of Part I of the Landlord and Tenant Act 1954;

(c) on the assumption that the tenant has no liability to carry out any repairs, maintenance or redecorations under the terms of the tenancy or Part I of the Landlord and Tenant Act 1954:

J

(d) on the assumption that the price be diminished by the extent to which the value of the house and premises has been increased by any improvement carried out by the tenant or his predecessors in title at their own expense;

(e) on the assumption that (subject to paragraph (a) above) the vendor was selling subject, in respect of rent-charges and other rents to which section 11 (2) below applies, to the same annual charge as the conveyance to the tenant is to be subject to, but the purchaser would otherwise be effectively exonerated until the termination of the tenancy from any liability or charge in respect of tenant's incumbrances; and

K

(f) on the assumption that (subject to paragraphs (a) and (b) above) the vendor was selling with and subject to the rights and burdens with and subject to which the conveyance to the tenant is to be made, and in particular with and subject to such permanent or extended rights and burdens as are to be created in order to give effect to section 10 below.' "

The words in italics are identical to those contained in section 9 (1) of the Leasehold Reform Act 1967, and I note here that by section 82 of the Housing Act 1969, there were inserted in section 9 (1): ". . . after the words 'a willing seller' the words '(with the tenant and members of his family who reside in the house not buying or seeking to buy).' . . ."

The tenant called Mr D J H Bliss MA (Cantab), holder of a certificate of proficiency in estate management, FRICS [partner in Catling, Brady & Bliss] and Mr A R Cook FRICS [partner in Gray Cook & Partners], both of whom practise in Cambridge. Mr Cook's evidence related mainly to comparables and improvements. The landlord called Mr M St J Hopper FRICS [partner in Gerald Eve & Co], who practises in London. The issues between the parties can readily be seen from the valuations submitted by Mr Bliss and Mr Hopper, which are as follows:

Mr Bliss (called by the tenant)

(1) *Term*			
Ground rent	£17.75		
YP 59 years at 9 per cent	11.042		
			£196.00
(2) *Reversion*			
Fair rent	£1,065.00		
Less repairs	220.50		
Net fair rent	844.50		
YP 25 years at 15 per cent	6.464		
		£5,458.85	
Deferred 59 years at 7 per cent		0.0184653	
			100.80
Vacant possession in 84 years		£37,000	
Deferred 84 years at 7 per cent		0.0034022	
			125.88
			422.68
Less value of improvements by lessee		£3,000	
Deferred 59 years at 7 per cent		0.0184653	
			55.40
			£367.28

Mr Hopper (called by the landlord)

(1) Value of lessor's interest exclusive of "marriage" value				£1,100
(2) Lessor's share of "marriage" value				
Agreed value of freehold in possession			£37,000	
Less for improvements			2,500	
(a) value of notional unencumbered freehold interest with vacant possession disregarding improvements			34,500	
(b) value of lessor's interest exclusive of "marriage" value		£1,100		
Value of leasehold interest	£30,000			
Less for improvements	2,000			
(c) value of lessee's interest exclusive of "marriage" value disregarding improvements		28,000		
			29,100	
(d) gain on "marriage" of interests disregarding improvements			5,400	
(e) lessor's share at 50 per cent				2,700
				£3,800

I consider these issues under the following headings:

1. Whether or not the sitting tenant should be considered as being in the market.
2. If so, how the amount of his bid should be ascertained: the value of the leasehold interest.
3. The value of the freehold subject to the tenancy, excluding the tenant's bid.
4. The amount to be deducted for improvements and the treatment of any such amount.

1. *The tenant's bid.*

Mr Bliss added nothing for this and the tenant urged me to disregard the sitting tenant's bid; to do so would give a fair value, he said, and he suggested that the price would then be on the lines on which a fair rent is approached. He pointed out that the statute does not refer to "marriage value," the concept which gives rise to the tenant's bid, and he submitted that this was a market philosophy which cannot be inserted into section 9 (1A) of the Leasehold Reform Act 1967 as amended. He considered that Parliament intended to exclude the tenant's bid. Alternatively, the tenant submitted that the amount of any marriage value was nil on the basis of Mr Bliss's and Mr Cook's evidence that since the Housing Act 1974 was passed leasehold values in respect of this sort of house had risen and had come right up against the freehold value.

Position as in "Custins" Case

Counsel for the landlord referred me to *Custins v Hearts of Oak Benefit Society* (1969) 209 EG 239 to support his submission that the sitting tenant should not be excluded from the market. The relevant words in section 9 (1) of the Leasehold Reform Act 1967 are identical with the words in section 9 (1A). The effect of section 82 of the Housing Act 1969 and the reasons for enacting it are set out in *Official Custodian of Charities and Others v Goldridge* (1973) 227 EG 1467, to which counsel referred me.

It is quite plain to me that the amendment contained in section 82 of the Housing Act 1969 was to section 9 (1) of the 1967 Act and that it cannot be read as amending section 9 (1A). The position here is, therefore, as it was when considered in *Custins* and I come to the conclusion that the sitting tenant should not be excluded from the market.

2. *The amount of the tenant's bid: the value of the leasehold interest.*

I reject the submissions made by the tenant and the opinions of Mr Bliss and Mr Cook that the amount of marriage value is nil, ie that the value of the freehold subject to the tenancy equals the value of the freehold in possession less the value of the leasehold interest. Mr Bliss sought to support this by some calculations, but I find these to be of no assistance to me. In assessing the value of the leasehold interest in the house to support this and in criticism of Mr Hopper's figure of £30,000, both these valuers had taken account of the tenant's right to enfranchise: it was not in dispute that this right had had the effect of increasing the values of leasehold interests in this category of property in the market in Cambridge. It is true that section 9 (1A) (a) requires the freehold subject to the tenancy to be valued without regard to the right to acquire the freehold and thus at first glance it could be argued that such an assumption should not be made when valuing the leasehold interest. However, when the price to be paid by the tenant for the freehold is found by considering *inter alia* the value of the leasehold interest, the same assumption must be made when finding the value of the leasehold interest, otherwise the two figures will not be on the same footing and the right to acquire will be reflected in the price to be paid contrary to section 9 (1A) (a).

Mr Hopper approaches the matter in this way. To the value of the freehold subject to the tenancy he adds half the difference between (1) the value of the freehold in possession (having deducted the value of the improvements) and (2) the value of the leasehold interest in the house (similarly reduced by the value of the improvements) plus the value of the freehold subject to the tenancy. The equal division between the parties was criticised by the tenant, but I was given no evidence as to any other proportion.

Counsel for the landlord referred me to *Siggs v Royal Arsenal Co-operative Society* (1971) 220 EG 39 at 47, to *Inland Revenue Commissioners v Clay* and *Inland Revenue Commissioners v Buchanan* (1914) 3 KB 466, to *Stokes v Cambridge City Council* (1961) 180 EG 839, and to the decision of this tribunal (Douglas Frank QC, president) in *Re SJC Construction Co Ltd's Application* (1974) 230 EG 906 in support of his submission that there should be an addition for tenant's bid and for an equal division of the marriage value.

Bargaining Powers Equally Balanced

I accept Mr Hopper's view that the bargaining powers of each party are equally balanced and I therefore agree that an equal division is appropriate in this instance. I also accept his whole approach, which seems to me to be entirely logical. I comment on the individual figures below.

I had little evidence from the tenant's experts as to the value of the leasehold interest disregarding the tenant's right

A to enfranchise. Mr Hopper supported his opinion by reference to the sales of six leasehold houses close by, the exteriors of which I inspected. I accept his figure of £30,000, although it was criticised by the tenant by reference to inflation.

3. *The value of the freehold subject to the tenancy, excluding the tenant's bid.*

Mr Bliss used 9 per cent in his calculation which he had taken from the decision of this tribunal in *Lead v J & L Estates* (1975) 236 EG 819. He supported this only by a letter from a firm of estate agents in Cambridge which set out particulars of two freehold ground rents which they B were offering for sale and which were secured on blocks of flats. He also referred to the return available from 3½ per cent War Loan unredeemable stock as being 14.58 per cent in February 1975. He then assumed that the tenant would remain in the house at the expiration of the ground lease at the agreed fair rent. He deducted outgoings from this, based on the statutory deductions from gross value to rateable value for rating purposes, to arrive at net income, which he capitalised at 15 per cent, being, as he said, an uncertain income, since the landlord was responsible for all repairs. He took a term of 25 years, which he estimated was the C average length of time a tenant paying a fair rent would remain in occupation. He deferred the result at 7 per cent and added the value of the freehold in possession as improved deferred 84 years, also at 7 per cent. I deal with improvements below.

Mr Hopper relied on four transactions, the facts of which were agreed, to support his figure of £1,100. These related to four houses fairly close to the reference house and showed prices from £550 to £900. They were cases where the purchasers had statutory rights to enfranchise at prices in accordance with section 9 (1) of the Leasehold Reform Act 1967 as amended by section 82 of the Housing Act 1969. The D leases range from 50 to 78 years and the ground rents from £7.50 to £12 per annum. The rateable values were from £426 to £451 (that of the reference house is £563) and the dates of sales were between May 1971 and May 1972, four different estate agents in Cambridge acting for the tenants. He compared and contrasted these four houses with the subject house but considered that a higher figure than £900 was justified because (1) the reference house was in a better situation, (2) values generally were higher in February 1975 than in May 1972, (3) the site of the reference house was bigger than three of these comparables and (4) because of the effect of the 50 years' extension of lease which had to be E assumed in these four cases under section 9 (1) (a) of the Leasehold Reform Act 1967. He considered the amount of the ground rent to be without significance.

Mr Bliss's approach was criticised by counsel for the landlord as relying too much on the money market (see *Gallagher Estates v Walker* (1973) 230 EG 359) as distinct from the land market (see *Finkel v Simon* (1974) 231 EG 329). It is true that Mr Bliss referred to the returns available from government stock and admitted in cross-examination that his calculation was an investment approach (apart from the freehold ground rents already referred to), but there is no reason why transactions in the land market should not be F analysed using the approach Mr Bliss sought to adopt (I refer to this again later). The fundamental weakness of his evidence lies in the fact that it was almost entirely unsupported by reference to comparable transactions. The information concerning the offer of freehold ground rents secured on blocks of flats in Cambridge is of no assistance to me, since the properties are not of the same type as the subject house. Mr Bliss's treatment of the reversion is open to the criticisms that the amount which he allowed for outgoings has no market basis and that he failed to justify the use of different capitalisation and deferment rates, treatment which is generally considered to be contrary to accepted principles and practice of valuation. His use of a high rate of interest G (15 per cent) on the grounds that the income is uncertain cannot be justified as, if the tenant failed to pay his rent, the landlord would have a possibility of obtaining vacant possession and thus of selling the house at a high price. In connection with Mr Bliss's assumption that the tenant would remain in the house at the end of the ground lease, section 9 (1A) (b) of the Leasehold Reform Act 1967 requires the assumption to be made that " . . . the tenant has the right to remain in possession . . . " not that the tenant will necessarily remain in possession. I note here that had Mr Bliss used 15 per cent in place of 15 and 7 per cent for the first part of his calculation of the reversion, he would have arrived H at £1: had he used 7 per cent for both parts he would have come to £180, which seems to me to be more realistic.

Landlord's Valuer's Approach Preferred

I prefer Mr Hopper's approach, since it is based on transactions in the land market in Cambridge which are sufficiently close in time. I agree with his reasons for taking a figure above £900. The amount is very much a matter of opinion, since a figure beyond a bracket is being estimated. Looking at the range of prices of the four comparables, it is obvious that Mr Hopper's opinion is, as counsel put it, " in the right J parish " and Mr Bliss's is not. However, there may be some elements of (1) the " Delaforce " effect, the anxiety to settle (*Delaforce v Evans* (1970) 215 EG 315), and (2) marriage value in the four settlements. The comparison made between the reference house and the other four may have been coloured by the improvements made to the former. In the light of my inspection and Mr Bliss's and Mr Cook's evidence, I come to the conclusion that Mr Hopper's figure of £1,100 is on the high side and that it should be £1,000.

4. *The improvements.*

The evidence as to the effect to be given to improvements K is paradoxical. The landlord's expert says £250 should be deducted: the tenant's £55.

Mr Cook's evidence was that the improvements added at least £3,000 to the value of the freehold in possession. Mr Hopper's opinion was £2,500. Counsel for the landlord at first submitted that only structural improvements should be considered as being relevant (he referred me to Schedule 8, para 1 (2) of the Housing Act 1974), but he did not pursue this, leaving it, as he said, for another day. I was given figures of historic costs and costs at February 1975 of the improvements carried out, but these are of little assistance: there is no doubt that it is " the value . . . by any improve- L ment . . . " which has to be considered. I accept Mr Cook's opinion, particularly in the light of my view of the subject property (I was provided with plans of the house as originally constructed and as now improved), and determine the amount of the improvements as £3,000 relative to the agreed value of the freehold in possession of £37,000.

Mr Hopper considered that the effect of the improvements on the value of leasehold interest was £2,000 (cp £2,500 for the freehold). I have already accepted Mr Hopper's evidence (see 2 above) that the leasehold interest in the subject house in February 1975 was £30,000 (improved). Mr Hopper pointed out that his figures of £2,500 and £2,000 for improvements M were in approximately the same proportion as £37,000 (the agreed value of the freehold in possession) and £30,000, the value of the leasehold interest.

The tenant submitted that " an improvement was an improvement " and if the effect on the value of the freehold in possession was £3,000, the effect on the value of the leasehold interest should also be £3,000.

It seems to me that the second must be lower than the first, as it is a wasting asset being tied to the length of the lease. Taking approximately the same proportion as Mr Hopper does, I find that the amount for the improvements (relative to the value of the leasehold interest) is £2,500 (cp £3,000 for the freehold). It will be seen that, since the difference between £3,000 and £2,500 equals the difference between £2,500 and £2,000, the net effect of substituting the two higher figures is nil.

As to the treatment of the value of the improvements, (d) provides that the price to be paid is to " be diminished by the extent to which the value of the house and premises has been increased. . . . " In cross-examination the tenant suggested that this meant that the value of the improvements related to the value of the freehold in possession should be deducted, but, as counsel for the landlord pointed out, this could in some circumstances, not difficult to imagine, result in the landlord paying the tenant to take the freehold off his hands! Counsel for the landlord submitted that the interest to be taken into account when considering the " value of the improvements " was the interest mentioned in (a), ie the freehold subject to the tenancy. Mr Bliss had taken this view and I agree. However, I am satisfied that Mr Hopper's approach, treating the house as being unimportant in all the constituent parts of his valuation, contains no error in principle, but I did not have the benefit of legal argument on behalf of the tenant.

Some Miscellaneous Points

I now comment on some miscellaneous points:

Neither party considered the open land at the rear of the subject house to have any substantial significance in view of its allocation in the town map.

With regard to section 9 (1A) (b), counsel for the landlord submitted that this had little effect, the tenant always having had the right to remain in possession; (c) was regarded as having little significance in these circumstances (it was suggested that with this type of house the tenant would keep it in repair, irrespective of his obligations, to maintain the value of his investment in it); and (e) (f) were not relevant here.

In criticism of Mr Bliss's calculations (set out on p 216), resulting in a value of £367.28, Mr Hooper produced a calculation (on the same familiar lines of a term and reversion).

Term		
Rent reserved	£17.75	
YP 59 years @ 9%	11.04	
		£196
Reversion		
Value of unencumbered freehold interest	37,000	
Less: 10% for the effect of tenant's potential right to remain in possession under Part I of the Landlord and Tenant Act 1954	3,700	
	33,300	
Defer 59 years @ 6%	0.032	
		1,065
		1,261
	Say	£1,250

It will be noticed that the figures for the term are the same as Mr Bliss's. If an approach on this basis is used, the result seems to me to be nearer the mark than Mr Bliss's calculation, but I think that the reversion should be to the value of £34,000 (£37,000 less £3,000 for improvements). If a deduction of 25 per cent is made for the right to remain, and using Mr Hopper's 6 per cent deferment, the result would be about £1,000. In this connection I agree with Mr Hopper's opinion that the landlord and tenant would negotiate and come to terms for a sale, but that factor must be excluded in arriving at 3 above.

Various general figures of house prices all over the country over a period of years were put to Mr Hopper in cross-examination together with figures for the same period for the retail price index and inflation. I derive no assistance from these figures: they are far too general.

In November 1975, Mr Cook acting on behalf of a purchaser of a leasehold house nearby and similar enough to the subject house to be of help, offered £3,250 for the freehold on his behalf. The purchaser had not occupied the house at all (let alone for the five years required to qualify) nor had the vendor served notice on the freeholder under the Leasehold Reform Act 1967 as amended by the Housing Act 1974, the rateable value being £613. The lease had 53 years to run (cp the subject house—59 years) at a ground rent of £12.40 per annum (cp the subject house—£17.75 per annum). This offer was refused. Counsel for the landlord stressed the closeness of this offer to the present case. Mr Hopper described it as a " back marker ": I agree that it is useful evidence.

Absurd Effects

The tenant stressed the sensitivity of the investment method of valuation in cases of remote reversions—a small change in the rate of interest used in reversion making a large change in the total value. He similarly criticised, and demonstrated both by calculations, a like sensitivity in Mr Hopper's approach to marriage value. Other calculations were submitted, one showing the absurd effect which Mr Hopper's figure of £1,100 had if the calculations were made in reverse under section 9 (1) of the Leasehold Reform Act 1967 as amended by the Housing Act 1969, ie on the basis of *Farr v Millersons Investments Ltd* (1971) 218 EG 1177, but I found this of no help at all. It only shows what absurd results can be obtained by applying blindly an inappropriate formula. The result of the calculation was to arrive at a value of £131,000 for the subject house, freehold in possession!

As I have indicated, I accept Mr Hopper's approach in principle, but my findings above require three figures in his valuation to be amended:

1. Value of lessor's interest exclusive of " marriage " value ..				£1,000
2. Lessor's share of " marriage " value				
(a) Agreed value of freehold in possession			£37,000	
Less improvements for value of			3,000	
			34,000	
(b) Value of lessor's interest exclusive of " marriage " value		£1,000		
(c) Value of leasehold interest ..	£30,000			
Less Value of improvements ..	2,500			
		27,500		
			28,500	
(d) gain on " marriage " of interest disregarding improvements			5,500	
(e) lessor's share at 50 per cent				2,750
Price to be paid for freehold				£3,750

I accordingly determine the sum to be paid by the tenant of 33 Barrow Road, Cambridge, for the acquisition of the freehold interest in accordance with the terms of section 9 (1A) of the Leasehold Reform Act 1967 to be £3,750.

Having read this decision, a sealed offer made by the tenant dated January 20 1976 and one made by the landlords dated February 5 1976 were opened. I find that both amounts are below the amount of my award. I heard submissions as to costs by counsel for the landlords: the tenant did not wish to comment. In these circumstances the tenant will pay the landlords their costs of the reference, such costs, failing agreement, to be taxed by the Registrar of the Lands Tribunal on the High Court Scale.

LANDS TRIBUNAL

RATING

A W & J PARKER LTD v LEICESTER CITY COUNCIL AND CULVERWELL (VO)

(LVC/514-515/1974)

(E C STRATHON FRICS)

January 30 1976

Estates Gazette April 3 1976

(1976) 238 EG 51

Inner-terrace butcher's shop in eight-shop pedestrian way linking two main shopping streets in Leicester—Zoning method of valuation—Rental evidence derived from inner-terrace comparables opposite appeal shop preferred to " assessment evidence " based on analysis of assessments of corner shops with principal frontages in the shopping streets—Ratepayers' appeal dismissed

B

Lord Colville (instructed by Owston & Co, of Leicester) appeared for the appellants; Mr C D Cochrane (instructed by the city attorney, Leicester City Council) for the first respondents. The second respondent, the valuation officer, appeared in person.

Giving his decision, MR STRATHON said: This consolidated appeal concerns the assessment of 2 Victoria Parade, Leicester, a ground-floor shop with a basement and three upper floors; it was entered in the 1973 valuation list at gross value £7,450. A local valuation court for the city and county borough of Leicester reduced the gross value to £7,300. From this decision W & J Parker Ltd (trading as W & L Stafford), the occupying ratepayers, appealed to the Lands Tribunal; the rating authority, the Leicester City Council, also appealed. The appeals were consolidated, by order dated January 14 1975, the ratepayers to be the appellants, the rating authority to be the first respondents and the valuation officer to be the second respondent.

C

Lord Colville for the ratepayers called evidence from Mr W G Simpson FRICS; Mr C Cochrane for the first respondents called Mr H M Wilks BSc FRICS; the valuation officer, Mr P N Culverwell ARICS, appeared in person.

D

Victoria Parade is an important pedestrian way which connects two shopping streets; at one end there is Gallowtree Gate, the primary shopping street of Leicester, and at the other end is Market Place, a wholly-covered popular shopping location. Victoria Parade is the most popular and the most used of four pedestrian ways which link these two shopping streets. Since the date of the ratepayer's proposal, certain structural changes have been made in shops in the parade, but at the relevant date it comprised the principal or return frontages of eight shops of which four are corner shops. No 2, a building about 150 years old, is one of two inner terrace shops on the north side of the parade; it has a frontage of 18 ft 6 in; it has been occupied for the purpose of a retail butcher's business for about 135 years. At the back of the building there is a goods access to Morley Parade, one of the four pedestrian connecting links.

E

The surveyor witnesses submitted valuations which for ease of reference are set out in comparative form in the schedule below:

F

The same method of valuation was used by the three surveyors, namely, the zoning method for the shop area and additional values for the basement and the three upper floors. The shop value zones adopted are those which prevail throughout the shopping areas of Leicester, namely, zone A, the first 15 ft of depth, zone B, the next 25 ft of depth and, finally, a remainder zone. But the witnesses differed in their approaches to their valuations. Mr Simpson adopted what I will hereafter call the assessment evidence approach, while the valuations of Mr Wilks and Mr Culverwell were derived from rental evidence.

G

Mr Simpson first analysed the assessments of the two corner shops on the same side as no 2, namely 27 Gallowtree Gate and 71 Market Place, both of which have a shop depth of 15 ft from their Victoria Parade frontages. Mr Simpson had learned from the valuation officer that, for the purpose of arriving at the gross values of the two shops, no 27 was zoned from the Gallowtree Gate frontage at values per sq ft of £20 for zone A and £10 for zone B, and no 71 was zoned from the Market Place frontage at £18 for zone A and £9 for zone B. By using the Victoria Parade frontages as bases from which to zone, Mr Simpson had found that on a devaluation of the gross values from this frontage zone A for no 27 would be represented by £15.20 per sq ft and for no 71 by £13.10 per sq ft. During the hearing it was accepted by Mr Simpson that a more correct analysis would be represented by £16.15 and £13.72 respectively. Relying upon his professional experience and his knowledge of the locality, Mr Simpson was of the opinion that by comparison with these two properties the value of no 2 should be less, and he had concluded that £12 per sq ft for zone A was a more correct value than £15.50 per sq ft, the value attributed to zone A of the appeal shop by Mr Wilks and Mr Culverwell. Mr Simpson said that his valuation of the upper floors was related to the assessments of office accommodation in the city. I will return later to what Mr Simpson had to say about the rental basis approach.

H

J

Very Few Rented Shops in Central Area

Mr Wilks said that there were very few rented shops in the central area of Leicester, but of the four inner terrace shops in Victoria Parade, rents were paid in respect of two of them, nos 1 and 3-5, both opposite the appeal shop. His devaluations of the rents, which he had adjusted for premiums paid on assignments, had produced, in terms of gross values per sq ft, zone A values of £17.80 for no 1 and £15.50 for no 3-5. He relied on this evidence for support for his valuation of £15.50 per sq ft for zone A of the appeal shop. He had valued the upper floors by comparison with the rental devaluation of the upper floors of the comparables, which were of similar quality.

K

Mr Culverwell also relied for support on the two rented comparables, nos 1 and 3-5. He also cited rents paid in respect of the two corner shops: (a) 27 Gallowtree Gate, which on devaluation produced an overall shop value of £17.10 per sq ft (the equivalent of zone A value if zoned from the Victoria Parade frontage) and (b) 69 Market Place,

Floor and Description	*Area* (*sq ft*)	*W. G. Simpson* (for the ratepayers: appellants) *Price* £	*Value* £	*H. M. Wilks* (for the rating authority: first respondents) *Price* £	*Value* £	*P. N. Culverwell* (valuation officer: second respondent) *Price* £	*Value* £
Ground floor							
Shop							
Zone A (15 ft) ...	264	12.00	3,168	15.50	4,092	15.50	4,092
„ B (25 ft) ...	285	6.00	1,710	7.75	2,209	7.75	2,208
Remainder ...	24	3.00	72	3.87	93	3.87	93
Pedestrian rear access ...				—	} 100		80
"Arch" advertisement ...				—			—
First floor							
Offices ...	263	0.60	158	1.10	289	1.10	289
Canteen ...	127	0.60	76	1.10	140	0.90	114
Second floor							
Stock ...	293	0.20	59	0.45	132	0.40	117
Staff ...	214	0.40	86	0.50	107	0.55	118
Third floor							
3 attics ...	313	—	25	0.20	63	0.15	47
Basement							
Preparation ...	500	0.40	200	0.40	200	0.45	225
Refrigerators ...			—		70		70
			5,554		7,495		7,453
		say G.V.	£5,550		7,450		7,450
						G.V. at	£7,300
						as determined by the local valuation court and accepted by the valuation officer.	

which produced on devaluation £18.50 per sq ft for zone A, whether zoned from the Market Place frontage or from the Victoria Parade frontage. Mr Culverwell explained that he had not appealed from the decision of the local valuation court reducing the gross value by £150 to £7,300 because, before the hearing by that court, he had offered, without prejudice, a similar reduction; he had made the offer to "grease the wheels," although he had felt there had been no justification for doing so, but having made it he concluded that the court's decision should be accepted by him.

I return to what Mr Simpson had to say about the rental evidence. In his view it was unreliable on two grounds: (a) the rented shops which have shallow depths are not physically similar to the appeal shop; the depth of no 1 is in part 7 ft 6 in and in part 14 ft and the depth of no 3-5 is in part 7 ft 2 in and in part 14 ft 3 in; such shops, Mr Simpson said, would attract specialist firms, whose bids of rent would be in a higher bracket than bids for deeper shops, (b) the rent-cum-premium of no 1 was a special bid by the occupier of the adjoining corner shop, 69 Market Place. Mr Simpson, however, produced a "check" valuation on the rental evidence and, by a devaluation process in which he had used depths for zone A of 7 ft 6 in, a gross value of £17.50 per sq ft for zone A of that depth had emerged. Using depths of 7 ft 6 in for each of zones A and B for the appeal shop, he had arrived at a total gross value of £5,612 by comparison with his primary valuation of £5,550.

I inspected the appeal hereditament and the comparables. I was told that in respect of some assessments there were outstanding appeals to the local valuation court and that an appeal to the Lands Tribunal was pending in respect of 27 Gallowtree Gate.

Skeleton in VO's Cupboard

On his assessment approach to value Mr Simpson had devalued by zoning from Victoria Parade frontages assessments which had been arrived at by zoning from the other frontages. He himself had had no experience of valuing shops by rezoning from secondary frontages. Mr Wilks illustrated that the results of such devaluations depend upon the relative proportions of the frontages and depths; thus zone A of 27 Gallowtree Gate from the Victoria Parade frontage comprised zones A and B from the Gallowtree Gate frontage. At the other end of Victoria Parade is 71 Market Place, which Mr Culverwell said had been zoned in error from that frontage whereas its dominant frontage is to Victoria Parade: this had resulted in an underassessment, which, he said, was a skeleton in his valuation cupboard. In my view the evidence adduced on the assessment approach is not sufficiently reliable on which to found the correct assessment of the appeal property, but I have yet to deal with valuations derived from the rental evidence.

Despite his criticism of the rental evidence, Mr Simpson had adopted it for the purpose of his "check" valuation and he was satisfied that, provided his method of devaluation was used, the evidence supported his opinion of value of the appeal property; indeed Mr Simpson said he regarded the rent of no 3-5 as the keystone of available primary evidence. But I find that having broken away from the pattern of zoning used throughout Leicester, there are no guidelines to Mr Simpson's selection of different zone depths and relative proportions of value as adopted in his valuation; as to the value of £17.50 per sq ft for zone A, Mr Simpson said "I plucked it from my experience," but I was told that in Leicester none of the surveyors had had experience in the valuation of shops which had involved the selection of a zone A depth of 7 ft 6 in. Mr Simpson had based his method on a passage from a decision of the Lands Tribunal in *Trevail (Valuation Officer) v C A Modes Ltd* and *Trevail (Valuation Officer) v Marks & Spencer Ltd* (1967) 202 EG 1175:

> "The second (observation) is that, just as depth affects remainder prices, so shallowness affects zone A prices: therefore it may be proper to 'gross up' the zone A price of a shop not having a depth adopted for zone A in that sector of shops."

I do not find this decision of assistance on the facts found in this case. In my opinion the method of devaluation and its subsequent use in the valuation process by Mr Wilks and Mr Culverwell, which conforms with the practice that prevails in Leicester, is to be preferred. It is also my view that the rental evidence approach to value in the instant case is more reliable than the assessment approach.

Taking the valuation of Mr Wilks for the first respondents, I accept his valuation of the shop, £6,394, which is derived from a gross value of £15.50 per sq ft for zone A. Mr Wilks was cross-examined at some length about the upper floors, but my inspection confirmed that they are broadly similar in quality to the rented properties and that the values attributed to them are fair by comparison. As to the basement, the valuations (of the floor area) of Mr Wilks and Mr Simpson are similar; Mr Wilks, however, adds £70 for the refrigerators, a value which I accept.

I therefore accept Mr Wilks' valuation, which supports a gross value of £7,450. It is not necessary to examine further the valuation of the valuation officer, who has accepted the decision of the local valuation court.

I therefore determine that the assessment of 2 Victoria Parade should stand in the valuation list at gross value £7,450, rateable value £6,180.

The appeal of the ratepayers is therefore dismissed and the appeal of the first respondents is allowed.

As to costs, the valuation officer had, at the hearing, asked that should the appeal be dismissed he should be awarded his costs in the sum of £150. At the reading of this decision I heard submissions by counsel for the ratepayers and the first respondents. My decision is the ratepayers will pay to the respondents only one set of costs, of which there shall be apportioned to the respondent valuation officer the sum of £150 and to the first respondents the remainder. The costs, if not agreed, are to be taxed by the Registrar of the Lands Tribunal on the High Court Scale.

PENNARD GOLF CLUB v RICHARDS (VO)

(LVC/448/1974)

(J H EMLYN JONES FRICS)

March 8 1976

Estates Gazette April 17 1976

(1976) 238 EG 199

Golf club at Swansea held to be in rateable occupation of golf course as well as clubhouse and buildings—Tribunal rejects contention that club's occupation is not exclusive because of serious interference with quiet enjoyment from grazing ponies under rights of common, extensive use of footpaths by public, and many acts of trespass—Burley Golf Club case distinguished—Exercise of commoners' rights a competing but subordinate occupation

John Diehl (instructed by George L Thomas, Nettleship & Co, of Swansea) appeared for the appellant ratepayers; Miss D M MacDonagh (representing Solicitor of Inland Revenue) for the respondent valuation officer.

Giving his decision Mr Emlyn Jones said: The appellant golf club, known as the Pennard Golf Club, occupies a golf course at Southgate on the Gower Peninsula. The land was formerly within the rating area of the Rural District of Gower but, following local government boundary changes, is now within the rating area of the City of Swansea. The club is an unincorporated members' club, so that, strictly speaking, the appeal should be lodged in the name of the trustees of the club, but no point is taken on this. The respondent is the valuation officer responsible for the valuation list for the City of Swansea and was responsible for the valuation of the subject hereditament entered in the valuation list with a rateable value of £1,125. Following a proposal made on behalf of the club on September 19 1973, the West Glamorgan Local Valuation Court determined an assessment of rateable value £1,040, their decision being given on October 1 1974. The appellants being aggrieved appeal to the Lands Tribunal. They accept that they are in rateable occupation of the clubhouse and premises but contend that there is no rateable occupation of the golf course and that, in any event, the assessment determined by the local valuation court is excessive.

Prior to the hearing the parties agreed alternative values as follows: If the whole constitutes a single rateable hereditament the correct assessment should be rateable value £970. If the golf course does not form part of a hereditament in the rateable occupation of the appellants, then the clubhouse and premises should be assessed at gross value £526, rateable value £410.

It will be seen, therefore, that the question at issue can be simply stated: is the golf club in rateable occupation of the golf course?

Pennard Golf Links, which occupy some 250 acres or thereabouts, lie on the southern coast of the Gower Peninsula some eight miles west of Swansea. It is an 18-hole course with clubhouse and a professional's shop; the total length of the course as shown on the club scoring card is 6,266 yds. It has many of the characteristics of a typical seaside links course and abuts at its south-western boundary on to seacliffs above Pobbles Beach. The golf course is traversed by some 16 public footpaths (of which two are also used as cart tracks), all appearing on the definitive map held by the West Glamorgan County Council. Additionally the links are included in the Register of Common Land held by the county council in accordance with the provisions of the Commons Registration Act 1965. Some 48 claims have been registered under the Act and although formal objection has been lodged by the club to all claims, the club recognise that many of these claims are genuine and perhaps as many as 30 will be difficult to defeat.

It appears that the game of golf has been played over the land from about 1896 and a club was formally constituted in 1908. Early documents of title were largely destroyed by enemy action in 1941, but it is not disputed that for many years the freehold of the land has been vested in a company known as Pennard Burrows Ltd. The shareholders of this company were originally persons who had advanced money to the club, but over the years all the shares have gradually been acquired by the trustees of the club. The company on various occasions in the past has granted a lease of the land to the trustees of the club, the last formal lease apparently running for a term of four years from March 25 1949, and since the expiry of that lease it appears that the trustees of the club have been holding over and currently pay a rent of £1 per annum to the company

Mr Diehl, for the appellants, called evidence from Mr C W Hutchinson MBE MA FRICS, a partner in Gerald Eve & Co, who, in addition to advising the club on rating matters, has been a member of the appellant club for 25 years and plays regularly over the course. I also heard evidence from the president of the club, Mr H E Davies, and from five other members including two past captains and the lady captain The evidence of these witnesses was directed to establish that there was serious interference with the quiet enjoyment of the golf course. This interference came from a number of sources. Firstly, as many as 50 ponies grazed over the links in exercise of their owners' rights of common. In order to protect the greens it was necessary to surround each green with a low single-strand post-and-wire fence. Secondly, the extensive use by members of the public of the footpaths. This was made worse because, although most of the footpaths followed the perimeter of the golf course (there were three shown on the definitive plan which cut across the middle of the course), the public did not stay on the paths, many of which were in any case ill-defined, but tended to wander at will over the course, the presence of the ruins of Pennard Castle and of an ancient church near the seventh green being a source of attraction. The third main category of disturbance came from an assorted collection of trespassers, marauders and unruly visitors.

The respondent valuation officer, Mr D G Richards FRICS, and a member of his staff who had also played golf over the course on various occasions, did not seriously challenge the evidence put forward on behalf of the club except in a matter of degree. Mr Richards pointed out that the incidents referred to by Mr Davies had to be measured against the length of time, 55 years, during which he had been a member of the club. I am satisfied, however, that in recent years the interference has become greater and I was particularly impressed by what was said regarding the route followed by the public in fairly large numbers on fine days in the summer months to and from Pobbles Beach along the edge of the 17th fairway.

Mr Diehl accepted that three of the four ingredients of rateable occupation as adopted in *John Laing & Son Ltd v Kingswood Assessment Committee* [1949] 1 KB 344, were satisfied: firstly, that there was actual occupation or possession by the club; secondly, that the possession was of some value to the club; thirdly, that it was not for too transient a period. But, said Mr Diehl, in the light of the evidence it was not possible to say that the occupation or possession was exclusive. He pointed to the occupation of the land by the commoners in exercise of their rights of grazing and the widespread use of the course made by members of the public. Mr Diehl relied in particular on *Peak (VO) v Burley Golf Club* and *Harding (VO) v Bramshaw Golf Club* [1960] 2 All ER 199, a case in which two golf clubs were held not to be in rateable occupation of land in the New Forest.

For the respondent valuation officer, Miss McDonagh referred to the leading cases on rateable occupation, *Westminster City Council v Southern Railway Co* [1936] AC 511, *Halkyn District Mines Drainage Co v Holywell Union Assessment Committee* [1895] AC 117 and also to Pimlico *Tramway Co v Greenwich Union* (1873) LR 9 QB 9 and *Margate Corporation v Pettman* (1912) 106 LT 104.

Quite clearly occupation in order to be rateable must be exclusive and that means that the person using the land has the right to prevent any other person from using it for the same purpose. Furthermore, the authorities indicate that in law two separate persons may be in exclusive occupation of the same land through the exercise of simultaneous, though different, rights (see, for example, the *Halkyn* case and *Lancashire Telephone Co v Manchester Overseers* (1884) 14 QBD 267).

In the present case it is contended for the appellants that the golf club are not in exclusive occupation of the golf course; firstly, because of the exercise of common rights, which in practice means the grazing of ponies on the land; secondly, by virtue of the exercise by the public of rights-of-way; and thirdly by what I understand is alleged to be occupation by the public as trespassers. Taking first of all the competing occupation by the commoners, were it not for the judgment in the Court of Appeal in the *Burley Golf Club* case to which I will return presently, I would have had no difficulty in deciding that this represented occupation of the land for a different purpose and although it represented an interference with the enjoyment of the land by the golf club it could not be said to interrupt the exclusiveness of the occupation of the golf course as a golf course. As to the exercise of the rights-of-way I need to look no further than the passage in the judgment of BLACKBURN J in the *Pimlico* case when he said at page 13:

"I also agree with what was decided in the case of *R v Jolliffe* that where a person merely enjoys a right to go across land in the sense of a right-of-way or wayleave, he is not in the occupation of the land and is not rateable."

It is true that in the *Halkyn* case the drainage company were held to be in rateable occupation of the tunnel which they used under the grant of an easement by the landowner; but in that case the rateable occupation by the company turned not on the fact that they made use of the tunnel under the grant of an easement but rather on the fact that they were in actual possession of the tunnel and that the use which they made of it amounted to occupation which was paramount, any rights retained by the landowner being subordinate.

With regard to the third matter, that is to say the many acts of trespass committed by members of the public, it is only necessary to observe that in rating law there is no such thing as rateable occupation by the public at large. As Lord Halsbury said in *Lambeth Overseers v London County Council* [1897] AC 625 at page 630: "The 'public' is not a rateable occupier; and I think that one sentence disposes of the case."

In the present case I do not consider that the members of the public can be said to be in occupation of the land over which they trespass, still less that they are in rateable occupation of it. If that be right then there can be no occupation to challenge the exclusiveness of the occupation of the golf club. The documents to title indicate that the company own the freehold of the land subject to the rights of the commoners and the rights-of-way. I am also satisfied—although the matter is not expressly set out in the original lease—that the club occupies the golf course subject to the same conditions. From the evidence it emerges quite clearly that the club fully recognise their right to exclude trespassing members of the public although they have given up trying to do so. Many years ago a ranger was employed at certain times during the week to keep the course clear. Nevertheless, members of the public are permitted to play golf only after payment of green fees and nearly £1,000 was taken in this way in each of the years 1972 and 1973. Over the years it appears that unauthorised golfers have been detected on about four occasions, but they were clearly recognised as trespassers. Indeed, in the 1936 lease between the company and the club trustees, the trustees covenanted, *inter alia*:

"7. To use the demised property as golf links and clubroom with the requisite ancillary buildings for the convenience of members of the club and their staff only or for such other games or recreations, if any, as may be sanctioned or approved from time to time in writing by the company."

I referred earlier to the *Burley Golf Club* case and I must now return to it in greater detail. In that case it was held in the Court of Appeal that the two golf clubs involved were not in rateable occupation of the golf courses, since they had no exclusive right of user by reason of the rights of the commoners and their inability to exclude the public (there was a further ground on which the findings were based in respect of the *Bramshaw Club* which has no relevance to the present appeal). But whereas in the present case the golf club occupy under a lease which grants them the exclusive right to play golf which they in fact exercise, in the *Burley* case the two golf courses were situated on Crown land within the New Forest and the clubs used each course under a licence from the Forestry Commissioners, who managed the land on behalf of the Crown. In the case of the *Burley Club* it was stated that the permission granted should not afford exclusive rights of user over the course and the evidence in both cases was that some members of the public who were not members of the club played golf over the courses but refused to pay and the clubs did nothing about it. Although Harman LJ appeared to regard the rights of the commoners who exercise rights of pasture over the course as evidence that no exclusive right existed for the benefit of the golf clubs, Pearce LJ said at page 204:

"In practice, the clubs do not prevent unauthorised persons playing on the courses. Nor, I think, do their licenses give them any right to do so as long as the Crown allows the public free access to this part of the forest. I cannot agree with the suggestion that the Crown by granting the licence are impliedly forbidding the public to play golf on the course or giving to the licensees an implied authority to prevent the public from doing so."

A and a little later in a reference to the members of the public who walk, picnic and park their caravans:

> "The club has no right to prevent them and although their presence on the course is directed to a different end from the purposes of the members, it does I think constitute some interference with the club's occupation for its particular purposes."

I think that the *Burley* case is to be distinguished on the ground that the clubs there made use of the golf courses under licences, one of which was expressly, and the other by implication, not exclusive. In my judgment the facts in the *Pimlico* case bear a closer relationship to the facts in the present case. In that case a tramway company under powers B contained in the Tramways Act 1870 laid down a tramway in a highway the soil of which was vested in the District Board. The company enjoyed the exclusive use of the tramway for carriages with flange wheels to run on the rail. By section 57 of the Act the company was not to acquire any right other than that of user of any road in which their tramways were laid; and by section 62 nothing in the Act was to abridge the right of the public to pass along or across every part of any road in which the tramways were laid. As Blackburn J said at page 14:

> "Clearer words can hardly be used than that for the purpose of C saying that the rails are laid down entirely to facilitate that purpose [the use of the rails for the movement of carriages with flange wheels] and for that purpose the promoters shall have the exclusive right of user."

Lush J said at page 15:

> "The tram rails occupy a portion of the soil; they are exclusively used by the tramway company for the purposes of the tramway, and that I think makes them occupiers of that portion of the soil. I do not think they are the less occupiers because the public still have the right of passing over the surface of their iron road. The road as a tramway is in their exclusive use and used for their exclusive benefit; therefore I agree in thinking they are occupiers, and must be held to be rateable." D

In the present case there is clearly some interferènce which disturbs the harmless pursuit of the golf ball around the golf course at Pennard Burrows, more particularly perhaps on fine summer weekends. Such interference in my opinion goes to matters of valuation and not to rateable occupation. In my judgment the occupation by the golf club through their E trustees of the golf course is exclusive for their purposes, just as the rails were held to be in the exclusive occupation of the Pimlico Tramway Company. No other person is in occupation of the golf course *qua* golf course and, in so far as the exercise of the commoners' rights may be said to be a competing occupation of the land, it is in my opinion clearly subordinate.

Accordingly, I hold that the appeal hereditament is in the rateable occupation of the appellants and since there is no dispute on the value I determine that the entry in the valuation list be amended to rateable value £970. The appeal is allowed F to that extent.

The appellants will pay the costs of the respondent valuation officer, such costs failing agreement to be taxed by the Registrar of the Lands Tribunal on scale 4 of the County Court Scale of Costs, less the amount of the setting-down fee.

GW00579863

Jock Phillips ONZM is a freelance historian based in Wellington, New Zealand.

Studying first at Victoria University of Wellington then Harvard, Phillips went on to teach American and New Zealand history at Victoria for 16 years, a time during which he also established the Stout Research Centre for New Zealand Studies.

He later became the Ministry for Culture and Heritage's Chief Historian, a position he held for 14 years. While there, he oversaw the development of the *New Zealand Historical Atlas*, initiated an oral history programme, and set up a Fellowship in Māori History. He also helped establish Te Ara: The Encyclopedia of New Zealand, acting as its General Editor from 2002 to 2011.

Phillips has taken on many governance roles throughout his career, including with the National Library Society, Fulbright New Zealand, the New Zealand Portrait Gallery, Victoria University of Wellington Council and Guardians/Kaitiaki of the Alexander Turnbull Library; he has served as Conceptual Leader for history exhibitions at Museum of New Zealand Te Papa Tongarewa, and is a trustee of Ngā Taonga Sound & Vision.

He has published 15 books on New Zealand history, the best known of which is *A Man's Country? The Image of the Pakeha Male — A History*. His most recent are *To the Memory*, a comprehensive illustrated history of New Zealand war memorials, and a memoir, *Making History: A New Zealand Story*.

Among Phillips' many accolades are the Royal Society of New Zealand's 2011 Pou Aronui Award, for service to the humanities, and the 2014 Prime Minister's Award for Literary Achievement.

Also by Jock Phillips

In the Light of the Past: Stained Glass Windows in New Zealand Houses (with Chris Maclean, 1983)

Biography in New Zealand (ed., 1985)

A Man's Country? The Image of the Pakeha Male — A History (1987, 1996)

Te Whenua, Te Iwi: The Land and the People (ed., 1987)

The Great Adventure: New Zealand Soldiers Describe the First World War (ed., with Nicholas Boyack and E. P. Malone, 1988)

New Worlds? The Comparative History of New Zealand and the United States (ed., 1989)

Towards 1990: Seven Leading Historians Examine Significant Aspects of New Zealand History (ed., 1989)

The Sorrow and the Pride: New Zealand War Memorials (with Chris Maclean, 1990)

Brief Encounter: American Forces and the New Zealand People, 1942–1945 — An Illustrated Essay (with Ellen Ellis, 1992)

Royal Summer: The Visit of Queen Elizabeth II and Prince Philip to New Zealand, 1953–54 (1993)

Going Public: The Changing Face of New Zealand History (ed., with Bronwyn Dalley, 2001)

Settlers: New Zealand Immigrants from England, Ireland and Scotland, 1800–1945 (with Terry Hearn, 2008)

Brothers in Arms: Gordon & Robin Harper in the Great War (ed., with Philip Harper and Susan Harper, 2015)

To the Memory: New Zealand's War Memorials (2016)

Making History: A New Zealand Story (2019)

A History of New Zealand in 100 Objects

Jock Phillips

PENGUIN BOOKS

PENGUIN

UK | USA | Canada | Ireland | Australia
India | New Zealand | South Africa | China

Penguin is an imprint of the Penguin Random House group of companies, whose addresses can be found at global.penguinrandomhouse.com.

First published by Penguin Random House New Zealand, 2022

1 3 5 7 9 10 8 6 4 2

Design by Carla Sy © Penguin Random House New Zealand
Author photograph by Robert Cross
Prepress by Soar Communications Group
Printed and bound in China by Toppan Leefung Printing Limited

A catalogue record for this book is available from the National Library of New Zealand.

ISBN 978-1-76104-721-3
eISBN 978-1-76104-722-0

penguin.co.nz

This book was written in the hope
that generations younger than
mine would find the history
of New Zealand exciting.

So, it is dedicated to:

My children — Jesse & Hester
My stepchildren — Laura, Eve, Julia & Lily
And my grandchildren — Elsha, Arlo,
Jade, Jack, Otis, George, Ocean, Neva,
Lachie, Leila & Tali.

Contents

Introduction — The touch of history

Like many Pākehā New Zealanders, I grew up with the belief that New Zealand had no real history. We were a 'young country' and the important events of the past happened in older civilisations across the seas. It took me a while to realise that we had a very ancient geological history, a human history going back at least seven centuries, and that we could not understand the society we had become unless we explored the journey over those years. I started to read and research New Zealand history, was inspired by two generations of fellow explorers of our past, and found the country's history full of rich characters and dramatic stories.

I eventually began thinking about writing a general history of our country. How to do so, in a way that was fresh and manageable for both author and reader alike? Inspiration came from listening on early morning radio to the gripping instalments of Neil MacGregor's *A History of the World in 100 Objects*. But it took the suggestion and encouragement from Penguin Random House's Jeremy Sherlock for me to be convinced that this might provide a way forward.

When helping to set up the day-one exhibitions at Te Papa, in the 1990s, I had learnt of the value of telling stories through objects. To embark on a general history of New Zealand through 100 objects seemed a demanding, but fruitful, prospect — an approach that might lead me into unexplored areas and illuminate the past in unexpected ways.

Naturally enough, I began with the history. I divided New Zealand's past into 11 periods — before 1800, 1800–39, and then nine 20-year blocks thereafter — and set out to find at least eight suitable objects for each period, with the remaining dozen slots available as seemed appropriate. I brainstormed the main events and developments within each period, and then looked for objects that might inform those themes. In other words, the book was initially driven by the story of New Zealand and its people rather than by letting the objects shout for their inclusion on the basis of their own intrinsic interest. Another historian with a different view of the past might have chosen 100 quite different objects.

It follows that the treatment is largely chronological (and, to an extent, it may help the reader to have read earlier stories in the sequence). I say 'largely chronological' because I quickly realised that, given limitations of space and the need for a variety of objects, there would be many occasions when I would have to reach far out of the particular period of an object to understand its significance.

For example, the wonderful Monck's Cave kurī leads to a discussion about the whole history of dogs and people in this country. The locomotive *Josephine* could only be understood if we traversed the wider history of railways, both here and in Great Britain. I looked for stories that represented the range of peoples who have lived in this country — Māori and Pākehā, the diverse cultures of different immigrant groups, rich and poor, women and men. I hoped to uncover this variety in unexpected places. One might anticipate that the story of Jimmy Hunter's All Blacks jersey would be about men and rugby, which it is, but the women who sewed the jersey also find an important place. The predominantly Māori stories throw light on aspects of our history that were new and at times very troubling to me.

Having set out with a general history in mind, however, I was soon captured by the objects. As I explored museums and read about their objects, I discovered they held such a range of stories and had so many human associations that the objects rapidly took centre stage. I began to realise that it was just as likely that people would dip into the book, attracted by particular titles and images, rather than doggedly start 18 million years ago with the central Otago crocodile jaw and end in

2020 with Ailys Tewnion's crocheted teddy bears.

I next focused on the specific criteria for my objects — the 'rules of the game', in MacGregor's words. Obviously, they did need to throw light on the general history, but they also needed to be fascinating in and of themselves. I quickly decided that although international museums hold many wonderful New Zealand objects, especially those taken during the early voyages of European discovery, I would confine my choice to objects held within New Zealand, and, in the hope that people might perhaps be inspired to go and look at them, I limited myself to objects in museums and public collections. I also decided that in order to highlight the physicality of objects, I would largely exclude paintings and documents. There are only two documents found here — the Treaty of Waitangi and the 1893 Women's Suffrage Petition. I decided that both had such importance to our history that a treatment of them as objects would provide an intriguing new perspective. To my surprise, I discovered that the Treaty of Waitangi was actually written on the skin of an Australian sheep!

I laid down several other definitions of objects. They had to be moveable, which excluded historic buildings or even statues and monuments, and they would not usually be natural history objects — such as shells, bones or eggs — unless they had a larger cultural significance or became part of an object with its own identity. I wanted to ensure that there was a range of social contexts explored, so they would not all be heirlooms of the rich and famous. We do have some objects, especially items of dress, of significant political leaders — William Wakefield's epaulets, Richard Seddon's coronation outfit, Helen Clark's trousers. But the little known also left objects with highly revealing stories — the anonymous eighteenth-century Māori woman who left behind her remarkable sewing kete; the bagpipes of an Irish publican, Paddy Galvin; the school uniform of Harold Pond, a Napier Tech pupil in the Hawke's Bay earthquake; the soccer ball that was a tribute to Tariq Omar, a victim of the Christchurch mosque shootings.

I was also keen that there be a range of geographical locations for the objects, and, even more, the stories that they told. The great museums — in Auckland, Wellington, Christchurch and Dunedin — are of course well represented; but I was able to travel to other parts of the country and search out objects from more distant corners. The sealskin purse from Solander Island in Foveaux Strait held at the Southland Museum; the waka huia found at Ruapekapeka and now at the museum at Waitangi; the kauri gum cathedral from the Kauri Museum; and Eve Bowes' flour bag bloomers from Shantytown on the West Coast — these are among my favourite items.

As I explored each object, I found myself asking similar questions: How does the object reveal a wider history? What is its provenance, and, if the museum does not know, what can we assume, and what can we find out from the family and other sources? What materials is it made from? Which led me to think about the wider history of materials — the book moves from objects of wood and stone to those made of metals and plastic. How does an object's design and purpose fit into the general history of that object? Thus, the story of the Parihaka plough throws a piercing light on the circumstances of that terrible moment in our history, but it also invites questions about the history of ploughs in New Zealand. The extraordinary career of the POLY 1 computer has to be seen in the context of the development of computers generally.

Above all, I began to realise that the stories of the individuals associated with objects were often as intriguing as the objects themselves. Amelia Haszard's sewing machine tells us about the Tarawera eruption of 1886 and the history of sewing machines; but, in the end, it is the brave resilience of this wife and mother that makes it worth the telling. Similarly, Tepaeru Tereora's tīvaevae quilt is a stunningly beautiful object, but it works as a way of communicating her impressive contribution as a migrant to New Zealand from the Cook Islands.

There are stories of violence and horror — the *Endeavour* cannons firing on waka in 1769, Les Adkin's baton that pummelled strikers in 1913, the Biko shields that tried to protect protesters against the Springbok tour in 1981. There are also stories of hope and domestic happiness — Emma Barker's delightful sewing box, Winston Reynolds' remarkable homemade Hokitika television set, the oldest working TV in the country.

In the end, *A History of New Zealand in 100 Objects* can be read in many ways — as a general history of the country, as providing insights into the physical objects that surround us, and as tales, standing on their own, about some of the extraordinary individuals, some famous, some unknown, who have shaped this country. My hope is that anyone who reads it will learn about our history; but, even more, they will have a good time along the way.

Jock Phillips
November 2022

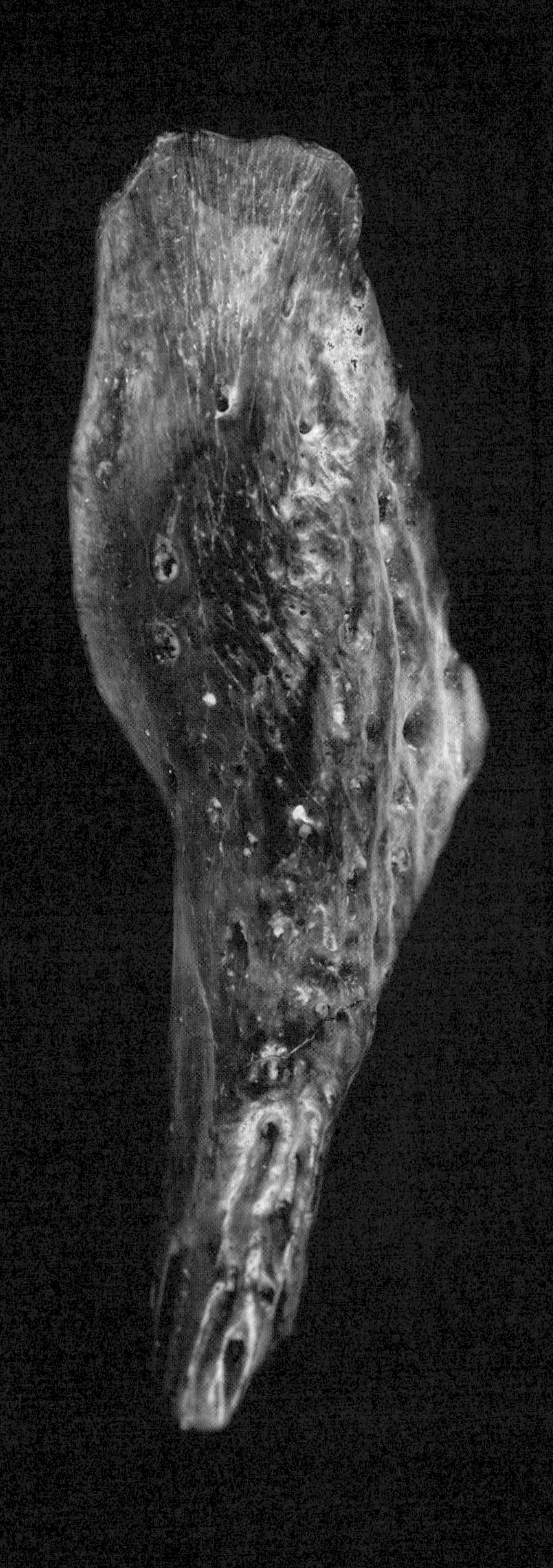

1.

Crocodile jaw

GEOLOGY MUSEUM, UNIVERSITY OF OTAGO

This grey, rather insignificant object, less than 7 centimetres long, may seem an unlikely choice as our first entry point to the history of New Zealand. Yet, when its discovery was announced in 1997, this object helped revolutionise the way we think of the country's history. We are looking at the jawbone of a prehistoric crocodile — or, more accurately, 'crocodilian' — the order which includes crocodiles, alligators and other ferocious creatures. This one was an inhabitant of a large inland lake in central Otago about 18 to 14 million years ago. The jaw is in the possession of the Geology Museum, University of Otago, but it is on public display in Otago Museum.

We think of New Zealand as a 'new country', a place without ancient history, a country which lacked large mammals and was nothing but a land of birds and trees. Yet our history goes back a long way — at least as far as 500 million years. We know that, because in 1948 Malcolm Simpson, a 14-year-old schoolboy on a trip up the Cobb Valley in north-west Nelson, chipped off a piece of limestone containing some indistinct, bug-like fossils. When the piece was looked at later it proved to be the oldest rock and evidence of life in the country, dated at 508 million years ago.

For the next 400 million years the New Zealand land mass was part of the ancient continent of Gondwana; but then about 80 million years ago we first asserted our independence. The southern land mass of Gondwana split apart and the large continent of Zealandia drifted away from Australia. It carried with it some unique plants and animals — dinosaurs, ancient reptiles called tuatara, perhaps the ancestors of moa, some very unusual insects like sandflies and the strange peripatus, or velvet worm.

Not all the inhabitants of this ark survived. Apart from birds, which are a kind of dinosaur, the remaining dinosaurs died out about 65 million years ago. This was probably as a result of a huge meteorite, 10 kilometres wide, which landed in the Yucatán Peninsula of Mexico and plunged the world into a wintry darkness for several months. At times, much of Zealandia sank beneath the seas and many other species of trees and birds disappeared.

But from about 20 million years ago those large parts of the land which had drowned began to re-emerge from the waves. It left large shallow lakes. One of these was in the central Otago area, a massive inland water system now called Lake Manuherikia. It covered some 5600 square kilometres — about nine times the size of Lake Taupō. At the time, the climate was much warmer than it is today — semi-tropical, in fact, with palm trees and eucalypts. It was here, close to St Bathans, that Mike Pole, researching fossil plants, discovered the fragment of a crocodilian jaw. It was the lower right jaw and its size suggests that the species was about 3 metres long — quite small compared to other members of the family. Since then, another 150 fragments have been discovered and the bones of a second species of crocodilian.

The ancient croc was not the only strange beast inhabiting this country at that time. Scientists have also found ferocious shark-toothed dolphins, a giant shark some 13 metres long with teeth the size of a fist, a terrestrial turtle, and a land-based mammal which palaeontologists have dubbed the 'waddling mouse'. Sharks, crocodiles and waddling mice — so much for the myth that ancient New Zealand was just a quiet haven for birds.

Where the crocodilians came from is not clear. They have been identified as probably belonging to a distinct group of crocodiles called mekosuchines, now universally extinct but once found in Australia, Fiji, New Caledonia and Vanuatu. Perhaps they became trapped here as Gondwana broke apart; perhaps they travelled here from Australia and the Pacific. Scientists now think that far more of our native birds and trees were migrants from across the seas carried by winds or the flotsam of the ocean. One might be sceptical that crocodile-like creatures could have swum that far. But in February 1970 the crew of a coastal trader did clearly see a large Australian crocodile some 300 miles from North Cape and over 800 miles from Brisbane. Perhaps this happened before, around 18 million years ago, or the crocodilians were indeed remnants from Gondwana itself. No one can say for certain.

What is known is that for over 500 million years the New Zealand land mass has been a changing world — sometimes hot, sometimes cold, sometimes under the sea, and it has been host to some very strange creatures. It was certainly a very different world from the land of trees and birds discovered by human beings a mere 700 years ago.

2.

Te Arawa taumata atua

AUCKLAND WAR MEMORIAL MUSEUM

This small, mysterious red basalt carving, a mere 34 centimetres in height, with its striking puzzled face and its three-fingered hands imploring heavenward, tells of the arrival of people to Aotearoa. Modern-day scientists, palaeontologists and archaeologists tell us that ancestors of the Māori reached these shores sometime in the late thirteenth century. But long before Pākehā scientists started digging and put on their lab coats, there was another source of knowledge about the first human migrations to these shores. These were the rich set of stories — oral histories if you will — which were told on marae up and down the country for hundreds of years. They were collected and written down by interested Europeans in the mid-nineteenth century, such as the missionary Richard Taylor, the doctor and administrator Edward Shortland and, above all, Governor George Grey, who relied on the work of his co-author Wiremu Te Rangikāheke, a leader in Te Arawa iwi around Rotorua, to produce in 1855 his famous *Polynesian Mythology*. Some of these stories contain magical and incredible incidents — such as people arriving in the country on the back of a whale. But they should not, and cannot, be simply dismissed. There is considerable consistency in the tales from different informants, despite iwi differences; and when you count the generations — the whakapapa — which they include, they reach back to almost exactly the date of arrival now posited by Pākehā researchers. Further, the stories tell us much about why people came to the new country and the values they brought with them. Thus, it is only appropriate, and helpful, to begin with a Māori account of the human arrival here.

This small red stone figure is our point of entry. It is said to have been carried from the homeland, Hawaiki, on board one of the arriving waka, the *Arawa* canoe. The figure was presented to George Grey during a visit to the Rotorua area in April 1866. He subsequently gave it to the Auckland Art Gallery who in turn lent it to the Auckland War Memorial Museum. Grey had come to Rotorua partly at the request of his old friend, Te Rangikāheke. Most of Te Arawa had supported the Crown during the wars against the Māori King and then against the radical Hauhau movement which followed, and in 1865 Te Arawa had put men into the field to assist the Crown in the Bay of Plenty. With their men away fighting, the crops had not been planted and the tribe was in dire straits. Grey came in part to respond to their calls and in part to request further assistance. While in Rotorua he went by canoe to Mokoia Island, that beautiful volcanic plug in the middle of Lake Rotorua which was the site of a famous love story involving Hinemoa and Tūtānekai. Some say that on the island Grey was presented with the stone god by Pango Ngāwene, a famous local tohunga (or priest), although another source suggests that Grey was given it at Ōhinemutu by the elderly chief, Ngahuru. There are also various stories about its name and origins. Some call it Matuatonga. William Mair, who was Grey's interpreter, denied that and claimed that the object had been brought from Taranaki by a war party and then buried on Mokoia. Others say that a much larger figure, some 2 metres high, was the one brought on the *Arawa* canoe, while other writers assert that this larger figure was but a copy of the original one brought on the canoe and presented to Grey. Certainly the larger icon is made of white rhyolite lava (pumice), the local rock of Mokoia; while red basalt is found in many places in the Pacific, but also in New Zealand; and the three fingers of the carving are of both Māori and Pacific style.

But in a world of oral history and memory, the exact provenance of the stone god is not really our primary concern. What remains important is that many in Te Arawa certainly believed that this small god had been brought on the *Arawa* canoe, and so it was invested with rich meanings and memories by generations of Arawa people. Their story of the canoe begins in Hawaiki, the fabled homeland, and the central figure was a spirited young man, Tamatekapua, the son of Houmaitawhiti and the brother of Whakatūria. The family had a pet dog which one day went missing. The brothers set out to find the dog and eventually went to a neighbouring village where the chief was

Uenuku. There they heard the dog howl, but could not see it anywhere. At last they realised that the howl was coming from the belly of one of Uenuku's villagers. The reason came out. Uenuku had been suffering from a pussing boil. The dog had eaten a discarded bandage; and this was viewed by Uenuku as akin to eating himself. So the dog was killed and consumed. Tamatekapua was furious and decided that he and his brother would take revenge on Uenuku by stealing breadfruit from his orchard, using stilts to disguise any tell-tale footprints. After many nights, the brothers were eventually caught, but through Tamatekapua's trickery, they each managed to escape. The result was a fierce conflict, with Uenuku's army storming the village where Tamatekapua, his brother and father lived. Casualties were heavy and Tamatekapua came to the decision that it was time to set sail for a new country. They carved a canoe, while just along the beach another canoe, *Tainui*, was also preparing to sail. Tamatekapua was farewelled by his father Houmaitawhiti, who told him to not follow Tū, the god of war, but to be peaceful: 'In war there is nothing but useless death.'

TAUMATA ATUA. AUCKLAND ART GALLERY TOI O TĀMAKI, GIFT OF SIR GEORGE GREY, 1887, ON LOAN TO AUCKLAND WAR MEMORIAL MUSEUM TĀMAKI PAENGA HIRA. G132.

As they were about to set off, Tamatekapua desired, for his protection and good navigation, to have Hawaiki's leading tohunga and priest, Ngātoroirangi (Ngātoro), as a passenger. But the tohunga was preparing to sail on the *Tainui* canoe. So Tamatekapua invited Ngātoro and his wife Kearoa to come on board and bless the *Arawa* canoe for the long journey. Immediately, Tamatekapua ordered the anchor to be lifted and the canoe set sail with Ngātoro and Kearoa on board. That was not the end of Tamatekapua's trickery. Kearoa was pretty, much fancied by Tamatekapua. One night, suspicious of Tamatekapua's intentions, Ngātoro came down from on deck where he had been observing the stars to keep the canoe on course. He realised that Tamatekapua had been sharing his berth with Kearoa. Ngātoro was furious. He called on the heavens for winds, and the canoe began to be sucked into a huge whirlpool, 'The Throat of Te Parata'. All feared for their lives. Finally, concerned about the women and children on board, the tohunga relented and the storm stilled. In some accounts the canoe's saviour was a large shark, an 'arawa', who thus gave his name to the canoe and its people.

The *Arawa* canoe reached New Zealand at Whangaparāoa near East Cape. The pōhutukawa was in full bloom. Seeing the scarlet blossom, a chief threw away his faded red-feather headdress and gathered the blooms to use instead — only to find that the blossom quickly wilted. Eventually they reached the headland of Maketū in the Bay of Plenty which Tamatekapua claimed to be 'the bridge of my nose'. The Arawa people had found their home.

What can we learn from this story about the migration of Māori to this land? First, that they came because of pressure over land resources. Growing populations competing over limited quantities of food seems a likely reason for people to head off overseas. People have always migrated when their stomachs are hungry. Second, the figure of Tamatekapua is the kind of hero who appears often in Māori tradition — the trickster and philanderer, a man who through fast thinking and sleight of hand gets himself and his followers out of difficult situations. Third, the Polynesian settlers who first reached these shores were amazing sailors and navigators. They read the stars and the pattern of waves to voyage thousands of miles across rough oceans. In this case they clearly survived a threatening storm, to arrive, like most of the canoes, on the North Island's east coast in early summer when the pōhutukawa was in full bloom. Finally, and most significantly, the people who came here were gardeners and farmers. This story begins in a breadfruit orchard. The stone god was itself a kūmara god — an icon placed in the fields at the time of planting to bring good crops. The early settlers from the Pacific brought with them certain foods for the new land — the kūmara most obviously (which had probably been retrieved from a voyage to South America), but also the gourd, the taro, the yam, the paper mulberry and a species of cabbage tree. Māori were from the start gardeners.

3.

Tairua pearl shell lure

AUCKLAND WAR MEMORIAL MUSEUM

Māori stories of the great canoe voyages have brought the first inhabitants from Hawaiki to Aotearoa. But where was Hawaiki, the ancestral homeland, and who were the first arrivals? To uncover this back story, let us turn to the diggings of archaeologists and particularly to one of their remarkable finds, this beautiful pearl shell lure, a mere 50 millimetres in length. The hole at the top was tied to a fishing line, while near the bottom are notches on either side which were used to attach an unbarbed bone hook. It was uncovered in 1964 in a dig led by Auckland archaeologist Roger Green, at the base of Paku, the hill guarding the entrance to Tairua Harbour on the Coromandel Peninsula. When examined by A. W. B. Powell, an Auckland Museum shell specialist, the lure turned out not to be formed of a New Zealand shell but of a black-lipped pearl oyster (*Pinctada margaritifera*), a tropical species. Here was the first object ever found to have been clearly brought from elsewhere. It is now on display at Auckland War Memorial Museum.

So where was the homeland? Similar pearl shell lures have been found in the Marquesas, a scattering of French-controlled islands in the far east of Polynesia, and Green initially claimed this to be the source; but it has subsequently been suggested that similar lures were also used in the Society Islands (which include Tahiti and Moorea) and the Cook Islands. The lure would have been trolled behind a canoe, its iridescent colours attracting bonito, a type of tuna. The lack of barbs would have allowed rapid catch and release of the fish into the canoe. The lure established solid evidence that New Zealand's first settlers came from the islands of East Polynesia. The dating has also been recently reconsidered and it is now thought the lure and its owner arrived between the late thirteenth and mid-fourteenth centuries — most likely between 1292 and 1332. Such dates fit perfectly with other evidence of the first arrivals to this country. The presence at the site of the bones of moa and other birds which have subsequently become extinct confirms this early dating.

The people who landed on the Coromandel shore came at the end of a series of very distant migrations. Over 3000 years ago, people from islands in the East and South China seas set sail eastwards and settled in the closest areas of the Pacific, in places like the Bismarck Islands and coastal New Guinea. They developed a distinctive style of pottery, Lapita, then set off eastwards again, first into the Solomon Islands and Vanuatu and then on to Fiji, Samoa and Tonga which they had reached by 2800 years ago. Then during the first millennium they explored even further eastwards into the volcanic islands of the East Pacific. Dates are uncertain but at least by 1000 AD a Polynesian community had emerged there on islands such as Tahiti, the Cooks and the Marquesas. They were a remarkable seafaring people able to cross vast distances by reading the clouds and the waves and watching closely the migratory patterns of birds and whales. They navigated as far as the Hawaiian Islands and Rapanui (Easter Island). These were the peoples who arrived on New Zealand shores, one of whom lost a lure on the Coromandel.

One might ask why these pioneering Polynesians brought a lure with them. Perhaps they had used it to fish on the voyage south to Aotearoa. Perhaps they expected to return one day to their homeland in the Cooks or French Polynesia. Certainly, Māori traditions tell of return voyages, and Mayor Island obsidian has been found at the Kermadec Islands, inviting speculation that the ancestors of Māori used the Kermadecs as stepping-stone points on the journey back. But there has never been any evidence found of such voyages on the home islands themselves. Was it simply that the lure had been a favourite and a highly successful way of catching fish, or was it perhaps a sentimental reminder of the sea world that they had left behind? There is no doubt that the earliest Polynesians were above all a sea people. Much of their food came from the sea or the rocky coasts; and they lived close to the shore. They spent much time in canoes fishing in lagoons or exploring islands, both close and distant.

When the Polynesians reached Aotearoa they must have been amazed at the scale of the landscape — the size of the mountains, the

extent of the inland forests and waterways. Yet they remained very much a sea people. Their earliest settlements were along the coast and for centuries they largely remained there. Their dominant foods were found by the sea — fish, seals and shellfish, not to mention seabirds like muttonbirds (tītī) and penguins; and they continued to use the sea as their primary highway as they criss-crossed water to hunting places or visit friends and relatives across harbours such as Tāmaki Makaurau (Auckland) or straits like Te Moana a Raukawa (Cook Strait) and Te Ara a Kiwa (Foveaux).

Many Māori traditions involve the sea. Māui fished up the North Island from his waka, the South Island. Te Ika a Māui, Māui's great fish, was seen as a giant stingray with Wellington Te Ūpoko o te Ika (the head of the fish). The legendary explorer Kupe was said to have come to New Zealand chasing a huge octopus which he eventually caught in Cook Strait.

Dependent on the sea for so much of their food, Māori were magnificent fishermen — I say 'men' because European explorers saw men do most of the fishing, while women collected shellfish from beaches and rocks. They developed impressive ways of catching fish — in place of the pearl shell lures they developed beautifully crafted lures using pāua shell. They came up with the idea of barbed hooks; and their nets amazed the Europeans when they finally reached New Zealand. On Cook's first voyage, Joseph Banks reported that Bay of Islands Māori laughed in scorn at the Europeans' fishing net and showed them one of their own, which Banks described as 'five fathoms' or 9 metres deep, and 400 to 500 fathoms or 700 to 900 metres long. Yet even that was small beside the 1.6km-long net made by Ngāti Pikiao in 1886. In the nineteenth century much early Māori trade with Europeans

TAIRUA LURE, MATAU. AUCKLAND WAR MEMORIAL MUSEUM TĀMAKI PAENGA HIRA. AU1785.

came through the sale of fish to the settlers, especially in Auckland. And when in the late twentieth century Māori were able to bring their traditional interests in fishing to the Waitangi Tribunal, a series of settlements, including 50 per cent ownership of the country's largest fishing company Sealord, ensured that about a third of the fishing industry was in Māori hands. Subsequently another taonga brought by our first settlers was announced — a shell tool made of another East Polynesian species which was found at Wairau Bar in the north of the South Island. But the full story of Wairau Bar and what it tells us about the country's earliest inhabitants is the focus of our next object.

Wairau Bar necklace

MARLBOROUGH HERITAGE TRUST

It was a hot Marlborough summer's day in January 1939. Jim Eyles had just turned 13 and he and his mate Billy were bored and looking for something to do. They were at Jim's family farm at Wairau Bar, and Jim remembered how his stepfather used to uncover moa bones when ploughing the family's paddock in front of the house. There was once a battered benzene can full of stone adzes which had sat under a tree. So the two lads decided to go hunting for 'Maori curios'. Jim took a shovel and decided to dig where some bones had recently been unearthed. For 20 minutes there was nothing of interest. Suddenly he saw a round hole — perhaps a rabbit burrow. He felt the hard edges — could it be an old gourd used by Māori to carry water? Jim carefully excavated it and his mother put the object in a pudding basin on the mantelpiece. When Jim's stepfather, Charlie Perano of the old Cook Strait whaling family, arrived home he immediately recognised the object as a very large moa egg. It had a small hole at one end and had been blown. It was over 19 centimetres long and 14 centimetres wide. The next day, in a cold southerly wind, Jim and his uncle went back digging. Before long, a brown skull emerged and around the neck below it were seven carved reels with a large hanging sperm whale tooth in the centre. There was much interest — the *Marlborough Express* quickly told the story; the local fish-and-chips shop displayed the finds in the window; and eventually the director of the Dominion Museum came across from Wellington and purchased the egg and necklace for the museum.

The interest was justified. At the time, some claimed that moa had died out before human beings had reached these shores; others believed that the 'moa-hunters' were a separate Melanesian people, the Moriori, who had been driven out by later Polynesian invaders and taken refuge on the Chatham Islands. These finds conclusively disproved such ideas. The seven-reeled necklace was quite different from later Māori ornaments, which were usually single pendants, often made of pounamu. But it was remarkably similar to necklaces found in Eastern Polynesia, from the Society Islands and the Marquesas. Later it was realised that the placement of the burials and the style of the adzes found there confirmed this cultural link. The inhabitants of Wairau Bar were clearly Polynesian, very early arrivals and distant ancestors of Māori. Latest research suggests that they had set up camp by the early fourteenth century. As for the blown egg found beside the skull, it showed that these people were clearly around at the time of the moa and regarded the birds with considerable awe. Subsequently, seven intact necklaces were found plus many individual reels. Sometimes, as with the first, and in our example from the Marlborough Heritage Trust, these were shaped from whale ivory, but many of the reels were carved moa bone. Most of these now dwell in Canterbury Museum from where Roger Duff, in large part from the Wairau Bar diggings, proceeded to tell a story of the moa-hunters of New Zealand. That is the story we now tell.

Moa were one of the great phenomena of New Zealand which attracted huge international interest when bones of the extinct birds were discovered in the mid-nineteenth century. They were ratites — relatives of the ostriches of Africa and the emus of Australia. There were probably nine species. They ranged from two giant moa, one on each island, to the relatively small bush moa under a metre high. The females of the largest moa grew up to 3 metres high and 250 kilograms in weight, while the males interestingly were only half that size. But all had small heads and poor eyesight, which must have made them vulnerable to hunting.

Moa were found throughout the country, but the most intense hunting appears to have been on the east coast of the South Island, where there was much scrub and small trees to browse. The most common place for moa-hunters to set up camp was at the river mouths of South Island rivers — such as Wairau Bar, or the Rakaia or Shag further south. Why the moa-hunters based themselves in such places is not entirely clear. The mouth of a river was often cold and windswept — not an inviting place to stay, especially in the probable hunting season of spring. Of course, river mouths offered plenty of driftwood for fires, and other foods were plentiful, such as seals, seabirds, fish and shellfish. But how did the river mouth assist the capture of moa? Roger Duff suggested that as at Wairau Bar the hunters

CENTRAL UNIT OF SPERM WHALE TOOTH, OTHER UNITS SHAPED FROM BONE OR SEA MAMMAL IVORY TO IMITATE WHALE TEETH C.1250 AD FROM TE POKOHIWI. SOME HIGHLY DEGRADED UNITS FROM THE SAME NECKLACE ARE NOT INCLUDED. 2006.138.0010 MARLBOROUGH HERITAGE TRUST – EYLES FAMILY TRUST COLLECTION

could drive the moa towards a natural cul-de-sac created by the sea on one side and the river and lagoon on the other. Others have argued that moa were frightened toward such places or into swamps by the lighting of fires. But such drives were not easy to control, and they did not flock in large numbers. Instead, moa might have been captured by set snares and then killed by harpoons or bird spears. It seems likely that males were caught when they were sitting on the eggs during the winter. Then they would often have their necks and feet removed and the large meaty legs were carried to a river, put on mōkihi (rafts) or canoes and floated down to the river mouths for cutting up and eating. At Wairau Bar there were few moa skull and foot bones found.

Wairau Bar was unusual because evidence of post holes and the burials suggests it was a permanent abode; while many of the hunting sites at river mouths were temporary butchery sites which could spread for as much as 100 hectares. What else do we know about the inhabitants of Wairau Bar? They were a tall people averaging 175 centimetres for the men and 163 centimetres for women. On average they died at the age of 28 and none lived beyond their early forties. Life was tough and short — but it does not seem to have been violent. There is no evidence of traumatic death and no fortifications surrounded Wairau Bar.

These people, our first inhabitants, were once called moa-hunters.

But they were clearly more than this. Moa provided a great feast, and the bones had considerable cultural significance for their community. But in day-to day life the moa-hunters probably consumed far more of other foods — seals (which could easily be clubbed along the beaches), fish, birds and shellfish, not to mention dogs. And despite the unique nature of their necklaces, archaeologists have come to recognise them as ancestors of the Māori, and Māori themselves have come to agree. When Jim Eyles first uncovered the skull at Wairau Bar his stepfather called in a local Rangitāne kaumātua, Maniapoto MacDonald, who decided, 'He's not one of ours'. However, the iwi complained to police about the removal of the kōiwi (bones) in 1946, and four years later took police to the site in the hope that the burials would not be disturbed and skeletons not removed to the museum. Some Pākehā rejected the lawful claims of the Rangitāne on the grounds that Wairau Bar moa-hunters were a different people. It took a long battle before, first, from 1964, digging was halted; and then in a moving ceremony in 2009 the kōiwi were brought back from the museum and reinterred. The idea that the Wairau Bar moa-hunters were among the earliest New Zealand ancestors of modern Māori had finally achieved full public and legal recognition.

5.

Monck's Cave kurī

CANTERBURY MUSEUM

Thank heavens for curious schoolboys! Like the Wairau Bar necklace, this intriguing wooden carving of a dog, no more than 8 centimetres high, was discovered by a young boy poking around in a cave system in Redcliffs, near Christchurch, the entrance to which had been blocked by a rock fall many centuries ago. Then in the course of building the Sumner Road, workers had dug into the loose gravel. The property was owned by J. S. Monck and one day in September 1889, Monck's son was out exploring when he noticed a small gap in the shingle. He managed to squeeze through and was encouraged by the sight of two eyes beaming at him. It was a stray cat who had got in first. But it was a dog, not a cat, that was the real find. The landslip had obviously occurred when the cave's inhabitants were away: only their possessions remained. There were adzes and fish hooks, and several wooden objects — a comb, a carved bailer, an outrigger, a painted paddle. There was also this carved wooden dog. Nothing like it had been seen before. It was not like traditional Māori wooden carving to be found on canoes or meeting houses. There were no spirals or ornamentation, just a charming, highly realistic dog with a hole through its curled tail. The bailer and painted paddle did show spiral carving, but they had been found on a higher ledge, and were thought to be of a later date. The dog was associated with bones of an extinct swan and of moa, and more significantly much moa eggshell, which suggested the inhabitants had been around when moa still roamed the country and their eggs could be had for breakfast.

'CARVED FIGURE OF MĀORI DOMESTIC DOG, KURĪ', MONCK'S CAVE, SUMNER, NEW ZEALAND, CANTERBURY MUSEUM, E158.356

Exactly why the dog had been carved remains a mystery. Early observers suggested that it was perhaps a child's toy; others that it was the blunt end of a paddle. Whatever its purpose, there was no doubt that the carving reflected the importance of the kurī to early Māori; and, apart from the ears which are missing, it provided a good idea of the dog which Māori had brought with them to Aotearoa. The carving has short stubby legs and a long body, notable characteristics of the kurī. With fur that was usually whitey yellow with occasional patches of black, the kurī had a sharp, fox-like face with strong jaws and sticky-up pointed ears. It did not bark but howled in a melancholy tone. Some early European observers thought that it was lazy and stupid. Yet the kurī was one of the two mammals (along with rats) the first arrivals carried with them to New Zealand, although the striking lack of genetic diversity suggests that the number of dogs brought was not large. Pigs and chickens, their usual fellow domestic species in the Pacific, did not make it, but kiore (rats) and kurī did and the dogs were so important that they survived the journey without being eaten.

So why were kurī so valuable for these early Māori? Perhaps, like modern dogs, kurī were companions, which might explain this slightly sentimental carving. In Māori tradition the explorer Kupe was said to have brought a pet kurī on his journeys; Tāneatua, the tohunga on the *Mataatua* canoe, had several; and Pāoa, captain of the *Horouta* canoe, was so sad at the loss of his kurī that he named the point of land Pākehā know as 'Young Nick's Head' Te Kurī a Pāoa. Their loss is also lamented in a number of waiata, and Māori would occasionally recount the whakapapa of their dogs alongside their owners. Yet many Māori sayings about kurī refer to them with some scorn; and in some tribal traditions Māui turns his brother-in-law Irawaru into a dog as a punishment for some offence, such as his laziness or greed. It is quite possible that kurī were valuable as hunting dogs. No early European observers saw them used this way, but the kurī's strong jaws would have been helpful in hunting ground birds such as moa, kiwi, pūkeko and weka. Certainly, traditional accounts describe kurī which were famous hunters, such as the kurī owned by Tara, son of the explorer Whātonga.

More certain was the value of the kurī's body. Their skins were cleaned and dried and used to make warm and waterproof clothing, while pieces of fur, especially the prized white fur, were woven into cloaks. Kahu kurī, dog-hair cloaks, had very high prestige, and dog hair often adorned the top of fighting weapons such as taiaha. Dog bone was carved into neck and ear pendants, while jaws and teeth became fish hooks. But undoubtedly the greatest value of kurī for Māori was as a source of food. We have already heard the story of Uenuku eating Tamatekapua's family dog (see page 18). There is no doubt that in a world where there was limited protein, dogs were a valuable source of nourishment. Joseph Banks reported that when he ate one in Tahiti, it was almost as tasty as 'an English lamb', which he put down to the fact that the dogs lived on vegetables. Archaeologists suggest that their diet in New Zealand was largely fish and vegetables; and there are few teeth marks on surviving bones of moa. But the good state of their teeth seems unlikely if they did not also eat meat; and James Belich has even suggested that dogs provided a kind of long-term storage of moa meat. Since people could not possibly eat all of the large birds when they remained fresh, it was fed to kurī, and the dogs were killed later for food. In the South Island it was said that kurī were castrated so they would fatten more quickly and make a succulent feed.

When the first European explorers reached New Zealand kurī were still plentiful around Māori settlements; but it appears they did not remain pure-bred for very long. William Colenso claimed that on his travels from 1834 he never saw a pure kurī. Instead they mated with escaped European dogs, and by mid-century packs of such mongrels roamed wild in the backblocks. When crossing what came to be called the Harper Pass between Canterbury and Westland in 1858, Leonard Harper described coming across packs of wild dogs and their howling at night. The last-known pure kurī is a somewhat bedraggled specimen now at Te Papa which was collected in the Catlins in 1876.

By then, dogs of other species had become essential companions and workers in New Zealand. Later explorers such as Charlie Douglas took dogs, no longer kurī, on their travels. For 20 years Douglas mapped the valleys of Westland with only a dog, which he always called Betsey Jane, for company. Other Europeans found other uses for dogs. Shepherds imported Border collies to round up sheep and keep their flocks within boundaries in the absence of fences. It was not long before sheep dogs became essential for mustering the high country. Sheep-dog trials emerged as a sport. Others developed breeds to help hunt down pigs. Māori having lost the kurī developed their own dependence on the imported breeds of dog. When the government introduced dog registration to help control the packs of stray and wild dogs, Māori objected to the payment of a 2s 6d registration fee and this almost led to war in the famous 'Dog Tax Rebellion' among Hokianga Ngāpuhi in 1898.

By the twentieth century, dogs had become so important to New Zealanders — Pākehā as well as Māori — that they entered national mythology. A sheep dog, simply called 'Dog', became the starring attraction of Murray Ball's cartoon strip and film *Footrot Flats*, and Lynley Dodd made Hairy Maclary a national icon with her children's books. By the end of the century, there were no fewer than 19 statues of dogs in public places, ranging from the much-photographed sheep-dog memorial at Tekapo to Hunterville's tribute to the huntaway breed and Te Aroha's bronze dachshund which also serves as a bike stand and drinking fountain. If our Monck's Cave kurī was indeed a sentimental tribute to man's best friend, it was but the beginning of a very long New Zealand tradition.

6.

Kahungunu hei tiki, Te Arawhiti

WHANGANUI REGIONAL MUSEUM

This beautiful hei tiki, in dark green pounamu with red sealing wax in the eyes, was a treasured taonga tuku iho handed down over generations, with its exact provenance lost to history. To approach Te Arawhiti we turn to the evidence found in Māori traditions, to the stories told in meeting houses or around a fire. In particular we must tell the story of a great love affair between Kahungunu, the eponymous ancestor of Ngāti Kahungunu, and Rongomaiwahine. Te Arawhiti was said to have been brought to Hawke's Bay by Kahungunu, who passed it on to Tauheikurī, the youngest daughter of Kahungunu and Rongomaiwahine. Tauheikurī reputedly wore it when a war party led by a famous warrior, Tūtāmure, had surrounded her people's pā at Maungakāhia on the Māhia Peninsula and she came out of the pā to plead for peace. Subsequently, it was handed on to her grand-nephew Tamaterangi, who is said to have always worn it into battle. The tiki was then passed through five generations to a chiefly woman, Te Arawhiti, and her son named it in her honour. In this way the tiki functioned as a keeper of whakapapa. Eventually the tiki was acquired in the early twentieth century by a Pākehā collector, W. E. Goffe, whose collection was sold in 1941 to the Whanganui Museum, where Te Arawhiti lives today. However, Tamaterangi's descendants, Ngāi Tamaterangi, who live along the Waiau River in the Wairoa district of northern Hawke's Bay, remain its kaitiaki (guardians).

To understand the fuller significance of Te Arawhiti let us begin with the journeys of one of the world's great lovers, Kahungunu. He was born and brought up way up north, on a pā near modern Kaitāia. He quickly established a reputation as a charismatic hard-working man skilled at the irrigation of gardens and the arts of carving and tattooing. He married and had three children. But he decided to head south, staying for a short time with his father at Tauranga; and then going further east and south. At each place he stopped — at Whakatāne, at Ōpōtiki, at Whāngārā around East Cape, at the Popoia pā in the Tūranga area (modern Gisborne) and at Whareongaonga — he married and had children before heading further on his journey. Some say that Kahungunu acquired the tiki in Whakatāne when he was offered it on the condition that he move on. Finally he heard reports of a stunning beauty, Rongomaiwahine, who lived at Nukutaurua on the Māhia Peninsula. Intrigued, Kahungunu set off, but arriving at Māhia he discovered that she was already married to Tamatakutai. He realised that if he was going to win her hand, he had to first win the support of her people and then disrupt her marriage. So he gathered huge quantities of fern root and rolled it into a bundle 'as high as man' which impressed the locals for whom this was a staple food. Then he climbed a hill and, watching seagulls (karoro) diving for fish, he practised holding his breath for as long as the birds were underwater. (The hill is still called Puke Karoro.) Then he used this new skill to dive for pāua. He filled several baskets and emerged from his final dive with his chest covered with the shellfish. A feast was in order and Kahungunu had the support of the people. Now to disrupt Rongomaiwahine's marriage. When the people feasted on the pāua, Kahungunu requested the roe, which is known for producing flatulence. That night he lay down beside the sleeping couple, and sure enough farted profusely. Mutual allegations and an argument ensued between husband and wife. The next morning Kahungunu took Tamatakutai surfing on a canoe. The waves were large, Kahungunu was steering and on one especially ferocious swell Tamatakutai fell out and was drowned. And so Kahungunu and Rongomaiwahine were married and settled for good at Maungakāhia on the northern part of Māhia Peninsula.

What does this story tell us? It reinforces the sense that pre-European Māori had a great respect for the 'trickster' male — the brave man who lived off his wits and won his battles through quick thinking and sleight of hand. There are many such figures in Māori tradition — from the legendary founding ancestor Māui, to Tamatekapua of the *Arawa* waka, to the historic figure Te Rauparaha.

Second, the story of Kahungunu's migration south and east points to a major development in early Māori society. It seems likely that when Polynesians first reached Aotearoa they settled along the coasts of both islands. Most found a home in the north of the country, where the climate was suited to a people who had come from the tropics, and where there was at first abundant food — moa in the forests and scrubland, and seals along the coasts. The early settlers fed well on this prey, and the 'protein boom' led to fast population growth. But the good times did not last. Moa were probably extinct within 200 years — by about 1500 — and the seals became confined to the coasts of the cold, windy and inaccessible Fiordland. In addition, it seems that the climate grew colder, wetter and windier, reducing crop yields. There was competition for resources. Many Māori turned to new areas where they hoped for good kai moana, birds in the forests and attractive lands for cultivation. They tended to move, like Kahungunu, towards the south and the east, away from the wind. As this competition for resources intensified, Māori came together in large groupings — small hapū of a few families, the original unit of settlement, joined other hapū into larger iwi controlling the resources of a substantial territory. So Kahungunu moved to the east coast of the North Island and around him an iwi, Ngāti Kahungunu, formed. There were also a series of migrations down to 'Te Ūpoko o te Ika' — the head of the fish or modern-day Wellington — and further south into the South Island. Ngāti Māmoe arrived in the south first, and then Ngāi Tahu from the North Island's east coast followed in the seventeenth and eighteenth centuries.

One consequence of North Island tribes moving into the south was a new importance attached to pounamu, or greenstone. When Māori first arrived, they quickly found sources of stone which were essential to making adzes and chisels for shaping wooden implements. Obsidian items from Mayor Island off the Bay of Plenty coast have been found in middens throughout the country. There was also argillite from the Nelson/Marlborough area. But early sites show little pounamu. At Wairau Bar, for example, there were only three nephrite adzes in burial chambers. A hard, deep green form of jade, pounamu, or nephrite, was only found in the South Island, which Māori called Te Wai Pounamu (Greenstone Waters). It came from the Arahura and Taramakau rivers of the central West Coast, and Lake Wakatipu in inland Otago. There was also a field of bowenite (a similar magnesium silicate) in Milford Sound. The hardness of the stone which could be ground to a fine edge, and its beauty gave pounamu great value to Māori. Once the North Island iwi had settled the south, good communications became established with relatives in the north. Pounamu could be carried by foot over the mountain passes of the South Island, worked at centres on the east coast and then transported by canoe as trade goods to the iwi of the north, becoming a treasured stone throughout Aotearoa.

At first pounamu was used simply to make tools — for adzes and chisels. Then Māori realised that it could be of value for weapons. Attached to a wooden handle, pounamu became a fearful tomahawk. Beautifully balanced mere were powerful implements to thrust at the chests of adversaries. Before long, mere became symbols of chieftainship. Such was its beauty and mana, that pounamu also became used for ornaments — to hang from the ears or tied with a flax string around the neck. Eventually, in one of those creative innovations for which early Māori society was so distinguished, used adzes became carved into hei tiki (hei meaning 'the neck', tiki meaning a human image). They became taonga of great personal and sentimental value, worn at first by men and then increasingly by women and handed down as precious items within families. They became associated with chiefly rank. Exactly when these developments happened is not clear. Pounamu hei tiki probably first appeared in the sixteenth century. By the time of European arrival in the eighteenth century, they were common and prestigious.

Exactly when Te Arawhiti was made is unknown. It was undoubtedly the work of a brilliant artist and is made of the prized dark green central West Coast kahurangi pounamu. The carving, especially of the ribs and prominent knobs on the knees, is outstanding, and the smooth surface suggests generations of handling. Nor does the presence of red sealing wax, a material which came with Europeans, imply that this was a post-European tiki. Often the wax was added to the eyes of an existing hei tiki after the original shell inlay had fallen out.

It is also uncertain what hei tiki represented. Some suggest they were symbols of fertility representing the human embryo, others that they were worn as mementos of ancestors who had themselves once worn the tiki. This was undoubtedly the case with Te Arawhiti. We can be sure that when the descendants of Kahungunu and Rongomaiwahine wore this splendid tiki, they recalled the great love of those two ancestors, who each gave their names to iwi which still flourish along the east coast of the North Island.

7.

Puketoi kete

OTAGO MUSEUM

The Polynesian settlers of Aotearoa left us objects made of the materials around them — rock, bone, shell and wood. All with the possible exception of wood, could be expected to survive over centuries. More transitory were the objects which Māori made from the herbaceous plants they discovered. This kete and its contents were almost wholly made from such fibres; so why have they survived? One reason is they were comparatively recent. This kete probably dates from about 1680 to 1730, two centuries or more after moa had become extinct, and indeed after the first brief contact by Abel Tasman and his crew. The second reason these items survived is that the kete was placed in a crevice inside a cave, near the Taieri River in central Otago, and the dry conditions perfectly preserved it.

The kete was discovered in July 1895 by a rabbiter who passed it on to David McKee Wright, an Irish-born shepherd working on Puketoi Station. Wright, a remarkable man, later found fame as an author of manly ballads about rural life. He announced the find of these 'Maori curios' in the local paper, the *Mount Ida Chronicle*, and showed the kete to Alexander Hamilton of the Dominion Museum, who wrote a detailed description. Later, Wright gave the kete to a Nelson collector, Frederick Knapp, from whom it was acquired by the Otago Museum in the 1920s. By then several items had disappeared — albatross and kākāpō feathers, and a bundle of cabbage-tree root — but enough remains to offer a superb sense of how Māori adapted to a new land.

The kete is really an ancient sewing kit, and Hamilton surmised that the kete and contents were 'probably the treasures of some industrious old Maori lady'. Were they indeed? The kete itself is made of four pieces (two ends and two sides folded under) of plaited flax, or harakeke. Harakeke was one of the wonder plants which Māori discovered in the new land. In their South Pacific homeland it was unknown for clothing and furnishing material. Polynesians depended on the paper mulberry tree which they beat into tapa cloth. The tree, which Māori called 'aute', was so valuable that it was brought to New Zealand by the first arrivals, but it never thrived in the cool climate. Because of that climate, early Māori had to find alternative ways to dress warmly. Harakeke provided an answer. Flax became so important that William Colenso reported one Māori pitying the English because it was not present in their homeland, declaring, 'I would not live in such a land as that.'

Māori sewers used harakeke in two forms — the flax was sometimes cut into even-sized strips for plaiting. But plaited flax was rather brittle, not comfortable for wearing. The alternative was to soak the flax, and then scrape it on both sides with a sharp implement such as a mussel shell to make muka, or string, for weaving. Polynesians had learnt finger weaving in the Pacific from making nets. They applied this skill to flax muka and wove large cloaks which fitted softly over the body. The kete itself was made from flax strips and then sewed with muka. There are also several bundles of flax fibre in the kete, two of them dyed black. There are some incomplete pieces of woven flax and a whitebait net also made from harakeke. Further, there are three large mussel shells, used to prepare the muka. It is almost certain that women would have done the work of preparing the muka and sewing it. We know this because in the 1790s Governor Philip Gidley King of New South Wales took two Māori chiefs to Norfolk Island to teach the locals how to prepare flax for sewing. It proved a poor choice, for these Māori men had no idea how to prepare the flax.

Probably the major use of flax muka was to make cloaks for warmth and protection from the rain. Cloaks were woven in harakeke and then thatched on the outside with other materials — flax leaves, cabbage-tree leaves, tussock, or the tomentum, or felt-like covering, of the leaves of celmisia, the spectacular mountain daisy found in the Southern Alps. There are four bundles of tomentum from about 1000 leaves of celmisia in the kete. This was a well-stocked sewing kit. Here were the materials for making cloaks; including feathers and even a few strips of dog skin for decoration.

Harakeke was not the only plant used for weaving. The kete also contains a bag, woven from kiekie, which grew as an epiphyte high on the

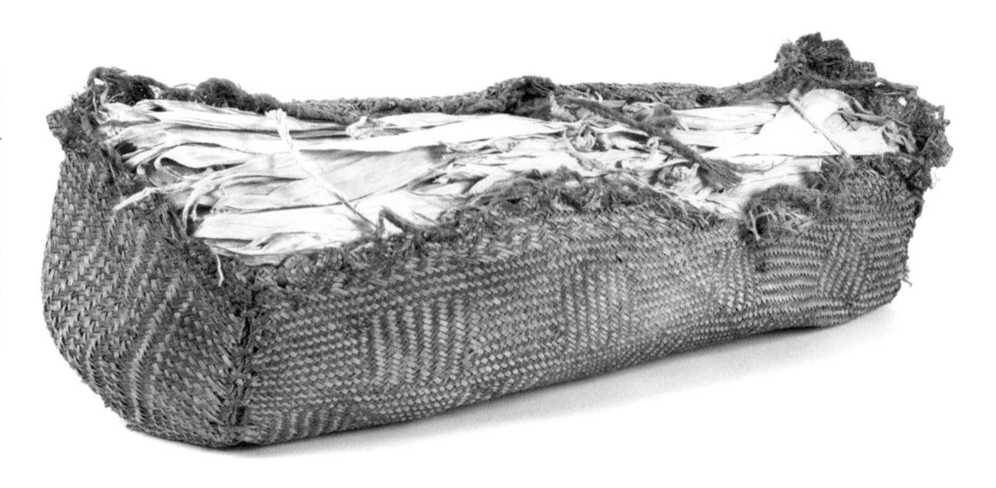

trunks of trees. Local Māori told Hamilton that kiekie bags (pūkoro) were normally used to strain poisonous seeds from tutu berries to create a safe and valuable red drink. But in this case the kiekie bag is carrying red ochre, a natural earth pigment, used to paint red on houses, people and clothing. Red ochre was considered a mark of status, the blood of Ranginui, the sky father. The kete also contained a large pāua shell with a flax handle used to hold the red ochre and the piece of an old mat for brushing on the colour.

These objects would have required time-consuming preparation. The kiekie leaves were collected, boiled and then left in the sun to bleach. For the black-dyed muka, hīnau bark was beaten and then the flax fibre soaked in it for at least 12 hours before being plunged in the dark mud found in swamps of kahikatea trees. Red ochre pigment was heated, and mixed with shark oil.

If the workers wanted a diversion while they slaved away, the kete offered something to chew. There is a small bag containing fragments of mīmiha, or ambergris, a black substance from whales picked up on beaches and used as a chewing gum to clean teeth. There were also two other varieties of chewing gum — one from the pittosporum tree and a second from the sticky leaves of celmisia.

Another plant used for weaving was the cabbage tree. Within the kete are two pairs of sandals (pāraerae) plaited from cabbage-tree leaves. These would have been worn on the long journeys which the people who left the Puketoi kete must have undertaken. For kiekie does not grow near central Otago, but is found after many hours of walking in the west and north of the South Island; and to collect the alpine celmisia would require a journey into the western mountains. There are also the mussel shells and mīmiha collected from the coast, perhaps a hundred kilometres away. The owners of the kete were frequent long-distance travellers. On the journeys the sandals would provide essential protection from the stony surfaces and matagouri prickles of central Otago.

It seems likely that these were a highly itinerant people. Perhaps with houses and gardens down on the coast, they would come up to central Otago to hunt for birds such as weka or to catch eels and lamprey. Perhaps they collected the flinty silcrete found locally to make adzes. The kete had presumably been hidden so that on their next journey was the wherewithal to make new clothing or sandals.

The kete suggests that over several hundred years, Māori developed an extraordinarily rich understanding of their natural environment, and clever ways of using it — whether scraping flax for muka, or manufacturing red and black dyes. The question remains whether these were indeed the items of a Māori woman, who in Hamilton's words, 'had been up to the alpine country to collect *Celmisia* tomentum for a mat for her lord and master'. One pair of pāraerae, which was well-used, seems to have been worn by a man; and there is some evidence that men occasionally made textiles. But because a pair of sandals fitted a man does not prove that it was made by him; and there is no doubt that women, more often than men, went out to gather the harakeke, kiekie and celmisia for their sewing, and then made the clothes, nets and kete.

Certainly, by the time Europeans reached New Zealand, Māori women were the great gatherers and weavers. Their skill in fashioning garments, largely from harakeke, and embellishing them with feathers, tomentum and dog hair, became legendary. Our Puketoi seamstress was primarily making rain cloaks to provide warmth and protection. But the kete originally contained a piece of kurī skin and feathers of an albatross and kākāpō. Were these also intended for a cloak? At the time of first European contact high-status cloaks worn by chiefs were predominantly kahu kurī with an outer layer of dog skin and fur. But one strip of dog skin would serve no more than a repair job. As for the feathers, those too raise questions. The famous feathered cloaks, kahu huruhuru, did not become fully established until the mid-nineteenth century. By that stage there was no doubt that the weaving of these taonga, one of this country's greatest contributions to civilisation was the work of women. The Puketoi sewing kete is a far-distant hint of a fine, predominantly female, tradition.

8.

Taiaha kura

MUSEUM OF NEW ZEALAND TE PAPA TONGAREWA

Apart from the fact that it was once in the collection of Walter Buller, the somewhat controversial Pākehā lawyer ornithologist, there is nothing particular known about the origins or ownership of this taiaha (fighting staff). Such implements are comparatively common in the museum collections of New Zealand. This is one of over 360 in Te Papa's collection. Very few come with any information about where they were made or even when, perhaps because many were acquired by Pākehā collectors in a dubious way or picked up from battlefields. Precisely because taiaha are so common they are worth exploring, for they are highly significant objects.

Let us begin looking at the taiaha from the top. There you can see an intricately carved tongue; or rather two tongues, for there are two faces and sets of eyes suggesting that the owner is all-seeing, alert for danger from any direction. The carving commands attention. We notice immediately that it features double spirals (rauponga). No spiral forms are to be found in those few wooden objects which survive from early Māori society — there are no spirals on the Monck's cave kurī, and famous early carvings such as the remarkable lintel unearthed from a Kaitāia swamp have notches not spirals. Further, there were few precedents for spiral carving among other Polynesian peoples. Where it came from is a matter of dispute. Some say the shape was taken from the coils of rope in the bottom of a canoe, others that it came from the

TAIAHA KURA, 1800–1850, NEW ZEALAND, MAKER UNKNOWN. GIFT OF W.L. BULLER, 1911. TE PAPA (ME002387)

coils of a lizard's tail or a seahorse, still others that it was the interlocking tongues of the manaia figure in Māori art. The koru of an unfurling fern is another popular idea. But whatever its origin, it could only have come about because Māori were able to carve into wood precisely and minutely. What conceivably made this possible was the widespread adoption of pounamu from about the sixteenth century. Pounamu was hard and chisels made from the stone could be filed to very sharp blades. Precise carving in forms such as the spiral became possible. From that time, magnificent Māori carving with flowing circular forms and spirals appeared on the prows of canoes, or the bargeboards of houses, and became a central feature of classic Māori culture.

Immediately below the carved tongue of the taiaha we see inlaid pāua eyes, and a collar of red (kura) feathers — hence taiaha kura. The red feathers were from the kākā parrot, and were traditionally associated with high rank. This was a chief's taiaha. Then come tassels of hair from a kurī. This tells us that that the taiaha must have been made before that dog died out — no later than the mid-nineteenth century but more probably in the eighteenth century. For taiaha like this were certainly a pre-European object and point to the centrality of war in classic pre-European Māori society. Along with patu, taiaha were the main fighting weapons for Māori in the centuries before Europeans arrived. Pre-European Māori had neither guns, nor bows and arrows. They did not often throw spears, but largely fought through close-quarters hand-to-hand combat. This was where taiaha came into their own. As we have already noticed, at the head (upoko) was a sharp carved wooden tongue (arero). The tongue was sometimes used for poking and thrusting, but that was never the main aim of the taiaha. Below was the body (tinana) of the weapon, about 1.5 to 2 metres in length, and this was gripped by the fighter and used to protect him from the blows of his enemy. The body gradually widened out to a flat blade, or rau, and this was the real killing zone of the weapon. The blade was used to thrust at the opponent and then strike him across the head. An experienced taiaha fighter used many feints and parries to try and catch his opponent off guard and allow the blade to descend with force. Taiaha had to be strong and resilient. Occasionally they were made of whalebone, but more often, as in this case, of a hard wood, such as mānuka or pūriri, which had been further hardened in a fire.

The many numbers of surviving taiaha emphasise the significance of war in Māori society during the several hundred years before Pākehā reached these shores. The reasons for this are well known. There was a growing population, and the initial boom came to an end with the wiping out of moa and other large birds. There was increasing competition for good sources of kai moana, forests well endowed with birds, fresh water for eeling, and suitable places for gardens. The result was fighting between groups. Northern tribes migrated into new areas further south which led to conflicts with those already living there. Not that competition for land and resources was always the spur for war. A well-known saying in Māori society was 'He wāhine, he whenua, e ngaro ai te tangata' ('Through women and land, men die'). The desire for mana, and the need to gain utu, payment to balance insult or injury, were also important. Chiefs emerged distinguished for their abilities in war; and the country, especially around the coast, became dotted with pā. It seems that building of pā began about 1500, and by the time of Cook's arrival there were an astonishing number — at least 6500, which works out at the rate of some 24 being constructed from earthworks each year. Some have suggested that these pā were primarily places to store food in cool pits in the earth, which was necessary as horticulture grew in importance. Undoubtedly most pā did provide such a role. But their location on hilltops with good views of the surrounding country points clearly to military purposes. At a time when there was growing competition for food, it paid to provide a stockaded protection for your larder; and of course, in the event of a siege, having food inside the pā made sense.

The need to fight successfully in order to survive led to other developments in 'classical' Māori society — the emergence of new weapons, such as the taiaha, the patu, or club, and the mere. So important did these become that both the mere pounamu and the taiaha became symbols of chieftainship. Chiefs carried a taiaha at their side when engaged in oratory and flourished them for emphasis.

That the arts of war had become so central to Māori society by the time European ships sailed over the horizon is of very great importance in New Zealand history. We all know Mao's saying, 'Political power grows out of the barrel of a gun'. The fact that Māori were so skilled in fighting became of great significance as the newcomers from Europe invaded this land. That Māori could fight provided them with a basis for respect and authority which would take a century to work through.

9.

Māori war trumpet or pūkāea

MUSEUM OF NEW ZEALAND TE PAPA TONGAREWA

We come now to the arrival of Europeans in Aotearoa, and we hit a problem: where's a suitable object? While there have been various suggestions that the Portuguese set foot on New Zealand soil, this has never been proved; and the first documented visit by Europeans came in December 1642 when two boats, the *Heemskerck* and the *Zeehaen*, under Abel Tasman's leadership, sailed up the west coast of the South Island and then anchored for two days in Golden Bay. The trouble is that there is no object left in New Zealand from that visit. Abel Tasman's crew never landed on the shore to leave anything behind. Instead, I am choosing a pre-European pūkāea, a trumpet used by Māori to signal over long distances and particularly to signal warlike intentions. For a war trumpet played a crucial role in the fateful encounter between Ngāti Tūmatakōkiri and Tasman's crew on that summer morning. Even then we are unsure whether the trumpet used was a pūtātara which was made from the shell of a conch (*Charonia lampas rubicunda*) to which was attached a carved wooden mouthpiece, or whether it was a pūkāea which was made from a long split length of mataī wood that was hollowed out and then bound together with the aerial roots of a kiekie vine. Pūkāea ranged from half a metre in length up to 2.5 metres. This example from Te Papa is about 1.4 metres. The flared end was formed like the petals of a flower. The pūtātara certainly produced a loud note that carried over long distances and it was occasionally played in a threatening situation. But the pūkāea gave out a deep trumpet sound also heard over a considerable distance. Although also played in peacetime, pūkāea seem to have been particularly used to signal during wartime to suggest that warriors were alert. Te Rangi Hīroa (Peter Buck) suggested it 'was essentially a war trumpet which was used to sound the alarm against attack and also used by troops during military campaigns'. So it seems probable that this was the instrument played by the tangata whenua on that December morning.

What happened was that two peoples, both creative, expansive and concerned about material welfare, came up against each other; and it was the trumpet which led to conflict. The tangata whenua, the Ngāti Tūmatakōkiri, were one of the many North Island tribes which moved south from the fifteenth century as populations grew amid competition for resources. Ousted from their original home, they had battled south down the Whanganui River and then followed other Kurahaupō tribes across Raukawa (Cook Strait) to Te Tau Ihu o Te Waka — the top of the south. Hassled out of the Sounds area by Ngāti Māmoe, and in conflict with Rangitāne and Ngāi Tahu, they had recently established themselves in the area west of D'Urville Island to Farewell Spit including Mohua (Golden Bay). Their new lands and coasts must have seemed attractive. In the hills behind were forests of kahikatea and miro supporting a rich bird population of tūī, kererū (pigeons), kākā, and ground birds like weka. The rivers abounded with tuna (eels), īnanga (whitebait) and kōura (freshwater crayfish) and the rocks along the coasts held pāua, mussels and oysters. As for the seas, there could be found hāpuku, flounder, blue cod and snapper. They were close to argillite stone in the Nelson hills which they used for tools and on a direct route to pounamu sources on the West Coast. The coastal flats and hillsides allowed flourishing gardens. Kūmara, gourds and taro were in late spring bloom. Not for the first or last time, Golden Bay must have seemed like paradise to the new settlers. It was worth holding on to; and the tribe were ready to defend their lands with fortified pā on the headlands.

The invaders, the Dutch sailors on board Tasman's two ships, were also from a restless group with the glint of material improvement in their eyes. They worked for the VOC, the Dutch East India Company. Holland at the time was at the forefront of European capitalism. Following Europe's discovery of new worlds in America and the East, the Dutch played a central role in trading the new goods, such as spices and precious metals, brought back by Spanish and Portuguese ships. Amsterdam became a thriving centre of European trade. Then in 1588 the Low Countries revolted against their Spanish rulers, and to secure their supply of goods from the new world the Dutch were forced to trade with the East Indies themselves. In 1602, to forestall competition among their shippers, the Dutch established a state-

PŪKĀEA, 1700–1850, NEW ZEALAND, MAKER UNKNOWN. TE PAPA (ME003950)

controlled East India Company based at Batavia, on the island of Java in present-day Indonesia. Abel Tasman worked for the company and in 1638 shifted his family to Batavia. By 1642 he was an experienced skipper. A truce with the Portuguese and English allowed the company to look further afield. Europeans had long believed that there must be a large 'Southland', to balance the continents of the northern hemisphere. The Dutch decided that peace gave them the opportunity to explore for the new land. The motivation was clear. The instructions penned for Tasman's voyage noted that in the past two centuries the kings of Castile and Portugal had discovered America, new areas of Africa and the East Indies — 'what inestimable riches, profitable tradings, useful exchanges, fine dominions, great might and powers, the said kings have brought to their kingdoms and crowns by this discovery'. Tasman's men were promised 'certain fruits of gain, and undying fame'. So the *Heemskerck* and *Zeehaen* came with trade goods to exchange for precious metals and spices; and were warned that 'the Southlands are peopled with very rough wild people, for which reason [you] must always be well armed and carefully on guard'. The two ships set off from Batavia, spent time in Mauritius and then sailed south and east. In November 1642 they reached Van Diemen's Land (Tasmania) and in early December headed further east. On 13 December they sighted 'a land uplifted high'. They were probably somewhere off Punakaiki. For four days they sailed up the West Coast looking for an anchorage to obtain fresh water. On 17 December the ships rounded Farewell Spit into Golden Bay. The next morning they sent two boats coastwards to explore the possibilities of water and of trade. By that evening the two ships were anchored about 2 kilometres offshore. As light fell in the summer evening four canoes approached no closer 'than a stonepiece's shot'. A warrior on one of the canoes blew 'many times on an instrument the sound of which was like a Moorish trumpet; we had one of our sailors (who could play somewhat on the Trumpet) blow back to them in answer'. This musical exchange was repeated. The Dutch had inadvertently accepted a challenge.

The following morning a canoe approached with 13 Māori who called out. It seems likely that the 'rough, loud' call was a haka. The Dutch crew tried to interest the locals by displaying trade goods — white linen and knives. The Māori were not interested. The expedition leaders decided that all the indications were friendly, so they moved the ships shoreward.

But they saw seven canoes setting off from shore. Most of the Dutch still believed they came as friends; but an officer on the *Heemskerck* was more cautious and sent a boat across to the *Zeehaen* with a warning to be alert. As the boat left the protection of the *Heemskerck*, the nearest canoe paddled quickly towards it. The Dutch boat was rammed, the quartermaster was clubbed with a 'long blunt pike' (presumably a taiaha) and knocked overboard, and three sailors were killed immediately and a fourth mortally wounded. The quartermaster and two others swam back to the ship.

Deciding that trade and watering were hopeless ventures on this coast, the Dutch weighed anchor and set sail, but not before firing several shots with their cannons. Tasman concluded 'that the detestable deed of these natives against four men of the *Zeehaen*'s crew, perpetrated this morning, must teach us to consider the inhabitants of this country as enemies'. The Dutch sailed north, attempted unsuccessfully to obtain water on the Three Kings Islands off the North Cape, and then departed.

The pūkāea played a key role in these sad events. By responding with their own trumpet call, the Dutch inadvertently accepted the military challenge. The local iwi prepared for battle. But why did Ngāti Tūmatakōkiri sound the trumpet in the first place? Anne Salmond argues that the locals viewed the pale-skinned Dutch as patupaiarehe, ghosts who were feared as likely to spirit away women and children. John Mitchell suggests that the Dutch had anchored near the cave of the taniwha Ngārara Huarau, and the locals feared their presence might inspire the taniwha to dangerous action. More likely their response was a simple act of defence. They had worked hard in recent years to win possession of this precious land. They had no intention of allowing another people, white or brown, to land calmly and occupy their forests and gardens. Sounding the pūkāea, with its strong warrior associations, would make clear that the whenua would be defended.

If that was the intention, then in large part Ngāti Tūmatakōkiri succeeded. Tasman and his men were frightened off. Tasman named the bay where they had stopped 'Murderers Bay' and their journals established a reputation for Māori as a dangerous people. Although the iwi would eventually be harried off their lands in Mohua, it was to be six generations, almost 127 years, before another European ship entered New Zealand's waters. The pūkāea had served its purpose.

10.

De Surville's anchor

MUSEUM OF NEW ZEALAND TE PAPA TONGAREWA

What was the first object left in New Zealand by people other than Māori? There are some intriguing answers to this question. All are metal objects — for metals, along with stones, are most likely to survive hundreds of years.

One possibility is the so-called Tamil bell. This was a bronze ship's bell, about 166 millimetres high and 153 millimetres in diameter, which first came to notice when the missionary William Colenso discovered it about 1836 being used to cook potatoes in a Māori village, probably near modern Whāngārei. Colenso learned that the bell had been found in the roots of a tree blown down in a storm. On the outside was an inscription which proved to be in archaic Tamil script from southern India. This dated the bell to about 1450. The inscription included the words 'Bell of the ship of Mohaideen Bakhsh'. Was New Zealand once visited by Tamil seafarers? Certainly, Tamil sailors, often in association with Muslim Arabs, did trade extensively across the Indian Ocean to Java, but they were never reported further east in Australia and the Pacific. Nor is there any other evidence of early Tamil visitors to New Zealand. Some have suggested the bell was brought by early Europeans, others that a ship was wrecked in the Indian Ocean and then carried by currents to New Zealand. A third suggestion is that the Tamil bell was carried as a souvenir from the East Indies to Spain, and later brought back to the Pacific by a Spanish caravel. This explanation links the bell to another object with claims to be the first non-Māori object — a 'Spanish' helmet. We know that the helmet was discovered in Wellington Harbour, and this must have been before 1904 when it was recorded in the collection of the Colonial Museum. The helmet has been dated to about 1580; and the suggestion is that it was lost overboard by Spanish visitors to New Zealand in the sixteenth century. The trouble is that it is very hard to sustain this romantic idea. It is not at all clear that the helmet is Spanish — the style might just as easily be Italian or English; and there is no evidence that it had been immersed in sea water for any length of time. Nor are there any other objects or remains of shipwrecks of a sixteenth-century Spanish origin.

The strongest claim for the first non-Māori object left here is one, or rather three, French anchors. A ship's anchor is not surprising as our first European object. It was a large piece of iron, the product of European industry. And it obviously came on a ship, which was for 150 years the main way Europe came to New Zealand. Further, metal anchors had origins deep in European culture. Wherever men have taken to the sea they have looked for ways to moor their boats. Where they had a permanent mooring they could attach their vessel to a tree or a wharf, and they also used large rocks with a hole at one end as long-term anchors. Māori used such stone anchors for their waka. But these devices did not work if you were highly mobile and wanted to be able to reuse your anchor in a new location. A large rock was simply too heavy for hauling aboard. So it is thought that the Chinese, and then the Greeks and the Romans, began to use anchors with arms attached to a shaft. At first these were wooden but eventually, especially with the emergence of blast furnaces to make wrought iron from the fifteenth century, iron anchors appeared. The grappling arms would dig in to the sand or mud at the bottom of the river or sea and hold the craft — not, as we shall see, that their holding qualities were always accident-proof. The iron anchors were effective and just light enough to be capable of being hauled aboard for reuse — although they were still heavy. This anchor is 1400 kilograms in weight. For several hundred years iron anchors had become an essential piece of equipment on European ships, and any vessel from that part of the world arriving in New Zealand waters would carry a set. Our example is one of a set of six.

The only surprise perhaps is that the first anchors left in New Zealand were French not British. That origin tells us how easily we might have become a French colony. For in the eighteenth century France was the most densely settled land in Europe, the centre of European intellectual life, and a country deeply interested in trading around the world, and competing with its old rivals, the English, in acquiring colonies and overseas interests.

ANCHOR, CIRCA 1760, FRANCE, MAKER UNKNOWN. GIFT OF MIKE BEARSLEY AND KELLY TARLTON, 1974. TE PAPA (NS000047/2)

The anchors were from the French ship *St Jean Baptiste*. The ship had been built largely on the initiative of Jean François Marie de Surville, a native of Brittany who had spent years trading with the French East India Company and had fought the British in the Seven Years' War (1756–63). The *St Jean Baptiste* was intended for the East India trade, but when in late 1768 rumours began to circulate that the British had discovered a rich South Sea Island, 'Davis Land', de Surville was sent as skipper of the boat to investigate. Supplies were inadequate and before long the emaciated crew were suffering from scurvy. De Surville, who possessed a précis of Tasman's journal, decided to head east and south to obtain fresh food and water in New Zealand.

On 12 December 1769 the crew saw 'the land of New Zealand, very high . . . very distant'. They were off the coast of Hokianga Harbour. They headed north and hit a fierce north-west storm. By an amazing coincidence as they were driven east, past North Cape on 16 December, James Cook's *Endeavour* must have passed them by, perhaps 50 miles away, blown northwards by the same gales.

The winds ceased temporarily and de Surville's ship headed south and anchored in a bay which James Cook had named 10 days before, 'Doubtless Bay'. For almost two weeks the *St Jean Baptiste* remained there, while de Surville's men landed, traded fish and provisions with the local Māori, and gathered wood, greens and fresh water. Their health improved. But on 27 December the weather turned again. A fierce wind, this time from the north-east, blew up and the sea became rough. One longboat and yawl managed to get back to the ship, their oarsmen encouraged by de Surville's bribe of a bottle of wine each; but as the wind rose the ship began to drift. It was dragging on its anchors. The ship headed towards the coast: 'we stared death in the face, seeing rocks along the length of the ship fit to make your hair stand on end'. A third anchor was dropped, and the ship briefly held. But then a cable to one anchor broke and the ship started drifting again. European anchor technology still had some progress to make! De Surville ordered the other anchors to be cut loose, hoisted the sails and the ship was saved just before it was wrecked. The loss of the three anchors was a fatal blow. They were left with only one. So, three days later, convinced that it was a country of storms, de Surville abandoned New Zealand and headed further east to South America where he drowned on a sandbar in Peru. James Cook meanwhile was safe on the other side of the island. He would proceed to circumnavigate both islands and claim New Zealand for the British Crown. We can only speculate what would have happened had the anchors held. Would de Surville have remained to claim Nouvelle Zélande for the French?

As for the anchors, they lay on the seabed of Doubtless Bay for 200 years. Then Kelly Tarlton, diver, conservationist and marine archaeologist, set out to find them. After consulting charts made by the French crew, he was towed underwater in a grid search of the area. The three anchors were found. Two were recovered. One is in the Far North Regional Museum in Kaitāia; the second now hangs on the high wall at the entrance to Te Papa Tongarewa the Museum of New Zealand. It reminds us how easily the country might have been French; and it is a fitting place for what is undoubtedly one of the first three European objects left in this country.

11.

Joseph Banks' or Daniel Solander's kōwhai specimen

MUSEUM OF NEW ZEALAND TE PAPA TONGAREWA

This 250-year-old specimen from a large-leaved flowering kōwhai tree (*Sophora tetraptera*) probably comes to us from the very first day, 8 October 1769, when Europeans, or indeed any non-Māori people, set foot on Aotearoa. It is one of 500 specimens in Te Papa from the *Endeavour* voyage and is a type of kōwhai once found only on the east coast of the North Island. The specimen was collected by either Joseph Banks or Daniel Solander, naturalists upon the British Admiralty ship. Banks and Solander had come ashore late afternoon with the ship's captain, James Cook, aboard two ship's boats, a pinnace and a yawl, on the east bank of the Tūranganui River near Tūranga, or Gisborne as it became. The yawl ferried the Europeans across the river leaving four young boys to guard it. The Europeans left nails and beads as a gesture of good will for the locals while Banks and Solander proceeded to botanise. Banks shot several ducks and then gathered 'a variety of curious plants in flower'. It was late spring and all along the banks of the river the brilliant yellow flowers of the kōwhai must have been in full bloom. The naturalists collected the leaves and flowers, brought them back to the *Endeavour* along with about 40 other plants new to their eyes, dried them, pressed them between pages of Milton's *Paradise Lost* and they were eventually carried back to London.

In one sense the day was indeed a case of 'paradise lost', for in his journal Joseph Banks makes no mention of collecting the plants. He focuses on the engagement with the tangata whenua, the people of the land. While the naturalists were exploring, four armed Māori men emerged from the bush. The young boys were scared. So they ran to the yawl and rowed it in panic across to the pinnace. There the coxswain fired muskets in the air to frighten the locals, who responded by brandishing their weapons. Provoked, the coxswain shot their leader, Te Maro, through the heart. The first day of meeting between Māori and Europeans had led, tragically, to the shedding of blood. It would be only the beginning of a series of bloody encounters around the country over the next few months.

Yet the tragedy of this incident should not obscure the gathering of plants which preceded it. Scientific exploration was at the very heart of the *Endeavour*'s purposes. The voyage was the inspiration of the Royal Society, founded in London in 1660 to further knowledge of the world. This was the age of the Enlightenment, when Europeans believed that a rational creator had laid down orderly rules in the natural world. It was science's mission to uncover those principles. The Royal Society had initially suggested to the King that a voyage go to Tahiti, as one of 77 stations around the world to observe the transit of Venus on 3 June 1769. This would help determine the exact distance of Earth from the sun. The King passed on the request to the Admiralty, who organised a ship.

The voyage had another purpose. Cartographers in Europe had long believed that the large land mass in the northern hemisphere must be balanced by a great continent in the south. So, in secret instructions from the Admiralty, Cook was told that following the observation of the transit, he was to proceed southwards 'to make discovery of the Continent', to explore it, 'to observe the Nature of the Soil, and the Products thereof' and to bring home 'such specimens of the Seeds of the Trees, Fruits and Grains as you may be able to collect' along with minerals.

When the *Endeavour* voyage was being planned, Joseph Banks suggested that he lead a scientific party on the journey. Banks was only 25, a recent member of the Royal Society; but he had a vision and gall. From a family of Lincolnshire landed gentry, he had gone to Eton and Harrow and one day while out contemplating nature decided that he preferred the study of plants and animals to Greek and Latin. Going to Oxford determined to study botany, he found no one there to teach it; so he organised a Cambridge botanist to come and offer classes. His proposition for a scientific role on the *Endeavour* came with a request for additional accommodation for another eight souls and a contribution of £10,000. His right-hand naturalist was Daniel Solander, a Swede, who was the favourite pupil of Carl Linnaeus, the genius from Sweden who had recently revolutionised natural science by developing a system of nomenclature by species and genus which still exists today. Solander's presence placed the *Endeavour* at the cutting edge of contemporary science. Banks and Solander's secretary was a Finn, Herman Spöring. Then there were two artists to record the natural history findings — Sydney Parkinson and Alexander Buchan, although the latter sadly died from an epileptic fit before the party reached New Zealand. Banks also brought along scientific equipment. The English naturalist John Ellis wrote to Linnaeus, 'No people ever went to sea better fitted out for the purpose of natural history, nor more elegantly.' Ellis went on to describe the equipment: 'They have got a fine library of natural history; they have all sorts of machines for catching and preserving insects; all kinds of nets, trawls, drags and hooks for coral fishing . . . All this is owing to you and your writings.' In his journal Banks described what would happen when they returned from shore — they would sit round the cabin table, Solander 'describing', himself 'journalizing'; and they would indicate to the artist Parkinson the features of each specimen while they were still fresh. Imagine their delight as they recorded the unique features of the kōwhai.

Yet the activities of Banks and his entourage were not disinterested science. The secret instructions to James Cook began with the statement that the discoveries 'will redound greatly to the Honour of this nation as a Maritime power, as well as to the dignity of the Crown of Great Britain, and may tend greatly to the advancement of the trade and navigation thereof'. By the eighteenth century, as Dutch naval power waned, Britain and France were battling each other across the globe for dominance in the world's trade and possessions. They fought one another in North America, in India and the West Indies; and with victory in the Seven Years' War Britain was eager to gain control over new areas of the world. In the next few years both countries sent voyages into the Pacific, 'an extension into peace of the rivalries of war'. The payoff would be trade and wealth. Botanising was not an abstract pursuit. Cook and Banks were looking for new natural products which could be sold; and they were interested in whether the new land would be worth settling by a European people.

The kōwhai was of little immediate value, but it was so striking that at the very least it might be worth adorning the gardens of the old world. Not that the adventures of this sample were over. The following year when the *Endeavour* ran aground on Great Barrier Reef, the botanical specimens kept in a hold deep in the ship's stern were immersed in sea water. Banks and Solander carried them ashore to be dried in the Queensland sun. When the ship returned to London, without Parkinson who had died in Batavia, the specimens were housed in Banks' house at 32 Soho Square where their fame became

LARGE-LEAVED KOWHAI, SOPHORA TETRAPTERA J.F.MILL., COLLECTED 8 OCTOBER 1769, NEW ZEALAND. TE PAPA (SP063797/A)

considerable. Eventually the herbarium collection was given to the British Museum which in turn sent some duplicate specimens back to Australia and New Zealand. This kōwhai specimen is in Te Papa.

It is worth reminding ourselves of the scientific impulse that lay behind the voyage of the *Endeavour*. Indeed, for some years, to those in Europe the *Endeavour*'s voyage was not Lieutenant Cook's but Mr Banks' exploration, just as for Māori it became the voyage of Tupaia the Tahitian navigator who always spoke to them. And it is worth reminding ourselves even more that scientific exploration was never a disinterested pursuit. The findings of science could tell future colonisers of the riches, the trade goods that might be found in the new world. The first Europeans to land came for knowledge, yes, but for trade and wealth even more. It no accident that in the portrait of Banks that was painted by Benjamin West in 1771, at his feet is a drawing of a flax plant, which was quickly determined to be the most valuable of the New Zealand species discovered.

12.

James Cook's cannon

MUSEUM OF NEW ZEALAND TE PAPA TONGAREWA

Despite the importance of James Cook's voyages to New Zealand and the time his ships spent in the country there are remarkably few objects held in New Zealand from those voyages. There are very large holdings of objects collected by Cook and his men scattered through the museums of the United Kingdom and Europe, but only a few of these have returned, either like the botanical specimens given back by the British Museum because they were duplicates, or several purchased at auction and sent home by expatriate Kiwis including a painted paddle, a wahaika (a striking weapon) and a waka huia. From the first voyage in 1769–70 there are possibly only two objects left behind in the southern hemisphere in our collections: a small medal, dated as 1761, found on a sand dune at Whāngārā on the East Coast and conceivably the medal hung around a chief's head on 19 October 1769; and a solid metal cannon from the *Endeavour* which is now held in Te Papa.

The provenance of the cannon is undoubted. It derives from the most terrifying moments on that first voyage. After leaving New Zealand the *Endeavour* crossed the Tasman and began exploring the east coast of Australia. By 11 June 1770 they were north-east of modern-day Cairns, creeping through the shoals of the Great Barrier Reef. That night, disaster struck. The artist Parkinson recorded in his journal: 'About eleven, the ship struck upon the rocks, and remained immoveable. We were, at this period, many thousand leagues from our native land (which we had left upward of two years) and on a barbarous coast, where, if the ship had been wrecked, and we had escaped the perils of the sea, we should have fallen into the rapacious hands of savages.' They had hit the coral at high tide, and although they threw over an anchor and pulled, the ship was firmly stuck. The only possible remedy was to lighten its load, so the men tossed over anything that was heavy — casks of rotting stores, ballast, barrels and the cannons. Eventually on the next night's high tide the *Endeavour* sailed free, and despite fears that a leak would sink the ship, the men eventually sailed it to the Endeavour River, where Cookstown is today, and temporarily repaired the vessel. Almost 200 years later an American team used a magnetometer and found metal objects on Endeavour Reef. Six cannons were recovered, and after treatment to conserve them, one was given to Te Papa in 1970.

That a cannon is one of the few objects in the country from Cook's first voyage is highly symbolic, for guns, and cannons in particular, were at the heart of the enterprise. Cook had participated in military operations during the Seven Years' War, and was fully aware of the value of armaments. He requested and obtained an extra four cannons on swivels when the ship was being outfitted to bring the total number to 12; and his request for 12 marines to take charge of military matters was finally granted. Cook came fully prepared for military actions. It is true that the Earl of Morton, president of the Royal Society, provided some salient advice: 'To exercise the utmost patience and forbearance with respect to the Natives . . . To . . . restrain the wanton use of Fire Arms. To have it still in view that sheding the blood of those people is a crime of the highest nature . . . No European Nation has a right to occupy any part of their country, or settle among them without their voluntary consent. Conquest over such people can give no just title; because they could never be the Agressors.'

Yet Cook was coming to claim new lands; and he wanted to protect his men and enforce their will on the locals. Within hours of his landing on 8 October 1769, Cook and his men had used their muskets and one man was dead. The next day the casualties were worse. When Māori appeared threatening, a musket was fired. But the locals grabbed the astronomer Charles Green's hangar sword; Joseph Banks fired his musket with small shot and hit at least one Māori. The sword was still not given up, so William Monkhouse, the surgeon, fired his musket using balls. At least three Māori were wounded. Later that day there was a further altercation on the water and four Māori were shot and killed. Banks conceded that this was 'the most disagreeable day My life has yet seen, black be the mark for it'. In the first 36 hours after Europeans landed in New Zealand at least nine Māori had been killed or wounded.

But the Europeans were starting to learn how to avoid such horrifying bloodshed. Banks recorded, 'we now despaired of making peace with men who were not to be frightened with our small arms'. So it was time to use the cannons. Four days later, on 13 October 1769, five canoes approached the *Endeavour* with people shouting and appearing threatening. A musket was fired over their heads in warning, but this seemed merely to encourage the threats. This time Cook brought up the cannons which were fired with a huge report using grape shot. The Māori left. So Cook and his officers began to enforce a hierarchy of military response. The Europeans carried two forms of guns — muskets and cannons — and both could be fired using small or grape shot, and alternatively lethal solid balls. The first response to signs of aggression was to fire muskets over the heads of Māori. If that did not work, then muskets were fired using small shot, which would hurt but not normally kill. The next stage was to fire muskets using balls. If these hit people, they would severely wound and probably kill. But as Cook learnt, 'Musquetary they never regarded unless they felt the effect but great guns they did . . .' So the next resort was the cannons. They could be used to warn initially by grape shot or firing overhead; or, as the ultimate deterrent, cannon balls would be lobbed, which either killed the locals or put holes in their waka, leading to drownings.

Behind this ghastly hierarchy of armaments were several considerations. The first was that Cook and his crew very quickly came to hold Māori in considerable respect for their military prowess. They realised that with their expert use of taiaha and mere, Māori were fearful opponents at close quarters. If it came to hand-to-hand fighting, Māori would clearly win. But they also noticed that Māori did not use what they called 'Omissive weapons . . . such as Bows and Arrows, Slings &c'. The only things they threw were stones and darts. Cook concluded that this gave the Europeans' possession of guns a huge military advantage — 'we have no weapon that is an equal match except a loaded Musquet'. He also learned that the threat of guns was as effective as their direct use. It was worthwhile firing the guns, and particularly the cannons with their loud explosions, to intimidate Māori. Banks noted, 'I am well convincd of, that till these warlike people have severely felt our superiority in the art of war they will never behave in a friendly manner'.

The second consideration was that the Europeans were determined to enforce their will on the locals despite the Earl of Morton's advice. They wanted to land in order to explore the new country, and discover its plants and animals, for this was potentially useful information for possible future exploitation and settlement. On 14 November 1769 Cook had no hesitation in taking formal possession of the place 'in the name of His Majesty'. The Europeans were also keen to trade with the Māori, partly to obtain their artefacts to bring back for scientific purposes and partly because they wanted the fish and produce that the Māori were often eager to exchange. The problem was that Cook's men simply assumed that only their ways of interacting were legitimate; and any acts which contravened European expectations were cause for the shooting of guns. Therefore when Māori approached the newcomers and performed a haka as a ritual challenge, which was their normal custom for any meeting, Cook and his men interpreted this as threatening behaviour. When Māori took goods from the Europeans on the basis that repayment would come later, Cook's men saw this as immoral stealing. There was a conflict of social mores and each time the European traditions were infringed, guns were fired.

The result was that whenever the *Endeavour* entered new waters, where the reputation of the cannons had not preceded them, muskets and especially cannons had to be fired. This was the case when they arrived in the Bay of Plenty, then again in the Coromandel, the Bay of Islands and finally when they reached Queen Charlotte Sound in January 1770. In all, there were no fewer than 19 days when there were armed confrontations using guns in the five months that Cook and his men were in New Zealand waters, and on 10 of those occasions the big guns were brought out and cannon balls fired. Three times the guns were fired directly at waka and holed them. How many Māori died as a result is not totally clear, but at least 10 deaths can be documented.

In sum, the cannon, looking strangely benign on its wooden wheels at Te Papa, tells us an important story about the invasion of New Zealand in 1769–70. European settlement of New Zealand was based on the barrel of a cannon.

CANNON, FROM HMB *ENDEAVOUR*., CIRCA 1750, ENGLAND, BY JOSEPH CHRISTOPHER, WALLY DOUGLAS. GIFT OF THE AUSTRALIAN GOVERNMENT, 1970. TE PAPA (DM000477)

13.

Solander Island sealskin purse

SOUTHLAND MUSEUM AND ART GALLERY

Imagine the scene. The men are huddled round a fire in a cave, the only protection from the wild winds and rain outside. They are on a small deserted island, Solander Island, about 1.5 kilometres long. The Māori name, Hautere, means 'flying wind', an accurate description. It is on the western end of Foveaux Strait, about 40 kilometres south of the Fiordland coast. The coasts are rocky and huge waves break continually. Behind are steep cliffs rising precipitously to a peak about 330 metres high. There are only a few small patches of flat land; only a couple of places where a boat might land. But sadly boats do not land. The five men, dropped there to hunt for seals, have been living here for over three years. They are clothed entirely in seal skins, which also constitute their bedding. They have almost given up hope of escape. So what can they do — apart from hope? They spend hours trying with little success to catch fish. They hunt for seabirds. They attempt to raise cabbages and potatoes, for one of them had brought seed, but the gales of sea spray put paid to such hopes. In the main they eat little else than seal flesh. But time hangs heavily. They try to cheer each other up. Perhaps, like a later group of sealers, they pass the time 'listening to the wonderful stories related by one of our party, of enchanted islands, haunted castles, and lovers' misfortunes'. Perhaps they sing. But it is conceivable that one of the party occupies his time sewing seal skins. Perhaps he makes leggings, or, thinking of a time when civilised pleasures will be possible again, he

HERITAGE STUDIOS FOR SOUTHLAND MUSEUM AND ART GALLERY

sews a purse — this fine object now living in Southland Museum.

Much of this is conjecture. The purse is certainly of seal skin and was found on Solander Island. We also know that five men were stranded there for years, four Europeans and an indigenous Australian. At least two of them, Thomas Williams and Michael McDonald, had been dropped there at the end of 1808 by the *Fox*, while a second group had arrived in 1810. So by the time of their rescue, in May 1813, which the Sydney newspapers attributed to 'nothing short of [that] divine interposition', some had lived on Hautere for over four and a half years. Whether these five included the purse-maker is uncertain; but since visits to that isolated windswept land were rare indeed, it seems likely.

The fact that sealers had been left on Solander Island in those years follows logically from the fickle fortunes of sealing, New Zealand's first European industry. Māori had looked on seals as a very significant prey for food and fish hooks ever since they had arrived in the country. However, by the time Cook reached these shores, seals were confined to the far south. During his stay in Dusky Sound he reported seals 'in great numbers, about this bay on the small rocks and isles near the sea coast'. After Cook's death in Hawaii his lieutenants noted that sea otter skins had a market in Canton for clothing. These two comments came to the attention of merchants in Port Jackson, later Sydney, which had been founded as a penal colony in 1788. The shippers who brought out convicts needed cargo for the return voyage. There were plenty of products which could fill the boats home from China, such as tea, silks and spices, but how to pay for them? One solution was to pick up seal skins and exchange them in Canton for Chinese goods. Cook had provided New Zealand with its first export. So in 1792 William Raven, master of the *Britannia*, decided to head 'to Dusky Bay to procure seals' skins for the Chinese market'. A gang of 11 was left there in October of that year, New Zealand's first non-Māori inhabitants, but they only obtained 4500 skins, which they considered disappointing. Instead they spent their time building the country's first ship. The experiment was not repeated; and the interest in hunting seals shifted to Bass Strait, which was closer to the Sydney merchants. But by 1803 those grounds were becoming depleted; so once again Cook's words came to mind and sealing gangs turned to Fiordland, Stewart Island and Foveaux Strait. By then the market for skins had increasingly moved to London, where Thomas Chapman had invented a process for separating the long guard hairs of fur seals from the valuable soft under-fur. The under-fur could be used to cover top hats much more cheaply than beaver. Prices rose on the London market to 10 shillings a skin. Such prices heightened the demand for seal skins.

Often seals would be hunted from whale boats. Sealers would land on exposed coasts, club the seals, skin them and leave the flesh to rot. But sometimes gangs would be left to hunt and be later picked up. The trouble for our Solander gangs was that in 1808 and 1809 when the men were dropped off the price of seal skin was high. But then the Chinese market became flooded with skins from South America; and in London there was a financial crisis and prices dropped precipitously to about 4 shillings a fur seal skin. Instead the Port Jackson merchants began to develop an interest in the oil of the elephant seal, which offered an odourless and smokeless fuel for lights and lubrication for machinery. So the focus shifted further south to the elephant seals of Macquarie and Campbell islands. The fur seal hunters on Solander Island were forgotten.

In the long term the Solander Islands were left to the winds and the seabirds. But the story of the sealers who were the country's earliest settlers from Europe is of significance. Cook's voyages had found and publicised certain items which Europe could use; and the first arrivals from Europe came to quarry these items. At first it was the seals which brought over 30 ships and men to the far south. Then, also in the 1790s, another sea quarry, the whales of the Pacific, attracted sea whalers to the far north of the country where they often stopped in for rest, recreation (mainly booze and sex) and supplies. Later Europeans would quarry timber for ships' masts and flax for naval rope. Of necessity these first arrivals came into contact with the indigenous people. In general Māori appear to have been welcoming to the sealers but inevitably, because the sealers were hunting prey which Māori also treasured, there was some conflict and there were at least three violent incidents and deaths in the far south of the country in the first decade of the century. One man, James Caddell, survived by living with the local iwi and marrying the chief's daughter.

These exploitative frontier occupations which were such an important part of the first settlement by Europeans brought a distinctive culture of trans-Tasman masculinity. In background, many were tough 'sea-rats', either footloose sailors or former convicts. Some had stowed away in Sydney. They endured extraordinarily challenging conditions and lived by killing animals. John Boultbee, a sealer during the brief revival of the industry in the 1820s, noted that the sealers' life had changed him 'from the delicate youth, to about as rough a piece of goods as ever weathered the wide world'. If the sealers were the first of those hard itinerant blokes that helped establish the pioneering masculine traditions in New Zealand, there is a certain charm that the sealing object we have chosen is not a 3-foot hardwood club used to bash seals over the head or a knife for skinning, but a delicate piece of sewing, a creation perhaps more often associated with the Māori women already in New Zealand or the European ones who would shortly follow.

TE PAHI MEDAL, 1806, NEW SOUTH WALES, JOHN AUSTIN. COLLECTION OF THE MUSEUM OF NEW ZEALAND TE PAPA TONGAREWA AND THE AUCKLAND WAR MEMORIAL MUSEUM TAMAKI PAENGA HIRA. PURCHASED 2014. TE PAPA (GH024416)

14.

Te Pahi's medal

MUSEUM OF NEW ZEALAND TE PAPA TONGAREWA

The wind was howling, the rain was falling sideways, but this did not deter a half-dozen expatriate Māori from a spirited haka outside Sydney's Intercontinental Hotel. Inside, a Sotheby's auction was beginning which the next day would see the hammer fall for a remarkable object, a silver medal, 45 millimetres in diameter, which was inscribed on one side with the words 'Presented By Governor King to TIPPAHEE a Chief of New Zealand during his Visit at Sydney NS Wales Jany 1806'. Who were Governor King and Tippahee, why was the medal given, and why were Māori protesting its sale? The answers tell us much about the economic activities which drew the first Europeans to these shores and the early relationships of Māori and Pākehā.

The story begins with Philip Gidley King on Norfolk Island in 1793. King was a naval officer who had accompanied Captain Arthur Phillip on the first fleet which established a convict settlement at Botany Bay near present-day Sydney. King was sent to Norfolk Island to establish a similar settlement, and became the Lieutenant-Governor. Noticing the abundant flax there and, having read Cook's accounts of flax in New Zealand, he arranged for two high-born Māori, Tuki-tahua and Huru-kokoti, to be captured off the Cavalli Islands in Northland and brought to Norfolk Island to teach flax-dressing. He quickly discovered that this was 'the peculiar occupation of the women, and as Woodoo is a warrior, and Tookee a Priest, they gave us to understand that dressing of flax, never made any part of their studies'. Yet Tuki and Huru lived with King for almost seven months and taught him some te reo. In November, King accompanied them back to their homeland, and when he left 'he never parted his mother with more regret than he did with those two men'. King became convinced that New Zealand would 'make a very eligible colony', of which he might one day be Governor; and he developed a huge admiration for Māori. He understood the crucial role played by gift-giving in that culture. Farewelling Tuki and Huru, King gave them clothes and tools; and, more importantly, several bushels of maize seed, wheat, potatoes, garden seeds and a dozen pigs. These taonga were widely distributed among Northland Māori and established good feelings towards King. Anne Salmond claims that, 'Much of the content of the term "Kawana" (governor) in Northland Māori derived from what people knew about Philip Gidley King.' Further, the pigs and potatoes firmly established these products in Northland.

While sealing attracted Europeans to the far south of the country, whaling brought them to the far north. Like seal oil, the oil of sperm whales was in demand for the bourgeois needs of urban Europe — as lubricant for machines and a smokeless fuel for cooking and lighting, while whale baleen was used in fashionable items like ladies' corsets and umbrellas. There were some whaling ships in New Zealand waters in the 1790s, but numbers rose when the East India Company's monopoly over trade was broken in 1798 and whalers developed a pattern of transporting convicts out and whale oil home. By 1801 Philip King, now Governor of New South Wales, reported six ships whaling off New Zealand; and from 1804 the numbers of whalers, both of British and American origin, rose fast. They stopped in the Bay of Islands for water, supplies and recreation.

This is when Te Pahi saw an opportunity. Te Pahi was a major Ngāpuhi chief, of Ngāti Torehina and Ngāti Rua hapū. He lived at Te Puna on an inlet in the north-west Bay of Islands, with his major pā on an offshore island, probably Mōtū Apo. It was a sheltered anchorage for visiting whalers and Te Pahi began to offer water, potatoes and other supplies. He saw the presence of European sailors as no threat. There were only a handful of Pākehā in the whole country; his was an overwhelmingly Māori world. But in exchange for supplying the whalers, Te Pahi could buy iron goods such as axes, spades and knives, which were of huge value in his community. This would help his people and increase his mana.

Te Pahi's role in supplying whalers brought him to the attention of Governor King. Keen on the economic development of New Zealand, King saw the provision of supplies as essential to the whaling industry. When in mid-1805 Te Pahi's son, Maa-Tara, joined the crew of a whaler and arrived in Port Jackson, King gave him tools and other gifts, plus three pigs and goats to recognise and encourage his father's hospitality. Later he sent 18 sows and two boars to Te Pahi from Norfolk Island.

In a culture where gifting was a major currency of relationships, Te Pahi decided to head to Port Jackson to thank the Governor personally. Despite adventures on route, including a threat to kidnap one of his four sons accompanying him, Te Pahi reached Sydney in November 1805 and laid at King's feet fine mats and a stone patu (club). The party stayed at Government House with King for three months and dined at his table. King was impressed by Te Pahi whose manners, he wrote, were those 'of a well bred Gentleman'. Te Pahi explored European society, examining techniques of carpentry, agriculture, spinning and weaving. Not that all European culture was to his liking. King reported that Te Pahi 'spared no pains to convince us that the customs of his country were in several instances better than ours'. Te Pahi was appalled to discover that a man was sentenced to hang for stealing pork, and pleaded that he come back to New Zealand 'where taking provisions was not accounted a crime'. He also had little respect for Aboriginal Australians. Te Pahi urged King to control some of the unscrupulous behaviour of the whalers, for both men wanted the exchange of goods in the Bay of Islands to flourish.

As Te Pahi was leaving in February 1806, King presented him with iron tools and utensils, a box of fruit trees, bricks and a prefabricated European house. Then as a final gesture King hung around Te Pahi's neck the silver medal which expressed his appreciation for his crucial role in New Zealand. It was probably made by an Irish silversmith, who had been transported to New South Wales for banknote forgery. Te Pahi was 'pleased and gratified'. Wearing taonga like pounamu hei tiki around the neck was a familiar tradition in Māori society; and he must have anticipated that it would carry huge mana at home. It appeared the beginning of a beautiful relationship.

Then things started to go awry. Many whalers were hard-bitten convicts or tough itinerant seamen. First Te Pahi's daughter and her new husband, a former convict, were kidnapped; then the whalers began to break the rules of exchange. On one occasion when asked for payment for the supply of pork and potatoes, they opened fire and weighed anchor. On another occasion Te Pahi was seized by a whaler for allegedly delivering only 19 instead of 20 baskets of potatoes (in fact the twentieth one had been stolen by a sailor). Then Te Pahi was flogged for a missing axe. He headed to Port Jackson to report the problems to the Governor. Things went from bad to worse. Travelling on the *Commerce* Te Pahi suggested that they call in at Whangaroa Harbour for provisions; but while there the captain dropped his watch into the harbour. The tangata whenua had been suspicious of the watch, and sure enough an epidemic broke out soon after. On the voyage across the Tasman Te Pahi fell ill and he arrived to a very changed world. His great mate, King, had been recalled; and the new Governor, William Bligh, had been deposed. Te Pahi was allowed to convalesce in Government House, but once recovered, he was forced to sleep under bridges or in the open. He returned after two months.

It was no better at home. Later the next year, in December 1809, the *Boyd*, on a voyage for kauri spas, put into Whangaroa Harbour. The captain and crew went ashore to inspect trees, were attacked and killed. Then the remaining Europeans were killed on board the ship which was accidentally set on fire. Some 70 Europeans died. The probable explanation was that on the voyage across, a young Whangaroa chief, Te Ara, who had joined the crew, fell ill and could not work. He was taunted and flogged. His people determined on utu, revenge. It was Te Pahi's misfortune that he arrived in Whangaroa soon after to deliver some fish and even apparently attempted to save some of the Europeans. But the whalers, perhaps confusing Te Pahi with a local Whangaroa chief Te Puhi, who was involved, decided Te Pahi was to blame. The latent racism of these hard whalers emerged. Te Pahi became the scapegoat, the treacherous cannibal. Groups of whalers took vengeance in two vicious attacks in March and April 1810. They destroyed his pā, killed some 60 of his people and Te Pahi himself was probably wounded, although one account suggests that he was killed. More likely Te Pahi fell several weeks later during fighting between his people and those from Whangaroa as a result of the *Boyd* affair.

As for the medal, it seems likely that it was stolen during the whalers' attack on Te Pahi's pā. Certainly, it disappeared from the record until mentioned in the will of a Dutch land surveyor, living in Australia, in 1899. The medal reappeared in 2014 when offered for sale by Sotheby's on behalf of a private vendor. Not surprisingly, Te Pahi's descendants were outraged. The evidence suggested that this was stolen property and should be returned to its rightful owners, Te Pahi's whānau, the peoples of Ngāti Torehina. But the auction went ahead despite protests and haka outside the Sydney auction rooms.

In the end there was some resolution. Te Papa and the Auckland Museum successfully bid for the medal, deciding to share its ownership but with a kaitiaki (guardianship) agreement with the descendants of Te Pahi. In December 2014, over 200 years since it had first been presented to Te Pahi, the medal returned to Aotearoa and was welcomed on to Te Pahi's tribal land at Te Puna Bay. The medal has become one of the great taonga of this land, a gentle reminder of a great chief and a moment in our history when Māori and a representative of the Crown held common interests and a mutual admiration.

15.

Hannah King's chair

AUCKLAND WAR MEMORIAL MUSEUM

It is a simple but rather elegant chair. The legs are surprisingly thin and the back shows three deft pieces of woodturning, perhaps done in England, more likely by a woodworker in Port Jackson (Sydney). Was he an ex-convict or a free settler? We do not know. We know that the chair was among the first pieces of European furniture to land in New Zealand. It is said the chair was used to lower Hannah King, seven months pregnant, from a ship, the *Active* when it anchored off Rangihoua in the Bay of Islands in December 1814. Hannah King and her fellow missionary wives, were among the earliest European women to set foot on this land. Only the convict mutineers Charlotte Badger and Catherine Hagerty are known to have preceded them. Further, Hannah and her husband John King are considered the first European family to stay and settle in New Zealand, which explains their self-description as the 'first' or 'founding' family. So this chair carries plenty of 'firsts' in Pākehā history. It also tells us about the central role of women in the missionary enterprise.

The back story begins with the principal chaplain in the New South Wales convict settlement, Samuel Marsden. Ever since he had heard about Tuki and Huru and then met Te Pahi in Port Jackson, Marsden had dreamed of a mission to New Zealand. He headed to England in 1807 to recruit missionaries, and returned on the *Ann* in August 1809 with two candidates, William Hall and John King. Also on the *Ann* was Ruatara, a Ngāpuhi leader and close relative of Te Pahi. He had lived most of his early

MRS HANNAH KING ARMCHAIR. AUCKLAND WAR MEMORIAL MUSEUM TĀMAKI PAENGA HIRA. 1965.78.258.

life at Te Pahi's home, Te Puna, and then headed off working on whaling boats in the hope of meeting George III. He did not meet the King and some whalers ill-treated him with beatings, little food and no pay. When Marsden met Ruatara on board ship he was depressed and vomiting blood. Marsden, and particularly his new recruit John King, befriended Ruatara, and became convinced that he might protect the missionaries at Rangihoua in the northern Bay of Islands. The *Boyd* incident delayed these plans and Ruatara stayed with Marsden for at least eight months, where he learnt English and techniques of western agriculture. Returning to Rangihoua, Ruatara became Te Pahi's acknowledged successor. By 1814 Marsden considered it time to begin the mission. A preliminary voyage including William Hall and a new recruit, Thomas Kendall, scouted out the scene in June and July 1814. Ruatara returned with the party. Five months later the missionary ship *Active* set sail again, and despite stormy weather, bouts of seasickness, and Ruatara's fears that welcoming the missionaries might herald an invasion from Europe, the missionary party anchored off Rangihoua on 22 December 1814. Besides Marsden and Ruatara, the party included the ship's captain Thomas Hansen, and the three missionaries,

Thomas Kendall, William Hall and John King, with their wives and children. King was originally a shoemaker from the Cotswolds in Oxfordshire. He had met and married Hannah Hansen, the captain's daughter, in 1812. They already had a son and Hannah was seven months pregnant. On that first day, only Marsden and his friend John Nicholas went ashore. The next day Nicholas reports that 'the wives of the missionaries' landed and paid a visit to Ruatara's wife Rahu. Whether Hannah King came by chair on that day or on Christmas Day when virtually the whole of the ship's complement landed to hear Marsden preach is not recorded. It seems likely that the chair (with or without Mrs King) reached dry land on Friday 13 January 1815 after a raupō hut 18 metres long by 4.5 metres wide had been built to accommodate the party and Marsden reports that Mr and Mrs King and their belongings landed.

The chair, and the accompanying furniture (which included a magnificent chest of drawers), were not brought out just to provide a sense of 'home'. They represented the essence of Marsden's missionary endeavours. Marsden never intended just to preach the Gospel. Māori had their own religious beliefs so why should they listen to the Bible? Rather, the first missionaries taught 'civilisation', for 'Civilisation offers a way for the Gospel'. The first male missionaries were not preachers. William Hall was a carpenter, Thomas Kendall a schoolmaster, and John King a ropemaker and shoemaker. Providing Māori with an example of civilised living was the goal. A chair was important because civilised people did not squat like Māori but sat on chairs. When John Nicholas first saw Ruatara's hut he was struck by the absence of 'domestic articles' — 'furniture there was none', he wrote with disapproval.

Women like Hannah King played a central role in the mission. In Britain there emerged a separation of home and work as middle-class men went out to the city, and women turned the home into a haven of moral purity. Men's sphere was public; women's was private. But on a mission station, the home was central to work. Women's role in representing civilised graces was not incidental but essential. Marsden insisted that the early missionaries should be married. Wives would hopefully prevent their husbands having sexual relations with indigenous women (a vain hope in the case of Thomas Kendall!). A wife could offer emotional comfort for her husband and provide children to anchor them in the new world. Hannah King gave birth to Thomas on 21 February, much to the astonishment of the locals whose normal practice was to give birth silently, crouching or kneeling in the open air, and who could not believe the loud groans that accompanied 'so trifling an affair as the delivery of a little infant'. She went on to have another 10 children in New Zealand.

Wives played many roles supporting their husbands. They were the spearheads of the civilising impulse. If their chairs taught Māori to sit, the example of their wardrobe and their role in teaching Māori to dress and sew were central aspects of 'civilisation'. In missionary eyes, civilisation meant transformation of the Māori body. They disapproved of tattoos and warmed to Ruatara because he had no moko. Kendall remembered that when he first 'saw the New Zealanders, dress'd as they were in their native raiment . . . I conceived they must indeed be sunk to the lowest pitch of human degradation'. On the first day of arrival at Rangihoua, Marsden presented Ruatara's wife with a red cotton gown and petticoat, as a gift from his wife. Hannah taught local girls how to sew, while her husband was pleased that on the Sabbath 'they are dressed in English clothing'. John King wrote to the Church Missionary Society with a list of materials ('30 yards of check . . . , 10 yards of black velvet . . . , 10 yards of Cambric Muslin') which Hannah could use in teaching sewing.

If missionary wives were essential in clothing Māori in western garb, they also tended the sick, and instructed Māori girls in reading and writing. John King wrote, 'Mrs King will have one or two Girls to instruct in readeing, writeing, sewing, making any sort of clothing to knit and spin. Above all, missionary wives modelled the civilised household. The Church Missionary Society told Henry Williams that he and his wife Marianne 'will exhibit to the Natives the instructive example of a happy Christian family'. Initially Hannah found this a challenge — in October 1814 John King reported, 'Our houses are made with flags, it will neither keep wind nor rain out, it has no chimney in it, the floor is Dirt, it is half over our shoes in watter [sic] when it rains.' But before long, Hannah had established a waterproof Christian home.

Compared to the role of women in introducing civilisation, the male missionaries, the ones who attract attention, had less to contribute. William Hall as a carpenter built housing which did provide an example, but as a shoemaker John King's service was not widely respected and his role in teaching Māori to weave rope from flax was never effective. What the locals craved was teaching about western agriculture, but the missionaries did not provide this and the steep slopes of Rangihoua were singularly unsuitable for farming. Further, the men did much to undermine the 'civilised' reputation of westerners. Kendall, Hall and King fought incessantly, and at one point a new arrival, Walter Hall, a blacksmith, fired a pistol at Kendall after he had advanced with a chisel. Neither was hurt, but it was hardly Christian behaviour!

Hannah King's chair tells us that the first long-term settlement by Europeans in New Zealand, the CMS mission at Rangihoua, gave women a far more significant role than perhaps credited. If the mission was intent on bringing civilisation, then it was women who carried the flame.

16.

Hongi Hika's gun

AUCKLAND WAR MEMORIAL MUSEUM

It was the high point of the year-long voyage, 13 November 1820. Dressed in European court clothes with a flax cloak over his shoulders, and with a spectacular full-face moko, Hongi Hika entered the lavish quarters of Carlton House. He was introduced to the King, George IV, bowed and uttered the famous words, 'How do you do Mr King George.' 'How do you do Mr King Shunghee,' came the reply. There was discussion about domestic things, with the King expostulating that 'a bad wife is rottenness to his bones'. Then responding to Hongi's evident interest in weapons, the King conducted him to his armoury and presented Hongi with a coat of chain mail, a helmet and several muskets. This gun, a British 'Brown Bess' muzzle-loading flintlock musket, was perhaps one of those, for a label attached to the gun in the Auckland War Memorial Museum reads 'Silver mounted fowling piece originally a flint-lock given to Maori chief Hongi by George IV, 1820'. I say 'perhaps', because Hongi's weapons have become encrusted with mythology. It was long believed that a helmet and breastplate in the Auckland Museum and another set of armour in Te Papa were among the items given to Hongi. But it is now clear that they were not Hongi's and his coat of chain mail was buried with him in 1828. The gun, which had been looked after following Hongi's death by Te Hikutū hapū of Ngāpuhi, does seem likely to have been one of the gifts (although Hongi also brought such a gun from the missionary John Butler in 1821, and by 1823 he claimed to own 15 such guns personally).

It certainly goes to the heart of why Hongi was in London on that day. For on that occasion it was also reported that Hongi asked King George whether he had told the missionaries not to supply him with muskets. King George replied that he neither knew the men nor had given such an order. Hongi must have felt free to use the muskets in future with impunity.

By 1820 Hongi Hika was in his late forties, a prominent Ngāpuhi chief based in the Bay of Islands. Like Ruatara, his nephew, he had originally seen major advantages in cultivating a relationship with the missionaries. He saw that a missionary presence would attract traders whom he might sell agricultural goods to and buy useful metal tools from. In 1814 he was introduced to Thomas Kendall and William Hall on their preliminary foray by Ruatara, and then accompanied all three back to Port Jackson, where Hongi met Samuel Marsden and learnt about agriculture. He returned home on the *Active*, and after Ruatara's death became the missionaries' main protector, encouraging them to open new missions on his lands at Kerikeri and Waimate North.

But unlike Ruatara, Hongi became primarily interested in how the trade created by the missionaries would bring in muskets. His interest in these weapons certainly went as far back as 1807 when Pōkaia led a group of Ngāpuhi hapū against Ngāti Whātua, an ancient enemy. At the battle of Moremonui, on modern-day Baylys Beach south of the Maunganui Bluff, Ngāpuhi had a few muskets. But taking advantage of the long time it took to load them, the Ngāti Whātua leader Murupaenga successfully ambushed Ngāpuhi. Hongi escaped but Pōkaia and two of Hongi's brothers were killed. Hongi determined to seek utu for the loss, and in subsequent encounters with northern tribes he learnt how to use muskets more effectively and realised the great advantage they posed when used in numbers. Māori were used to hand-to-hand fighting, but muskets allowed killing from a distance of some 50 metres. Hongi determined to increase his armoury, despite the opposition of some

missionaries, especially Marsden, to supplying him with guns.

So in 1820 he found a way to get to England and meet the King. The specific reason was that the mission schoolmaster, Thomas Kendall, had been working to compile a Māori dictionary. It was arranged that he would bring Hongi Hika and a young companion chief, Waikato, to Cambridge to work with the leading orthographer Professor Samuel Lee. But Hongi always saw guns as the main purpose of his visit. Captain R. A. Cruise met Hongi two days before he travelled and reported that there were efforts to stop him going for fear of what might happen in his absence. Hongi answered that he would die if he did not go, '"that if he once got to England, he was certain of getting 12 muskets, and a double-barrelled gun", which latter article', Cruise continued, 'in the opinion of a New Zealander, exceeds in value all other earthly possessions'. In England Hongi proceeded to meet the King and receive his gift of guns (including probably this weapon) and armour, and also the reassurance that the missionaries' opposition to supplying guns did not have royal support. It was not the only gift on that journey. When Hongi and Waikato returned home and stopped at Port Jackson he decided to sell most of his gifts and use the money to purchase muskets. It seems likely that the price of such guns at Sydney had fallen as the market became flooded with the end of the Napoleonic War. Certainly, he came back home with over 300 muskets. His armoury full, the way was open for Hongi Hika to use his military dominance to exact utu on his traditional enemies.

So began what has become known as the Musket Wars. Hongi's monopoly of muskets was not the only cause. Taua, or raids, had long been a traditional way of resolving disputes in Māori society. And muskets were not always effective. Many of them were second-rate 'trade' muskets, rather than the military ones which Hongi had probably purchased in Sydney, and they were often highly inaccurate, using inferior powder and stones or bits of metal as shot. They misfired frequently, with their barrels sometimes exploding, and their shots were

HAKIMANA, FLINTLOCK MUSKET WHICH BELONGED TO HONGI HIKA. AUCKLAND WAR MEMORIAL MUSEUM TĀMAKI PAENGA HIRA. A7101.

slow, allowing, as at Moremonui, time to take evasive action between the flash and the arrival of the bullet. Nor were muskets the only change which made warfare more effective. The arrival of potatoes, a more productive crop than kūmara, broke the old tradition that only men could work in kūmara fields, allowing men to get away for some time leaving wives and children to work the fields; and the agricultural revolution which came with iron tools, pigs and seeds for wheat and potatoes gave Māori the wherewithal to purchase guns. As trade increased, the price of a musket fell from about 150 baskets of potatoes and eight pigs in 1814 to 70 'buckets' of potatoes and two pigs in 1822 (although sometimes the sexual services of women were the currency of exchange). New agricultural patterns increased the need for slaves, which helped motivate warfare. Potatoes and guns went hand in hand.

But muskets were crucial. They produced devastating losses — as many as 20,000 deaths is the most common estimate for the whole Musket Wars (from about 1821 to 1840); and for about five years they gave Hongi an effective monopoly of guns which freed him to take vengeance on his people's traditional enemies. So for three years, every summer from 1821, Hongi and his men set off south on raids which lasted several months. He would take this gun, his favourite, which he called 'Patu iwi' ('Slay the people'). In 1821 Ngāti Pāoa at Mokoia (near Mt Wellington/Panmure in Auckland) and then Ngāti Maru, a tribe based near Thames, suffered hugely; the next year came a major campaign against the Waikato tribes when over 1500 died; and in 1823, after an epic journey which involved portaging their canoes from the coast to Lake Rotorua, there was a fierce attack on Te Arawa tribes gathered on Mokoia Island. Hongi was not interested in permanent land acquisition — he took utu and left — but his forays often led to major movements of population.

In 1825 Hongi's greatest desire for revenge was largely quenched when his forces defeated Ngāti Whātua at Te Ika-a-Ranganui. Again Hongi used Patu iwi in the battle. He returned to pursue more local ventures. But Hongi's luck was turning. Other tribes began to acquire muskets and the arms race evened up. Further, in 1827 he decided to attack the people across the hills at Whangaroa, partly as long-pent-up utu for the events of the *Boyd* affair and partly to assert rights inherited from his father. On most occasions Hongi used to wear the chain mail and helmet he had received from George IV at Carlton House into battle; but on this occasion he did not and a ball from a musket passed through his chest. To acute injury was added the sadness that came with the death of his blind senior wife, Turikatuku, two days later. A gentle family man, he had cared for her and she had been his great confidante. Two years before, he had lost his son, also in war. Hongi lingered on before finally dying in March 1828, a victim to the weapon which he above all had introduced to Aotearoa.

17.

Te Rauparaha's mere, Tuhiwai

MUSEUM OF NEW ZEALAND TE PAPA TONGAREWA

If our previous object, a double-barrelled gun, was a clear representation of war and symbolises the beginnings of the Musket Wars, Tuhiwai, a beautiful pounamu mere, is an expression of peace and signifies the end of those wars. Mere and other pounamu artefacts were traditionally given to cement peace agreements; and according to Te Papa who now care for the mere it was given to Te Rauparaha in 1843 by the Ngāi Tahu chief Te Mātenga Taiaroa as just such an agreement. However, in her biography of Te Rauparaha, Patricia Burns cites Mātene Te Whiwhi who claimed that the mere was taken from Ngāi Tahu by Te Rauparaha's nephew, Te Rangihaeata, after the siege of Kaiapoi pā in 1831 and that it represented 'their mana over the land'. There is a famous lithograph by Richard Oliver which shows Te Rangihaeata holding Tuhiwai. Yet, if we go back to Burns' original source, we discover that it was more likely to have been obtained after an attack on Kaikōura in 1828. But Burns also acknowledges that Ngāti Toa have long believed the mere was received from Taiaroa in exchange for a waka taua (war canoe), and she also notes that there was an exchange of 'greenstone weapons' with Te Rauparaha as part of a peace treaty in 1843.

There can be no doubt that this mere, over 1 kilogram in weight and over 400 millimetres in length, is one of the finest examples of pre-European craftsmanship in the country, and that it came to Ngāti Toa through their fierce battles with Ngāi Tahu in the 1830s. My guess is

OLIVER, RICHARD ALDWORTH, 1811-1889 :TE NANGIHAETA. NEW ZEALAND CHIEF. CAPT OLIVER DELT. DICKINSON & CO. LITH. [LONDON, 1852]. OLIVER, RICHARD ALDWORTH, 1811-1889 :A SERIES OF LITHOGRAPHIC DRAWINGS FROM SKETCHES IN NEW ZEALAND. LONDON, DICKINSON & CO, [1852]. REF: PUBL-0032-1. ALEXANDER TURNBULL LIBRARY, WELLINGTON, NEW ZEALAND.

that there were two Tuhiwai, one grabbed in battle and the other given in peace. Which of these is the Te Papa Tuhiwai is not clear (although Mātene Te Whiwhi does allege that the mere given in peace was presented by Te Rauparaha to George Grey, whereas the Te Papa mere came from the Wineera family). Ngāti Toa and Te Papa certainly believe that Tuhiwai represented a reconciliation of Te Rauparaha and Taiaroa, two fascinating and highly significant leaders.

Let's begin with Te Rauparaha's story. He was born in south Kāwhia probably in the late 1760s of Tainui heritage — his father of Ngāti Toarangatira, his mother of Ngāti Raukawa. He was a wild young man who received a full-faced moko early (minus one spiral in his lower cheek). He was never tall — about 170 centimetres — and he was never regarded as handsome, with his 'aquiline', some would call beaky, nose. He had six toes on one foot. Nor was he a rangatira of the highest lineage. But he became one of the most powerful and impressive leaders Aotearoa has seen, a brilliant military strategist and author of the nation's most famous haka, 'Ka Mate', penned while he was suffering the ultimate indignity of hiding in a pit beneath a woman's genitals. What drove Te Rauparaha was his realisation in his middle age that his people were suffering too much from the constant warfare with the Waikato tribes who were Ngāti Toa's neighbours in Kāwhia.

In 1819–20, when Te Rauparaha was already in his fifties, he joined Ngāpuhi led by Tāmati Wāka Nene and Patuone in a taua, a great expedition, to the south-west of the North Island. It is said that twice he saw European ships in Cook Strait. Te Rauparaha knew what this meant. If there were ships, there was the possibility of trade and buying muskets. Wāka Nene said to him, 'Oh Raha, your home! Take it, so that you may be near the Pākehā. The people sailing there in the ship are helpful people. If you take this as your home, you will become a great chief; if the Pākehā give you muskets, you will be able to defeat the tribes here.' At that stage only the Bay of Islands tribes had such opportunities and Te Rauparaha had begun to understand what control of guns meant. He also realised that he would be close to the South Island, the home of the precious pounamu which he much craved, and that the land around the Kāpiti district was fertile. Neither the Kāpiti area nor the great harbour of Wellington, Whanganui a Tara, seemed to be heavily peopled. He conceived of a new home for his people who, he hoped, could live peacefully alongside the tangata whenua. But it was the chance to acquire guns easily and so gain authority and power that was the real drive.

In 1821 Te Rauparaha led his people, women and children, along with warriors, south. Near the Mōkau River they encountered Ngāti Maniapoto, so the women were dressed as chiefs and stood by lit fires to suggest they were a large force — hence the name of this migration, Te Heke Tahutahuahi, migrants who lit fires. They stayed with Ngāti Mutunga in Taranaki for some months recovering and then set off further south. The next journey was even more exhausting and difficult and came to be called Te Heke Tātarāmoa — the bramble bush migration. They finally reached the Kāpiti Coast in 1822 and settled on a bend of the Waikawa River. Despite Te Rauparaha's hope for a peaceful integration, these hopes were misplaced. It was not until a series of battles with the local tribes, Muaūpoko, Ngāti Apa and Rangitāne, that Ngāti Toa established themselves. Te Rauparaha set up a home on Kāpiti Island and his mother's people, Ngāti Raukawa, came down from the north.

From the mid-1820s, Te Rauparaha's vision started to flower. On arrival in 1822 it is said that Ngāti Toa had no more than four muskets. But once he had conquered the area Te Rauparaha encouraged his people to raise potatoes and pigs, and gather the abundant flax for trade with the Pākehā. From the late 1820s, whalers began to base themselves along the coasts including from 1832 on Kāpiti itself. European ships increased in number. Te Rauparaha could now do for himself what Hongi Hika had done in the Bay of Islands — control trade with the Pākehā and arm himself with guns. And just like Hongi, Te Pehi Kupe, another Ngāti Toa chief, boarded a ship bound for London, and returned with muskets purchased in Sydney from the proceeds of his English gifts. By late 1830, Te Rauparaha claimed to have 2000 muskets on Kāpiti.

Now with a monopoly of guns behind him Te Rauparaha could look south across Cook Strait where Ngāi Tahu had been weakened by an internecine feud called Kaihuānga — eat relations. For over a decade from 1827 there were a series of fights in the South Island involving Te Rauparaha. Sometimes it was in response to fearsome challenges — in one case he was threatened with having his head beaten by a patu aruhe (fern root pounder); in another he was told that he would have his belly ripped open by a barracouta's tooth (niho mangā).

There were some unusual and dramatic moments — such as when Te Rauparaha smuggled 140 men into the European ship the *Elizabeth* to attack Whangaroa (Akaroa); or when his siege of Kaiapoi pā was only successful because the wind changed direction and set fire to the palisades; or when he was ambushed while snaring ducks on Lake Grassmere. Eventually the conflict even reached the far south of the country before Te Pūoho, one of Te Rauparaha's allies, was defeated in deepest Southland at Tuturau in 1837.

It was now time for peace. So we must introduce our other party, Te Mātenga Taiaroa. Although born at Te Waihora (Lake Ellesmere), he made his home at Ōtākou on the Otago Peninsula. Even as a young man, Te Mātenga established a reputation for peace-making. Despite provocations from a sealing gang in 1823, he encouraged traders to call at Ōtākou; and when the Ngāi Tahu civil war broke out Taiaroa warned relatives on the opposing side and shouted, 'Escape! Fly for your lives!' In the late 1820s he reportedly visited Te Rauparaha on Kāpiti and the two men got on well. Then, during the siege of Kaiapōhia, Taiaroa, who had been visiting relatives, was apparently allowed to leave unharmed by Te Rauparaha. In return it seems that during the surprise attack at Lake Grassmere Taiaroa allowed Te Rauparaha to escape.

So it is hardly surprising that the two men were able to come together for peace negotiations, and the mere was handed over as an expression of this peace between Ngāi Tahu and Ngāti Toa. Exactly when this happened is not clear. Perhaps it was in the late 1830s. Ngāti Toa tradition is that Tuhiwai was subsequently used by Te Rangihaeata to execute Captain Arthur Wakefield at Tuamarina in 1843 when in the Wairau affair Te Rangihaeata's wife was killed. Ngāti Toa had gone there to defend their land from illegal surveying. Other sources suggest that the peace and the handing over of pounamu weapons occurred after the Wairau incident when Taiaroa sought out Te Rauparaha with the possibility of an alliance against the Pākehā. What is clear is that by the time the mere was exchanged the Musket Wars were coming to an end. Guns had spread through the Māori world and a balance of power had been established. The killings and the migrations were over; and a new challenge faced Māori — the mass settlement of the Pākehā.

18.

Betty Guard's comb

MUSEUM OF NEW ZEALAND TE PAPA TONGAREWA

It is now time to return to the few Pākehā who lived in New Zealand in the first four decades of the nineteenth century. There were sealers, and traders and missionaries; but the major economic activity among Europeans in those years was whaling. From the turn of the century there had been whaling ships visiting New Zealand from Britain, the United States and eventually France. They hunted sperm whales out at sea. But in 1827, so the story goes, Jacky Guard, a tough former convict who had been sent to Australia for stealing a five-shilling quilt, was on a sealing expedition and found his boat driven into Tory Channel off Cook Strait in a storm. He landed at Te Awaiti and the next day while climbing the hill nearby to search out his position saw a number of whales. They were not sperm, but 'right' whales. The oil of right whales was not quite so valuable as sperm whale oil, but duties on colonial oil had been reduced in 1825, and right whales had a regular migration route from the subantarctic islands to the tropics passing along New Zealand's east coast. Right whales also provided baleen, which hangs in fringes around a whale's mouth and was valuable for making corsets and whips. So a new form of whaling — from the shore — became possible. Whether Guard began shore-whaling immediately in 1827 at Te Awaiti is in dispute but certainly he had joined the game by the following year and in 1829 he moved his operation to Kākāpō Bay in Port Underwood. Eventually the Guard family donated their collection of historic objects to Te Papa. But what object should we choose from the collection to tell our shore-whaling story?

The collection does not include a whaling boat, the light double-ended clinker-built boats which normally carried six men; but there is a harpoon which the whalers would throw into the whale's side usually with a line up to 300 metres long. Once the harpoon was attached, there would be a 'Nantucket sleigh ride' when the crew moved to the back of the boat and held on. Eventually the whale would tire, and the steersman would plunge a lance into the whale's heart. There is just such a lance in the collection. The whale, now dead, would be towed to shore, and its carcass stripped of blubber for boiling up into oil. There is a mincing tool, a blubber spade, and a scraper in the collection. We could have used any of these precious whaling objects.

Instead we have chosen this delicate tortoiseshell comb, or rather part of the comb, for its teeth are missing. Just why there are no teeth goes to the heart of the story and allows us to explore what the coming of such Pākehā meant for Māori–Pākehā relations in New Zealand. The comb belonged to Jacky Guard's wife, Betty. There is an oft-repeated story that as a 12-year-old Betty had been a passenger on a voyage with Guard to the Marlborough Sounds in 1827. It is more likely that she reached New Zealand in 1830 aged 15 after marrying him in Sydney. She allegedly became the first European woman to settle in the South Island, and their first son, John, became the first Pākehā born there.

In 1834 the family sailed to Sydney for supplies and set off back home on the *Harriet* in April. Sixteen days out the ship was caught in a southerly storm and driven ashore just south of Cape Egmont, Taranaki. The ship's party made it safely to shore; but the local Ngāti Ruanui, with the assistance of two sailors who had deserted, attacked the party. Exactly why this happened is unclear — perhaps because under the tradition of muru the occupants of a shipwrecked vessel were liable to be plundered for offending Tangaroa, god of the sea; or perhaps, as James Busby suggested, because the crew had robbed the Māori when they came to sell potatoes. In any case, Betty and Jacky Guard, their two children and some sailors were captured; but in the process Betty was twice attacked with a tomahawk which would have split open her head if it had not been for the protection offered by this comb. The teeth of the comb are said to have been buried in her skull for the rest of her life. It is interesting that Betty was wearing a comb. This was not a comb for brushing hair, but one which sat at the top of the hair to hold it in place. Tortoiseshell combs like this had become highly fashionable for women in both European and American circles in the 1830s, so despite her distance from metropoles, Betty had kept up with the fashions. Further, it seems probable that the tortoiseshell comb came from China, for at

HAIR COMB, CIRCA 1830, MAKER UNKNOWN. GIFT OF THE H.A. GUARD AND GUARD FAMILY. TE PAPA (GH003415)

this time the sealing trade had opened up commercial relationships with that country. Usually such combs had teeth of up to 100 millimetres in length; so it is hard to imagine that very much of the teeth remained in her head as the legend claims.

That was just the beginning of the story. The accounts of Betty's subsequent treatment vary. Sydney newspapers claimed she was stripped naked and would have been killed if a chief's wife had not intervened; other reports say she was treated well and protected for five months by a chief, Ōaoiti. Jacky, meanwhile, was released after two weeks with the understanding that he would return with a cask of gunpowder as a ransom for the captives. He eventually made it back to Sydney but there changed his intentions. Instead he returned with several military officers and 65 soldiers of the 50th Regiment in a man-o'-war, the *Alligator*, and a schooner. They managed to recapture Betty and after some delay young John Guard; but the process required a fierce bombing of the coastal pā which ended symbolically with the soldiers of the 50th playing football with the severed head of a Ruanui chief. This was the very first occasion on which British soldiers had fired their guns against Māori; and the incident did not pass without comment. W. B. Marshall, the surgeon on the *Alligator*, was horrified at the behaviour of the soldiers and was especially critical of Guard. He recalled asking Guard how Māori should be civilised, and Guard responded, 'Shoot them to be sure. A musket ball for every New

Zealander is the only way to civilize their country.' As for Betty, she too did not escape suspicion, with Edward Markham suggesting she became Ōaoiti's mistress and was brought to bed with twins and 'they were rather dark'.

In many respects this conflict between the country's first shore whaler and Māori was somewhat unusual. Guard himself was said to have an excellent working relationship with Te Rauparaha, in whose lands he had established his station and with whom he signed a deed of sale. He considered Te Rauparaha his protector and in 1833 and again in 1834 spent weeks sheltering from Ngāi Tahu on Kāpiti Island. His son, John, grew up speaking fluent te reo and later married a Māori woman. In other shore-whaling stations, especially on Banks Peninsula, at Waikouaiti and Otago Harbour and in Southland, there was extensive interaction with the tangata whenua. Some 40 per cent of shore whalers were Māori (over half in Otago). Many Māori women intermarried with whalers and were hired to work around the houses or in the surrounding gardens.

Yet the fierce interaction of Betty and Jacky Guard on the Taranaki coast came to be seen by authorities, especially in London, as a classic example of the disorder and violence which was coming to characterise early New Zealand; and it helped to encourage moves towards the establishment of British legal authority that would lead in the end to the Treaty of Waitangi. Betty Guard's comb tells a very large story.

19.

Bishop Pompallier's printing press

HERITAGE NEW ZEALAND POUHERE TAONGA

In December 1834, having returned to Sydney following the bombing of the Taranaki pā, Jacky Guard joined the *Blackbird* for a sealing venture. But first the ship had another task — to take William Colenso and a printing press to Paihia for the Church Missionary Society. That press no longer survives; but six and a half years later, in June 1841, another press landed on the beach at Kororāreka and was hauled up by ropes 'to the measured rhythm of the sailor's cry'. That press is now the oldest surviving press in the country and is displayed in working order at the Pompallier Mission at Russell (formerly Kororāreka). This wonderful large metal object tells much about the central role of printing in New Zealand of the 1830s and 1840s.

The arrival of the printing press resulted from the coming of Roman Catholicism to New Zealand. Catholicism arrived in the person of Jean Baptiste François Pompallier, born in Lyons in 1801 and appointed Bishop of Western Oceania in 1836. He came with one priest and one brother of the Society of Mary in January 1838. With extraordinary energy Pompallier set up no fewer than nine separate missions by 1841, mostly around Northland but extending as far south as the French settlement of Akaroa in Canterbury and as far east as Tauranga. Pompallier realised that if the Catholic Church was going to compete for Māori adherents with the Anglicans and Wesleyans, he needed a printing press. So when Jean François Yvert, a schoolmaster, offered his services to the Society

GRANT SHEEHAN FOR HERITAGE NEW ZEALAND POUHERE TAONGA

of Mary as a printer, he was given a short course in printing and went to Paris to buy the press and accompany it to New Zealand. As the handsome brass label tells us, the press was a 'Gaveaux', made in Paris in 1840. It is a heavy iron object which is interesting since, from their invention by Johannes Gutenberg, about 1440 printing presses had been made of wood. It was not until the early 1800s that the Earl of Stanhope developed an iron press. French manufacturers picked up the idea and Gaveaux was one of the leading makers. The press was only the beginning, for as Father Antoine Garin wrote in early 1842, 'We have a press but no room to set it up in.' The mission set to work to build a printery from rammed-earth walls, and the house included not only a room for printing, but also space for a tannery so that the books could be bound in leather. On this press, over 30,000 volumes were produced in the 1840s, the vast majority in the Māori language. Today you can see the whole operation at the restored Pompallier Mission.

Pompallier's recognition of the importance of printing books in Māori followed the lessons learnt by his competitors, the Anglican Church Missionary Society (CMS) and the Wesleyans. When Samuel Marsden's CMS missionaries arrived at Rangihoua in 1814, they believed the way to convert Māori to Christianity was through civilised arts, especially crafts. The first missionaries were William Hall a carpenter, John King a ropemaker, and Thomas Kendall a schoolmaster. The problem was that Māori learnt the crafts and the value of trade with Europe but ignored the holy message. By 1822 the CMS reported that the New Zealand mission was 'fruitless' and 'in confusion'. There had not been a single baptism. With the arrival of Henry Williams at Paihia the focus shifted towards literacy and evangelising. Progress in conversions was still slow — the first conversion was not until September 1825, and by 1830 fewer than 10 Māori had been baptised. The missionaries came to believe that religious texts in Māori were essential tools of conversion. Christianity, after all, was a religion based largely on the written word, and the sacred text of the Bible was at its centre.

Beginning with the work of Thomas Kendall and his journey to Samuel Lee in Cambridge with Hongi Hika and Waikato, a system of spelling te reo Māori in Roman letters — an orthography — became established. Then translations of religious texts into te reo began, first by James Shepherd in 1822, and then by William Williams following his arrival that year. In 1827 Richard Davis took some key biblical texts in te reo to Sydney for printing. The pamphlet was only 31 pages long, and only 400 copies were printed; but it was a start. Three years later William Yate brought back a simple press from Sydney to the Bay of Islands plus a 15-year-old assistant. But both workers and press were inadequate and only a few hymns and a catechism were produced before the experiment was abandoned. However, the realisation remained that the printed word was a key to converting Māori. So in 1834 William Colenso, a Cornishman trained as a printer, was sent out to the Paihia mission by the CMS with a Stanhope press. It took time to get the press operating — the heavy press had to be ferried ashore on two canoes, and when unpacked Colenso found, to his distress, no printing furniture, no galleys, no cases for the lead, no leads, no inking table or composing stick, and, most seriously, no printing paper. But with ingenuity he made do, making an inking table from pōhutukawa planks, and an imposing stone from local basalt blocks. Over five years from January 1835 Colenso's press produced 74,100 books of four pages or more — some 3.5 million pages of text in te reo Māori which included a complete New Testament in 1837, 11,000 copies of the Psalms and 6000 prayer books. The Wesleyans followed suit and set up a press at Mangungu, Hokianga, in 1836.

The impact of this outpouring of religious texts in te reo Māori was huge. Māori responded with enthusiasm to the new form of communication. An Anglican missionary recalled that when visiting remote Māori settlements, 'Their cry was the same as in almost every place we staid at, — Books, books, "E mate ana matou i te pukapuka kore". We are ill for want of books.' Printed editions of biblical tracts quickly ran out. Colenso could not bind them fast enough. Māori gladly offered a pig in return for a copy of the 1837 New Testament. Māori enthusiasm reflected their recognition of the power that came with this new form of communication. Because nearly every printed text was religious, this gave the churches an effective monopoly of information. Not surprisingly, as Christian texts became available in Māori the number of adherents rose. By 1838 the CMS claimed that there were 2176 Māori in their congregations and there were 10 Anglican missions in the North Island. As slaves returned home from capture in the Bay of Islands in the Musket Wars, they carried their new faith and their interest in literacy further south.

So it was obvious to Bishop Pompallier that the success of Catholicism depended on printing texts in te reo Māori. The earlier productions were specifically designed to communicate Roman Catholic truths. The very first was *Ako Marama Te Hahi Katorika Romana — Clear teaching of the Roman Catholic Church*; while the largest work printed at Kororāreka was *Ko te ako me te karakia o te Hahi Katorika Romana — Teaching and Prayers of the Roman Catholic Church*. The press had its effect. By 1846 over 4000 Māori had been baptised into Catholicism throughout the country. Conversion was not the only effect of the missionaries' printing efforts. Biblical language became part of Māori culture and Old Testament stories became significant in Māori understandings of the world. Beginning with the first prophetic religion, sparked off by Papahurihia in the Bay of Islands and the Hokianga in the 1830s and 1840s, Māori found inspiration in Old Testament stories which they combined with Māori tradition. Thinking of themselves as Jews, a chosen people, who had a promised land, they found in biblical texts a route to a restored Māori authority.

In 1846 Pompallier returned to Europe for four years, and in his absence the Northern War led to an increasing resentment against Europeans. The northern mission virtually collapsed. When he returned in 1850, Pompallier moved to Auckland. That year this Gaveaux press which had served the Church so well was packed up. But that was not the end of its historical significance. In 1857 the Māori King asked Pompallier for a printing press, and at some point the Gaveaux was sent. Certainly from 1891 it was used to print the important Kingite newspaper *Te Paki o Matariki o te Kauanganui*. So to its crucial role in the history of Catholicism and Māori literacy in New Zealand was added another significant contribution. The Gaveaux press deserves its status as among the country's most important historical artefacts.

20.

Flag of the United Tribes

MUSEUM OF NEW ZEALAND TE PAPA TONGAREWA

It is somewhat ironical that this, the earliest surviving flag of the United Tribes of New Zealand, originally approved by 25 northern chiefs in 1834, was raised on Petone Beach on 30 September 1839 by members of the New Zealand Company ship *Tory*. It is ironical because the flag was originally accepted by Māori as an assertion of their sovereign independence, a claim which the New Zealand Company through its dubious land purchases and poor respect for Māori rights implicitly failed to honour. It says something about the company's disrespect for the flag's meaning that the 1839 version was incorrect — it has four stars with six points, rather than the eight points of the original, and there is no white border between the smaller red St George's Cross and the blue background.

The immediate need for such a flag came about because in November 1830 Sydney customs officials seized a ship, *Sir George Murray*, which had been built at Te Hōreke shipyards in the Hokianga. New South Wales operated under British law which required all ships to sail under a flag. At that time New Zealand was not a British colony, so the ship could not use the Union Jack, and although two rangatira on board hoisted a Māori cloak up the mast, this was not deemed acceptable. Māori had quickly accepted the symbolic significance of flags. In the 1820s, Māori villages flew red or white flags on Sundays to signal their observance of the Sabbath. They used white flags to signal a truce during the Musket Wars; and illustrations of large hākari at the time show flags flying on top of the stages built for the feasts. But it took more than three years after the Sydney seizure before the United Tribes flag was adopted.

To understand their motives we must briefly canvass Māori ideas about their relationship with the British. On two occasions, at Mercury Bay and again at Queen Charlotte Sound, James Cook claimed to take 'formal possession of the place in the name of His Majesty', but neither Māori nor other international powers gave these acts the slightest credence. However, as contacts between northern Māori and the British unfolded in the early nineteenth century, some significant Māori leaders began to treasure contact with the British. The British offered economic opportunities, and British officials in New South Wales seemed to provide protection against foreign invasion and some policing of the rowdy whalers and early settlers who were often ex-convicts. Several times Governors of New South Wales issued proclamations against the kidnapping of Māori and seeking to force ships to sign bonds of good behaviour. None of these measures were very effective, but Māori continued to look to the British Crown for protection. In 1820 Hongi Hika travelled to England, partly to obtain arms, but in equal part to meet the King of England.

Then in 1831 a rumour that a French warship was intending to annex New Zealand raised the stakes higher. Māori long remembered their experiences with the French explorer Marion du Fresne in 1772. He had been killed in the Bay of Islands after an apparent infringement of local tapu and his followers took reprisal by killing some 250 tangata whenua. Māori feared that the 'tribe of Marion' had returned and looked to the British King for protection. Rāwiri Taiwhanga at Paihia sought to hoist a British flag as a warning to the French; and 13 rangatira signed a petition asking the King to become 'a friend and guardian of these islands'. The response was the appointment of James Busby as a Resident in the Bay of Islands.

The Māori request was not an invitation for direct British rule. The British had implicitly accepted Māori independent sovereignty and Busby had no authority to enforce any laws he might promulgate. But he hoped that the leading northern chiefs might establish a system of governance and impose effective law in the country. Soon after his arrival in 1833, when he was greeted by some 600 Māori and 50 Pākehā, Busby decided that the crisis over the lack of flags for New Zealand-built ships offered an opportunity to encourage this development. He consulted with Governor Bourke in Sydney who approved and sent a suggested flag with four horizontal blue bars on a white background and the Union Jack in the top-left corner. The missionaries were not impressed with the design, criticising its 'total absence of red, a colour to which the New Zealanders are particularly partial and which they are accustomed to consider as indicative of rank'. For Māori, red implied rangatiratanga, or

NEW ZEALAND COMPANY / UNITED TRIBES FLAG, 1839, MAKER UNKNOWN. GIFT OF ANDREW HAGGERTY RICHARD GILLESPIE, 1967. TE PAPA (GH002925)

chieftainship. Instead the missionary Henry Williams ordered that three alternatives be made in Sydney and these were delivered on 9 March 1834 aboard HMS *Alligator*, the ship which later that year would be involved in bombing Betty Guard's captors on the Taranaki coast.

Eleven days later on 20 March 1834, 25 chiefs and hundreds of supporters turned up at a marquee in front of Busby's new home at Waitangi. The three flags flew on poles outside the tent. After a ludicrous performance when each chief was separated from his followers by being forced to creep under a rope, 12 chiefs voted for a flag which they already knew. It was the Church Missionary Society flag which would have flown above the mission at Paihia. It incorporated the flag of the Anglican diocese of New South Wales imposed on the naval white ensign. The winning flag, now the United Tribes flag, was hoisted up the main flagstaff alongside the Union Jack and 21 guns were fired. An embarrassing scene followed where the Europeans were invited for a meal, and the Māori were offered 'a thin paste made of flour and water'. Yet Busby did honour the flag. It was sent to the United Kingdom and received the King's approval, which Busby recognised as an acknowledgement of 'the sovereignty of the chiefs of New Zealand in their collective capacity'. Māori had proclaimed their independent status while in effect being recognised and protected by the great sovereign in London.

It was in this spirit that 19 months later, on 28 October 1835, leading northern chiefs once more gathered at Busby's place at Waitangi, yet again in response to fears of a French takeover. This time it was the news that Charles Philippe de Thierry was about to take up his 1820 purchase of land in the Hokianga from Hongi Hika and establish an independent state. In response, the chiefs put to paper the concerns that had lain behind their acceptance of the flag, and 34 of them (later supplemented by another 18) signed He Whakaputanga o te Rangatiratanga o Nu Tireni (the Declaration of Independence of the United Tribes of New Zealand). There were four articles. The first declared the independence of the country and an independent state (using in Māori the word 'rangatiratanga'); the second declared that sovereign power (using the word 'kingitanga') lay with the United Tribes; the third stated that the confederation of chiefs would meet each autumn to frame laws; and the last said that a copy would be sent to the King and thanked him for recognising the flag. What is of interest is that the words for an independent state were 'rangatiratanga' and those for sovereign power were 'kingitanga' and 'mana' — translations which had real importance when the Treaty of Waitangi came to be signed four and a half years later. As for the assembly of chiefs, that never met. Māori were not interested in a western-style parliament. They already recognised a system of governance by leading chiefs responsible to their hapū. What did interest them was recognition of their authority as chiefs and their independent status — by the declaration no less than by the flag.

The United Tribes flag had an interesting later history. Subsequently, Māori with a strong commitment to an independent Māori nation flew the flag. The Ngāi Tahu chief Tūhawaiki did so on Ruapuke Island at the eastern end of Foveaux Strait, and when Māori gathered at Pūkawa on Lake Taupō to consider choosing a Māori King in 1856 the flag was flown at the top of the flagpole. The Kotahitanga (unity) movement in the early 1900s also flew the flag; and it has been used by protesting Māori on many occasions since the 1980s. Pākehā too used the flag but with very different meanings. As already noted, it was hoisted at Petone Beach on 'an immense flag-staff' to claim the New Zealand Company's presence in Aotearoa, an act that was hardly in the spirit of Māori sovereignty. Lieutenant-Governor William Hobson saw this flag as a challenge to his authority. So, after the signing of the Treaty of Waitangi, he sent an armed party to Petone and the flag was lowered on 3 June 1840 to be replaced by the Union Jack. Later, in 1858, the flag, as modified by the New Zealand Company, was adopted by the Shaw Savill Line as a trademark. Shaw Savill were responsible for transporting many thousands of people from the United Kingdom to settle in New Zealand. That a company dedicated to assisting this latter-day invasion adopted the flag suggests how far the vision that lay behind the 1834 United Tribes flag had come to be subverted.

21.

William Wakefield's epaulets

TE WAKA HUIA O NGĀ TAONGA TUKU IHO | WELLINGTON MUSEUM

'What is an epaulet?' you ask. Unless you are a guard at Buckingham Palace or an expert in military uniform, you may not know that epaulets were ornamental shoulder pieces worn by military officers to display their rank. These epaulets in striking gold plate have dazzling embossed silver insets showing a crown which refers to the rank of colonel and the insignia of the 1st Reina Isabel Lancers, the British Auxiliary Legion, Spain. The epaulets once belonged to William Wakefield. The cabinet in which these epaulets are displayed today at the Wellington Museum also includes the naval uniform of William's brother, Arthur, and were given to the Wellington Historical and Early Settlers Association in 1938 by the very respectable Mrs Swinfen Jervis.

William Wakefield has often been known as 'the founding father of Wellington', but apart from a monument at the Basin Reserve where it is enjoyed only by a few bored cricket watchers, there are few physical reminders of his founding role. Even the *Dictionary of New Zealand Biography* does not include him among its 3000 entries. Yet the man's career tells us much about the impulse which brought thousands of British settlers to New Zealand from 1840; and his epaulets are deeply revealing of the man's character. William Wakefield was awarded the epaulets when serving as a mercenary on the Iberian Peninsula. He had signed up in 1832 and served for 20 months to help Dom Pedro, the Emperor of Brazil, restore his young daughter to the throne of Portugal.

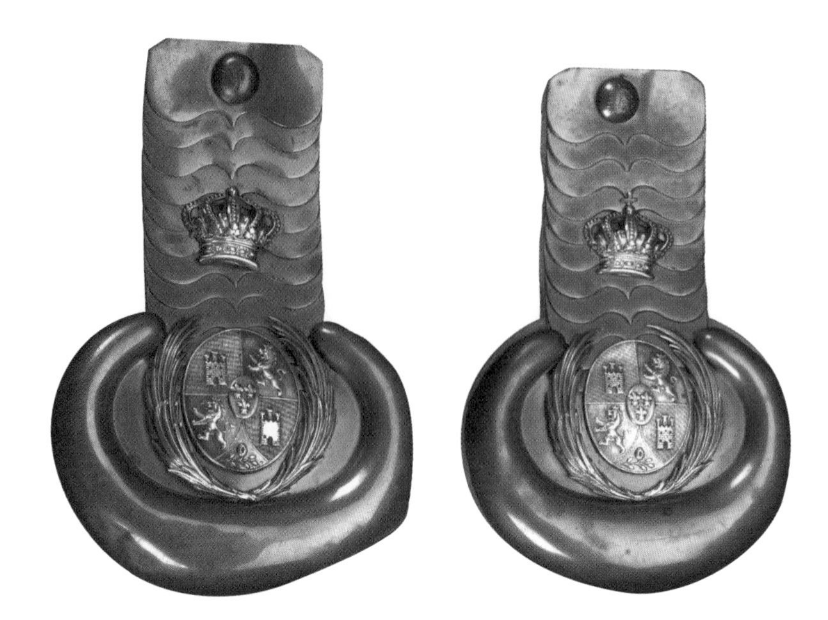

COLLECTION OF WELLINGTON MUSEUMS TRUST

Wakefield was made an Officer of the Tower and Sword. Then when a similar conflict broke out in Spain to restore another young queen to her throne, William joined the British Auxiliary Legion in 1835. He became known for his 'gallantry' and 'almost reckless' courage, and returned to England in 1838 as a full colonel and a Knight of San Fernando 1st class.

The interesting questions are: why did William Wakefield join these Iberian adventures, and what relevance do they have to his later career in Wellington? The answer to the first goes to the heart of the Wakefield legend. The Wakefields came from a respectable background. William's grandmother was a philanthropic Quaker and his father an unsuccessful city businessman and Essex landowner. William's oldest brother, Edward Gibbon, was an unscrupulous intriguer who aspired after money and influence. He targeted Ellen Turner, the daughter of a rich trader who had a parliamentary seat under his influence. As the prosecutor said, Edward hoped 'by the hand of the daughter, to pick the father's pocket'. Ellen was only 15, still at school; and Edward determined to abduct her and force her into marriage. Edward roped in his younger brother William to help. Ellen was uncooperative and in 1827 the two brothers were sentenced to three years in prison.

The brothers reacted differently to their crime. Edward Gibbon read classical economics and mused over 'the condition of England' — its poverty and overpopulation. From prison he wrote *A Letter from Sydney* (1829). Challenging the idea that colonies were an unnecessary drain on capital, he suggested they offered a solution to England's problems. The key was charging a sufficient price for land there. This would prevent labourers squatting on distant land and ensure a labour supply for capitalists. The colonial economy would flourish, providing a use for Britain's surplus capital, a market for their manufactures, and a supply of food grains for the domestic consumer. It would also offer opportunities for labourers whose migration would solve domestic poverty and overpopulation. Further, the sufficient price of land would prevent dispersal and encourage 'taste, science, morals, manners, abstract politics'. A hierarchical civilised society would emerge in the colonies. This brilliant vision had a major influence on the development of New Zealand.

William took longer adjusting to the ignominy. Once released he was at a loose end. Becoming a mercenary offered a different culture and activity. He flourished, finding the army life invigorating which became central to his identity and view of the world. He remained 'Colonel Wakefield' henceforth. But what was he to do once the Spanish adventure was over? His older brother offered an answer.

Edward Gibbon was determined to try out his colonisation ideas. An attempt in South Australia was not to his satisfaction. In 1837 a New Zealand Association was established to promote colonies in New Zealand under Wakefield's principles. The Colonial Office was not impressed. With close links to the Church Missionary Society, it was determined to protect Māori rights. The number of Europeans settling in New Zealand rose from about 300 to 2000 during the 1830s. The possibility of a Wakefield colony upped the concerns. William Hobson was appointed a consul to New Zealand in December 1838 and the prospect loomed that the British Crown would seek sovereignty and insist on an exclusive right to purchase Māori land. Hearing this, the New Zealand Company (as the association had been renamed) acted fast. Once more Edward asked William to do his bidding. William should command the company's ship, the *Tory*, and rush to New Zealand to buy land, explore the country and prepare for settlement. William obtained testimonials as to his fitness in maintaining order and discipline amongst men from those who knew of his military career. On 5 May 1839 the *Tory* left London, with William commanding; Edward Chaffers, once master of Charles Darwin's *Beagle*, as captain; Ernst Dieffenbach as scientist; Charles Heaphy as draughtsman. They expected a fertile pastoral and empty England in the south seas. Arriving at Queen Charlotte Sound this was not quite what they found. No matter, land had to be bought urgently.

The *Tory* headed to the largest harbour nearby, Port Nicholson, on the North Island's south-west corner. Wakefield, the military officer, took charge. Within a week of arriving on 20 September 1839 Wakefield had gained 16 signatures of Māori chiefs on a deed purporting to sell the company 160,000 acres of land. The transaction was highly questionable. The deeds were in English, and the explanations to the chiefs were in the hands of a whaler, Dicky Barrett, who spoke a crude form of 'pidgin Māori'. The boundaries were inadequately defined, and it was not at all clear, given that the harbour had seen recent changes of occupation by Māori tribes, that the appropriate chiefs had the authority to sell land. While Māori were promised one tenth of the land as reserves, they had no choice as to which land. Their holdings were 'pepper-potted' so they would live amid Europeans and become assimilated. It seems likely that the signatories, the local Te Āti Awa chiefs, thought they were offering use rights to a small number of white people with whom they could trade. The invasion of several thousand settlers in January 1840 was a shock. Not surprisingly, local iwi removed surveyors' pegs where they ignored traditional burial grounds, pā sites and gardens. The British authorities were not pleased. With the signing of a treaty at Waitangi on 6 February and then in Port Nicholson on 29 April, the provision of the Treaty that all land had to be bought and sold by the government came into operation. Land Claims Commissioner William Spain led a long inquiry into the original purchases. Throughout, Colonel Wakefield was stubborn in his contempt for the process and determined to hold on to the land. Eventually in 1844 he agreed to pay £1500 for some 67,000 acres. The year before, the Colonel's real instincts had come out. Arthur Wakefield, two years older than William, had been shoulder-tapped by Edward Gibbon to lead another Wakefield settlement in Nelson in 1842. Finding it impossible to satisfy company promises of farming land, the Nelson settlers looked to the Wairau Plains near modern Blenheim. The Ngāti Toa chiefs, Te Rauparaha and Te Rangihaeata, warned that these lands had not been sold. When the settlers set out to survey the Wairau in June 1843, they were again warned and were attacked. Arthur was killed. The Colonel was incensed and his Iberian experience came to the fore. He wanted the Ngāti Toa chiefs brought to justice, formed a volunteer corps, held daily drills and set out to lead an elite cavalry.

The next year, 1844, when there were raids by Māori on Hutt Valley farms, once more William called out a militia of 220. They drilled using muskets brought out as potential trade goods for land purchases. Redoubts were built at Te Aro, Thorndon and Kaiwharawhara. Two years later when Governor George Grey arrived, he found support from Wakefield. After Māori had burnt settlers' homes in retaliation for their failure to recognise Māori occupation rights, Grey, egged on by Wakefield, declared martial law, building a redoubt at Paremata and a military road to Porirua. When fighting broke out in the Hutt, and the Governor arrested

Te Rauparaha, William again called out the militia and issued arms.

Finally in 1847 when Māori attacked the settlement in Whanganui, William joined Grey's military party. William was bitterly disappointed that the fight ended so quickly. He reported: 'The affair ended in nothing but a few shots fired . . . and I must say I was a good deal vexed at returning to the stockade so soon — particularly as I had gone to the extent of rigging myself in one of Tommy Pedder's blue worsted shirts and of carrying 2 barrelled gun, with ammunition enough to exterminate all the natives . . .'

It would be unfair to claim that the 'father of Wellington' was always keen to start shooting. He had a huge task getting the young settlement under way; but he does seem to have come alive when military duties called. On other occasions his contemporaries commented on his lack of energy and human warmth. A withdrawn person, he spent much time in the local baths, which has led one historian to suggest he may have been gay. But he always showed speed, decisiveness and engagement when there was a chance to carry arms. There was a final example of this zeal. His doctor Isaac Featherston wrote an editorial in a Wellington paper criticising the fact that land he had purchased from London was actually a swamp. He accused the company, and by implication Wakefield, of being a thief. The Colonel felt his honour impugned. He immediately challenged the doctor to a duel. Featherston fired first and missed; whereupon Wakefield fired in the air 'with the comment that he would not shoot a man who had seven daughters'.

Eighteen months later in September 1848 William Wakefield died. Over a thousand mourners followed the coffin to his grave. If Wellington's citizens over 170 years later are inclined to forget him, they could do worse than take a look at his epaulets, for they speak of a time when the man's military style and character was forged and which came to the fore only too often in his years as 'father of the city'. William Wakefield was always called, and deserves to be remembered, as 'the Colonel'.

Ko Wikitoria te Kuini o Ingarani [illegible] Rangatira [illegible]

[illegible]

Ko te tuatahi

[illegible]

Ko te tuarua

[illegible]

Ko te tuatoru

[illegible]

W Hobson Consul & Lieutenant Governor

[illegible]

Ko nga Rangatira o te wakaminenga

[illegible]

Tawata [illegible]

Na te [illegible]

Ko Dade [illegible] Piter

[illegible]

Hau[illegible]

[illegible] Tahua

Daniel [illegible]

Wiremu [illegible]

[illegible] March 1840

Inoka

[illegible]naki

[illegible] Moka

Awa[illegible]

[illegible]

[illegible] Jow[illegible]

[illegible]

22.

The Waitangi sheet of the Treaty of Waitangi

ARCHIVES NEW ZEALAND

We come to the most famous, some may say infamous, event in New Zealand history, and we need an object to tell the story. Perhaps we might have chosen the pounamu mere which the Hokianga chief Patuone presented to Lieutenant-Governor Hobson after signing the Treaty of Waitangi on 6 February 1840; but he was representing a particular iwi and cannot stand for the feelings of others. Perhaps we might have chosen a quill pen from the signing, but these are all in private, not public, hands. Instead we have chosen the piece of parchment which was laid out on the table above a Union Jack and signed with their marks by 45 chiefs. But this is a document, not an object, you say. It is a document with words seared into our history but it is also a fascinating object. Let's focus on the Treaty from that perspective.

But first we must explain the origins of the Treaty, which lie in two previous stories — the Declaration of Independence and the New Zealand Company. The declaration, endorsed by British policy-makers, proclaimed Māori sovereignty in Aotearoa and the organiser of the declaration, James Busby, envisaged that the United Chiefs would become a supreme legislature for the country. The chiefs, however, were happy to continue ruling as chiefs responsible to their own hapū. They never met subsequently as a body. Both Busby and the missionaries became concerned that without such authority New Zealand was becoming a frontier of Pākehā disorder and crime which would eventually hurt Māori. The local merchants in Kororāreka petitioned for relief from this disorder. Busby and the missionaries believed the increasing European population was purchasing Māori land in shady deals. Hearings in the British Parliament heightened this concern. Supporters of the missionaries demanded action to protect Māori. There was also a fear that the French might establish a colony in the Pacific. What finally determined action was the New Zealand Company — its establishment as the New Zealand Association in 1837 and then two years later William Wakefield setting off in the *Tory* to purchase large areas of land for settlement. If colonisation was going to happen, the supporters of aboriginal rights argued, it should be done in an orderly way which might protect those rights.

At first William Hobson understood that acquiring sovereignty in a part of New Zealand would be sufficient; but as the ambitions of the New Zealand Company became clear he set out to acquire the whole country for Britain. He was asked to ensure that Māori handed it over by consent. Britain was used to signing treaties with indigenous tribes, and Hobson's draft was modelled on such treaties. The initial draft was in English and after a preamble which stated that a treaty was necessary because of 'the rapid extension of Emigration both from Europe and Australia' and the need to 'avert the evil consequences . . . to the Native population and to Her subjects', the first article ceded sovereignty to the Queen of England. The second clause, largely the work of James Busby, confirmed to the chiefs and tribes 'the full, exclusive, and undisturbed possession of their Lands and Estates, Forests, Fisheries, and other properties'; and stated that the Crown had an exclusive right of pre-emption over lands. The third article extended to Māori 'Royal protection' and granted them 'the Rights and Privileges of British Subjects'. That was the English version.

But the Waitangi sheet on which the chiefs affixed their signs on 6 February 1840 was in Māori. The English text had been translated on the late afternoon and evening of 4 February by the missionary Henry Williams and after the meeting on the 5th the Māori text was inscribed onto parchment by another missionary, Richard Taylor. It was the Māori version that was signed; and it is the Māori meaning of the words that explains why many chiefs were ready to sign. For a start, the first article ceding sovereignty to the Queen used the word 'kawanatanga' for sovereignty. But in the Declaration of Independence sovereignty had been written as 'kingitanga'. 'Kawanatanga' by contrast was derived from the English word 'governor' and would have been understood as implying the kind of role that the Governor of New South Wales exercised. Since the time of Te Pahi those governors had sought to restrain the lawlessness of British subjects in New Zealand. Presumably

Māori thought that 'kawanatanga' did not imply a loss of sovereignty but an expectation that a governor would keep order among the unruly Europeans in New Zealand. On the other hand, the second article confirmed in Māori eyes that their 'rangatiratanga', their chieftainship, was being recognised and upheld. In the debates on 5 February which preceded the signings a number of chiefs made it clear that they read the Treaty as implying no loss of their status to the Governor.

In sum, Māori chiefs saw in the Treaty a confirmation of the Governor's role in controlling Pākehā, and a protection of their own authority and lands. They also hoped for a safeguard against invasion from other European powers, especially the French. They wanted to keep the Pākehā around, because Pākehā brought trade, guns and material benefits. They did not expect an invasion of white people to their country.

In this spirit the chiefs placed their marks on a large piece of parchment containing the Māori words of the Treaty. Here we must begin to look at the sheet as an object. What is parchment? It is the skin of an animal, which has been soaked to remove the blood and muscle, dehaired in a lime bath, then dried and stretched. It seems likely that Richard Taylor found some parchment at the mission house at Paihia; and that it had been made in Sydney. An 1835 dictionary of commerce reported that parchment sheets were for sale in Sydney at 1d to 2d each and that '[a]t present there is at least one parchment-maker in Sydney; two could not support themselves by their trade.' Although several animals could be used for parchment, a recent genetic analysis of the Waitangi sheet suggests that the skin came from a ewe (a female sheep). The writing was done with a quill pen, most likely a goose feather also purchased in Sydney. The ink was derived from tannin found in the galls (apple-like growths) which were found on the trunks of an oak tree. The ink sat on top of the parchment rather than sinking in it. Oaks, geese and sheep — the Treaty was an object created by distinctly European materials, almost certainly grown and made in Australia.

Beneath the text of the Treaty are signatures of the chiefs. A scribe has written them in script and the chiefs have added a mark which is usually an image of their moko. The first invited by Hobson to sign was Hōne Heke, who on the previous day was the first to speak in favour of the Treaty — or at least that was how European observers interpreted his words. His mark can be seen fourth below the text — perhaps for reasons of seniority three other chiefs made sure their marks were placed above Hōne's. On that first day probably 45 signed, three of whom were women. Four days later another six chiefs signed; and on 12 February at Mangungu in the Hokianga, in an even larger gathering than at Waitangi, another 60 signed. Eventually, after another signing in Kaitāia on 28

April, there were over 200 chiefly marks on the Waitangi sheet. At some stage there was not space to hold all the names, so an additional piece of parchment was stitched on the bottom. But on 1 March Hobson had fallen ill with a stroke so he delegated to others the task of taking the words of the Treaty to other parts of the country. Another seven sheets, all but one on paper, rather than parchment, were taken for signature elsewhere and one of these, a printed sheet in English, was signed by 39 chiefs at Waikato Heads and Manukau. A ninth sheet, in both Māori and English, was sent with only Hobson's signature to the Colonial Office.

What happened to the Waitangi sheet of the Treaty? It was placed in an iron box, and, when the nation's capital moved from Russell to Auckland in 1841, the box and Treaty were taken to a wooden government building at Official Bay. Early one morning that year a fire broke out in the building. The records clerk, George Elliott, raced up the hill, put a handkerchief over his face against the smoke, unlocked the box and rescued all the Treaty sheets. They were transferred to the Colonial Secretary's safe. When Wellington became the capital in 1865 the Treaty moved south and was obviously in reasonable shape in 1877 when the government published a facsimile copy. Then the Waitangi sheet, indeed the very existence of a treaty, seems to have been forgotten — at least by Pākehā. When Thomas Hocken rediscovered the Waitangi Treaty sheet in the basement of the old government building in 1908, water had got in and crumbled the parchment. More seriously, the parchment, being the skin of a sheep, had proved attractive to rodents, which had nibbled at the edges. The paper treaties, not such attractive fare, remained almost untouched.

There was further damage when the Waitangi sheet was taken by train back to the Bay of Islands to be displayed in full sunlight for the centennial of the Treaty in February 1940. The threat of Japanese invasion was another challenge and the Waitangi sheet was carted off to Masterton to spend the war years leaning up against a wall in the Public Trust Office. Gradually there was a revival of interest in our most precious object. In 1961 the Waitangi sheet was put on public display for some 17 years in the Alexander Turnbull Library. But concerns about security and the state of the sheet remained. It was withdrawn from sight and placed in a vault in the Reserve Bank. Then in 1991 a Constitution Room was established in the new National Archives Building in Wellington with the Treaty sheets having pride of place. They remained there until 2017 when a beautifully designed exhibition space was created in the National Library to show off the Treaty sheets alongside the Declaration of Independence and the 1993 Women's Suffrage Petition. Today you can creep into a dark room press a button, and the Waitangi sheet is lit in all its glory before your eyes.

23.

Ruapekapeka waka huia

TE KŌNGAHU MUSEUM OF WAITANGI

This waka huia was found in the Ruapekapeka pā near Kawakawa in Northland and kept as loot by a British soldier in January 1846. It now belongs to the Waitangi National Trust. Waka huia were small beautifully carved boxes. They were originally designed to hold feathers such as the rare and highly prized black and white huia tail-feather (hence the name) which would be worn in their hair by leading rangatira on special occasions. But the waka huia also came to house other precious ornaments such as pounamu ear pendants and tiki. Waka huia were often suspended from the ceiling to allow the carving underneath to be displayed. So, in addition to holding precious taonga, waka huia were taonga in themselves. The discovery of this waka huia at Ruapekapeka has a double significance. It helps tell us something about the nature of the pā; and it may assist in explaining the course of the battle there.

First, we need to explain why Bay of Islands Māori and British redcoats were fighting at Ruapekapeka high up in the Tapuaeharuru Range some 20 kilometres from the sea. The story begins with Hōne Heke Pōkai's disappointment at the results of the Treaty of Waitangi. Hōne, believing that the Treaty would ensure a continuing prosperous trading relationship with Europeans under the protection of the Governor, had been the first chief to sign the parchment on 6 February 1840. But the hopes of him and his people were quickly dashed. In 1841 the capital was moved from Ōkiato near Kororāreka to Auckland. There was a sharp downturn in trade, shipping revenue now went to the government and the felling of kauri was banned, leading Heke to believe that his rangatiratanga was being undermined. In a symbolic protest, he cut down the flagstaff on Maiki Hill above Kororāreka. Three further times the flagstaff was replaced, and three further times Heke had it felled. On the last occasion he arranged for his ally, the leading Ngāpuhi chief Te Ruki Kawiti, to carry out a diversion in Kororāreka and allow his warriors to surprise the blockhouse which had been built to protect the flagstaff. In the confusion, a British guard dropped ash from his pipe into the town's gunpowder magazine, which exploded spectacularly. All hell broke loose. There was looting and bombarding of the town and the north was at war. The British sent regular troops, the 58th Regiment, to Auckland and then up to the Bay of Islands.

Expectations of a quick British victory were dashed. With Kawiti taking charge of the defences, Lieutenant-Colonel Hulme's men were defeated at Heke's pā Puketutu. Then more impressively Colonel Henry Despard with over 1300 men plus 400 Māori warriors brought by Wāka Nene failed in their assault of Kawiti's pā at Ōhaeawai on 1 July 1845, losing 40 killed and 70 wounded. The British were determined to teach the Māori a lesson. Kawiti responded by creating a defensive pā at Ruapekapeka. The site itself had no strategic importance. Its value was that it was a magnet to attract the British troops, keen to avenge their defeats. Kawiti believed that the

COLLECTION OF 'WAITANGI NATIONAL TRUST, WNT1966/1/5'

British would exhaust themselves just getting their forces there. So it proved. The British were forced to clamber up 300 metres of altitude and 30 kilometres of steep bush-covered hills and across rivers to approach the site, manoeuvring over 25,000 kilograms of artillery including three huge 32-pound cannons. It took them three weeks. When the 1100 men finally arrived at Ruapekapeka, they discovered a pā brilliantly designed by Kawiti to resist their artillery.

The pā, about 100 metres long and 70 wide, was surrounded by two palisades. The outer one, the pekerangi, was several metres high and had a flax covering which could absorb the battering of musket and some cannon fire. About a metre further in was a taller palisade, about 6 metres high, of pūriri trunks lined with wooden branches, and behind it was a 2-metre-deep trench. The defenders could stand in the trench and shoot their muskets through loopholes in the first barrier and underneath the outer barrier. The outside walls were irregular with many flanking angles to allow crossfire along the face of the walls. This was brilliant engineering.

But the real genius was inside the pā and here the waka huia starts to tell its story. Kawiti realised that his people may have to be there for some time and they needed to be kept safe. He constructed in the soil 2-metre-deep rua, or pits, which were roofed by wood, rammed earth and stones to become artillery-proof. On the floor were soft fern and flax mats, and store pits of kūmara, dried eel and preserved pigeons. Here the Māori defenders could sleep and live for some time. We know there were women and children there who were employed cooking and packing cartridges. This is why there was a waka huia at Ruapekapeka. This was not just a temporary battlefield sortie. Ruapckapeka became for a month or so a place where Māori lived; and rangatira would bring their precious objects such as their waka huia of taonga which they might want to wear on special occasions.

The second lesson the waka huia can perhaps tell us is what happened at the end of the encounter. By New Year's Day 1846 the British had finally made it up the hills and the cannons were in place. For 10 days they pounded the pā while Kawiti's people sheltered in their bunkers. Eventually on 10 January a small breach appeared; and the next day, 11 January, a Sunday, soldiers approached the pā and discovered that it appeared deserted apart from Kawiti himself and a dozen warriors relaxing on the far side. Firing commenced and Kawiti and his men retreated outside the pā into the bush behind. Some of the British soldiers followed into the bush whereupon they were attacked by Ngāpuhi. According to James Cowan, 12 British soldiers were killed and 30 wounded. But Kawiti and his people did not return to the pā and the British claimed it — although this did not represent, in any sense, a triumphant victory.

There are two explanations for this sequence of events. One is that 11 January was Sunday and that the defenders of Ruapekapeka had left the pā to hold prayers. Kawiti stayed behind to hold the fort because he was one of the few unbelievers. Recent historians have been sceptical of this theory. They state that there had been no previous observances of the Sabbath during the conflict (and they cite Ōhaeawai as a prime example). Instead, influenced by James Belich's view, they suggest that Ngāpuhi had deliberately abandoned the pā and were waiting outside to ambush the British. Kawiti's role had been to entice them to follow his men into the bush. There were reports of Māori leaving the pā in the days before 11 January.

What does the fact that a waka huia was left behind in the bunkers tell us about these two interpretations of the Ruapekapeka battle? If the people had left to attend prayers, then they might quite reasonably have left such taonga as a waka huia behind. For they would have intended returning to reclaim such treasures. But if they had planned to abandon the site and entice the British into an ambush, then it seems strange they would have left taonga behind. Further, the Sabbath interpretation was not a post facto view. Several observers of the event at the time, including one from the Māori side, did claim that the camp had retreated for prayers. In any case, the oft-cited example of Ōhaeawai as evidence that Sunday was not observed is unconvincing. The major actions at Ōhaeawai occurred on a Monday and Tuesday.

One should not see the waka huia as conclusive evidence. It seems likely that Māori were preparing an ambush because pūriri defences were found among the trees. And there is a convincing report of Heke arguing when he joined the forces on 10 January that a retreat and an ambush in the bush was a good strategy. What the presence of the waka huia may imply is that while an ambush was planned, it was not expected to occur on the Sabbath, but subsequently. Hence the taonga was left behind on that Sunday 11 January; and the next day some of Kawiti's men were seen prowling around the destroyed pā. Were they perhaps looking for special possessions? For our sakes, perhaps, we must be grateful that the waka huia was left, so that it encourages us to reflect on one of the most remarkable battles of the New Zealand Wars.

24.

Millstones from Pātea

AOTEA UTANGANUI – MUSEUM OF SOUTH TARANAKI

The traveller who passes through Pātea today is inevitably triggered, by the sight of the huge abandoned freezing works, to think about the social costs of this fallen industry. But if you stop at the Museum of South Taranaki on the main street you will find evidence of another abandoned industry which once, like the freezing works, was the dominant economic force in the area. There are two millstones from a flour mill which from 1848 through the early 1850s ground wheat at Mokoia, a little further north on the road to Hāwera, for Ngāti Ruanui. The stones are a reminder of a time when Māori were the economic powerhouse of the country, or at least the North Island.

From the earliest arrival of Pākehā on these shores Māori had seen the economic possibilities of trade. At first it was simply exchanging fish for iron implements; but the need for muskets led to more expansive ambitions. Māori farmed pigs and grew potatoes and then sold them to whalers and sealers. Soon flax and timber were supplied to traders. The services of women also entered the cross-cultural marketplace. It took longer for wheat to become a favoured item. The Ngāpuhi chief Ruatara had brought back wheat seeds from Sydney in 1813, but the locals, used to the way potatoes and kūmara grew, dug up the roots looking for the grain; and it was not till the following year that Ruatara used a hand grinder to make flour and then cooked a cake in a frying pan. Slowly Māori came to like the new food — 'stirabout', a boiled mix of flour and

"MILLSTONES FROM PĀTEA" AOTEA UTANGANUI MUSEUM COLLECTIONS

water, and even better 'thunder and lightning' which included treacle, became popular.

It took a while before Māori traded flour rather than ate it. But the arrival of thousands of Pākehā settlers in the 1840s offered a market. Māori became the chief entrepreneurs of the age, supplying large quantities of food to the new arrivals. They began to bring vegetables such as potatoes and pumpkins, fruit such as melons, and pigs in canoes to Auckland. By 1853 there were 50 Māori canoes beached at Mechanics Bay; and Māori also bought schooners to carry their produce to market. Increasingly flour became a major item. The first water-powered mill had been built at Waimate by the missionaries in 1834; and Māori soon picked up on the innovation. Assistance came from Governor George Grey who, after the Northern War ended in 1846, came to believe that Māori economic development within the European economy was the key to a peaceful future. Many Pākehā observers saw the use of the plough by Māori to turn wilderness into wheat fields as the very essence of progress and civilisation. Grey provided assistance in the form of loans and waterwheels and stones for mills; and at times Donald McLean, chief land purchaser, offered a flour mill alongside land purchases. Missionaries too sought to curry favour with iwi by supplying equipment and even knowledgeable millers for new mills. So in the late 1840s and early 1850s flour mills spread rapidly among tribes in the Waikato, the Ōtaki area and in the south Taranaki/Whanganui region. Chiefs came to regard the ownership of a mill as a mark of mana. By 1855 there were over 50 Māori flour mills in operation. There is no doubt that by then such mills were essential to supplying bread to the Pākehā townspeople; and flour was being exported across the Tasman.

In the south Taranaki district around Pātea travellers reported acres of wheat fields growing on Māori land. Richard Taylor passing very close to Mokoia expressed his pleasure that the 'cultivations at Ōhangai are extensive and had a very pleasing appearance'. The first mill in the area was completed in 1847. The Mokoia stones were from one finished the next year. Ngāti Ruanui saw a profitable future in the growing of wheat and the selling of flour.

Building such a mill was not a straightforward process. A site beside a flowing river was essential; and then a waterwheel had to be brought in, along with two millstones, which most often were of quartz material and imported from Australia. One imagines that, like the stones for a similar mill at Waitōtara, the stones might have been rolled from Whanganui port along the beach with a large branch in the centre. But there are also records of stones being carried on waka up rivers — 32 waka no less in the case of the millstones still in place at the Kawana Kerei mill at Matahiwi on the Whanganui River. That mill was named after Governor George Grey who gifted the millstones. On one occasion two stones were so heavy they overturned a waka; and swimmers had to dive 9 metres to the bottom of a river, tie a rope to the stones and pull them to shore.

The stones had to be heavy to crush the tough outer husk of the wheat grains and release the flour. The stones were about 1.2 metres in diameter, normally made from quartz and bound by an iron hoop. In the Pātea example however, the rock is one piece of volcanic rock, and so did not need a metal surround. The bed stone at the bottom was fixed while the runner stone above turned round via a shaft geared to the waterwheel. The carved furrows on the stone allowed the grain fed from above to work its way, eventually as flour, to the skirt of the stones and then through a chute to a sack below. It was ancient technology but greeted enthusiastically within the Māori communities of the period.

The mill would usually be operated by a Pākehā miller who leased the equipment. Often the millers married into the local iwi. A good example was Richard Pestell, who began as a miller with Ngāti Ruanui in south Taranaki, perhaps at the Mokoia mill. In 1854 he left and took over at the Kawana mill from 1854. He married into the local iwi, became known as Wiremu Petara and adopted a Māori 'way of living'. He always wore a blanket over his shoulders.

By the mid-1850s the Māori flour mills of the North Island and the wheat fields that supplied them were the most dynamic part of the New Zealand economy. But the good times did not last. In 1850 Māori were still the majority population, especially in the North Island, so their economic dominance was unsurprising. But by 1858 Māori had become a minority and, as the European economy grew, some Māori were attracted into the individualistic wage system, weakening the economic authority of chiefs. In 1853 George Grey, whose 'flour and sugar policy' had encouraged mills, left New Zealand. Māori water mills could not compete with new steam-driven mills in Auckland and Napier; and many tribal lands began to suffer from soil exhaustion. More seriously, the end of the Victorian gold rush reduced demand and there was an influx of wheat from Tasmania, Chile and soon after from the South Island. By 1869, 83 per cent of New Zealand wheat was grown in Canterbury and Otago. Nor were Māori well suited to the new bonanza — wool farming. Much of the suitable pastoral land had been sold and Māori had no experience of tending sheep. As wheat growing and milling declined, Māori in those areas, such as the Waikato and Taranaki, where it had once flourished, faced a serious loss of land as the colonial government set out to grab it for settlers. Māori became alienated from the European economy and instead put their energies into resisting land sales and promoting the Māori King.

However, we should not forget that for almost a decade from the mid-1840s to the mid-1850s Māori commercial endeavour was the driving economic force of the country. The millstones in Pātea are an important symbolic reminder.

25.

John Buchanan's table

TOITŪ OTAGO SETTLERS MUSEUM

At first sight there is nothing much of interest about this crude rectangular table with its three irregular planks as a top and the ill-fashioned legs held together with wooden cleats. This was not the work of a master furniture-maker. But that is precisely why the table is of interest. It was made by 48-year-old John Buchanan soon after his arrival in Dunedin on one of the Otago Association's founding boats, the *Philip Laing*, in 1848. Along with his wife, Margaret, and his two young daughters, Isabella and Jeannie, Buchanan was a steerage passenger not able to bring many precious items from home. On arrival he had to fend for himself and make his own furniture. So he put together this table from items at hand — the three pine planks were cut from the bunk on which he had slept on the voyage out. The initials JB are scratched on them. The legs were mānuka taken from the bush behind Buchanan's home in Maclaggan Street about half a kilometre up the hill from the centre of Dunedin. There are still a few bits of bark attached to the legs. Scottish pine from the *Philip Laing* and Dunedin mānuka — a perfect symbol of the migration experience.

What brought John Buchanan and his family to cold, muddy and isolated Dunedin in 1848? His migration had its roots in a religious schism in Scotland's Presbyterian Church in the 1840s. The specific issue was whether gentry or the church congregation should be able to appoint the minister of a parish. But what lay behind this was a conflict between those who were content with a gentle liberal church and those who aspired to a more forthright fire-and-brimstone evangelical religion. Eventually, with the Great Disruption of 1843, 450 Presbyterian ministers left and founded the Free Church. Some of the Free Church people came up with the idea of establishing a city on a hill, a perfect commonwealth overseas. Scotland was suffering from rapid population growth, an influx of Irish, and the dire social consequences of industrialisation and agricultural capitalism. Captain Cargill and two ministers, the Reverend Thomas Burns and Dr Andrew Aldcorn, looked to the new world to create a stable pre-industrial community. Here they came into contact with Edward Gibbon Wakefield's New Zealand Company. Wakefield's dream of a similar society in the antipodes had led to settlements in Wellington, Taranaki, Nelson and Whanganui; but the company had been disappointed by the social disorder of these places. Perhaps what might bind the new community was religion. So was born the idea of a Scottish Free Church settlement under the Otago Association in Dunedin, and an Anglican settlement in Canterbury.

The company granted the association land in Otago and in 1848 the *John Wycliffe* with 97 passengers led by Cargill and the *Philip Laing* with 247 under Burns set out. There were many ways in which John Buchanan was typical of the working people who occupied the dormitory-like steerage quarters (there were only 36 paying cabin passengers). Like

COLLECTION OF TOITŪ OTAGO SETTLERS MUSEUM

most of Dunedin's early Scots settlers, the Buchanan family were not Highlanders but came from the Lowlands, from Kirkintilloch, in Dunbartonshire, about 16 kilometres north-east of Glasgow. Among New Zealand's Scots settlers up to 1852, almost three quarters came from the Lowlands around Edinburgh and Glasgow. Buchanan was a hand weaver, and like many Scots he had found his traditional job become increasingly irrelevant as textile manufacturing moved into factories with the invention of power looms. Many of New Zealand's earlier settlers were forced onto the boat when their hand crafts were no longer needed. The Buchanan family had been Lowlands weavers in the area north-east of Glasgow for generations. This was unlike many other early Scots migrants who had previously moved, with thousands of others, from rural areas closer to the urban centres. For such people migration to the other side of the globe was not the first time they had moved home. It must have been a bigger wrench for John and Margaret.

Buchanan was typical of Dunedin's early Scots immigrants in that he came with his wife and daughters. His brother Thomas was also on board. Wakefield settlements consistently looked to attract families, since Wakefield believed that a surplus of single men would be an invitation to drunkenness and frontier wildness. The family was the essence of civilisation. There were 92 children and only 30 single passengers on the *Philip Laing*.

Yet another way in which Buchanan was representative of the first Scots arrivals was his religious faith. The Otago Association was looking for settlers 'imbued with the principles and habits of Scottish piety', and about half of the Scots migrants to Dunedin up to 1852 were, like the Buchanans, members of the Free Church. The colony's first minister, who sailed on the *Philip Laing*, was Thomas Burns, nephew of the famous nationalist poet Robert Burns, and he conducted services twice a day on board. Soon after arrival, Burns organised the erection of a wooden church where he held two services every Sunday and preached a gospel of discipline, hard work and piety. Buchanan clearly became close to the minister for he took a job working with Burns' son, Arthur, across the harbour at Grants Brae; and he served for many years as the beadle, or church official, at Burns' church and rang the town bell four times a day.

Finally, Buchanan was a good representative of Dunedin's founding Scots settlers in his ability to 'make do'. We have already seen evidence of this trait in his functional table; but not far from that display in Toitū Otago Settlers Museum is a reconstruction of the home he fashioned in 'Squatters' Gully' on Maclaggan Street about 18 months after his arrival. John obtained from Captain Cargill a free site for a house in the bush, and then, as he wrote in a letter home:

> After great perseverance and difficulty, I have at length succeeded in erecting a house single-handed with the exception of two half days which I got from Alexander Watson. The house consists of one apartment about 12 feet square, the sides upwards of 7 feet high are posted, the posts being about a yard asunder, are wattled across and clayed betwixt the wattles. It is roofed with grass and I have a clay chimney attached to it. I have got a garden cleared out of about 24 yards by 30 yards. It consists of potatoes and cabbages which are looking well, also peas and a few garden seeds but they are not looking so well.

Buchanan spent 14 years on Maclaggan Street before moving to Great King Street. He became one of the town's first bird watchers, and as a settler, keen to make a home in the new world, he had a particular interest in native birds. He died aged 80 in 1880.

There has been much mythologising about the do-it-yourself Kiwi ingenuity tradition, which has become more of a fiction in the modern urban world; but there can be little doubt that the first European settlers to this country were forced by circumstance — the lack of specialised craftspeople and the absence of commercial infrastructure — to furnish their own needs from what was close to hand. John Buchanan's reconstructed hut at Toitū, and even more his handmade table, are as good an example of such ingenuity as we can find; and they also point to the life of a man who exemplified so much of Dunedin's first settlers from Lowland Scotland in the late 1840s.

26.

Emma Barker's sewing box

CANTERBURY MUSEUM

This charming work or sewing box, only about 170 millimetres wide and high, now sits in the Canterbury Museum displayed in a reconstructed V-shaped tent house where Emma Barker spent some of her earliest days in Canterbury. But the box arrived there after a long and perilous voyage and several adventures, which tells us much about Canterbury's early European settlers.

The journey began when Emma and her husband Alfred Barker left their home in Rugby, in the Midlands not far from Birmingham where Alfred's family had been merchants, and boarded the *Charlotte Jane*, one of the 'first four ships' sent out by the Canterbury Association to Lyttelton in 1850. Alfred was the ship's doctor, responsible for the health and moral discipline of the 150 prospective colonists on board. Consequently, Alfred and Emma travelled as cabin passengers and could bring plenty of luggage including this work box; but in other ways they were typical of the migrants which the association wanted to recruit for Canterbury. Inspired by Edward Gibbon Wakefield's vision, the association was looking for Anglican families. Both aspects were important. Wakefield believed that a civilised colony would only eventuate if as many women as men migrated. 'As respects morals and manners,' he wrote, 'it is of little importance what colonial fathers are, in comparison with what the mothers are.' Wakefield and the association were especially keen on young married couples. Alfred was 31, Emma 30, and they brought with them three children, one only six months old. Emma was pregnant with her fourth. Young married couples with children were two thirds of the *Charlotte Jane*'s passengers. There were 30 married couples and 53 children on board, with only 26 unmarried men and 12 unmarried women.

As regards religion, the Canterbury Association was looking to establish a model Anglican settlement, and again the Barkers fitted well. Both were pious Anglicans, and Alfred's father had been notable for his efforts to convert Jews to Anglicanism. Three of Alfred's four brothers were Anglican ministers. Emma wrote to her mother on arrival in Lyttelton, 'What I feel we ought to aim for first is to get schools and churches etc as soon as possible.' John Robert Godley, founder of the Canterbury settlement, who had been educated in high church Anglicanism at Christ Church College, Oxford, would have been well pleased.

Once the Barkers reached Lyttelton on 16 December 1850, the work box still had some journeying to do. Alfred decided that they would set up home across the hills in Christchurch. There was no road to the plains so a man was hired to carry the goods by boat from Lyttelton across the Sumner bar. But this proved too treacherous so, as Alfred described, the luggage was landed 'on the naked rocks some miles away from the only house there'. The boxes were undamaged, and 'after an immensity of trouble' Alfred succeeded in getting them within the town boundary, but 'landed on the swampy banks' where they remained for many weeks. It was not until Alfred had constructed a tent from a ship's sail, and laid down a wooden floor, that Emma and the children joined him and at least some of the luggage was unpacked. Whether this included the work box is not certain; but eventually, once Alfred with assistance from the few carpenters available had built a house with wood frames and sod filling, then the work box would certainly have been unpacked and come into its own.

Why was it important that Emma and Alfred go to such lengths to bring such an object with them? The first reason was that the box, with its image of an Olde English cottage, was a reminder of the old country. Both had close family at 'Home' — indeed Emma's sister had married Alfred's brother; and their letters attest to how important these relationships were. Emma knew that once she boarded the *Charlotte Jane* she would never see her English family or friends again. Further, the whole migration experience was distinctly unsettling, especially for women. The voyage out, in this case just a few hours short of 100 days, was trying and stressful. Emma describes in letters home the early days of seasickness which turned them into 'a set of poor weak-looking mortals', then the exhausting heat of the tropics, and the fierce cold

'MRS EMMA BARKER SEWING BOX', CANTERBURY PILGRIMS & EARLY SETTLERS ASSOCIATION INC COLLECTION, CANTERBURY MUSEUM, 1949.148.806

and mountainous seas of the final beat across the southern waters. There was sickness — Alfred, as doctor, records the deaths of three children on board, and Emma's children had boils and diarrhoea — and there were rodents and cockroaches.

Arrival brought excitement but also new challenges. The promised houses had not been built; so while Emma lived for a time on board ship and then stayed in a pub at Sumner, Alfred constructed the tent house. Although he humorously called it 'Studding Sail Hall', the house was cramped and when the south-westerly rains hit, it flooded profusely. Even when this 'Gypsy life' was over and their first house was completed, there were continual trials of a colonial variety — such as the fact that the 'vagabonds' from the pub next door would keep the family awake with their drunken debauchery. It was only natural that Emma and Alfred would turn for consolation to symbols of home. They had brought with them on board a cat and kitten, and although the kitten was lost overboard, the cat survived. Alfred wrote of 'poor old puss who keeps purring around me and reminding me of old times'. Plants were another way to remember the old country, and although the case of plants he had brought with the family 'was smashed to pieces on the jetty at Lyttelton' and the seeds mostly rotted, Alfred ordered replacements from home.

The community also encouraged a sense of Englishness, most obviously in the names which were bestowed on the new land. Māori nomenclature was replaced by the town of 'Christchurch', the province of 'Canterbury', the 'Avon' River, and 'Worcester' Street on which the Barkers built their new home. So the image of the English cottage on the work box served like these reminders of Olde England to keep the alienation and unsettlement of migration at bay.

The work box itself contained Emma's sewing equipment — needles, scissors and thread. This too tells a story of colonial women. In the old world, sewing such as embroidery was one of the expected accomplishments, along with singing and drawing. A sampler was often stitched to show off such skills. But once women arrived on the colonial frontier, sewing became a more essential task. Whether it was knitting a garment from fresh, or patching and mending worn clothes, or putting together quilts and curtains from used material, colonial women had to sew. Imported clothes and fabrics were expensive, and not always available. This applied not only to working women, but also middle-class women like Emma. The colonial middle class simply found it harder to obtain servants to carry out housework and domestic chores like mending. This was partly because of the shortage of unmarried women, who tended to be quickly married off. Although Emma was never without some domestic help, most often a married couple, she did complain about the lack of servants. The day after her arrival in December 1850 she wrote to her father, 'I fear our servants besides being great plagues will be very expensive and that those may think themselves lucky who can secure a man and his wife for £50 and their keep. I trust that the merciful God will guide us through the trying scenes which are always the lot of colonists . . .' It is likely, therefore, that Emma's work box took on additional value in New Zealand.

Emma did not have a long life in New Zealand. While Alfred developed a thriving practice as a doctor, she had a child every two years, but in 1858, soon after the birth of her eighth, she caught a chill and died. Alfred was devastated and consoled himself by throwing his energies into becoming one of the colony's finest photographers.

Today in a reconstruction of the Barkers' first V-shaped tent you can still see her work box. It is a reminder of the way in which New Zealand's early Pākehā settlers looked for a way to remember the old country, and also of the crucial role which sewing played in an isolated frontier world where servants were few and expensive. It also reminds us of a fine Anglican whose journal and letters evoke the unsettling experience of being a woman on the colonial frontier.

27.

Te Reko's cast-iron pot

SOUTHLAND MUSEUM AND ART GALLERY

At one level the story of this iron pot concerns the search for land for sheep farmers in the South Island in the 1850s; but at another it is about the enjoyments and perils of eating in nineteenth-century Aotearoa.

Let's begin with sheep. For a few years the Pākehā settlers who had come out to New Zealand with the Wakefield companies had little interest in sheep-farming. They clustered around the ports where they had arrived hoping to grow crops and grains. But the drive to find an economic basis for the settlements soon led them to look for additional land (with fateful consequences in the case of the Wairau where 22 Europeans and at least four Māori died) and they learnt from the Australian example that the pastoral farming of sheep offered a saleable export in wool. The sheep's meat, mutton, proved of little value for it rotted too fast to be carried any distance. Usually the meat was simply boiled down for tallow. In contrast, wool was readily transported, and there were cloth manufacturers in Britain willing to pay for it. In 1843 and 1844, four men imported 1600 sheep from Australia, and then Charles Clifford and Frederick Weld drove 900 sheep around the coast from Wellington into the Wairarapa. But it was on the South Island's east coast that sheep-based pastoral farming really took off and made fortunes.

This squat but heavy iron cooking pot owes its fame in part to the drive to find land for sheep in that part of the world.

The pot was given to a Ngāi Tahu chief, Te Reko, based at Tūtūrau in Southland by Nathanael Chalmers, an Otago farmer, in return for guiding his exploration of inland Otago. Born about 1803, Te Reko had already entered recorded history as a young man when he had taken part in Ngāi Tahu's defence of Kaiapoi pā against Te Rauparaha's forces in 1828. He had subsequently fought against Te Pūoho, Te Rauparaha's Ngāti Tama ally, in his famous raid down the West Coast before his defeat close to Te Reko's home at Tūtūrau in late 1836 or early 1837. Having defeated the Māori enemy, Te Reko was, however, prepared to cooperate with the Pākehā. Like many Ngāi Tahu, he hoped that the arrival of Europeans would provide economic opportunities and trade for his people. First, there was the small matter of land. If pastoral farming was to flourish, then Europeans wanted to control it. When the Scots settlers arrived in Dunedin, they did so on the basis of a relatively confined coastal purchase of the Otago Peninsula. But George Grey was keen to acquire more. So in 1848 no less than 20 million acres, one third of the country's land area, was bought by the Crown in the Kemp purchase for a mere £2000, plus a promise, scandalously not fulfilled,

HERITAGE STUDIOS FOR SOUTHLAND MUSEUM AND ART GALLERY

to protect traditional Māori food sites and reserve sufficient land for their own prosperity. Only 6359 acres, of the 20 million, was reserved for these purposes. Then in the early 1850s William Mantell was sent south to negotiate the sale of much of Southland. A meeting is recorded between Te Reko and Mantell in late 1851 in which Te Reko agreed to sell but in return for a reserve of 287 acres. Eventually on 17 August 1853 the whole Murihiku block, some 7 million acres, was sold for a paltry £2600.

It was shortly after this that Nathanael Chalmers met Te Reko and presented him with the pot. Chalmers, born on the Isle of Bute in Scotland, had migrated, still a teenager, to Dunedin in 1849. Ignoring Cargill's hope that the population would remain concentrated and therefore civilised, Chalmers and his brother took up land at Kaihiku near Balclutha. It was tough country and Nathanael headed to the Victoria goldfields before returning accompanied by the first flock of sheep landed in Southland. Whether it was a vision of sheep-farming which inspired him is not certain; but soon after, and indeed just a month after the Murihuku block had been sold, Chalmers visited Te Reko and convinced him to lead an expedition further north into the inland parts of Otago, areas which had already been sold in the huge Kemp purchase. The price for Te Reko's guidance was this three-legged cast-iron pot.

Why was Te Reko keen for the pot? From the earliest days of European contact, iron goods had been highly desired by Māori. Ploughs, chisels and of course guns were in hot demand. But in other respects Te Reko showed himself to be technologically a traditionalist. Harriet Beattie records that in the previous year Te Reko had been seen hollowing out a canoe with a stone adze. Apparently he scorned the use of matches, and another observer recorded that he 'had a light from his firesticks nearly as quickly as I did. He rubbed a pointed stick in a groove in another stick.' But he welcomed the iron pot enthusiastically. One must presume that he saw it as an easy way to do 'boil-ups' over an open fire; and it was sufficiently large that it could feed the large numbers of whānau to whom he wished to offer manaakitanga (hospitality) and so increase his own mana. One wonders what the pot would have contained — kai moana, such as mussels or kōura (crayfish), from Southland's south coast; mutton-birds, or tītī, from the Tītī islands off Rakiura (Stewart Island); perhaps ground birds such as weka. We can be sure that it would have been a tasty stew bubbling over the fire.

So with the promise of the pot Te Reko accepted the deal and, with a Māori companion, Kaikōura, set off north with Chalmers. Intending to live off the land, they carried no provisions. While Chalmers brought a gun, the two Māori carried eel spears and a fish hook. They walked the whole way in pāraerae, sandals woven from flax; and when his boots failed, Chalmers was forced to follow suit.

The party of three headed off on foot from Tūtūrau along the banks of the Mataura River. They took a right near Nokomai into the Nokomai River valley, and then over the hills to the Nevis Valley. They dined on duck and eels, and reached the Kawarau, which they crossed by jumping over the famous rock bridge. They headed upstream and Chalmers set his eyes on Lake Wakatipu. Then it was back downstream to the Cromwell Flats and the Clutha. Again the party headed upstream and Chalmers became the first Pākehā to see Lakes Wānaka and Hāwea. Te Reko then offered to lead Chalmers further north as far as the Waitaki River. But the off-the-land diet had not suited Nathanael — or perhaps they carried some cooked meat from the iron pot which had gone bad. For whatever reason, Chalmers was wracked by chronic diarrhoea, and yearned for home. The quickest way back was by the river. So the Māori built a mōkī, a raft of flax, and the party raced through the Cromwell Gorge and down to the Otago coast. Fifty-seven years later, in 1910, Nathanael Chalmers recalled, 'I shall never forget the "race" through the gorge . . . my heart was literally in my mouth, but these two old men seemed to care nothing for the current.'

Chalmers forgot neither the race, nor the rich land he had seen. Within three years the Otago province had lowered the price of land to encourage settlement. Prospective runholders like John Chubbin also sought directions from Te Reko who used a stick to draw a map of the Mataura River in sand, 'the streams being represented by hollows and the mountains by little mounds of sand'. Otago appointed John Turnbull Thomson as the provincial surveyor. He set off to survey the interior and once more, like Chalmers and Chubbin, went to the oracle of geographical wisdom, Te Reko. There was no need for the gift of a pot this time; but again Te Reko proved a valuable guide and there is a famous painting of him paddling Thomson across the Mataura in January 1857.

Within a year or so sheep farmers and their animals were 'flocking' westwards; and Otago, like Canterbury, was becoming one huge sheep farm. Three to four million acres of land had been applied for. The iron pot had cooked a very valuable stew.

28.

Colonel Robert Wynyard's epergne

AUCKLAND WAR MEMORIAL MUSEUM

This beautiful object, made in 1858 by London silversmiths Stephen Smith and William Nicholson, may seem an unlikely entry in a collection about New Zealand history. For an epergne was an artefact of European aristocratic taste. Epergnes emerged in France in the eighteenth century as elaborate decorations placed in the centre of grand dining tables beneath the candelabras. They were intended to bear fine sweets or delicate fruits for the third, or dessert, course. Epergnes subsequently became fashionable among the British bourgeoisie.

Yet there are obvious New Zealand elements in this particular epergne. For a start, the object is kept in the Auckland Museum. And the finely worked decorations include a Māori man, woman and child, a soldier of the 58th Regiment who fought in the New Zealand Wars, and a delicate punga as the high point of the design. Further, if we look closely, we can see an inscription which begins: 'Presented to Colonel Robert Henry Wynyard C.B. of Her Majesty's 58th Regiment on his departure from New Zealand by a number of the inhabitants of the City and Province of Auckland . . .' So this aristocratic table piece certainly has New Zealand associations and the story behind it, ironically, concerns the coming, not of aristocratic rule to the country, but of New Zealand democracy.

That story begins with the establishment of formal British authority over the country with the signing of the Treaty of Waitangi. When Māori signed the Treaty, the other party was the Crown, and Māori assumed that the exercise of royal power would be in the hands of the Governor, the figurehead whom they had learnt to trust. So in the early 1840s while the Governor was assisted by an executive council of officials, it was essentially he who called the shots. This did not please the increasing numbers of British settlers, especially those brought out by the New Zealand Company, who felt that the early Governors, especially Robert FitzRoy and George Grey, were too sympathetic to Māori and restrained their own ambitions for cheap land. In the southern areas, especially Nelson and Wellington, there was agitation for 'self-rule', which meant, of course, rule by white male settlers.

The British government was thinking along similar lines influenced by a report written by Lord Durham following rebellions in Canada. Durham, assisted by Edward Gibbon Wakefield, had suggested that in the colonies the executive government should work in unison with those who had a majority in an elected legislature. Government would be 'responsible' to an elected assembly. So in 1846 the British Secretary of State for the Colonies, Earl Grey, instructed his namesake but no relation, George Grey, to gradually introduce a system of self-government. There would be elected local councils; provincial assemblies with members elected by those councils; and a General Assembly which would include representatives chosen from the provincial assemblies. It was a significant step towards colonial self-rule, if not full democracy. The charter also authorised the establishment of separate Māori districts to be governed by their own methods. But George Grey was wary of the proposal. At a time when there were some 100,000 Māori in the country and a mere 13,000 Europeans, he feared that the

THE WYNYARD TESTIMONIAL EPERGNE. AUCKLAND WAR MEMORIAL MUSEUM TĀMAKI PAENGA HIRA. S963.

new constitution would be used by the settlers to oppress the indigenous people. Grey postponed self-government and instead divided the country into two provinces, New Munster (Wellington and the South Island) and New Ulster (the rest of the North Island). When looking round for a person who could be his lieutenant in the northern province of New Ulster, he chose Robert Wynyard.

Wynyard was born in 1802 at the royal castle of Windsor where his mother was lady-in-waiting to the Queen. Robert followed his father by becoming a military officer and helped to keep the peace in Ireland. Eventually Wynyard was asked to command the 58th Regiment which was posted to Sydney in 1844 and soon after sent to New Zealand. There he was involved in the fight against Hōne Heke and Kawiti and took part in the storming of Ruapekapeka pā. Then in 1851 he was given command of all the official military forces in the country; and soon after was appointed by Grey as Lieutenant-Governor of New Ulster. But Grey conceded few powers to his lieutenant, and the proposed elected assembly was never called. Pākehā settlers, especially in the southern areas, continued to press for self-government. Constitutional associations sprang up to press this cause. Something had to give.

Finally in 1852 the British Parliament recognised colonial aspirations. A New Zealand Constitution Act was passed which gave the country 'representative government', by which was meant that a majority of the legislature was elected. The franchise was still property-based, but for the time it was very democratic and included a large part of the Pākehā male population. Māori were given the vote but, since the suffrage was based on property holding, they were disadvantaged because Māori land was communally held. Six provinces were established, and voters could choose the provincial councils, with the possibility of electing the superintendent of each province. But representative government was only half of the settlers' ambitions. They also wanted 'responsible' government so that the executive, where day-to-day decisions were made, would be drawn from the elected legislature. In both Wellington where Isaac Featherston was superintendent and Canterbury where James FitzGerald held the post, the two superintendents made moves to ensure that their government worked 'responsibly' with the support of the elected councils.

It was a different situation in Auckland. There, our friend Robert Wynyard was elected as superintendent, and he did not act in this way. His election was somewhat of a surprise because his opponent, William Brown, the merchant partner of John Logan Campbell, had long argued for democratic aspirations against the Governor. With the support of the *Southern Cross* newspaper Brown's 'Progress Party' stood for settler aspirations — especially cheap available Māori land. But Wynyard was able to win for two reasons. Since Auckland was the nation's capital at this stage, there was a large group of government servants who gave him support. Secondly, the southern frontier of Auckland had been settled by 'Fencibles', ex-soldiers who were given land to provide a military protection for the province. These ex-military men gave their fellow soldier, Wynyard, solid backing. But when it came to 'responsible' government, Wynyard had no intention of handing over his executive authority to elected representatives.

The situation became even more fraught. Wynyard was elected Superintendent in July 1853. But George Grey left for England six months later and Wynyard found himself, not only Superintendent, but also acting Governor. This provided even more ammunition for those who saw Wynyard as an autocratic ruler, hostile to democratic impulses. The allegations became fiercer when New Zealand's first Parliament, elected under the 1852 constitution, met in May 1854. The assembly immediately passed a resolution, moved by Edward Gibbon Wakefield, calling for responsible government. Wynyard responded that he had no authority to do so and would need to consult with his imperial masters. There followed an unruly period of conflict and eventually deadlock, until at last in August 1855 Wynyard was forced by instructions from Britain to agree to an executive formed from elected members. He had no chance to see in the new democratic form of government, for the next month, the new Governor, Thomas Gore Brown, arrived. Wynyard retreated to his military duties and in 1858 the 58th Regiment was recalled to England where Wynyard died in 1864.

Just before Wynyard left, 'a number of inhabitants of the City and province of Auckland' presented him with 300 sovereigns in appreciation 'of the abilities, integrity and urbanity' of the man. Wynyard chose to spend the money on this silver epergne which is inscribed with these appreciative remarks from his followers, many of whom were probably military pensioners or officials. The object was in every sense an appropriate symbol of Wynyard's role during his years in New Zealand. He was among the last opponents and resisters of the coming of colonial democracy, so an object which breathed aristocratic pretension was wholly fitting. The symbols represented on the epergne express well Wynyard's sense of New Zealand. There is the figure of a soldier, and romantic images of punga and a Māori family. What is not to be found among the silver embellishments is any representation of the 'new' New Zealand, the ambitious white settlers, whose desire for cheap land and plentiful trade had helped to fuel the colonial drive for democracy.

29.

Wiremu Kīngi's tauihu

PUKE ARIKI

This beautifully carved tauihu, the prow of a waka, has already received considerable public attention. There have been two recent publications about the taonga, to both of which I am highly indebted; and it was placed beside the figurehead of HMS *Eclipse* at the entrance to Puke Ariki's path-breaking exhibition on the Taranaki War in 2010, *Te Ahi Kā Roa, Te Ahi Kātoro Taranaki War 1860–2010: Our legacy — our challenge*.

Both objects were entirely appropriate. The carved wooden figure of a woman from the prow of HMS *Eclipse* symbolised the heavy involvement of British forces in our civil war. The *Eclipse* was a wooden gun vessel, launched in 1860, which arrived in New Plymouth in May 1863 with British officers and men of the 40th and 70th Regiments to fight in the Taranaki War. Over the next couple of years the ship was used to carry troops and armaments to various war theatres. Its cannons fired on Tātaraimaka Redoubt in May 1863 and also on the earthworks at Katikara on 4 June 1863 when Governor George Grey observed from its decks. Later the ship was involved in carrying soldiers and equipment to the Waikato War, where its seamen took part in attacking the Māori defences at Rangiriri in November 1863. The next year men from the *Eclipse* were among those killed in the attack on Gate Pa and in 1865 the ship was sent to Ōpōtiki to investigate the killing of Carl Völkner. So the ship and the men that it carried were in many ways veterans of the military conflicts of the early 1860s and a fitting symbol of the involvement of British forces, not only in the Taranaki War, but in the New Zealand Wars generally.

As for the tauihu, this too fully justified its position at the entrance to the exhibition. It represented well the concerns of Te Āti Awa which led to the outbreak of the Taranaki War. A tauihu was an ornately carved figurehead which was lashed to the prow of a canoe. Its carving usually involved a large expenditure of time and energy, and frequently, as here, consisted of two large scroll designs (pītau) and a prominent figure in front with protruding tongue and arms. The figurehead assisted Māori navigators in setting a course ahead. Tauihu were so valuable that they were frequently moved to new canoes after an older one had been destroyed. That seems to have been the case with this tauihu. Its origins lie with the migration of many Te Āti Awa from their home in Taranaki to Waikanae. When Ngāti Toa led by Te Rauparaha decided to leave Kāwhia and head south in the early 1820s, they were joined by some of their Taranaki kin, especially Ngāti Mutunga, Ngāti Tama and Te Āti Awa. Although they had journeyed by foot to Kāpiti the new arrivals realised they would need canoes to defend their new possessions and assert their authority on the top of the South Island. So Ngāti Toa invited their Arawa relations, Ngāti Tuara, to help build and carve new waka. It seems likely that this tauihu was carved by Ngāti Tuara with input from Te Āti Awa, for the Taranaki carver Hemi Sundgren argues that, while its spirals and powerful head were based on the conventions of North Island carving, there were also elements of Te Āti Awa design. Whether the tauihu was the figurehead for a waka taua taking warriors to the South Island we do not know. What seems more certain is that at some stage it was removed and attached to a new, much smaller waka called *Kaupahanui* which in April 1848 carried the Te Āti Awa leader Wiremu Kīngi Te Rangitake, along with two other men, eight women and three children, back to Taranaki and eventually into a confrontation with the Governor that led to the outbreak of the Taranaki War. Both the *Eclipse* figurehead and the *Kaupahanui* tauihu were carrying combatants to battle.

Why did Wiremu Kīngi and his people return to their ancestral lands near Waitara? When Kīngi had first settled on the Waikanae coast he had been converted to Christianity and showed considerable support for the Pākehā settlers. He had drawn his moko on a copy of the Treaty of Waitangi brought to Waikanae in 1840. But in 1841–43 about a thousand British settlers arrived in Taranaki as settlers with the Plymouth Company, an offshoot of the New Zealand Company. They came in the naïve belief that the company had purchased 20 million acres in the southern part of the North Island including Taranaki.

A78.128 TAUIHU, COLLECTION OF PUKE ARIKI, NEW PLYMOUTH

The government's land commissioner, William Spain, dismissed this purchase but in 1844 he did recognise a second purchase of 60,000 acres which included Te Āti Awa land at Waitara. Wiremu Kīngi was incensed at the threatened loss of ancestral land. He wrote to Governor FitzRoy, 'Waitara shall not be given up; the men to whom it belongs will hold it for themselves.' FitzRoy listened and refused to enforce Spain's award. So far so good, from Kīngi's perspective. But the New Plymouth settlers were hungry for farming land, and in Britain the Secretary of State William Gladstone heard their pleas. FitzRoy was replaced by George Grey; and in 1847 he authorised the reacquisition for European use of the 60,000 acres. It was this threat which induced Kīngi to head north back to his ancestral home. In a flotilla involving one sailing ship, four whaling boats and 44 canoes, 587 people set sailed. The *Kaupahanui* with this tauihu at front and Kīngi on board was in the fleet. Kīngi and his people settled on the south bank of the Waitara.

There for some years Kīngi's community flourished. By 1854 they had 35 ploughs, 300 cattle and 150 horses and were selling £6000 of produce each year. But the desire for land among the New Plymouth settlers grew stronger as the population increased. In 1859 Governor Thomas Gore Browne purchased 243 hectares on the south bank of the Waitara River from a junior Te Āti Awa chief, Te Teira Mānuka, who claimed sole rights to the land. Kīngi was not interested in individual rights. As principal chief of Te Āti Awa he saw himself as protector of the whole tribe. He wrote to the Governor, 'I will not agree to our bedroom being sold (I mean Waitara here), for this bed belongs to the whole of us . . . none of this land will be given to you, never, never, not till I die.' The Governor for his part was determined to assert his authority. He believed that the Māori King movement was behind the opposition and that British sovereignty had to be upheld. The stakes had risen on both sides to matters of high principle.

In November 1859 a first instalment of £100 was paid to Te Teira and when surveyors appeared Te Āti Awa peacefully obstructed them. On 17 March 1860 the first shots were fired in the Taranaki War when the British attacked Te Kohia pā. The tauihu had brought one party to battle, and before long the figurehead of the *Eclipse* would bring British reinforcements. For some six years, between 1860 and 1866, there was spasmodic fighting in Taranaki. On a conservative estimate several hundred Māori died, and almost 1.2 million acres of Māori land was confiscated.

The later history is mixed. Wiremu Kīngi left Taranaki to stay with Rewi Maniapoto at Kihikihi and then for 12 years with Ngāti Maru in 'strict seclusion' in the bush between Waitara and Whanganui. In 1872 he joined Te Whiti-o-Rongomai's community at Parihaka and he died in 1882. As for the tauihu, according to Kelvin Day of Puke Ariki, it first resurfaced in 1883 when it was gifted to the New Plymouth Boys' High School museum by William Lawrence, who kept a shoe shop in Waitara and was a collector of 'old Maori curios'. How he came by the precious prow is unclear, but the damage to one of the spirals suggests that it went through tough times. In 1901 the tauihu was transferred to the newly established Taranaki Museum, where under the museum's new identity as Puke Ariki, it remains today, as a striking memorial to the origins of the Taranaki War.

30.

James Quedley's red coatee

AUCKLAND WAR MEMORIAL MUSEUM

For over 200 years, from the seventeenth to the nineteenth centuries, the red coat was the symbol of British soldiers. Even in the late Middle Ages red, derived from St George's Cross, was the mark of an English soldier; and following Parliament's use of red uniforms in the English Civil War, red coats were adopted by the King's army during the Restoration. From then on red jackets were universal in the British army. This may seem an odd choice, since, to modern eyes used to camouflaged army wear, red appears a dangerously conspicuous colour. But in early modern European warfare the smoke of cannons quickly hid infantry from the eyes of the enemy while red helped British soldiers identify their comrades. There is no truth to the old belief that red was designed to hide bloodstains. Red coats were used in European wars whether against Napoleon at the turn of the nineteenth century or in Crimea in the 1850s. They were prominent in repressive actions in Ireland, and were worn around the Empire, notably in Africa and India, to keep colonial populations in check.

Red coats signalled the arrival of full-time British soldiery and it tells us much about the New Zealand Wars of the mid-nineteenth century that a number are found in New Zealand museum collections. Some 18,000 'redcoats' served here between 1840 and 1870, and they played a major role in establishing European dominance and the massive transfer of Māori land into the hands of white settlers. Although by 1860 there were

SHELL JACKET OF PRIVATE JAMES QUEDLEY, 14TH REGIMENT OF FOOT 2ND BATTALION. AUCKLAND WAR MEMORIAL MUSEUM TĀMAKI PAENGA HIRA. U101.

more Europeans than Māori in New Zealand, the Europeans were not able, unlike Māori who raised about 4000 fighters for the war, to depend upon their own population to carry out their military ambitions. True, there were about 3000 volunteers and militia drawn from the white male community; but additional fighters had to be brought in — up to 4000 were military settlers, largely recruited in Australia, who were given the promise of land in return for fighting; and then there were the redcoats. At the high point during the Waikato War in 1863–64 there were over 10,000 professional British soldiers here. This was a small proportion of the over 200,000 redcoats spread around the Empire, but they were a major contributor to establishing white hegemony in Aotearoa.

This particular jacket, or coatee, was worn by James Quedley. He was fairly typical of the redcoats who established the 'pax Britannica' in the Empire. He was not of Irish origin as were about a third of the common soldiers in the British army in the 1860s, and over half of those who came to New Zealand. But he was from a family of rural labourers, a favoured background. 'There are no men so good soldiers as the man who comes from the plough,' a sergeant-major told a royal commission in 1835. Quedley had been born in 1828 at Cutcombe, Somerset, in the west of England. At the age of 17 he 'took the shilling' (referring to the daily pay of a private) and joined the 55th Regiment of Foot serving in Ireland, Gibraltar and at the Crimean War. It was a tough life. Living conditions were often uncomfortable; there was constant danger; officers (who had purchased their commissions) could be erratic and authoritarian, and occasionally flogged their men; and among fellow soldiers there were high levels of illiteracy, violence, desertion and drunkenness. James Bodell's memoir of being a private in the New Zealand Wars notes how often the men got 'locussed', or drunk. Yet the pay, once board and lodging are taken into account, was about what a rural labourer might earn.

Quedley, like most recruits, joined the army for life, and in March 1859 he transferred to the 2nd battalion of the 14th Foot Regiment, known colloquially as 'the old and bold'. In November the next year he sailed from Ireland with the regiment to Auckland. His wife Mary and their children accompanied him. Several months later Quedley found himself assisting in building Pratt's famous sap (a trench) at Pukerangiora, in the last significant action of the initial Taranaki War. The regiment remained in New Zealand, and on 12 July 1863 it was called on to begin the invasion of the Waikato when 380 men of the regiment, including Quedley, crossed the Mangatāwhiri River.

This act was the culmination of several years of growing tension between the Governors of New Zealand (Gore Brown and then George Grey) and Waikato Māori. The background to the invasion was that in

1856 a group of central North Island Māori had gathered at Pūkawa on the shores of Lake Taupō to establish a Māori King. Their goal was not to challenge the authority of Queen Victoria but rather, by imitating the British sovereign, strengthen Māori ability to control their own affairs. A powerful Waikato chief, Pōtatau Te Wherowhero, was elected Māori King and supporters placed the mana of their land under the King. Pākehā and especially Governor Browne and his successor, George Grey, feared that the Kīngitanga would halt the settlement of Europeans in the rich lands of the central North Island. Their suspicions were strengthened when Waikato went to the aid of Wiremu Kīngi and Te Āti Awa in the Taranaki War. Grey, who arrived as Governor in September 1861, was determined to break the power of the Kīngitanga. He ordered the building of the Great South Road from Auckland to the Waikato River and arranged for armour-plated river steamers to enter the river.

By mid-1863 Grey and his forces were ready. Using the pretext of the Waikato involvement in Taranaki and the false allegation of a plot to attack Auckland and massacre its people, Grey issued a proclamation to Māori living in the Waikato. This warned that military posts would be established in the Waikato and those who resisted would lose the protections of the Treaty of Waitangi and have their lands confiscated. The model was Ireland where the British had used the confiscation of land and occupation to enforce their will on the Irish Catholics. The proclamation was dated 11 July 1863, but there is strong evidence it did not reach the Waikato until 15 July. By then it was too late, for on 12 July the 14th Regiment, including James Quedley, crossed the border, the Mangatāwhiri River, and the Waikato War had begun.

Over the next few months Quedley was involved in a series of fierce engagements. Five days after crossing into Waikato territory there was a successful attack by Quedley's battalion on Māori positions at Koheroa where they used bayonets and killed some 14 Māori. On 1 November British forces achieved a 'bloodless victory' at Meremere when Māori abandoned their pā following bombing from two river steamers. Thirteen miles upriver, however, the Kīngitanga forces had constructed an impressive pā at Rangiriri between the Waikato River and Lake Waikare. The pā featured high parapets protected by a ditch over five metres deep. On 20 November the British set out to storm it. The 65th Regiment led the charge, but were hampered by their ladders being too short. The 14th Regiment, presumably with Quedley among them, then joined in. They also failed with heavy loss of life. Eventually the attack was suspended. The next morning a white flag was seen being hoisted by Māori. It was interpreted as a sign of surrender. Whether this was actually the case has long been a matter of dispute, with some claiming that the flag was in response to one on a British steamer, and indicated a desire to talk terms. Whatever the case, the British interpreted the flag as surrender and proceeded to capture many Waikato prisoners.

This was not the end of the Waikato War. There was a nasty attack on the civilian settlement of Rangiaowhia, and from 30 March to 2 April 1864 at a hastily prepared pā at Ōrākau, the Kīngitanga force of about 300 was surrounded. When they made a rush to escape many were hunted down and killed. War came to an end in the Waikato, although the Kīngitanga remained loath to accept the sovereignty of the Queen.

There is little evidence that Quedley was present at Ōrākau. But he was probably there at Meremere and Rangiriri. The question remains as to whether he was wearing this red coat on those occasions. It seems a little improbable. This coat is a short shell jacket, tight-fitting and only extending down to the waist. Such coats had been introduced in the 1830s to make bush fighting easier than with the full dress uniform which was much looser, bulkier and extended down to the knees. A distinction emerged between the formal dress parade coats and the 'undress' or working coats. The Auckland Museum also has a longer red dress coat which belonged to Quedley who must have worn it on parade and on formal occasions. But even the shorter 'undress' red coat must have been a bit uncomfortable in the rough and tumble of warfare. There was also a shortage of red cloth. So it appears that a light, loose-fitting, dark blue smock shirt was introduced for combat in the New Zealand Wars. Colonel J. E. Alexander, who commanded the 14th Regiment, wrote: 'the skirts of the men's great coats were cut off to enable them to wear them in skirmishing in the bush and scrub. This plan I did not think well of, and afterwards, when preparing some of the 14th Regiment for fighting, I gave them blue smocks, over which the great coat was worn . . .' It seems likely that this blue shirt, rather than the red shell jacket, was Quedley's uniform at Meremere and Rangiriri in 1863. Yet the presence of two red coats at Auckland Museum once owned by Private James Quedley is a powerful symbol of the role of the British imperial army in bringing the Queen's sovereignty to New Zealand, not to mention the confiscation of Māori land which followed.

As for James Quedley, at the end of the New Zealand Wars he went with the 14th Regiment back to Melbourne, where he was soon discharged as medically unfit. He returned to New Zealand and settled in Auckland as a military pensioner with a land grant of 50 acres in Pakuranga. He died in 1918 aged 90 and his son donated the jackets to the Auckland Museum some years later where they remind us of the important place of 'redcoats' in New Zealand's colonial past.

31.

Moutoa flag

WHANGANUI REGIONAL MUSEUM

Māori were New Zealand's first enthusiasts for flags. They chose the country's first national flag and the importance they attached to flags encouraged Hōne Heke to cut down the flagpole at Russell. Later Māori, such as the Kīngitanga and Te Kooti's supporters, used flags as symbols of resistance. This remarkable and unusually large flag (originally over 2.5 metres wide) obviously has strong European elements. The background is fine white silk; a crude Union Jack adorns the top left; in the centre is a gold and red crown above two hands, one brown, one white, grasped in friendship, and the central images are surrounded by classical laurel leaves of victory and the word 'Moutoa'. The flag was conceived and executed by Europeans, but was presented to Māori for whom it became a significant symbol.

The story begins with the emergence of the Pai Mārire movement, or Hauhauism from 1862. The founder of Pai Mārire, Te Ua Haumēne, was a former Wesleyan preacher. Seeing Māori fighting Pākehā and losing their land, Te Ua claimed the Archangel Gabriel had visited him promising to unify Māori and drive out the Europeans. Te Ua saw Māori as Old Testament Jews seeking restoration of their homeland. Pai Mārire means 'goodness and peace'; and Te Ua insisted that his vision would occur peacefully. But his followers, especially in Taranaki, encouraged a warrior response. Their rituals included dancing around niu poles bedecked with ropes and, yes, large flags which communicated messages from

God conveyed on the wind (hau) — hence their common name Hauhau. In 1864 the Hauhau spread their message to other areas, carrying the dried heads of European soldiers killed in Taranaki. In April of that year Mātene Rangitauira brought the Pai Mārire message to his kin at Pipiriki on the Whanganui River, where the people, seeing the Taranaki war, feared future loss of land and authority. Tōtara poles with flags bearing Pai Mārire symbols appeared.

Mātene Rangitauira indicated that his followers would come down the river and attack Whanganui township. This was not welcome news to the hapū of the lower river. Those at Pūtiki, just across the river from the township, were strong Christians, dependent on the Pākehā economy. They treasured and benefited from their relationship with Europeans. Māori further upriver, although sympathetic to the Kīngitanga, were affronted that their mana over the river might be infringed and feared that a Pai Mārire presence would draw government reprisals. They would not allow the river to become a highway for military purposes. Ngāti Hau, Ngāti Pamoana and the lower-river hapū gathered at Rānana to resist the Pai Mārire forces. Their chiefs, Mete Kīngi and Haimona Hiroti, sent a message to Mātene that they would challenge the upriver Māori on Moutoa Island. The date set was 14 May 1864.

Moutoa was about a kilometre north of Rānana and about 70 kilometres north of Whanganui township. It was a kilometre-long island in the middle of the river, mostly shingle and sand and a few scrubby trees. Tradition said it was a chip from Taranaki as he fled westwards after angering Tongariro by trying to abduct his wife Pīhanga.

The battle was a formal, almost ritualistic, encounter lasting less than half an hour, and involving only about 200 warriors on either side. It was almost like a football match, with the Pai Mārire fighters occupying the northern part of the island, the downriver Māori the southern part, and supporters watching from each bank. Yet it was a dramatic and bloody encounter. The lower-river men were almost driven into the water when Tamehana Te Aewa refused to give way. Singlehandedly, he killed five enemies and although a bullet shattered his knee, he delayed the Hauhau sufficiently that Haimona Hiroti was able to rally the men and turn the tide. The Hauhau lost some 50 men, including their leader Mātene. The lower Whanganui tribes lost 15 warriors.

When news reached Whanganui, the sense of relief among the Pākehā population was huge. Within two months of the battle the Wellington Provincial Council decided that 'in recognition of their patriotic services a suitable monument be erected in the town of Wanganui, sacred to the memory of those friendly natives who lost their lives in defence of their European fellow settlers'. Even earlier, indeed one week after the

PARTINGTON, WILLIAM HENRY THOMAS, 1854-1940. MAORI MEN ALONGSIDE THE MOUTOA FLAG. TOURIST AND PUBLICITY. REF: 1/2-012420-G. ALEXANDER TURNBULL LIBRARY, WELLINGTON, NEW ZEALAND

battle, Anne Logan, wife of the officer commanding imperial forces in Whanganui, issued a circular in which she 'begs the Ladies of Wanganui will join her presenting a flag to the friendly and loyal natives who so gallantly fought at Moutoa for the protection of them and their homes. Also in memory of the brave fellows who fell in the conflict.' The women, largely drawn from significant commercial families in Whanganui, raised £33.8.0 towards the materials for the flag and collected funds to distribute clothing and food to the families of those who had fought. The local magistrate, John White, suggested a design. The women drew on their sewing and embroidery skills and their stocks of gold ribbon, green velvet and white silk from cast-off fine clothing to complete it.

The flag was finished by February 1865 and plans were drawn up for a presentation. But then there was further fighting between the upper- and lower-river Māori. At Ōhoutahi, near Hiruhārama (Jerusalem), a leading chief from Pūtiki, Hone Wiremu Hīpango, was killed. His body was brought down to Pūtiki and a funeral was held on 27 February. The newly finished flag was laid on his coffin and accompanied the procession to his burial on the hill above Pūtiki. Five days later Anne Logan made the formal presentation to Mete Kīngi at the Market Place (later Pākaitore, or Moutoa Gardens).

Europeans had to wait for the first public appearance of the Moutoa flag. It came at the unveiling of the Moutoa memorial, New Zealand's first war memorial, at the Market Place on 26 December 1865. As the name suggests, the site was where upriver Māori gathered to moor waka and trade with the locals. On that afternoon the flag was gathered from Mrs Logan's residence by Māori warriors, brought to the monument site and placed, presumably face down, on the top of the memorial. The provincial superintendent, Isaac Featherston, spoke to the crowd of over 2000 including over 500 Māori from the southern North Island. He suggested that the memorial and flag were intended to honour the deeds of heroic men, but also to stimulate future efforts by Māori in support of the Crown. He looked forward to the time when 'Mrs Colonel Logan and the ladies of Wanganui [might] inscribe upon the "Flag of Moutoa" the names of other victories greater and more decisive in their results than those they have already won'. Featherston then pulled the flag off the memorial to reveal an image of a weeping woman. The English inscription was 'To the memory of those brave men who fell at Moutoa 14 May 1864 in defence of law and order against fanaticism and barbarism', and on the side were the names of 15 Māori who died fighting the Pai Mārire forces along with lay brother Euloge who was hit as a spectator. The flag was unfurled, to three cheers for 'Mrs Colonel Logan and the ladies of Wanganui'.

The flag was not forgotten. Four years later, in May 1869, when Queen Victoria's son, Alfred, Duke of Edinburgh, visited New Zealand on this country's first royal tour, Major Kemp (Keepa Te Rangihiwinui), another local leader who fought for the Crown, led some Whanganui Māori to Auckland and presented the Duke with the Moutoa flag, inviting him to 'present it back to us to prove our loyalty and allegiance to the Queen'. Prince Alfred, wisely recognising its significance to the community, politely did so. Then in 1897, during Queen Victoria's diamond jubilee celebrations, Whanganui Māori again paraded the flag; and four years later the Moutoa flag was taken to the Māori reception at the Rotorua racecourse for the next royal visitors, the Duke and Duchess of Cornwall and York. When the Pioneer (Maori) Battalion were welcomed back at Pūtiki pā after World War I they were honoured with the display of the Moutoa flag.

In 1945 the family looking after the flag, the descendants of Mete Kīngi, loaned it to the Whanganui Museum. There it remains today, a continuing symbol of the fact that, while many nineteenth-century Māori found the European invasion of the country oppressive, there were some who formed close associations — economic, religious and political — with the Pākehā community and, for their own reasons, not necessarily loyalty to the Crown, gave their lives protecting that community.

32.

Kereopa's bailer

AUCKLAND WAR MEMORIAL MUSEUM

If the Moutoa flag was given by Pākehā to Māori to soothe feelings after one fierce conflict in the New Zealand Wars, this bailer was given by a Māori to a Pākehā in 1872, to assist healing following another violent encounter in those wars. The giver was the brother of Kereopa Te Rau, hanged for murder; the recipient was Governor George Bowen. It was a highly symbolic gift. Māori bailers (tīheru or tatā), with their unique forward-projecting handles to prevent wrist strain, were widely used in waka. Two people would face one another, bailing in turn in a rhythm, and sing waiata calling upon the gods to convey the waka 'to the sheltered home where sleep is sound'. It was presumably this sentiment which encouraged Kereopa's brother to give the bailer to the Governor following the judgment on Kereopa to be hanged for murder — 'Take you this baler and bale out everything who is evil, so that henceforth we may paddle in the canoe of Queen Victoria.'

The gift of the bailer was an attempt to ease intense feelings among the Pākehā following the hanging of an Anglican minister, Carl Sylvius Völkner, in Ōpōtiki on 2 March 1865. It was believed Māori acted under the influence of Pai Mārire. What horrified the Pākehā population was not that Völkner was the first Anglican minister to be killed, but the manner of his death. Völkner was captured, held captive, taken to a willow tree near his church and hanged. His body was taken down, his blood drunk from a chalice in a bizarre reference to the Christian communion, his head cut off and then Kereopa removed his eyes and swallowed each in turn — one representing the Queen and the other, Parliament. Völkner's body was then thrown down a latrine. This horrific vision of a cannibal act was amplified by racist fears and became seen among Europeans as the purest expression of the 'savagery and barbarism' (to quote the Moutoa monument) of the Hauhau.

How to explain this awful act? The victim, Völkner, was born in Kassel, Germany, in 1819 and was sent to New Zealand in 1849 by the North Germany Missionary Society to work as a Lutheran missionary in Taranaki. As a German, Völkner was among the largest non-British migrant group in nineteenth-century New Zealand; but he aspired to a closer relationship with the British. In 1852 he offered to serve the Anglican Church Missionary Society and two years later married a wealthy Englishwoman, Emma Lanfear, daughter and sister of Anglican ministers. In 1857 Völkner was naturalised and in 1861 ordained as an Anglican priest. He moved to Ōpōtiki, partly because of his wife's health. There he organised the local Whakatōhea iwi to build a church, Hiona. When war broke out in the Waikato, Völkner was keen that local Māori stay out, believing they would suffer from fighting the Crown. More important, because of his German origins, he wanted to reaffirm his British loyalty. While the local Catholic priest, Father Garavel, served as an intermediary between Waikato and Whakatōhea, Völkner took another route. He became, in effect, a government spy, writing to Governor George Grey in 'private communications' detailing the Kingite preparations for war in the Bay of Plenty and even sending a plan of Rangiaowhia pā. Whether it was these letters that roused the suspicions of Whakatōhea or not, they certainly regarded him as a government informer. From February 1864 when the iwi committed to assisting the Māori King, Völkner was viewed as a traitor. When local Māori held a posthumous trial after his death, it was not the letters but Völkner's four trips back and forth to Auckland, where his wife was living in 1864–65, which confirmed suspicions of him as a spy. He was seen as working for the Governor.

Other factors heightened animosity towards the Anglican minister. In April 1864 a leading Whakatōhea chief, Te Aporotanga, was killed near Matatā by Te Arawa, and Whakatōhea were furious that the Governor, and by extension his supporter Völkner, had not taken action against Te Arawa. It was also believed, incorrectly, that Father Garavel, the Catholic priest and their Whakatōhea supporter, had been taken to Auckland and hanged. In addition, from mid-1964 there were epidemics of typhoid and measles and at least 200 local Māori had died; while the disruptions of war caused considerable economic distress. By February 1865 feelings were already strong against Völkner. Then, on the 25th, Kereopa Te Rau and

TIHERU, BAILER. AUCKLAND WAR MEMORIAL MUSEUM TĀMAKI PAENGA HIRA. 60.

Pātara Raukatauri arrived as Pai Mārire emissaries from Te Ua Haumēne. Kereopa was from a Te Arawa iwi, Ngāti Rangiwewehi. He had been baptised a Catholic and worked as a policeman in Auckland, but in the 1860s he joined the King's forces in Waikato. It is said that his wife and two daughters were killed during the British attack on the peaceful Waikato village of Rangiaowhia in February 1864. They were among the victims of a whare fire lit while those inside were at worship. The next day he saw his sister killed at Hairini. Perhaps these personal tragedies made the Pai Mārire vision of a land freed of the Pākehā particularly meaningful for Kereopa. He joined the new faith soon after. Te Ua sent him and Pātara eastwards to preach the faith in peace. Arriving with about 40 followers at Ōpōtiki, Kereopa found their teachings well received. Before long, there was a niu pole erected and Pai Mārire dancing around it.

On 1 March Völkner returned by boat from Auckland. He and his fellow missionary Thomas Grace were held and that night it was decided that Völkner should be hanged. Kereopa was apparently not present when the decision was taken, and he claimed later that he resisted the decision, although subsequently supported it. He was probably present at the hanging, but he consistently denied playing a leading role, and there are no reliable witnesses to the contrary. There is no question that Kereopa swallowed Völkner's eyes. This was seen by Māori as a customary insult, a way of expressing contempt for a despised enemy — appropriate treatment for a traitor who represented a land-taking church. Kereopa explained to Sister Mary Aubert on the night before his execution that someone had thrown Völkner's head through the church window, 'and when I saw the eyes of my enemy like that, all the Maori in me rose up. I could not help myself. I seized them and swallowed them.' How far he was responsible for Völkner's killing is more questionable.

These were nuances of little concern to the Governor and the Pākehā population. They wanted revenge. In September a force landed at Ōpōtiki and swept through the district killing 58 Whakatōhea people; 211,060 acres were confiscated of which Whakatōhea lost 144,930. Five members of Whakatōhea were hanged for Völkner's murder, including Mokomoko who was later shown not to have been present on that occasion. In 1992 he was pardoned and in 2013 legislation passed, restoring his character and reputation.

But Kereopa was the government's real target. Pai Mārire was seen as the primary cause of the murder, and Kereopa its instigator. He had to be punished. Kereopa was pursued, but escaped capture narrowly on two occasions, taking refuge in the Urewera. There for five years, colonial troops and Māori supporters under Rāpata Wahawaha hunted for Kereopa and his fellow Māori nationalist Te Kooti Arikirangi. Eventually in September 1871 Tūhoe became tired of being invaded and handed over Kereopa to Rāpata. He was taken to Napier and in December 1871 was convicted and sentenced to hang. There were major concerns about the trial. Kereopa was not allowed to summon witnesses for his defence; and those testifying for the Crown were granted immunity from prosecution, even those who played a major role in the event. Evidence that Kereopa had not instigated the murder, was simply ignored. The Crown wanted a sacrificial victim, someone who could tie the murder to the hated Pai Mārire faith. The witnesses in turn wanted to offload their responsibility onto a man from another iwi.

Kereopa's conviction did not pass without questioning. William Colenso, in Napier, wrote a furious letter to the newspaper protesting Kereopa's innocence. Sister Aubert spent the last night before his execution on 5 March 1872 giving him comfort. His brother Rāwiri who had been with him when he was captured in the Urewera visited him in prison. Rāwiri sought to rise above the immediate pain by giving the Governor the bailer. If he hoped that this gesture would ensure safe sailing on the waka carrying Māori and Pākehā, it was a vain gesture. The memory of Völkner's killing lived long and not painlessly. But in 2014 a Treaty of Waitangi settlement with Ngāti Rangiwewehi provided a statutory pardon for Kereopa Te Rau. Perhaps at last the waka could sail without fear of sinking.

33.

Paddy Galvin's pipes

LAKES DISTRICT MUSEUM AND ART GALLERY

While the North Island of the 1860s was convulsed by war, the South was transformed by the gold rushes. So what object do we choose to tell the story of this hugely significant event? If you look at images of miners on the road to the diggings, they inevitably carry a billy and a pannikin, or metal mug. Both were essential for brewing the black tea which kept the miners alive through the searing cold of central Otago and the penetrating dampness of the West Coast goldfields. Then there is the equipment used by miners to find gold from rivers — a wide metal pan for sorting the gold specks from the sand, or a shovel for digging out the shingle, or even a rocker or cradle. This was a wooden box into which the miner poured water over the sand with one hand while he rocked the cradle with the other to separate out the golden grains. But while these items survive, few are endowed with particular stories or associated with interesting individuals.

Instead we have chosen Paddy Galvin's Irish bagpipes, which tells us something about one of the cultures brought by the miners. Paddy Galvin was born in 1840 in County Clare in western Ireland, an area with a heavy Catholic population and a county known for traditional Irish music. In the years of Paddy's childhood County Clare suffered from the severe potato famine. Millions died and thousands set off for new lives overseas. In 1857 Paddy followed, migrating to Victoria where he carted goods from Geelong to the goldfields. By the early 1860s the Victorian fields were losing their lustre; so in 1862 he joined thousands from those fields and headed to the newly discovered Dunstan field on the Clutha River in Otago.

This was not the first gold discovery in New Zealand. There had been brief rushes to the Coromandel and Collingwood; and occasional reports of gold in Otago. But the very 'proper' leader of the Dunedin Presbyterian community, Captain Cargill, feared that a gold rush would lead to 'mischievous results', so such reports were suppressed. However, in 1861 Gabriel Read, a well-educated Tasmanian with considerable experience on both the Californian and Victorian goldfields, followed up a suggestion from a Hindu Anglo-Indian, Edward Peters, to look for gold on the Tuapeka River near present-day Lawrence. On 23 May 1861 Read found gold 'shining like the stars in Orion on a dark and frosty night'. He claimed the £500 reward recently promised by the provincial council and the rush was on. There was a stampede from Dunedin and diggers flocked to Melbourne to catch ships for Otago. By August there were over 2000 diggers under canvas at Gabriel's Gully.

The fine pickings did not last. By the end of the year there was a steady outflow back across the Tasman. Then in August 1862 an Irishman, Christopher Reilly, and an American, Horatio Hartley, formerly mining mates in California and British Columbia, deposited 87 pounds (39 kilograms) of gold on the counter of the Treasury in Dunedin. They had found it using a cradle on the banks of the Clutha in the Dunstan Gorge (which ran from modern-day Cromwell to Clyde).

This time Paddy was interested. He joined the new exodus of miners from Victoria and headed for the Dunstan. The newcomers were a polyglot group. One observer wrote, 'There are people here on the gold fields from all parts of the globe, there are men here as black as your kettle to as white as snow.' There were Americans, Canadians, Scandinavians, Italians, Germans and eventually, Chinese. Among those from the UK there were many Scots and English, and about a third were Irish like Paddy. When the gold rush moved to the West Coast in 1865 the proportion of Irish rose to almost half. Among the Otago miners those from County Clare constituted the highest numbers from Ireland except for the Ulster folk. So the goldfields along with the settlement of ex-soldiers brought a distinct Irish and Catholic community.

Like many newcomers to the goldfields Paddy did not come as a miner. He worked in Victoria servicing the miners. The service he provided was their leisure-time enjoyments. Being a miner was tough. The climate in central Otago was not easy — hot summers, bitter winters and heavy snowfalls, spring floods from melting snow which swept away miners' tents and possessions as in 1863. Drowning was common. There was little wood for fires. Food, especially meat and vegetables, was

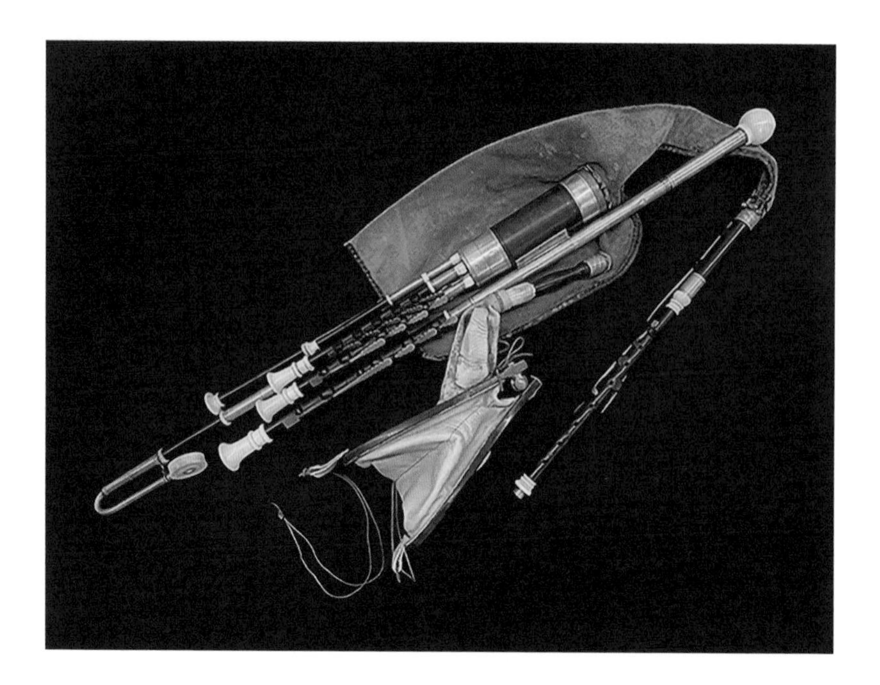

COURTESY OF THE LAKES DISTRICT MUSEUM REF#X0023

expensive and hard to find. Panning or digging for gold was physically demanding, and monotonous. Disease was rampant.

The miners found respite in communal enjoyments — singing and yarning with their mates around a fire. But their passion, once they had sold their gold and had money in their pockets, was heading to the pub on Saturday afternoon and Sunday and going on a 'spree'. Drinking was the cure for pain and sorrows. When in July 1861 the Weatherston brothers found gold in the next valley over from Gabriel's Gully, 'fourteen hotels, dancing saloons and numerous gambling dens' quickly sprang up. The *Evening Post* reported in 1865, 'Everybody who does not dig sells grog and everybody who digs drinks copiously.' This is where Paddy Galvin found his role. He first tried selling items such as drapery goods at various sites along the Clyde. But by 1864 he had purchased the Sportsman's Arms Hotel in Alexandra and then the Union Hotel there. It did not fully work out to Paddy's satisfaction. He abandoned the licence when he discovered that he was not allowed to play a musical instrument in his house. That was no good for Paddy because he had become a legendary player of Irish pipes; and singing and enjoying music was central to the miners' leisure. He purchased the Shingle Creek Hotel also on the Clutha where he could play to his heart's, and the miners', content. The Irish, or uilleann, pipes he played were rather different from the well-known Scottish bagpipes. Instead of blowing into a bag, Paddy inflated the instrument by a set of bellows strapped around his waist and right arm. It meant that he could sing while he played; and the sound flowing forth was quieter and sweeter in tone than the boisterous Scots bagpipes. He normally played while sitting down. The set of pipes owned by Paddy Galvin and now in Arrowtown's Lakes District Museum is said to have been made by Michael Egan, a famed maker of Irish pipes. Egan migrated from Ireland to Liverpool, like many Irish, and then moved to New York. Where and when Paddy Galvin obtained the pipes is unknown; but their quality attests to how important the playing of pipes was to him. He became a legendary figure among the diggers of central Otago.

In later years Paddy moved to the Cardrona Valley, which had also seen a gold rush. He eventually built a rammed-earth house just uphill from the famous and still surviving Cardrona Hotel. Paddy spent many years until his death in 1929 playing and singing his pipes at the hotel. It was said that he could fall down drunk but never miss a note. Paddy married another migrant from County Clare, and today the grim southerlies which sweep up the Cardrona Valley are known as 'Ma Galvin's breath'. We are fortunate that the remarkable set of pipes and such sayings live on to remind us of the music and drinking which kept miners, especially those from Ireland, content against the rigours of the goldfields.

34.

Chinese li-ding scales

TOITŪ OTAGO SETTLERS MUSEUM

There are a number of objects, excavated or preciously handed down, that point to the distinctive material and social culture of the Chinese men who inhabited the New Zealand goldfields in the 1860s and 1870s. They include beautiful ceramic dishes, joss sticks and packapoo tickets, and also this delicate 'li-ding', or portable set of scales. At least two li-ding survive, one in the Lakes District Museum at Arrowtown and this set in Toitū in Dunedin. They throw light on a community which lived apart and developed its own unique way of life.

What brought the Chinese down to the 'New Gold Mountain' of central Otago, Southland and the West Coast? Until the 1860s, there had been very few Chinese people in New Zealand. Appo Hocton had jumped ship in 1842 and became a successful carter in Nelson; and in 1853 Edward Jerningham Wakefield had promoted a plan, suggested by his uncle Edward Gibbon, to bring in Chinese as cheap labour. The idea was stoutly resisted by working men and never got off the ground. But by 1865 the leading merchants of Dunedin, who had done well from the rush, became concerned that the Otago goldfields had run their course. Quantities of gold were declining; and miners were either heading back to Victoria, or moving to new fields on the West Coast. The Dunedin Chamber of Commerce, believing the Chinese to be 'temperate, frugal and well-behaved', sought a guarantee that they would be afforded legal protection and encouraged the superintendent to issue a formal invitation. There were already thousands of Chinese on the diggings of California, Victoria and British Columbia, so they responded quickly. The first group arrived in late 1865 and eventually up to 8000 came. Some arrived from across the Tasman, but nearly all were originally from the counties around the south China city of Guangdon (Canton). What pushed them south was partly civil and ethnic strife in the region, and partly serious economic decline. Canton had been a prosperous port, with a monopoly on the distribution of opium which the British East India Company had sold to the Chinese in return for tea, silk and porcelain. When the Chinese authorities restricted imports of the drug, the British responded with force. The two Opium Wars (in 1839–42 and 1856–60) forced open the opium trade to other ports and Canton suffered acutely. Peasants in rice-growing areas around the port looked to other options and gold became their first choice.

Let's not forget opium. It is central to our story. If we look closely at these exquisite scales, we immediately wonder why they were brought and for what they were used. The obvious explanation is that they were for weighing gold. This is certainly possible. Living as a highly separate community, the Chinese miners sometimes sold their gold, not to European bankers or merchants, but to their own local storekeepers. Conceivably these scales were used by a Chinese storekeeper to measure out the findings of a Chinese miner or, more likely, the winnings of a small group, perhaps four or five in number, who often worked together under a headman. But the surviving gold scales of banks and European merchants are much larger. One suspects that these delicate weighing machines were too modest for a decent haul of gold. It is just as likely that the scales were used by storekeepers to measure out opium.

Most of the opium came in from Victoria or from China as solid brown cakes, arriving in 'packets' of eight large tins with each containing 20 koons, or small tins. Many koons have been found in central Otago excavations. When it came to be smoked in larger Chinese settlements, storekeepers offered 'dens' with little rooms. There would be a brass tray and a set of scales, and one of the miners would buy a pennyweight that would be weighed out and placed on the tray. He would pick up a bamboo pipe, about 2 feet long with a ceramic bowl at the end. He would break off a piece of opium, light it, wait for the opium to burn and take two or three draws, often sharing the pipe with a mate. The process would be repeated until the purchased opium was all gone; or until one of the miners asked for another piece to be cut off and weighed. The effect was feelings of quiet euphoria followed by a sense of restful comfort lasting several hours.

Opium use was not unknown in European New Zealand. Occasional

dignitaries such as Sir George Grey partook of the drug, and it was usually a central ingredient in laudanum taken by sickly women and even given on occasion to infants. But the use was not extensive and customs figures show a dramatic increase in imports from 0.85 grams per head in 1867 (when there were already over 1000 Chinese miners) to 3.56 in 1871. So it was the miners who really brought the habit from their homeland. Estimates of its use among miners vary from Alexander Don's figure of about two in ten to Young Hee who claimed in 1899 that 800 of the 900 Chinese on the West Coast smoked it, with 270 regular smokers. It seems likely that a relatively small number were addicts because the drug took a toll. It undermined initiative, affected the diet and induced a pallid, listless look. Regular use became very expensive. More likely most Chinese were occasional users — on Saturday afternoons or Sunday.

The interesting question is why opium was so popular among Chinese miners. In part it was a habit they brought with them. Opium was introduced to the Canton area by the British who had brought it from Bengal. The drug was initially swallowed, not smoked, and primarily used to heighten sexual pleasure. By the mid-nineteenth century it was well-established in the country. The miners were keeping up a homeland custom.

There are other factors. Nearly all Chinese miners came as 'sojourners' not settlers. While the women stayed home looking after the shop or land and spun silk, the males went overseas to the diggings to earn good money for a year or so and then return. They either remitted their earnings home or brought the money back with them. As an exclusively male community, isolated from European society, the Chinese must have missed their families, especially those who had left behind wives and children. There were only a handful of women among the 4000 Chinese in New Zealand. Opium was arguably a way of compensating for the miners' emotional pain and sublimating their sexual urges. Alexander Don reported Sham Paak of Adams Gully saying, 'We Chinese are not accompanied by our wives, and the younger of us find opium very useful in repressing our passions.'

Opium may also have been a way to escape a tough physical life quite unlike the world at home. The rugged rocky landscape of central Otago and the dripping-wet forests and frightening rivers of the West Coast were very different from the flat rice paddies in the counties around Canton. The harsh climate and rough terrain brought injury and death. There was much sickness. Not all were successful prospectors and their accustomed foods of rice, pork and tea were expensive. Suicides among the Chinese were not uncommon. While other miners drowned their pains in alcohol and the comradery of the pub, few Chinese took that route, although peach brandy was an occasional indulgence. Further, they must have sensed some hostility among their European neighbours. The mere mention of possible Chinese miners in their area in 1857 led the citizens of Nelson to express concern about their 'filthy and immoral habits'. When the Chinese reached central Otago, there were scuffles with white miners over claims and miners' rights. Digger Matthew Wright wrote that the Chinese 'stirred up our little camp as much as if we had been informed that a Bengal tiger was in the vicinity'; and in a notorious case in Naseby in 1857 Ah Pack had his pigtail hacked from his head and he was rolled about town in a wooden barrel. Such hostility can be exaggerated. To be fair the *Otago Witness* claimed in 1868 that the Chinese had generally 'been received with an entire absence of ill-feeling — in some cases with cordiality'. However, there is no doubt that the two communities kept very much apart; and when a Chinese miner felt lonely or exhausted or homesick for his family, he would turn to the closest escape and visit the den for a spot of gambling and opium.

The story of the Chinese and opium has an impressive ending. During the 1880s and 1890s, New Zealand, following the lead of Australian states and with support from working people, restricted migration from China. In 1881 a poll tax was imposed on Chinese immigrants of £10 and there was to be only one Chinese passenger for each 10 tons of a ship's cargo. Fifteen years later the figures were raised to £100 and 200 tons of cargo. There was growing anti-Chinese prejudice. A young West Coast MP, Richard Seddon, distinguished himself by claiming that 'there is about the same distinction between a European and a Chinaman as that between a Chinaman and a monkey'. Anti-Chinese and White Race Leagues appeared. But there were already some 4000 Chinese in the country who were anxious that their wives join them. So they sought to challenge negative aspersions about their race by adjusting their habits. Their principal target was opium smoking. Initially in 1888 they set out to reform themselves. Then they turned to Parliament for legislation; 542 Chinese — mainly Christian people on the West Coast, led by a young law clerk, Young Hee — petitioned for opium imports to be ended. Hee played on fears that the habit might spread to the wider community. Further petitions followed from Chinese in other parts of the country. In 1901 the smoking of opium and imports of the drug in a form suitable for smoking were prohibited. However, in a revealing provision which showed how deep ran suspicion of the Chinese, police did not need a warrant to search a Chinese house, but did to search a residence of anyone else. From their arrival in numbers in the 1860s to the turn of the century the Chinese were always 'different'. These scales help reveal the nature of that difference.

35.

Timaru life-saving rocket

SOUTH CANTERBURY MUSEUM

It was late afternoon on Saturday 24 May 1869. For two days huge waves had rolled in from the south-east onto Timaru's exposed beach. Out in the harbour roadstead two barques, the 450-ton *Collingwood*, which had been loading wheat, and the smaller (180-ton) *Susan Jane*, on a coal run from Newcastle, Australia, were trusting to their anchors. Strangely, there was not a breath of wind. As darkness fell, there was a brief puff of wind and a signal from shore that now was the time for the sailing ships to head out to sea and to safety. The *Susan Jane* hauled up her anchors, but then the wind eased. She tried to drop the anchors but they would not hold; and as she drifted she fouled the *Collingwood*'s anchor cable. The *Collingwood* found herself heading towards shore and soon after 5 am she grounded on the Waimātaitai Lagoon. Waiting on the beach were a number of uniformed men, the 'Volunteer Rocket Brigade', as they were known. They fired a rocket towards the ailing ship, but the harbourmaster Mr Mills was standing too close and was burned about the face and hand. A second rocket was fired and this time it passed right over the vessel. The line was seized by the crew on board. Quickly a larger rope was sent across, and a life buoy with canvas breeches (pants) attached was hauled to the ship. Before long, the officers, the crew and seven others working on the cargo were winched safely ashore — 17 men in all.

Focus now turned to the *Susan Jane*. She had put down her anchors and had all her sails aloft in the hope of being able to sail out to sea, but she continued drifting towards shore watched by hundreds of people along the cliffs. Eventually it was decided to slip the cables and the barque surged towards the beach. Once more the rocket crew were on hand. They fired a line between her masts, the life buoy was rigged, and the crew were hauled to shore with the exception of the cook and the mate who fell into the sea, but were safely rescued. Twenty-six men had been saved.

This was the first occasion on which this unusual object, Boxer's rocket life-saving apparatus, now on display in Timaru's museum, had been used to save life. The apparatus had a number of components. There was a tripod on which rested the metal rocket which carried a light hemp line. In training, the rocket men found that they could shoot the line up to three quarters of a mile, although it was normally fired at shorter distances. When the men on ship recovered the line they pulled it and brought aboard another heavier rope, the whip, and a light pulley and tail block. On the block were instructions to attach the pulley and block to a mast and tie the ends of the whip together. Back at shore the rocket men tied an even stronger rope, the hawser, to the whip line and this was hauled out to the ship by the rocket men and attached to the mast. The breeches buoy was hung from the hawser and tied to the whip rope. This allowed the men manning the rocket to pull on the whip so that a person sitting in the breeches could be carried to shore. At the same time a second team manipulated the tripod to ensure that the hawser kept the passenger above the waves.

SOUTH CANTERBURY MUSEUM

The apparatus had been purchased in 1867 for a cost of £160 including the rockets. Over the next 15 years it was involved in a number of stirring rescues. Indeed, newspaper reports suggest that the crews of no fewer than 10 ships wrecked at Timaru, from 1869 to the last rescue from the *City of Cashmere* in January 1882, were brought ashore using the rocket. In all, 104 people were carried to safety on the system's breeches. There were also a number saved using the apparatus down the coast at Ōamaru. To be fair, the 10 rescues represented under half of the 28 ships wrecked at Timaru from 1865 to 1890; and they were not always successful. When four ships became beached in September 1878 the rocket lines sent across to the *Melrose* either went too high or became fouled; and the crews of all four ships except for two men swam ashore holding bits of wreckage. Yet the rocket crew were repeatedly praised for their work. When the brig *Layard* was wrecked in June 1870 the nine crew were brought ashore in only nine and a half minutes; and when the last of the five rescued from the *Pelican* reached dry land in June 1879, the rocket men were given a 'good hearty and well-deserved cheer' from bystanders. However, in the tragic wreck of the *Ben Venue* and the *City of Perth* on 14 May 1882, which is still remembered in Timaru, rockets could not be used. But the rocket crew helped man the lifeboats and were among the nine who lost their lives when the boats capsized.

Several questions remain from this story of the rocket apparatus. The first is why were there so many shipwrecks in New Zealand, and Timaru in particular, in the mid-nineteenth century? In Timaru's case there was no protected port, but only a beach, and initially ships had to berth exposed to the open sea with fatal consequences when huge waves rolled in from the south-east. There were similar beach harbours at Ōamaru, Napier and New Plymouth. Other harbours shared particular difficulties, such as the river ports of Greymouth, Westport, Gisborne and Whanganui and the bar harbours of the north, the Hokianga, Kaipara and the Manukau (the site of New Zealand's worst wreck in 1863 when the *Orpheus*, bringing men and supplies to the New Zealand Wars, was caught on the sandbar and 189 people died). Apart from these 'graveyard harbours' there were generic problems. For a long time the coasts were poorly charted and until the 1860s lighthouses were few; and of course before roads and railways, most colonial settlements were on the coast rather than inland. The greatest problem was, as in the Timaru examples, that these were sailing ships exposed to New Zealand's fierce winds and impossible to manage when there was no wind at all.

The second question is why, faced with such difficulties, did ships continue to be used? Very simply, nineteenth-century New Zealand was a dependent colony reliant upon sea transport for the migration of its people and its economic welfare. From the beginning of European settlement non-Māori people had sought to sell goods overseas and had found the need to import things to continue the lifestyle they had been used to on those islands on the other side of the globe. The sea was the essential highway of economic and cultural survival. The Timaru wrecks point up this dependence. The *Collingwood* had been loading wheat to take to Australia for sale; the *Susan Jane* had come to Timaru to drop off coal, sourced from the mines of Newcastle, for use in homes and local factories. Even when goods or people needed to be transported to other parts of the country they were nearly always carried by coastal ships. Railways and metalled roads were still in the future. Early parliaments in Auckland could not begin work until the coastal sailing ships bringing the members had docked.

So the rocket apparatus was not just a 'nice to have'. It was a piece of equipment that serviced a major economic process. Fortunately, the need for such a device declined quickly. From the mid-1870s steamships increasingly replaced sailing ships on both passenger and freight routes. This made the ability to steam out to sea for protection much easier. The beach harbours such as Timaru built protective breakwaters to guard the ships in port from huge breakers. After various false starts a solid breakwater began construction in 1878 and by the 1880s was providing an excellent refuge. In addition the port acquired a sturdy steam-powered tug which could go to the aid of any vessel in distress.

After 1882 the rocket apparatus was never again fired in urgent necessity. For a time the Volunteer Rocket Brigade continued to meet. The brigade had initially functioned as one of those exclusively male uniformed groups that characterised nineteenth-century New Zealand — like military volunteers or friendly societies, and eventually rugby teams, where mateship and comradery was as important as their explicit role. Indeed, their members were apparently recruited from the old Volunteer Battery of Artillery and the rowing club. The brigade met monthly to carry out drill, although their membership also included honorary members whose function was to raise money. But after 1884 there are no more reports of their meetings. Three years later they participated among a number of other male groups in the parade to celebrate Queen Victoria's golden jubilee. Then they virtually disappear from history. In 1900 when a piece of a Timaru wreck was recovered, their services were recalled in the local paper under the heading 'A Relic of the Past'; and four years later it was conceded that few of its members still survived. What does survive is the rocket launcher itself, bearing witness to the central role of shipping and the port in the economic and cultural life of South Canterbury in the 1860s and 1870s.

36.

The locomotive *Josephine*

TOITŪ OTAGO SETTLERS MUSEUM

The coming of steam power and metal construction, the revolution which had brought some degree of safety to ships at sea, also brought a revolution on the land with the arrival in New Zealand of steam-powered railways. The 1870s were the crucial take-off for this form of land transport here and holding pride of place at Toitū is *Josephine*, the earliest surviving rail engine from those founding days of New Zealand railways. *Josephine* is our biggest object. It is an extraordinary-looking engine, a kind of push-me/pull-you with chimneys and a barrel at each end and the driver's compartment and central firebox in the middle. The only difference between each end is the handsome red and gold signage which proclaims 'Josephine' at one end and 'Fairlie patent', referring to the type of locomotive, at the other.

Although a treasured historical artefact in New Zealand, *Josephine* is a comparative latecomer internationally. In the United Kingdom as early as 1830, Robert Stephenson's 'Rocket' had begun operating on the Liverpool and Manchester Railway; and by 1870, trains were common in the old country. Many immigrants to New Zealand would have begun their journey by travelling to the departure port by train. But here people continued for a long time to use their own legs or those of a horse for short trips and cartage on land, and sail in ships for longer journeys. The continued importance of the sea meant that the first steam-powered railways in New Zealand were intended to take people and goods to and from ports. The very first steam train (as distinct from the horse-drawn Dun Mountain railway in Nelson) was the *Pilgrim*, which began running between Christchurch and the wharf at Ferrymead in December 1863. Almost exactly four years later Christchurch finally solved the problem of its distance from the port of Lyttelton with the first train through the new tunnel. The same year saw Invercargill linked to its port of Bluff with a 27-kilometre iron track. *Josephine* was brought to New Zealand in 1872 for the same purpose — to link Dunedin with Port Chalmers.

The initiative for the Dunedin–Port Chalmers line was a private one, and the company, Proudfoot, Oliver and Ulph, chose Robert Fairlie's firm to design two engines which were then built at the Vulcan Foundry in Lancashire. The engines were unusual in that all the driving wheels (four at each end) were mounted on swivelling bogies intended to cope with the steep grades and severe corners in New Zealand. The two engines were named *Josephine* and *Rose* by Richard Oliver, one of the company's partners. They were brought out to Port Chalmers in pieces and then assembled, all 28 tons of them, on the wharf. Accompanying the locomotives on board was Jack Thomas from Bristol. It was he who drove the *Josephine*, after a celebratory whistle, on her first trial run to Sawyers Bay on 10 September 1872; and again on the first goods journey a week later when the train brought three hogsheads of beer from Burke's brewery into Dunedin. The engine was then used to help construct the rest of the 13-kilometre line, which was formally opened by the Governor, Sir George Bowen, on 31 December 1872 with a return journey from Dunedin to Port Chalmers and a luncheon. Sadly the hall was too small to admit ladies, who were instead given seats in the train's carriages to await the return of their well-fed men!

COLLECTION OF TOITŪ OTAGO SETTLERS MUSEUM

Originally the line had been surveyed as 4 feet 8½ inches wide; but then it was decided to conform with the 1870 Railways Act so the gauge was built as 3 foot 6 inches. *Josephine* became the first train to run using these dimensions in New Zealand. In this respect she really was the starting point of a new era. For the 1870 Act signalled the effective beginning of railways in the country. At that date there were only 46 miles of rail, all in the South Island. But the Colonial Treasurer Julius Vogel, a London Jew who had worked on the Victorian goldfields, and then came to Dunedin where he started and edited the *Otago Daily Times*, had a vision of a transformed New Zealand in which the railways played a central role. He proposed that the government borrow £10 million by 1876 to fund 1000 miles of railway and bring from the United Kingdom thousands of immigrants to work on the railways and exploit the development which the iron pathway opened up. He envisaged railways connecting farms and businesses to the ports, thus encouraging a huge growth in exports. They would also allow the exploitation of lands in the North Island confiscated from Māori. The railways in frontier New Zealand became the engine of growth as had happened in North America. As part of this scheme the government decided to investigate the most suitable gauge. It was decided that shortage of capital would handicap the construction of tunnels, earthworks and bridges, which might be needed because New Zealand was hilly and with many rivers and gorges. Instead it was suggested that routes be circuitous to avoid obstacles. A narrow 3 foot 6 inch gauge was a way of coping with the resulting steep inclines and tortuous curves. In this way the *Josephine* with its swivelling bogies and narrow gauge was consistent with the new vision.

Within four months of the Port Chalmers line opening, the government had purchased it from its private developers and handed it over to the Otago province. Three years later it was absorbed into the central government's national system. Vogel's vision in the 1870 Act very much came to fruition. The railway did connect farms and mines to the ports facilitating the export of wool and gold, and opening up formerly Māori lands for settlement especially in the North Island. As Te Kooti said at the time, 'the whistling God of the Pakeha . . . a ngarara . . . belching flames, and smoke' helped destroy the Māori grip on their lands. The state became involved not only in running the railways, but also through the Public Works Department building the lines, and through new workshops servicing the engines and rolling stock. Thousands of immigrants were brought in to build the lines. Cities, now connected to their hinterland, grew; and by uniting the country in a large-scale system the railways helped create a nation. Indeed, the unwillingness of people in Canterbury and Otago with provincial loyalties to allow the central government to reserve millions of acres along rail routes as a guarantee for Crown loans was one of the factors that convinced Vogel and his supporters to push for the abolition of the provinces — a measure that finally happened in 1876.

By 1880 there were 1200 miles of rail track in the country, three quarters in the South Island. In 1878 Christchurch was linked to Dunedin by rail, and the next year to Invercargill. In the north, by 1875 a line had gone from Auckland to Mercer and the lands of the Waikato beckoned; while Wellington sought to exploit the grasslands of the Wairarapa with the famous Rimutaka Incline opened in 1878. It was not until 1908 that a main trunk finally connected Auckland and Wellington.

The *Josephine* had its place in these later developments. With the abolition of the provinces she became part of the national network, and while her partner *Rose* was damaged irrevocably in an accident in 1878, *Josephine* had another (somewhat ignominious) moment of significance. When the first train went from Christchurch to Dunedin in September 1878, she was pulled initially by a Yankee 'K' engine; but at Ōamaru *Josephine* was added to 'double bank' the train. While the guests banqueted, there was a fierce argument as to whether the newer American locomotive or *Josephine* should lead; but the driver of the Yankee held back and *Josephine* made hard work of the load. Eventually at Seacliff she was cut free and left behind. Her design with two driving mechanisms on swivel bogies made it very difficult to ensure that the steam and exhaust pipes remained intact, and 'Old Joss' as she was known, became regarded as 'flighty' and unreliable.

Such problems did not mark her end. In 1883 *Josephine* went north to work on lines around Whanganui under the name of E-175; and then 17 years later she was taken over by the Public Works Department, given a new name as PWD 504, and used in railway construction. She made a contribution to the building of sections of lines in central Otago, and then from 1903 at Mangaweka she worked on the North Island main trunk line. In 1917 she was sold to the Otago Iron Rolling Mills as scrap, but only survived out of sentiment, until one of the company's boilers collapsed and *Josephine* was hauled in to supply steam. Even then her days were not over. She became an exhibit at Dunedin's 1925–26 New Zealand and South Seas International Exhibition, and at the end was presented to the Otago Early Settlers' Association. Placed outside the museum, the locomotive began to rust and deteriorate until in 1966 a public appeal raised money to restore her. Dunedin's Hillside railway workshops, one of the operations set up by the 1870s vision, restored her and returned her to the Early Settlers, who today give her pride of place at Toitū. She stands there embodying so much of the history which brought the railways to New Zealand.

37.

Elisabet Engebretsdatter's hair embroidery

DANNEVIRKE GALLERY OF HISTORY

With gold returns declining in the South Island and the North still suffering from a decade of war, the development of a train system was one leg of Julius Vogel's 'Think Big' vision to restore New Zealand's growth. The other leg was a massive increase in immigration. This would provide labour to build the train tracks and to exploit the new lands confiscated from Māori and opened up by the railways. This unusual piece of hair embroidery, proudly displayed at the Dannevirke Gallery of History, speaks eloquently to the new migration of the 1870s.

The embroidery was the work of Elisabet Engebretsdatter, who with her husband, Bernt Johannessen, was among the very first group of 18 Scandinavian families to be brought to New Zealand. Born in 1844 in Nordli in central Norway, about 500 kilometres north of Oslo, Elisabet had been engaged to Bernt, who was working on a farm, for three years and was expecting her first child. Conditions in rural Norway were tough and there had been a steady stream of about 12,000 migrants a year leaving for the United States where many settled in the Midwest.

In August and September 1870 Isaac Featherston, soon to become New Zealand's first Agent-General in London and tasked with recruiting migrants for Vogel's scheme, toured Norway, Sweden and Denmark. A former Danish premier, Bishop Ditlev Monrad, had shown the potential of Scandinavian settlement when he arrived at Kārere (near Longburn) in the Manawatū in 1866 and his sons and friends were effective in clearing the bush. Assisted by Monrad, Featherston sold the idea of migration to New Zealand to people in Scandinavian countries. He had in mind that they could help clear the huge 70-mile bush in southern Hawke's Bay and northern Wairarapa, and also the forests of Manawatū. Having cut down trees, they would lay down rail tracks and open areas for farming. Featherston was enthusiastic about the suitability of Scandinavians, especially Norwegians, for this task. He noted that Norwegians were chiefly agricultural labourers or farmers, but were also excellent axemen and sawyers, and because of the long winters they were skilled in various trades such as carpentry. They were well educated, 'extremely honest, frugal, and industrious'; and 'in habits, manners, and customs resemble very closely our own countrymen, especially the Scotch'. In sum they seemed the perfect migrants for opening up the southern North Island; and this judgement was supported at home. The *Wellington Independent*, pointing to the contribution of the Vikings to Anglo-Saxon culture, noted, 'What the hardy Norsemen of the 8th, 9th, and 10th centuries did for our forefathers at that end of the world, our Danish, Norwegian, and Swedish contemporaries promise to do for the "Great Britain of the South" in the 19th and 20th centuries.'

Featherston's visit must have attracted the interest of Elisabet and Bernt, keen to improve their lot. On 3 October 1870 they were married in a Lutheran church; and two days later sailed for Gravesend. There they boarded the *Celaeno* along with another 51 people, all but four of whom were from Norway. After 124 days at sea they arrived in Wellington in February 1871, were temporarily housed in the Mt Cook barracks, and then taken by boat to Foxton and up the Manawatū River. Their boat ran aground, but the children and supplies were loaded in drays and the adults walked to Monrad's old stamping ground of Kārere. There, just at the south-west end of Papaioea, which would become Palmerston North, Bernt and Elisabet took up 40 acres. Bernt worked 8 am to 5 pm helping to build roads and laying the rail track from Palmerston to Foxton, and in his 'spare' time he built a slab hut, cleared his land and began farming. Elisabet kept chooks and a cow which wore a bell so Elisabet could quickly find it in the bush for milking. She made all the family's clothes and would walk into Palmerston North for supplies. Their first-born, Johan, had been born on the *Celaeno*, and another nine children followed in the following years. The family was not always successful. The land was prone to flooding and eventually Bernt left his original block and took up another 40 acres just north-west of Palmerston on the Rangitīkei line.

The family was sufficiently content, however, that in 1875 Bernt became the first Scandinavian to be naturalised and at the same time

DANNEVIRKE MUSEUM

he Anglicised his name to Johansen. It cannot have been easy leaving behind memories and family in far-off Norway. To some extent the adjustment was made easier by further migrations from Norway and other parts of Scandinavia. By 1878 there were 1213 people born in Norway living in New Zealand, along with 1162 born in Sweden and 2225 in Denmark. Some of these were in the Manawatū but more were over the Tararua Hills in southern Hawke's Bay (at Dannevirke and Norsewood) or northern Wairarapa near Eketāhuna. There was also the Lutheran Church to give comfort and fellowship. But Elisabet especially must have felt a degree of nostalgia for the traditions and people left behind across the other side of the globe.

Perhaps this hair embroidery was one response. In Norway the long winters had encouraged women to spend time in sewing and handcrafts. Featherston had reported that, 'During the long winter months, the women employed themselves in spinning and weaving', and assert 'with no little pride, that every article of clothing worn by the men, from the knitted woollen shirt to the warm blue frieze coats, was entirely of home manufacture'. Using a person's hair for embroidery had become a common practice in Nordic countries, especially in the nineteenth century. A Swedish proverb runs that 'rings and bracelets of hair increase love'. So in part Elisabet was keeping alive an old Scandinavian tradition. Hair jewellery was particularly used to remember those who had died.

It seems likely that the occasion for this work was the death of one of her children, Mary, in 1886 — hence the date. It is probable that some of Mary's hair was used in the work. Also visible are the initials of other members of the family — 'J' for Bernt Johannessen and 'E' for Elisabet, the two parents, which sit immediately below the central cross; on the left, 'J' for Johan, the boy born on the *Celaeno*, and 'L' for Lauritz; above the date, 'A' for Ann and 'O' for Ole who died in infancy; and on the right, 'E' for Embret, 'M' for Mary (who died as an infant) and two others who were also named Ole and Mary, after the dead children (it was Norwegian practice to reuse a name when an infant died). The number of children, who eventually totalled 10, and the high incidence of infant deaths were characteristic of colonial New Zealand families.

Both Bernt and Elisabet had long and ultimately successful lives in the Manawatū. In 1903 Bernt set up in business, 'B. Johansen and Son', as a wood and coal merchant on Main Street East of Palmerston North. He died in 1924 aged 80. Elisabet outlasted him by four years, dying in 1928 aged 84. The two were buried beside each other in the Lutheran section of the Terrace End Cemetery in Palmerston North. As for Elisabet's hair embroidery, in the belief that it was a seed mosaic, the object was destined for disposal at the Dannevirke dump. But a Danish visitor suggested that in fact the material was human hair; so instead the object was saved, restored, and today continues to express one Norwegian/New Zealand family's feelings about death and migration.

38.

Ōkaramio blacksmith bellows

HAVELOCK MUSEUM

These blacksmith's bellows, today held in the Havelock Museum, are huge — over 2 metres long and over a metre wide. They once belonged in the blacksmith's forge at Ōkaramio, about 12 kilometres south of Havelock on the main road to Blenheim. It seems probable that they were proudly acquired in the 1880s by the local blacksmith at the time, Tom Maxted. We can tell the approximate date because there is a copper label on the top of the bellows bag which reads: 'William Alldays Patent . . . Alldays & Onions Ltd Sole Makers Birmingham, England'. We know that William Allday and Sons merged with John C. Onions in 1885, and was then reorganised as Alldays and Onions Pneumatic Engineering Co. in 1889. The 1880s date is suggested too by painted words also on the bellows: 'Medals Each time of Exhibiting . . . London 1880'.

Why did Tom Maxted and the blacksmiths who followed him require such a huge piece of equipment? The problem was that iron needs to become red hot before it can be shaped into a tool or horseshoe of the right size, and simply placing the metal into a coke fire does not achieve that heat. So a bellows was used to direct air to the centre of the fire and create a spot hot enough to allow the iron to be shaped. The smith would pump the bellows with one hand and manipulate the iron with tongs in the other. When it was glowing red (or even white if he wanted to weld a piece of farm equipment), the smith would lift the metal object from the fire and take it to his anvil for working. To protect himself he would wear a thick leather apron. If as was most likely he was shoeing horses, he would first have removed the old shoes and trimmed the hooves. Horses' hooves are adapted for grazing on grass, but working horses and those which carried riders often worked on rough roads. Without protection the hooves would split. But hooves, like a person's nails, keep growing, so every six to eight weeks the blacksmith needed to file them down and fit new shoes. Once he had shaped the shoes on the anvil, the blacksmith would plunge them into water to cool, although they were often put on quite hot and would singe a little as they fitted onto the horse's hoof and were nailed tight. Thus the blacksmith's shop was the car mechanic's of the nineteenth century — only more so, because a six-week turnaround meant frequent visits there for people using horses and the shop often became, especially in the cold of winter, a warm centre for community gossip.

What gave these bellows and the blacksmiths who used them such importance was that horse-power was the key form of harnessed energy in the nineteenth century. Horses, in the form of one stallion and two mares, first landed in New Zealand at Rangihoua in the Bay of Islands on 23 December 1814, to the amazement of the locals. When Marsden

HAVELOCK MUSEUM

mounted a horse and rode it up the beach, J. L. Nicholas recalled that, 'To see a man seated on the back of such an animal, they thought the strangest thing in nature; and following him with staring eyes, they believed at the moment that he was more than mortal.' From then on, by Māori and Pākehā alike, horses came to be regarded as essential to the development of the country. They became so important that prices for a good horse rose rapidly. By 1856 a saddle horse cost £30, a cart horse £50. They were brought in from Australia, and some from as far away as the United Kingdom. By 1861 there were over 28,000; by 1881 over 160,000 — or about one horse for every three people.

As at Ōkaramio, the major use of horses was in farming. In that area of Marlborough, sheep-farming with some dairying was the main form of agriculture in the late nineteenth century. Horses would be used by the farmers to help round up sheep and drive them to the shearing shed; or would be employed to haul hay for feeding out to the dairy cows in winter. In other parts of the country such as the Canterbury Plains from the early 1880s, there was considerable growing of arable crops, especially wheat. This called for ploughing. Until mid-century, oxen had often been preferred for such a purpose. They were cheaper to feed than horses, less prone to disease and injury, and could eventually be sold for beef. But with their greater speed and mobility, draught horses had taken over by the 1860s. The A&P (agricultural and pastoral) associations and shows encouraged good breeding and held ploughing matches to refine farmers' skills. Horses became essential for ploughing obviously, but also for harrowing, drilling and harvesting. They were also used frequently for carting — manure around the farm, milk to the factory, wheat and wool to the port or railway station.

So the big bellows in the Ōkaramio blacksmith shop were crucial to the activities of local farms in keeping the horses working properly and helping fix farm equipment such as ploughs or harrows. But blacksmiths also serviced another crucial nineteenth-century role for horses — in the transport industry. The original Ōkaramio blacksmith, Tom Maxted, sold the operation to Edward Hart in 1888. Hart had originally learnt his trade in Brightwater where he was a mate of Tom and Henry Newman. From the early 1860s, Cobb and Co. coaches pulled by horses had begun operations in Otago, and such services spread throughout the country. In 1891 the Newman brothers followed suit and began a coach service from Nelson to Blenheim, deciding that Ōkaramio would be a good place to change the horses on the route. They built a stable there and asked their old friend Edward Hart to shoe their horses. He charged them 6 shillings for each horse. Perhaps on the basis of such custom, Hart began to do well. In 1900 he built a new wooden smithy and outside the door planted a chestnut tree, as a nice nod to the old Henry Longfellow poem 'The Village Blacksmith', which runs, 'Under a spreading chestnut-tree / The village smithy stands; / The smith, a mighty man is he, / With large and sinewy hands'.

Edward Hart's gesture was a characteristic act of rural nostalgia for village life in the homeland across the sea; but horses also had a hugely important place in the city, a long way from rural Ōkaramio. There, horses became essential for local transport. Aided by the development of mechanical springs from the 1870s, elegant landaus and four-wheeled ladies' phaetons appeared. Only the rich could afford their own stables and coach houses, but other people travelled around the city in four-wheeled hackney cabs or two-wheeled hansoms. In bigger centres like Christchurch, Nelson and Wellington, horse-drawn omnibuses emerged. There were also of course many delivery carts in the city used by butchers, bakers and milkmen. With every horse producing about 6 to 7 tons of manure a year, the cities were forced to hire an army of street cleaners to remove the droppings.

Finally, horses were also used in nineteenth-century New Zealand for recreation, occasionally for games of polo by the affluent or show jumping at the local A&P shows, but more significantly for horse-racing. As early as the 1840s, European settlers would commemorate their anniversary days with horse races; and from the 1880s, especially following the introduction of the totalisator in that decade, there was a huge growth in horse-racing. The 'races' became central to community life in New Zealand in the twentieth century.

Horse-power would not remain so important. The spread of train transport, then the automobile, and the invention of tractors, meant that from the First World War horses began to disappear from both the cities and rural areas of New Zealand. In 1918 there had been over 380,000 horses in the country; 40 years later the number was under a third of this figure — some 123,000. As for the Ōkaramio blacksmith, the shop was sold to Ern Nicholas in 1912, who in turn sold it in the 1940s, and it was closed soon after. The need for a smithy in Ōkaramio had passed; and the bellows which had given over 50 years of service in repairing farm equipment and shoeing horses were needed no more. We are fortunate they were given to the Havelock Museum as a powerful reminder of the central role once played by the horse in rural New Zealand.

39.

Parihaka plough/parau

PUKE ARIKI

The 1880s opened with one of the most shameful episodes in New Zealand history — the invasion on 5 November 1881 of the peaceful Māori community of Parihaka in Taranaki. To represent this event we cannot do better than this sturdy iron parau, or plough, now in possession of Puke Ariki in New Plymouth. Information from the museum tells us that the parau 'was said to have been used to plough as part of the Parihaka passive resistance' and that it once marked the burial site of a Parihaka ploughman who was buried at Kawau pā urupā (burial ground).

Ploughing began in New Zealand when John Butler, a missionary at Kerikeri, recorded in his diary: 'On the morning of Wednesday the 3rd of May 1820 the agricultural Plough was for the first time put into the Land of New Zealand.' He hoped that the anniversary of the day would be kept 'by ages yet unborn'. Butler's plough was wooden and drawn by a team of six bullocks. Wooden ploughs continued to be used especially by Māori in their work growing wheat for market in the 1840s and 1850s. But from the 1860s iron ploughs began to be imported or increasingly made in New Zealand; and they were drawn by horses rather than bullocks. In the South Island, heavy iron ploughs with multi-furrows made possible the boom in grain-growing in the 1870s and 1880s. So this simple plough pulled by horses is exactly the type of plough which the Parihaka ploughmen would have used. Reports from that area at that time record ploughing with horses with about four people for each horse.

Whether or not this plough was actually used in the Parihaka ploughing, it was certainly believed to be the case and this object has become heavily encrusted with symbolic Parihaka meanings. It was the centrepiece of the famous exhibition *Parihaka: The Art of Passive Resistance* in 2000 at the City Gallery in Wellington which brought the Parihaka story to wider consciousness, and it had a similar central position in the Taonga Māori gallery at Puke Ariki. The plough was put on display in front of the table where Te Huanga ō Rongo (the deed of reconciliation between the Crown and the people of Parihaka) was signed on 9 June 2017. This plough deserves its status as a taonga with deep Parihaka associations.

Why was a plough, a pre-eminent object of European civilisation, so symbolic of this event? The story begins with the wars in Taranaki. After war had broken out over the Waitara 'purchase' in 1860, a truce was agreed the following year. But in May 1863 the truce was broken when Crown troops occupied land at Ōmata and Tātaraimaka and in response Māori ambushed British troops who were on Māori land at Ōakura. From that point warfare spluttered on with varying degrees of intensity until 1869 when Tītokowaru's forces, who, as James Belich noted, 'had brought the colony to its knees', abandoned the great warrior at Tauranga-ika near Waitōtara.

As part of their strategy, the New Zealand government decided on a policy of confiscating Māori land. The idea was that the threat of losing land would deter Māori, and that once confiscated the land could be settled by Europeans and a frontier peace imposed. Huge areas of the Waikato, the Bay of Plenty and Taranaki were declared confiscated. In Taranaki the area was 1.275 million acres (slightly more than in the Waikato). But the Act establishing the process stated that the land needed to be settled by Europeans within a specified time for the confiscation to apply. Further, those who had remained loyal during the conflict were to be compensated with the return of lands, while other Māori were to be given reserves so that they would not remain destitute. None of these things happened. The land was declared confiscated but for over a decade it was not occupied by Europeans and Māori did not receive back the lands they expected.

Meanwhile, two Māori leaders, Tohu Kākahi and Te Whiti-o-Rongomai, having been pushed out of their coastal home at Wārea near Cape Egmont and then repeatedly attacked at other settlements, decided in 1866 to move inland and establish a Māori community at a site about halfway between the mountain and the sea. It was obviously on confiscated land. Both men were of Te Āti Awa and Taranaki tribal background, both had studied the Bible and had

A95.932 PARAU, COLLECTION OF PUKE ARIKI, NEW PLYMOUTH

strong spiritual gifts. Both had been sympathetic to the Māori King, but had come to accept that fighting provided no solution and were happy to work in partnership with Pākehā so long as they retained much of their traditional land. Neither took part in Tītokowaru's war. Accepting that Europeans would settle south of the Waingongoro River, they saw their new settlement of Parihaka as a self-sufficient Māori community which would develop its own economy and way of life, and be recognised by Pākehā. Before long, disillusioned Māori from throughout Taranaki — those whose land had been confiscated, kūpapa (allies of the Pākehā) who had never received compensation or who were tired of the war — gravitated to Parihaka. By the late 1870s there were several thousand Māori living there and the place was regarded as the most prosperous and peaceful Māori community in the country. For over a decade the people of Parihaka controlled the area of central Taranaki from Waingongoro to Hangaatahua and operated a toll gate on the road through it. But in 1878, with growing numbers of migrants pouring into the country, the government decided to enact the confiscation. In July 1878 surveyors crossed the Waingongoro River. At first the Parihaka community was relaxed, believing that Māori reserves would also be marked out. But when in early 1879 road-makers cut through Tītokowaru's grass-seed crop and burial grounds and the Native Minister refused to discuss reserves, Te Whiti and Tohu had had enough. They believed, quite correctly (as the Waitangi Tribunal has shown), that confiscation was not valid after a decade of Māori occupation and they were angry that there had been no provision for Māori reserves. Te Whiti told the surveyor James Mackay, 'you wish to take the whole of the blanket and leave me naked'. So in May 1879 Te Whiti ordered 20 unarmed men to plough his moko into land at Ōakura, the very spot where the second Taranaki War had begun. Within weeks the ploughing proceeded on settler-occupied lands which had been confiscated on the coast between Pukearuhe and Hāwera. The act of ploughing was full of symbolic meaning. The tool that was used was not a traditional Māori tool such as the digging kō. The plough was a European implement, which demonstrated a message to both peoples about ownership. Ngāti Mutunga ploughed Mutunga land, Ngāti Ruanui their land. Ploughing picked up on Donald McLean's suggestion to Māori after the war ended, 'Let your future fighting be with the soil . . . return to the land not as strangers but as children of the soil.' The act was, as Te Whiti said, 'ploughing the belly of the Government', but it was also a deliberate act of peace, a turning of swords into ploughshares as the missionaries had urged.

This highly symbolic act did not go unnoticed. Within weeks, on 29 June 1879, George Grey, now a minister not the Governor, ordered that the ploughmen be arrested. By 6 July, 105 were in custody; by early August there were 170 imprisoned in Wellington and 25 in Dunedin. Eventually over 400 ploughmen, among the most distinguished of the Parihaka community, were in prison. The men were detained without trial, for as the Minister for Native Affairs John Bryce said, Magna Carta and habeas corpus were 'mere legal technicalities' of interest to lawyers not statesmen.

By mid-1880, with the prisons bulging, the government tried new tactics. They surveyed and offered for sale the seaward Parihaka block, the most fertile area around Parihaka and land which Te Whiti's Taranaki ancestors had cultivated for centuries. The government also began to build a road from Hāwera to New Plymouth which would pass through Parihaka land. Soldiers were ordered to cut through crops recently planted and break down the fences which surrounded them. This time Te Whiti and Tohu's response was not to send out the plough, for this land was already in cultivation. Rather, they sent out fencers to calmly and peacefully make repairs. They too were arrested and by September over 200 were in prison, also held without trial.

There is nothing so annoying for an aggressor than to find his opponent responding with the calm peacefulness of the plough rather than in violence. Whether it was this, or a growing sense that Te Whiti and Tohu were dangerous fanatics, but by November 1881 the government and especially the Native Minister John Bryce were determined to assert authority. Te Whiti knew what was in store, but insisted, 'Though the lions rage still I am for peace.' Almost 1000 volunteers around the country and over 600 Armed Constabulary were summoned, and a cannon was hauled to the top of the hill, Purepo, which overlooked Parihaka township. On 5 November 1881, soldiers marched into Parihaka led by John Bryce with a sabre and full military uniform on a white charger. They were greeted by 2000 Māori sitting on the marae in their best clothes. The children were singing waiata, the women were greeting the soldiers with loaves of warm bread. The leaders were arrested. Tohu noted, 'We looked for peace and we find war'; Te Whiti was charged with being a 'wicked, malicious, seditious and evil-disposed person'. Women were raped, the township of Parihaka was destroyed, its taonga were looted, and the scorched-earth programme extended to the cultivations around.

It is true that eventually reserves were provided for Taranaki Māori, but ironically these were vested in the hands of the Maori Trustee who usually offered the lands in perpetual leases to European farmers. The wealthy soil of Taranaki passed from Māori to Pākehā. But protest by the plough was not forgotten. In 1886 ploughmen were found working fields farmed by settlers near Pātea and once more Te Whiti, Tohu and Tītokowaru were arrested. Eleven years later, 92 Māori were sentenced for ploughing land near Hāwera. New Zealand has not invented many forms of civil action which warrant the attention of the world; but the peaceful act of ploughing as a form of civil disobedience is one that deserves to be remembered. The Parihaka parau tills very rich soil.

40.

Totara Estate killing knives

HERITAGE NEW ZEALAND POUHERE TAONGA

Cutting the throat of a sheep with a knife was once a coming of age for Kiwi blokes, and it was certainly for almost a century an essential first step in one of New Zealand's most important sources of wealth — the export of refrigerated lamb and mutton to the United Kingdom. This trade began on 6 December 1881 when six butchers at Totara Estate, near Ōamaru in North Otago, began slaughtering sheep. Some of these knives, imported from Sheffield, the great British home of knife-making, were used to kill the sheep, skin them, disembowel them, behead them, and finally cut away the surplus fat. It was a highly skilled job and the butchers prepared about 40 sheep each day, making a daily total of 240. The carcasses were then taken by a farm cart to the local station, loaded onto a train with a central block of ice, and transported to Port Chalmers. There they were cooled, sewn into a calico bag and taken to the lower hold of the sailing ship *Dunedin* where they were frozen and eventually transported to London. Thus began one of the most important economic revolutions in New Zealand history.

The origins of this trade lay in the need to find a new economic use for sheep products. During the 1850s and 1860s the sheep flocks of New Zealand, and especially the South Island, had expanded fast in response to the international demand for wool. But over the next decade the price of wool fell in part because the new process for making worsted cloth could not handle the short fine merino wool. Farmers also found that there was a limited local demand for the meat of merino sheep which was tough and of poor taste. The usual response was either to drive old sheep over a cliff, or into pits for burial, or to boil up the flesh for tallow used in making soap and candles, but this offered poor returns. There were some experiments at salting and canning meat but neither was well received overseas. Yet there was a potential market in the United Kingdom. The British regarded meat-eating, and especially the consumption of mutton, as one of the distinctive characteristics of their civilisation. The Sunday roast was a central ritual of their culture. As the British population rose, people worried about the 'meat famine'.

So how to get New Zealand's surplus sheep meat to the British housewife? The Americans had chilled meat and brought it across the Atlantic cooled by ice, but the long sea voyage from New Zealand through the tropics made this impossible. The solution was refrigerated shipping using a steam-powered freezer unit which worked by compressing air. In 1877 there had occurred the first successful long-distance shipment of frozen meat — from Buenos Aires to France — but the real precedent that awoke New Zealand interest was the sailing of SS *Strathleven* from Sydney to London in 1879–80.

The 'Strathleven experiment' attracted the attention of William Soltau Davidson, the general manager of the New Zealand and Australia Land Company, which held over 3 million acres in the South Island. From a Scottish banking family, Davidson had worked on estates in Canterbury and Otago in the 1860s and 1870s, and as he rose in the company he became concerned about the lack of profitable outlets for the land's produce. Refrigerated shipping seemed a solution and he determined to try it. From his Scottish base he convinced the Albion Line to add a refrigerated hold to a passenger sailing ship, the *Dunedin*. Davidson chose a coal-powered Bell Coleman freezing plant which cooled the hold to 22 degrees Celsius below the outside temperature. It made for a very strange-looking boat with a large funnel for the refrigeration placed between the sails. Davidson delegated to the company's New Zealand superintendent, Thomas Brydone, also a Scot by background, the New Zealand side of the experiment. Brydone built slaughtering facilities at the Totara Estate and organised the transport of the carcasses to Port Chalmers.

The loading was not entirely smooth — at one point the crankshaft of the refrigeration machine broke down, spoiling over 600 carcasses, but finally on 14 February 1882 the loading was finished and the *Dunedin* sailed the next day. Davidson had travelled there to see the ship off; and he beat the ship back, presumably travelling by steamer, to greet it in London. The voyage lasted 98 days but was not without drama. Sparks from the refrigeration machinery set the sails alight, and on another

TOTARA ESTATE. IMAGE BY JOCK PHILLIPS.

occasion ducts for cold air became blocked and the captain had to crawl down to clear the ice and almost froze to death. In the refrigeration chamber were 4331 mutton, 598 lamb, and 26 pork carcasses. To Davidson's delight they were in good condition on arrival and their healthy colour, particularly by comparison with the dark-coloured meat from Australia, attracted good prices at Smithfield market. The company made a profit of £4700 on the voyage. It took some years before the trade was firmly established, but by 1895 New Zealand was exporting 2.3 million sheep carcasses, and by 1910, 5.8 million. By then there were 28 freezing works in the country. Increasingly, lamb rather than mutton became the chosen export. As for the *Dunedin*, it was lost in the Southern Ocean on its tenth voyage to London in 1890.

The consequences of a profitable refrigerated meat trade were huge. Small farms became economically viable, and helped encourage the Liberal government to break up the large estates and purchase several million acres of Māori land. In the South Island, tussock was ploughed under and replaced with European grasses. In the North, sheep-farming spread to the hill country where bush was cleared for pasture. Blood-and-bone, a product of freezing works, fertilised the soil. As meat became a profitable element of sheep-farming, there was encouragement for the development of new breeds which were suitable for both wool production and good lamb meat and mutton. There had been some experiments in cross-breeding in the South Island before 1882, encouraged by William Davidson; but now they became more widespread. Near Ōamaru on his Corriedale estate, James Little led the experiments and by cross-breeding merino ewes and Lincoln rams he evolved the Corriedale flocks which matured early to produce tasty meat and excellent long wool. Unlike the pure merino wool, Corriedale fleeces were suitable for the fashionable worsted cloth. Elsewhere the Romney was imported from southern England to produce quick-maturing meat carcasses. With an export trade in both wool and sheep meat, flocks grew — from a national total of just under 13 million in 1881 to over 25 million by the First World War.

The sheep trade to the United Kingdom brought increasing affluence from the turn of the century and helped lock New Zealand more firmly into an economic and cultural relationship with the old world. From 1900 to the 1950s over 80 per cent of the country's exports by value went to the United Kingdom. Furthermore, the use of refrigerated shipping attracted the interest of dairy farmers. Indeed, Davidson and Brydone had seen these possibilities for the export of butter and cheese and had encouraged the development of the nation's first dairy factory at Edendale in Southland. But that is another story.

The story told by these knives now adorning the walls of the killing shed at Totara Estate is of the beginnings of refrigerated meat exports, a revolution which transformed New Zealand's economy and its relationship with the world.

COLLECTION OF TOITŪ OTAGO SETTLERS MUSEUM

41.

Gaelic Society targe

TOITŪ OTAGO SETTLERS MUSEUM

This striking round shield, or targe, was a symbol of the founding in 1881 of the Gaelic Society in Otago. The society had been established to encourage the perpetuation of Scottish Highland culture in New Zealand, of which the most important element was the Gaelic language. A targe was chosen because it was an important piece of battle equipment for Highlands infantry from about the thirteenth century to the Battle of Culloden in 1746. The targe, of which the diminutive gives us the common term 'target', was normally about 500 millimetres in diameter, with a concave face. It was made of two pieces of wood laid crossways, and was covered by leather which was fixed by brass studs often, as in this case, forming a design. At the back were two handles, one adjustable with buckles for the left forearm and the other gripped by the left hand. The targe was the main form of defence against swords and lances, but when a metal point was attached to its centre it could also become an effective offensive weapon.

The coming of muskets really spelled the end of the targe and the moment of truth was at the Battle of Culloden. This was the culminating act of the Jacobite rebellion. Bonnie Prince Charlie had led a movement to replace the Hanoverian King, George II, with a restored Stuart monarchy. The Jacobite infantry went into battle carrying targes, but they were no match for the gunfire of the English forces; and soon after the Scots outlawed their use in future battles.

The Jacobites did not all fit the stereotype of being Catholics from the Scottish Highlands. They included many Episcopalians and some Lowlanders. But the mythology developed that the battle was the final defeat of Highland independence; and the targe became a stirring symbol of Highland culture. This targe was never used in war. It was a purely symbolic shield, made in Dunedin by a local craftsman and decorated appropriately. Around the border are images of the thistle of Scotland, in the centre there are St Andrew's Crosses, and in between are words in Gaelic: 'Comunn Ghaidhealach, N.Z. 1881 Far Am Faigheadh Coiggreach Baigh' — 'Gaelic Society, N.Z. 1881 where the stranger would receive kindness'.

Why was a Gaelic Society founded in Dunedin in 1881 marked by this unusual article of war? One might think it was because there were large numbers of Gaelic-speaking people from the Highlands living there at the time who wished to preserve their heritage. Certainly, when the first notice of the intended society appeared in the *Otago Daily Times* on 5 January 1881, the claim was that 240 gentlemen had indicated their interest, and just over two weeks later 450 names had been sent in. The founding meeting stated that the objects included 'to foster and perpetuate the Gaelic language', and 'to encourage the cultivation of Gaelic literature and music'. Their rules laid down that members 'must be Scottish Highlanders or descendants of Scottish Highlanders, possessing an acquaintance with and a desire to improve their knowledge of the Gaelic language'.

Yet there were not in fact many migrants from the Highlands in Otago, let alone in New Zealand generally. True, the Scots were disproportionately represented among the non-Māori settlers of nineteenth-century New Zealand, comprising about a quarter of those arriving from the United Kingdom from 1840 to the turn of the century, and Otago and Southland were overwhelmingly more Scottish than other regions. But those from the Highlands were few. Only about 13 per cent of New Zealand's Scots came from the Highlands; and in Otago the proportion from that region was even lower. A large majority, over three fifths, came from the Lowlands of Scotland, from the urban industrial communities in and around Glasgow and Edinburgh. In 1872 only about 2000 people spoke Gaelic in Otago from a Scottish-born population of 21,400.

So it is not surprising that the major purpose of the society, to preserve the Gaelic language, was a vain endeavour. For those not born in the Highlands it was a big ask to encourage young New Zealanders to learn Gaelic. Those able to follow the language became progressively fewer in the south; and elsewhere, even in isolated communities such as among the Reverend McLeod's followers at Waipū in Northland or among the Turakina Highland families, the speaking of Gaelic largely disappeared by the First World War.

But Highland influence in New Zealand was significantly greater than this might suggest. If few came directly from that area, Highland culture remained an important folk memory. Many who stepped on the boat in the Lowlands were descended from Highland stock. In the late eighteenth and early nineteenth centuries there had been a major economic revolution in the Highlands. In place of the traditional arable pastures farmed in a semi-communal way by tenants, landlords began farming sheep. This required less labour than in the past, and it meant enclosing the fields with fences. Landlords began to push tenants off the land in the notorious 'clearances'. There were other pressures on the people of the Highlands — it was a time of growing population and opportunities for women in spinning and weaving were threatened by the emergence of textile manufacturing. The potato famine of the 1840s, which so afflicted the Irish, also spread to the Highlands. Highlanders had little choice. They packed their bags and headed away — some onto ships heading towards the United States, Canada and Australia, and others into the factories of the Lowlands. While few of New Zealand's Scots came direct from the Highlands, more were descendants of families who had moved south a generation or so before.

So when Scots people in New Zealand wanted to recall their origins and encourage a sense of Scots identity, they turned to the culture of the Highlands. The Gaelic Society was one expression of this romantic impulse. What handicapped its growth was the focus upon the language. But the society encouraged other Highland traditions which had a wider appeal. There were monthly concerts of piping and Highland dancing, and those who contributed to the music were given the honorary title of 'bards'.

Such traditions had already been taken up by other institutions. In 1862 the first Caledonian Society was established in Dunedin. Over the next quarter-century societies were established throughout Otago and Southland, in Timaru, and also in the southern North Island, especially the heavily Scots community of Turakina. The major activity of the Caledonian Societies was to promote Highland games — piping, dancing and such competitions as throwing the caber. The wearing of kilts was encouraged. While the games gave Scots a sense of home, they also attracted interest from the wider community. The most popular activities were those which had a broader appeal, such as wrestling, foot races and jumping. In the New Year of 1879 some 33,000 people attended games in Invercargill, Dunedin, Ōamaru, Timaru and Wellington. Piping began to attract support from the wider community. Growing out of the Caledonian Society, the Invercargill Pipe Band was formed in 1896 and attracted young New Zealanders of many backgrounds. The old Scots game of golf, which had been established in New Zealand by Scots migrants in Dunedin in 1871, took off among the wider community from the 1890s. By the First World War there were 73 men's clubs and 20 ladies' clubs in the country. As for the old Scots custom of Hogmanay, while some precise rituals such as 'first-footing' (in which the first person to step into a house in the New Year was celebrated) disappeared, the Scots tradition of revelry on New Year's Eve became a generally accepted part of the New Zealand calendar; and New Zealanders came to have two holidays — the English Christmas and the Scots New Year.

One other impact of the Gaelic community is best expressed by the career of Sir John McKenzie. Born in eastern Ross-shire in the Highlands in 1839, the five-year-old John had witnessed crofters and their families huddling in a graveyard after being evicted from their lands by an enclosing landlord. John came to New Zealand in 1860 and eventually, after becoming a successful north Otago farmer, entered politics, and set out to ensure that clearances could never occur in his new country. As Minister of Lands and of Agriculture in the 1890s Liberal government, he promoted policies such as a graduated land tax, the lease-in-perpetuity, and the 'busting up' through compulsory purchase of large estates — policies designed to foster widespread landholding and create an Arcadian vision of the family farm. Sadly this vision also involved extensive purchase of Māori land for use by Pākehā family farmers. Highland dreams were satisfied while Māori dreams were broken. Despite this Highland heritage, John McKenzie broke with the language of his youth. He and his wife Ann had been brought up speaking Gaelic, but in New Zealand McKenzie spoke English in the family home, diligently practised making speeches in English, and used Gaelic in public only in swearing asides. McKenzie became Chief of the Gaelic Society in 1894 until his death in 1901, but when installed he excused himself from responding in his mother tongue because 'he was more accustomed to speaking in English'.

The lesson from these examples is clear. While the Scots tried to keep alive homeland traditions to affirm their identity in a new land, those traditions which survived were the ones which could be adopted easily by people of other cultural traditions — playing music such as the bagpipes or games like golf, singing Auld Lang Syne as the New Year arrived, preventing large landholdings. But narrower traditions, such as speaking Gaelic, were much harder to sustain. The Gaelic Society soldiered on with a diminishing number of Gaelic speakers and put its energies into other Highland traditions. Finally in 2006 the society closed its doors. Its shield was handed over to Toitū where today it reminds us of the varying successes and failures of Scots Highland traditions in New Zealand.

42.

Amelia Haszard's sewing machine

BURIED VILLAGE OF TE WAIROA

In the early hours of a cold winter's morning, 10 June 1886, occurred New Zealand's greatest natural event of the century, the eruption of Mt Tarawera. The explosions were heard as far south as Christchurch and as far north as Auckland. At least 108 people died, perhaps up to 150. Among those who died were Māori and Pākehā at the tourist village of Te Wairoa some 10 kilometres from the mountain. Te Wairoa was covered in over a metre of mud and ash and today the museum at the 'Buried Village' holds items recovered from the explosion. There are no fewer than four sewing machines.

The number is unsurprising. Hand sewing was a central pursuit of women, Māori and Pākehā alike. Early missionary women believed teaching Māori girls European sewing skills was essential for mending and making clothes, and as part of the 'civilising process'. In the early 1800s, needles and thread were the tools of the trade. Hand sewing was time-consuming. But from the 1860s sewing machines became available. Efficient machines had been developed in the United States by a tailor's apprentice, Elias Howe, and then more successfully by a machinist, Isaac Singer, and dress patterns appeared in household guides such as Mrs Beeton's. So we would expect to find sewing machines at Te Wairoa. To be fair, one machine, found 4 metres up in the fork of a tree, was apparently placed there in the early 1900s; but the other three were survivors of the eruption, as can be seen from the layer of volcanic mud which encrusts them.

The machine featured here was found in the schoolhouse and belonged to Amelia Haszard, wife of the schoolmaster Charles Haszard. It seems likely that in addition to sewing clothes for her five children, Amelia taught needlework to the girls in the school. The shape of the machine suggests it was an American lock-stitch 'home' shuttle sewing machine. Advertisements for this brand appeared in Auckland newspapers every second day in the early 1880s. They were sold for £4, which at about $680 in today's terms was a considerable investment. It was a hand-turned machine rather than the more expensive foot-treadle type. It is also revealing that an iron was found close to the machine, for irons were an essential accompaniment of neat sewing.

Amelia was originally from Prince Edward Island in Canada and in 1859 came to New Zealand where she married her cousin Charles, a teacher. The couple had lived in the schoolhouse at Te Wairoa for some nine years. They had come to a Māori community of about 120 people in a considerable state of transformation. Te Wairoa had been established in 1852 by an American Anglican missionary, Seymour Mills Spencer, as a model Māori community like an English village with a schoolhouse and a flour mill. Each whare had a fenced garden. Spencer left the area when Te Kooti briefly threatened, and the focus of Te Wairoa became tourism. It was only 7.5 kilometres from the famous Pink and White Terraces, hailed as the 'eighth wonder of the world', two sets of crystallised silica which flowed down in a series of beautiful pools. The White Terraces, Te Tarata, were larger, some 240 metres wide and fan-shaped, descending gradually from a boiling lake. The water was an azure blue and the terraces themselves formed a delicate fretwork. The Pink Terraces, Ōtūkapuarangi, were smaller but the colours were beautifully subtle and they provided superb bathing with the water hotter at the top so visitors could choose their desired temperature. From the 1870s the terraces attracted many international tourists from English royalty to American moguls. The visitors would travel by coach

BURIED VILLAGE MUSEUM. IMAGE BY JOCK PHILLIPS.

from Tauranga or by train and coach from Auckland, stay a night at Ōhinemutu, and then coach on to Te Wairoa. There two hotels had been built, the two-storeyed Rotomahana Hotel and the temperance Terrace Hotel. In the evening the visitors were entertained with haka in the Hinemihi wharepuni. The next morning they travelled by whale boat across Lake Tarawera to Te Ariki landing, walked with bathing gear and packed lunch to Lake Rotomahana, then travelled by canoe to the terraces with two famous Māori women guides, Sophia Hinerangi and Kate Middlemass.

International tourism brought both wealth and disease to Te Wairoa, and Charles Haszard added other roles to his teaching. He promoted temperance, warning the locals against alcohol, and dispensed medicines. Here he came into conflict with the local tohunga, Tūhoto Ariki, who offered very different traditional remedies.

In mid-1886 Charles and Amelia were experiencing a tough autumn. Levels of sickness were high, and Amelia herself had fallen ill. Then on 31 May came reports of strange sightings by a tourist party led by Sophia. They apparently witnessed unusual surges on Lake Tarawera followed by a 'phantom' war canoe with 13 warriors which sailed before their eyes. But the night of 9 June promised happier times. The family were hosting two surveyors as guests, John Blythe and Harry Lundius. It was Amelia's birthday, one shared with Lundius, and they held a lively celebration around the fire in the schoolhouse. The five children, Clara (aged 21 and an assistant to her father in the school), Ina (15), Adolphus (10), Edna (6), Mona (4) and a nephew Charles (5) crowded around Blythe reading *Tom Sawyer*.

By midnight all were asleep. But at 1.15 am Clara was woken by an earthquake. The rumblings continued and by 2 am all were up. Columns of dark smoke and huge flames appeared above Mt Tarawera. Charles stood on the verandah and exclaimed, 'What a grand sight! — should we live a hundred years we shall never again see its equal.' There were fountains of glowing scoria from a line of fire, a 17-kilometre-long rift along the mountain, clouds of ash and shafts of lightning. Amelia was worried. She persuaded the party to move next door away from the bedrooms at the front of the schoolhouse and to the drawing room in the corrugated-iron annex which would be stronger. In the room where they had celebrated birthdays hours earlier, they lit a fire and sang hymns around the organ. About 3 am, rocks started falling on the roof and the quakes became stronger. Amelia pushed her chair against a chest of drawers with her younger children around her. Then the roof collapsed. Clara had just raced to the bulging outside wall to brace it. She was hemmed in by mud and plunged in darkness. Eventually she felt the hand of Lundius and soon after that of Blythe. The three were able to climb out and spent the rest of the night in the hen-house which astonishingly still stood.

The eruption continued through the night but tailed off just as first light appeared. There was over a metre of ash and mud covering everything. Sixty-three people were sheltering in Sophia Hinerangi's whare including all but one of the guests from Rotomahana Hotel. Elsewhere in the village 11 Māori had died. Clara believed her whole family were dead, but going into the house Lundius released Ina who had been trapped inside. Then in mid-morning two rescuers from Rotorua arrived and began digging at the schoolhouse. Pulling a woollen wrapper they saw fingers moving. It was Amelia Haszard. She had been buried for seven hours. Roof timber had pinned her leg, but also kept the mud off her face. Later she told her story. When the roof fell, Charles was killed instantly. Amelia held her youngest child Mona in her arms, but eventually mud smothered the girl. Close by was Adolphus who comforted his mother with the words, 'Mama, I will die with you', and indeed was dead soon after. Another daughter, Edna, also huddled close by and eventually died. Amelia waited with three dead children beside her. When finally freed, Amelia craved a drink, but there was no drinkable water in Te Wairoa, and on the uncomfortable journey by stretcher to Rotorua she had to make do with brandy and port wine from the hotel — not a happy solution for a dedicated teetotaller.

Amelia was not the only person from Te Wairoa dug out of the volcanic debris. Three days later the tohunga Tūhoto, allegedly aged 100, was excavated from his house after being buried for 104 hours. He died three weeks later. Elsewhere the Pink and White Terraces had disappeared and several villages which served visitors to the terraces, Mourā and Te Ariki, were wiped out and their inhabitants killed.

As for Amelia, she was granted two years' payment of her husband's salary and joined her daughter Clara who became the schoolmistress at Te Waotu school near Arapuni in the Waikato. Amelia became the school's sewing mistress. She died in Auckland aged 82 in March 1925. Her sewing machine lay covered in ash and mud for over 40 years before being excavated in the early 1930s. By that time hand-powered sewing machines were relics of the past. New Zealand women still sewed clothes for their families, but either used treadles or electricity to power their machines. Amelia's machine went on display at the Buried Village, a reminder of a resilient and bereft woman and her love of sewing.

43.

1893 Women's suffrage petition

ARCHIVES NEW ZEALAND

We arrive at one of the seminal events of New Zealand history, the achievement in 1893 of suffrage for adult women in the national legislature, the first time this had been won in the world. Suffrage was made possible by the huge petition which preceded the legislative Act. That petition, cared for by Archives New Zealand and now on display in the National Library in Wellington, is a document, but it is also a remarkable object.

For a start, there is its size. When rolled up the diameter is almost 50 centimetres. It is thick because there are no fewer than 546 individual sheets glued together, and the sheets contain 25,519 signatures of women throughout the country (plus a few men). When another 12 petitions presented separately are added, the total number is 31,872, representing 22 per cent of the women aged 21 or over in the country. It was the largest petition in either Australia or New Zealand up to that time; and was so weighty that it is said Sir John Hall had to carry it to the House of Representatives in a wheelbarrow. When he reached the House, according to one story, he unrolled it with the help of a messenger along the floor and it hit the far wall with a thud. According to another, it was then rolled back to cover the floor twice. Members were deeply impressed by its scale.

The 1893 petition was the third large petition that New Zealand women had presented in pursuit of women's suffrage (there were over

10,000 signatures in 1891 and over 20,000 in 1892). The questions these petitions invite are: what motivated women to spend such enormous energy on behalf of women's suffrage, and why were New Zealand women able to win this important right before women of other countries?

There were two impulses behind the suffrage campaign. The first was the liberal tradition of equal rights, the view that women as intelligent human beings, many paying taxes, had as much right as men to choose who governed them. This view had been articulated during the Enlightenment by Mary Wollstonecraft and then by John Stuart Mill, who in 1866 presented the British Parliament with a petition for women's rights and wrote a famous tract, 'The Subjection of Women'. Mill influenced two significant New Zealand feminists with the name Mary Ann(e) — Mary Anne Müller in Nelson and Mary Ann Colclough (or 'Polly

Plum') in Auckland, both of whom argued for women's suffrage in the late 1860s and early 1870s. The case became stronger once all adult males received the vote in 1879 (all Māori men had received it in 1867 when Māori seats were established). Equal rights feminists won New Zealand women access to secondary and higher education in the 1870s and property rights within marriage in 1884. But efforts to win the vote failed in 1878 and 1879.

The second argument motivating the suffrage fight was women's distinctive role. Many women saw themselves as primarily mothers and protectors of the family who had particular responsibilities for society's morality. This was significant in a frontier society such as New Zealand with more adult men than women. With many men unmarried, habits of drinking, gambling, sexual vice and fighting developed among the

footloose male community. Those places where women first won the vote, notably the American states of Wyoming, Colorado and Utah, all had a large imbalance of the genders. Women's vote would bring morality and sobriety to the community. Sir John Hall, the supporter of women's suffrage in the House, said it 'will increase the influence of the settler and family man, as against the loafing single man'. A crucial expectation was that the women's vote would restrict alcohol consumption. The pub was the centre of single male culture in nineteenth-century New Zealand. Men working on the frontier would come to town to spend their earnings at the bar. Fisticuffs and gambling would follow. For married women, a drinking husband caused distress — he spent household earnings at the bar and on returning home domestic violence was likely. Women concerned about men's drinking habits were directed towards the vote by Mary Leavitt of the American organisation the Women's Christian Temperance Union (WCTU). Founded in 1874, the WCTU argued that the vote would protect women and children from the effects of liquor. Mary Leavitt arrived in early 1885, travelled the length of the country, and 10 branches followed.

The women drawn to the WCTU were largely from the non-conformist churches, Methodists, Congregationalists, Presbyterians. Among these was a well-educated middle-class mother of one, Kate Sheppard. Her father had been a Scottish lawyer and Kate arrived with her parents at Lyttelton in 1869 aged 20. She married a Christchurch gentleman, Walter Sheppard, became involved in the YWCA, was a member of the Congregational church and taught in the Sunday school. Joining the WCTU, Kate helped organise a petition against barmaids, but when the parliamentary committee refused to make a recommendation she was outraged and took up the cause of women's suffrage. In February 1887 she became national superintendent for the franchise. She was a brilliant organiser and used many tactics — letters to newspapers, pamphlets, columns in temperance journals, public meetings. But the major tactic was the petition. There were international precedents — British Chartists had organised a petition of over a million signatures asking for universal manhood suffrage after the 1832 Reform Bill, and the WCTU often organised petitions. Indeed in 1893 they presented a 'monster petition', claimed to be 'the largest petition in the history of the world', seeking the abolition of traffic in alcohol and opium. It was displayed at the Chicago World's Fair.

The major reason behind the suffrage petition was to show that New Zealand women wanted the vote. The WCTU had sent in small petitions in 1886 and 1887, and while the House had supported women's franchise, this was overturned by a procedural sleight of hand. Newspapers questioned whether women themselves were keen. *The Lyttelton Times* wrote, 'Women will have to give some sign of a wish for the franchise before they get it. At present they are so entirely quiescent in this part of the world that it is generally and easily assumed that they have no such wish.' When in 1890 a women's suffrage Bill again failed to become law, the chief sponsor, Sir John Hall, suggested a petition.

The following year saw the first mass petition. Through a Riccarton choral group, Kate Sheppard had developed a close friendship with her later husband, William Smith, who was a printer. He printed the petition forms. By June 1891 some 7700 signatures had been collected and the petitions were presented. Once again a female suffrage Bill received a majority in the lower house, but was narrowly defeated in the upper house. The WCTU tried again in 1892. This time over 20,000 women signed, but the upshot was the same. The ministry and lower house were too scared to defeat the Bill lest it cost them in a subsequent election if women did get the vote. But following their own prejudices and heavy lobbying from the liquor industry they had it rejected in the upper house through an impractical amendment for postal voting.

Depressed about the effectiveness of petitioning, Kate had doubts about a third major effort; but Hall believed the strategy worthwhile, arguing that the number of signatures had to be even greater lest it appeared women were getting cold feet. Kate set out to organise a new petition. This time the opening prayer did not argue the case for women's suffrage, but simply pointed to the failure to pass the law despite both houses affirming 'the justice of the claim and the expediency of granting it'. She reached out beyond the WCTU to other women. In Dunedin, Harriet Morrison called on the Tailoresses' Union to give a hand. Women's Franchise Leagues were established throughout the country calling on women of all faiths, including Anglicans and Catholics. In Christchurch the Canterbury Women's Institute was established, also aimed at non-Temperance women.

All these organisations became involved in canvassing. Once again William Smith printed the forms, each one with the prayer at the head. Once again there were two petitions, one for each chamber of Parliament. Kate sent the forms out before Christmas 1892 in the hope of capturing women on holiday. The organisations set to work. At a meeting of the Christchurch WCTU Kate unrolled a map of Christchurch and members took responsibility to canvass door to door. In other centres women set up tables in busy places such as railway stations. Women signed in pens and pencils, even crayons. A few signed with an 'X' and the canvasser filled in the name. In Whanganui the local paper reported that the petition had been left at shops, a milliner's, a chemist's and confectioner's, for women to sign. The result was signatures from around the country. Places with well-organised women's groups, such as Dunedin and Christchurch, were well represented. Some communities with few signatures in the main petition were better represented in one of the 12 supporting petitions.

Most sent their completed forms back to Kate Sheppard in Christchurch. There she laid them on her kitchen floor, prepared some glue by boiling down the skin, bones and hooves of animals, and then pasted one form on top of another using the prayer at the top of each form as the backing glued section for each page. Legend has it that she wrapped the glued sheets around a broom handle. Eventually her hard work was done and on 5 July 1893 she sent the huge petition to Wellington for Hall to unroll along the floor of Parliament.

Such an effort was hard to resist. Richard Seddon, now in charge of the government following the death of John Ballance, wanted, as Hall explained, 'to defeat women's suffrage, and he is very cunning'. The liquor lobby tried to upstage the women's petition by organising a counter-petition; but its 5000 signatures were shown to be the result of fraud and the use of paid canvassers. Seddon's manipulations to defeat the Bill while seeming to support it failed. Two members of the upper house, appalled by Seddon's hypocrisy, voted for the measure and it passed in that house by two votes. Even then came petitions to the Governor not to sign. On 19 September Richard Seddon telegrammed Kate Sheppard that the Governor had signed the Bill into law.

Subsequently, both William Smith and Kate Sheppard herself wrote accounts to counter the view 'that the franchise was conferred upon the women of this dominion without any effort on their part'. If the 1893 petition shows anything, it is the truth of this statement. It was the extraordinary organisational efforts of women to demonstrate that women in New Zealand really wanted the vote that brought about its eventual success. That is the meaning of this remarkable object.

44.

Meri Te Tai Mangakāhia's parliamentary chest

AUCKLAND WAR MEMORIAL MUSEUM

This handsome chest of inlaid wood, about half a metre wide, reads on its top, 'Votes For Women'. It tells us that alongside the 1893 petition there is a quite different women's suffrage story in New Zealand. This chest is said to have been a 'parliamentary chest' used to hold important documents; and the chest commemorated the effort, by Māori women, also begun in 1893, to win for themselves the vote and representation in the Māori parliament.

The 'Māori parliament', you say. What was that? The institution only lasted about a decade — from 1892 to 1902 — and it emerged out of the efforts of Māori to cope with their disastrous loss of economic resources and authority in the decades after the New Zealand Wars. As the Pākehā population rose and the Māori population, crippled by disease, fell (to no more than 40,000 by 1890), there was a sustained effort by the European community to obtain land still held by Māori and use it for sheep and dairy farming. The setting up of the Native Land Court in 1865, which allowed for the individualisation of land ownership, facilitated the selling by individual Māori. By 1890, Māori had lost almost all the South Island and only about 40 per cent of land in the North remained in their hands. Even then much of it was leased to Europeans. True, there were four Māori members of Parliament, but that was a paltry number in a house of 74 and consequently they could not achieve much. Māori had to find other ways of retaining their land and independent self-government. The King movement based in the Waikato and Te Whiti's settlement at Parihaka were significant attempts. Elsewhere there were efforts such as the Repudiation movement in Hawke's Bay where the aim was to relitigate land sales. The setting up of a Māori parliament, or Te Kotahitanga (unity), was another important effort.

The idea emerged from the annual intertribal gatherings at Waitangi each year and in 1889 it was proposed that there be a 'Maori union of Waitangi' to administer Māori affairs. This was not accepted by the Parliament in Wellington so eventually it was decided in 1892 to set up a separate parliament without formal endorsement from the government. The participants were the leading tribes from the south, east and north of the North Island, many from areas like Hawke's Bay, Whanganui and the Wairarapa where there was long-established Pākehā settlement. Some had been involved in the Repudiation movement, others were former Māori MPs. Māori were invited to sign one of eight parchments pledging that 'the Native race . . . are to combine as one'. Women were invited to sign a separate parchment. By 1893 there were 21,900 signatures (on a similar scale to the suffrage petition). The parliament had 96 elected members with electoral districts based on tribal loyalties, and there was an upper house of 44 elected by the lower house. For 11 years the Kotahitanga parliament met annually, moving around the North Island. An attempt to win legal acceptance from the national Parliament failed with the proposal not even being debated in 1893, while the next year members got up and walked out so that the lack of a quorum prevented debate continuing. Not even Bills for women's suffrage had been treated with such contempt.

With that background let us return to the parliamentary box. On the top are two names, Meri Te Tai Mangakāhia and Hāmiora Mangakāhia. Meri comes first partly because at the meeting of the Māori parliament at Waipatu marae in Hastings in 1893 she was invited by the speaker of the lower house to speak to her motion that women be given the right to vote and stand as members of Te Kotahitanga. Meri was born at Panguru in the Hokianga in 1868 as the daughter of a leading Te Rarawa chief. She was educated at St Mary's Convent in Auckland and married Hāmiora Mangakāhia who was from Ngāti Whanaunga on the Coromandel. Together they built a homestead on Hāmiora's land at Whangapoua on the Coromandel Peninsula. The box shows two images of the property, with the homestead beautifully displayed on the inside cover. So, in one respect, the box is a tribute to both husband and wife, and their family home. Meri was Hāmiora's third wife, and the couple had four children. The box was ordered for them by their close friends Hēnare and Ākenehi Tōmoana.

NEW ZEALAND WOMEN'S TEMPERANCE UNION KOTAHITANGA PARLIAMENTARY CHEST PRESENTED TO MERI TE TAI MANGAKĀHIA. AUCKLAND WAR MEMORIAL MUSEUM TĀMAKI PAENGA HIRA. 2006.98.1.

Both Hāmiora and Hēnare had been assessors in the Land Court; and both had played prominent roles in the Māori parliament. Hāmiora had been elected premier of Te Kotahitanga at the previous session, and although he was no longer premier he had great mana in that community. Hēnare, a former member of Parliament, had been speaker of Te Kotahitanga, and was the host at the Waipatu gathering in 1893. Ākenehi, like Meri, was a third wife and bore Hēnare 13 children. She was a good friend of Meri and a supporter of her campaign. In one sense the box is an expression of friendship.

But votes for women, the phrase in bold letters along the base of the box's cover, was clearly a major focus. It is interesting that along each side of the top are the letters WCTU. Was temperance, represented by the Women's Christian Temperance Union, also an important influence for Meri and Ākenehi's campaign for women's place in the Māori parliament as it had been for Kate Sheppard? It seems probable that Meri had heard about the women's suffrage campaign and petition, although it is striking that her motion came before the debate in the New Zealand Parliament. It is possible too that she may have had contact with the WCTU but there is no solid evidence of this. When she spoke on why Te Kotahitanga should give women the right to vote and stand as members, she did not mention the problem of alcohol. Māori men had been given the right to vote for Parliament when Māori seats were introduced in 1867 — 12 years before all European men were given that right. So Māori were well used to the idea of voting. But when Meri spoke at Te Kotahitanga on 18 May 1893 her explicit reasons were all about land, the issue that had motivated the Māori parliament in the first place. She noted that many women were widowed and held land, that many were knowledgeable about land matters while their husbands were not, and that some women had elderly fathers and had taken over management of the land. She might also have noted that although Māori women did not usually speak on the marae, some had signed the Treaty of Waitangi, many had signed petitions about land to the New Zealand Parliament, and others had taken cases to the Land Court. There is no question that land, rather than temperance and moral reform, was the major focus of Māori women activists in the period.

So, one wonders why the letters WCTU are so prominently displayed on the box. It is possible that this was because it was the work of a Pākehā cabinetmaking firm, Andrews and Sons from Wellington, and that they decorated it in native woods with the words which they assumed had motivated Meri as they had motivated Kate Sheppard and Pākehā women. It is suggestive that the box has no Māori designs or carving and contains no words, apart from names, in te reo, which seems unusual if Hēnare and Ākenehi had specified its design and wording in detail.

Despite the close friendship of Meri and Ākenehi, reflected in the box, Ākenehi actually spoke against giving Māori women their rights at the 1993 Te Kotahitanga. Her reasons apparently were that this should await the Māori parliament receiving full recognition from the New Zealand Parliament. However, the matter did not die. At the next gathering of Te Kotahitanga at Pākirikiri in Gisborne in 1894, women's suffrage was again discussed. Women were authorised to set up their own committees, and there was a great flowering of such komiti wāhine over the subsequent few years. Finally, in 1897 when Te Kotahitanga met at Pāpāwai, suffrage of Māori women in their own parliament was finally granted. The victory came very late because within five years the last meeting of the Māori parliament had been held.

There is a final element of the box worth noting. On the front is a small oval brass plate which reads 'To Connie Cherry, Farewell 20th January 1927 Caranderrk Station, Healesville'. The back story is that the box was held by Meri's family until her death from influenza in 1920, and then somehow ended up in Queensland and was used as a gift to Connie. Eventually it returned to New Zealand in 2006 when purchased at auction by Auckland Museum. Today the box is once more treasured as a tribute to a brave and impressive woman whose battle for suffrage in the Māori parliament in 1893 points up yet again how similar, yet different, were the concerns of Māori and Pākehā in late nineteenth-century New Zealand.

45.

H. B. Lusk's cricket bat

NEW ZEALAND CRICKET MUSEUM

Looked at with fresh eyes this is a strange object — a hunk of wood with a handle attached to it. Not a weapon of war, but a piece of equipment for the noble English game of cricket. Although it lives today in the Cricket Museum at Wellington's Basin Reserve, we don't know much about this bat, neither its precise date nor the person who obviously cherished it on the evidence of the binding around the blade. What can we find out?

We begin with the date. Cricket arrived in New Zealand with the English, a pastime born in rural England like horse-racing, hunting with hounds, and village entertainments like sack races and chasing the greasy pig. But unlike these other pursuits, cricket had an ethical dimension as the essence of English civilisation. The first recorded game here was supervised by the Anglican missionary Henry Williams at Paihia in December 1832. Three years later Charles Darwin saw young Māori playing cricket at the Waimate mission station. British soldiers played it during the New Zealand Wars. The first provincial game (between Auckland and Wellington) was in 1860.

It seems unlikely our bat was from these early days. Until the nineteenth century, bowling was underarm (like Trevor Chappell's bowl in the infamous 1981 game!), so bats were curved like hockey sticks. Eventually bowlers began looping the ball and bats developed parallel edges like this one. But they remained bottom heavy. Then in 1864 came over-arm bowling. Balls bounced and came at speed, so bats became lighter to allow the batter to adjust quickly. Springs were introduced in the handle to prevent jarring. In the 1880s Charles Richardson invented a handle with springs which was inserted into the willow blade through a tapered splice. The taper prevented the bat splitting at the join. Our bat has a spliced handle, so it must date at least from then. Further, it was made by John Wisden & Co. John Wisden, a former cricketer who gave his name to the annual cricketing almanac, started a sports goods business which, after his death in 1884, became a major international brand. The bat must be later than this. On the other hand, it is unlikely to date from the twentieth century because in the 1890s bat makers switched from heavy heart willow in the blade to sap willow. This made the bat lighter. Heavy willow bats were dark brown in colour and weighed up to over 2 kilograms. Later bats were paler and weighed about 1 kilogram. This bat is heavy and dark in colour. So it is unlikely to date from after 1900. The museum's suggestion of 1895 seems plausible.

The date is relevant to the mystery of the bat's owner. It was given to the museum as belonging to H. B. Lusk. The trouble is there were two H. B. Lusks, both prominent cricketers. The first was Hugh Butler Lusk who was born in 1866, began playing cricket for Auckland in 1889, and then Hawke's Bay from 1891 to 1909. He played for New Zealand from 1896 to 1902 and was vice-captain of the New Zealand team on their first overseas tour to Australia in 1898–99. The other Lusk was his cousin, Harold Butler, born 11 years later in 1877. He also started his first-class career for Auckland, in 1899, and played for various provinces until 1919, including one season for New Zealand (in 1909–10). Given that Hugh played earlier and included the mid-1890s it seems probable the bat was his.

Whichever Lusk owned the bat, their lives tell us much about cricket in nineteenth-century New Zealand. They were both grandsons of Robert Baillie Lusk who had come out in 1849 from Greenock, north-east of Glasgow. Don't be fooled by this Scottish origin. The Lusk family had strong English connections, were Anglicans, and Robert came as a surveyor for the New Zealand Company. Among his sons were Major Daniel Lusk who achieved fame commanding companies of the Forest Rifle Volunteers in the New Zealand Wars; and his younger brother, Hugh Hart Lusk, who became a barrister and member of the Auckland provincial government. These were the fathers of our cricketers, Harold and Hugh. The two brothers married sisters, both daughters of Captain William Butler, so Hugh and Harold were cousins twice over — through both parents. William Butler was the son of an Anglican Dorset clergyman who became a rich merchant at Doubtless Bay. There he fathered the mothers of Hugh and Harold. Through their mothers the

Lusk boys had a strong English inheritance; and this helps explain their interest in cricket.

For colonial New Zealand saw cricket as a way of keeping alive English civilisation. Settlers constantly feared giving in to the frontier and losing civilised graces. Playing cricket offered assurance that English moral values were secure. 'The lessons cricket teaches are just those features which distinguish the British character in every department of life,' editorialised the *Otago Daily Times* in 1884. The game was strongest in Anglican Christchurch, and always a bit suspect in Scots Dunedin, even more the Irish West Coast. Hugh Butler cemented this Anglophile heritage when he married the daughter of W. L. Rees who was part of the Grace family, W. G. Grace being the most famous English cricketer of the century. When an English team visited in 1877, Rees 'hoped that the visit of the cricketers . . . would help to strengthen the ties which bound England to her children in these far distant regions'.

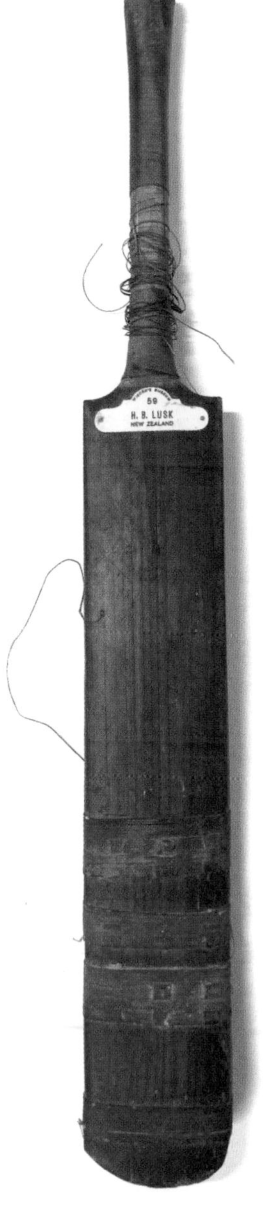

NZ CRICKET MUSEUM

The Lusk story also points to the importance of the elite secondary schools in encouraging the game. Harold Lusk became a schoolmaster at Anglican Christ's College in Christchurch, and eventually became headmaster of the Auckland Anglican school King's College. Elite boys' schools, whether privately owned like Christ's and King's or state schools like Wellington and Nelson Colleges, were not very popular in the late nineteenth century. Only about 3 per cent of boys attended them. Yet they had a disproportionate influence on cricket. Headmasters like C. C. Corfe at Christ's College and Joseph Firth at Wellington College put energy into promoting cricket as essential to English moral standards. About a third of the provincial cricketers of Auckland and Canterbury came from such schools.

The game was dominated by the urban middle class. Hugh Lusk became a barrister and solicitor. Harold Lusk, a headmaster of King's College. In England the social elite enjoyed the game, but in rural areas working men participated in village cricket. The stereotype was of batsmen as classy, while the fast bowlers were blacksmiths or labourers. In first-class cricket there was a distinction between paid professionals and gentlemen amateurs. In New Zealand the urban middle class had time and money to play games, while working men were initially handicapped by the lack of a universal Saturday half-holiday. There were some cricket teams of working people, such as butchers and bakers and a Dunedin team of the Hillside railway workshops, but in general it was urban professionals who played cricket and dominated representative teams. Even the rural elite were inhibited by the lack of easy travel options and did not have available local grounds. It took resources to prepare grounds with English grasses and mowed wickets. The first five presidents of the New Zealand cricket council formed in 1894 were lawyers.

Finally, the Lusks were Pākehā and male. Although missionaries encouraged Māori to play cricket, most lived in rural areas, while cricket, was an urban game. Concerning gender, the Victorian view was that women's primary role was as mothers and an active life might threaten their reproductive capacity. Less active games such as golf or croquet were acceptable for women; cricket, or heaven forbid, rugby, much less so.

A question remains: why, given its early start in colonial New Zealand, did cricket not become the national game as occurred in Australia or India? The first rugby game was not played until 1870, almost 40 years after the first cricket game. Yet by 1905 rugby was clearly the national game. This had largely come about because of international competition in which New Zealand rugby players more than held their own. But in cricket, while there were as many as 22 visits from English and Australian teams from 1864 until 1914, the local teams were so poor that before 1902 New Zealand teams did not field the same number of players as the visitors. Usually it was 11 against 15 or 22 locals, who even then were usually beaten. The reason was that, as Lusk's story suggests, cricket here was played by a select group — those with English affinities, elite secondary education, and from the urban middle class. In many parts of New Zealand, cricket was a foreign pursuit — among Otago's Scots, among North Island's Māori, among the West Coast Irish, among rural labourers and city working-class men, and of course among women. So the bat of H. B. Lusk, whichever Lusk it was, tells a larger story about how by the 1890s New Zealand was a diverse community of many subcultures. Cricket flourished in just one subculture.

46.

Northland gum cathedral

THE KAURI MUSEUM

While people in the south and west of the South Island were still hunting for gold in the 1890s, men (and a few women) in the far north of the North Island were pursuing another form of gold — kauri gum. This remarkable work of folk art, a model cathedral, 44 centimetres high and made of over 600 pieces of kauri gum, throws a golden light on the distinctive culture and way of life of the Northland gumfields.

What initially drew the diggers to the far north was the golden resin, known as 'gum', which oozed from the damaged parts of huge kauri trees. After the trees had died, the gum remained in the ground, at times up to 20 feet below the surface of swamps. Māori had used the gum for chewing and lit it for torches, and they would burn it and mix the ash with oils or the fat of dogs to produce the blue/black colouring which was then rubbed into lacerated flesh to produce tattoos and moko. Early Europeans were intrigued by the gum and it is said that in 1836 about 20 tons were shipped to London, but then declared worthless and dumped in the Thames. However, it was not long before a novel use was found for the new gold. Kauri gum became a valuable ingredient in varnishes. It had a much lower melting point than the resins previously used, so heating it to mix with oils was very much safer, and in addition the gum held its colour much better than alternative resins. From the 1840s, high-quality varnishes, used particularly on wooden furniture, came to depend on kauri gum. From that time there was a flourishing export trade, and by the 1890s some 8000 tons a year were leaving New Zealand for London, bringing in about £500,000, more than was earned by coal and about half that earned by gold.

A substantial number of diggers scoured the Northland swamps for kauri gum — some observers said there were 2000 people involved, many as part-time workers, while the 1893 commission suggested there were about 7000 full-time. By then it was hard work finding and recovering the gum. The first pieces, called 'bold gum', were easy pickings, lying on the surface or barely embedded in the soil. But by the end of the century those days were long gone. The best finds were old pieces of gum left deep in swamps when the kauri died from wet feet, leaving the gum behind. To find good gum the digger had to put on rubber boots and wallow around in the thick mud. He would use a heavy spade but, when the gum was really deep, pulled out a gum spear up to 8 metres long to feel for pieces deep down. Some of the diggers used a 'joker' — four rings of metal wrapped round the spear blade which helped the spear penetrate the mud. Having located the gum, they would use a long piece of galvanised pipe with a steel toe welded to form a hook. This would allow them to retrieve the gum and place it in a bucket or kerosene tin before transferring it to their pīkau, a pack made from a grain sack, to be carried back to camp.

There they lived in sod whare thatched with nīkau or raupō reeds, or in canvas tents, trying to find some protection from the Northland rain. It was not an easy or pleasant existence. Often the diggers felt they were exploited by the local storeman, who both sold them supplies and bought their gum; and as increasing numbers of diggers arrived from across the seas, mainly Dalmatia on the Adriatic coast, there was resentment that their systematic methods and work in groups rather than as nomadic individuals left thin pickings for others and lowered the price of the gum. They were accused of being 'birds of passage' who sent their earnings home and soon returned to Croatia themselves. Others found the landscape a depressing wasteland of mānuka scrub and mud. Jane Mander, one of the great novelists of the gumfields, described the scene as 'sprawling, undulating waste. Bare and black . . . the earth was shrivelled in permanent exhaustion.' William Satchell, another gumfield novelist, called it 'the land of the lost'.

One way the diggers brought some colour and inspiration to their lives was through carving gum. When they came back at night they needed to scrape the gum with a jackknife to remove the dirt and make it presentable for purchase. It was not long before scraping the dirt morphed into carving works of art. Just as sailors or sealers would occupy their spare time carving whale teeth into scrimshaw, or even getting a mate to give them a tattoo, so the gumdiggers occupied themselves at night, or when it was raining, or on Sundays, with their form of folk art. They would work slowly carving the gum, and once the carving was complete, would polish it with emery paper to bring out the golden glow. Their subject matter was often the typical designs of scrimshaw or tattoos — hearts or anchors, the kind of subject matter also found on nineteenth-century tombstones. Interestingly, given the fact that many Māori worked on the fields, there are few Māori designs. Often religious subjects seem to have been an inspiration. Some of the diggers were of an Irish Catholic background; and from the early 1890s many came from Dalmatia bringing their Catholic faith with them. By 1893 there were over 500 Dalmatian diggers. So among their subjects were golden crosses, or models of a Bible.

This cathedral was an exceptional piece of religious imagery. It has four towers, like a Romanesque cathedral, a cross (or perhaps star) at the centre of its façade, and a beautiful round orb in the middle. At the rear

THE KAURI MUSEUM. IMAGE BY JOCK PHILLIPS.

is a mirror to reflect the glorious colour. It is said to have taken almost 10 years and was completed in 1896. The 600 pieces were carved and then stuck together by being melted with a flame. The artist was Allen Addis, probably of English background. We know little about him. Whether he began the work on the gumfields, whether he had extensive contact with the Dalmatian community, or even what his religion was, are questions unable to be answered. What is certain is that the original artists, the first people to develop the craft of gum sculpture, were the gumdiggers of the north.

From about 1910 the kauri gum trade began a slow decline. Although the discovery that it could be used in linoleum kept up some demand, by the 1930s the amount exported was well under half what it had been at the turn of the century, and the arrival of synthetic lacquers spelled the end of the trade. The Dalmatian diggers intermarried with the local Māori and became leaders in the wine industry; other diggers took up land as growers of kūmara or as dairy farmers. The tough days of the gumdiggers were forgotten. We are fortunate that the Kauri Museum in Matakohe keeps their memory alive, and in extraordinary works of art such as this provides a tribute to the creativity of working people.

47.

Westport District Gold Miners' Union banner

MUSEUM OF NEW ZEALAND TE PAPA TONGAREWA

This beautiful banner, 4 metres square, and now held in Te Papa with a superb copy in the Coaltown Museum in Westport, tells us much about the history of gold-mining and of labour unions in New Zealand. It was purchased for £40 in 1899 by the Westport District Gold Miners' Industrial Union of Workers from the famous makers of banners, George Tutill, whose signature and address (83 City Rd., London) are found in small white letters on the bottom right.

The first surprising aspect of the banner is that it was gold miners, not coal miners, who ordered it, for it is the coal miners who entered New Zealand mythology as one of New Zealand's great leaders of industrial unionism; and a gold miners' union seems unexpected — we think of West Coast gold miners as largely self-employed fossickers who worked with their mates in small groups. Why did they need a union? It is true that after the discovery of gold on the West Coast in 1865, those who came, from central Otago or even Victoria, in search of the shining metal tended to be individuals or a couple of mates, joined later by small groups of Chinese. But by the 1890s things had changed. As James McGowan, Minister of Mines, wrote in his annual mining report: 'I find much of the ground that in early days gave rich returns to the individual miner has now been so thoroughly worked that larger combinations of men and money are necessary in order to obtain the gold at greater depth.'

Three forms of industrial mining took off — crushing quartz such as at Reefton on the West Coast or Waihī in the North Island; dredging rivers, which was common in central Otago; and complex water races and sluices. It was this last form which was significant on the West Coast. Indeed, there were no fewer than 3285 'European' alluvial miners and 804 Chinese in Westland and Nelson provinces (the latter included West Coast communities of Īnangahua, Ahaura, Charleston, Westport and Lyell) in 1900 as compared with 804 quartz miners, nearly all of whom were at Reefton. It was a large sluicing operation which sparked this union. A German company, the General Exploration Company, began sluicing at Fairdown just outside Westport. By 1897 it had constructed over 7000 metres of water races and around 3000 metres of tunnels; 218 men were employed. But the miners were dissatisfied with their pay, which was significantly less than other craftsmen like carpenters or blacksmiths; and they experienced a high level of accidents. In 1899, for example, there were no fewer than 16 fatal accidents in the alluvial mining sector, considerably more than in the coal mines or quartz mining. So in May 1897 a group of miners met and decided to form a union. The next year a branch was established at Addisons Flat where the company had also begun development; and soon after a meeting was held at the Westport skating rink and the Fairdown miners' union became the Westport District Gold Miners' Industrial Union of Workers. A banner was ordered.

As for the coal miners, before the turn of the century gold-mining was a far more important economic activity both on the West Coast and in New Zealand than coal-mining. In 1899 the value of gold was over three times the value of coal produced; and according to the 1896 census there were 18,583 people engaged in mining in New Zealand, but only about 1500 of these were mining coal by comparison with over 9000 alluvial gold miners and almost 4000 mining quartz for gold. Yet despite their relatively small numbers, coal miners were important in the early history of unions on the West Coast. John Lomas, who had arrived with 50 other miners from Barnsley in Yorkshire, set up a union at the new coal mine at Denniston, high up on the plateau west of Westport, in 1884. When the Fairdown gold miners' union was formed, they did so in the front room of Frank Beirne of the Denniston coal miners' union, and James Patz, secretary of that union, was present to provide guidance. He and John Foster, president of the Denniston union, led the Westport meeting in 1898. So coal miners were important to the story of our union.

Let's return to the banner. Why did the union want a banner? Union banners go back to the origins of unions in Britain. In the early nineteenth century when unions first appeared they were groupings of skilled craft workers. They did work for better wages, but they were primarily organisations to promote the prestige and welfare of their fellow craftsmen. They provided support in times of illness or unemployment, rather like friendly societies. George Tutill began sewing regalia and eventually banners for friendly societies. Indeed, some of the regalia of New Zealand friendly societies like the Foresters or Oddfellows were Tutill's work. The banners would be carried during local parades on anniversary days or communal celebrations. The unions of skilled craftsmen not surprisingly picked up the practice and also wanted to carry banners on such days. This served to display their role in the local community. Tutill developed a trade sewing silk banners for independent groups of craft workers. The first unions in New Zealand were of this nature — groups of skilled workers organised locally and committed to such aims as better wages but also the eight-hour day and the prestige of their craft. Then in the 1880s economic depression hit. Australian influences led to the unionisation of unskilled men, such as shearers and seamen. Increasingly, unions saw themselves representing a class, the wage-workers, not just a particular craft. They sought to build bridges between the working class and encourage a socialist revolution. John Lomas organised an inclusive labour union in Westport. There were international links and even women became drawn into union activity through the tailoresses. This movement reached heights with the creation of the Maritime Council, New Zealand's first colony-wide union. It then fell in the maritime strike of the 1890s when unskilled men, especially maritime and transport workers, went on strike in support of colleagues in Australia. The strike failed and the council collapsed.

But the political rise of working men had an impact. The new Liberal government of the 1890s passed a series of factory Acts, set up a Department of Labour, and eventually under William Pember Reeves passed the Industrial Conciliation and Arbitration Act which was a lifeline for the union movement. Workers in any industry could form a union, apply for registration, and in any dispute employers were required to enter into conciliation and then arbitration. There was a large increase in unions — from 75 in mid-1896 to 202 in 1901. The very first union to be so registered was the Denniston miners, and the Westport Gold Miners' Union was established under the provisions of the Act. The Act led to a revival of small craft unions; but the vision of a united working class, involving unskilled workers, which had first become evident in 1890 remained very much alive.

This is clearly reflected in our banner. For a start, the gold-mining element of the union's origin is rather hidden. All we get are the initials

TRADE UNION BANNER, 1899, LONDON, BY GEORGE TUTILL & CO., LTD. GIFT OF THE CENTRAL AMALGAMATED WORKERS' UNION, 2009. TE PAPA (PC004649/1-7)

'DGMI'. Instead what is spelled out are the words 'Union of Workers'. The union made strenuous efforts to expand its community. As early as 1898 John Foster had offered to enrol all workers in Westport in the gold miners' union; and eventually in 1908 it would become the Westport General Labourers' and Mechanics' Industrial Union of Workers. The same impulse is reflected in the imagery of the banner. True, there is a gold miner represented no fewer than three times — twice on the front and once on the back, dressed in his red shirt, with shovel and pick and rocks behind him. But in each scene he is reaching out to other workers. On the back he is shaking hands with a blacksmith, symbolising a comradeship with industrial workers. In the front he is paired with an agricultural labourer holding his staff; and in the middle he shakes hands with a Māori man. This sense of a common kinship with all working and oppressed peoples is expressed even more strongly in the words. On four occasions the word 'unite' or 'unity' is found, including the famous phrase 'United we stand, divided we fall'.

The clear message is the strength of working people acting together. This is the vision of 1890 class unity which would become more pronounced in the 1912–13 strikes.

Rather than being carried in community parades like friendly society banners or those of the old craft unions, this banner was marched through Westport streets on Labour Day parades behind a band and with enthusiastic followers. Labour Day was first held on 28 October 1890 to celebrate Samuel Parnell's success in winning an eight-hour day in 1840. The day was initiated by the Maritime Council of maritime and transport unions who held it on the anniversary of their establishment. At that stage the council was gearing up for the maritime strike. So Labour Day expressed the unity of all working people. In 1899 the day became a public holiday. In Westport the gold miners' union led the call for a demonstration on Labour Day, in part, no doubt, to display their new banner.

There are several other symbols worthy of remark. On the back are masts of a ship and a Scottish thistle — reminders that many who led the union were recent migrants from the old world. Their loyalty to the British Empire is expressed on the front of the banner with the Union Jack and lion of the United Kingdom wrapped in the New Zealand ensign; while the phrase 'Advance New Zealand' is held by three women who could well be two young figures of Zealandia with an older Britannia behind. The beehive may suggest that the migrants travelled to 'a land of milk and honey' — a traditional Masonic symbol also represented by the all-seeing eye on the front. Masonic symbols had long appeared on union banners, indicating their close relationship with such fraternal orders.

So the banner carries with it the varied traditions that lay behind New Zealand unionism — the craft tradition which brought unionism to New Zealand and found expression in the development of union banners in the first place; and the millennial vision first unveiled in 1890 of a world where unions working together created a socialist commonwealth.

48.

George Bradford's bandolier

NATIONAL ARMY MUSEUM TE MATA TOA

On 28 December 1899, four days before the twentieth century began, Trooper George Bradford died of wounds in a Boer hospital in South Africa. He thus became the first New Zealand soldier to die in service in a foreign country and was indeed the first soldier from any of Britain's colonies to die in the South African War. He began a tragic tradition which would become hugely important for New Zealand in the new century. This leather bandolier designed to hold rifle bullets was allegedly worn crossways over his chest (from left shoulder to right waist) by Bradford on the fateful day, 10 days earlier, when he was shot. It subsequently became the badge of office for the Dominion President of the South African War Veterans' Association from 1920 to 1980 with each president identified by a silver bullet.

Why were Bradford and the 214 other members of the first contingent fighting in South Africa in December 1899? The specific cause of the war was that the Boers, descended from Dutch settlers, wanted to preserve the independence of their territories in Transvaal and the Orange Free State, while the British were determined to regain control and protect the rights of 'Uitlanders' (foreigners, mainly British citizens) in Boer areas. New Zealand's involvement was explained by a heightening loyalty to the British Empire. Most Pākehā were either British migrants or children of migrants, and with the development of the frozen meat and butter trade, there was a growing economic relationship with Britain. Imperial rivalries led to a sense of the superiority of the British 'race'. New Zealanders were notably enthusiastic towards the golden and diamond jubilees of Queen Victoria's reign in 1887 and 1897. So when Britain went to war in South Africa, New Zealanders were keen to play their part. There were 40,000 people in the streets accompanied by eight brass and pipe bands to farewell Bradford and his fellow troopers when they sailed from Wellington on 21 October 1899, only eight days after war had been declared. It was the anniversary of Nelson's victory at Trafalgar. Governor Lord Ranfurly was sure the men 'would prove themselves no unworthy descendants of that dauntless Island race whose colours wave over them'. Premier Richard Seddon promised that the country would 'fight for one flag, one Queen, one tongue, and for one country — Britain'. The ship carrying the contingent raced across the Indian Ocean to be the first colonial force to reach the scene.

The first contingent was drawn from a few permanent soldiers plus volunteers, who were the local territorials. Originally, to express colonial patriotism, it was hoped that all members would be New Zealand-born. This proved impossible; not helped by the ruling from the mother country that the conflict was 'a white man's war' and Māori were not welcome. George Bradford, an immediate volunteer, was not a New Zealander by birth. He had been born in 1870 in Sussex, not far from where the Battle of Hastings had been fought 800 years before. His father ran a pub. George showed an early interest in the military life, joined the Grenadier Guards for six years and actually served in Africa. Then, apparently for health reasons, he came to New Zealand in 1895 and became a miner on the upper Thames goldfields. As the movement for volunteers grew, Bradford responded. In 1897, once a volunteer regiment was organised in Paeroa where he lived, Bradford offered his services. He became the Regimental Sergeant-Major of the Ohinemuri Rifles No. 1 in early 1899 and was described as a most effective drill instructor. The local Thames paper was surprised he had not been made an officer, but on the voyage to Africa he used his experience to drill the men on board ship. The only issue was that the contingenters were a mounted force. This meant that while they did not fight from horses like cavalry, they rode on horses to the action. All recruits were supposed to be experienced riders, but Bradford admitted, 'I can't ride worth a cent; but, I suppose, if I squeeze in between two other fellows, I shall manage to stick on.' Despite this he enlisted as a farrier. Bradford was unmarried, and it was reported from Africa that when the soldiers sat down to write home, he stated that 'he had neither kith nor kin to write to'.

Bradford died in the first real engagement faced by the New Zealanders. The Boers went on the offensive with the outbreak of war, cutting the railway linking the Cape with Rhodesia and besieging the towns of Ladysmith, Mafeking and Kimberley. They also occupied the northern Cape Colony town of Colesberg. The New Zealanders were sent there to join Sir John French's harassing force. The landscape was open with no trees and many kopjes (hillocks) strewn with rocks. On 18 December 1899 the New Zealanders were despatched to check out Jasfontein farm, near Colesberg, where the Boers were suspected to be hiding. The farm was deserted, but Boers were seen in the vicinity and New Zealand horsemen advanced to investigate. They came under fire and in the retreat of some 150 metres back to the farm over rocky ground, Bradford was shot. He apparently placed his hand on his groin and cried out 'I've got one' before falling off the horse and gashing his head. With blood flowing from his head, it was presumed he was dead. Later a party was sent out to retrieve his body but discovered that Bradford had only been wounded and taken prisoner. It was 10 days before he died, apparently of meningitis, in a Boer hospital.

Because the newspapers initially reported Bradford's death, then that he was only wounded, and finally his death, there was ample opportunity for New Zealand correspondents and editorial writers to reflect on 'the first man down'. The descriptions of his death were accompanied by reports of several other incidents. The first was during the Jasfontein farm affray when a New Zealand trooper returned to recover a mate whose horse had been shot, an act illustrated in the *Weekly Press*. The second was a dramatic engagement in mid-January 1900 when the Boers attacked a kopje being held by 70 New Zealanders and 30 Yorks men. The Boers almost reached the top when the New Zealanders responded with

a bayonet charge and forced them downhill. The kopje was renamed 'New Zealand Hill'. These incidents led to much colonial pride — 'The story of their gallant deeds . . . must send a glow of pride through the heart of every colonist,' wrote the *Evening Star*. The papers relished the praise bestowed on 'our boys'; but also smugly noted that those involved in the 'hot incident' when Bradford was injured did not 'indulge in self-glorification'.

Glorification was confined to the people at home, where there was much mythologising about the qualities of the New Zealand trooper. For a start, the papers were pleased that the colonial soldier was being taken seriously. The fear that they were there simply for sentimental reasons and condemned to garrison duty was replaced by a recognition that 'our boys' were the real thing — 'well-mounted, first-class horsemen, good shots'. The papers even elevated the colonials into superior soldiers of empire. Brought up on the colonial frontier, they were physically tougher than the urban Brits; they were good riders, and they showed adaptability and initiative by comparison with the overdisciplined class-bound Tommies. If there was any criticism of Bradford's wounding and death, it was that his mates had not scooped him up and won a Victoria Cross as a result — which was strangely insightful because the one VC that the New Zealanders did win in South Africa was gained by W. J. Hardham for just such an action. The growing enthusiasm for New Zealand's military achievements was reflected in a huge increase in such activities at home. The volunteers rose from 7000 in 1899 to 17,000 by mid-1901; and there was an even more dramatic increase in school cadets, from just over 2000 in 1897 to over 15,000 ten years later.

As for Bradford, his status as the first colonial to die in a foreign

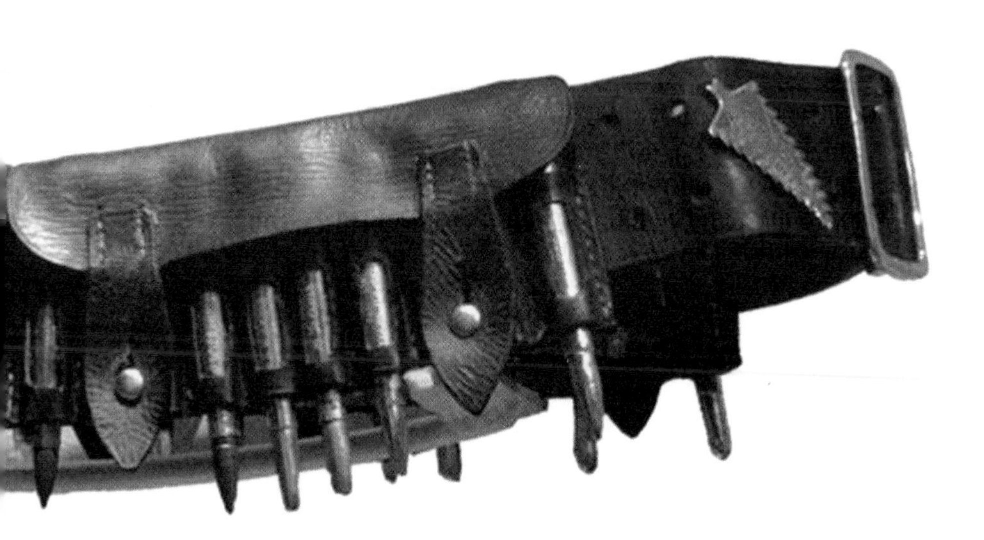

1981.350 NATIONAL ARMY MUSEUM NZ

war was not forgotten. Within a week of his death, in Auckland 'The Collosophone' was advertising a musical offering, 'The Soldier's Farewell' accompanied by 'limelight' pictures of the late Trooper Bradford; while in Nelson patriotic badges with an image of Bradford went on sale. A month after his death Paeroa held a memorial service; in June his old unit the Ohinemuri Rifles offered a Bradford Memorial Belt as a prize for shooting; and four months later the Thames Naval Volunteers had a portrait of Bradford for their competitive shoot. Early the next year a meeting was held to consider a permanent memorial to Bradford and in 1902 Richard Seddon unveiled an impressive memorial fountain on Tuikairangi/Primrose Hill, a promontory above Paeroa.

It took longer before the returned veterans glorified his memory. There were various returned soldiers' groups quickly established, such as the First New Zealand Mounted Rifles Association formed in 1901 and the Southland Returned Troopers Association. It was not until 1920 that the South African War Veterans Association was created for former members of the 10 contingents. It is not entirely clear how George Bradford's bandolier was acquired by the association as the badge of office for their president. But it was a wholly appropriate symbol. The South African War established a new tradition that New Zealand men were good at fighting in Britain's imperial wars; and George Bradford's death marked the real beginning of such a view. Over the next century over 30,000 young New Zealanders paid with their lives in upholding the tradition.

The South African War Veterans Association held its final meeting in 1973 and from 1974 the bandolier hung in the RSA clubrooms in Cambridge. When the association was dissolved in 1980 the bandolier was given to the National Army Museum at Waiōuru where it resides today.

49.

Richard Seddon's coronation coatee

MUSEUM OF NEW ZEALAND TE PAPA TONGAREWA

This amazing wool velour coatee with its spectacular gold braid and nine gilt buttons, each enriched by the royal crest, was worn by Richard Seddon to the coronation of Edward VII, on 9 August 1902 (after a delay of two weeks while the King recovered from appendicitis). His uniform also included black woollen breeches, white stockings, black slip-on shoes with gold buckles, and capping it all off was his black bicorne hat topped with white fur. The clothes were made by Hill Brothers, a very respectable military tailor based in Old Bond Street in the heart of London's classy West End. The outfit was the Privy Councillor's First Class dress uniform as specified by an official *Schedule of Civil Uniforms*. The Civil Uniform First Class had gold braid no more than 4½ inches wide, whereas Second Class Councillors had to make do with only 4 inches!

Richard Seddon must have looked impressive in the clothes, since he was a very big man — about 6 feet tall and nearly 20 stone in weight, which in part reflected his enjoyment of fine food and alcohol. There is little doubt that Seddon loved wearing it. He had always taken great pains with his public appearance, often choosing a frock-coat with a flower in the buttonhole. As Grand Master of the New Zealand Lodge he enjoyed dressing up, and being photographed, in the full Masonic garb. His wife, Louisa, who accompanied him to the coronation, did not let the side down, wearing 'a magnificent violet velvet gown, artistically trimmed with iridescent embroidered blonde and pale mauve chiffon'.

It was not only wearing splendid clothes which gave Seddon pleasure on the coronation day. He always enjoyed royal occasions, from the time in his childhood when Queen Victoria visited the local estate, near his home in Lancashire. When the Queen celebrated her diamond jubilee in 1897, Seddon set off to take part in the full commemorations; and four years later he enthusiastically arranged a visit to New Zealand by the Duke and Duchess of Cornwall and York. Seddon ensured he met the royal couple in Auckland, travelled with them to Rotorua and was at their sides in Wellington, Christchurch and Dunedin.

What must have given Seddon pleasure during the coronation celebrations was that the British press considered him the 'lion' of the

COATEE, CIVIL UNIFORM, FIRST CLASS [FULL DRESS], 1897, ENGLAND, BY HILL BROTHERS.GIFT OF DAME ELIZABETH KNOX GILMER, 1955. TE PAPA (PC000765)

coronation, the charismatic colonial star; and their views were reported widely in New Zealand where newspapers reproduced 'home tributes'. Apart from Peter Fraser at the founding of the United Nations in 1945, or David Lange at the Oxford Union in 1985, or even Jacinda Ardern after the mosque shootings in 2019, never has a New Zealand prime minister (a term used by Seddon in place of 'premier') had such overseas fan-interest. Wearing these clothes Seddon must have looked in the mirror and seen himself at the high point of his career — the crowning of 'King Dick' as well as Edward VII.

It is revealing to examine the story told about Seddon at this time. It explains much about both his overseas image and the enormous political hold he gained at home. There were two elements to his image. The first was as 'a man of the people'. It would be untrue to claim that Seddon made his way from abject poverty. Both his parents were school teachers near St Helens in Lancashire and he himself gained an engineer's certificate. But his English career was marked by disputes and illness; and he eventually sailed for the Victorian goldfields, and then, in 1866, to the West Coast. At the Waimea diggings he sluiced for gold before achieving greater success opening a store and a pub at Kūmara. He acquired a reputation for settling disputes with his fists. In 1870 he was elected to the Arahura Road Board and nine years later to Parliament where he became a spokesman for the gold miners. In 1891 he joined the Liberals' ministry under John Ballance and when Ballance died in 1893 he cleverly outwitted his rivals, William Pember Reeves and Robert Stout, to become premier. So there was enough in this story to evoke an image of the poor boy made good, the hero who had pulled himself up from a working man to 'King Dick' of the colony (a phrase first widely used in 1894). He enriched that image with his style, his vigorous electioneering, and his habit of dropping his 'aitches'.

What cemented Seddon's reputation as the friend of working people was his politics. He was never a socialist, far from it. But he did seek to make New Zealand a haven for working people. Early in his career he gave staunch support to McKenzie's plans to break up land monopolies and provide opportunities and loans for small farmers. Then he endorsed Pember Reeves' plans to protect the wage-earners with factory acts and the industrial conciliation and arbitration system which he believed would ensure fair wages and thriving unions. When Reeves began to propose more radical measures he was promptly despatched to London as the Agent-General; but Seddon continued to promote the arbitration system as a cure for inequality and poverty. But without a doubt the policy which Seddon claimed as his major contribution was the introduction of old-age pensions. Aware that the generation who had migrated from the old world in the 1870s were coming to the end of their lives, he was keen to reward and recognise them for their contribution to the colony. Unquestionably, Seddon deserves credit for the passage of the measure. He studied the question intently and pushed it through Parliament despite strenuous opposition in both houses. Once implemented, he proclaimed its value whenever he went overseas, along with industrial arbitration, housing for workers and, interestingly, women's suffrage, which he had originally opposed. He presented these measures as New Zealand's contribution to the world, the 'social laboratory'. So it was easy for Seddon himself and for journalists to see him as a working man made good and friend of working people.

Seddon's second image, at the coronation in 1902, was as an ardent imperialist. He was a noisy enthusiast for the territorial expansion and superiority of the British race. At home this meant working to exclude the Chinese and expanding New Zealand's own empire into the Pacific. He did not succeed in acquiring Samoa, but the Cook Islands and Niue became part of the New Zealand 'empire' in 1901. He became a tub-thumping supporter of the Empire's involvement in the South African War, farewelling the New Zealand contingents, welcoming them home, and following the action with pins on a big map of South Africa on the wall of his Thorndon house. En route to the coronation he stopped in South Africa to lecture Kitchener on how to fight the war, and Lord Milner on how to ensure a tough peace. Once in London he promoted imperial preference in trade and urged regular imperial conferences. He was seen by the British press as the 'arch-imperialist'.

So there is no doubt that Seddon would have enjoyed putting on this garb for the coronation with all that the gilt implied about his status in the Empire. Yet it is revealing that while other imperial politicians accepted royal honours on the occasion of the coronation, Seddon did not. There was much trumpeting of the fact that he had turned down a possible knighthood or baronetcy. Seddon could enjoy dressing up as Privy Councillor First Class, but he would never undermine his reputation as a friend of the working man. The title 'King Dick' — or 'King Richard', as some preferred — was enough.

50.

Christchurch 1906 exhibition waharoa

MUSEUM OF NEW ZEALAND TE PAPA TONGAREWA

This impressive carved waharoa, or entranceway, 6.7 metres high, with 5.5-metre side posts, now stands opposite the entrance desk at Te Papa. It was originally carved by Ngāti Tarāwhai artists from Rotorua, Neke Kapua and his two sons, Eramiha and Tene, to stand at the entrance of the inner pā at the Christchurch International Exhibition, from 1 November 1906 to 15 April 1907.

The exhibition, which was held on 5.7 hectares of Hagley Park and attracted almost two million visitors (nearly twice the population of the country), was another initiative of Richard Seddon. Returning from his success at the coronation of Edward VII, Seddon was keen to further promote the wonders of 'God's Own'. In 1903 and 1904, he called for an international exhibition as a way of proclaiming 'that New Zealand was a great country'. He laid the foundation stone in December 1905.

Sure enough, the exhibition did promote national identity. The main building, entered beneath two towers, held displays of the country's prosperity to illustrate what James Cowan called the 'material progress of New Zealand since it was first redeemed from barbarism by the white man'. Paintings of 'beautiful New Zealand' and walls of deer antler heads and mounted trout to promote the angler's and hunter's 'El Dorado'. The Department of Labour court showed off the healthy working conditions of the 'social laboratory of the world' in contrast to the sweated labour in the old world, while the exhibition events included a military tournament and axemen's competition highlighting New Zealand as 'a man's country'. The large British court emphasised New Zealand's continuing loyalty to the mother country. The exhibition also promoted New Zealand as 'Maoriland'. At the entrance was a large sign, 'Haere Mai'; the certificates awarded to successful displays were rich in Māori motifs; there was much selling of 'greenstone' trinkets and tikis in the stalls; and above all, besides the artificial lake (which became Lake Victoria) was a large pā. What explains this view among some Pākehā that New Zealand was 'Maoriland'? The context was that Māori loss of land from confiscation during and after the New Zealand Wars, and the continued pressure to sell especially under Seddon's Liberal government, had led to growing Māori impoverishment. European disease had a serious impact. By 1896 the Māori population, at least 100,000 when Pākehā arrived, was some 42,000. Many Pākehā described the Māori as 'a dying race'. In response, some intellectuals began to argue that Māori culture should be preserved before it was too late. In 1892 a Polynesian Society was formed largely for this purpose. Museums expanded their collections of Māori art and artefacts. In Wellington, Augustus Hamilton, an English immigrant, and director of the Colonial (later Dominion) Museum, replaced the natural history exhibits in the main hall with Māori art and authored an influential book on *The Art and Workmanship of the Maori Race*. Other intellectuals, taking their lead in part from American interest in native Americans, suggested Māori culture and history could provide a young colonial society with an instant set of traditions. Painters like Charles Goldie, writers like Arthur Adams and journalist/historians like James Cowan turned to Māori subjects. In addition,

WAHAROA, CIRCA 1906, WELLINGTON, NEKE KAPUA. COMMISSIONED 1905. TE PAPA (ME001771/1-3)

tourist promoters saw Māori culture as adding a uniquely exotic element to New Zealand's scenic attractions. The most influential person in the tourist business was T. E. Donne, the first general manager of the new Department of Tourist and Health Resorts.

The people involved in the Maoriland movement were central in creating a Māori pā at the exhibition. Donne had visited the 1904 St Louis Exposition, where native peoples had been grouped into 'living villages', and returned determined to repeat the idea. To implement this vision, James Carroll, the first Māori Minister of Native Affairs, appointed Augustus Hamilton to take charge. As early as 1899, Hamilton had suggested an 'olden time' model pā at Rotorua peopled by Māori in traditional dress and without any 'pakeha pots, pigs and petticoats'. For help on the Christchurch pā project, Hamilton turned to John Macmillan Brown, a Canterbury professor, who had written dubiously about the migration of Māori, James Cowan who was the official recorder of the exhibition, and James McDonald who had worked with Cowan at Donne's Department of Tourist and Health Resorts and became an assistant at Hamilton's museum. On the pā committee there were only two Māori — J. H. W. Uru, a Ngāi Tahu rangatira, and the young doctor Peter Buck, later the famous anthropologist Te Rangi Hīroa.

Not surprisingly, what emerged from this largely Pākehā group was a pā that presented Māori as 'semi-barbaric' and was 'located in a timeless romanticised never-never land of the European imagination'. The pā was positioned at the far end of the grounds, beside the amusement park. In a sense the pā was an entertainment, for which you paid sixpence to enter. Hamilton was insistent that the Māori in residence at the pā had to wear traditional dress, especially piupiu. They lived in whare and took part in pursuits such as mock warfare.

The pā was divided by a stockade into two areas, an outer and inner pā, with the waharoa at the entrance to the inner pā. All visitors passed under its central arch. Much of the carving at the pā was a polyglot collection of various pieces. Some were older museum carvings. Other pieces came from the proposed village at Rotorua, from Whanganui River, and from Pūniho in Taranaki. Two poupou outside the meeting house had come from a burial ground at Lake Rotoiti. But for most of the new carving Donne and Hamilton turned to Ngāti Tarāwhai carvers. This was a Te Arawa iwi based around Okataina, who as allies of the Crown had retained much land and a thriving tradition of tohunga whakairo (master carvers). In particular, Tene Waitere and his brother-in-law Neke Kapua became well-known to Europeans. Donne had asked the pair to carve a small model house in 1905 and this was obtained for the exhibition.

Wanting more carving for the exhibition, Hamilton brought Neke

Kapua and his two sons, Eramiha and Tene, to the colonial museum. There Hamilton gave instructions suggesting they copy pieces in the museum or from his book of Māori art. For the gateway he pointed to a sketch of a waharoa at Maketū drawn by Horatio Robley; and he purchased a large piece of tōtara from the Taupo Totara Timber Company for £11 15s. The carvers listened but made their own adjustments. Unlike the Maketū waharoa, this one has two supporting pillars, each with manaia figures at the base and two ancestors above. The central archway has a beaked manaia immediately above the ground and two huge ancestors towering in the centre. Hamilton was possibly not wholly enthusiastic about the resulting work, because, according to Roger Blackley, the central ancestor under which visitors passed originally included male genitalia. Today it appears as visibly emasculated. The three carvers came to the exhibition and remained on site for the whole time. To entertain visitors they were required to wear piupiu and cloaks, store their food in a pātaka, cook in traditional ovens, and take part in haka and ceremonies of welcome.

If being an 'old-time Māori' for the enjoyment of Pākehā wanting symbols of national identity sounds an uncomfortable role, at least the carvers were rewarded. They were paid 10 shillings a day and board and travel were covered; and the three men continued to find work from Pākehā patrons after the fair. Further, Neke Kapua appears to have accepted the aims of the pā, arguing that 'our works of ancient times have been brought here, so that the peoples of the earth may know that the Maori is still living'.

Although the pā was named Te Araiteuru after a famous South Island canoe, the local iwi, Ngāi Tahu, played little part in the fair, perhaps because of continuing bitterness over land issues and a sense of being upstaged by North Island tribes, especially Te Arawa who sent several large contingents. Other Māori were enthusiastic. Accounts of the exhibition in the Gisborne Anglican Māori paper *Te Pipiwharauroa* suggest considerable excitement that Māori were being included, and they were especially positive about a visit to the pā from Cook Islanders who brought with them the adze Te Aumapa which was said to have carved the *Tākitimu* canoe. As so often in the past and in the future, Māori reacted creatively to opportunities; and their sense of the pā's meaning was quite different from that of the Pākehā promoters or visitors.

Yet this should not disguise the fact that the primary driver of the Māori pā at the Christchurch exhibition, and the reason that this impressive waharoa exists today, is the desire of New Zealand Europeans for a sense of identity. For a few short years, when Māori had been largely stripped of their land and resources, Pākehā could enjoy an identification with the 'old-time Māori' who provided them with a romantic history.

51.

Jimmy Hunter's All Blacks jersey

NEW ZEALAND RUGBY MUSEUM

This jersey, a precious object at Palmerston North's Rugby Museum, has many threads — male culture, the evolution of rugby, and (yet again!) the development of New Zealand nationalism under Richard Seddon. But let's start at an unexpected place, the history of women. It was women who sewed this striking woollen jersey in the factory of J. Stubbs Hosiery of Palmerston North. This business began after Mary Alice Stubbs and her husband John migrated from England in 1880. While John tried farming, the couple imported a Griswold hand-operated sock knitting machine in 1884, initially for the family's needs. The socks became popular and soon Mary Alice had a thriving cottage industry. In 1889 they recognised the opportunity and the family opened a business as a hosiery manufacturer. In 1905, to attract more women employees, they moved to Newtown in Wellington and won a contract to supply jerseys for the New Zealand rugby team.

Sewing clothes in factories was a growing occupation for women. For much of the nineteenth century there were more Pākehā men than women, so single women were snapped up as colonial wives and 'helpmeets'. Among those employed, the most common occupation was domestic service in affluent homes. But by the 1900s the gender balance evened up, and cities were growing. Single women, and some married women, looked for work as dressmakers in small factories or sweatshops. In 1906, 21,000 women in New Zealand were domestic servants, but manufacturing clothing, the second occupation for women, had over 14,000.

It also tells a story about New Zealand men's love of rugby. How did New Zealand men learn to love rugby? Football began as a rough melee in English villages. Concern about public order led to the suppression of such games, but they lived on in English public schools, where masters decided to impose discipline and rules and introduced a referee. Anxious that, as Britain urbanised, public school gentlemen might become soft, teachers saw football as training manly, muscular Christians.

Rugby came to New Zealand in 1870 when Charles Monro, an old boy at Sherborne College, initiated the first game at Nelson College. At the time, soccer and Australian rules were also played. Rugby's advantage was that the old boys of English and local secondary schools had time and money to travel for regional games. Additionally, rugby was a rough physical game, appealing to men brought up with a tough pioneering heritage. It took off in parts of the world, like Welsh mining communities, where masculine values emphasised physical strength. Rugby was easy to organise. All that was needed was a paddock and a pig's bladder, and the coming of a Saturday half-holiday in 1894 helped. Further, following the South African War, rugby was viewed, in Thomas Ellison's words, as 'the good, manly and soldier-making game', training men to serve the Empire on the battlefield. In the new century the game became a central ritual of boys' secondary schools. In 1902 the Governor, Lord Ranfurly, gave a shield for interprovincial competition, and over the next two

1905 ALL BLACKS JERSEY OF JIMMY HUNTER, NEW ZEALAND RUGBY MUSEUM

years there was the first international test against Australia and the first home test against a Great Britain team at Athletic Park before 20,000 spectators. Then in 1905 the first national team set off for Britain; and Mary Alice Stubbs provided their jerseys.

The next question is why a black jersey with a silver fern? Thomas Ellison successfully proposed at the first AGM of the New Zealand Rugby Union in 1893 that the New Zealand colours be 'Black Jersey with Silver Fernleaf, Black Cap with Silver Monogram, White Knickerbockers and Black Stockings'. Ellison, of Ngāi Tahu and Te Āti Awa, was born at Ōtākou on the Otago Peninsula. His father had come south in search of gold and discovered a claim on the Shotover River now called Maori Point. Ellison eventually became a distinguished Wellington-based lawyer fighting for Ngāi Tahu land rights, but his passion was rugby. In 1888 Ellison was chosen for the 'Natives' rugby team, of whom all but five were Māori. The first New Zealand team to tour Great Britain, they were away for 15 months playing 107 matches. New Zealand's first international jersey on a New South Wales tour in 1884 had been dark blue. But Joe Warbrick, one of the Natives' organisers, decided they should play in black, partly to show off the silver fern. Another path-breaking initiative of the team was to perform a haka before every game. When on his return Ellison was named captain of the first official representative New Zealand team in 1893, he recalled the colours of the Natives team and the black jersey became accepted. Black knickerbockers replaced white in 1901, perhaps because this made for easier laundering!

As for the silver fern, Joe Warbrick had chosen this on the basis of a whakataukī: 'Mate atu he toa, ara mai he toa / Mate atu he tetakura, are mai he tetakura' — When one warrior dies, another takes his place / When one fern dies, another emerges'. It was an image of teamwork. Pākehā too had adopted the fern as a national symbol. In the 1860s settlers sent albums of pressed ferns to families at 'home'; and the cabinetmaker Anton Seuffert used ferns for intricate marquetry. When in 1890 Pākehā organised Native Associations to promote national identity, they adopted the silver fern as their symbol.

The colour black had a further significance. It gave birth to the team's name as the 'All Blacks'. A star of the team, Billy Wallace, claimed that after one game, an English newspaper intended describing the team as 'all backs' but a typographical error changed the words to 'all blacks'. This seems implausible, since it was common to describe a team by reference to their jersey colour. Indeed, after the tour's first game a newspaper noted, 'The All Blacks, as they are styled by reason of their sable and unrelieved costume . . .'

Yet Wallace's argument the team was viewed as 'all backs' had a measure of truth. What attracted enthusiasm in England was that while rugby had traditionally been a rough game of scrummaging forwards, the 'All Blacks' amazed by their passing and running. Of the backs, two shone out — the fullback Billy Wallace and the tiny 165-centimetre-high second five-eighths from Hāwera, Jimmy Hunter. He scored 44 tries in 24 games, a feat never equalled. Twice he scored five tries in a game. After one display the *Daily Mail* wrote 'The electric Hunter . . . cut and wriggled his way through an almost solid phalanx of opponents, feinting, dodging, swerving.' It is apt, should we say 'fitting', that this jersey, sewn precisely for his small frame, belonged to Jimmy Hunter, the 'little wonder'.

The black jersey, the silver fern, and the 'All Blacks' thus became established in 1905. The jersey and team became central to national mythology. This was in part because of their phenomenal success. They won 34 of 35 games (only beaten 3–0 in a controversial 'test' against Wales), scoring 976 points to 59. It was also because English observers heaped praise on the All Blacks. A recent British parliamentary committee investigating the 'physical deterioration' of the race exposed by the poor physical standards of recruits for the South African War had blamed their urban lifestyle. British observers interpreted the All Blacks as showing an 'ethnological fact' that on the colonial frontier British manhood was improved. New Zealand males were the hope of the Empire and the Anglo-Saxon race. The All Blacks were judged superior because they had been toughened on the pioneering frontier. There was much mythology here — the team's huge victories were in part because they arrived early in the season before the locals were match fit, and they played English teams weakened by the withdrawal of the tough Northern players for professional game rugby league. Furthermore, many All Blacks were not country types at all, but from New Zealand cities.

Yet the mythology was important and New Zealand politicians exploited it. The Agent-General, Pember Reeves, travelled with the team, appealing for immigrants and emphasising the country's healthy conditions. Richard Seddon inevitably got into the act. He arranged for the Post Office to post results outside newspaper offices and was so enthusiastic about the team's projection of national identity that he rewarded them with an 'American picnic' — a trip through North America to display their superior manhood. When the team arrived back in early 1906, Seddon, nicknamed 'minister of football', escorted the players to the wharf amid 10,000 cheering Aucklanders. The 'All Blacks' came to personify New Zealand manhood — men who were versatile, modest, good team members, and trained on the frontier to be tough and courageous. In sum, perfect 'sons of the Empire'.

The All Blacks jersey, in this case the exquisite work of women seamstresses, would henceforth carry the hopes and pride of a nation.

52.

The *Dundonald* coracle

CANTERBURY MUSEUM

In December 1907 New Zealand headlines were dominated by the 'thrilling story' of the survival and rescue of sailors from the *Dundonald*, which had been wrecked on Disappointment Island in the Auckland Islands. Large crowds greeted the men at Bluff, and the Bluff brass band performed an open-air concert in their aid. The survivors were given money from the Shipwreck Society, newspapers issued special editions with photographs of the castaways' 'Thrilling Experiences', and items, notably this 'coracle', were displayed in Invercargill, Dunedin, Timaru and Christchurch to raise funds for the men. The coracle became one of Canterbury Museum's most popular exhibits and is still on display today.

The Auckland Islands, 460 kilometres due south of Fiordland, are a wet and windy place sitting in the teeth of the Furious Fifties westerly gales. They became part of New Zealand territory in 1863. Long before then, human beings had arrived. Polynesians settled briefly on these subantarctic islands at the time that they colonised New Zealand, and then in the early 1800s the hunt for seals brought men south for a brief period. In the 1840s two hardy groups settled. After their conquest of the Chatham Islands some Ngāti Mutunga came south with Moriori slaves; and 200 British settlers were lured there by the unrealistic promises of Enderbys, a London whaling company, to establish a whaling base and agricultural community at Port Ross. But the acid soil and lack of sun made growing crops impossible and the whalers returned empty-handed. After several years of misery both Māori and the Enderby community abandoned the Auckland Islands.

For the next half-century the only humans who endured the islands' terrible climate were castaways. The islands were on the grand circle route from southern Australia to Europe via Cape Horn. With inadequate charts and poor visibility, seven ships foundered on their rocky coasts. After the first two sets of castaways in 1864, the New Zealand government established depots with emergency supplies (including three-piece woollen suits for warmth!) and put up finger posts dotted around the main island which pointed in the direction of the depots.

The *Dundonald* was the last ship to strand men in the Aucklands. Although by 1907 steamers were carrying men and goods to Europe, the *Dundonald* was a four-masted sailing barque on a voyage from Sydney to London with 28 men on board. On 7 March in thick fog they suddenly found themselves facing 'black, frowning, threatening cliffs'. Before long, the ship crashed against the rocks and some managed to clamber onto the cliff face and to safety. There were 16 survivors who soon became 15 when the mate died. 'We were on a bleak, barren island about three miles long and two wide. No trace of water, no sign of life . . . All silence, all mountain, and scrub and loneliness.' They were on Disappointment Island, a name given by an early unsuccessful sealer. There were six miles of stormy seas to the main Auckland Island, which the men quickly labelled 'the island of dreams'. The island was thick with nesting mollymawks which they grabbed and ate raw, 'tearing their warm flesh with our teeth, like wild beasts'. They supplemented this diet with seaweed and seal meat. Landing with little clothing, they returned to the wreck and recovered canvas from the sails, which they sewed into clothes using the bones of mollymawks as needles. They made blankets from mollymawk feathers and huts from branches and grass thatch. They managed to save a box of matches and kept a fire burning day and night.

But as the mollymawks fledged and left, the men became more desperate. Their only hope was to cross the perilous waters to the main island and find a depot. But how to build a boat? There were no tall trees on the island and the wood from the wreck had drifted away. Eventually they discovered a grove of hebe bushes, *Veronica elliptica*. It was a hard wood, which had grown in fantastic shapes, curled and twisted by the wind. To avoid blunting their precious knives, the men took off the knobbly bits, lashing branches together using strands of rope salvaged from the wreck. Then they carefully stitched canvas from the sails to cover the frame. For paddles they sewed canvas between

'DUNDONALD CORACLE', JANE USHER PHOTOGRAPH, CANTERBURY MUSEUM, 1952.137.2

forked pieces of hebe. When the coracle was done, they found that with three men on board, not shifting position, there were only a few inches of freeboard. The men could not move. One bailed continuously, while two paddled. They waited until the sea was dead calm. Eventually at the end of July such a day dawned and three men set off. They reached the main Auckland Island, but were unable to clamber through the tortuous rātā forest to reach the depot far to the north. Defeated, the trio returned.

But the depots remained their only chance of survival. The men set to work to build two new coracles. Once more, the gnarly hebe was the frame; but the canvas was all used, so the men unpicked their clothes and stitched bits of canvas together for a new covering. One new coracle was smashed on the rocks when tested. Then the winter gales and driving rain made a voyage to the main island impossible. There was an hour of excitement when they saw a ship approaching, but hopes were dashed when it turned away. They became so ravenous they tore up grass and chewed like cattle. For weeks they waited for the winds to calm. Finally, with the sea still not flat, four men set off on 7 October. Down to their last matches on the island, they carried fire in a peat sod. They reached the shore but waves wrecked their coracle and doused the fire. They were down to two matches. Only one lit — for a split second, then faded. Forced to exist on raw seal meat, they set off northwards through impenetrable bush, becoming wet, cold and ravenous. Then in the darkness they came across a finger post: '4 Miles to Provision Depot'. A strenuous night's tramp and they reached the depot. It had been looted but there were enough ship's biscuits and tinned meat to make the doings for a stew. More importantly, the shelter contained a boat.

They set off to rescue their colleagues but the effort of rowing some 20 miles was too great. They returned to fashion a sail from canvas and this time succeeded. Seven men were dropped at the landing spot immediately opposite the island to walk to the depot and eight returned in the boat. Then all 15 waited for rescue. On 16 November a steamer, the *Hinemoa*, arrived. It was carrying 22 scientists, nearly all the great New Zealand naturalists of the age, including Leonard Cockayne, George Hudson and H. B. Kirk. They were on an expedition initiated by the Philosophical Institute of Canterbury to document the subantarctic islands. After heading further south to explore Campbell Island, the *Hinemoa* returned to retrieve the castaways. But before heading back home the party returned to Disappointment Island to retrieve the body of the mate for burial in a small graveyard at Port Ross established by Enderby settlers. When the party reached Disappointment Island the first thing they saw on the landing beach was the original coracle, since the two later ones had been wrecked by the sea. Looking at the frame of the boat, the scientists were amazed that it could ever have reached the main island. They immediately said 'it ought to be taken back and put in one of our museums, so we gave it to them right away and in the Christchurch museum, New Zealand, it is shown to this day'.

Interestingly, the scientists, believing at the time in a doctrine of the evolution of mankind, took a particular view of the coracle. Leonard Cockayne stated that the value of the survivors' artefacts was 'not because of the daring and heroic deeds connected with them, but . . . they showed how man, when he was thrown into absolutely primitive conditions, in order to obtain sustenance and the necessaries of living once more became exactly a man of the stone age . . . [A]s such the curios were of high ethnological value.' Others, however, saw in the coracle a tribute to British pluck. The Christchurch *Star* waxed lyrical at 'the heroism and defiance of the members of the crew' and the *Dominion* pointed out that when the *Hinemoa* approached the landing in front of the depot 'three hearty cheers left no doubt as to the nationality of the castaways'. Although the 15 survivors included two Norwegians, a Chilean, two Russians and a German, the lesson for New Zealanders was the pluck of the British race. It would not be the last time in the first two decades of the new century when such sentiments were evoked.

53.

Dannevirke Plunket scales

DANNEVIRKE GALLERY OF HISTORY

During the nineteenth century two devices became essential to capitalist society — the clock to measure the hours people worked, and scales to weigh products exactly. Both devices entered the home to facilitate cooking. Recipes with exact measurements and timings became the housewife's friend. Then in the early twentieth century the clock and scales became central to raising children in New Zealand through the efforts of the Plunket Society. Plunket scales became an archetypal object and there are many in New Zealand museums. In later years, Plunket scales became sturdy metal objects, over a metre high, painted cream and with a large basin for the baby; but before then there was no standard design and household scales were adapted for weighing babies. Auckland Museum has American cast-iron scales and a wicker basket for the baby, and Timaru has brass balance scales with round weights placed on one side and a large metal pan on the other. We have chosen the Dannevirke example because it shows the adaptation from kitchen to nursery. These are Salter scales, made by the descendants of Richard Salter, who invented the spring scale in 1770. There is a clock-like face and at the top would normally be a large pan for measuring flour or sugar to make scones and puddings. Instead, a soft canvas hammock is attached by a metal frame. It could be easily carried into mothers' homes or left in the Dannevirke Plunket rooms for visits.

Why did scales, not to mention the clock, become so central to baby-care in New Zealand from the early 1900s? The context was that, as city populations rose, married women changed from being active helpmeets around the farm to becoming full-time mothers and wives dedicated to protecting the morality and welfare of the family. There was also a striking decline in the number of children per family. Among married women in 1880 the average number of children born was 6.5; by 1913 the average had almost halved to 3.3. Each child became more precious. Not everyone was enthusiastic about the decline in numbers. Some saw it as race suicide. Was the Anglo-Saxon race failing before the rising tides of the other races, especially Asian people? Was the British Empire fading for lack of white men? Concern about the falling birth rate led to a society for the protection of women and children, the registration of midwives and the setting up of St Helen's maternity homes (named after Seddon's birthplace!).

Enter in this crisis Frederic Truby King. Although born to a well-off Taranaki family, Truby was a sickly child. He suffered from the typical diseases that produced high infant mortality — diarrhoea, pneumonia, pleurisy and tuberculosis. Two of his four siblings died young and King himself was somewhat of a hunchback and lost the sight in one eye from TB. After initially following his father into banking, King became interested in medicine and set off to Edinburgh where he graduated as the most distinguished student. There he married Bella, an academically successful daughter of a jeweller. Returning to New Zealand, Truby became superintendent of Wellington Hospital and then took charge of the largest mental asylum in the colony, a Scots baronial edifice at Seacliff on the coast north of Dunedin. There is much to admire about King's role at Seacliff. He expanded the hospital's farm and flower garden, providing inmates with fresh air and exercise, good food and blooms in their wards. He established a fishing station at Karitāne, 6 kilometres away, providing inmates with further work and nutritious food. But he also became a good friend of one inmate, the Old Etonian Lionel Terry, who in 1905 had shot a Chinese man, Joe Kum Yung, to raise concern about the 'Yellow Peril'. In the years 1904 to 1907 King's interests came together. From the farming and gardening ventures he developed ideas about correct feeding, and when in 1904 he and Bella adopted a young baby girl, his attention turned to how good motherhood could contribute to alleviating the 'suicide' of the Anglo-Saxon race and the levels of insanity and crime in New Zealand society.

He first emphasised the importance of breast-feeding. Where that was not possible, as in the case of Mary, he recommended 'humanised milk' — milk that approximated mother's milk. Success would be measured on the scales. Then he turned to the development of a child's character. The aim was for children to restrain their impulses and to obey moral dictates. They should not be 'spoiled', lest selfishness and immorality followed. Parents were not to give in to a child's cries. The infant and toddler must be given regular routines in feeding, bowel movements, washing and sleeping. All was governed by the clock — ironical because King himself was notoriously disorganised and only kept to schedule by the meticulous attention of Bella! However, King's gospel laid down: '10 o'clock in the morning [must not] pass without getting baby's bowels to move'. Only through regular routine would children have the inner discipline to succeed in the battles of life — in the marketplace, on the sports field or in the wars of Empire. Without such 'character', children would end up in prisons, asylums or in charitable institutions.

The focus was on boys, future warriors and businessmen. For women the country's destiny depended on their becoming superior mothers. Separate education for boys and girls was central to King's vision. In part he and his supporters were reacting against the feminist movement which claimed political rights and occupational opportunities for women. In part he was reacting against the growing employment of women in urban factories and offices. While boys needed 'a cold bath in the morning . . . and plenty of open-air exercise', women should avoid too much academic work which would make them flat-chested and take their energies away from reproduction and motherhood. In good motherhood lay the destiny of the Empire.

King was a charismatic proponent of his views and in May 1907 he spoke to a crowded meeting in the Dunedin Town Hall. The result was

DANNEVIRKE MUSEUM

the Society for the Promotion of the Health of Women and Children, with a committee of wives of prominent lawyers and businessmen. The Governor-General's wife, Lady Plunket, became patroness of the society and gave it her name. With her support five branches were set up in 1908, and two years later Truby King delivered over 100 lectures on a nationwide speaking tour. Another 60 new branches sprang up. Among these was probably the branch at Dannevirke from which our scales derive. In 1909 it was reported that the Napier Plunket nurse with 'a weighing machine' had been visited by babies from as far away as Dannevirke, and in 1912 the town had a committee for a visit from King. Certainly by the late 1920s there were Plunket rooms at Dannevirke.

Plunket rooms were sanctuaries to motherhood for early twentieth-century women, who, with fewer children, and less reliance on nursemaids, must have been anxious that their children were blooming. The regular weighing of babies on these scales provided reassurance that their infant was 'doing well'. One must acknowledge too that the Plunket routines for feeding, sleeping and defecation by the clock, designed to give young New Zealanders 'character' and 'discipline', also offered mothers 'opportunities for sleep, rest, housework, outings, exercise and recreation'.

In 1908 Bella and Truby, under the pseudonym 'Hygeia', began a regular newspaper column, 'Our Babies'; and two years later the society published King's *Feeding and Care of Baby*, which sold 20,000 copies in five years. The government commissioned *Baby's First Month*, which was given to all new mothers. Karitane hospitals were established to care for sickly infants, and a scheme was set up to train 'Lady Plunket nurses' who could visit mothers at home to give advice and, of course, weigh the babies. King became the Director-General of Child Welfare based in Wellington in 1921 and spent much time overseas promoting 'the New Zealand way' of rearing babies. By 1931 the Plunket Society was claiming credit for the lowest infant mortality rate in the world (which conveniently ignored Māori). Plunket nurses were making 200,000 calls to mothers while 600,000 mothers and babies visited Plunket rooms. Fifteen years later 85 per cent of all New Zealand infants were 'Plunket babies', whose weight gains were carefully inscribed into their Plunket book.

As for Truby King, he was knighted in 1925 and in 1957, after his death, became the first New Zealand citizen to feature on the country's postage stamps. He had become a national icon, and the Plunket way had become central to the early development of New Zealanders. Plunket scales deserve their place in any history of the evolution of the country and its people.

54.

Hurleyville dairy cooperative cheese crate

AOTEA UTANGANUI — MUSEUM OF SOUTH TARANAKI

Blocks of cheddar cheese, cow cockies in gumboots, Anchor butter — the dairy industry has contributed hugely to New Zealand's identity, not to mention to its wealth. It was in the first decade of the twentieth century that dairying really took off. Frozen carcasses of lambs and ewes had been the first exports opened up by refrigerated shipping; butter and cheese followed. In 1890 dairy products earned only some 2 per cent of our export receipts. By 1912 the figure had risen to about 18 per cent — virtually the same as frozen meat.

The new 'gold' was not usually refrigerated, but to withstand passing through the heat of the tropics and prevent the butter and cheese going off, it needed cooled temperatures, just above freezing. Dairy produce travelling to the British consumer was packed in wooden crates — the butter in square boxes, the cheese in 12-sided crates. They were made of kahikatea, the native white pine, because that wood did not taint the food. This crate end, on display in the museum at Pātea, was like many thousands shipped to the British market. At each end was a stencilled marking with the characteristic New Zealand fern leaf, the weight and the description of its contents. There were some complaints from New Zealand's 'Dairy Commissioner' that the markings were not entirely satisfactory, especially the fact that the weight was often scrawled in lead pencil which rubbed off and there was no marking to distinguished so-called white cheese from coloured. But no changes were made. What is somewhat distinct in this case was the word 'XMAS' at the top of the crate. This is not a Christmas greeting or an indirect date stamp, but refers to the name of the exporters, Messrs Lovell and Christmas. They were the agents who in 1912 had been consigned the season's output for sixpence a crate by the Hurleyville Cooperative Dairy Company. Hurleyville is about 10 kilometres north and inland from Pātea in south Taranaki, and its story provides an excellent case study of the developments which made the huge rise in dairy exports possible. For a start, there was the land. Like much of the North Island bush country which became the home for dairying in the Waikato, Taranaki and the Bay of Plenty, Hurleyville was originally land confiscated from Māori after the New Zealand Wars. When in this case the confiscation was not followed up by immediate possession of the land by Europeans, Māori moved back on, and in 1875 and 1877 two blocks totalling over 26,000 acres were 'purchased' from Māori essentially through the use of a gratuity or bribe at the bargain rate of about 8 acres per pound. From 1878, the Crown sold the land off to European settlers, who soon replaced the Māori names of Ōtoia and Ōpaku with Hurleyville, which recognised one of the early settler families, the Hurleys.

The settlers set about making a living, as elsewhere in the North Island at this time, beginning by clearing the bush. Once the trees were gone, they planted introduced English grasses, and for a time the district made a precarious living from the selling of cocksfoot grass seed. From the beginning the early families would have had a few cows, probably of the shorthorn breed which had been brought to New Zealand by Samuel Marsden. The housewife probably milked them by hand in the paddock sitting on a stool or upturned bucket. Then she would leave the milk to stand, pour off the cream on the top, and hand-churn it into butter. In most places the handmade butter would be bartered for other groceries with the local shopkeeper.

Sales of large quantities of butter and cheese were not possible because they would turn rancid. But refrigeration, or rather cooled shipping, changed all that. The first milk factories appeared in the 1880s, and by the early 1890s the Hurleyville settlers saw prospects for commercial dairying. John Sawyer, the Chief Dairy Instructor, reported giving a lecture at Hurleyville in 1892 on 'Cooperative dairying as an economic factor in the prosperity of New Zealand agriculture'. Men began to replace women in charge of the cows. They increased their herds, built simple walk-through milking sheds, and then in May 1893 at a public meeting one Saturday evening in the local schoolhouse 'the Hurleyville folks' decided to set up a dairy factory on cooperative lines. They were following a practice that had first begun on the Otago

"HURLEYVILLE CHEESE CRATE" AOTEA UTANGANUI MUSEUM COLLECTIONS

Peninsula in 1871. Local farmers would band together and own a dairy factory collectively. By the end of the 1890s, 40 per cent of the 150 dairy factories in the country were owned by farmers.

In Hurleyville's case 926 shares at £1 each were bought by the locals and this was enough to fund the building of a wooden factory. Its opening was celebrated with a concert and dance. Within two years the Hurleyville Cooperative was winning prizes for its cheese at the Egmont A&P Association show and claiming to be 'champion in the North Island'. The factory, which also made butter, was fed by about 20 dairy farmers and was 'within cooee of the railway station'. The farmers invested their proceeds in more cows, which were increasingly Jerseys not shorthorns, and before long mechanised milking machines and sheds with a concrete floor. By 1908 some two thirds of Taranaki's sheds had concrete floors instead of mud.

Such developments were happening all over Taranaki as dairy factories sprang up. By 1897 two thirds of the country's butter exports sailed out of New Plymouth or Waitara. But the process was also happening in the Waikato, Bay of Plenty and north Auckland. In 1920 there were 600 dairy factories, 85 per cent of which were farmers' cooperatives. There were changes in the system. Originally the farmers would take their milk to the factory in 20-litre metal cans using horse-drawn carts. There they might meet their mates for a short gossip. But there were issues. In Hurleyville, correspondents wrote to the local newspaper complaining about the muddy state of the roads which made it difficult to transport the milk to the factory. So in many places a creamery was set up close by where the farmer could have his milk separated into cream, which went on to the factory, and skimmed milk, which he took home to feed his pigs. Eventually even taking the bulk milk to the creamery was too arduous and dairy farmers began to buy Alfa Laval or Lister separators which allowed them to separate the butterfat from the skimmed milk in their milking shed. Then only the cream had to be carted to the factory.

There is one more matter to contemplate while looking at this crate which proudly proclaims its content as 'N.Z. CHEDDAR CHEESE'. Cheddar was originally a form of hard cheese that had developed in the Somerset village of Cheddar, probably as early as the twelfth century. Although many early European settlers to Taranaki came from the same area of south-west England as the cheese, it seems unlikely that they brought the making of cheddar with them. It is more plausible that it came via Australia where one of the sons of Joseph Harding brought his father's modern and industrial way of making cheddar to the new country. Certainly, from the mid-nineteenth century the mass production of cheddar took off, and cheddar became the most popular variety for the British consumer. It was this demand that explained the rule of cheddar cheese in New Zealand.

Despite the pride which the people of Taranaki came to have in their land's ability to feed 'Her Majesty the Cow' (a phrase spoken by the Minister of Agriculture, Thomas Macdonald, at Hāwera's 'Dominion Dairy Show' in 1910), there was never any development of specialty cheeses or distinct local flavours. From the time when locals got together to build dairy factories, they had their eyes on export to the British housewife. They wanted to satisfy her taste buds, so they spent little time in developing new kinds of cheeses with unusual tastes. Even if they had tried, specialty soft cheeses perished quickly and were not up to the long sea journey, while cheddar actually improved on the seas. It was not until 1951 when the New Zealand Co-operative Rennet Company at Eltham released a blue vein in silver wrapping that New Zealanders developed a taste for a different style of cheese. Until then the industrial production of bulk cheddar which could sit neatly in these crates and travel to Britain was the focus of the Taranaki cow cocky.

The New Zealand fern-leaf brand stencilled a little crudely on these kahikatea planks became the symbol of both salty New Zealand butter and unadorned New Zealand cheddar which were together used in sandwiches by working people throughout Britain. The two products brought to places like Taranaki and Waikato enough money that they could fund, for the men at least, a distinctive lifestyle — hard work in the sheds morning and night, with the occasional beer and game of rugby at the weekend.

55.

Leslie Adkin's baton

MUSEUM OF NEW ZEALAND TE PAPA TONGAREWA

On 7 November 1913 George Leslie Adkin, a 25-year-old Levin farmer, was riding up Tory Street in Wellington with his troop of fellow special constables. Bystanders cheered the specials who replied with 'Maori war cries'. Then, to quote his diary, 'the fun began — some of our troops (the Levin "boys") batoned ruffians using foul and abusive language. No stones were thrown so I confined myself to hitting pretty smartly across the shoulders and got four — a fifth who swung a bag of tools at me got a good crack on the wrist — grand total of five.' Leslie Adkin had obtained a baton, made of hardwood and about half a metre long, through his old mate from Wellington College, fellow special and Levin dentist, Bernard Freyberg, who would later become a national war hero and Governor-General of New Zealand. When the 'fun' was over, Adkin carved five notches representing his five victims into the handle along with the date ('7/11/13') and his initials (GLA). Thanks to Anthony Dreaver the baton now resides in Te Papa.

Leslie Adkin was a well-educated, intelligent farmer who would subsequently win distinction as a geologist, naturalist and ethnologist. How did such a man become a 'skullcracker', one of 'Massey's Cossacks', in the most violent civil disturbance in New Zealand since the New Zealand wars? The explanation goes back to the 1890s when both small farmers and working people supported the Liberal government. Small farmers benefited from the break-up of large estates, the purchase of Māori land and the provision of loans; workers benefited from the protection of the new Department of Labour and the Industrial Conciliation and Arbitration Act, which gave unions recognition through registration. Both groups slowly lost faith in the Liberals and by 1913 were mortal enemies.

Let's begin with the unions. Despite prosperity in the new century, real wages did not rise significantly. Unionists believed the conciliation and arbitration system had been captured by employers. For 12 years after 1894 New Zealand was a 'country without strikes'; but in 1906 Auckland tramwaymen struck successfully. Two years later the stakes were raised when at the Blackball coal mine Pat Hickey, Paddy Webb and five others were dismissed for claiming a half-hour rather than 15 minutes for 'crib time' (lunch break). The miners went on strike, and after three months the company gave in. Strikes seemed to work. Success in Blackball inspired the New Zealand Federation of Miners, which the next year became the New Zealand Federation of Labor. The spelling of 'Labor' was American, which in part reflected the influence of Hickey, who had worked with the 'Wobbly' Western Federation of Miners.

Hickey, Webb, Bob Semple and their supporters had a vision, derived from overseas, of a class war which would introduce industrial rule by workers. The federation motto was 'The world's wealth for the world's workers'. The strike, not arbitration, became their weapon. Before long, the shearers, the Auckland brewers under M. J. Savage, the Auckland labourers under Peter Fraser and the watersiders joined. By 1912 the 'Red Feds', as they became known, had 15,000 members. In response employers encouraged workers to set up alternative unions under the arbitration system. This was first tried at the country's largest gold mine at Waihī. When the Red Fed union struck in response, strikers and their families were shunted out of town and a striker, Frederick Evans, was killed.

By 1913 the militant anti-arbitration unions tried to establish an enlarged United Federation of Labour, which employers and the new Massey government were determined to thwart. Tensions came to a head in two minor strikes in October 1913 — at the Huntly coal mines and a second on the Wellington waterfront where watersiders were dismissed for holding a stop-work meeting to discuss an issue faced by shipwrights. The shipowners wanted to force the watersiders and Red Fed unions back under the arbitration system. The Red Feds were equally

SPECIAL CONSTABLE'S BATON, 1913, MAKER UNKNOWN. GIFT OF ANTHONY DREAVER, 2016. TE PAPA (GH024824)

determined not to give in. They succeeded in blocking 'scab' workers from the wharves. Wharfies throughout the country gave support, as did seamen and miners. The situation on Wellington and Auckland streets became tense.

With the police unable to maintain order, the government initially turned to the army. But the chief of staff, respecting the British tradition of 'no standing armies', was unwilling. Instead 'special constables' were suggested. Where to obtain them? Some local professional men enlisted as foot specials. But the major source for mounted specials was the farming community.

Why would farmers respond and why had they abandoned their alliance with working men under the Liberals? One factor was that the Liberal government refused to change leasehold land into freehold. Further, farmers became hostile to unions attempting to unionise farm workers. In 1899 a group of small farmers, took a leaf from the working men's book, and created a Farmers' Union. The Farmers' Union put pressure on the Liberals, especially over freehold land, and by 1912 with the backing of the Farmers' Union many farmers abandoned the Liberals and supported a new party, Reform, led by 'Farmer Bill' Massey. They cheered when Massey's government took office in mid-1912, and were enthusiastic about the fierce police response to the Waihī strike. In 1913 farmers feared that union troubles on the wharves would mean their butter and cheese turning rancid or their frozen meat rotting. In late October the export trade peak was several months off; but farmers anticipated the worst.

When the army suggested 'special constables', farmers were an obvious constituency. In 1909 the Defence Act had replaced volunteers with regional territorial units. In the Wellington area territorial regiments were used to recruit the specials, but the Farmers Union was also important. In Adkin's case he was attracted by a public meeting in the Levin Druids' Hall called by Herbert Richards, a close supporter of the Farmers' Union's founding president, James Glennie Wilson. Farmers on their horses constituted many of those who came to Wellington, and then Auckland, in late October and early November 1913. Adkin loaded his horse Fanny, newly shod, onto a railway wagon at Levin station along with 24 other steeds. Of the 1131 mounted specials in Wellington, most came from the Wairarapa, Taranaki, Hawke's Bay and Manawatū. Dairy farmers from south Taranaki were well represented. There were others — Freyberg, Adkin's closest mate whom he described as 'a great chap', was a dentist; and Adkin befriended a fellow special, Elsdon Best, the prominent ethnologist.

Initially the Wellington specials were housed in the Post and Telegraph building. But on 30 October there was a violent altercation when revolvers were allegedly fired in Post Office Square, so when Adkin arrived on

2 November the specials were in the army drill hall on Buckle Street, Mt Cook. Locals in the working-class Mt Cook area were not happy, and when Adkin arrived the troops were greeted with hooting and stones. Later, order was maintained with machine guns at each end of Buckle Street and the specials charging on their horses. Tensions were high. Abuse flowed. The strikers called the specials 'cow-spankers', 'teat-pullers', 'country bumpkins', not to mention Edward Tregear's dismissal of them as 'outcast scum'. The sentiments of the specials were well expressed in the words carved by a Taranaki famer, Hugh White, into his baton, 'Society Combats Anarchy Brutality Syndicalism'. Given such feelings it was fortunate that batons, not guns, were the weapons of choice. Of course batons could hurt — especially when wielded under the instructions of Colonel Chaffey: 'Let the first charge be a lesson to the workers of New Zealand. Pick your man and put force behind your blow.' The baton became the specials' symbol, featuring in cartoons including one in the *Free Lance* showing a figure of Zealandia handing a baton to a farmer with the words: 'Mob rule threatens my towns . . . Take this baton and uphold law and order.' This was class war — employers and farmers versus workers.

In the short term the employers and farmers won. On 5 November the specials cleared the Wellington wharves of strikers and allowed strike-breakers to load the ships. Auckland wharves were seized three days later. Although a call for a general strike followed, the uptake was unimpressive. On 22 December the watersiders returned to work. Leslie Adkin went back to his Levin farm, after collecting £8 for 16 days' 'service'. The crisis cemented the alliance of employers and farmers behind Bill Massey's Reform Party. They would remain for over a century the core constituency of Reform and the National Party that succeeded it. As for the workers, many Red Feds hoped that revolution would come, not through party politics, but through workers taking over industry directly. Yet just months before the wharf battles, Paddy Webb, coal miner and unionist, was elected Member of Parliament for Grey. Radical unionists began looking to parliamentary politics to remedy their problems. The Labour Party was formed in 1916 and 19 years later it would take office as the first Labour government, which included many who had fought in 1913. That division between workers and farmers which weakened the Liberals had long-term consequences.

Within a year many who had learnt to wield batons among the specials would find themselves in a very different war. In Wellington the commander of the specials, in effect if not name, was Andrew Russell, who would lead the New Zealand Division on the Western Front; while in Auckland the foot specials were commanded by Arthur Plugge, who led the Auckland Regiment at Gallipoli.

This baton carries a very long story.

56.

Edmond Malone's lemon-squeezer

PUKE ARIKI

On 16 October 1914 New Zealand soldiers sailed out of Wellington Harbour to head for the conflict that became known at the time as the 'Great War'. During the next four years over 100,000 New Zealand men and 550 army nurses served the Empire overseas, and over 18,000 died in the most horrific experience for Pākehā New Zealanders, and for many Māori, since colonisation. The Great War became a defining experience for the country and because so many served overseas, ways of identifying New Zealanders became significant. The fern leaf was used on army uniforms, but as a name for the country's soldiers it never took off. The kiwi became a distinctive symbol most famously expressed in the large kiwi carved into the hill above Sling Camp in Wiltshire, but it was only later used as a collective name. Often the men referred to themselves as 'Anzacs' or 'diggers' but these were terms shared with Australians. Arguably the most enduring symbol of national identity for the New Zealand soldier became the lemon-squeezer hat. It appeared as a symbol on war memorials throughout the country in the 1920s, and two decades later the lemon-squeezer was enlisted to recruit men for another world war. This lemon-squeezer, now at Puke Ariki in New Plymouth, is a fitting example for it was owned by Edmond Malone, a Taranaki boy and son of William G. Malone, who was probably responsible for the adoption of this headgear.

William Malone had migrated from London in 1880, joined the Armed Constabulary and taken part in the attack on Parihaka the next year.

PA2009.038 HAT, LEMON SQUEEZER, 1914–1918, EDMOND LEO MALONE, MARY MALONE, COLLECTION OF PUKE ARIKI, NEW PLYMOUTH

He worked as both a farmer and lawyer near Stratford in Taranaki, but his real passion was the military. Convinced that war with Germany was inevitable, he studied military history, and prepared himself for the coming conflict by rationing his food, sleeping on a military stretcher and keeping fit. He also played a leading role in local military endeavours. During the South African War he organised the Stratford Rifle Volunteers, and after the military reforms of 1909 took command of the local territorial force, the 11th Taranaki Rifles Regiment. It was in this context that Malone was allegedly responsible for the lemon-squeezer.

The common story is that at the annual camp in 1911 Malone introduced the lemon-squeezer as the regimental hat. Until then the normal headdress was a soft felt hat, turned up on one side, as worn by the New Zealanders in South Africa. The lemon-squeezer was the same hat, with a flattened brim and four dents on each side creating a peak. This seemed appropriate because its shape mirrored the outline of Mt Taranaki; this can be seen in the regimental badge fixed to the front of Edmond Malone's lemon-squeezer which shows the mountain echoing the shape of the hat. When on the outbreak of war Malone was appointed to command the Wellington Battalion, the lemon-squeezer was extended to the whole battalion.

This sequence of events has been challenged. The granddaughter of Hubert Hart, Merrilyn Bartram, claims that he had admired its use by the South African constabulary while on service in the South African War. Just before the Great War, Hart was a company commander in the Ruahine Regiment, so he took a hat, gave the peak four dents and the style was adopted by the whole regiment. Then when war broke out Hart is said to have convinced Malone, despite his scepticism, to apply the style to the entire Wellington battalion. Colin Andrews has yet another story. He claims that from their organisation in 1911 the Taranaki Regiment had worn undented slouch felt hats like other regiments. But at the divisional camp at Takapau in April–May 1914, there had been torrential rain creating miserable cold conditions, and the men were delayed in camp for an additional four days without extra pay. They rushed the headquarters, shouting abuse and throwing stones, and as a concession Malone allowed the men to change their slouch hats to a lemon-squeezer so that it afforded better protection from the rain.

Both these versions suggest that Malone was unwilling in his acceptance of the lemon-squeezer; but there are several photographs taken before 1914 showing Malone himself wearing a military hat in that style. It may not only have been the echo of Mt Taranaki which appealed to Malone. The lemon-squeezer had also been the favoured style of Robert Baden-Powell who used it for the Scout movement. The Scouts expressed that belief in British manliness and imperial pluck which Malone himself accepted. So it seems fair to credit Malone with the lemon-squeezer and to recognise that while it carried a distinctive local, and eventually national, message, its origins were firmly within the tradition of the Empire.

Photographs confirm that when the Wellington Infantry Battalion left New Zealand heading for Egypt, they were wearing lemon-squeezers. Members of other units did not. The 'lemon-squeezer battalion' was distinctive in other ways. As its commander William Malone was a real martinet. His diary of the voyage to Egypt and then the five months spent camped at Zeitoun just outside Cairo show a man determined to enforce military discipline and British imperial values. As the ship left Wellington he refused permission for the men to be excused drill to see the harbour heads and ordered the band to stop playing because that was 'too cheap for anything'. He was determined to avoid 'slackness and demoralisation'. In Egypt he worked his men mercilessly.

Once the men landed on the steep slopes of Gallipoli on 25 April 1915 and were faced with the ordeal of fighting a stubborn Turkish enemy, who held the higher slopes and were defending their homeland, things changed. Faced with the constant danger of snipers above, and lice and dysentery in the trenches below, parade ground uniforms were minor problems. Photographs suggest that all the New Zealanders, the Wellington Battalion included, wore a mix of styles. Many at first wore the 'Tommy hat', a flat-topped cap with a peak in front. Others wore pith helmets. Increasingly to protect themselves from the summer sun they wore the felt hat pulled without dents onto their heads. As for Malone, he changed his view of his men. He became convinced that the hard discipline of training had paid off and they were 'grand fellows . . . Nothing better in the world.'

After Malone spent most of June and July 1915 'straightening things out' at Quinn's Post, a weak point on the front where the Turks were only some 50 metres away, his Wellington Battalion took part in the attack of early August on Chunuk Bair. After fierce fighting he led his men to the top, the high ridge on the Gallipoli Peninsula from where they could look down on both sides. The men hoped it was the key to victory, but there was little support. Malone and his men stuck it out all that fateful day of Sunday 8 August. In the late afternoon Malone stood up to survey the situation. A shell fired by a British destroyer burst over his trench and he fell dead. Only 76 of the 700-strong Wellington Battalion were not killed or wounded on that day. The summit was eventually retaken by the Turks and the Gallipoli invasion was doomed.

Edmond was wounded on Gallipoli, as was his elder brother Terry. But when the New Zealanders withdrew from Turkey and sailed to Europe, Edmond rejoined them, this time as a second lieutenant in his father's old Wellington Battalion. The New Zealanders were no longer part of the Australian and New Zealand Division, but had become a national unit, the New Zealand Division. In helping to create a distinctive identity for the new division, it was decided that the lemon-squeezer would become the official headdress. Order 135 of the New Zealand Infantry Division in September 1916 was: 'The slouch hat will in future be worn by all ranks as follows — brim horizontal, crown peaked.'

As for Edmond, he soon found that trench warfare on the Western Front was no picnic after Gallipoli. The New Zealanders had exchanged heat and rocky hillsides for the mud and constant bombardment of France and Flanders. The long-drawn-out trench warfare on the front took a terrible toll. Almost 12,500 New Zealanders lost their lives, two thirds of the country's total war dead, and thousands were injured in that theatre. Edmond was one who paid a fearful price. He was injured three times. The first was in early June 1916 north of the line near Armentières. While recuperating in England he fell in love and married an English nurse, Mary Brocklehurst. The second occasion was the following autumn in the awful offensive of October 1917 at Passchendaele when the rain did not stop and the Germans' barbed wire and concrete pillboxes thwarted the attempted advances. Edmond was shot in the shoulder and was awarded a Military Cross 'For conspicuous gallantry and devotion to duty'. Once more he returned to the front. This time, in March 1918 when the New Zealanders were resisting the Germans' spring offensive, he was again shot, and died 10 days later.

Edmond Malone's lemon-squeezer tells us two stories about the impact of the Great War on the distant small community that was New Zealand in 1914–18. One is a personal story about a family who believed wholeheartedly in the imperial cause, and paid for their commitment with two tragic deaths. The second is of a nation finding a way to mark their distinctiveness as a people. The lemon-squeezer came to be treasured as one expression of that sense of nationhood that the war helped to create.

57.

Margaret Cruickshank's gold watch

WAIMATE MUSEUM AND ARCHIVES

On 13 February 1913 Dr Margaret Cruickshank was presented with this beautiful gold watch and chain plus a purse containing 100 sovereigns at a public meeting in Waimate. She had been a doctor there for almost 16 years, and was given these expressions of admiration by 'her many Waimate friends', as the inscription on the back reads, before leaving for a trip to the old country. The gold watch, now held in the Waimate Museum, opens up several fascinating stories.

Let's begin with the watch itself. Portable watches were made possible by the invention of a mainspring in the fifteenth century, and within a century they were fashionable items for European gentlemen. Gold watches were often given to mark special occasions, not necessarily retirement. Such was the case with Margaret Cruickshank's watch.

Watches remained items of luxury in New Zealand until the end of the nineteenth century. For both Māori and Pākehā, time was measured by the rising and going down of the sun, or in towns by the firing of a cannon at midday. Then came a greater need for precision. When the telegraph system imposed Wellington time on the country in 1868 to enforce consistency in manning their offices, a national time was established by parliamentary resolution. New Zealand was the first country to do so. By the end of the century larger cities called for greater attention to exact time. Precise hours and minutes were important for train timetables and events like concerts or sports games. Hours of work and the opening and closing of pubs and shops were prescribed. Health too required exactitude — Truby King laid down 'feeding by the clock', while doctors and nurses measured heartbeats by a watch. Wristwatches did not come into general use until made popular by the physical conditions of trench warfare in the First World War. So this watch was both a mark of esteem, a precious heirloom which carried associations of gentility and popular regard, but also a reminder of the central place of the clock in modern life and medicine.

The second issue is what induced the citizens of Waimate to present a gold watch to Margaret Cruickshank? Margaret and her twin sister Christina were born at Palmerston (south) on New Year's Day 1873, daughters of an engineer who had come out to the Otago goldfields. Ten years later their mother fell ill and died and the two girls alternated between attending school and looking after their five younger siblings. They both went to Otago Girls' High School, sharing the prize as dux, and, perhaps inspired by that school's pioneering commitment to female education, both proceeded to Otago University. While Christina studied the arts eventually becoming principal of Wanganui Girls' College, Margaret followed her friend Emily Siedeberg to Otago Medical School. They were pioneers in a profession where women were not welcomed. There is one story, later downplayed by Emily Siedeberg, that in the dissecting room the young male students threw flesh at her. Margaret persisted and in 1897 followed Emily as the second woman to graduate from a New Zealand university in medicine, capped in the same ceremony as Ethel Benjamin, the first woman law graduate. That same year she

WAIMATE MUSEUM

became an assistant to Dr Horace Barclay at Waimate, and thus became the first New Zealand woman to practise as a general practitioner. She lived with the Barclays until her death.

Margaret Cruickshank's arrival was not wholly welcome. The *Bruce Herald* headlined its story 'Licensed to Kill' but the Waimate paper grudgingly accepted her, conceding, 'The day of the lady doctor, like the women's franchise and the labour legislation, seems to be upon us.' However, she rapidly won the admiration of the local community for 'her largeness of heart and well known generosity'. She apparently combined a fine understanding of medicine with enormous personal sympathy for the sufferings of her patients. The Presbyterian minister recalled that her strong Christian faith involved 'washing dishes when she saw there was a need . . . pulling off her coat and scrubbing floors when the sick one could not do it'. She was known as a special friend to the poor, and did not worry if they could not pay. So, when in 1913 she decided to take a year's leave to extend her medical learning in London, Edinburgh and Dublin, the community expressed their esteem with a gold watch and 100 sovereigns. What is interesting about the watch is that it is like a nurse's watch upside down with the 12 o'clock at the bottom. This allows it to be read from above when pinned to clothing, thus freeing the hands to take the pulse. One wonders if Margaret's gender encouraged the community to conflate her role as both doctor and nurse!

Not long after she had returned from Europe, the Great War broke out. Dr Barclay departed to serve, and Margaret was left to take on his full caseload and share responsibility as superintendent of the Waimate Hospital. She organised the local Red Cross, and was so popular that she was Waimate's candidate in a Queen carnival to raise money for voluntary war support. Then her duties became especially onerous. At the end of October 1918 as New Zealanders eagerly anticipated the coming peace in Europe, people started dying in Auckland from a highly contagious new form of influenza which appeared to hit adults in the prime of life and turned bodies of the victims dark purple and black. The flu appeared first among soldiers at camp in the American Midwest. At this stage it was comparatively mild and no one was concerned. But, perhaps because of the atrocious conditions and the effect of mustard gas, the flu mutated among soldiers on the Western Front. Many died from the pneumonia triggered by the illness. As the soldiers returned home they carried the flu around the world.

At the time, it was believed that the terrible illness was brought to New Zealand by the RMS *Niagara*, which arrived on 12 October and was said to have been excused from quarantine because it carried Prime Minister Bill Massey and his deputy Joseph Ward back from a wartime London conference. The evidence does not fully support this conjecture, but certainly by 29 October Auckland was in the full throes of the epidemic. Then it spread south. By Christmas, when the plague was over, between a half and a third of all New Zealanders had caught it. Over 6000 Pākehā (six in every thousand) and over 2000 Māori (at the appalling rate of 42 a thousand) died.

In the emergency, shops, offices, factories, schools, hotels and theatres were closed and temporary hospitals were set up in church halls and racecourse grandstands. With a third of the country's doctors overseas on war service, medical personnel were stretched. Many cures and preventatives were tried — medicinal doses of whisky and brandy, kerosene sprinkled on sugar, and the inhalation of zinc sulphate — but they usually caused more illness than they cured. Many places were organised into blocks and those voluntary organisations which had been set up to provide support during the war — the Red Cross and patriotic societies — were summoned back to work. They visited every house in each block providing food, bedding and childcare or sponging patients with raging temperatures.

Waimate saw signs of the flu in the second week of November as people returned from the Christchurch Show, the Addington races and celebrations of the Armistice. By 15 November there were five bad cases in hospital and it was decided to erect marquees to take the expected influx. Shops were closed at 3 pm to encourage people to take health-giving exercise in the open air. Margaret Cruickshank worked at a fierce pace. When her driver fell ill, she visited sick families by bicycle, and in addition to dispensing medicine would lend a helping hand by feeding the baby or even milking the cow. On 21 November it was reported that Dr Cruickshank had caught the flu and a week later she died aged 45. She was one of 14 doctors around the country who succumbed, and one of 17 people who died in Waimate town, a rate of death close to double that nationwide.

The community was so bereft by the loss that they lined the streets to honour her cortège as it passed and within days a memorial was suggested. After debate it was decided to erect a marble statue in her honour. The sculptor, William Trethewey, heavily involved in carving war memorials at the time, imported an 8-ton piece of marble for the task, and then displayed clay models of Margaret Cruickshank in a Waimate shop window so that he could get the likeness right. The fine statue, 2.7 metres high, was unveiled in Waimate in 1923. The inscription reads, 'The Beloved Physician / Faithful unto Death'. Her gold watch turns out to be just a precursor to Waimate's larger tribute to New Zealand's first woman doctor.

58.

Tainui mere, a princely gift

MUSEUM OF NEW ZEALAND TE PAPA TONGAREWA

If you like royalty, then this story about these two beautiful pounamu mere is for you. They were presented on behalf of the Māori King to the dashing young Prince of Wales at Rotorua in 1920. They continued to be held by the Duke of Windsor, as the Prince became after his abdication as Edward VIII, at his villa in Paris. When the Duchess died in 1986 the estate was leased by the Egyptian owner of Harrods, Mohamed Al-Fayed, who purchased the contents. These were expected to be auctioned in late 1997, but three days before, Al-Fayed's son Dodi and Diana, Princess of Wales, were killed in a car crash so the auction was postponed. When the auction was held in 1998 Tainui lobbied to have the mere returned. They were bought by an anonymous New Zealand collector who paid eight times their estimated value. Eventually they came to Te Papa.

Behind this short story is a richer tale. The mere were a central symbol in a long courting of British royalty by Tainui and the Māori King movement. That movement emerged in the 1850s because some Māori had met Queen Victoria in England and saw the strength that derived from a royal house. They decided to set up a Māori King, not in opposition to the Queen, but in imitation — to provide one powerful voice of Māori interests. The tribes of the *Tainui* canoe, especially Waikato, responded to the idea and the Waikato leader Pōtatau Te Wherowhero was anointed by Wiremu Tāmehana, the King-maker, at Ngāruawāhia in 1859. The King movement continued to seek a close relationship with the British Crown partly because Victoria was the other signatory to the Treaty of Waitangi and also because they wished for the Queen's support in their concerns about New Zealand government action. They hoped royal acceptance would bring recognition within New Zealand and the return of confiscated land. The result was a long courting of Victoria and her family by the Kīngitanga.

When Prince Alfred, the Duke of Edinburgh, and second son of Queen Victoria, paid three visits to New Zealand in 1869–71, the first time a member of the British royal family had set foot in the country, there were complicated negotiations to have the Duke visit the new King, Tāwhiao, at Ngāruawāhia. Eventually the efforts fell through. However, Tainui remained committed to the idea that a meeting with the Queen could bring recognition and a resolution of their grievances, especially loss of land. In 1884 King Tāwhiao travelled to London in the vain hope of seeing the Queen. Then in 1901 came another chance. The Queen's grandson, the Duke of Cornwall and York, paid a visit in June. A major gathering of Māori at Rotorua was planned. But the Māori King wanted to meet royalty on his own ground. Tainui insisted that it was against tradition to meet a distinguished guest for the first time on land other than their own, and they had a long enmity with Te Arawa, the Rotorua hosts. The King offered to meet the Duke either in Auckland where they had traditional land or at Ngāruawāhia. Suggestions that Tainui send out waka to greet the Duke and Duchess when they arrived in Auckland were spurned by the government, who insisted on just one Māori gathering. Tainui did not attend the huge Rotorua hui.

The dream of a royal meeting persisted. In 1914, just before the outbreak of war, the fourth Māori King, Te Rata, travelled to London and was eventually granted an audience with George V, on the condition that no grievances be raised. Then in 1920 came a third royal visit. The visitor was Edward, the Prince of Wales, heart-throb of the Empire. The major reason for Edward's visit was to thank New Zealand for its service to the Empire in the Great War. Edward, himself an officer in the war, arrived in Auckland to a large banner: 'Welcome to the Digger Prince from New Zealand Diggers'. He wrote later, 'Returned soldiers and shrieking people and school children are all that I remember of my visit.' Māori had sent some 2000 soldiers to serve first in the Native Contingent on Gallipoli and then in the Pioneer Battalion on the Western Front. As in 1901, one major gathering was planned for Māori at Rotorua, where the Prince could inspect Māori veterans, and attend another large reception at Arawa Park racecourse.

Most Waikato supporters of the Māori King had refused to volunteer for the war and, when conscription was selectively imposed on the tribe, some had been arrested. But they remained committed to meeting and achieving recognition from the royal visitor. As in 1901 Tainui and the Māori King were unhappy about attending the Rotorua hui. Once again they invited the Duke to Ngāruawāhia, hoping to greet the Prince and accompany him to Rotorua. About 1500 Tainui supporters gathered at the local station with stacks of food ready. The Prince's train slowed; Māori clad in mats held a placard, 'Haeremai', and the kuia waved greenery and sang waiata; but the train did not stop. Rumours that Waikato would sit on the line did not eventuate.

The gathering at Rotorua was huge — over 5000 Māori, most of whom, transported there free by train, camped on the racecourse. The 'Young Maori Party', that group of reforming leaders who had made a major impact on politics from the turn of the century, were present. Peter Buck (Te Rangi Hīroa), a former MP from Taranaki, used his medical background and wartime experience to oversee the camp. Māui Pōmare, another doctor and now government minister, welcomed the Prince to the gathering and guided him through the ceremonies. Apirana Ngata, also an MP, was in charge of the canteen and led Ngāti Porou in a fierce haka. The Prince was not impressed — he wrote to his mistress, 'I've had such a terrible day of Moaries [sic] & all their comic stunts . . . a reception which lasted three hours.' But he did recall that 'they gave me some fine presents when all the hakas & poi dances were over'.

The Tainui tribes stuck to their principles. 30 Tainui came to Rotorua 'in a private capacity' and lived in quarters independent of the camp. They did not take part in haka or poi. The Tainui group were led by Tupu Taingakawa. Of Ngāti Haua, Taingakawa was the son of Wiremu Tāmihana Te Waharoa, the famous King-maker. From the 1890s Taingakawa was the tumuaki, in effect the Prime Minister, of the King movement and firmly committed to Māori self-government. He had initiated and led the delegation to King George V in 1914.

The Tainui presence, although 'unofficial', did not go unnoticed. The Prince, keen to reciprocate the meeting in 1914 with his father, apparently enquired whether the King Te Rata or Tupu Taingakawa were present and requested that they come to receive a message from the King. When Tainui found out that the invitation was direct from the Prince and not from the Government or Te Arawa, they agreed to meet at the Grand Hotel. Expressing a hope that royalty might support upholding the principles of the Treaty of Waitangi, Taingakawa presented the Prince with two mere. The shortest and heaviest one was 'Kauwhata', which Taingakawa handed over with the handle pointing to the Prince, signifying that it should remain the Prince's property forever. The second mere, the thinner and taller one, 'Wehiwehi', was presented with the handle pointed away from the Prince suggesting that it should eventually be returned to the tribe.

As for the mere, they were undoubtedly old. A mere pounamu had spiritual qualities and was the main symbol of chieftainship. To give it was a sign of good faith. Their origins are less clear. Taingakawa, keen to emphasise the Kīngitanga heritage, stated when presenting 'Kauwhata' that it belonged to his ancestor, Tāmihana; and that 'Wehiwehi' was the battle mere of the first Māori King, Te Wherowhero who used it to slay 20 enemy in a battle at Ōkoki in Taranaki in 1821. But the names of the mere refer to two famous Tainui ancestors, Kauwhata the father and Wehi Wehi his son, who were the eponymous ancestors of two hapū based on the west side of Maungatautari. Both hapū had suffered grievously from the operations of the Native Land Court and had travelled to Rotorua with the mere to have the Prince initiate investigations of their land loss. But they had not been allowed to make their own representations to the Prince personally or present their ancient mere so Taingakawa did so on their behalf. Perhaps the gift of 'Wehiwehi' with the handle pointing to the Tainui donors, suggested that it might eventually be returned along with their land.

While not all stories have a happy ending, this one does have some resolution. The King movement's gesture at Rotorua did not immediately establish full relationships between the Māori King and British royalty. On visits by members of the royal family to New Zealand in 1927 and 1934, appeals for a welcome at Ngāruawāhia fell on deaf ears. Finally, on 30 December 1953, the royal car turned into Tūrangawaewae marae and the first reigning British monarch to come to New Zealand, Queen Elizabeth II, was given a traditional welcome and led to the whare Mahinarangi. As for the mere, the anonymous New Zealand collector who bought them at auction was persuaded in 2001 to sell them to Te Papa for $143,000; and they are held today by the museum, which shares guardianship of the taonga with Ngāti Kauwhata and Ngāti Wehi Wehi.

59.

ATCO motor mower

MOTAT | MUSEUM OF TRANSPORT AND TECHNOLOGY

On the first Saturday of December in 1923 the 'famous ATCO motor mower' was demonstrated in Victoria Square, Christchurch. We do not know the size of the audience, but by the 1920s New Zealanders would have been interested in easier ways of mowing their lawns. From 1922 the ATCO motor mower was frequently advertised as 'the most efficient mower on earth'. There are two of these mowers in New Zealand museums — one at MOTAT in Auckland and one at Gore.

The idea of mowing grass for smooth lawns was European. It began with a French eighteenth-century fashion for formal gardens and was carried to Britain's aristocratic estates. Then in 1830 a Gloucestershire engineer, Edwin Beard Bunning, invented a mechanical mower with blades around a central cylinder. At first the mowers were pulled by horses, sometimes fitted with leather boots to protect the grass; but from the 1880s hand-pushed mowers were marketed in the USA and Britain. The first machine-driven mowers, using internal combustion engines appeared at the turn of the century. The ATCO model was the invention of Charles Vernon Pugh, who had to mow grass around his father's Birmingham factory. The company made small metal goods such as bicycles or items for armaments, but once the ATCO mower went into production in 1921 it rapidly became the firm's signature product. The ATCO used a Villiers two-stroke petrol engine. Both examples have the name 'CHAS H PUGH L^{TD} BIRMINGHAM, ENG.' inscribed along the mower's metal handles, and 'THE VILLIERS ENGINEERING COMPANY LIMITED WOLVERHAMPTON' on the flywheel.

The arrival of the ATCO mower was perfectly timed for New Zealand's needs. British colonists had introduced English grasses to replace indigenous tussocks as food for farm animals. Economic reasons were supplemented by sentimental ideas that an English lawn represented British civilisation. The rich developed mown lawns in gardens, providing a place for croquet and tennis. Town pioneers set aside parks and public gardens for entertainment and as a memory of 'home'. Auckland had its Domain, Christchurch its Hagley Park.

But city life was regarded by colonials with some suspicion. The crowded slums of Britain were one of the 'diseases' which the new society sought to avoid; and much nineteenth-century migration propaganda promised immigrants the lure of owning a piece of land. Those recruited were often farm labourers rather than city slum-dwellers. But inevitably New Zealand towns became cities. The 1911 census claimed that for the first time more non-Māori New Zealanders were living in towns and cities than in rural areas, and by the 1920s two thirds were urban dwellers. This was a source of some concern. Early New Zealand towns and cities had poor infrastructure — roads muddy and littered with rubbish and horse manure. The view of the city as unhealthy was reinforced by the 1918 influenza pandemic, which hit urban dwellers hard. City life was seen as undermining the pioneering outdoor life which was so central to national

CHAS. H PUGH LIMITED. CIRCA 1920S. LAWN MOWER [ATCO], 1967.112. THE MUSEUM OF TRANSPORT AND TECHNOLOGY (MOTAT).

identity. Urban life might bring physical degeneracy and crime.

One solution was to clean up the city with drainage and sewerage systems, and municipal rubbish collections. Another was to limit the city's 'immoral' attractions. It became an offence for prostitutes to solicit for business; and during the Great War six o'clock closing of pubs was introduced which became permanent after the war. Shopping hours were also limited. But the key development, proclaimed in a town planning conference in 1919, was promoting the suburb as not just for the elite but for the middle and working class. The suburb was a place where New Zealanders could enjoy the healthy life of a garden and lawn, while working in the city. Once the pubs closed their doors at 6 pm the workers could return to their moral suburban haven. The promise of every family owning land could be fulfilled in the suburb rather than on the farm.

Two other forces created the 1920s suburb. One was a revolution in transport — not the building of tar-sealed roads and private cars (although they helped), but the development of tram services, especially in the four main centres. From the 1880s these were horse-drawn, and property speculators took the opportunity to subdivide land on city fringes — areas like Mt Eden in Auckland, St Albans in Christchurch and Newtown in Wellington. Then in the 1900s electrical trams were introduced to the big cities and also Invercargill, Whanganui, New Plymouth, Napier and Gisborne. By 1926 Auckland had over 80 kilometres of tramlines, and suburban housing had developed along these lines. Christchurch had 10 spokes going out from Cathedral Square. The result was the suburbs of Riccarton, Fendalton and Papanui in the west; St Albans, New Brighton and Sumner in the east; Cashmere and Addington in the south. People came to town to work, do the shopping, watch the movies or sports, and then, when the pubs and shops closed, get on the trams and head to their quarter acre suburban paradise. By the 1920s only a quarter of city-dwellers lived in the inner city.

The second development behind the suburb was the Reform Party government headed by William Massey. Although 'Farmer Bill' fought the radical labour movement in 1912 and 1913 and was a jingoistic imperialist, he courted the support of working people. At the end of the Great War, New Zealand society was highly unstable. Men returned from war injured or psychologically damaged. There were strikes, a growing Labour Party and a Bolshevik revolution abroad. There was conflict between Catholics and Protestants. Massey wanted to settle down society and gain the confidence of hard-working New Zealanders. In that tradition of conservatism carried on by Keith Holyoake, Robert Muldoon and Jim Bolger, Massey happily used the state for populist ends. Under his rule there was a 40 per cent increase in the state's share of GDP. His government raised old-age pensions, added a widows' pension and family benefit, and increased spending on education to open up secondary education to most New Zealanders. But his most important act was using the State Advances Department to lend money on easy terms so that workers could purchase a suburban home. 'If you wanted to knock Bolshevistic notions and revolutionary socialistic notions out of men,' he said, 'you give them a stake in the country — something to lose, something to take pride in.' Caring for and mowing your piece of land was an alternative to revolution.

So in the 1920s suburbs proliferated. The style of house changed with their location. Inner-city working cottages had a small living room in front, kitchen behind, while lavatory and wash-house were outhouses in a tiny back section. The suburbs had streets of California bungalows with a central passage, larger living and kitchen/dining rooms and a water closet inside, connected to a sewerage system. In front was a small garden with a neat lawn and welcoming flowers, while at the back was the full glory of the quarter acre — a lawn for games, a vegetable garden, perhaps a shed. Within this universe a peculiar gender separation took place. The women, no longer in paid employment although receiving the family benefit, looked after the children, cooked and cleaned. The men left on the tram for town and returned when the pubs closed; and during the weekend they did odd jobs to the house, grew vegetables in the back garden, and of course mowed the lawn. The bloke lived by his *Yates Garden Guide*, the woman by the *Edmonds Cookery Book*.

How many of these suburban blokes were proud owners of an ATCO motor mower is uncertain. Some newspaper advertisements aim at the do-it-yourself gardener with the promise that an ATCO mower 'gives you more time for real gardening'. But the cheapest version cost at least £40 (just over $4000 in today's terms and over a fifth of the average annual income) so few suburban house-owners could have afforded one. Most would have used a simple push mower. More likely such mowers were used by sporting clubs, while the local councils preparing sports pitches may have purchased them. Some advertisements for the ATCO mower targeted 'Bowling Clubs, Golf Clubs, Tennis Clubs, Cricket Clubs and everyone else with a large lawn'. Certainly, as suburbs expanded, new parks emerged. In Christchurch the nine parks owned by the council in 1914 became 22 by 1939. Membership in sports clubs flourished. Bowls saw a big increase in membership once the three separate associations amalgamated in 1913.

Whether it was providing a beautifully manicured back lawn to embellish the floral border, or high-quality cricket or bowls pitches for the playing of weekend sport, the ATCO motor mower was at the heart of that 1920s revolution which took Kiwi families away from the dangerous inner city to the sanctity of the suburb.

60.

Rudall Hayward's camera

AUCKLAND WAR MEMORIAL MUSEUM

This handsome-looking camera in its smart wooden case was used by Rudall Hayward, the great New Zealand movie director, for silent films from 1925 until about 1930. It was used in movies which tell us much about New Zealand of the 1920s.

Rudall Hayward's career developed alongside the rise of moving film. He arrived from Wolverhampton in England aged four in 1905, accompanying his family who were members of a touring entertainment troupe. Moving pictures were just beginning. Without electricity, film was projected using an incandescent stick of lime (hence 'into the limelight'). Several early films were made covering significant public moments such as the departure of the second contingent to the South African War in 1900 and the tour of the Duke and Duchess of York in 1901. But for popular entertainment the main offerings were melodramatic theatre, vaudeville music or black-face minstrel shows presented by travelling companies. This was the Haywards' market. Their company, the Brescians, including three Hayward brothers and four Italian sisters from Brescia, offered costumed concerts; but they also showed films. These were a success. Reluctantly the Haywards took over theatres for film-only shows. In 1909 they opened a dedicated picture house at Auckland's Royal Albert Hall and the next year erected the country's first purpose-built cinema in Dixon Street, Wellington. Rudall, aged 10, got his first job hand-cranking the bottom spool of a projector.

MOTION PICTURE CAMERA, RUDALL HAYWARD. AUCKLAND WAR MEMORIAL MUSEUM TĀMAKI PAENGA HIRA. 1974.103.1.

The 'flicks' (referring to the staccato moving image) took off. By 1912 the Haywards owned 33 theatres. Five years later over half a million New Zealanders attended the movies each week. By 1928 there were 612 cinemas in towns and cities and 359 country circuits. At first most New Zealand films were government documentaries, but some independent film-makers, including the Frenchman Gaston Méliès, turned to Māori subjects, drawing on Māori performance skills and the exotic location of Rotorua. Many early films were an extension of the tourist industry.

Throughout the 1920s films were dominated by Hollywood productions, and crowds flocked to the stars — Charlie Chaplin, Mary Pickford, Laurel and Hardy. Alongside phonographs, films brought a powerful American influence to New Zealand. Always a cultural nationalist, Hayward determined to provide alternatives to Hollywood dreams. American hits were of two types — comic romances and swashbuckling 'cowboys and Indians' westerns. Hayward offered local versions of each.

His first venture, made in 1920, was a romantic comedy, *The Bloke from Freeman's Bay*. Hayward admitted the film was terrible and his uncle offered him 50 pounds to destroy it and protect the family name. Instead Hayward plastered the town with posters and attracted a sell-out audience. There was obviously a thirst for local films. At the end of the 1920s Hayward returned to this recipe of local romantic comedies, when he directed 23 two-reel comedies with such titles as *Natalie of Napier* or *Betty of Blenheim*. Using the same plot — a love element, a chase, and a rescue — he attracted large crowds as 'extras'. A week later the film was shown and people flocked in to see themselves on the big screen.

But Hayward had more serious ambitions, in the form of Kiwi westerns. To shoot proficiently, he needed a good camera. This camera is the one he used. The outside label reads 'Empire Cinematograph Camera No. 3' manufactured by W. Butcher & Sons Ltd, London. Butcher & Sons were certainly camera-makers, established in 1913. Inside the camera is another label, 'Prestwich Manfg. Co.'; Prestwich was another noted London maker of silent cameras. To add further complexity the camera lens was a Newman-Sinclair model. Newman-Sinclair was yet another British firm of camera-makers noted for their cameras' light metal bodies. The best assumption is that Hayward put together this camera using parts from three models to produce one that was light and easy to use in demanding conditions. Brought up in the film industry Hayward was a 'jack-of-all-the-cine-trades', and we know he was capable of do-it-yourself construction because in the 1930s, when Hollywood control blocked access to sound cameras, Hayward made one himself.

Hayward certainly had a workable camera by 1925 when he began work on his first Kiwi western, *Rewi's Last Stand*. There were two inspirations for the film — the first was Hayward's desire to provide a New Zealand version of a Hollywood western. The second was the writings of James Cowan. Cowan was a popular historian and journalist who had grown up near the site of the battle of Ōrākau, the culminating conflict of the Waikato War. He was influenced by those 'Maoriland' intellectuals early in the century who saw Māori mythology and history as a basis for New Zealand culture. In the first chapter of his history of the New Zealand Wars (1922), Cowan claimed that the New Zealand historian 'must turn to the story of the white conquest in America for the record of human endeavour that most closely approaches the early annals of these Islands'. These echoes of the American frontier drew Hayward to Cowan's work and inspired his films. Reading Cowan's work, Hayward realised that here 'was material for film plays just as exciting and colourful as any Hollywood western'. Hayward used Cowan's account of the battle of Ōrākau to provide the basis for *Rewi's Last Stand*. He followed this up two years later, also using this camera, with *The Te Kooti Trail* which recounts Te Kooti's raid on Jean Guerrin's flour mill near Whakatāne.

We do not need to recount the plots of these films, especially since little of the first remains; but it is worth exploring the main messages. Much in the two films echoes conventional American westerns — the portrayal of heroic pioneering by white settlers; stirring battle scenes, horseback chases and sieges; and softer romantic love stories. There were significant differences from the American model though. Several love stories cross racial boundaries with romance between white men and Māori women, in striking contrast to American models particularly Griffith's *Birth of a Nation* which saw miscegenation as a major crime after the civil war. Māori battlefield chivalry and courage are given extensive treatment especially in the battle of Ōrākau. Māori language (in subtitles) is used at times. Further, Hayward insisted that Māori actors play Māori roles, having found distasteful an experience of rubbing red ochre onto the legs of an Australian actress to make her look Māori. He even tried to find descendants to take roles. This was not always possible, and for the 1925 version of *Rewi's Last Stand* Te Puea turned down his request to use Waikato warriors in the Ōrākau scenes. He had to call on Te Arawa, who had fought for the Crown in the 1860s! When remaking the film in a sound version in 1940, Hayward did film around the original site and used supporters of the Kīngitanga.

Hayward was ahead of his time in his understanding of Māori and his sensitivity to the dynamics of particular iwi. He learnt to speak te reo and eventually married Ramai Te Miha, who had been the star of the second *Rewi* film. But the ideology in his films, that out of the conflict of war and the chivalry of Māori, a deep respect between the two peoples emerged, became a central myth for New Zealanders, white and Māori. During the Great War, Māori had fought alongside Pākehā and anti-Māori genetic racism, as distinct from cultural prejudice, was not common. In the 1920s the Māori population was only about 60,000 (under 5 per cent of the total) and most lived away from white settlement in rural areas. There was not huge contact between the peoples, so Pākehā could afford to be benevolent and smug about their race relations.

Although the 1920s saw several investigations of Māori land loss and some minimal compensation paid, most white New Zealanders were unaware of the confiscations or loss of land through the Land Court. When James Cowan wrote about the loss of Māori land in the Waikato for his official centennial publication *Settlers and Pioneers*, the chapter was removed. White New Zealanders of the interwar years did not want to be confronted. Instead Hayward's films offered a competing vision of two peoples, happy to accept mixed marriages, and believing that the New Zealand Wars had led to 'a mutual respect'. Here were the origins of the view that New Zealand had 'the best race relations in the world'. It is unfair to impute all this to Hayward, but certainly his 1920s films, shot on this camera, helped create a central New Zealand myth.

61.

The thermette

PUKE ARIKI

During the January summer holidays in 2019 Radio New Zealand invited listeners for suggestions of those objects which could tell New Zealand history. There was an enthusiastic response. The striking characteristic of the objects was the number associated with eating and drinking, and the most popular was the thermette. Let's follow the poll and explore what this celebrated item of so-called 'Kiwi ingenuity' tells us about our social history. Strangely, despite the public's fascination with the item, there are few thermettes on display in our nation's museums. The solitary representative I have found is a rather battered example from the 1970s on display at Puke Ariki.

The 1970s are over 40 years after John Ashley Hart invented the thermette. Born in London, Hart migrated to New Zealand with his family in 1902 aged 15. He served and was wounded on both Gallipoli and the Western Front. Both his father and later his father-in-law were electrical engineers, so not surprisingly John began work for the plumbing and electrical engineering firm of A. & T. Burt selling electrical goods in the Manawatū. By 1929 when he apparently invented the thermette he had moved to Auckland and set up his own firm selling electrical equipment. The thermette was one of 32 inventions that he patented, but by far the most successful. By the end of the 1930s he had sold over 8000.

The idea was simple. The metal base provided a fireplace, open at one side. Fitting neatly above the base was a cylindrical container for water with a chimney running through the centre. You could feed the fire anything flammable — sticks, paper, dried grass, even petrol — and the fire below and the hot chimney would heat water remarkably quickly. The long chimney caused the smoke to rise rapidly and drew in air to feed the flames. Hart promised in his advertisements that the thermette would produce 12 cups of tea in five minutes, and 'the stronger the wind the quicker it boils', for the wind would create a burst of hot flame and embers. The original models were offered in blue, green or orange tin for 12s sixpence or a more expensive version in copper. Our Puke Ariki example in red with a painted stylised flame and a slogan 'PREVENT FOREST FIRES' is very much a later model.

In New Zealand legend the thermette was a classic example of 'Kiwi ingenuity'. Yet this is in part myth, for the idea of heating liquids in a sleeve around a central chimney has international precedents such as samovar tea urns from eastern Europe and Russia, or Chinese hot pots. Its design was similar to the 'volcano kettles' made by Irish gypsy tinsmiths and manufactured by the Kelly Kettle Company in the early twentieth century. Whether Hart saw examples of these devices during his war service is unknown, but he certainly had the concept by 1928 when he gave the idea to George Marris & Co. of Birmingham who registered a design for a volcano kettle.

Whatever the precise origins of the thermette, John Hart had a fine understanding of how his new device could fit into New Zealand culture. For a start, it was intended to provide the means for making a 'cup of tea'. Tea-drinking entered British culture at the very time when Europeans reached New Zealand. Indeed, one motivation for the sealing trade was to sell seal skins to China to pay for imports of tea. During the later nineteenth century tea plantations developed in India and Ceylon (Sri Lanka) and the temperance movement promoted tea-drinking as an alternative to alcohol. Tea houses were more

PA2008.069 THERMETTE, CIRCA 1970S, HARCO SALES (1972) LTD, COLLECTION OF PUKE ARIKI, NEW PLYMOUTH respectable gathering places than pubs. By the 1870s Australia and New Zealand had the highest consumption of tea per head in the world — almost three times that of the UK. As the twentieth century opened, tea kiosks developed in public gardens, tea rooms in department stores, and 'smoko' for a fag and cup of tea became a central ritual for manual workers.

So the thermette catered to a hugely important New Zealand taste. This was particularly the case for those whose jobs involved moving away from base — post and telegraph men, roading gangs, foresters, even drovers. But the Kiwi family was the target market. Behind its use was the mythology of the pioneering heritage. As New Zealand society became increasingly urban, or rather suburban, in the interwar years, the frontier tradition could be kept alive by weekend and holiday pursuits — fishing, camping under canvas, or picnicking by a stream. Here the thermette came into its own. It is noteworthy that Hart advertised the thermette widely in December and January as families planned summer holidays. In December 1930, the first year when these advertisements appeared, the *NZ Truth* ad claimed, '"Camping out" is growing more popular every year for the annual Christmas holidays — the open air, the freedom from restraint of everyday living make an irresistible appeal to the city folk'. Tapping into memories of the campfire, the thermette was often described as a 'Picnic Billy'. Camping was not the only occasion for the thermette. There was also the 'picnic, bach, beach, farm'. It was aimed especially at men — 'a wonderful Xmas present for Dad'. The 'farmer' and the 'sportsman' would especially enjoy the device.

A major factor in the rise of the thermette at this time was the arrival of the motor car. The thermette was not something that your outdoors bloke would consider taking when he went hunting or tramping on foot. Too heavy and bulky for a pack, it had to be carried in the boot of the family car; it came into its own when families started taking the car for a Sunday drive to a 'beauty spot' in the bush, or drove to a campground or even to a river for a spot of fishing. An advertisement in the *New Zealand Herald* in November 1930 stated that it was 'A boon to every Motorist' as well as 'Camper, Bach Owner and Farmer'. The thermette was 'boiled alongside your camp or car'. One West Coaster remembered, 'My in-laws kept one in the boot of the car in case they wanted a cuppa when they went on a trip'; and another recalled, 'We lived for ours on long road trips from Haast to Motueka and Nelson. We would buy fish'n'chips, then park up on the side of the roads for a cuppa and food and let the dogs have a run.'

Cars had appeared in New Zealand in 1898 when they were permitted to operate provided they were lit at night and did not go faster than 20 kilometres per hour. For a quarter of a century they were very expensive and owned only by the rich. As late as 1925 when vehicle registration was introduced, the *Yearbook* noted that the exceptional increase in vehicle imports was surprising, 'considering that the motor car is still regarded as an item of luxury and not always as a necessity'. Many roads were rough muddy tracks, very few tar-sealed. There were some 70,000 cars in 1925. But by 1930, when John Hart introduced the thermette, the numbers had more than doubled to 154,000; and, with one car for every 10 people, New Zealand had the second highest level of vehicle ownership in the world after the United States. Cheaper American models — the Model T Ford replaced in 1927 by the Model A — brought car ownership to farming and urban middle-class people. It was perhaps unfortunate from the point of view of sales of the thermette that numbers of cars then faltered in the Depression years, but they increased in the late 1930s to reach over 200,000 in 1940.

The Second World War opened up two new opportunities for John Hart's invention. The story is that an army sergeant, who had purchased a thermette just before the war, suggested to his officers that it could be valuable for the army. When war broke out in 1939 the army asked Hart to waive his patent to allow them to make their own thermettes which were subsequently issued to mobile units. It is said they came into their own especially from late 1941 when New Zealand soldiers arrived in North Africa from Greece and Crete to pursue Rommel's army across the desert sands to Tunisia. The thermette was dubbed the 'Benghazi boiler' after a town in Libya where the New Zealanders had fought from late 1942 to mid-1943. It is also claimed that the Germans and Italians who occupied sites previously used by the Kiwis were puzzled by the round circles of scorched earth caused by Benghazi boil-ups. Nor was North Africa the only site for the thermette, if Hart's advertisements are believed. In 1945 they claimed it was used 'in the dark wet jungles' of the Pacific. Families were urged to send a thermette in parcels to their boys overseas. The other wartime market for the device was on the home front. People were urged to buy a thermette in case of a gas shortage or an electricity power cut. Families could brew up using a thermette in front of a candle.

By 1960 there were over half a million cars registered — and with increasing numbers of roads sealed in tar rather than gravel, the Sunday drive and picnic by a stream became a common experience. The thermette came into its own. New colours were introduced and a legend formed. It would not be until the opening of coffee shops at scenic sites and the arrival of natural gas boilers that the thermette began to move from an essential item of Kiwi families to an item of national mythology, no longer widely used but remembered fondly.

62.

Harold Pond's Napier Technical School uniform

MTG HAWKE'S BAY

For two days the wind had blown and the sea had been turbulent; but 3 February 1931 dawned with still blue skies — a typically hot and sultry Hawke's Bay summer day. It was the first day back at school, Harold Pond's third year at Napier Technical School, and he had just enjoyed the morning break catching up with mates. Now it was back to lessons on the second floor of the handsome old brick school building. The school of 350 would shortly move into new premises under construction.

Suddenly, at 10.46, just after class resumed, the floor heaved upwards, and the building trembled. There was a cascade of bricks and heavy roof beams fell as the outer walls crumbled. A girl was thrown across the room and through an opening in the wall. The noise was deafening, the air thick with dust. It was hard to breathe. The shaking stopped and the only sounds were cries of terror and moans of pain. Some pupils tried to pull themselves out from under the bricks and beams. A few raced for the stairs. Then it started again, this time more violently, and the floor fell. When it stopped, two and a half minutes after the first quake began, Harold found himself at ground level. The building had collapsed. Harold was dazed, his head bleeding, his body covered under a pile of rubble. Around him his classmates were sobbing and crying for help. Already boys were trying to rescue their mates. But Harold could not move. He was trapped up to his neck. His whole body screamed with pain. He could only wait and hope.

Harold Pond had been caught in the most deadly natural disaster in New Zealand's recorded history. Far below, tension built where the Pacific plate collided with the Australian. On 3 February it was released in a 7.8-magnitude quake which thrust the land upwards about 3 metres along a 100-kilometre fault. The centre was only 15 kilometres north of Napier.

Immediately following the quake, men from the nearby railway workshops rushed to Napier Tech and using crowbars, shovels and axes freed many pupils. After an hour or so Harold himself was rescued. He had been found buried under a pile of bricks and his shirt was cut open to free him. He was badly cut and bruised and taken to Nelson Park where he was bandaged around the head and on both legs. Not all his schoolmates were so lucky. There were six classes of some 30 each on the second floor when the quake hit. Remarkably, only nine died — eight boys and one girl. Several boys were killed, in 'gallant deeds of heroism', when they returned to the building attempting rescues and a beam fell. The other fatalities were those who had made it to the stairwell after the first shock and were killed when it collapsed. Only one died from Harold's class.

The school was one of the technical schools established in New Zealand after 1905. Compulsory secular free education had been established under the Education Act of 1877. All children aged seven to 13 had to attend. Secondary schools were few, attended by far more boys than girls, and because they charged fees, only children of affluent families attended. In the 1900s a reforming Inspector-General, George Hogburn, extended the leaving age to 14 and in 1903 introduced a certificate of proficiency in the last year of primary school which provided free places at secondary schools. The numbers going to post-primary schools rose fast and by 1928 over half went on from primary school. The established secondary schools, modelled on English public schools, were highly academic, offering literature, history and the classics but little science. In 1905, technical schools were established for urban working-class children to learn trades and office skills. The boys learnt woodwork, metalwork and book-keeping; the girls sewing, cooking and homecraft. There was a clear class distinction between the technical and the secondary schools — boys from the elite schools would chant, 'With the hob-nailed boots and the unwashed neck, they don't come here, they go to the Tech.' Some district high schools (with secondary education added to a primary school) were also set up.

It was not surprising that Harold had chosen the Napier Technical School. His family had been among the many UK migrants to New Zealand in the decade after the Great War. His father had worked at the Devonport dockyard, and was brought out in 1926 as an engineer by the Napier factory of William Plowman, which made jellies, jams and vinegar. Harold was 10 at the time. Following his father's example, Harold

HAWKES BAY MUSEUM

decided that he also wanted to be an engineer, so the technical school was the obvious option.

It was unfortunate for Harold on the day of the earthquake that he chose the tech. At several primary schools the quake hit during the morning break, and while some were bowled over in the playground, few were injured. One child at the Central School broke a leg. At Napier Boys', the middle-class school, the first week was always set aside for 'Military Barracks', so the boys were outside practising being soldiers. They too were in a relatively safe environment and teachers urged them to run away from falling buildings. The assembly hall where the whole school had been gathered minutes earlier collapsed in a heap. At the district high school in Hastings, the senior boys were initiating the third-formers by dipping them in a horse trough and covering them with blue bags (used in laundry) so they too were on the playing fields.

If most schoolchildren mercifully escaped death, many others did not: 161 people died in Napier, 93 in Hastings, two further north in Wairoa — 256 in all. Some were killed by falling masonry in the streets, 11 staff and nurses lost their lives when the recently completed nurses' home collapsed, and 15 died in an old men's home in Taradale (although a 91-year-old was pulled from the rubble three days later and commented that he could have done with a gallon of beer). The damage to the buildings of Napier and Hastings was huge; and within minutes of the quake the air in Napier became acrid with smoke as fire broke out in three chemist's shops where Bunsen burners were sealing prescriptions with wax. The fire burned for a day and a half, covering 11 blocks of the central city. Over 500 aftershocks rattled the ruins and unsettled nerves. In the chaos, looting broke out.

It was fortunate that a British warship attached to the New Zealand navy, the HMS *Veronica*, had arrived at Napier port on the morning of the quake. Immediately following the first shock the *Veronica* contacted the Devonport naval base and two light cruisers about to depart for naval exercises in Australia, the *Dunedin* and *Diamede*, were sent south, arriving early on 4 February. Here was a group of 1000 strong young men, with safe accommodation and used to working in teams. The naval men worked tirelessly searching for survivors amid the ruins, setting up food depots, patrolling the streets for looters, and helping to fight the fire.

Medical attention was an immediate need. The hospital had suffered severe damage and was unusable so patients were treated in open areas, such as the botanical gardens and the racecourse. Nelson Park where Harold had his wounds dressed and bandaged was a cricket ground just south of the centre. Those who had lost their homes congregated there and tents were put up. In late morning a good friend of Harold's, Ron Liddell, called at his home to see if Harold was safe. The Pond home, a wooden turn-of-the-century cottage just west of the Bluff Hill, was unharmed, but Harold's parents knew nothing of his fate and were worried. Ron headed back to town and found his mate recovering in a tent at the park.

With no running water, a shortage of food and the ongoing dangers of fire and aftershocks, Napier was not a healthy environment. The authorities encouraged women and children to head out of town. The Nelson Park camp became the organising centre for evacuation and within two weeks 6700 people had left, some by ship, some by car and others by train. Harold and his family were evacuated to New Plymouth.

Eventually they returned to Napier to pick up their lives. Harold's school was now destroyed. For some weeks classes were held in army tents on Clive Square around the corner from where the school had once been. But only 115 of the 350 pupils turned up, and it was decided that the cost of rebuilding was too great. Despite much lobbying and local opposition the Education Department decided to amalgamate Napier Tech with the two local high schools, Napier Boys' and Girls'. Teachers went reluctantly to other technical schools around the country, and many pupils decided to leave. This included Harold. He was soon to turn 16. He still hoped to become an engineer but in the depths of the Depression opportunities were few. Napier was faced by the urgent need to rebuild. The temporary Tin Town erected in Clive Square was replaced with remarkable speed by shops and offices built in the 11 blocks decimated by fire. Within two years the low-rise, concrete-reinforced city with its striking art deco decoration had emerged. Harold took advantage of this building boom and became a fibrous plasterer, which turned into his lifetime occupation.

As for the uniform which Harold wore on his last dramatic day at school — the shirt with its cut in the front, the neat cap and the woollen regulation shorts — he carefully folded them, put them in a plastic bag and placed them in his wardrobe. Although Harold, like many earthquake survivors, rarely talked about the event, the uniform was clearly central to his identity. The bag of clothes was found by his family after his death in 2003, and now lives in the Hawke's Bay Museum.

63.

Eva Bowes' flour bag bloomers

SHANTYTOWN HERITAGE PARK

There had been warnings beforehand — nasty downturns in 1921–22 and 1926–27. This was to be expected. As a country with a small domestic market, New Zealand was heavily dependent on the exports of its agricultural goods. Small changes in their prices could bring severe consequences. But the Great Depression of the 1930s was of quite another order, although it took a little time before the full impact on New Zealanders' lives was felt. The trigger was the Wall Street crash in the United States in 1929 which led to lower demand and prices for New Zealand's agricultural exports. Receipts from wool dived — in the 1931–32 season the returns were almost a third of three years before. Total export earnings fell over the same period by 40 per cent. Faced by lower taxes and customs revenue, the Reform government of George Forbes, with a flinty Dunedin Minister of Finance, Downie Stewart, in charge, set out to balance the books and slashed public spending. Public-sector wages were reduced on two occasions by 10 per cent, public works were reduced, old-age and war pensions were cut, family allowances introduced by Massey's government in 1926 were abolished. Not surprisingly, unemployment rose. Initially there was no state provision for the unemployed, but in 1930 relief payments were introduced and those without work began to register — almost 2500 in February; over 11,000 by the end of the year — and then the downturn really hit. By October 1931 there were over 50,000 registered unemployed; by July 1933 the figures reached

OWNED AND POSSIBLY MADE BY EVA BOWES NEE KERR OF BLACKBALL. DONATED TO SHANTYTOWN HERITAGE PARK BY EVA'S GREAT GRAND-DAUGHTER VALDA RHODES.

80,000 — and this did not include women, or young people, or Māori. If they were counted, about 40 per cent of the workforce was out of work. Forbes introduced a policy of no pay without work, and from 1933 his government required men to join camps in the countryside to qualify for relief. Forbes' partner in government from late 1931, Gordon Coates of the United (formerly Liberal) Party, was notoriously alleged to have told the unemployed 'to eat grass'.

The human consequences of this extreme loss of income were severe. Some could no longer afford housing and were forced to crowd in with their relatives or even sleep rough. Many found it hard to obtain food, queuing for soup for themselves or small boxes of food for their families. One person remembered standing patiently in line at the council chambers from 7.30 in the morning until 2.30 in the afternoon, just for 12 shillings and six pence worth of food — bread, tea and butter. 'Silver beet, mashed spuds and a poached egg was a feast.' Many protested. Those without work set up an Unemployed Workers' Union which marched on Parliament to argue for decent relief; and in 1932 there was breaking of windows along the streets of Auckland, Wellington and Dunedin.

Others tried to eke out their small incomes by 'making do'. They lined their floors with potato sacks to keep out the draughts. They made soap by boiling fat and caustic soda on a kerosene tin over an outside fire. Many wore clothes that were hand-me-downs, or sewed new ones with cast-offs. They carried their food in old sugarbags, which gave a name to the era, 'the sugarbag years'. The museums of New Zealand are full of examples of this Depression make-do and ingenuity. There are pieces of furniture constructed of recycled cheese boxes or packing-case timber by the blokes, and floor rugs from scraps of worn-out clothing, and quilts and assorted garments from patchwork, made by women. There are even beautiful embroidered bags from sacking. For the kids, there are golliwog dolls from rags and toys from wooden cotton reels.

This object is one such examples of 'making do'. The elegant women's bloomers were carefully sewn with frilly bottoms from a pair of flour bags. They come from the West Coast, where they now reside at Greymouth's Shantytown. The museum's donation form states that the labelling on the flour sacks dates the object to early 1930, and that the bloomers were made by Eva Bowes, whose maiden name was Kerr, of Blackball. We have not traced an Eva Bowes in that town, but there was an Eva Kerr living on the Main Street on the 1928 electoral roll. There were four other Kerrs living there; three were miners and one a fireman. It is reasonable to believe that the maker of these bloomers was from a Blackball mining family. If so, it is understandable why Eva would have tried to save money by making clothes from a flour bag.

For Blackball was a solid working-class community which fell on tough times in the Depression. Situated on a misty plateau inland from Greymouth, the community owes its existence to the Blackball Coal Company which was set up in 1889 to supply cheap coal to the Blackball Shipping Line. The mine began in 1892; but conditions were never great for workers. The hours were long (10 hours a day), pay was low and the physical environment was dangerous. In 1908 discontent had spilled over into a famous strike when Pat Hickey led a demand for a half-hour crib time to have lunch. The success of the strike attracted other radicals and in the 1920s the town became the home of both the United Mine Workers and the Communist Party, with Blackball resident Angus McLagan secretary of both. But as the Depression hit, work in the mine dried up. Shifts were reduced from three to two and then only one. Single men lost out completely, and married men received work only three days a week. Then came proposals for a new shaft using water sluicing to be worked by a small group of miners who would be paid, not wages, but a contracted price. The response was a bitter five-month strike. The union's allowance ran out and eventually a few men returned to work.

Francis Bennett, who took up a position as the local doctor in 1928, initially found Blackball a dispiriting place. On arrival he wrote to his wife about 'the soot, the sulphurous smell, the lack of trees, of paint . . . it was just a place where people existed', and he decided to stay just long enough to earn some money and leave. But in the end the spirit of the people won him over. He judged his five years there 'the happiest of our lives'. Bennett described the miners' monotonous work with constant heavy shovelling; and as a doctor he was only too aware of the dangers to their health. The rock falls and explosions brought invalidity and premature death. In addition, there was the 'early heart failure, silicosis, pneumonia, chronic bronchitis, emphysema and premature aging'. A local newspaper, the *Hokitika Guardian*, reported on the mine's dangers at that time — in February 1930 a miner was hit by a fall of rock and died four months later; in September a long-time miner was struck by some uncoupled trucks and killed. There was also a bad flood across Blackball's main road. Times were tough in Blackball in 1930, so it was not surprising that Eva Bowes found solace and saved money for the household by sewing this pair of bloomers.

The bloomers were made from two 50-pound bags of Peerless flour. The bags were soft, preventing any chafing from the trousers or skirts worn over them, and would have been readily at hand because flour was one of the few essential items for the thrifty housewife during a time of depression. She could save money by making her own bread or even scones — no doubt with the aid of the Edmonds cookbook, first published in 1908 as the *Sure to Rise Cookery Book*. She would have bought Peerless flour because that brand was heavily promoted in West Coast newspapers. The flour was milled from Canterbury wheat by D. H. Brown & Sons at their Moorhouse Avenue roller mills in Christchurch, not very far from the Edmonds baking powder factory. Brown's mill had a railway siding close to the main tracks so flour was easily transported to the West Coast by rail through the Ōtira tunnel which had opened in 1923. Peerless was promoted as 'the best household flour' especially suited 'to make those distinctive scones. You know the sort, the "more-ish" variety, the praises of which hubby is always singing.' The brand was used at local shows for competitions for a plate of homemade scones. In 1918, cinemas even showed a film about the milling of Peerless flour and at the Saturday sessions children were offered a free ration of 'kiddies buns' made with it.

Eva Bowes' bloomers seem at first sight a simple and charming object. Yet they allow us to approach the hard-working world of a conscientious housewife during the years of the Great Depression. They lead us to the scones and bread which women cooked to eke out a limited diet; and they suggest how many were able to make do, clothing themselves and their children at a time when money was always short and all women had was their ingenuity and a needle and thread.

64.

Photo of Michael Joseph Savage

MUSEUM OF NEW ZEALAND TE PAPA TONGAREWA

This object is not one unique heirloom. There are hundreds, and once were thousands, of them — carefully framed copies of this particular photograph of Michael Joseph Savage, hung on living-room walls or sitting on mantelpieces of working people's homes across the country. The image was reproduced in many forms: a black-and-white drawing; in a print on enamel signed by the man himself; and in an embroidery done in 1939 by Henrietta Watkins, a Picton housewife, who added a silver fern for good measure. It has become obligatory for every Labour Party leader to display it in their offices. No other New Zealand political image has attained such devotion. There is the famous shot of Whina Cooper, mokopuna in hand, leaving Cape Rēinga on the 1975 Land March, and Jacinda Ardern in a black hijab after the 2019 mosque shootings. But while these have been reproduced often, this portrait acquired a unique status and popularity. There were hundreds of photographs of Savage taken in the 1930s, but only this one, carefully shot in the studio with mood lighting in early 1936 by the Wellington society photographer Spencer Digby, achieved immortality.

What explains the photograph's unique status? The image shows not a heroic strong leader, of the kind found at that time in Hitler's Germany or Stalin's Russia. We see a small gentle face, cocked to the left, with forgiving eyes looking downward directly at the viewer behind round frameless glasses. The mouth's subtle smile evokes happiness and warmth. The feeling is not of a know-it-all genius or strong man, but of a kindly uncle at the bedside. The reassuring image was perfect for the time. By 1935 many New Zealanders had endured five years of anxiety — declining wages, unemployment, families separated with those on relief forced to go to work camps, the fear of ill health. In the angry autumn of 1932 some unemployed had taken to the streets and attacked shopfronts; but most New Zealanders simply endured, struggling 'to make ends meet'. At this moment cometh the man — Savage represented the benevolent saint who would take care.

Obviously this perception was mythology — Savage could be tough and intolerant. But his background in part explains the messages found in this face. He was born in 1872 in the Australian bush, Ned Kelly country, of an Irish family. There he knew grief and poverty. His mother died when he was five. Michael was cared for by his sister Rose and spent much time looking after his crippled brother Joe. In 1891 both Rose and Joe died, and in sorrow Savage added Joseph to his name. His father's farm was too small and unsuited to crops. Droughts were severe. Michael took a job working 90 hours a week in a wine and spirits store. In subsequent years he was a labourer and ditch-digger in the Riverina district of southern New South Wales and a gold miner in Victoria. Pay was poor, hours long, conditions dangerous. Savage found refuge in unions and labour politics, and inspiration in the writings of the Americans Henry George and Edward Bellamy who offered a vision of a cooperative egalitarian society. Eventually in 1907, aged 35, and inspired by Seddon's reforms, he followed his mate Paddy Webb across the Tasman. He was one of over 20,000 of his countrymen who in that decade settled in 'Maoriland' (as the Aussies called New Zealand). He landed with a shilling in his pocket which he immediately spent on a shave.

It was not long before Savage made his mark in both unions and politics. He joined the revolutionary industrial union the 'Red Feds', and was a leader in Auckland in the 1913 strike. After failing to win success for the Socialist Party he joined the Labour Party when it was formally established in 1916, and by 1919 had become the MP for Auckland West. Throughout the 1920s and early 1930s he was a leading voice for Labour. He had two major interests — the economic system where he spoke for higher wages and cheap loans; and a social security system of decent pensions for all and a free universal health system.

Savage became the voice of humanitarianism in the Labour Party, contrasting with the more severe and doctrinaire socialist principles of his leader Harry Holland. In 1933 at the funeral of the Māori King Te Rata, Holland collapsed and died. Savage succeeded as the party leader. His chance for power came at the election in 1935. The party's policies

PHOTOGRAPHIC PORTRAIT OF MICHAEL JOSEPH SAVAGE, 1936, WELLINGTON, BY SPENCER DIGBY, THE WEEKLY NEWS. PURCHASED 1994. TE PAPA (GH005501)

were for a guaranteed income for workers and farmers, education from kindergarten to the grave, and Savage's signature commitment to health and superannuation — in sum 'everything necessary to make a "home" and "home life" in the best sense of the meaning of those terms'. Savage never married and spent his years in New Zealand living with the family of Albert and Elizabeth French where he was a favourite of Aruba their daughter. The lack of his own family allowed Savage to become the benevolent 'uncle' to the nation.

The 1935 election marked the beginning of a new political and economic order, but it was not a total triumph for the Labour Party. Although they received 53 of the 80 seats, they won only 45 per cent of the vote and the right wing was handicapped by two smaller alternative parties which took votes away from the new National Party (formed by the union of Reform and United). When Savage formed his ministry, which included four other Australian-born unionists, the initial photographs portrayed him as a serious man of action — smartly dressed, eyes looking into the distance, papers in hand. Quickly the new government showed its stripes — with Christmas bonuses for the unemployed and recipients of charitable aid, an increase in pensions and a Christmas holiday for relief workers. Not surprisingly, Savage's image needed to change. Fears about a Labour government needed to be met by reassurance. On the eve of the new Parliament in March 1936 the *Weekly News* published the classic Savage photo. The *Weekly News* was the *Herald*'s weekly summation of news and, like other weeklies such as the *Free Lance*, it offered a central glossy section of black-and-white photos. In the days before television weeklies gave New Zealanders a visual illustration of the news, they were examined devotedly. The *Weekly News* of 25 March 1936 opened its glossy section of images with the full-page photo of Savage under the heading 'New Zealand's first Labour Prime Minister'. Beneath was a caption: 'This portrait of the prime minister was specially taken for "The Weekly News" by Mr Spencer Digby of Wellington.' Spencer Digby had established a business of society photography on Lambton Quay in Wellington and his images were known for their use of photo-flood and spotlighting in the studio. His Savage portrait became by far his most famous work.

In the next two years the Labour government implemented some of its promises — a higher minimum wage, compulsory unionism, and the restoration of conciliation and arbitration. Dairy farmers received guaranteed prices, pensions were increased, the Reserve Bank was nationalised to allow better public control over credit, and a major state housing programme was begun. But the heart of Savage's programme was the Social Security Act. In one piece of legislation came universal superannuation for the elderly, a free national health system, family allowances, and emergency benefits and pensions for widows, orphans, the sick and the unemployed. Many of these provisions already existed in some form. Now they were regularised and made more generous. The cost of benefits doubled over the next year. Savage cleverly delayed the implementation of social security until after the election of 1938, so the election became a referendum on the policy. In this context, the 1936 portrait of Savage, the kindly uncle looking after his national family, came into its own. The image was used on huge hoardings around the country alongside the slogan, 'Now Then! To ensure Social Justice and Security vote Labour'. It was shown during the interval at picture theatres and in newspaper advertisements.

Although born Catholic, Savage was never a practising one, but increasingly he drew on religious sentiments suggesting that his humanitarian aims were 'applied Christianity' — 'aimed at establishing the Kingdom of God upon Earth'. His followers saw him as nothing less than a saint, even Christ himself. This canonisation grew after 1938 when Savage fell ill and eventually died during the Labour Party conference in March 1940. Over the last years he was bitterly attacked by John A

Lee on the grounds that his economic policies were conservative and his health a handicap. Lee became the Judas Iscariot to the Labour Christ. This adoration of Savage reached a climax at his death. The *Weekly News*, the *Free Lance*, and the Labour Party weekly, *The Standard*, all reproduced in full page the famous Spencer Digby portrait. It seems likely that it was after this that Labour Party stalwarts and those who felt grateful to Savage for social security cut out the glossy photograph from the weeklies and framed it in gilt. A copy was placed on the top of Savage's casket when lying in state in the vestibule of Parliament for two days to be visited by 50,000 people; and then when it travelled by train to Auckland, stopping 20 times along the main trunk as crowds gathered to the hymns 'Abide with me' and 'Lead kindly light'. Eventually at Takaparawhau/Bastion Point a memorial was built above his grave on which was carved in stone a replica of the famous portrait above the words, 'Michael Joseph Savage 1872–1940. He loved his fellow men.'

The life of the famous photograph continued long after Savage's death. Michael Bassett (named after the man) remembered that, in his childhood, there were few working-class homes in Auckland without the framed, sometimes tinted, photograph of Savage. By the time Bassett himself became a Labour candidate in 1966 the photos were disappearing and he went looking for one. Eventually a woman in Devonport contacted him. 'It was a simple, Catholic, working class home. It was dark, and had become rather musty. There he was in the hall next to the crucifix. She picked him down, kissed him, and told me that Savage was the nearest thing to Christ in her life.' The Devonport woman was far from alone in this belief. Spencer Digby's photograph was one element that assisted and expressed that canonisation.

MURCHISON MUSEUM

65.

Pacific Burleigh radio

MURCHISON MUSEUM

The interwar years saw the greatest revolution in material objects of any decades in New Zealand history. Two major advances in power, first developed in the late nineteenth century — electricity and the combustion engine — came into their own. Within a few years New Zealanders had access to motor cars, aeroplanes, refrigerators, gramophones and radios. This imposing radio, a Pacific Burleigh from 1935, is a striking example of this technological bonanza.

The idea of transmitting sound over long distance was an exciting prospect. True, there were telephones in New Zealand since the 1880s but these were for private calls, not music and voice available to all. There were also gramophones from the turn of the century, although they were not popular until the 1920s, and were recorded, not live sound. Understandably, Clive Drummond in Wellington was hugely excited when in November 1921 he heard 'Come into the garden, Maud' arriving from the transmitter of Robert Jack, Otago Professor of Physics. It was only a year after the first broadcast in the United States, from station KDKA in Pittsburgh. Radio had been used from the early 1900s to transmit Morse code. The transmission of sound was a breakthrough.

In the years after that first broadcast, technical developments cleared the path towards our 1935 radio console. Those who listened to Jack's broadcasts did so on earphones using crystal sets involving a coil and a 'cat's whisker'. The first important change was vacuum cylinders, or valves, to receive the sound; but these were heavy on power and depended on bulky batteries which needed replenishing. However, from the late 1920s electricity, previously used for street lighting and factories, became available in homes. Reliable 'all-electric' radio receivers could be 'plugged in'. Electricity permitted built-in turntables in radio stations, replacing the microphone hanging precariously above a wind-up gramophone. The final development in the path to our radio was good speakers. Until about 1925, listening on earphones was a solitary, anti-social experience. Then radios began to have separate, large free-standing cone speakers shaped like a trumpet. From 1929 the Americans adopted moving coil speakers which were housed as part of the radio within a large console. The technical path to the Pacific Burleigh console radio was clear.

This radio also required an entrepreneur who was up with these technical developments. That person arrived from Russia in 1926 under the name William Markoff. He changed his name to Marks and got a job repairing electric meters for the Wellington Municipal Electricity Department. Then in 1930 Marks started a business making amplifiers and transforming imported American radios from 110 volts to the local 230. The next year, realising the demand, he started making his own radios, called 'Courtenays' because they were distributed by a Courtenay Place firm, Stewart Hardware. In 1932 the company became Radio Corporation (NZ) Ltd which, with a staff of 50, produced 500 radios a month. The firm competed successfully with American imports and soon became one of three major radio manufacturers in the country. They even exported to Australia. William Marks died in 1946, and in 1958 the firm sold out to Pye, which wound it up in 1982.

With the arrival of efficient radio receivers, sales took off. From 1924, New Zealanders had to purchase a radio licence at a cost of £1.10s. Under 3000 licences in that year gradually increased to about 50,000 in 1930 from a population of 1.5 million. Then numbers rose. By 1935, when the Pacific Burleigh was released, over 150,000 had bought a licence; and the following year half the country's households had a radio, which was fourth in the world. In 1940 over 85 per cent of households had radios.

Some New Zealanders bought small mantel radios, but many splurged out and paid up to £20 (about $2400 in value today) for large console models like the Pacific Burleigh. These had large speakers which produced excellent sound. The console radios initially had mock antique designs, but by 1935 the Burleigh was graced with stylish art deco features expressing modernity. It was designed to hold pride of place in the family living room. An advertisement for a Pacific console radio in 1936 promised, beneath an image of a three-seater sofa and two armchairs, 'Enjoy the winter at home . . . Now is the time you can appreciate the cosiness of your home — with the home fire burning — a PACIFIC RADIO bringing you the news from any part of the world.' The 1930s radio was inextricably part of family domesticity.

As the advertisement promised, part of radio's appeal was allowing the family to hear the news. Events such as the Napier earthquake and its aftermath, or the feats of aviators — the tragic and unsuccessful attempt by Moncrieff and Hood to cross the Tasman in 1928 and the successful crossing by Kingsford Smith later that year — were closely followed on radio. The staple fare was music, and from the beginning there was a battle between the 'high-brows' preferring the classics and the 'low-brows' wanting popular American music. This contrast in tastes was reflected in the distinction between the national 'YA' stations and the 'B' stations. This became stronger after 1936 when James Shelley directing the non-commercial stations promoted a BBC-style radio rather than a vaudeville style. For those wanting vaudeville, Colin Scrimgeour was appointed to direct the commercial stations. But even national stations began offering 'light' and 'dance' music. Both types of station were initially restricted to playing only a quarter of their music through records so there was much live music broadcast. However, increasingly records and American popular culture grew in importance, especially on the 'B' stations.

Another attraction of radio was sports broadcasts, despite the nervousness of some codes about losing spectators. During the 1920s rugby became hugely popular and most provinces built grandstands to accommodate growing crowds. The first live rugby broadcast was a club game in Christchurch in 1926. The following year there was commentary on a famous Ranfurly Shield game, 'the battle of Solway'. Hawke's Bay, having held the shield since 1922, had finally lost it to Wairarapa. Five weeks later, in a game noted because a player on each side was sent off and drunken spectators invaded the ground, Hawke's Bay won on the scoreboard only to lose when a player was declared ineligible. When the All Blacks toured Britain in 1935, Pacific Radios were advertised: 'Hear the All Blacks play — Every detail of the game direct from English fields to your own home.' Racing and wrestling commentaries became popular.

While the whole family might listen to music and even sport, radio also catered for particular groups. In the mornings when the bloke was away, a women's session on the ZBs was dominated from 1936 by the famous 'Aunt Daisy' Basham. In late afternoons when the kids returned from school there was the 'Children's Hour', inevitably hosted by an 'Aunt' or an 'Uncle'. In the evenings, musical offerings were supplemented on the non-commercial stations by live talks and plays. Before long, domestic serials from overseas arrived, usually on the commercial stations, although the famous Australian serial *Dad and Dave from Snake Gully* was cornered by a non-commercial station.

As listening to the radio became popular, its potential for political communication became obvious. International examples ranged from the avuncular fireside chats of President Roosevelt to the conscious use of radio by the Nazi leader Goebbels as a major weapon of propaganda. In New Zealand from the start government had controlled the airwaves. They did not own the stations but, for security purposes, kept a close eye on content. As radio's popularity radio grew, governments became more involved. In 1932 the Coalition government replaced the private broadcasting company with a government agency which set out to suppress controversy. Talks by a visiting Indian philosopher and the Social Crediter Major Douglas were censored. The 'B' stations were told to avoid controversial topics, but in 1934 a Methodist minister, Colin Scrimgeour, known as 'Uncle Scrim', with his radio church, the 'Fellowship of the Friendly Road', took over 1ZB and in his Sunday-evening 'Man in the Street' sessions he expressed sympathy for those suffering in the Depression. Michael Savage had always been a radio enthusiast, especially for comedies and wrestling, and became a supporter of 'Scrim'. Then in 1935 Scrim's final broadcast on the eve of the election was jammed. There was an uproar.

Savage promised the new government would support broadcasting and restore advertising. Instead they set up a national broadcasting service under the academic James Shelley and purchased the 'B' stations to operate under Colin Scrimgeour. Savage believed that radio offered an opportunity to break the power of the newspapers, which he considered unfairly conservative. Consequently, the broadcasting of Parliament was set up, and to some surprise it became popular. Janet Frame remembered that her mother in Ōamaru listened religiously, 'and as each speaker finished, she praised the "goodies" and criticised the "baddies"'. This did not mean that the radio became a free and fair site for political discussion. From 1937 the national stations broadcast an official news service produced in the Prime Minister's Department and broadcast without change. The government banned controversial material and, apart from in Parliament, there was no criticism of government on the airwaves.

Large console radios like the Pacific Burleigh made a major contribution to the interest and enjoyment of people in the 1930s. They made the family home an important site for entertainment in the tough years of the Depression. They strengthened the impact of American popular culture. They also facilitated a direct communication between government and people which, especially in World War II, brought New Zealand closer to being a totalitarian country than ever before.

66.

Te Winika waka

WAIKATO MUSEUM TE WHARE TAONGA O WAIKATO

Fittingly it has its own gallery. You walk down a ramp beside it, examining the intricately carved taurapa (sternpost) decked with white puhi (feathers). You look down at the hull, 20 metres long, enough to sit 36 paddlers, and protected in the base by flax matting, and finally just in front of the tauihu (prow) you look out at the powerful Waikato River. Thus you experience *Te Winika*, the great Tainui waka taua, which has pride of place at Waikato Museum Te Whare Taonga o Waikato.

Its history goes back over 180 years to 1838 when Te Wherowhero, later the first Māori King, commissioned tribes on the west coast near the mouth of the Waikato River (Ngāti Tipa, Ngāti Maru and Ngāti Māhanga) to go looking for a tōtara to build a war canoe. They found a suitable tree and high up where the trunk forked discovered an orchid, te winikā (*Dendrobium cunninghami*). When the tree was felled and the huge waka taua carved, they gave it the name *Te Winika*. The waka's first role was to go north to protect Auckland in 1844 from Ngāpuhi during the Northern War. Then it returned to the Waikato River where it became a frequent sight and even carried Governor George Grey during a visit in January 1863. But six months later Grey sent men across the Mangatāwhiri River to invade the lands of the Waikato. Realising the strategic significance of the river, the invaders set out to destroy Māori waka, and it is said, although this is not proven, that *Te Winika* was pulled apart by Gustavus von Tempsky and his Forest Rangers. Von Tempsky became known as

1972/88/1 – TE WINIKA, WAKA TAUA, COLLECTION OF WAIKATOMUSEUM TE WHARE TAONGA O WAIKATO

'Wawāhi waka — the breaker of canoes'. All that was left of *Te Winika* was the central part of the hull, which lay in mud near Port Waikato and was used by locals as a hurdle to jump over on their horses.

There the remains of the waka would have lain, slowly to rot, had not a hugely impressive woman, Te Puea Hērangi, entered the scene. Te Puea was born in 1883, the granddaughter of Tāwhiao Te Wherowhero, the second Māori King, and niece of Mahuta the third King. It was Mahuta who recognised her leadership abilities and brought her back to the Waikato in 1910 to help the Kīngitanga. From that time, through the interwar years and particularly during the reign of the fourth King, Te Rata (1912–33), she became the effective leader of the movement. Her first major contribution was to convince the Waikato peoples to vote into Parliament Māui Pōmare, one of the Young Maori Party, that group of educated Māori committed to bringing European knowledge, medicine and education to their people while also upholding the best of Māori traditions. She then established her mana by breaking with Pōmare and supporting her people to resist conscription in the Great War. But the real focus of her life was to take a community with low esteem, poverty, little regular work, disease and the loss of land from confiscation, and restore the identity and prosperity of the King movement. This involved moving the community back to its traditional home at Ngāruawāhia on the banks of the Waikato where a swampy land covered in gorse and blackberry was transformed into a proud home, Tūrangawaewae, 'a place to stand'.

From the beginning, cultural expression was central to Te Puea's vision of restoring dignity and identity. In 1922 she set up Te Pou o Mangatawhiri (TPM) which toured much of the North Island performing poi dances, waiata and haka. Among those attracted by this initiative was another member of the Young Maori Party, Apirana Ngata, who as a leader of Ngāti Porou, member of Parliament and Cabinet minister, not only encouraged cultural initiatives but also facilitated her attempt to obtain and develop productive lands. Another major cultural initiative was her vision to bring back carving traditions. She brought in Te Arawa carvers to train those from Waikato. Their first major project was a new carved house, which was opened with great ceremony in 1929. Ngata provided the name, Māhinarangi, the ancestress who had united Waikato and East Coast tribes through her marriage with Tūrongo. At the opening hui, Te Puea saluted Ngata, 'You found Waikato dejected; now they are hopeful'.

If a concert party and a new carved house were the first major cultural expressions of Te Puea's vision, the third was to restore large waka to the river. The Waikato were above all a river people. They found food in the river in the form of tuna (eels), kōura (crayfish) and waterfowl, they used its waters for irrigation, and it was the main highway through their lands. As Michael King wrote, 'the life of the river became inseparable from the life of the people, and each took the name of the other'. But by the 1930s waka had disappeared from the river. The sight and sounds of a team of paddlers chanting and moving in unison had inspired Te Puea as a child. She was determined to restore them and saw the project as nationally significant. By 1936, thoughts were turning to the forthcoming centennial of the Treaty of Waitangi in 1940. Te Puea presented the idea of creating seven new large waka, representing the seven founding canoes. The idea was taken up with enthusiasm.

Now Te Puea had to deliver. She remembered the hull of *Te Winika* lying at Port Waikato. This could form the skeleton of the first waka taua. In May 1936 Te Puea arranged for it to be brought to Tūrangawaewae. It was called onto the marae. She summoned a man in his nineties, Rānui Maupakanga, skilled in traditional carving, to supervise the work. Beneath Rānui was a team of Waikato craftsmen led by Piri Poutapu who had been one of Te Puea's many adopted children. Recognising his skill, in 1929 Te Puea had sent Poutapu to study carving with Te Arawa at Ōhinemutu where he worked closely with the brothers John and Pine Taiapa. Now with Rānui's support, Poutapu led the waka carving.

The first task was to obtain two new tōtara trunks for the additional sections of the waka. They were found in deep bush at Waingaro (some 15 kilometres west of Ngāruawāhia). After felling they were hauled foot by foot over 5 kilometres of bush country by 30 young men using traditional hauling chants. Then it was time for the splicing of the sections. In Poutapu's memory, this brought a moment of doubt. 'Rānui abandoned his blanket on the ground, shifted to the fire and pulled out a bundle of sticks from the embers. Then he went to the hulks and traced the outline for the dovetail joints without measurements. The joints were cut according to the markings and the hulls slotted together perfectly.' The joints were caulked using the traditional mixture of the seeding head of the raupō (bulrush) mixed with fat.

When the carving began, Poutapu followed Rānui's advice but he also visited the great waka taua *Te Toki-a-Tāpiri* in Auckland Museum to examine how it had been carved in 1836. One of Poutapu's team at Tūrangawaeawae, and the best pupil he ever had, was a young carver from Ngāti Raukawa who worked on the magnificent bow section. Inia Te Wīata would later win fame as one of the country's finest opera singers.

In early 1938 the final carvings were fitted to *Te Winika* and it was painted blood red. The waka was completed just in time for its first great moment on 18 March. It was a gloriously fine day with an audience of 5000. The Governor-General Lord Galway boarded *Te Winika* at Ngāruawāhia, its carving embellished with a headdress of plumes, and was paddled upstream by 36 chanting warriors to reach Tūrangawaeawae. There he invested Te Puea with a CBE and opened a second large carved house, Tūrongo, a house for the King and companion for Māhinarangi. The *Auckland Star* described the occasion as 'the greatest day the Waikato, indeed Maori New Zealand, has seen for years'.

Te Winika performed on important occasions. Two more waka taua were carved, *Tūmanako* and *Te Rangatahi*, although the dream of seven for the centennial was never completed, and in the end Te Puea boycotted the celebrations; but from 1942 an annual regatta at Tūrangawaewae became a significant event when the waka taua raced.

Te Puea died in 1952 recognised as a great New Zealander, and 20 years later *Te Winika* returned to the waters of Waitematā Harbour when it took part in the Auckland Regatta. But the waka was starting to show its age; so Piri Poutapu agreed to lead the carving of a new waka taua, *Tāheretikitiki II*, which was launched as part of the fiftieth celebrations for Tūrangawaewae marae. That evening the Māori Queen, Te Arikinui Te Atairangikaahu, announced that she would give *Te Winika* to the people of Hamilton. On 8 July 1973 *Te Winika* was paddled upstream from Tūrangawaewae to Hamilton where it was hauled up onto the grass. Finally in 1986 the waka was scrupulously restored, its red paint removed to reveal the full glory of the carving, and *Te Winika* made its final journey into the new gallery at Waikato Museum with a view over the Waikato River. It is a resting place suitable for a pre-eminent taonga.

67.

Centennial exhibition souvenir ashtray

MUSEUM OF NEW ZEALAND TE PAPA TONGAREWA

Despite the war that arrived in September 1939, New Zealand's centennial in 1940 was a big event. There were large ceremonies at Waitangi, Petone and Akaroa. There were innumerable parades of horses and bullock drays and people dressed up in colonial costume. Christchurch's procession was 10 miles long. There were no fewer than 256 centennial monuments; and significant cultural accomplishments — an art exhibition, concerts, a centennial film, *A Hundred Crowded Years*, a 30-part pictorial magazine, *Making New Zealand*, and a series of book-length surveys. But the biggest centennial event was the exhibition held on the windswept sand dunes of Rongotai, Wellington, beside the airport. It ran for six months, from November 1939 to May 1940, and attracted over 2.6 million visitors from the country's population of 1.6 million. The exhibition's 'Playland' pulled in over 2.8 million to ride on the Cyclone roller coaster and dodgem track and gaze at 'freaks' in the 'odditorium' such as Mexican Rose, the 'world's fattest woman'.

Apart from a coloured fountain in Kelburn Park and the statue of Kupe, now turned into bronze on the Wellington waterfront, there are few reminders in the landscape of this extraordinary public event. But there are many souvenirs, plates, teaspoons, cups, salt and pepper shakers, and ashtrays, found in museums and second-hand shops. Our ashtray is one of these. Souvenirs, small items designed to remind

ASHTRAY, CIRCA 1989–1940, JAPAN, MANUFACTURER UNKNOWN. PURCHASED 1995. TE PAPA (GH004471)

people of a happy visit to a special place, emerged as commercial products in the nineteenth century. It was the World's Fairs that established their place. Most of the 27 million people attending the Chicago Exposition in 1893 left with a small remembrance trinket. It was hardly surprising that the centennial exhibition spawned such items. This striking ashtray allows us to read the meaning of the centennial exhibition and anniversary, and also invites questions about the existence of ashtrays in twentieth-century New Zealand.

At the top of the ashtray in capitals is the name 'NEW ZEALAND'. When a centennial exhibition was first promoted by the Mayor of Wellington, T. C. A. Hislop, in 1936, he envisaged an 'International Wellington Exhibition'. But a 1928 convention limited the term 'International' to exhibitions endorsed in Paris; and Joe Heenan, the energetic Under-secretary of Internal Affairs, opposed a Wellington focus because he wanted the centennial to cultivate 'the national spirit'. So the exhibition became the 'New Zealand Centennial Exhibition, 1940'. The anniversary year was intended to imbue a sense of nationalism and the unity forged by a powerful state. But this formulation hid some major problems which only became fully revealed in the years ahead.

For a start, the souvenir ashtray shows no sign of Māori input or design. Did not the centennial mark 100 years since the Treaty of Waitangi? A twenty-first-century New Zealander would expect a focus on the relations of Māori and Pākehā and on the obligations implied by the Treaty. True, on 6 February 1940 there was a ceremony and re-enactment at Waitangi, and a Whare Runanga was opened to complement the Treaty House. But the re-enactment was one of many such events in 1940 which usually involved the arrival of the Pākehā. The Whare Runanga was a major enterprise, but W. E. Parry, the minister responsible for the centennial, anticipated its opening as 'no occasion for mourning an alien conquest, but an occasion for rejoicing'. In the event, the Kīngitanga boycotted the hui on the grounds that the confiscation of Waikato land had not been settled and the King had not been exempt from having to register under the Social Security Act; while Ngāpuhi displayed red blankets in protest at loss of land. Speaking on the occasion, Apirana Ngata asked what in the year of the centennial did Māori see: 'Lands gone, the powers of the chief crumbled in the dust, Maori culture scattered — broken.' Few Pākehā listened. Ngata's positive comments on the occasion were played up in the press and the protests largely ignored. Most Pākehā saw the centennial of the Treaty as representing the beginnings of British settlement and organised government. To the extent that Māori were remembered it was, as Cheviot Bell, president of the Founders' Society said, 'to mark . . . the free entry one hundred years ago of the Maori race into the great privilege of membership of the Commonwealth of peoples that we are proud to call the British Empire.'

If the centennial did not mark Treaty obligations, its primary message is found in the soaring tower at the centre of our ashtray. The centennial celebrated 100 years of material progress since the beginning of British settlement. The tower became a central symbol, universally found on souvenirs of the centennial. Almost 50 metres high, the tower's 'massive lines and well-proportioned height symbolised the progress and ambition of the young nation' as the official history noted. The exhibition was designed by Edmund Anscombe, an enthusiast for World's Fairs who had worked as a builder at the 1904 St Louis Exhibition and then designed the New Zealand and South Seas International Exhibition at his home town of Dunedin in 1925. For 1940, Anscombe tried to evoke a sense of speed and modernity by using an art deco streamline style. The tapering tower with its four central ribs gave a sense of soaring heavenwards into the future. The crest was marked by the proud scarlet-lettered figures 1940; and at the base, vaguely suggested on the ashtray, was a frieze of material progress going from horses and bullocks to tractors and cars.

Shown on the ashtray behind the tower are searchlights piercing the sky. Electricity, the wonder energy, which had transformed New Zealand in the interwar years, was everywhere at the exhibition — over 37,000 lights representing over a million watts. At night, buildings were floodlit and the tower illuminated in bright red.

The counterpoint of this paean to material progress was the centennial's second great theme — praise of the heroic pioneers. The foundation stone for the exhibition described the event as 'commemorating the dauntless courage of our pioneer men and women'. In front of the tower were large sculptures of a pioneer man and a pioneer woman. The noble pioneers were honoured repeatedly in 1940 — in pioneer memorials, and in the centennial film *One Hundred Crowded Years* which portrayed the struggles of a pioneering family voyaging out to a distant land and carving a home from the bush; followed by a cavalcade of progress as New Zealand went from 'savagery to civilisation in one hundred crowded years'.

If the central tower on the ashtray evokes material progress, what about the fern leaf on either side? This was not just the obligatory nod to a recognised national symbol. The centennial also expressed pride in the country's natural beauty. Beneath the exhibition's Dominion Court with its display of cities and transport was a model of the Waitomo Caves. There was a scheme for schoolchildren to plant native trees so that New Zealanders might form 'strong enduring friendships with the forests'. Yet few at the time saw any contradiction between the country's enthusiasm for material progress and the worship of the scenic wonderland of the Pacific. New Zealanders in 1940 blithely assumed they could have both.

There is another contradiction slightly hidden on this ashtray. In front of the tower are flagpoles; but, although not visible, the flags were Union Jacks. While the centennial promoted a national spirit, the mother country played a large part in the national definition. The first building at the exhibition was the United Kingdom Court, and almost every public ceremony saw speeches by the Governor-General or the British government's special representative, Lord Willingdon. As he left, Willingdon expressed delight that 'I have found New Zealand as British as ever before'.

So there were multiple contradictions in the centennial, between the Treaty and Māori claims for rights, between material progress and the country's beautiful environment, and between nationalism and continued colonial subservience. A few people pointed these out — Māori who challenged the country to live up to the words of the Treaty; and some intellectuals who expressed discomfort at the puffery and contradictions of the centennial. Frank Sargeson's centennial short story, 'The making of a New Zealander', portrayed a recent immigrant, a Dalmatian, trying to feel at home in puritan materialistic New Zealand; and Denis Glover wrote later, 'In the year of centennial splendours / there were fireworks and decorated cars / And pungas drooping from verandas / But no one remembered our failures.'

There was another conflict hidden in 1940. Our souvenir is an ashtray, designed to sit, perhaps alongside a centennial cigarette box, on a respectable living-room table. Hospitality in the 1940s and 1950s assumed that people smoked cigarettes. Smoking tobacco had come to New Zealand with the first settlers, but during the nineteenth century it was almost always pipe-smoking, predominantly partaken by men, especially those working in outdoor situations like gold miners or soldiers. Māori took up smoking pipes, including some Māori women. Then in the 1880s machines were developed in the United States to manufacture cigarettes and in the new century cigarette smoking became increasingly popular — although often, until the 1950s, in the form of 'roll-your-owns'. In the First World War soldiers welcomed free cigarettes to help calm anxiety; while in the interwar years advertising increasingly targeted women. Images of glamorous film stars with cigarette in hand encouraged women to see cigarettes as a necessary accompaniment for leisure-time enjoyments. Doctors recommended smoking as an aid to frayed nerves. By 1940 the average New Zealand adult was consuming 2.5 kilograms of tobacco a year. It was not until 1945 that the New Zealand Health Department issued its first warning against smoking; and by the 1950s increasing levels of lung cancer became obvious.

This small ashtray evokes the optimism and self-satisfaction of New Zealand's centennial in 1940. But it hides many contradictions and problems behind the certainties of 1940. Over the next half-century many of these issues would become more obvious and much social energy would go into responding to what was revealed.

68.

The Pope's accordion

NATIONAL ARMY MUSEUM TE MATA TOA

At 9.30 pm on 3 September 1939, New Zealand declared war on Germany — coinciding with Britain's declaration in London. But this time, despite Savage's famous phrase, 'Where Britain goes, we go', we spoke for ourselves, not bound, as in 1914, by the British declaration on behalf of the Empire. Over the next five years almost 105,000 men and women sailed overseas with the 2nd New Zealand Expeditionary Force (2NZEF). Their experience can in large part be recalled by the story of this piano accordion, given to Private Clifford Ewing by the Pope at an Italian prisoner-of-war camp in 1943.

Cliff Ewing, 'Spike' as he was known, was born in 1916. His father was a farmer at Five Forks about 20 kilometres north-west of Ōamaru. In 1939 Cliff was a farmhand and enlisted immediately. Less than a month later he went by train to Burnham to join the D (Otago) Company of the 20th Battalion, the last of the three battalions which comprised the first echelon of the 2NZEF. Its commander was Howard Kippenberger. With cast-off equipment from the Great War, the men did basic drill, and in the evenings there were sing-songs in the Salvation Army tent. The 20th, the first unit to leave, sailed from Lyttelton on 5 January 1940. They headed for Maadi Camp, just outside Cairo. The next year was spent training in harsh desert conditions.

In March 1941, over a year after their arrival, the New Zealanders were sent to the front lines in Greece. Characteristically, on arrival they entertained locals with a concert. But instruments were quickly replaced by weapons. Germany invaded from the north and the New Zealanders, including Private Ewing, were sent to hold a defensive line west of Mt Olympus. The Germans, controlling the air, were formidable, and after six weeks the New Zealanders were forced to evacuate to Crete. There, while waiting for the expected German landings, the men were again entertained with music. The recently formed Kiwi Concert Party provided a vaudeville show involving violins, brass, double bass, drums and, inevitably, an accordion. The men did not long enjoy the sun and vineyards of the Hania area. On 20 May, warned by Brigadier Freyberg's knowledge from the Ultra code that an invasion was imminent, the men hunkered down beneath a fierce bombing and watched thousands of paratroopers floating down on their parachutes, and gliders heading for the Maleme airfield. By the next day the Germans controlled the airfield, and on the 22nd the battalion including Private Ewing made an unsuccessful attempt to recapture it.

The loss of the airfield effectively meant the loss of Crete; but for Cliff Ewing the drama did not end. On the 25th the Germans briefly captured the town of Galatas and the battalion was in danger of being cut off. Cliff became the runner for D Company, sprinting through the grapevines with orders for the rest of the battalion to withdraw. He described his experience: 'I ran as fast as my legs would carry me. On my left a crowd of Germans were squatting in the oats, spraying bullets about . . . I dashed across the track, hurdled a stone wall, and raced up a grape-vine slope to meet LT O'Callaghan. I gave him the orders . . .' But the battle was lost and Private Ewing had to endure the long march over the White Mountains to the south coast village of Sphakia where he was among those lucky enough to be evacuated. Many were left behind. The New Zealanders lost a third of their strength, wounded, killed or taken prisoner.

If war had meant for 'Spike' Ewing long periods of boring training and two ignominious retreats, it did not improve. He returned to Egypt and intense training in the desert, scorching by day, fiercely cold at night. Rommel's German troops were pushing east. In Operation Crusader in November 1941 the Kiwis were asked to relieve the besieged Libyan city of Tobruk. The 20th Battalion was ordered to capture the hill of Belhamed, an escarpment close to Tobruk. Ewing's company went forward with cries of 'Otago' and the attack succeeded. For almost six days the men, cold, hungry and damp, sought cover in a rocky exposed position. Hopes soared when they believed South African tanks were advancing to relieve them; then dashed as the tanks turned out to be a German panzer division. The Germans arrived with the words, 'Stick your hands up, Tommy. The war for you is over.' Cliff Ewing was among 354

1982.876 NATIONAL ARMY MUSEUM NZ

men of the 20th Battalion who became prisoners of war. Some 8800 New Zealanders were captured in the war — 2042 in Operation Crusader, 1856 in Greece, 2180 in Crete, over 1800 the next year in North Africa, and a few taken in the more successful advances across North Africa to Tunisia and up Italy in the last years of the war.

Following capture, Private Ewing and his fellows were held in overcrowded transit camps. Sanitation was poor, and food and water inadequate. Many had dysentery. Eventually, while those caught in Greece and Crete were sent to camps in Germany, the North African POWs went to Italy. Ewing went first to a temporary camp in southern Italy, and then in February 1942 to Campo PG 52 at Chiavari in the north near Genoa. For about a year Ewing was one of 950 prisoners, 30 to a hut. Then in early 1943 he was sent to a work camp in the north-east corner of Italy near Trieste, Campo PG 107 at Torviscosa. In September 1943 Italy gave up the fight, and the Italian prisoners were moved to Germany. Ewing went to Stalag XVIIA at Wolfsberg in eastern Austria.

The physical conditions in these camps were not atrocious — crowded sleeping arrangements with no privacy, and at Chiavari the huts leaked and lacked heat, but at least they had blankets and straw palliasses for sleeping. There were two special torments. The first was food. 'You talked food and dreamt food,' remembered one POW. In Italian camps meals were a small slice of bread, macaroni and a tiny triangle of cheese; in German camps it was watery soup, boiled cabbage, sauerkraut or swede, and one seventh of a small loaf. The men longed for the Red Cross parcels with the tin of pressed meat, chocolate and jam.

The other torment was boredom. There were hours of monotony standing in line to be counted three times a day. One of Ewing's fellow prisoners at Chiavari wrote in his diary: 'If ever there was a lazier, more useless, hungry, and at times hopelessly boring life than that of a prisoner of war in a foreign country, I cannot imagine what or where it could be.' How did the men cope? They read, especially in Stalag VIIIA which had a library of 15,000 volumes. They played games — chess and bridge in winter, soccer, baseball, even cricket in summer. They gambled with two-up and mock race meetings. They performed plays. But in every camp it was music that kept them going. There were concerts, such as one at Torviscosa devoted to Māori waiata and haka; and there were endless singalongs. In this respect accordions where one POW could play and the others sing were a favourite. A New Zealand woman who worked with the Red Cross in Geneva supplying the camps remembered: 'They all wanted accordions. It was accordions, accordions, accordions.' However, it was not the Red Cross that supplied Cliff Ewing with his accordion. One day when at Torviscosa, a papal delegation arrived. The Pope, Pius XII, had an ambivalent relationship with the Fascists and Nazis and has been accused of doing insufficient to prevent the Holocaust. However, he wanted to give gifts to POWs in Italy. On this occasion the men were presented a clock, some stamp albums and two accordions. The guards grabbed one of the accordions; but Cliff, known as a good pianist, was given charge of the other and began to play for his mates. When a short time later Italy capitulated and the prisoners were sent to Germany, the senior Allied officer in the camp said to him, 'You take that accordion with you and look after it. That is an order!' Cliff followed orders and when the prisoners moved to Stalag XVIIIA the accordion came with him.

As the war ended, the Germans moved the POWs further west to avoid capture by the Russians. Some POWs in camps in east Germany and Poland marched west in January 1945 in the depths of a vicious snowy winter. Their marches were long, painful torments. Cliff's journey west from Stalag XVIIIA in April 1945 was shorter but still arduous. The accordion was regarded as essential, so it was slung on a pole and carried by two POWs. In May, after being evacuated by American troops, Cliff brought the accordion to England and then back by boat to his home at Five Forks. There it lay in poor condition for many years until 1982, when after talking to his surviving POW mates, he donated it to the Army Museum at Waiōuru. The accordion sits proudly on display there, a symbol of the music which helped the 100,000 men of 2NZEF survive the trials and boredoms of a dangerous, fast-moving war.

69.

Victory cot blanket

AUCKLAND WAR MEMORIAL MUSEUM

This remarkable cot blanket was completed in 1945 by a woman in Thames, given to her neighbour years later stuffed in a pillowcase, and now resides as a special treasure in the Auckland War Memorial Museum. The knitting of the blanket and the iconography embroidered on it tell us much about the home front during World War II.

The blanket is an extraordinary piece of knitting, some 1.7 metres long and over 71 centimetres wide. One can but imagine the huge needles that must have been used, for there is no evidence that smaller pieces were sewn together. The knitting is traditional stocking and garter stitch with an amazing range of colours. Some images are part of the knitting, and further images and words have been embroidered on top. Knitting had originated in North Africa, spread to Europe, and became popular in Britain in the seventeenth century when English wool was used for stockings. In the early nineteenth century, middle-class women took up the craft and it was brought to New Zealand by missionary women. At the end of the century knitted jerseys, based on printed pattern books, became popular. During the Great War, inspired by Lady Liverpool's knitting book, women produced socks, balaclavas and gloves for men at the front. In August 1916 over 130,000 knitted articles were produced. Many women and their daughters learned to knit, creating a knitting boom during the interwar years. Women's organisations such as the Women's Institute and the Women's

Division of the Farmers' Union encouraged knitting; and local A&P shows offered prizes. Working-class women supplemented the family income through knitting.

When another war broke out, once more the needles came out. 'Men must fight and women must knit' was the slogan; and from July 1940 patterns for knitted items were provided in *Comforts for Men in the Armed Forces*. The National Patriotic Council collected the work of the Red Cross, the Women's Institutes and the Lady Galway Guild and sent over 1.2 million knitted items to men overseas including to POW camps. Other women sold tea cosies and children's clothes locally to raise funds for food parcels. Wartime knitting was not without problems. There were shortages of wool, leading on one occasion to a crowd jostling outside a shop after supplies had arrived. From October 1942 wool was rationed. Many were forced to unravel old jerseys. By the time peace arrived our Thames woman must have been well practised in knitting. Did she perhaps unravel old jerseys to get the strips of bright colour?

We should not assume, however, that knitting was either the main or most important role for women in wartime. Indeed, at one level, World War II, unlike the first war, represented a breakthrough in women's social position. The assumption that war service was a purely male responsibility was rejected, with all three services recruiting women. Both the army WAACs and the air-force WAAFs numbered over 4000 by 1943, alongside 500 in the navy's Wrens; 78,000 women served in the Women's War Service Auxiliary (WWSA) helping in canteens, camps, hospitals and alongside the Home Guard. More women entered paid employment. From 1942, childless women were conscripted into work, and by 1945 the number of women in the workforce had risen by over 50,000. They moved into factory work and the public service and took on new roles as drivers, tram conductors and herd testers. Yet this turned out to be a false dawn. Women's work was regarded as a temporary expedient, acceptable while the 'boys' were away, but not once they returned; and in many jobs gender roles remained — in butcher's shops men handled the knives and carcasses, the women dealt with the customers.

Furthermore, men's return placed a special emphasis upon women's role as the caring centre of the family. When the men came back, often injured or traumatised by their experiences, it was women's role to provide psychological support and unquestioning love. A famous advertisement for blankets at the time showed a woman and a soldier kissing and the words 'Back home for keeps'. Having children was seen as New Zealand's best form of defence against possible future invasions. In 1944 the family benefit was increased and in 1946 made universal.

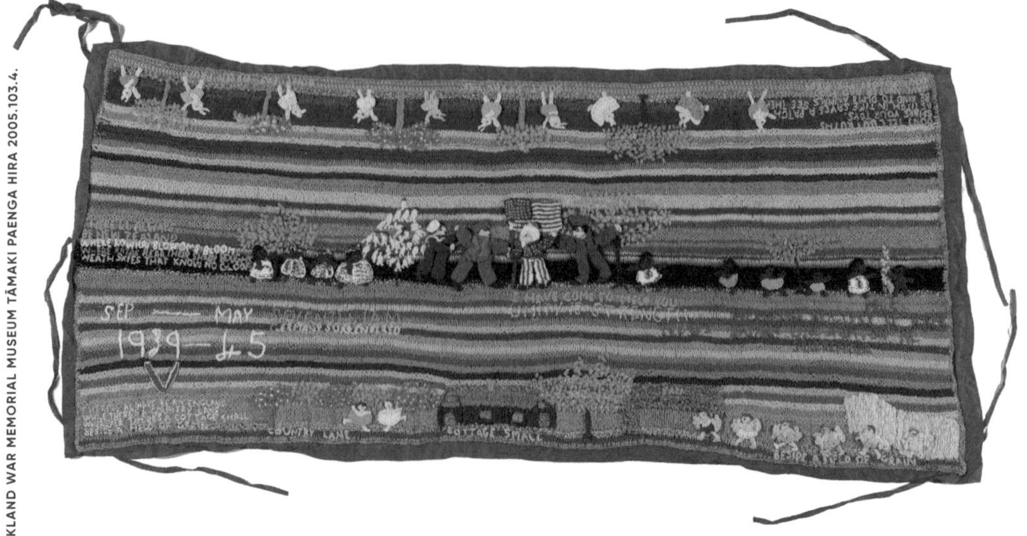

VICTORY KNITTED COT BLANKET. AUCKLAND WAR MEMORIAL MUSEUM TĀMAKI PAENGA HIRA 2005.103.4.

A parliamentary inquiry into 'race suicide' concluded that the 'best immigrant was the New Zealand-born child'. This explains why a blanket for a baby's cot was such an appropriate piece of knitting to celebrate the coming of peace. Victory meant that women should return to the home and have children.

The images and words so deftly embroidered on the knitting also throw fascinating light upon the wider home-front experience. Let's begin with the large V for Victory in red with the precise dates and times in white: 'May seventh 1am Germany surrendered' and 'Wednesday August 15 1945 11 am: Japan's unconditional surrender'. The knitter is correct with the timing (in New Zealand) of Japan's surrender; but not so Germany's. In fact Germany surrendered at 2.41 am Greenwich Mean Time on 7 May which was early afternoon New Zealand time. The news arrived in New Zealand for the morning papers on 8 May, but Walter Nash, acting Prime Minister, insisted that celebrations wait until Winston Churchill's official announcement. This did occur at 1 am in New Zealand, but on 9 May not the 7th. The confusion of times throws light on New Zealanders' respect for regulation at home during the war years. One might have thought that news of peace in Europe on 8 May would lead to a spontaneous outburst of joy. But apart from Dunedin, where the coincidence of a student capping parade on that day lit the fuse, there was remarkable respect for government orders. Most people reported to work, the pubs were full but quiet, and people in the street were surprisingly subdued. It was not until the next day, following Walter Nash's instructions, that the bells and sirens sounded and people danced and kissed strangers in the streets. This suggests the extraordinary control by the state on New Zealanders during the war. From 23 July 1940 New Zealand became the first country outside Britain to introduce conscription and within two years half the eligible men were in uniform. All men aged up to 59 and women 18 to 40 were liable to be 'manpowered' into specific occupations. Prices, rents and wages were controlled, and many goods, from petrol to eggs, could only be bought through tickets in ration books. New Zealanders even had to buy 'austerity clothing' — no cuffs for men's trousers, no pleats for women. New Zealand has never been so regulated and the respect shown for government authority was perfectly illustrated by the response to VE and VJ Days.

Along the bottom of the blanket is a tribute to mother England quoting 'There'll always be an England' from a famous patriotic song sung by the British 'forces sweetheart', Vera Lynn, evoking a nostalgia for rural England which is supported by images of a cottage and a field of grain. Such romanticism about rural England was not surprising. The 1936 census showed that over 98 per cent of the non-Māori population were descended from British or Irish people and almost 10 per cent had been born in England. Huge loyalty to the mother country remained. Vera Lynn's recording was widely played on the radio. New Zealanders fought in North Africa as part of the British Eighth Army; and when the war in Africa was over and Japan entered the war, New Zealand chose to keep 'our boys' in Europe defending Britain rather than bring them home to defend New Zealand from the Japanese as did the Australians.

True, the blanket includes a romantic tribute to New Zealand: 'There'll always be New Zealand / where kowhai blossoms bloom / where ratas rear their shady boughs / neath skies that know no gloom', but the local images occupy about a third of the space accorded Olde England. There is a group of Māori women dressed traditionally as a romantic addition to beautiful New Zealand but this greatly understates the role of Māori in the war. Learning from the contested conscription of Waikato in the Great War, Peter Fraser resisted conscripting Māori; but led by Apirana Ngata and the four Māori MPs, Māori were keen to make a major military contribution. A Māori battalion, with the companies organised on tribal lines, won a superb reputation in Africa and Italy as a fighting force at the cost of 642 deaths. At home, too, Māori were actively involved. In June 1942 a Maori War Effort Organisation was set up which over the next two years represented a distinctive effort at Māori political leadership and autonomy. It helped recruiting Māori, encouraged primary production and directed Māori into essential workplaces. The wartime reality of Māori experience was very different from their diminished image on the blanket.

In the middle of the blanket above the words 'I have come to help you / Unity is strength' stands Uncle Sam holding the American flag. Between 15 June 1942, six months after the Japanese attack on Pearl Harbor which brought war to the Pacific, and mid-1944, there were always between 15,000 and 45,000 American soldiers and marines stationed in New Zealand. They came partly to shore up New Zealand defences. (By early 1942 Hong Kong and Singapore had fallen to the Japanese and Darwin was bombed. A Japanese invasion seemed possible.) They also came to train men for the Pacific war and provide rest and recreation for those who had suffered from the gruesome conflict in the tropics. The 'American invasion' was of particular importance to women. With sons or husbands fighting overseas, many women provided welcoming home visits for the Yankees. Inevitably the good manners and generosity of the visitors impressed and almost 1500 women married Americans. Many New Zealanders, especially dry cleaners, taxi drivers and florist's shops, would not quickly forget the 'invasion'. Uncle Sam deserves a place at the centre of this blanket.

Finally, at the top of the blanket and only readable from that side, is a frieze of dancing rabbits based on Beatrix Potter's famous English book *The Tale of Peter Rabbit*. Alongside the rabbits are moral slogans for children: 'Don't tell untruths', 'Share your toys', 'Don't steal, ask'. The frieze takes us back to the larger context. Along with celebrating the end of war, this cot blanket was intended for use in raising the next generation. After the adventures of war, women had another national responsibility. The beginning of the baby boom was just around the corner.

70.

Chip and Rona Bailey's typewriter

MUSEUM OF NEW ZEALAND TE PAPA TONGAREWA

On 30 November 1949 the National Party was elected to office on a promise of 'freedom' — freedom from wartime restrictions and rationing. Yet within less than 15 months, in February 1951, the new government imposed some of the most severe restrictions on New Zealanders' freedom ever seen. This typewriter, now held at Te Papa, was a prime weapon in the struggle to retain freedom of thought during the 1951 waterfront 'lockout' and associated strikes.

The background was that in the late 1940s shortages had fuelled increases in the cost of living. The watersiders, whose work was dirty, dangerous and erratic, fought for higher wages and in May 1950 received a 6 per cent rise as a 'final' settlement. Then in early 1951 the Arbitration Court provided for a 15 per cent general wage order, but the shipowners agreed to pass on only the 9 per cent difference. On 13 February 1951 the watersiders decided, in protest, that they would work a strict 40-hour week and not do overtime. The government stepped in, locked the wharf gates and refused to allow entry to the Waterside Workers' Union members. They asserted that the dispute would end only when each port had new unions.

For 151 days the dispute continued. There were heightened feelings on both sides. The watersiders looked back to the traditions of 1890 and 1913 strikes, but were deserted by their expected allies, the Federation of Labour and the Labour Party. On the other side, the shipping companies, keen to reduce costs, wanted a showdown and were backed by a new government with strong allegiances to farmers and business people. They promised to defeat militant unions. It was the Cold War, and in mid-1950 following the 'fall' of China, a hot war had broken out in Korea. In the United States Joe McCarthy unleashed a frenzy of anti-communism which had echoes here.

In this environment Sid Holland's National Party had no difficulty imposing an astonishing level of repression. Drawing on the Public Safety Conservation Act passed in the Depression, the government declared a state of emergency and issued various regulations. The lockout was defined as a strike, and it became an offence to be a party to a strike, or to contribute to 'strikers' or their families, including children. Equally serious were attacks on freedom of expression. It also became an offence to encourage a strike, to picket, hold processions or meetings, display posters, or publish any 'statement, advertisement, or other matter, in support of the strike'. The normal methods of democratic communication were outlawed. Newspapers could not report the watersiders' point of view and the government set up a 'stink factory', an information section, to craft anti-watersider bulletins for the local radio. Freedom of expression had fallen victim to the Cold War.

The watersiders and supporters were forced to go underground to present their viewpoint and sustain their comrades' morale. They decided on regular bulletins. This is where the typewriter played its central role. It was owned by Chip and Rona Bailey. In February 1951 Chip was 29 years old, a former watersider and former member of the Communist Party, driving taxis in Wellington. Once the lockout began, he volunteered to write the bulletins, and put together 52 issues. They covered news of the lockout, notices of meetings and assistance to families, messages of support, and scurrilous material about 'scabs' and the government. Chip typed the bulletins onto wax sheets.

His 'Good Companion' typewriter, produced by Imperial Typewriters in Leicester, was named after a famous novel by J. B. Priestley, *The Good Companions*. It became a popular portable typewriter in the 1930s. Chip's model has the royal insignia following its appointment as supplier to King George V. Typewriters dated from the 1870s, and in the early twentieth century became increasingly important for commercial letters and reports. In New Zealand cities, women found growing employment as typists. By 1951 the number of women in their traditional occupation, as domestic or personal servants, was under 26,000, while those in administration who would have used typewriters rose to over 45,000. So it was interesting that Chip did the typing because one might have expected his wife Rona to do so. But Rona had another role.

Once Chip had typed the bulletins, his mate, Max Bollinger, a freezing worker, took over. The freezing workers, like seamen and coal miners, struck in protest at the watersiders' treatment. Max embellished the sheets with magnificent cartoons. It was difficult scratching images onto wax, but Max's drawings of scabs as 'rats' and caricatures of police became legendary icons of the dispute. The wax sheets were taken secretly to another location and run off in their thousands on a Gestetner machine. To ensure the Gestetner was not found by police, the heavy machine was moved from house to house. The paper used for the bulletins was supplied, ironically, by the government, smuggled in by sympathetic public servants.

The printed bulletins then had to be distributed. Now Rona took over, working with Kay Bollinger, Max's spouse. Rona had an interesting background. Daughter of a successful Gisborne businessman, she followed interests in physical education and dance to the United States in the late 1930s where she became committed to left-wing movements. After a brief marriage to a fellow communist and later prominent economist, Ron Meek, she met Chip through radical circles and they married. She worked in the Physical Welfare Branch of the Department of Internal Affairs which had been set up under a Labour Party minister, Bill Parry, to encourage New Zealanders' involvement in physical activity. Throughout the lockout Rona worked, earning sufficient to pay the owner of Chip's taxi £20 a week and allow him time for the bulletins. Along with looking after her two-year-old daughter, Rona spent the early morning hours, from about 2 am to 5 am, distributing bulletins in her old Riley car. An elaborate decoy system was

TYPEWRITER, CIRCA 1950, ENGLAND, BY IMPERIAL TYPEWRITER COMPANY.GIFT OF MEG BAILEY, 2007. TE PAPA (GH011698)

devised. Suspicious-looking men in big coats would ostentatiously try to avoid the police. The police would take the bait while Rona and friends smuggled the bulletins in for distribution. About 10 to 15,000 copies were distributed to sympathisers each time — some 650,000 bulletins and 400,000 other broadsheets were given out.

The police determined to track down the Gestetner and the typewriter. Under the regulations they could enter and search a house without a permit. One evening Rona was at their Vivian Street flat with her sleeping two-year-old. The doorbell rang and there were two policemen. One was the infamous 'Call me Dave' Patterson, referring to his cheery manner. There was nothing cheery about his manner that evening. Despite Rona's pleas the police stormed into her daughter's bedroom and ransacked the house, strewing papers everywhere in search of the machines. They never found the typewriter. Put away in its appropriately red case, it was hidden above the safe in the kitchen and covered by a wooden panel. The police believed, however, they had found the Gestetner: an old printing machine parked in the bedroom. When this proved not to be the offending item, Rona and Chip had to pay £14 for an unregistered machine!

Max Bollinger's home in Vogelmorn was also raided by 'Call me Dave', and Kay too tried to protect their sleeping kids. The support Rona and Kay gave to their striking husbands was typical. Because the lockout meant a worker's whole family suffered, women became crucial. Few distributed pamphlets like Rona, but they played other roles. Some, like Kay, brought in funds; others distributed clothes, toys and food supplied by sympathetic farmers; and at every port a women's committee strengthened women's determination to 'stay solid'.

In the end, despite the bulletins, and women's loyalty, the watersiders gradually lost support. After 151 days they admitted defeat. It marked, for several decades, the end of radical unionism. Soon after, the National government called a snap election and received a ringing electoral endorsement. National held power, barring a single one-term Labour government, for the next 21 years. Legislation made permanent many restrictions imposed in 1951. The number of days lost to strikes plummeted until 1968. This small typewriter signalled, not the beginning of a workers' revolution, but the last hurrah of radical unionism. As for Rona and Chip, they remained police targets. Many years later, after Chip had died prematurely, and Rona put her energies into new causes such as opposing the Vietnam War and the Springbok tour, she discovered a listening post and a wire in her new house at Robieson Street at the top of Mt Victoria. If 'Call me Dave' was no longer ransacking her place to search for an offending typewriter, the New Zealand police still regarded Rona with suspicion. The 'red' typewriter was not forgotten.

71.

Jim Bradley's flagon case

MATAURA MUSEUM

This case, and its two flagons hidden inside, was designed to carry beer from a pub and it became a staple item for any 'real Kiwi bloke' in the years after World War II. The origins of this object take us into the history of beer-drinking in New Zealand and especially the impact of the two world wars.

Beer had been a staple of the British diet in the centuries before migrants came to New Zealand. It was both a food since it was made of grain and an essential safe drink at a time when water was often contaminated and milk quickly went 'off'. But alcohol was unknown to Māori and beer was a bulky item that was expensive to carry far, so until the end of the 1860s it was not commonly drunk in New Zealand. Men in frontier communities and new towns would come to the pub as a place of warmth and convivial company, and predominantly drink fortified wines like port or sherry or, more often, spirits, especially brandy and rum. Although the consumption of alcohol per head was lower than in the old country, the pattern of drinking was that men would come to the pub and go 'on the spree' — 'melt their cheque' and get very drunk, with much gambling and fighting as a result. From the 1870s breweries sprang up and beer-drinking rose.

Horrified by the social consequences of such drinking, and keen to protect the welfare of wives from drunken husbands, evangelical Protestant churches like Methodists and Presbyterians and activist women organised a temperance campaign to restrict opportunities for drinking. Wine could only be sold for consumption in hotels, and then in 1881 its sale from vineyards was allowed but only in quantities of at least 2 gallons (9.1 litres). To carry the wine away ceramic flagons were developed. Other more wide-ranging prohibitions on the consumption of alcohol followed — no food or barmaids in the pub, no Sunday drinking, new licences were restricted, the age of entry raised to 21, windows of pubs were frosted to cut out the sights inside, and 12 of the 76 electorates voted themselves entirely dry. In 1919 in two referenda New Zealand only just avoided complete prohibition. For our story the most important restriction arrived in 1917 when, as much for economic reasons of 'national efficiency' as moral, the pubs were closed at six o'clock following an Australian precedent, and the next year this became permanent by one vote in Parliament. Men were kicked out of the pubs once serving stopped at the stroke of six. That was the crucial measure that created the flagon case, but there were several other developments which were essential.

World War II brought the next important change. In 1942, as a rationing exercise to save grain, the government lowered the alcohol content of beer, and to compensate additional sugar was added. At the

MT2017.12 FROM THE COLLECTION OF THE MATAURA & DISTRICTS HISTORICAL SOCIETY – MATAURA MUSEUM.

turn of the century New Zealand had already changed its style of beer from flavoursome English pale ales to German-style lager; and this now became watery and weak. In order to get drunk, New Zealanders had to drink large quantities of the lager.

The war itself encouraged many men serving overseas to develop a pattern of drinking on their time away from the front. For many veterans, drinking beer had become a regular habit. In 1949 another referendum confirmed the six o-'clock closing of pubs. The breweries realised that the weaker beer could be sold at the same price for lowered costs of malt and hops, and also reduced taxes which were levied on the basis of alcohol content. So, when the beer regulations were lifted in 1949, the breweries resisted returning to stronger beers.

The conditions were now in place for the notorious 'six o'clock swill' which astonished, and horrified, visitors and especially the large number of British migrants to New Zealand in the late 1940s and 1950s. For one frantic hour between the end of work at 5 pm and the end of beer sales at 6 pm, men consumed at incredible speed before going home to the suburbs. It was a disgusting sight — an exclusively male space (women were not allowed to work there and their attendance as customers was frowned on), crowded to the gunnels with no comfortable seating, just small tables for standing, and concrete or linoleum on the floor or, if you were lucky, a soggy carpet to soak up the spills. It was more evocative of a rugby scrum than a genteel savouring. By 1957 in terms of alcohol the quantity of beer drunk had risen more than four times the 1933 figure, and in terms of volumes of liquid the figure would have been even higher. Because the emphasis was on quantity and speed of drinking, beer taps in pubs were replaced by plastic hoses connected to a 2000-litre tank in the basement, jugs were introduced, and to deliver such huge quantities beer tankers were developed in 1948. The very first, an Austin K4, is now on display at MOTAT in Auckland. The use of tankers was promoted by the emergence of a nationwide duopoly of beer production. New Zealand Breweries and Dominion Breweries systematically bought up local breweries and established their control of New Zealand's weak, sweet, watery beer. The development of the continuous fermentation technique in 1958 only added to the industrial scale of the operations.

But what happened when the bell rang at 6 pm? If the Kiwi blokes wished to continue drinking, they needed to carry away enough of this low-alcohol beer to keep their mates happy around the kitchen table. The flagon became the essential piece of equipment, a way of avoiding the restrictions of six o-clock closing. It would be filled at the hose, or swapped for a prefilled and corked one. The flagon was such a loved item that it acquired various nicknames — a 'half g', a 'rigger', a 'bluey', a 'goon', even a 'Peter'. Of course, it did not look good to be seen carrying large flagons of beer home in the tram or bus. New Zealand remained sufficiently influenced by the Protestant crusade to retain a cult of respectability. So the contents of the flagons were discreetly hidden in a flagon case. The case also allowed beer to be carried modestly to rugby games or put in the car along with the thermette (see page 263) for picnics or fishing expeditions. The only requirement was that the beer had to be drunk fast, for within 24 hours flagon beer was usually flat. But that was not normally a problem for the Kiwi bloke of the 1950s!

This particular flagon case and flagons was originally owned by Jim Bradley, who with his wife Maud owned a taxi business in Mataura, Southland. There are four flagons, two of which fitted inside the case, and they have metal lids in green marked with the words 'Draught ale bottled by Mataura Lic. Trust'. This raises another interesting aspect of New Zealand's alcohol story. By 1908, 12 electorates had voted themselves dry, but as prohibition sentiment waned and habits of drinking alcohol were revived, those electorates eventually lost their 'dry' status. In the early 1940s, led by Rex Mason, Minister of Justice, and supported by Peter Fraser, the Labour government promoted the idea that if liquor were sold it should benefit the community, not the avaricious profit-focused breweries. So communities were given the option of establishing licensing trusts. They had a monopoly of liquor sales in their area and used the profits for community assets. Invercargill was the first in 1944; and eventually 30 became operational. They often flourished in areas that had previously been dry electorates. One such place was Mataura, which the strong Scots Presbyterian community voted dry in 1902. In 1955 the electorate decided to set up the Mataura Licensing Trust which in 2020 still had a monopoly on liquor sales in the community.

If the trust remained, the days of the 'half g' were long gone. In 1967, New Zealanders voted to end six o'clock closing, and the extension of hours allowed a more civilised style of drinking in which women were allowed into the pubs. New Zealanders began to change their drinking tastes — wine sales took off from the 1970s and a fashion for spirits-based RTDs (ready-to-drink) began in the 1990s. By the 2000s the amount of alcohol consumed in beer was not much larger than that consumed in the form of wine or spirits. Furthermore, beer was increasingly craft beer, and much of it from 1999 was sold in supermarkets. Whereas until the 1980s only about 20 per cent of beer sold in New Zealand was in bottles or cans, by 2011 the figure was two thirds. The age of the flagon and the watery beer that it once held was long over.

72.

Godfrey Bowen's handpiece

RICHIE GOULD MUSEUM

It was a typically hot summer's day. By mid-morning 6 January 1953 there were over 200 cars parked in the paddock, and some 2500 people crammed into the large red wood and corrugated-iron woolshed on the Akers estate at Ōpiki, about 15 kilometres south-west of Palmerston North. The crowd had each paid five shillings to watch a 31-year-old shearer, Godfrey Bowen, attempt to beat Percy de Malmanche's record, 10 years old, of shearing 409 ewes in a nine-hour day.

Bowen had begun at 5 am. He was aiming for 475 sheep to smash the record. For the first five hours he was exactly on schedule with hourly tallies of between 51 and 53 ewes. As each tally was announced the crowd cheered. But in the afternoon heat, things got tough. The crowd had filled the temporary grandstand inside and people were perched on every rafter. A non-smoker himself, Bowen was struggling with the overheated air polluted with tobacco smoke. There was one tiny window for fresh air and spectators craned their heads through it, blocking the breeze. 'The air could have been cut with a knife,' he recalled later. In addition to shearing the sheep with 13 months' growth, Bowen had to catch them in the stall nearby. Exhausted, struggling to breathe and suffering attacks of biliousness, he was near collapse. At each hourly break he sprawled on a seat, consuming black tea. He had abandoned the standard black 'Jacky Howe' woollen singlet for a light T-shirt.

However, the huge crowd, not to mention the National Film Unit who were shooting proceedings and the BBC who were giving a radio commentary, could not be disappointed. Bowen struggled on, paced by a local woman shearer, Julie Schwamm. Although he failed to reach the 475 target, he established a new record of 456 Cheviot-Romney cross ewes, the record confirmed by three Justices of the Peace. The £404 raised from the spectators was enough to pay off the mortgage on the local interdenominational church. As for the shearer himself, he had weighed in at 5 am at 14 stone 11 pounds, and weighed out at the end of the day nine pounds lighter!

This Wolseley 10 handpiece was the one used by Godfrey Bowen to achieve the record on that day in 1953. It was cleaned and put away. Bowen found it again over 30 years later and gave the handpiece to Richie Gould, a shearing acquaintance, who now gives it pride of place in his private museum at Pleasant Point in South Canterbury alongside an account of the day from Bowen himself.

Originally sheep were shorn using hand blades. Some merino sheep are still shorn that way. But handpieces date back to the early years of shearing in New Zealand. Sheep became valuable for their wool from the mid-1840s, and in the 1850s a handpiece, using a cutter sweeping back and forth against a toothed comb, was invented in the United States. This required a source of power, so for several decades shearing continued to be done out in the open, usually on tarpaulins and with hand shears. However, in the 1880s the first machine shearing was demonstrated successfully in Australia, and Frederick Wolseley developed an effective powered handpiece which was attached to a down-tube from an overhead gear. By the 1890s some large stations were shearing in woolsheds with machines driven by a steam-powered traction engine. The first electric machines were seen in New Zealand in 1909.

Shearing gangs emerged who travelled from station to station. They included 'rouseabouts' ('rousies' for short) to pick up the shorn wool, table hands to sort

RICHIE GOULD'S MUSEUM, PLEASANT POINT. IMAGE BY JOCK PHILLIPS.

it, and 'pressers' to operate the wool press. The fast shearers, or 'guns', were paid by the number shorn, not time, so there was a premium placed on speed. Before long, A&P shows began to hold shearing competitions. Famous shearers emerged — such as the Ngāti Porou 'gun' Raihania Rimitiriu — in the interwar years. In addition to improved equipment, shearers developed new techniques for shearing fast. They learnt to hold the sheep between their knees, and from about 1905 a technique emerged to lay the sheep down for a long 'blow', a series of strokes with the handpiece along the sheep's back. The handpiece was refined with the development in the 1930s of a high-tension nut which made the pressure of the cutter on the comb fully adjustable. The nut can be seen clearly on Bowen's Wolseley 10 handpiece.

An improved technique also made Godfrey Bowen's record possible. In this he owed much to his older brother Ivan. The Bowen brothers were sons of a man who had farmed in Hawke's Bay before establishing a sawmill in Te Puke. Ivan and another brother, Eion, established themselves as shearing contractors and Godfrey joined them shortly before World War II. Because shearing was essential work, he was exempt from military service. Ivan developed a new style of shearing which used the full width of the comb and refined the sequence of 'blows'. Godfrey picked up the technique and used it to achieve his world record.

Bowen's achievement in January 1953 received extraordinary publicity in the press, not to mention radio and film. One wonders why it was so hailed. Part of the explanation was that it was a 'Kiwi' achievement, and New Zealanders in the 1950s hungered for international recognition. It was also partly that through the A&P show competitions, shearing had been transformed in the public mind from a job to a sport, and at Ōpiki on that day it was very definitely a spectator sport. Interestingly, in the media Bowen was presented, largely fictitiously, as a 'hobby' shearer on the grounds that he prepared the accounts for his father's sawmilling business. He was thus a noble 'amateur' of British sporting tradition. Like other contemporary sporting heroes, Bowen was described as a family man with two small children, and it was noted that he was a dedicated Christian. One *Evening Post* editorial, using sport as a measure of the nation's fitness and reflecting continuing racial anxiety, noted Yvette Williams' gold medal in the Olympic long jump the previous year and Brian Sutcliffe's 385 runs in a Plunket Shield cricket match, and claimed that Godfrey Bowen's feat showed 'that in New Zealand, so far from deteriorating, we are breeding a race which will bear comparison with the best qualities of its predecessors'.

Behind the public fascination with Bowen's achievement lay a larger reality. Wool had become the centrepiece of the New Zealand economy.

From the halcyon days of the 1870s when wool comprised up to 60 per cent of New Zealand's export receipts, its value had fallen, and in the 1930s wool constituted only about 20 per cent of the country's overseas earnings. It was significantly less valuable than frozen meat or butter. In 1945 wool constituted under 16 per cent of our export value. But the following year things improved. Prices and sales rose. In 1951 wool prices spiked as the United States sought to clothe its forces in warm wool garments for winter in the Korean War. That year, half of New Zealand's export earnings came from wool, and, while prices then dropped, throughout the 1950s and early 1960s wool represented 30 to 40 per cent of New Zealand's overseas income, significantly more than earned by meat or dairy products. The sheep farmer was king, and wool the golden fleece.

Behind this economic success was the New Zealand Wool Board. Established in 1944, the Wool Board worked assiduously to promote wool, increase production in New Zealand, and smooth out prices. The publicity garnered by Godfrey Bowen in breaking the record was too good an opportunity for the board to ignore. Within months they had enlisted his aid. Bowen became 'chief instructor' of the board's new shearing service, providing demonstrations in woolsheds and country halls throughout rural New Zealand, training regional instructors in the Bowen technique, and teaching courses at Massey and Lincoln agricultural colleges. He produced a much-used book on shearing, *Wool Away*, in 1955. So effective was Bowen in his role as 'Mr Wool' that he attracted international attention. He gave performances around the world, including the Soviet Union where Nikita Khrushchev made him a 'Hero of Socialist Labour' in 1963!

Not that shearing was just a skilled occupation, essential to the economic welfare of New Zealanders in those years. It remained for Bowen in large part a sport and an entertainment. He encouraged people in the Wairarapa to establish a major competition, the Golden Shears, at Masterton. At the first event in 1961, Godfrey was pipped to victory by his older brother Ivan. This was not the first time that Ivan had outdone his younger sibling. Ivan had helped pace Godfrey during his record-breaking run in January 1953, and they often sheared beside each other, sheep for sheep. Later that year it was agreed that Ivan should be given the opportunity to equal his brother's record. Unknown to Ivan, those marshalling the sheep added one extra ewe, so to his embarrassment and surprise Ivan ended up with the new record of 457. Soon after, Godfrey had a consolation prize. After sustained public pressure following his January record, it was agreed that Godfrey could demonstrate shearing for Queen Elizabeth II during her visit to New Zealand at the end of the year. Appropriately, Godfrey Bowen received a royal honour, MBE, in 1960.

73.

Crown Lynn coronation mug

TE TOI UKU – CROWN LYNN AND CLAYWORKS MUSEUM

This handsome mug was made by the New Zealand pottery firm Crown Lynn in 1953 as a souvenir for the coronation of Queen Elizabeth II. One thousand cups were made. The mug invites two revealing stories — New Zealanders' fascination with royalty in the 1950s and early 1960s, and the rise and fall of Crown Lynn Potteries. Both tell us much about New Zealand's infatuation with the 'Old Country' during those years.

The mug was described as a 'loving cup' — not because it expressed a love of the Queen, but because its two handles supposedly allowed two people to use it. It is large, 115 millimetres high, 100 millimetres wide at the rim and a full 190 millimetres if you include the handles which are gilded with a representation of the British lion. On one side is an imported transfer, found on cups by other makers, which shows the Queen's coat of arms, and the words 'Queen Elizabeth II Coronation Souvenir June 2nd 1953'. On the other side, 'ERII' is inscribed in gold and black. This was not the only coronation cup made by Crown Lynn and was one of many New Zealand souvenirs of the occasion. There were golden spoons, fold-outs of the coronation procession, and metal coaches and horses for kids.

The coronation of the young Queen, still in her late twenties and with two small children, was a major occasion in New Zealand. In the evenings before the big event crowds thronged city streets to gawk at the floodlit decorations, the bunting and royal emblems picked out in electricity.

LOVING CUP – QUEEN ELIZABETH II CORONATION, CROWN LYNN POTTERIES LIMITED. TE TOI UKU CROWN LYNN AND CLAYWORKS. 2008.1.261.

There were coronation balls, and on the actual day processions of floats, bands and marching soldiers. Since the exact hour of the coronation in New Zealand was at night, it was marked with bonfires and fireworks, the firing of guns, and the pealing of bells. Auckland bookings for the coronation films *A Queen is Crowned* and *Elizabeth is Queen* broke records. The coronation gained lustre when, on its eve, news arrived that Edmund Hillary had climbed Mt Everest. This was seen as New Zealand's special gift to the Queen.

Six months later this public enthusiasm was far eclipsed when from 23 December 1953 to 30 January 1954 the Queen and Prince Philip toured New Zealand. At least two in every three, perhaps three in every four, New Zealanders dressed up in their best, took along periscopes and their cameras, and Union Jacks to wave, and waited for that unforgettable moment when the royal couple drove past. The social energy that went into this event was remarkable — facades painted, milk bottles given commemorative tops, sheep dyed in the patriotic colours of red, white and blue, and huge floral arches made, one in the shape of Mt Everest. Every schoolchild received a commemorative medal and was offered free train travel to see the Queen.

How to explain this extraordinary enthusiasm? It was partly that this was the first occasion a reigning monarch had set foot in New Zealand — a visit much anticipated after planned trips by George VI were cancelled in 1940, 1949 and 1952. It was partly that the Queen was a media idol, a 'young and radiant' princess 'of fairy-tale'. As a young mother she personified the family values of post-war New Zealand. It was also that New Zealanders had a special love of Britain at that time. In explaining the public response, the *Hawera Star* wrote: 'The great upsurge of feeling owed something, too, to unexpressed feelings of relief from the tension of the war days when New Zealand was drawn so close to Britain that listening to the voices of her leaders became part of the daily routine of living.' The Queen visited partly to thank New Zealanders for their war service to the mother country. The nature of the population reinforced these sentiments. New Zealanders in 1953 were overwhelmingly of British descent. Of some two million people in 1951, only 115,000 were Māori (6 per cent of the population), many of whom had shown their devotion to the Empire in war. Only 30,000, 1.5 per cent, were born outside the Commonwealth. Each year about 10,000 assisted migrants arrived from England and Scotland. Only the several thousand Dutch settlers added a non-British element. Naturally, the Union Jack, not the New Zealand flag, was waved by Her Majesty's loyal subjects in 1953–54.

The mug was the product of Crown Lynn Potteries, and here too English influence was profound. Crown Lynn had its origins in the Clark family's brick and pipe company, in West Auckland. Begun in the 1890s, it was called Ambrico from 1925. Cargo restrictions during the Second World War opened up markets for household pottery. The firm supplied crockery to the American troops in New Zealand and then won a contract with the Railways Department. The sturdy white mugs marked NZR used at railway cafeterias became a New Zealand icon. Trying to smash them against a tunnel wall was a national sport. In 1948 Tom Clark, head of the firm, decided that, as affluence returned and New Zealanders moved into suburban homes, there was an opportunity to offer high-quality domestic dinnerware. The problem was that New Zealanders expected to eat off plates made in the old country. Clark decided that a new name with 'home' connotations was essential — hence Crown (with its royal associations) and Lynn (from the factory's locality at New Lynn). With imports still flooding in, Crown Lynn explored specialty lines. The firm made jugs to accompany McAlpine refrigerators for example. They also saw a market in souvenirs, such as for the 1950 British Empire Games in Auckland. The coronation mugs were another such effort.

But the mass production of dinner sets was Crown Lynn's goal. To achieve this, Clark believed he needed expertise from Stoke-on-Trent, the home of Wedgwood and the British potteries. He had to make Crown Lynn ware look similar in quality and style to British imports. He advertised in Stoke papers, visited the area, imported equipment and recruited skilled workers. In the late 1940s seven skilled workers arrived; and another 10 or so in the 1950s. Three of these English workers were largely responsible for the coronation mug — partly designed by Peter Cooke, hand thrown by Ernest Shufflebotham and decorated by Doris Bird. Furthermore, Crown

Lynn style had to suggest English traditions. Tom Clark originally stamped his dinnerware 'Made in New Zealand'. But this put off buyers. Instead he labelled the pottery 'British' or back-stamped it 'Regal Potteries'. He named sets with English associations — 'Ascot and Wentworth ware', 'Covent Garden British', 'Lido British', 'Sylvia Rose'. The motifs were northern hemisphere floral designs, such as 'Autumn Splendour' and 'Fashion Rose'.

Admittedly Clark aspired to interesting local designs. In the late 1940s a Wharetana range of Māori designs was launched, but it was a limited run aimed at overseas tourists. In the early 1950s Clark hired several artists from Europe: two Dutchmen, Frank Carpay and Daniel Steenstra, who both produced dynamic work; and a Czech, Mirek Smíšek, who added a striking line called Bohemia ware. None were successful commercially. Clark recalled that when he showed buyers Carpay's work, 'You would see them shuddering because it wasn't British.' Carpay waged 'a war against the rosebuds', without success. After three years he was let go.

From the late 1950s things began to change. In 1958 Walter Nash's one-term Labour government was faced by a serious crisis in the balance of payments. In the notorious 'Black Budget' new taxes on cigarettes, petrol and booze were introduced, and import controls on foreign manufactures were reimposed. Crown Lynn now had protection from British imports. Another help was the next tour by the Queen in 1963. This time she visited the Crown Lynn factory at New Lynn and although the Queen was not impressed by the ornate urn specially designed for her, Tom Clark believed that the visit 'deodorised the whole stink of the past'. She convinced New Zealanders that local dinnerware was respectable and could foot it with that from Stoke-on-Trent.

In the mid-1960s, with protection and royal acceptance, Crown Lynn was booming, and soon employed over 500 workers producing over 15 million articles each year. Most New Zealanders were eating off Crown Lynn china. Clark's efforts to encourage local designs began to pay off. Some designers used fern leaves and koru patterns. The Air New Zealand dinner service was embellished with kōwhaiwhai. Local idioms were never hugely popular but they represented a new cultural nationalism.

The good times did not last. In 1984 a Labour government was elected and this time, also facing an economic crisis, they lowered prices for consumers by ending import duties. Cheap dinnerware flooded into the country from Asian factories. Crown Lynn could not compete and in 1989 the factory was closed. Today Crown Lynn products can only be found in second-hand shops and in the pantry cupboards of older New Zealanders. We remain fortunate that a few of the firm's souvenir works from 1953 survive to remind us of the love of royalty and of the old country in 1950s New Zealand.

74.

Happy Families card game

PUKE ARIKI

Picture the scene. A wet weekend in the late 1950s. No shops are open at weekends; there is no television to watch. An open fire is roaring and there is a nice woollen carpet to lie on. It's time for the kids to play a game and there's a wide choice. Perhaps the old stand-by 'Snakes 'n' Ladders', or 'Tour of New Zealand', a local board and dice game introduced a couple of years before (in 1956). Let's play the one that Mum has just got from the local Four Square grocery, 'Happy Families'. The cards are dealt; and Mary begins. 'Have you', she asks John, 'Mrs Creamoata?' Reluctantly John hands over the card and Mary puts down four cards, Mr, Mrs, Master and Miss Creamoata. 'I have a happy family!' she gloats.

Happy Families was an archetypal New Zealand game of the decade. There were other favourites in those years — the Buzzy Bee for toddlers was invented in the 1930s but took off when Hec and John Ramsey started mass production in 1948. Young boys had Dinky cars and train sets; young girls had plastic dolls. But Happy Families was universally popular. This set comes from Puke Ariki. Happy Families was an old English game, but the Kiwi version was devised by the Four Square grocery cooperative in the late 1950s. Judging by the brands featured, it was after the launch of Jojo jelly in 1958, and before the sale of Crest in 1959.

Four Square was established in the mid-1920s as a cooperative company to distribute goods to independent grocery stores. The

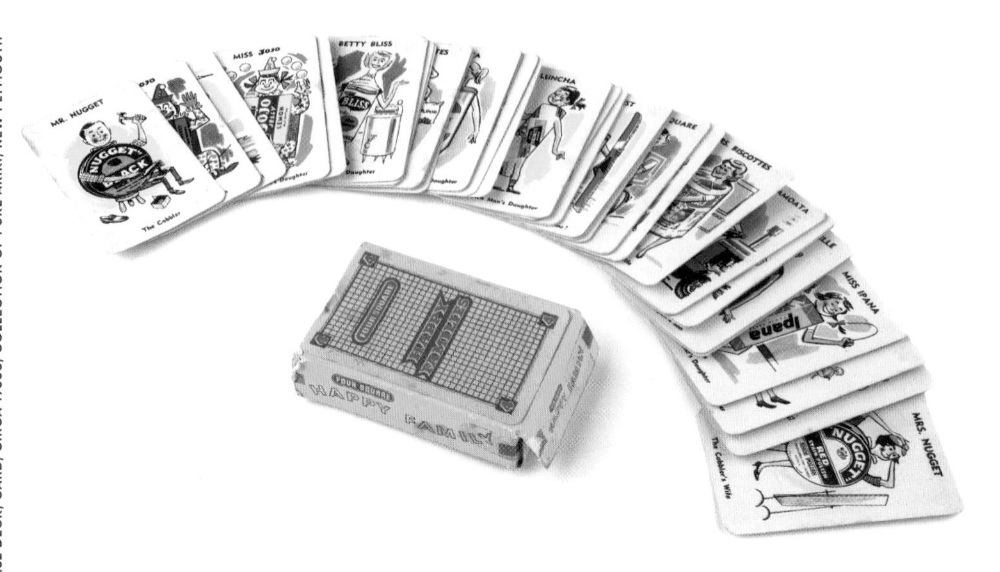

TM1998.52 DECK, CARD, CIRCA 1950S, COLLECTION OF PUKE ARIKI, NEW PLYMOUTH

cooperative had two distinctive icons. One was the four-square symbol which allegedly arose from a doodle around the date drawn by the company secretary John Heaton Barker. The other icon, developed in the 1950s, was 'Cheeky Charlie', the Four Square storeman with the cheery face and a big thumbs up. Four Square took off in the post-war years. In 1935 they had 285 grocery stores; by 1956 there were over 1000. Housewives of the 1950s usually visited the grocer about three times a week; and following the introduction of self-service from 1948 they could wheel a trolley and help themselves rather than ask for items individually. Supermarkets did not appear until the 1960s. So the Four Square store was an important presence. Many families pinned the Four Square Christmas calendar of beautiful New Zealand on their walls.

There are 48 cards in the game, 12 families, each with a father, mother, daughter and son. Each family represents a popular product — from Nugget shoe polish to Creamoata oats. Every character is smiling broadly (they were 'happy' after all), and they reflect 1950s gender roles. Thus most 'Mr's are in paid employment. Mr Nugget is a cobbler, Mr Ipana a dentist, Mr Biscottes a baker. But not the women, who are defined as the 'wife' of the cobbler or dentist or baker. Mrs Nugget is putting on a hat in front of the mirror; Mrs Biscottes is making a cake. Even some of the Masters, the male children, work for pay, but none of the Misses. Master Four Square delivers groceries, while his sister plays on a scooter. Master Creamoata, or Sergeant Dan (a reference to the product's mascot), works on a tractor, while Miss Creamoata plays with a pet lamb. Even where the father is not seen working, his role is clearly demarcated from his wife's. Mr Gregg reads the paper, while his wife makes pudding; Mr Bliss cleans the car, Mrs Bliss cleans the dishes. Even the kids have distinctions — Master Ipana plays cricket, Miss Ipana just skips.

These gender differences certainly reflected social realities. Once peace arrived in 1945, women were urged to go back home and have children. Fears of 'race suicide' and an 'Asian invasion' led to public policies to encourage families to have more children, along with assisted migration from the United Kingdom. In 1946 the family benefit, paid to mothers, became universal, and the next year the baby boom began with a record number of births. Men's pay rates were intended to cover the living costs of a wife and children, while women's pay rates were lower, encouraging mothers to return home and procreate. The number of children under the age of 15 rose from about 450,000 in 1945 to almost 800,000 in 1961.

Most women married, gave up paid work and became full-time housewives and mothers. Throughout the 1950s women's average age of marriage was under 23 and over four in every five women aged 25 to 34 were married. Once married, fewer than one in 10 in 1951 remained in paid employment, the lowest in the western world. Children too were no longer in paid work, with schooling becoming compulsory from the ages of six to 15 in 1944. There was a huge expansion in secondary schools. Secondary school pupils increased from under 40,000 in 1943 to some 150,000 in 1963. At the other end of childhood, government assistance from 1947 led to more three- and four-year-olds attending playcentres and kindergartens.

Several factors made this baby boom possible. One was the level of affluence. With exports of wool, meat and dairy products still gratefully received in the UK, and domestic manufacturing protected, unemployment was low and incomes the second highest in the world. Families could survive on one working income. In addition, government policies and funding facilitated a burst of suburban housing. There was a major increase in new state houses built, and to assist families into their own homes, State Advances provided cheap loans and guaranteed mortgages. Families could capitalise family benefits. From 1950 to 1965 there were on average about 20,000 new homes built each year and the total number increased by about 50 per cent. Those living in boarding houses and rental accommodation declined as New Zealanders moved into their own homes. Rental dwellings fell from about 35 per cent in 1945 to under a quarter in 1961.

Many of these family homes were in the suburbs around cities. In South Auckland and especially on the North Shore (once the Auckland

Harbour Bridge opened in 1959), there were acres of new housing, as there were in Tawa and Porirua north of Wellington, in the Hutt Valley and in the outskirts of Christchurch. In Auckland and Wellington, suburban trains took the men home to families, and bus routes proliferated. The number of New Zealanders in urban areas increased by half between 1951 and 1966 and the Auckland urban area reached half a million. The suburban quarter-acre pavlova paradise (more usually an eighth of an acre!) became a realisable dream for thousands.

These 1950s family homes were also better furnished. During the decade new 'mod cons' arrived — electric stoves replaced coal ranges, refrigerators and electric washing machines became standard, inside flush toilets replaced outside 'loos'. Wool carpets were laid on wooden floors and to clean them electric vacuum cleaners became widespread. The single-family suburban home, with Dad at work, Mum the efficient housewife, and kids at school or playing sport at the weekend, was the site for New Zealand's 'Happy Families'. This was as true of the British immigrants who flooded in, and the Māori who came to the city for work, as of the Pākehā. Some houses funded by the Department of Maori Affairs were deliberately 'pepper-potted' through the suburbs to enforce the suburban family norm on Māori too.

The families in our game highlight characteristics of this suburban life. Nugget shoe polish, Bliss cleaning fluid and Ipana toothpaste suggest that the family and home were expected to be hygienic and well scrubbed. Creamoata for the morning porridge, Crest cans of peas, Betta peanut butter and Luncha grease-proof paper suggest the housewife was expected to provide healthy meals and pack school lunches with nutritious sandwiches. Not that there were no treats — Jojo jelly and Gregg's instant puddings provided quickly made desserts. The busy housewife herself could always take a break with Rawakelle tea.

While the promotion of a happy suburban family was widespread in 1950s New Zealand, not everyone found the ideal satisfactory. Many Māori joined the rush to the suburbs but some preferred to remain close to their home marae in rural New Zealand, and not all were comfortable with the nuclear family structure assumed by the two- or three-bedroom suburban home. Many wanted to be surrounded by their wider whānau which 'pepper-potting' did not recognise. Gay New Zealanders suffered from the heterosexual assumptions of the 'Happy Families' model. Lesbians kept their relationships from public view, although there was much titillation about the lesbian aspects of the infamous Parker-Hulme murder in Christchurch in 1954. Gay men too hid from the police, for homosexual activities between males were still illegal. Ten years after the Parker–Hulme case, also in Christchurch, six youths who beat up a gay man in Hagley Park were acquitted of manslaughter by a jury. Some women found the prescriptions of the 'Happy Families' model limiting; but when Phoebe Meikle penned an article in *Landfall* critiquing the extreme gender distinctions of 1950s New Zealand, she chose to hide her identity behind an androgynous pseudonym, Leslie M. Hall.

Finally, there were some young New Zealanders, the very children at the centre of the ideal, who, as they matured, found the moral repressions of 'Happy Families' limiting. They wished to experiment with sexuality and to develop their own culture centred on the milk bar and rock 'n' roll. But when the wider society heard about this there was a national outcry. The government sponsored an inquiry into the alleged goings-on of teenagers in the Hutt Valley, and the resulting Mazengarb report was sent to every household in the country in November 1954.

While most New Zealanders accepted the 'Happy Families' ideal of the 1950s, the understated rebellions from women, gay people and Māori were a foretaste of things to come. The baby boomers who played Happy Families in the 1950s would play a very different game when they reached adulthood in the late 1960s.

75.

Winston Reynolds' television set

HOKITIKA MUSEUM

In 2011 when digital television was about to come to New Zealand the Going Digital team had a competition to find the country's oldest working television set. The winner, to everyone's astonishment, was a kitset television owned by 92-year-old Elva Reynolds in Hokitika. That seemed extraordinary given that the West Coast was about the last area of the country to officially receive a television signal.

The person who made the set was Winston Reynolds. He came from a well-established West Coast family. His grandfather, James Reynolds, a Methodist lay preacher from Cornwall, had been one of many from that area who migrated in the nineteenth century in search of gold. He was sufficiently successful to buy land at Waitaha almost 40 kilometres south of Hokitika. Winston's father took over farming, but Winston, an only child, was not keen on that life. His passion was radio and sound. He trained as a radio serviceman in Wellington, and, in the air force in World War II, specialised in radio communication. But there was not much work in that line on the West Coast, so he earned a living at the McKays Creek power station on the Kanieri River. After work and at weekends he pursued his hobby, repairing old radios in his 'shack' or setting up a sound system to play music and provide a microphone at public events. He built a recording machine to cut acetate records and recorded the family singing around the piano.

Through his contacts with overseas radio hams, Winston heard about television and in 1958 managed to obtain parts for a kitset television which he then assembled. At the time, there was no organised television transmission in New Zealand. The country was far behind developments overseas. Both Britain and the United States had introduced public television before World War II. Australia had its first stations operating by 1956. In New Zealand various government committees had been studying the medium since 1949. There were some closed-circuit demonstrations from 1951, and in the mid-1950s both Pye and Bell displayed television at shows. But governments were put off by the expense and the mountainous terrain which inhibited getting signals to many areas. Eventually Walter Nash's Labour government committed itself and the first official transmission, for a mere two hours, came on 1 June 1960. It was only receivable in Auckland; and only for two nights a week. A year later, on 1 June 1961, Christchurch received its first television transmission, followed by Wellington. But few received the signal at that stage. There were under 5000 sets in the country and the licence fee was £6.10.0 (almost $300 in 2020). Dunedin had a station the following year. But the West Coast, isolated from east and north by the Southern Alps, had to wait until 1975 to obtain a satisfactory signal.

So, having built his television set in 1958, what did Winston Reynolds watch? He erected a huge aerial above his house in Hampden Street, Hokitika to obtain a snowy signal with erratic sound from a Sydney station across the Tasman. Then in 1964 Winston discovered that, by locating his set on the Hokitika wharf, it was possible, in good conditions, to receive a signal from Mt Sugarloaf transmitting the Christchurch station. There were only two other places on the coast where this occurred and one required TV addicts to clamber up to a high point above Denniston with a homemade aerial. Winston was by now a public-spirited local councillor, so he decided to make his television a social service. He set up the 'Hokitika and Districts Televiewers' Society', erected a 70-foot aerial on the wharf, and residents of Hokitika were free to come down, sit on hay bales and banana boxes in the old wharf building and enjoy Christchurch television (even if occasionally interrupted by Australian signals!). One Hokitika resident, a child at the time, remembers watching Peter Snell winning a gold medal in the 1500 metres at the Tokyo Olympics in 1964. Perhaps deservedly, Winston Reynolds was elected Mayor of Hokitika and died in office in 1982 after 26 years in local government.

While the West Coast had a long delay in enjoying television, it took some time after 1960 before most New Zealanders watched the box. There was delay in some regions relaying the signal. There was suspicion from those who suggested the 'goggle box' would lead to 'square eyes' and rot the brain. More important was the expense — the steep licence fee and the cost of sets which, made locally in a protected industry, were usually £200–£300 (about $8000–$12,000 in today's money!). However, by 1965 there were 300,000 sets and over half a million by 1967. By then the television had become the centre of domestic life, with the furniture in living rooms rearranged for the best viewing. Adult New Zealanders were watching two to three hours of television a day, about 15 per cent of their waking hours.

What would Winston Reynolds' TV fans have watched on the Hokitika wharf? In one sense it would have seemed familiar. The box followed the movies, and radio, by emphasising the cultural dominance of the old country and of Hollywood. On the first night of television in Auckland in June 1960 the offerings included two programmes from Britain including *The Adventures of Robin Hood*, and two series from the United States. The local content was an interview with a visiting English ballerina and a performance by the Howard Morrison Quartet. In 1965 only a quarter of the programmes were local with the rest divided equally between American and British programmes. No doubt the Hokitika wharf watchers would have followed closely the adventures of *Peyton Place* which ran from 1964 to 1969 and *Coronation Street* which first aired in 1960 and of course is still running.

HOKITIKA MUSEUM

Until 1969 there was no national network and the four main centre stations had their own schedules. News items were flown to each for showing in turn. On the night of the *Wahine*'s sinking in April 1968 Christchurch's Channel 3, keen to get footage for their news bulletin, sent a reporter up to Kaikōura. There he picked up the Wellington signal and filmed it from the screen before rushing back to Christchurch! So Winston's television viewers would have enjoyed, alongside the Hollywood and BBC programmes, some Christchurch content. Especially in the folksy and parochial *Town and Around* programmes, covering local events and personalities, which played in the early evening.

In 1969 the news became networked and from 1975 there were two national channels, TV One and TV Two, available throughout the country. Both channels carried advertisements but there remained a public service ethic; and this facilitated some distinctive homemade programmes which, despite the continued importance of overseas (especially American) material, became central to an evolving national identity. At the very moment when most people on the West Coast first received television in 1972, the medium was starting to define the national conversation. Since there was only one channel, and after 1975, only two, and most households by that time had sets, much of the population shared the viewing experience. People would talk about what they had seen around the tea trolley next day. The West Coast became part of a monolithic national media world. The best expression of this was the news programmes — the networked news and current affairs slots like *Compass* (1964–69) and *Gallery* (1968–74). Interviews with politicians were followed closely and election campaigns were fought in the television studio rather than on the hustings. Those who had a 'television presence', such as Norman Kirk with his gravelly voice and silver hair (at least by 1972), and Rob Muldoon with his air of belligerent authority, did well. Those who were not telegenic, such as Bill Rowling with his soft speaking and mild manner, failed to attract the voters.

The two-channel state monopoly of television led to other widely watched programmes which helped define a New Zealand world view. From 1966 Winston Reynolds' TV fans would have enjoyed the country's longest running show, *A Country Calendar*, which strengthened a sense of the country's rural mythology at a time when it was becoming overwhelmingly urban. This was reinforced both by the popular *A Dog's Show* in which shepherds put sheep-dogs through their paces and the skits of Fred Dagg with his gumboots and black singlet. His local brand of comedy was followed up by *Lynn of Tawa* (Ginette McDonald) and David McPhail and John Gadsby chatting in a pub in *A Week of It*. There was some impressive discovery of Māori history in Michael King's path-breaking *Tangata Whenua* series and the epic drama *The Governor* about George Grey, both aired after the West Coast received a good signal. There was a local soap opera, *Close to Home*, and a legendary popular music show, *C'mon*, which brought local performers like Dinah Lee and Ray Columbus to the fore.

After initial resistance by rugby authorities to the televising of games, from the early 1970s there was coverage of significant national sporting events. The most dramatic moment came on the opening day of the 1974 Commonwealth Games in Christchurch when Richard Taylor won a gold medal in the 5000 metres. This was televised in glorious colour which had been introduced specifically in time for the games. There were local children's shows like *Spot On* and distinctive Kiwi game shows such as *Top Town* and *It's in the Bag* with the ebullient host Selwyn Toogood, which were two of Winston Reynolds' favourites. *Koha* and *Te Karere* introduced the first programmes using the Māori language. In retrospect, the years when Winston Reynolds and his mates began watching television from the late 1960s through to the late 1980s were a halcyon period. Never before or since did one communication medium so define the stories that New Zealanders talked about. Telethons gripped the nation's attention.

The situation did not last. In 1989 Television New Zealand lost its monopoly and became a state-owned enterprise, no longer supported by a licence fee, and reliant on commercial advertising. A private channel, TV3, was launched that year and the Sky network followed a year later. Satellite technology allowed for international feeds 24 hours a day. In the 2000s the coming of the internet and then of digital television in 2012 opened up a huge range of competing channels and streaming programmes like Netflix. New Zealanders became part of a global audience, with the same programmes and news shared from Auckland to Arkansas. To be fair, a new agency, New Zealand On Air, was established in 1989, to support local programming, Māori Television was introduced in 2004, and even Television New Zealand was given 'social objectives' which were enforced with varying commitment. There were some significant local programmes still made, often by independent producers. *Shortland Street* from 1992 became the closest we have ever had to a national soap. But in general, diversity, choice, and the ever-growing impact of American media undermined that national conversation around the box, which was so characteristic of New Zealand in the years when Winston Reynolds worked his magic to bring television to the people of Hokitika.

76.

Mt Eden gallows

AUCKLAND WAR MEMORIAL MUSEUM

Stored in the basement of Auckland War Memorial Museum is the most macabre object in New Zealand. It is a set of wooden planks, made of heart kauri, and steel uprights along with metal plates, braces and various fittings. Known colloquially as 'the Meccano set' with reference to a popular children's construction toy of the early twentieth century, and marked with letters and numbers to allow quick assembly, it is a set of gallows. It was used to hang every New Zealander, 16 of them, sentenced to death for murder since 1924. Nearby are boxes containing the ropes, block and tackle, even a black hangman's cap, used for these gruesome events. The story of these objects' transition from a central tool of justice to a museum object hidden away in a basement makes for a fascinating and revealing story of the 'civilising process' in New Zealand.

Death by hanging became a part of New Zealand life with the formal establishment of British legal authority in 1840. It was a mandatory sentence for those convicted of murder, treason and piracy, unless the Executive Council granted mercy. The first hanging, in March 1842, was Maketū Wharetōtara, the 17-year-old son of a Ngāpuhi chief. Over the next 93 years, until 1935, another 77 people were hanged; almost a third (24 judging purely by their names), like Maketū, were Māori. They included Kereopa Te Rau, hanged at Napier gaol, and five others like him who went to their death for their part in the killing of Carl Sylvius Völkner (see pages 140–43). One of these, Mokomoko, went to the gallows pleading his innocence in the immortal words, 'Tangohia mai te taura i taku kakī, kia waiata au i taku waiata' — 'Take the rope from my neck that I might sing my song'. He has subsequently been pardoned. There was only one woman hanged, the notorious 'baby-farmer' Minnie Dean in 1895.

The first eight hangings, until 1858, were held in public and large crowds attended. In that year it was prescribed that hangings should occur only within the gaol. This did not stop, for a time, nosy spectators climbing the slopes of Mt Eden to peer down at the awful events within. Each prison faced with an execution had to build a temporary scaffold; and they were not always fit for purpose. A number of hangings were sadly bungled in the nineteenth century, on one occasion when wood was warped by rain and the trapdoor failed to open fully.

So in the early 1920s it was decided to commission the railway workshops in the Hutt Valley to construct a reusable scaffold which could be disassembled and sent by rail in boxes between Wellington and Auckland whenever an execution was planned at either Mt Crawford or Mt Eden. Hence the 'Meccano set' came into being, 2.4 metres square, 3 metres high, and painted silver. New efficient equipment for hanging went along with the ending of other gruesome customs such as the tolling of the prison bell and the flying of a black flag on the day of an execution.There was also increasing use of a reprieve by the Executive Council. From 1920 to 1935, 26 men were convicted of murder but only half were eventually executed. Then in 1935 the new Labour government was elected. The Labour Party had held to a policy of opposing capital punishment since 1918. The Deputy Prime Minister Peter Fraser, who always had some authoritarian instincts, opposed its abolition; but all convicted murderers were given mercy. Then in 1941, while Fraser, by now Prime Minister, was overseas, the government abolished death by hanging, for all crimes except treason. Flogging was also abolished. Sadly that was not the end of the story. The National Party had consistently supported hanging, and emboldened by a flurry of eight convictions for murder in 1950 the new National government reintroduced the death penalty. The Meccano set was brought out of storage, and a request sent to Wandsworth prison in England for new fittings such as two new rope nooses and a block and tackle.

Over the next few years, until the last hanging in 1957, eight men were taken to the gallows. The scaffold was set up in a small enclosed yard in the punishment block, at the east end, of Mt Eden prison. The yard was covered in a wire mesh over which tarpaulins were spread to hide the view from above. The length of the rope was dependent on the man's weight — if the man was light and the rope not long enough, he could suffer a protracted and painful death as occurred on one terrible occasion in the 1930s. So once the sentence of death was confirmed, the victim was weighed every day, to ensure he did not know when hanging was imminent. On the fatal day he would be brought to a cell close by the gallows. Before 1935 other prisoners would hear his march and respond in traditional fashion with the shouting of obscenities and the banging of chamber pots; so in the 1950s a movie was scheduled for the prisoners to divert their attention and then the route was laid with seagrass matting to muffle the footsteps. The prisoner was offered a powerful sedative of phenobarbitone and morphine. When the sheriff summoned the prisoner, prison staff pinioned his arms, led him up the 17 steps to the top of the scaffold, pinioned his legs, and placed a white linen hood over his head and the rope around his neck. The hangman's only role was to pull the lever that released the trapdoor. For this role he was paid £50 a time. Sometimes the metallic clang of the trapdoor was heard and incited noisy banging and calls from other prisoners. It was an hour before the man's body was recovered.

There were always a few spectators allowed — the press, officials, the police, a doctor. They were offered beer and whisky to strengthen their moral fortitude — the only occasion that alcohol was allowed in Mt Eden prison. Those more closely involved were naturally upset. The first sheriff involved in the 1950s hangings fell sick with a duodenal ulcer the day after a hanging and eventually resigned; his successor became morose and withdrawn after four executions and resigned before the last execution of Walter Bolton in 1957; and the man who presided over that event immediately took time off, suffering from nervous exhaustion.

If the officials pointed up the inhumanity of the death penalty, there was also the fear of executing innocent men. This was especially the case with the last hanging, Walter Bolton, accused of poisoning his wife. He pleaded innocence and there was some evidence that arsenic from sheep dip might have penetrated his farm's water supply. He went to his death on 18 February 1957.

Unsurprisingly, when a Labour government was elected in 1957, they refused to allow the death penalty and took refuge in the Executive Council's privilege of offering mercy for any death sentence passed. But the time for abolition had arrived; and interestingly it came under Keith Holyoake's National government. The Minister of Justice was Ralph Hanan, an Invercargill lawyer and war veteran, who became one of the country's great legal reformers. He was responsible for the establishment of an Ombudsman, and also for the Law Revision Commission which eventually became the hugely valuable Law Commission. Robert Muldoon remembered that 'he had a mind like a corkscrew and a puckish sense of humour'. Hanan had always opposed the death penalty; but in

1961, as Minister of Justice, he was obliged to introduce a revised Crimes Bill which included the death penalty for three types of murder. However, Hanan indicated that the member for New Plymouth, Ernie Aderman, a clergyman and National Party MP, would move an amendment abolishing the death penalty and that, although he had introduced the Bill, he would support the abolition amendment on conscience grounds.

The leading proponent for capital punishment was the Deputy Prime Minister, Jack Marshall, whose major argument was its deterrent effect; and he was supported by older MPs who, having previously voted to reintroduce the death penalty, would not change their votes. Hanan circulated masses of literature on the subject to younger MPs. In the House, Muldoon claimed that he had never experienced such a dramatic mood, with one columnist describing it as 'Parliament's finest hour'. Hanan delivered several powerful speeches, arguing that the experience of the previous three decades showed no correlation between the number of murders and whether or not the death penalty applied. He also noted that there was a danger of error in executing innocent people and that the hangings had been 'a severe ordeal for prison officers and sheriffs'. Finally, he appealed to Christian and moral principles.

When the amendment was moved, the Labour Party supported it and were joined by 10 from the National Party. As expected, they included Hanan and Aderman and also six young MPs elected in 1960, Robert Muldoon and some of his fellow 'Young Turks', Bert Walker and Duncan MacIntyre, among them. Brian Talboys, elected in 1957, joined them. These were men who would reshape the political style and direction of the National Party in the 1970s. As for the Prime Minister, Keith Holyoake, unlike other party leaders in the House and despite the importance of the matter, he did not speak other than to oppose sending the Bill to select committee. Ever the unideological pragmatist and easy-as-it-goes politician, Holyoake stood aside from the debate. He was never one for high moral principles. It is said that he was keen that if the death penalty was to be abolished it should be by a convincing margin. He waited until it became evident that this would be the case and voted against the abolition in part because he had supported the death penalty in 1950. The vote for abolition was 41 to 30.

In this way the death penalty for murder was abolished in New Zealand and 28 years later, in 1989, it was also abolished for treason. At last the 'Meccano set' no longer needed to be held in readiness, and it was able to find a permanent home in a museum, a grisly reminder of a penalty which belongs in less civilised times.

77.

Margaret Sparrow's contraceptive pills

MUSEUM OF NEW ZEALAND TE PAPA TONGAREWA

This packet of Anovlar contraceptive pills, just 21 tabs, appears small and insignificant. But it is among the most important objects featured, since from the 1960s the contraceptive pill revolutionised the lives of New Zealand women.

Pills were hardly foreign objects for women in the early 1960s. They were the keepers and the dispensers of pills. Their favourite pills included aspirin, a remedy for headaches and pain; and penicillin or antibiotics which from their first use in the early 1940s became the main response to bacterial infections, from pneumonia to ear and throat infections. Also common in the housewife's medical kit were pills to treat children who suffered from worms or car-sickness; and also Valium, of which 40 million doses were prescribed in 1969 New Zealand, many to housewives suffering from depression or 'suburban neurosis'.

But a pill for contraception, that was something new. Until then contraception was a fraught subject and methods were crude and unreliable. At the end of the nineteenth century, New Zealanders experienced a demographic revolution as women bore fewer children in part through reducing sexual behaviour. As families had fewer children, anxieties rose about 'race suicide' — the fear that New Zealand would be overrun with faster-breeding races, especially 'Asian hordes'. A Dominion Population Committee in 1946 described contraceptives as destroying 'the moral stamina of the nation'. Women using contraception were regarded as

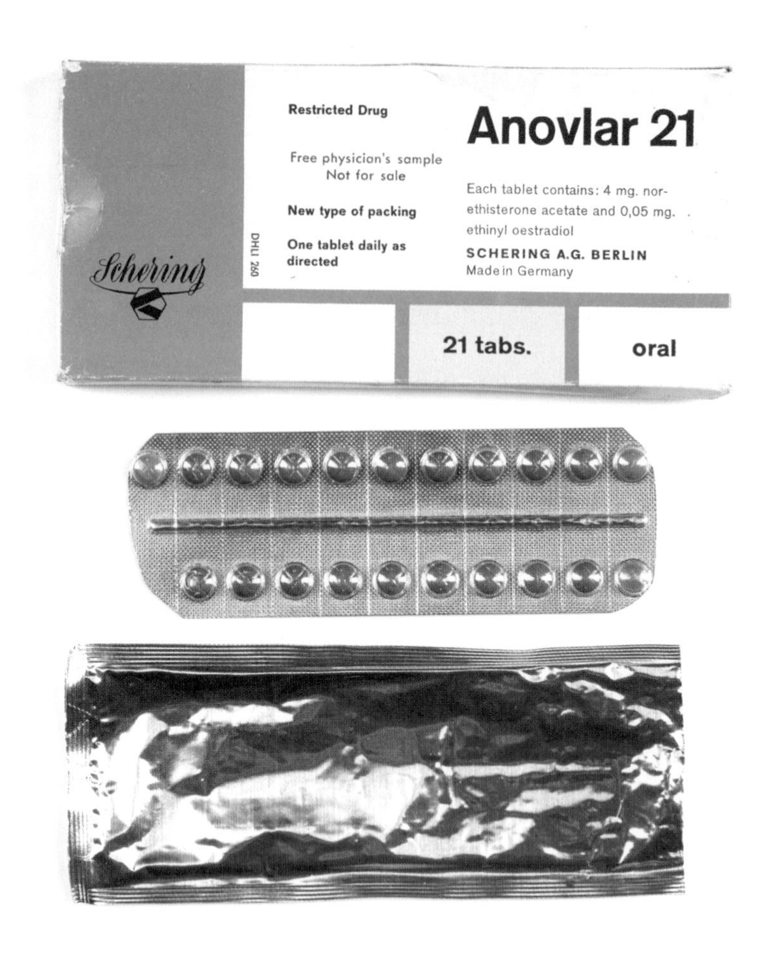

ANOVLAR 21 CONTRACEPTIVE PILL, CIRCA 1965, GERMANY, SCHERING AG. GIFT OF DAME MARGARET SPARROW, 2011. TE PAPA (GH022147)

'selfish'. Doctors were loath to prescribe contraceptive methods.

In addition, the available methods were unsatisfactory. The most common was sexual abstinence which did little to aid in happy marriages. Since rape in marriage was not a crime, women had little power to resist their husbands' sexual advances yet had to carry the consequences. Coitus interruptus, withdrawal, was common but unreliable, and also harmful to mutual pleasure. Some women used douches after the sex act, but this was ineffective. The most common mechanical methods were condoms and diaphragms. Condoms were imported by ship and it was believed that the journey through the tropics perished the rubber, while those made of tougher rubber were uncomfortable. Condoms also had a questionable reputation, being associated with prostitution. 'Good women' did not encourage their use and they were sold discreetly in plain packaging. As for diaphragms, they normally required fitting by doctors, nearly all male, who were often unwilling to assist.

In the absence of effective contraception, a common remedy was abortion. Indeed in the 1930s about one in five pregnancies ended in an abortion which was an illicit matter, dangerous to the woman's life. Some women used pills, such as Dr Bonjean's Female Pills; others hid in cars in the darkness to avoid detection while an abortionist inserted a catheter; 223 women died of septic abortions from 1927 to 1935. Another response was for women to carry their child, only to lose the infant immediately by adoption. Under the system of closed stranger adoption the baby was placed in the hands of a suitable nuclear family away from the 'sexually promiscuous' mother. The child was denied access to information about their birth parents. It was usually heart-breaking for the mother, not to mention the child.

Given this history, New Zealand women welcomed the contraceptive pill with enthusiasm. The pill contained progestin and oestrogen hormones which shut down ovulation so no egg was released. The one week's gap in taking the pill allowed a menstrual bleed to occur. The pill was effective and easy to take. There was no messy 'mucking around', fiddling with condoms or diaphragms, in the middle of the sexual act.

The packet of pills shown here was collected by Margaret Sparrow who gave them to Te Papa. Her story illustrates the experiences which led New Zealand women to adopt the pill so readily. The story begins in 1956. Margaret, originally from Inglewood, was 21, recently married and studying like her husband to be a doctor at the University of Otago. She became pregnant, but was unwilling to interrupt her studies. She realised the medical profession would not help; but her husband knew about the Christchurch chemist George Bettle. The couple wrote to him and obtained an inky black elixir which arrived in a brown-paper parcel. Margaret had no idea whether it would work or was safe. She took the mixture anyway and it worked — just like 'a heavy period'. 'Anticlimactic, really,' she recalled. But she never forgot the feeling of being trapped. Two years later, despite using a diaphragm, Margaret Sparrow became pregnant again. She had her first child, and then a second in 1960; but the following year, 'the pill came along . . . for me my whole life changed because that was the first time that I ever really experienced good fertility control. I was one of the first women in New Zealand to use the pill.' She had obtained the pills from her husband, who in the course of general practice training had received sample packets. The packet she tried was an Anovlar.

In 1969 Sparrow became a doctor with the student health service at Victoria University. She was constantly asked about contraception and abortion by female students but the doctor in charge followed the ethical guidelines of the New Zealand Medical Association that contraceptive information not be given to unmarried persons. Eventually Sparrow attended a conference of the Family Planning Association. This had been established in 1936 by Elsie Freeman (later Locke) and set up clinics in the main centres in the 1950s. Sparrow was inspired. She promptly set up a contraception display at the student clinic, and, despite her boss's discomfort, it remained. Margaret Sparrow's experiences informed her later life's work to improve fertility control for women in New Zealand. She pioneered vasectomies for men for Family Planning and use of the morning-after pill, and for over 30 years fronted the crusade for more liberal abortion laws. She also set up a company to import an abortion pill which allowed women to have medical, not surgical, abortions. She became a Dame for her lifelong efforts. Over the years, Margaret Sparrow collected many contraceptives which she presented to Te Papa in 2011. This packet of German-made Anovlar, the very brand she herself first used in 1961, is part of her Te Papa donation.

The contraceptive pill was not without health risks. The early pills with high doses of oestrogen often brought on weight gain and could cause headaches, nausea and bleeding. Occasionally there were more serious effects such as blood clots, strokes and even heart attacks. Despite these health risks, the pill's relatively high cost, the continued opposition of some doctors, and the refusal of a few Catholic pharmacists to fulfil prescriptions, the uptake was remarkable. By 1965, only four years after the pill reached New Zealand, it was claimed 40 per cent of married women of fertile age were using it; and 17 brands were sold. Two years later another survey of all women of reproductive age claimed 20 per cent were on it, compared with only 3 per cent in the United Kingdom. By 1975 the figure had risen to 35 per cent. The reliability of the pill and its separation from the sex act won support, and even doctors found the pill an easy solution to contraceptive anxieties.

The pill had two major impacts on New Zealand women. First, by separating sex from reproduction it encouraged their sexual liberation while also creating the expectation that women should be sexually available. Traditionally New Zealanders regarded sex as an act only within marriage, and sexual activity outside marriage created considerable public alarm. This was best expressed in 1954 with the public horror about sexual liaisons in Lower Hutt milk bars, which led to a public committee on moral delinquency in young people. It became an offence to sell contraceptives to those under 16. But the earlier age of sexual maturity as a result of good food, and baby boomers' challenge to their parents' values, could not stop the change. Ex-nuptial births, usually to teenage girls, rose from 11.67 per 1000 Pākehā women in 1945 to 24.14 in 1961. The pill built on, and encouraged, this revolution.

A generation coming of age in the 1960s, less anxious about economic pressures and surrounded by sexual images in the media, took advantage of the new contraceptive methods. No longer having to worry about getting pregnant was a psychological breakthrough in sexual freedom. Of course the downside of this was that a new expectation was imposed on young women to fulfil their male partner's sexual desires. Fear of pregnancy no longer worked as a reason to reject a sexual advance.

The second impact of the pill was in enabling women to better plan their lives and careers. The year the pill arrived in New Zealand, 1961, was the peak year of the baby boom with the average woman giving birth to 4.3 children. A woman could expect long-drawn-out years as a full-time mother. The pill allowed young women to delay children after marriage so they could complete their education and plan a career. They could have a successful sex life, fewer children and a fulfilling work life. The number of children per married woman fell to 2.1 in 1978, under half of what it had been 17 years before. On the other hand, the numbers of married women in the workforce increased steadily. In 1951 fewer than one in 10 married women were in paid work; but in 1981 close to a half (46.7 per cent) were contributing directly to the household income and they represented over half of the female workforce.

This was a huge revolution in the structure of the family and the sense of worth felt by New Zealand women. It would be unfair to ascribe this change just to the pill. There were other major factors. A group of activist women, imbued with feminist ideas from an international feminist movement, put huge amounts of energy into opening doors for other women. There were certain significant advances. The scene was set by the early 1960s when equal pay was won in the public sector and the tax status of married women was changed to that of individuals. Then the emergence of a local feminist movement, inspired in part by international writers like Betty Friedan and visits by others such as Germaine Greer in 1971, led to further action. In 1972 equal pay was extended to the private sector, and in 1973 there was a select committee on women's rights. Ideological and institutional changes were crucial, but they were in part made possible by women's ability to control their fertility.

It was not all plain sailing. There were strong battles in the 1970s, not wholly successful, to ensure that when contraception still failed, the possibility of abortion remained. 'Our bodies, ourselves', women's right to control their own bodies, became a major slogan of the 1970s. But behind that slogan lay the revolution that had brought the pill to New Zealand women — and in that battle Margaret Sparrow had played an enormous role.

78.

Dunedin Committee on Vietnam protest banner

HOCKEN COLLECTIONS – UARE TAOKA O HĀKENA, UNIVERSITY OF OTAGO

This beautifully painted and very large (1200 millimetres by 800 millimetres) calico banner was the work of the Dunedin Committee on Vietnam, and now resides in the Hocken Collections in Dunedin. The banner was one of many thousands carried in the streets or held up at demonstrations throughout New Zealand from the mid-1960s.

Until then protest banners were not a common sight in the country. Ornate banners, such as the Westport gold miners' banner (see page 207), had been carried in parades for a long time to identify groups, but banners or placards of protest had been comparatively rare. Members of the Christian Pacifist Society marched in Wellington in 1940 wearing sandwich boards with slogans. There were some simple wooden placards with bold words carried by those on the Easter anti-nuclear marches from Featherston to Wellington in 1961–64; and also a few on the 'No Maoris, No Tour' protests in 1960, but these were isolated examples. Several factors made banners more common from the mid-1960s. Most obvious was the rise in civil protest. There were more marches, and bystanders needed to be informed what the marches stood for. There were often counter-demonstrations which wanted to differentiate the points of view. The advent of television made visual messaging hugely important. But it was protest about New Zealand's involvement in the Vietnam War which made banners, and indeed marches, commonplace and established modes of protest which became frequent in the last third of the twentieth century.

The Vietnam War was a civil conflict that became an international proxy battle between communism and 'democracy'. In 1954 Ho Chi Minh's Communist Vietminh defeated the French colonial rulers at Dien Bien Phu. The country was divided with the expectation that it would be reunited after nationwide elections. The elections never happened; and the Americans, increasingly seeing the world through Cold War eyes and made anxious by the fall of China in 1949 and the Korean War, regarded South Vietnam as the front line of a battle against communism. As the conflict in Vietnam intensified the Americans became directly involved. From 1965 American military involvement ramped up and there was pressure on Australia and New Zealand to follow suit.

Two factors pushed New Zealand's involvement. One was Cold War thinking that led many to believe in 'forward defence' — the idea that communism would sweep down through South-East Asia, Singapore and eventually to Australia and New Zealand. The argument was that we had to fight the 'commies' on the Mekong or we would be fighting them on the Waimakiriri. The second factor was New Zealand's sense of dependence upon the US for its security, which had been signalled by the signing of the ANZUS treaty in 1951. With Britain no longer prepared to be our shield we had to curry favour with the alternative protector, the US. New Zealand flew the flag in Vietnam for long-term protection.

The person who ultimately decided to get involved was the Prime Minister, Keith Holyoake. He accepted the doctrines of collective security and 'forward defence' in Asia, but wished to avoid controversy and domestic unrest. Very reluctantly in May 1965 he agreed to send New Zealand military to Vietnam, but the support was minimal. At first there was only an artillery battery of 120 men, rising in 1968 to a peak of 540, compared with over half a million Americans and over 7500 Australians. New Zealand's was a token gesture and Holyoake avoided the domestic distress caused by conscription as occurred in Australia and the United States.

If Holyoake went only 'part of the way with LBJ' and barely satisfied American government opinion, this did not appease domestic opposition. Soon after Henry Cabot Lodge visited Wellington to ask for assistance in April 1965, groups opposed to involvement were formed — Committees of Vietnam in Wellington, Christchurch, Palmerston North, Nelson and Dunedin. In Auckland there were several committees. At first those involved were the old left — not the Labour Party, because as Warren Freer noted, the one thing all 80 MPs agreed on was 'the desire to discourage the southward march of communism through Asia'. Rather, it was old-time socialists or sympathisers of the Communist Party and pacifists, especially those who had been active in the Campaign for Nuclear Disarmament. Quickly the anti-war movement attracted a new group drawn from the

DUNEDIN COMMITTEE ON VIETNAM BANNER, C.1960S. NEW ZEALAND CAMPAIGN FOR NUCLEAR DISARMAMENT – SOUTHLAND BRANCH RECORDS 85-095, HOCKEN COLLECTIONS UARE TAOKA O HĀKENA, UNIVERSITY OF OTAGO.

university. Initially this was academics challenging the crude jargon of the anti-communist crusade and arguing for an independent foreign policy. In Dunedin, where this banner emerged, academics were especially prominent in the local Committee on Vietnam. Their presence led to an outburst of pamphlets and written media analysing the war.

Before long, the anti-war movement attracted a younger group. These were 'baby boomers', born after 1945 and growing up in the affluence and Cold War politics of the 1950s. With a different life experience from their parents, they had different attitudes and political styles. Even before the 1965 anti-war protests, students had marched in the streets for improved bursaries. Now, they challenged our involvement in the war. There were sit-ins, some chained themselves to pillars, and banners and chanted slogans became the style on marches. Not all students became anti-war. Eric Herd, a Dunedin anti-war professor, remembered that lawyers, commerce students, dentists and medicos were predominantly in favour of involvement. At Canterbury in 1965 only 34 per cent of the students opposed the government's policy. But many arts students took to the streets; and the student newspapers became vocal in their anti-war stance.

Visits by representatives of participating countries became a focus for protest because they highlighted the lack of independence which critics claimed was the signal failure of New Zealand's Vietnam policy. Lodge's visit in 1965 had triggered the anti-Vietnam committees; and although Lyndon Johnson's visit in 1966 was widely regarded as a public relations success because crowds cheered the first visit by a sitting US President, there were noisy demonstrations with over 1500 outside Government House. The next year 2000 protested outside Parliament against Johnson's envoys Clark Clifford and General Maxwell Taylor. In a characteristic act of the new politics, students carried a coffin 'in memory of New Zealand democracy'.

It was not surprising that the visit of Air Vice Marshal Nguyen Cao Ky in January 1967 to thank the country for its support stimulated protests and banner-making. Ky had become Prime Minister of the republic of Vietnam in a military coup replacing a civilian government in 1965. From a privileged background, Ky was known for his flamboyant style and high living. His pearl-handled chrome-plated pistols were notorious. To the anti-war movement he personified the undemocratic character of South Vietnam. Ky had closed newspapers and allegedly set up posts for executions in Saigon's main square. The anti-war movement aimed to greet the man appropriately. In Auckland a special Ky Protest Committee was formed and decided on a slogan, 'Ky No'. In Wellington the Committee on Vietnam developed a song, 'No, No, Kiwi'. The committee secretary, Jeremy Lowe, sewed three Viet Cong flags. In Dunedin the committee prepared this large banner and decided on a Hitler theme. Hitler was of course in 1965 the ultimate symbol of anti-democratic fascist tyranny; and Air Vice Marshal Ky had been reported as saying to a London reporter, 'People ask me who my heroes are. I have only one — Hitler . . . but the situation here is so desperate now that one man would not be enough. We need four or five Hitlers in Vietnam.' So the Dunedin committee designed a banner which showed Ky with a Hitler moustache and characteristic lock of hair on his forehead, and the slogan 'Vietnam's Mini-Hitler'. In place of the swastika on his chest is a US dollar.

Ky did not visit Dunedin, and the banner was probably used at Christchurch airport when he arrived in New Zealand on 23 January 1967 to about 1500 demonstrators and '50 or 60 banners of varying degrees of unfriendliness'. The Hitler iconography of the banner was supported by chants of 'Sieg Heil' and 'Fascist Ky'. With the 'mobilisations' of 1970 and 1971 anti-war protest became a mass movement drawing support from younger New Zealanders and students. The 30 April 1971 mobilisation saw 32,000 in the streets, at that time the largest demonstration in the country's history. New leaders such as Tim Shadbolt with his Auckland University Society for the Active Prevention of Cruelty to Politically Apathetic Humans developed ingenious modes of resistance alongside their ingenious names. The methods of protest developed in opposition to New Zealand's involvement in the Vietnam War were picked up by other causes — Māori and women's rights, environmental causes, anti-apartheid sport and the anti-nuclear movement. The next two decades would see 'banner politics' transform political and social debate in New Zealand.

79.

Save Manapouri Campaign share certificate

ALEXANDER TURNBULL LIBRARY, NATIONAL LIBRARY OF NEW ZEALAND

On 1 June 1970 the Save Manapouri Campaign offered 30,000 50-cent shares as a contribution to their campaign. The issue coincided with an issue of preferential shares by Comalco, the multinational consortium which had an agreement to use power from Lake Manapōuri to process aluminium at a smelter, Tīwai Point. The plans included raising the levels of two lakes, Manapōuri and Te Anau. Comalco's shares had been offered to leading politicians and journalists. The issue of Save Manapouri shares to the 'New Zealand Public' came one week after the presentation of a parliamentary petition against raising the lakes which contained 264,907 signatures, then the largest petition in our history, announcing the arrival of a major environmental movement.

From the nineteenth century, there had been isolated voices defending New Zealand's flora and fauna from economic development. In 1923 the Native Bird Protection Society was founded which became the Forest and Bird Protection Society in 1948. Others worked to protect the country's natural beauty. Scenery preservation societies emerged and reserves were established. National parks followed. The Milford Track became famed as the most beautiful walk in the world. So there was always a consciousness of 'beautiful New Zealand'. But these were quiet voices drowned out by the rush to carve out farms and cities from the bush.

Electricity brought a major challenge. The country had abundant opportunities for hydro-electric generation. At first power stations were small and local. Reefton was the first place to have a municipal electricity system in the southern hemisphere in 1888. During the 1920s, as electricity use began for domestic lighting, heating and cooking, larger hydro-electric schemes were built in isolated areas such as Waikaremoana. From the 1950s, huge housing growth and new electrical appliances, from refrigerators to vacuum cleaners, increased electricity demand dramatically. By 1966 over 80 per cent of households had electrical water heating and used electricity, not gas, for cooking. There were new power stations on the Waikato, Clutha and Waitaki rivers. While in the 1950s 60 per cent of electricity was consumed in the home (compared with a quarter in the USA), increasingly it was used to power factories in protected industries like car assembly or clothing.

However, the idea that New Zealand electrical power itself might become a source of foreign exchange was new. The motivation was a belief that the economy was too dependent upon traditional agricultural exports — wool, meat and dairy. If the UK entered the European common market, New Zealand's farm exports might be challenged. Industries drawing on the country's electricity were promoted — the Glenbrook steel plant, pulp and paper mills at Kinleith and Kawerau. An aluminium smelter was another option. In the post-war years aluminium use dramatically increased. A light, strong metal, aluminium was used in aircraft and domestic implements. In 1956 a large bauxite deposit was found on Cape York Peninsula in northern Australia. To transform the ore into aluminium called for an electrolytic process requiring huge amounts of electricity. Consolidated Zinc believed they had found a cheap source of power at Lake Manapōuri. In January 1960 Comalco, as the firm was now called, was given an exclusive right to develop and use Manapōuri power for 99 years. Two years later, the government agreed to develop the power station with Comalco using the power for the Tiwai Point smelter.

The power station was a major engineering and construction project. Water would be taken from Manapōuri's west arm in penstocks 200 metres down to a large powerhouse carved out of rock 100 metres long and 40 metres high. After passing through the turbines the water would flow into a 10 kilometre long tunnel tailrace to Deep Cove in Doubtful Sound. Building the station was a treacherous and difficult operation. Eight million man-hours were spent on the job and 18 people died.

The environmental controversy hinged on a relatively small issue. The long-distance fall of water from the west arm to Deep Cove would produce 600 megawatts of power, equal to one sixth of all power generated in New Zealand in 1970. But the agreement with Comalco was that Lake Manapōuri be raised by 8.4 metres. This would only increase the power by an additional 60 MW (about 10 per cent) and provide storage in dry

Save Manapouri Campaign
(ASSETS:- NATURAL)
50c ONE SHARE 50c
CAPITAL: 30,000 Ordinary Shares @ 50c
SHAREHOLDERS: New Zealand Public
Nº 17187
PERPETUAL DIVIDEND : Retention of Lake Manapouri in its natural state
Dated this 1st day of June, 1970
David G. Kember, Secretary.
This Certificate acknowledges your donation to the National Save Manapouri Campaign

SAVE MANAPOURI CAMPAIGN: ONE SHARE 50C. BY KEMBER DAVID G. 1ST OF JUNE 1970. [EPHEMERA OF OCTAVO SIZE CONCERNING THE ENVIRONMENT AND ENVIRONMENTAL PROTECTION] COLLECTION. REF: EPH-A-ENVIRONMENT-1970-01. ALEXANDER TURNBULL LIBRARY, WELLINGTON, NEW ZEALAND.

spells. Unraised, the lake would deliver Comalco's needs for 96 per cent of the time. The gains seemed small. The costs of raising the lake were huge. With 44 islands and a shoreline clothed in bush, Manapōuri was regarded as 'the most beautiful of the southern lakes', quoting the 1960 *Yearbook*. As part of the Fiordland National Park, the lake was assumed to be free of development. The visual damage would be severe. Islands would disappear, trees would become unsightly skeletons, beaches would be drowned.

From the beginning, there was opposition to raising the lake. Forest and Bird submitted a parliamentary petition in 1960. Three years later they combined with the Scenery Preservation Society on a second petition. Then new activists emerged. An early critic was Les Hutchins, who ran a tourist operation on the lake and believed the scheme would compromise Fiordland's tourist appeal. He joined other Southlanders who met in the home of an Invercargill city councillor, Norman Jones, in October 1969 to form the Southland Save Manapouri Committee. The committee chair was Ron McLean, a local farmer with a deep love for Fiordland. Ron joined his daughter on a national roadshow which produced 19 branches of the Save Manapouri Committee. Doctors were attracted to the movement and in Wellington the battle was led by a prominent epidemiologist, Ian Prior. A leading botanist, John Salmond, and a distinguished Otago botanist, Alan Mark, joined the battle. These were not rabble rousers, but were well-established 'professional men'.

Their ingenuity was impressive. The issuing of shares arose because the group wanted funds to prepare submissions to the commission of inquiry which Keith Holyoake was forced to establish. The idea of shares was a deliberate reference to the threat of 'big business' to the the environment, and coincided with Comalco's share offer. The language on the certificate was clever — the shareholders were the 'New Zealand Public' and the 'PERPETUAL DIVIDEND' was 'Retention of Lake Manapouri in its natural state'. The certificates were designed by two prominent Wellington artists, both commited to the country's heritage. One was Roy Cowan, a well-known potter, and son of the popular historian James Cowan. The other was John Drawbridge who had been part of the legendary attempt by the photographer Brian Brake to film an ascent of Mt Aspiring in 1949. Later Drawbridge went to Britain and returned as a distinguished printmaker. The same year, Roy Cowan and John Drawbridge made murals for the New Zealand pavilion at the 1970 World Exposition at Osaka.

This movement inspired a generation who had learnt to march against the Vietnam War. To discredit the huge petition, the Minister of Works and Electricity, Percy Allen, claimed that two thirds of the signatories were children too young to vote. That was wild speculation but undoubtedly many younger people did sign. In November 1970 there was a large march of the Youth for Manapouri Action Committee. Photos show males with beards and long hair. Eighteen months later, anticipating the 1972 election, the Values Party emerged. Led by baby-boom activists, the party was the world's first 'Green' party.

The Labour Party, led by Norman Kirk, was forced to listen to the new voices. Kirk pledged in the 1972 election not to raise lake levels, and after winning the election appointed critics of the Manapōuri scheme including Ron McLean and Alan Mark as Guardians of Lakes Manapōuri, Monowai and Te Anau. The younger activists were inspired by the Manapouri campaign to further environmental fights. On 4 July 1975, in a characteristic ritual of the new politics, a group of young environmentalists signed the 'Maruia Declaration' while gathered around a campfire on the snow-covered banks of the Maruia River. The declaration called for 'the logging of virgin forests to be phased out' and declared that 'the wholesale burning of indigenous forests and wildlife has no place in a civilised country'. Ecology Action groups sprang up, and environmentalism became a powerful new concept. Some set out to save indigenous forests and birds; others to save the planet through an anti-nuclear movement. Over the next 15 years, these ideals became centre stage in New Zealand. Large-scale milling of indigenous forests was phased out; greater funds and energy went into protecting native birds; and both Norman Kirk, and then David Lange, led campaigns against French and United States' nuclear ambitions in the Pacific.

There is little doubt that the Save Manapouri Campaign represented a turning point in the New Zealand value system. This share certificate is a fitting symbol for the birth of a new consciousness.

80.

Ngati Poneke record *Aku Mahi*

MUSEUM OF NEW ZEALAND TE PAPA TONGAREWA

This old item of technology, a gramophone record, was produced by EMI (New Zealand) in 1973 and features songs by the Wellington culture group Ngati Poneke. The appearance of this object at that time throws light on one of the most significant social changes of twentieth-century New Zealand, the migration of Māori to the city.

There had always been a strong tradition of migration by Māori. In the first great migration they travelled thousands of kilometres across the ocean to reach Aotearoa, and after arrival many tribal groups moved some distance to find suitable places of settlement. Some heke, such as those led by Te Rauparaha from Kāwhia to the Kāpiti Coast, continued even after Pākehā arrived. Once settled, iwi often travelled for seasonal hunting and gathering of food. Yet for many years the city was primarily Pākehā space; and while some Māori travelled to the city to attend schools or for sporting or ceremonial occasions, few — well under 20 per cent of their population — lived long term in the city until World War II.

The war began a significant movement to the city. Many single Māori women and those men not in the armed forces were 'manpowered' to essential industries in the city. By 1945 over a quarter of Māori (26 per cent) were living in places of over 5000 people, although they comprised a small percentage of city populations. Then in the 1950s Māori families began to move to the bright lights and this became a flood in the next two decades. The overwhelming push factor was the loss of Māori land;

NGATI PONEKE RECORD, AKU MAHI, UNIVERSAL MUSIC NEW ZEALAND

with much of that remaining leased to Pākehā farmers. Seasonal work as agricultural labourers was erratic and poorly paid. A major pull factor was the growth of industry in the city. Places like freezing works, or industries protected by tariffs such as car assembly plants and clothing factories, were crying out for workers. Money was a major drive, but so were the movies and the buzz. 'Employment and enjoyment' summed it up. Government promoted the migration, believing it assisted the 'assimilation' of Māori to Pākehā habits and wanting Māori employed in the expanding economy. From 1956 to 1966 the proportion of Māori in the city rose from 35 per cent to 62 per cent. Ten years later it was over three quarters of all Māori.

Adjusting to the city was not easy for rural Māori. A few had to learn to speak English. Most left behind large, comforting extended families for a nuclear family situation or boarding as single people. They had to learn managing city finances — cheque books, rents, mortgages, hire purchase. There were problems of debt and the easy attraction of alcohol. Some withdrew into loneliness. Housing was a major issue. At first Māori concentrated in the inner city — Freemans Bay or Ponsonby in Auckland, Newtown in Wellington. Often there was overcrowding and lack of hygiene. Department of Maori Affairs houses were 'pepper-potted' among Pākehā families, but this produced isolation from whānau and culture. In the 1960s mass housing was built in South Auckland, especially Ōtara and Te Atatū, while in Wellington Māori found themselves in distant suburbs in Wainuiomata and Porirua.

The challenge was to provide cultural sanctuaries where Māori felt comfortable, were supported by their own people and could speak their own language. Some churches stepped up. In Wellington in the 1930s Māori Anglicans met monthly at St Thomas's Church in Newtown for services conducted by the Ōtaki minister who travelled in by bike and train. After the service, young people remained, initially to practise choral singing, but this was soon followed by waiata, poi and haka. From 1937 the group called themselves 'Ngati Poneke Young Maori Club'. The name, Ngati Poneke, referring to the English name of Wellington, Port Nicholson, had been coined by Sir Apirana Ngata the year before. The Maori Purposes Fund had agreed to support a new marae at Waitara in memory of Māui Pōmare, and young Māori were brought to Wellington to do the carving and tukutuku work. Ngata encouraged the young people to gather after work for action songs. He called them Ngati Poneke and they held a concert in the Wellington Town Hall in May 1936. The St Thomas's group followed their namesakes on stage and also held an inaugural concert at the same venue in May 1938. Annual concerts followed.

The Ngati Poneke cultural group became a focus for young Māori in the city. An early member, Paul Potiki, recalled of their weekly practice, 'Monday nights then became very, very rewarding nights for young people starved of Māoriness.' Vera Morgan remembered, 'It was like a home away from home for me, being with my own people'; while Mihipeka Edwards said, 'Ngāti Pōneke was our tūrangawaewae, our rock and strength, our protection. Without it we would have gone around like people with no heads. We'd have been lost.'

Cultural groups like Ngati Poneke emerged in other cities. In Auckland they were initially based at Auckland Maori Community Centre, opened in 1949 in an old army barracks next to Victoria Park. The attraction was not just the music and dance. Ngati Poneke and other such groups were providing a place to stand, functioning almost like a marae. It was natural that would evolve into fully-fledged urban iwi, open to all tribes, and with their own wharenui. After 20 years' planning, Ngati Poneke opened a meeting house at Pipitea in Thorndon in 1980. In the same year in West Auckland a similar pan-tribal house, Hoani Waititi, was opened, followed by Nga Hau e Wha in Christchurch and Araiteuru in Dunedin. Eventually a host of other marae opened in New Zealand cities — some were 'taura here' (binding ropes), a gathering of urban-based members of a particular iwi, such as Tūhoe or Waikato; others were found at schools and universities.

Another crucial institution which assisted city Māori was the Maori

Women's Welfare League. Here too Ngati Poneke played a role, for it was in its hall rooms in the Red Cross Building in Wellington in 1951 that the league was established. Te Puea Hērangi became patroness, and Whina Cooper was elected inaugural president. By 1956 there were 300 branches and 4000 members.

Despite Ngati Poneke's other contributions, its core activity remained waiata and haka. During the war Ngati Poneke greeted distinguished visitors such as the US President's wife, Eleanor Roosevelt, entertained crews of visiting ships and in 1946 welcomed home the Maori Battalion. The club's annual town hall concerts were regularly filled to capacity, with the audiences including Pākehā as well as Māori. By 1973 the club's performances were well known.

It was hardly surprising that in the early 1970s a recording company, EMI (New Zealand), saw the opportunity to record their performances on vinyl. EMI dates to the beginnings of gramophone records when Emile Berliner founded the Gramophone Company in 1897 in London. This became EMI (Electric and Musical Industries) in 1931 and in New Zealand the firm distributed imported records under the HMV (His Master's Voice) label. Gramophone records took off in the 1920s, at first used on wind-up gramophones and then electric radiograms, but it was not until 1949 that the firm made records in New Zealand and did local recordings. Two musical revolutions led to a huge increase in the sale of HMV records — the rock 'n' roll revolution symbolised by Elvis Presley in the 1950s and then from 1963 the Beatles' Liverpool sound. The Beatles visited New Zealand in 1964 to adoring crowds in the streets and packed theatre audiences. They were followed by the Rolling Stones whose records EMI and HMV also distributed. Local stars Lee Grant and Maria Dallas, both promoted on television, were recorded on HMV.

So in 1973 when the HMV name was retired and EMI (New Zealand) was established with a warehouse and recording studio in Lower Hutt, they looked round for local talent. HMV had already recorded two Ngati Poneke records — in January 1964 with the revealing title, aimed at Pākehā, *Songs of the Maori*, and again in 1967. They were successful, so in 1973 came a third record.

The initiative came from Wiremu (Bill) Kerekere. Originally from Te Aitanga-a-Māhaki just south of Gisborne, Kerekere was a musician, songwriter and leader of the Waihirere Maori Club. He worked with Leo Fowler on the local radio station, so when in 1964 Fowler came to Wellington to form the Māori programme section of the New Zealand Broadcasting Corporation, Bill Kerekere came as his assistant and took over the section within two years. As a migrant from country to city, Bill joined Ngati Poneke and encouraged new waiata as well as traditional ones. Of the 13 songs on the record, only four were traditional in origin. The remainder were modern compositions. Four were the work of Bill Kerekere himself. Another four were composed by Kohine Ponika who had been born in the Urewera centre of Rūātoki with both Tūhoe and Ngāti Porou lineage. Encouraged by Apirana Ngata, she would compose songs at the kitchen table and then get her son and eight adopted children to learn the words and melodies. Without being able to read a word of music, she became one of the country's great composers.

The title song, 'Aku Mahi' (translated as 'My Work') was written by Kohine Ponika and the actions were orchestrated by Bill Kerekere. It was a creative partnership. The song won a competition for a new waiata held at the first Polynesian Festival (which eventually became Te Matatini) held in Rotorua in 1972. 'Aku Mahi' was conceived as an invocation to Māori pride and identity. It ends with the following:

Ki āku mahi o āku tīpuna e Ka haka tēnā, ka poi tēnā He mahi a ringaringa a e Te hiki taku mere, i taku taiaha Te mana taku ihi e,	In the activities, of my ancestors, the haka, the poi and action song. I raise my greenstone blade, and my fighting staff The spirituality energises me.
Takahia, takahia Kia whakarongo ai ngā iwi. Pupuritia, pupuritia, pupuritia A tāua mahi e.	Stamp your feet, stamp your feet so that everyone can hear you. Hold on, hold on, hold on to these customs.

The words express perfectly the inspiration behind Ngati Poneke and this record — through waiata and haka, Māori could find strength against the challenges of the city.

81.

Te Rōpū o te Matakite (Māori Land March) pou whenua

TE KŌNGAHU MUSEUM OF WAITANGI

A pou whenua is a wooden marker traditionally used by Māori to mark the boundaries of their land. This fine pou whenua was carved from tōtara by Whina Cooper's son-in-law, Moka Puru, and carried some 1100 kilometres at the head of the Māori Land March from Te Hāpua in the far north to Parliament Grounds in Wellington from 14 September to 13 October 1975.

The march was a powerful protest about Māori loss of land. Under the Treaty of Waitangi, the chiefs and tribes of New Zealand had been guaranteed the 'full, exclusive, and undisturbed possession of their Lands and Estates'. It had proved a false promise. Subsequently, Māori lost much land through dubious sales. Huge areas were confiscated during and after the New Zealand Wars. The establishment of the Native Land Court in 1865 allowed for the individualisation of title and subsequent sale; and then a series of laws saw additional lands lost — especially through the Public Works and Town and Country Planning Acts. By 1965, of the original 66 million acres, Māori ownership had fallen to 4 million acres. Then in 1967 the Maori Affairs Amendment Act facilitated further purchase of 'uneconomic' parcels of Māori land. The Act was seen as 'the last land grab'. There was growing discontent among Māori, many of whom had been driven to the city or into rural poverty. By 1975 only 2.5 million acres were left in Māori ownership. Many Māori argued that the economic loss of land was only part of the cost. Te whenua, the land, is also the word for the placenta, the organ that feeds growing life; and Māori regard land as central to their culture and to tribal and hapū identity.

The idea of expressing this discontent through a march brought together several groups. There were younger urban-based Māori who had formed Nga Tamatoa in Auckland and the Maori Organisation on Human Rights in Wellington. They especially campaigned for the protection and encouragement of te reo Māori. Out of their protests came Maori Language Day which soon became a week. There were also older Māori with strong links to the rural heart of different iwi, some of whom had moved to the city and were active in the national organisations created in the 1950s and 1960s. In the organising committee of the Land March, alongside Syd Jackson and Titewhai Harawira of Nga Tamatoa were Graham Latimer of the New Zealand Maori Council, Mira Szászy of the Maori Women's Welfare League and Ranginui Walker of the Auckland District Maori Council. But the inspiring leader of the older generation with strong rural roots was the 79-year-old Whina Cooper. Born and brought up as the daughter of a Te Rarawa chief in the Catholic area of the north Hokianga, Whina had shown remarkable leadership — stopping the draining of local mudflats by a Pākehā farmer and leading land development schemes in the area. She even became the first president of the local rugby union. In 1951 she joined the urban migration to Auckland and almost immediately became the founding president of the Maori Women's Welfare League.

Unsurprisingly, it was at the home of Whina's son, Joseph, that a new organisation was formed in February 1975, Te Rōpū o te Matakite — those with foresight. A month later, at Te Puea marae in Māngere, it was decided to proceed with a march. A march was not a traditional Māori form of protest — acts of civil disobedience, such as at Parihaka, or occupations were more typical. The idea of a march derived from American examples, both the civil rights marches to Washington and Selma, and the 'Trail of Broken Treaties' of indigenous Americans to Washington in 1972. Closer to home, Pākehā groups had marched through city streets to protest such issues as the Vietnam War, student bursaries, and women's rights. Further, a march through the North Island would emphasise the nationwide character of the issue. The march would stop at a different marae each night, allowing the issues to be discussed and local iwi to express support.

What was different about the Land March from its urban equivalent was the absence of banners or painted slogans (although the phrase 'Not one more acre' was often spoken). Instead the march was led by the pou whenua to which was attached a large white flag with the words

TAME ITI HOLDING POU WHENUA, ACCOMPANIED BY WHINA COOPER, LEADING MAORI LAND MARCH ALONG HAMILTON STREET. HEINEGG, CHRISTIAN F, 1940- :PHOTOGRAPHS OF THE MAORI LAND MARCH. REF: 35MM-87527-2-F. ALEXANDER TURNBULL LIBRARY, WELLINGTON, NEW ZEALAND. EXTRACT FROM POEM BY HONE TUWHARE, 'PAPA-TU-A-NUKU (EARTH MOTHER)'. *SMALL HOLES IN THE SILENCE*, 2016.

'Te Roopu o te Matakite'. The pou was intricately carved. At its head was a lizard with its tail embracing another lizard. Lizards were creatures which, as the carver Moka Puru said, had 'been on the earth for generations'. Beneath the lizards were a series of manaia heads and then a representation of tears, symbolising the old people crying for their land. Finally, there were carved lines with notches to express whakapapa, the idea that land belonged to people over generations. The pou was to be kept off the ground for the whole journey to signify the huge area of land which had been lost and could not be marked. For much of the march the pou was carried by Cyril Chapman, a 20 year-old whose maternal grandparents had been jailed for returning to their ancestral land in the 1950s. Young leaders of local iwi were invited to carry it through their lands — Joe Hawke, Ngāti Whātua leader, carried the pou across the Auckland Harbour Bridge; Tame Iti carried it through Hamilton. An image of the pou was used to separate names and addresses on the petition of support.

Marchers set off from Spirits Bay and Te Hāpua on 14 September, the first day of Maori Language Week, and were cheered by a message of support from Matt Rata, who came from Te Hāpua, and, as Minister of Maori Affairs, had felt somewhat undermined by the march proposal. As the 50 or so marchers walked, they sang songs to keep their spirits up, one written by Whina Cooper. But there were many blisters and sore limbs, and some older people such as Whina herself travelled at times in an accompanying bus. Hone Tuwhare, the distinguished poet who walked the distance, expressed the sentiments perfectly:

> We are stroking, caressing the spine
>   of the land.
> We are massaging the ricked
>   back of the land
> With our sore but ever-loving feet.
>   Hell, she loves it!

At the end of the day, after legs had been massaged and the company well fed, there was serious talk in the wharenui. The rangatira were invited to sign a memorial of rights addressed to the Prime Minister, and others to sign a petition, which described the march as 'a climax to over a hundred and fifty years of frustration and anger over the continuing alienation of their lands'. By the time the hīkoi reached Wellington on a rainy 13 October there were over 5000 marchers and 60,000 signatures on the petition. The petition and memorial were presented to Prime Minister Bill Rowling. Three days before, his government had passed an Act to establish the Waitangi Tribunal, which, following amendment in 1984 to allow historic claims, would have a major impact on the recognition of Māori land claims.

There was some immediate disappointment for Whina, because a 60-strong splinter group of Te Rōpū set up a camp on Parliament's steps — until it was evicted by the new Robert Muldoon government on Christmas Eve. But the march had more substantial impacts. Two people on the march, Joe Hawke and Eva Rickard, led major occupations over land over the next few years. On 5 January 1977 Joe Hawke and Ngāti Whātua began an occupation of Bastion Point to protest plans by Muldoon's government to use land there for housing. The land had been originally taken for defence purposes and represented the last remaining Ngāti Whātua land in Auckland. After 506 days Ngāti Whātua were evicted by over 600 police and army, but in 1987 a Waitangi Tribunal report recommended the return of Bastion Point to the tribe. Eva Rickard was involved in a 1978 occupation of tribal land at Raglan which had been taken during the Second World War for an airfield and then leased to the local golf club. Rickard and fellow protesters from Tainui Awhiro were arrested on the ninth hole, but here again the land was eventually returned.

As for the pou whenua, it was brought out again in 2004 to lead a hīkoi to Parliament to protest the Crown's attempt to take over through legislation Māori rights to the foreshore and seabed. Once again protest had an impact, with a new law in 2011 which gave iwi a mandate to seek customary rights through the courts.

The pou had one final outing. On 3 February 2020 George Puru, son of the carver, carried the pou while escorting the Prime Minister, Jacinda Ardern, at the unveiling of a memorial statue of Whina Cooper at her home community of Panguru in the northern Hokianga. The statue was based on the famous photograph of Whina setting out from Te Hāpua on the Land March holding hands with her three-year-old mokopuna. The pou now resides at Te Kōngahu Museum of Waitangi — until perhaps another moment, significant to Māori, again calls it out.

82.

Tepaeru Tereora's tīvaevae

MUSEUM OF NEW ZEALAND TE PAPA TONGAREWA

This story concerns the experience of Pasifika people in New Zealand, but it really begins far away on the beautiful tropical atoll of Manihiki, with its palm trees, exuberant flowers and coral beaches. The atoll is tiny — just 4 square kilometres of land and only 500 people. It is one of the northern Cook Islands atolls, almost 1300 kilometres north of the main island of Rarotonga. There in 1934 Tepaeru Tereora was born. She is the designer, and joint maker, of this stunning tīvaevae which is now one of the treasures of Te Papa and which evokes some of the bright colours of her birthplace.

In 1944, in the first of many migrations, Tepaeru moved with her sister to Rarotonga to further her education, and the next year joined the only girls' organisation on the island, the Avarua Girl Guides. Eight years later, at the age of 18, she travelled to New Zealand for the first time to attend a Guides jamboree at Marton in the Manawatū. Fortunately, Guides was not the only element of female culture to which the young woman was exposed. Both her birth mother and Mama Inē who brought her up were experienced in making tīvaevae, but at first Tepaeru was not interested. It was only when she began helping by cutting up paper patterns that her interest was aroused.

The exact origins of tīvaevae are uncertain, but it is generally agreed that the art developed from patchwork quilting introduced by early Christians in the 1820s — perhaps the wives of London Missionary

TIVAEVAE, 1980, COOK ISLANDS, MERE TEPAERU TEREORA. PURCHASED 2002. TE PAPA (FE011717/1)

Society missionaries, perhaps French nuns bringing Catholicism. The art quickly developed distinctive local forms and began to reflect the vibrant colours, the flowers and the plants of the islands. Four types of tīvaevae evolved — tīvaevae taorei which was the most complex, being a patchwork of small pieces of coloured fabric cut into shapes; the tīvaevae manu, probably the most simple, which was patterns sewn onto a plain background; tīvaevae tataura which started out like the manu but had detailed embroidery stitched onto the patterns; and the tīvaevae tuitui tataura which consisted of embroidered squares sewn together. Tepaeru's one shown here is a tīvaevae tataura. Sometimes women worked on the tīvaevae by themselves but more often they did so in groups, singing and chatting as they sewed. The finished quilts became items of great value associated with important moments in the family — given away at baptisms, covering chairs at weddings, or when a boy first had his hair cut (an important male rite of passage), even thrown over the casket at funerals. The tīvaevae are 'stitched with love', as Tepaeru said.

Through her family and local women Tepaeru eventually learnt to love tīvaevae and become a ta'unga, a woman who could design as well as cut and sew the magnificent quilts. Her first real opportunity to help inspire their creation came in the mid-1960s. The Cook Islands had been annexed by New Zealand in 1901 and power concentrated in the hands of a somewhat authoritarian Resident Commissioner. But in 1965, on the back of over 20 years of agitation by the Cook Islands Progressive Association, the islands were given self-government and independence from New Zealand. The first Premier, Albert Henry, was committed to strengthening national culture. Tepaeru Tereora was hired as a Woman Interest Officer to travel round the islands establishing a women's federation and teaching women domestic crafts such as cooking, weaving, setting up smokeless stoves, and above all sewing tīvaevae. She became skilled at designing tīvaevae by using pieces of folded paper. Her designs were based on the natural world — flowers, birds, coral and fish.

Then in 1969 Tepaeru left Rarotonga for Auckland and the next year became a welfare officer with the Department of Maori and Island Affairs based in Wellington which became her permanent home. Since the Cook Islands had been New Zealand territory for so long, its people were New Zealand citizens with free right of entry. There had been some young single migrants in the 1950s, with men working largely as agricultural labourers and the unmarried girls as domestic help. In 1966 there were 8663 Cook Islanders living in New Zealand. Then as New Zealand industry boomed in the 1960s, creating a shortage of low-skilled labour in factories, especially in Auckland, employers looked to the Pacific. Many came in from Western Samoa and Tonga; but also some from the

Cook Islands, especially after the opening of Rarotonga International Airport in 1973. Often they were hopeful of improved education for their children. Many settled in Ponsonby in Auckland, or Newtown and then Porirua in Wellington. By 1976 the numbers of people describing themselves as Cook Islanders by ethnicity had risen by 10,000 to 18,610. This made them easily the largest Pacific Islands group in New Zealand after Samoans (27,876).

As a welfare officer, Tepaeru confronted the social issues facing the Cook Islands and Pasifika community, and there was much to do. It was hard for them adjusting to the cold winds of New Zealand. This was particularly the case after 1973 when the terms of trade turned against New Zealand and unemployment rose. There was growing resentment against Pacific people, especially in Auckland. As New Zealand citizens, Cook Islanders had every right to be in the country; but the Samoans and Tongans had entered on three-month visas since 1964, and from 1967 there had been annual quotas. While the economy boomed, these regulations were not enforced. Suddenly the situation changed, unemployment rose and Pacific Islanders were blamed. From 1974 there were dawn raids on overstayers. Further, in the 1975 election the National Party used Hanna-Barbera cartoons, one of which portrayed the influx of 'brown people' from the Pacific Islands as angry and violent. While Samoans and Tongans were a particular target, many Pasifika people including Cook Islanders were made to feel uncomfortable.

Tepaeru worked on many levels helping the Cook Islanders adjust successfully to New Zealand. She took some inspiration from the revival of Māori culture. Drawing on the cultural and linguistic links between New Zealand and Cook Islands Māori, she joined the Maori Women's Welfare League and became involved in the beginnings of the early childhood language movement Te Kōhanga Reo. Following this example she founded in 1984 Te Punanga o Te Reo Kuki Airini Inc., a Cook Islands language nest, which began in the garage of Tepaeru's home in Berhampore, Wellington with two children. Encouraged by her brother, Kauraka Kauraka, a well-known Cook Islands poet, she also attempted to keep the Cook Islands language alive by writing several children's books, one of which was in the distinctive language of her home island, Manihiki. In 1976 she was a founding member of the Pacific Allied (Women's) Council Inspires Faith (In) Ideals Concerning All Inc. (PACIFICA) and eventually became its president. The organisation was committed to helping Pacific families find work and adjust to city living. There was also her involvement in Aotearoa Te Moana Nui a Kiwa Weavers, a group of Māori and Pacific women interested in sharing skills in weaving through regular hui. Once more, Tepaeru learnt and then taught new sewing skills and once again she showed characteristic leadership by joining the committee.

Throughout this time Tepaeru continued to work at weekends with the Mamas, her fellow Cook Islands women, in sewing tīvaevae, which was important both for their own sense of identity and to recognise significant family events. The Newtown PASIFICA branch met every Saturday when Tepaeru taught tīvaevae to the Cook Islands women. This particular tīvaevae was completed over a week by four Cook Islands women led by Tepaeru in 1980. They started in the middle and worked out to the four corners, gossiping and singing as they sewed. It is over 2 metres wide and 2.5 metres long. Against a stunning yellow cotton background there is a central pattern of eight water lilies each beautifully embroidered in several colours. Around the outside is a gay garland of buds, flowers and leaves. By the 2000s, over 60,000 people living in New Zealand described themselves as Cook Islanders, while there were under 15,000 living back home in the Cooks — only about a thousand of whom were in the northern Cooks, Tepaeru's place of origin. So it was of huge importance to that community that the bright colours and distinctive shapes of the foliage back home were kept alive in beautiful tīvaevae such as this. It may be a work of art but it is also a symbol of the cultural diversity and creativity that arrived in New Zealand from the Pacific in the 1960s and 1970s.

83.

Montana Blenheimer wine cask

MARLBOROUGH MUSEUM

In autumn 1973 four men motored around the Wairau Plains in Marlborough looking for land. There was John Marris, a land agent; Frank Yukich, son of a west Auckland Croatian winemaker, and founder of Montana Wines; Frank's brother Maté; and Wayne Thomas, a plant scientist in charge of viticulture for Montana. Over 10 days they purchased nine sheep farms at an average price of under $500 an acre.

Although a century earlier David Herd had a small vineyard in the area, by 1973 that venture was forgotten. Most regarded Marlborough as too cold for wine. But Frank Yukich had visions of a New Zealand wine empire. He purchased land in Gisborne and the Waikato; and Wayne Thomas convinced him of Marlborough's potential. Confidently, Yukich paid the 10 per cent deposits from his own pocket; so it was a shock when the company's board rejected the purchases. Only after Wayne Thomas convinced fellow scientists from California to write in support did the board agree.

The holdings were in the name of a legal entity, 'Cloudy Bay'; and over the winter there were plantings of müller-thurgau and cabernet sauvignon and a symbolic blessing following Croatian custom. There were problems. The Mayor of Blenheim convinced Montana that 'Cloudy Bay' was too negative a name for successful wine; and over summer 80 per cent of the vines, some of which had been planted upside down, died from drought. But within four years a new industrial-scale winery was built, the first grapes of a new variety, sauvignon blanc, were planted, and the first wines made.

Montana decided to market some of the müller-thurgau wines in a cardboard cask. The idea of a cask originated with a South Australian wine-maker, Thomas Angove. In 1965 he had watched a person safely carrying battery acid safely in a plastic bladder protected by corrugated cardboard. He realised the same could be used for wine. Originally the plastic bag was taken out of the box, cut in the corner and resealed with a peg, but by the early 1970s the dispenser tap had become universal. Cask wine was dismissed by many as 'chateau cardboard'; but for those wanting to drink in quantity or have wine always accessible in the fridge it was a boon. The cask signified wine joining beer as a mass-consumption alcohol. The company pleased the local mayor and referenced the grape's German origins by calling their flask wine 'Blenheimer'.

The development of high-volume cheap cask wine from Marlborough carried two significant stories. The first looked back to a transformation in New Zealanders' patterns of consuming alcohol; the second looked forward to the emergence of wine, especially Marlborough wine, as a major export industry.

Looking back, most early inhabitants of New Zealand were not from wine-drinking cultures. Māori had not known alcohol, and settlers from Britain and Ireland were drinkers of beer or spirits. Some early non-Māori such as Thomas Busby saw a potential for growing wine grapes, but those who actually pursued the craft were continental Europeans such as French Catholic missionaries in Hawke's Bay, or Croatians who had originally migrated to dig kauri gum. In West Auckland there was a cluster of European wine-makers in the early twentieth century. But they found little market for table wines and concentrated on fortified wines such as sherry or port. As late as 1962, 88 per cent of the wine was 'fortified', not 'table', wine. Instead of European wine grapes they used hybrids which could also be sold as table grapes.

Wine-making was further handicapped by the prohibition movement which made it illegal to sell alcohol with food, and did not permit sales in fewer than a dozen bottles. The fear that the country might become 'dry' made investments in wine-growing highly risky. William Massey dismissed wine as 'a degrading, demoralising and sometimes maddening drink'. Little wonder the Croatian wine-makers supplemented their incomes with orcharding and cows.

Then things changed. People travelling overseas developed a wine taste, especially in Europe, or even Australia where Italian and Greek migrants had brought Mediterranean habits. Some New Zealanders aspired for the urban culture of older civilisations. The six o'clock swill was boorish. The law responded. From 1957, vineyards could sell single bottles of wine and wine shops were given licences. Three years later the first restaurants were able to sell wine, and from 1976 BYO was allowed.

The emergence of a local wine market encouraged new varieties and new areas of grape-growing. Instead of hybrid vines, specialised wine grapes were planted. The two initially favoured were those planted by Montana in Marlborough — müller-thurgau and cabernet sauvignon. Müller-thurgau, or riesling-sylvaner, was a high-cropping grape which made a German-style white suitable for unsophisticated quaffing. It was perfect for cask wine. From the later 1970s new varieties emerged, especially sauvignon blanc, a dry white wine, and pinot noir, the Burgundy red. The transition was aided in the mid-1980s when, after several years of overproduction and poor sales, Roger Douglas, Minister of Finance, was convinced, against his 'new right' instincts, to subsidise a 'vine pull'. The aim was to prevent a wiane glut, but the effect was to replace müller-thurgau grapes with more sophisticated varieties.

There was also expansion into new areas — Gisborne, the Wairarapa, north Canterbury and central Otago. But it was Marlborough which developed huge plantings. The area's appeal was less the soils, so much as the climate. Days of sunshine and warmth followed by cool nights brought intense flavours and a distinctive grassiness.

The change in land use in Marlborough was aided by the fact that initially land was cheap. Until the 1970s the major economic activity was sheep-farming. But in 1966 the wool industry hit a crisis as synthetics led to a fall in international demand. The Wool Board was forced to purchase thousands of bales. Demand never fully recovered. With Britain's entry to the European Economic Community, the secure market for sheep meat also came into question. Wine provided an alternative use for Marlborough land.

With traditional exports facing rocky times, could fine wines become a new export? This was not Montana's initial focus — müller-thurgau wines were packaged in casks for locals. The export drive was made possible by others who followed Montana's lead and purchased land in Marlborough. They were smaller,

MARLBOROUGH MUSEUM

family wine-makers. One was an Irishman, Ernie Hunter, who planted sauvignon blanc grapes on a new vineyard near Renwick. In 1986 Hunter's Fumé Blanc (a name for oaked sauvignon blanc) was voted as the most popular wine at the Sunday Times Wine Club Vintage Festival in London. Hunter's success, accompanied by clever promotion at New Zealand House made Marlborough sauvignon an international hit. Close to Hunter's, a vineyard was established by David Hohnen from Western Australia. In 1986, the same year as Hunter's won London accolades, the new vineyard produced its sauvignon blanc. It achieved international fame under a name which might have astonished the former Mayor of Blenheim — Cloudy Bay.

In 1985 about 800,000 litres of wine was exported — 1.3 per cent of the total production. Within three years the quantity had risen to almost 3 million litres, and by 2020 over 300 million litres of New Zealand wines were exported — over 85 per cent of wine production. By then wine exports ranked tenth in overseas income earned. Montana was exporting to over 30 countries and had been bought by Allied Domecq, the world's second largest wine and spirits firm. Production was industrial-style with large stainless-steel tanks; and Marlborough wine-makers brought wine-making into a new age by replacing corks with screw caps.

Revealingly, as Marlborough, developed a major export in wine, the Blenheimer cask came into question. In 2004 Marlborough wine growers tried to prevent Marlborough cask wine sales in British supermarkets for fear that it would devalue their premium image. Twelve years later vintners promoted a change in the name of the region's city, Blenheim, to 'Marlborough City', because they believed Blenheimer wine muddied the region's reputation for high-quality wines. Even Montana, with export markets in mind, abandoned its own name in 2010 and became Brancott, because Americans associated Montana with an American state. Marlborough was downplayed because of its echoes of a cigarette brand, Marlboro.

The Montana Blenheimer cask, which had once expressed the early promise of Marlborough wine-making, had become, in the new world of high-quality exports, everything wine-makers rejected. The rise and fall of the Blenheimer cask reflected dramatic changes in tastes and the emergence of an important new export.

84.

Biko shield

AUCKLAND WAR MEMORIAL MUSEUM

'Bi-ko's foot-steps. Bi-ko's foot-steps.' The Biko Squad advanced to the chant. It was like a scene from the Middle Ages — bicycle helmets, chests protected by cardboard tubing, and these impressive green shields, most made of plywood, others from plastic corned-beef containers. The squad was on Sandringham Road, 400 metres from Auckland's Eden Park. Between them and the ground were rolls of barbed wire and a perimeter of jumbo bins and shipping containers. In a 10-metre gap between the bins were other New Zealanders — the police Green Squad with metal helmets, visors before their eyes, large metal, not plywood, shields, and in their right hands an ominous baton, the infamous Monadnock PR24. These American batons became known derisively by protesters as 'Minto bars', after John Minto, a leader of the anti-apartheid group HART, and a marshal with Biko Squad that day.

The squad came up to police lines eight across, three deep. A marshal called to the police, 'Green Squad, lay down your batons.' The batons remained poised. Back to the marshal, 'Advance! Bi-ko. Bi-ko. Biko Squad, brace yourselves . . . At the count of three we're going through them.' The Biko Squad inched forward. 'One . . . Two.' They eyeballed the police, then stopped. The Biko shields, 1.2 metres high, 60 centimetres wide, provided reassuring protection. They might also be weapons. But not this time. 'Relax, and fall back,' came the order. Four times over the next hour the Biko Squad advanced; four times they halted. It was a strangely unnerving ritual — young New Zealanders prepared to get hurt campaigning against apartheid, up against police protecting other New Zealanders' desire to play rugby with South Africans.

The Biko Squad was one of three, each about 2000 strong — the Tutu Squad of older people or church groups, keen to avoid confrontation; the Patu Squad, attracting Māori and lesbian women and prepared for violent interactions; and Biko, largely students, ready for a disciplined encounter as on Sandringham Road. Elsewhere on that day, 5 September 1981, when the Springboks played Auckland, protesters broke police lines and were batoned to the ground. A week later, in the final game of the tour, police and protesters returned to Eden Park. Once more these shields were used by the Biko Squad in nerve-racking confrontations on Sandringham Road. This time the police were less tolerant. They ripped away helmets and shields and baton-charged. The squad were left bleeding on the road. Forty Biko shields were lost. Elsewhere violence was worse, notoriously when eight Artists Against Apartheid members in fancy dress — five clowns, two rabbits and a bumblebee — were pummelled to the ground. One suffered a burst eardrum and injuries requiring 12 stitches.

What brought New Zealand to this extraordinary moment when homemade plywood shields faced police batons? It was a battle between love of rugby and concern about racism. From 1905 rugby had been central to New Zealand identity and South Africa was 'the old enemy'. Contests with the Springboks were epic — Trevor Richards remembered the 1956 series as 'like the Farmers' Christmas parade going on for three months'. But South Africa was a society governed by racial distinctions, and in 1948 an apartheid system was established. Gradually, disquiet about its racism affected New Zealand. The first issue was South Africa's refusal to accept Māori in All Black teams. There were murmurings in 1928 when the legendary George Nepia was not selected; then in 1949 the unrest became more vocal. When the 1960 team excluded Māori, opposition was heated, and eventually Keith Holyoake's government accepted the principle of 'No Maoris, No Tour'. In 1970 a tour went ahead including Māori. Now the battle lines shifted. The issue was not the place of Māori in New Zealand teams, but of blacks in South Africa. Sport became a way to oppose apartheid.

New Zealand was increasingly divided, as Norman Kirk cancelled the 1973 Springbok tour for fear of civil disruption and then from 1975 Robert Muldoon campaigned for provincial support by backing sporting tours. National embarrassment followed with an African walkout at the 1976 Montreal Olympic Games because of an All Black tour to South Africa, and fierce opposition from the Commonwealth which imposed the Gleneagles Agreement on a grumpy Muldoon. Meanwhile, the old New Zealand of war heroes, and white 'kith and kin', argued for 'keeping politics out of sport' ('KEEPOOS' in Tim Shadbolt's words!).

Opposing them was educated urban New Zealand, which through foreign travel and reading accepted cultural diversity. Many were young, in generational revolt against their parents' values, and often students. HART (Halt All Racist Tours) had been founded in 1969 at the New Zealand University Students' Association's Easter Council and was led by veterans of student politics — such as Trevor Richards and Mike Law. They were imbued by a vision of a world free of racism. Many in the Biko Squad had a student background, including the marshals, Heather Worth, John Minto and Fuimaono Norman Tuiasau.

Heather Worth was among many women protesters. It was not only apartheid's evils that motivated women. They also opposed the centrality of rugby in New Zealand culture. One Wellington female protester said, 'I have for years resented the dominance that rugby has in the homes, schools and society in general. It's time that a few other values took over from bloody rugby.' Thirty-three of the 37 marshals in the Patu Squad were women. Others were Christians. When the South Africans arrived they were met at the airport by 60 staff and students from St John's Theological College, and among those who stormed the field at Hamilton on the second game were three St John's students carrying a wooden cross.

Young Māori were well represented. They saw parallels between apartheid and racist policies in New Zealand. The Patu Squad was heavily Māori, with Nga Tamatoa radicals such as Rebecca Evans and Donna Awatere amongst its leaders. The squad carried a banner with Ngāti Maniapoto's famous words from Ōrākau: 'Ka whawhai tonu matou, ake, ake, ake!' — 'We shall fight on for ever, ever and ever!' The leading marshal of the Biko Squad, Fuimaono Norman Tuiasau, was a law student of Samoan background who was active in the Polynesian Panthers, a radical Pacific Islands group.

BIKO SQUAD SHIELD, 1981. AUCKLAND WAR MEMORIAL MUSEUM TĀMAKI PAENGA HIRA. 2002.55.1.

Why did protesters carry shields? Photographs from the Springboks' first game in Gisborne show protesters wearing woollen hats, not helmets, and no shields in sight. The next scheduled game, Springboks against Waikato, at Hamilton changed the situation. Five thousand protesters marched to Rugby Park. A radical section, in 'Operation Everest', pulled down the fence, and raced onto the field. As the crowd became abusive, the 400 or so protesters, aware the game was beaming live to South Africa, chanted 'The whole world is watching'. The 500 police had no choice. The game was called off. While rugby supporters inflicted nasty beatings on protesters as they left, the police felt humiliated. Determined it would not be repeated, they turned to the helmets and batons.

The first display of this new philosophy was on a peaceful march up Wellington's Molesworth Street on the night of Diana and Charles' wedding. To their horror, marchers saw police blocking passage and batoning the unprotected front rows. From then on, rolls of barbed wire circled rugby grounds and armed ritual became the format. The police came to every game with helmets, visors and batons, to prevent pitch invasions. The anti-apartheid campaigners developed a strategy of probing the park perimeter. To protect themselves from both police and angry rugby supporters they wore motorbike helmets, cricket boxes, cardboard chest protectors and carried shields. The shields had two functions — they provided protection and offered a surface for anti-apartheid slogans.

These shields are inscribed in black with 'BIKO'. Steve Biko was a young black nationalist student who led the black South African Students' Organisation. 'Black is beautiful' was his favoured slogan. In 1973 he was banned from public appearances and four years later arrested. He was beaten to death in his cell by state security officers and his funeral attracted 20,000 mourners. As a charismatic fighter for black rights, Biko became a symbol of protest. Much was made of the fact the last test was 12 September 1981, the anniversary of Biko's death in 1977. Posters inviting protesters to Eden Park on that day portrayed Steve Biko's face, and when the marchers gathered at the start of the day, Prain sang Peter Gabriel's song 'Biko'. The Artists Against Apartheid created a banner inscribed with, 'BIKO', which was carried across Eden Park by eight balloons. Biko was the dominant anti-apartheid symbol. The bold black letters on these shields express this perfectly.

In retrospect, given the feeling between anti-apartheid protesters on one side, and police and rugby supporters on the other, it seems extraordinary there were no deaths. We remain thankful that police carried batons, not guns, and the protesters held only plywood batons. The conflict became a ritual, not outright war. Our shields tell us much about the limits of this most divisive social conflict.

85.

Nuclear-free badges

MUSEUM OF NEW ZEALAND TE PAPA TONGAREWA

This hessian strip holding 38 anti-nuclear badges was used by the New Zealand Peace Foundation to sell badges for $1.50 each in the 1980s. The sentiments found on the badges reflect the ideas which made the anti-nuclear movement so important to New Zealand identity in those years.

The idea of displaying a badge goes back to the European Middle Ages when pilgrims wore pewter badges to show their faith. Subsequently, badges identified membership of a group or organisation. The idea that they could promote a political movement emerges in the nineteenth century with the coming of democratic politics. In the USA, from 1824, buttons identified support for presidential candidates, while British movements such as anti-slavery and women's suffrage adopted metal badges. There is no evidence of New Zealand suffrage movement badges, partly because it was not until 1896, after the 1893 achievement of suffrage, that a patent was issued for cheaply made celluloid-covered button badges. As identity politics emerged in New Zealand from the 1960s, protest badges became essential tools of communication. They were cheap, easy to make, and could be worn by either gender on any form of clothing. They allowed the wearer to announce their political identity and promote their cause with logos and slogans. Badges were characteristic items of 'baby boom' politics.

The peace foundation which sold these badges had been established in

DISPLAY OF BADGES, 1980S, NEW ZEALAND, VARIOUS MANUFACTURERS. GIFT OF THE PEACE FOUNDATIONDISARMAMENT AND SECURITY CENTRE, 2009. TE PAPA (GH020698)

1975 by a small group, with Quaker connections, who wished to promote peace education. But the commitment to peace expressed in these badges goes back long before 1975. There were Polynesian traditions of non-violent resistance in Aotearoa, displayed most famously in the nineteenth century by Moriori in the Chatham Islands and the Parihaka community in Taranaki. In both world wars, some with a strong Christian or socialist commitment refused conscription. They were imprisoned, or, in the case of the infamous 14, forcibly transported to the front line. It is the anti-nuclear movement with its symbol of outstretched despairing fingers against the circle of the world which is best represented in our badges.

No fewer than 10 badges feature the CND symbol, which was invented in 1958 when the Campaign for Nuclear Disarmament (CND) was launched in Britain. By that stage New Zealand was a willing partner in the nuclear arms race. Even during World War II the country sent scientists to the United States to help in the Manhattan Project, and from the late 1940s, bomb-testing came to the Pacific, with New Zealand's blessing. The National Party tried to develop 'heavy water' for use in the nuclear industry, and New Zealand made no complaints when the United States tested bombs at Bikini Atoll and the British exploded devices in northern Australia, and at Christmas Island in the Pacific. New Zealand even supplied two frigates to assist weather reporting for the Christmas Island tests. By signing the ANZUS agreement with the United States in 1954, New Zealand implicitly accepted protection under the nuclear umbrella.

Initially there was little domestic opposition. But in 1961, inspired by the CND marches in the UK to, and then from, the Aldermaston nuclear weapons facility, a few New Zealanders, concerned about the nuclear arms race, began annual marches over the 52 miles (the distance from Aldermaston to London) from Featherston to Wellington. In 1963 the New Zealand CND movement supported the Australian Labor Party suggestion for a nuclear-free South Pacific with a petition of over 80,000 signatures.

Despite such numbers, it was the French decision to start nuclear bomb-testing at Moruroa Atoll in French Polynesia which awoke widespread concern. The French wanted a nuclear capacity independent of the USA and Britain, and the loss of Algeria meant their Sahara testing site was unavailable. Moruroa was about around 1600km from the Cook Islands and over 4200km from New Zealand. Tests began in 1966 and continued over several years. Some went awry and rising levels of radioactivity were detected in New Zealand milk. The National government protested politely but opposed vociferous opposition because, with Britain joining the European Economic Community (EEC), they knew France could exercise a veto over future access.

It was left to private people to express resistance. In 1972 a group,

Peace Media, sent a flotilla of small boats into the Moruroa testing zone. There were several inspirations for this. There was the growing environmental movement which found strongest expression in the campaign against raising Lake Manapōuri. Young activists had a vision of New Zealand as a country free of the industrial and military poisons of the old world. The environmental impulse is reflected in the green colour of some badges, the flowering pōhutukawa on the 'Bread not Bombs' badge, and the pastoral landscape of the Kumeu Peace Group badge. There is also the irony of the 'Save the Humans' badge, a reference to the movement to save whales. Second, there were people who had become convinced of the need for an independent New Zealand foreign policy through opposing the Vietnam War. One who boarded a protest vessel was Barry Mitcalfe, previously chair of the Wellington Committee on Vietnam. Third, there were some, especially Māori, who saw the fight against French testing as representing the movement to reclaim dignity for the South Pacific's indigenous people. Another who boarded a vessel was the Māori leader and Labour politician Matiu Rata.

By 1973 the situation had changed. Negotiations for Britain's entry to the EEC were over; and Norman Kirk's Labour Party won the 1972 election on a platform of vigorous opposition to French testing. Following the election, Kirk sent his deputy to Paris to argue against the tests and New Zealand joined Australia in submitting a case to the International Court of Justice. More dramatically, Kirk despatched the frigate HMNZS *Otago*, followed by HMNZS *Canterbury*, into the test zone with a Cabinet minister, Fraser Colman, on board. The French exploded their tests; but Kirk won attention abroad and huge plaudits at home. New Zealanders felt pride that their small nation had spoken against the old world's nuclear ambitions. In 1974 France announced that future tests would be underground.

Now the focus of the anti-nuclear movement changed. In 1974 Norman Kirk died and the next year Robert Muldoon was elected Prime Minister of a new National Party government. As on the issue of playing sport with South Africa, Muldoon was keen to cock a snook at the baby-boom activists on the nuclear issue. He insisted that as a loyal member of ANZUS, the country must accept nuclear arms. He lifted a ban on nuclear warships and urged the United States to send nuclear-powered and -armed ships here. He wanted 'to show the protest movement who was in control'. So when the United States accepted Muldoon's invitation the peace movement took to their boats, hoping to prevent the warships and submarines berthing. The USS *Truxtun* in August 1976 and the USS *Longbeach* in October were met by peace flotillas of yachts, dinghies, pleasure craft, even surfboard riders. When a nuclear-powered submarine, USS *Haddo*, visited in January 1979, a protester managed to dance on its nose. Visits continued through the early 1980s and some local bodies responded by declaring themselves nuclear-free.

By 1984, three of the four major parties, Labour, Social Credit and the New Zealand Party, were opposed to the visit of nuclear-armed or -powered boats. In June, Richard Prebble introduced a Nuclear Free New Zealand Bill, and Marilyn Waring, a National MP, announced her support. Robert Muldoon responded by calling a snap election. He was happy to fight for a nuclear-welcoming New Zealand. In the event, the Labour Party headed by David Lange won handsomely. Lange was committed to refusing nuclear-armed ships, but seemed evasive on the issue of nuclear-powered vessels. It seems likely the invitation to argue for a nuclear-free world at the Oxford Union nerved him and he decided that neither nuclear-powered nor -armed vessels were welcome. 'An old diesel rustbucket', the USS *Buchanan*, was refused a visit. Both the United States and the United Kingdom were decidedly unhappy; and while the USA did not formally expel New Zealand from ANZUS, they refused to admit the country to military exercises. David Lange duly won international plaudits at the Oxford Union by a brilliant performance outpointing Jerry Falwell. Lange harnessed the country's environmental mood and the burgeoning nationalism to the anti-nuclear cause.

The Oxford debate occurred in March 1985. Four months later a Greenpeace boat, the *Rainbow Warrior*, docked in Auckland Harbour, was blown up and a crew member killed. Two weeks later, after an international whodunit, two French secret service agents were arrested. The French government had been trying to rid itself of the annoying protests at Moruroa. The agents were convicted of manslaughter. New Zealand received an apology and $13 million from France; but France had the last laugh when the agents were released early, decorated and promoted. Several badges refer to these events. Whether the 'Stop French Testing' badge with its tricolour colours derives from 1984 or earlier is not clear; but certainly other similar badges encouraged a boycott of French goods. There are also badges which draw on rainbow colours, most effectively the Greenpeace badge with the words 'You Can't Sink a Rainbow'.

Over the next few years, the country took further anti-nuclear steps. New Zealand and seven other South Pacific nations signed a South Pacific Nuclear Free Zone treaty; and in 1987 a nuclear-free Bill was passed. The United States downgraded New Zealand's status from 'ally' to 'friend'. Nuclear-free New Zealand became a widely accepted national principle which even the National Party accepted when back in office in 1990. This transformation in public attitudes is well reflected in these 38 badges. Who can say how much their wearing by committed activists contributed to changing the minds and aspirations of fellow New Zealanders?

86.

Save Our Post Office poster

MUSEUM OF NEW ZEALAND TE PAPA TONGAREWA

This poster was one of four designed and produced by the Wellington Media Collective for the New Zealand Post Office Union in 1987. The poster offers an insight into three powerful value systems which battled for influence in the mid-1980s: the welfare state, new right economics, and the youthful counterculture.

The first value system was the belief in a strong, benevolent state. This vision lay behind the growth of the state-run Post Office system defended by these posters. Post offices arrived with British rule in 1840, and as European settlement expanded, hundreds of post offices to collect and deliver mail were started by both provincial and central governments. From the 1860s the Post Office added services — a savings bank in 1867; telegraphic communication from 1881; telephone services from the end of the century. By 1900 there were over 1700 post offices found in every small community. The Postmaster-General was an important cabinet post. During the next half century the state increased the Post Office's role — it registered births, marriages and deaths; issued television and fishing licences; registered cars; paid pensions; enrolled voters. The Post Office Savings Bank lent for housing and encouraged children to learn saving habits through the Schools Savings Scheme. By 1964 The Post Office employed some 26,500 people. Post offices were a centre of community life, representing the protective state.

One believer in the benevolent welfare state was Robert Muldoon,

'OUR LOCAL POST OFFICE...WHERE WILL WE STAND WITHOUT IT?' POSTER, 1986, NEW ZEALAND,WELLINGTON MEDIA COLLECTIVE. PURCHASED 2015. TE PAPA (GH024617)
DESIGN AND ILLUSTRATION ©1986 SHARON MURDOCH/WELLINGTON MEDIA COLLECTIVE.ONE OF FOUR POSTERS PRODUCED FOR THE NEW ZEALAND POST OFFICE UNION.

National Party Prime Minister and Minister of Finance from 1975 to 1984. When he came to power the economy was stretched. Wool prices collapsed in 1966, and then in 1973, Great Britain joined the EEC (European Economic Community) and oil prices reached record levels. The balance of payments was under severe strain. Unemployment and inflation rose fast. Growth virtually ceased and New Zealand went from being the world's sixth most wealthy country per head in 1965 to nineteenth in 1980. Muldoon responded in the tradition of the protective state. He made superannuation more generous, froze wages and prices, added new export subsidies, imposed import controls and promoted a 'Think Big' set of schemes led by major government investment. The result was slight economic growth, but also criticism of Muldoon's 'welfare state' philosophy.

The chief critics were a younger generation who questioned the strong state. This is our second world view. Some believers in economic liberalism, like the politicians Roger Douglas, Richard Prebble and Michael Bassett, had been Labour Party supporters since childhood but believed the welfare state created a culture of dependency, producing people unable to take entrepreneurial risks. David Lange was initially their public leader, and became Prime Minister after a tipsy Robert Muldoon called a snap election in 1984. Until Labour's re-election in 1987, when he aired doubts, Lange was the charismatic champion of the free-marketeers. They found support in the business community and among neo-liberal economists at the Reserve Bank and in Treasury. Overseas Margaret Thatcher and Ronald Reagan were both preaching a gospel of downsizing the state and trusting market forces.

The free-market reforms, known as 'Rogernomics' after the Minister of Finance Roger Douglas, began with deregulating the finance market. The dollar was devalued and then floated. To provide an incentive for risk-taking, higher levels of taxation were reduced, paid for by a Goods and Services Tax and the closing of tax loopholes. Export subsidies and tariff protections were removed. The result was greater consumer choice, but also unemployment in traditional export sectors like farming and in protected industries like clothing and car assembly.

Central to the reformers' vision was bringing business 'efficiency' to the state. This involved two processes. The first meant turning arms of the state into state-owned enterprises (SOEs) operating on business lines and paying a dividend to government. The Electricity Department became Electricorp, State Coal Mines became Coal Corp. Within a week of the first SOEs being established in April 1987, 4732 workers went into voluntary redundancy and eventually 19,133 people lost their jobs. The second process, begun by the Labour government and pursued in the 90s by Jim Bolger's National government, was selling off government activities to the 'efficient' private sector.

The Post Office was transformed by both processes. In April 1987 it was spilt into three SOEs — New Zealand Post to deal with mail, Telecom for telecommunications, Postbank for the savings bank. Many were angry at this attack on a hallowed institution. At Waipū in Northland, Post Office staff stamped 'Waipu Post Office Will Not Close' on every piece of mail. People were concerned, by the loss of jobs, but even more by the end of a community service.

The threat of this break-up inspired this poster. Sponsored by the Post Office Union, the posters were the work of the Wellington Media Collective. They represented the third significant cultural force in 1980s New Zealand. The Media Collective were an archetypal expression of the baby-boomer counterculture of those years. For a start, the collective was set up, not as a business to make money, but as a collective to share media skills with community groups — 'We will work with you, but not for you' was their slogan. Initially the members worked after-hours and charged only for the cost of materials and equipment. They supported the identity movements of the 1970s and 1980s, preparing posters and graphic material for the anti-nuclear movement, opposition to the 1981 Springbok tour, promoting childcare, Waitangi protest and land rights. One of their earliest posters was for a United Women's Convention in Hamilton in 1979, an important gathering of feminists. Dave Kent, who was in the collective from its start in 1978 to its winding-up in 1997, explained his background: 'A mix of my friends, the lively political debates and activism at university, opposition to the Vietnam War, and Rona Bailey and her circle of friends/activists'. Apart from the last point, it could have been replicated by many leaders of the political counterculture. Rona Bailey was Dave Kent's cousin, and her home hosted the first gathering of the collective. She provided a connection with the radical Marxist tradition which was famously expressed in Dave Kent's poster of Karl Marx as a black-singleted Fred Dagg.

While the collective produced many kinds of publication, the poster was their stock-in-trade; for this was how the countercultural political movements got their message across.

This poster was the work of Sharon Murdoch, who had joined the collective in 1985, bringing a strong feminist perspective. The poster shows five people including a Māori woman and her son having a rug pulled out from under them by the strong arms of a corporate gentleman in a dark suit. The text notes that the aim of the Post Office reorganisation is 'profit before service'. It claims that many post offices would be closed and some services become unavailable. Then in recalling the old Post Office, the text says, 'If you use your local Post Office to do banking, collect benefits, buy stamps, pay bills or collect mail' this will affect you. Murdoch was responsible for a second poster showing the line to a telephone being cut; while Dave Kent also prepared two posters. One showed a piggy bank and the words, 'When is a people's bank not a people's bank?', and the answer, 'When it's carved up for profit before service!' The other portrayed six pairs of hands posting letters in a post box which was being pulled away by, of course, a suited businessman. Probably because the union paid, the posters were not screen-printed, like most Media Collective work, but offset printed using strong primary colours to speak of formal government authority.

For all their graphic impact, the posters changed nothing. Within three years New Zealand Post closed 432 post offices while both Postbank and Telecom were sold to private owners. As for the Media Collective, the need to earn a living and pay mortgages turned the group towards becoming a paying business. Technology changed and by the mid-1990s computer graphics were replacing handmade design. The collective ceased trading in September 1997. By then, state-owned enterprises were firmly established, and while some such as the railways, Air New Zealand and Postbank had been sold, it was becoming clear that the state might have to buy them back, or in Postbank's case establish another state bank, Kiwibank. The posters speak of a moment in the 1980s when the old belief in the benevolent state jostled with the new ideas of Rogernomics and the countercultural vision of the baby boomers.

87.

POLY 1 personal computer

MOTAT | MUSEUM OF TRANSPORT AND TECHNOLOGY

With its smooth, moulded plastic surround holding a screen and keyboard all-in-one, the POLY 1 was by far the most elegant micro-computer around. It came in six colours including the cream one shown here from MOTAT, and it even had handles at each side for ease of carrying. When turned on, the screen was a delicate mauve. With full colour graphics and 64 kilobytes of memory it was among the most technically adept. For a time, in the early 1980s, the New Zealand-conceived and -built POLY 1 looked as if it was leading the world in the advancing field of micro and educational computers. Silicon Valley were worried.

The development of the POLY came at a crucial moment in the history of computers. Exactly when computers began is a cause of debate. Some say the abacus used by Greek and Romans constitutes the beginnings; others point to Charles Babbage's Analytical Engine in England of the 1820s. New Zealand can claim a contribution with George Julius inventing the totalisator for betting on horse races in the 1910s. Electronic computers really emerged in the 1940s driven by military needs. In the mid-1970s they were mostly large mainframes using punched cards for the storing of data and with separate machines and operators for sorting and collating, not to mention printing. Such equipment was used in large organisations — in government such as Treasury or by law enforcement authorities which set up the Wanganui Computer Centre in 1976; in

POLYCORP NEW ZEALAND LIMITED. 1981. COMPUTER [POLY1], 1999.2. THE MUSEUM OF TRANSPORT AND TECHNOLOGY (MOTAT).

banking where by 1969 all the trading banks were involved in Databank Systems Ltd to read and process magnetic strips on cheques; and in some businesses which used computers to substitute for clerical work in doing accounts or paying bills. There were also a few mini-computers, used largely for controlling manufacturing processes. Where computers were not found was in secretarial word-processing, or in schools and homes. Opportunities beckoned.

Internationally, the first steps towards micro-computers, or personal computers as they were eventually called, came in the mid-1970s. In 1975 the Altair 8800 appeared using a microprocessor. The size of a thumbnail, microprocessors replaced the need for separate integrated-circuit chips for each function. Two years later the first 'trinity' of micro-computers hit the market — Steve Jobs and Stephen Wozniak's Apple II which had a colour display when linked to a television; the Commodore Pet; and Radio Shack's TRS 80 which had a separate monitor. The same year a micro-computing club was set up in New Zealand and in 1980 came the first exhibition of the new small computers. Most interest at this stage came from those with a passion for games. Quickly other overseas offerings appeared, even though the Labour government in 1975 introduced a 40 per cent sales tax on such machines. In 1981 came the IBM PC (Personal Computer) and the next year the Commodore 64 which would become the best-selling PC of all time. There was so much interest internationally in the new machines that *Time* magazine named home computers the 'person of the year' for 1982. So in the early 1980s there was interest in personal computers, yet there was uncertainty as to which machines would rule — the Apple II, the IBM PC perhaps, and there were offerings from the BBC, Amiga, Osborne and Kaypro. The opportunity was open for a New Zealand entrant to the race. Here the POLY 1 made its pitch.

The initial conception of the POLY micro-computer came from two teachers in electronic engineering at Wellington Polytechnic, Neil Scott and Paul Bryant. They were primarily interested in designing a small computer for use in schools, and attracted the support of the Minister of Education, Merv Wellington. The ministry promised $10 million to produce 16 POLY 1s for every school in the country. To fulfil this order Scott and Bryant turned to the Lower Hutt firm Progeni, the brainchild of Perce Harpham. Harpham had initially worked for Dulux Paints, but on a work trip to the United Kingdom he became excited about computers; and with the UK having just agreed to join the European common market, he realised that New Zealand needed new exports. In 1968 he established the country's first software company, SPL, which became Progeni.

Progeni had considerable success. They designed a system for Ron Jarden, former All Black and sharebroker; provided software for the Wanganui law enforcement system; and worked on a GIS (Geographic Information System) for the Department of Lands and Survey. So, when the Development Finance Corporation approached Progeni to provide the software and experience to produce the POLY, Harpham readily agreed. Progeni put up $250,000 to get the work started. The government indicated they would buy 1000 machines each year for five years. It took a remarkable eight months to go from the polytechnic prototype to 50 working machines. In the Christmas of 1980–81, 60 teachers gave up their holidays to develop course material; 30 suppliers made parts. In May 1981 a lengthy evaluation in classroom trials before three colleges and the Department of Education showed all expectations had been met.

The prospects looked alluring. The POLY 1 was faster, better designed, and with its full colour display was superior for school use to any international competitor. Its memory was four times the size of the Apple II; and its development was 18 months ahead of the BBC Micro which would dominate the educational market in the UK. Because it was an all-in-one unit, teachers only had to worry about two cords — one for power, and one for the network. The latter connected the machines to a central server, the Proteus, which networked the content to 32 computers, in different classrooms. In turn Proteus could be a gateway to a national mainframe, so that excellent software could be distributed to every school in the country. It was a brilliant concept which promised to make New Zealand an international leader. An export market beckoned.

Inevitably there were grumbles. Some teachers found the carrying handles made the machines a bit wide and difficult to get through doors. And it was unfortunate that the initial name, POLYwog (Poly after its institution of origin, and polywog which was a tadpole), was seen as racist. But what largely killed the POLY 1 was colonial cringe. There was a sense that anything technological must be better if it came from overseas. Further, the New Zealand government was coming under pressure to ensure a fair commercial playing field, and open up opportunities to other players. The big international outfits saw the threat. They pressured the government. Apple took direct action. In their first educational deal outside the United States, Apple offered every school one free Apple II and reduced the price from $4812 to $1200. A small local start-up did not have the resources to compete in a global market. In homes around the country, IBM's PCs were becoming increasingly dominant. Before long the government had reneged on its earlier agreements to purchase the POLY 1, with the Minister of Foreign Affairs Warren Cooper saying that he 'could see no reason why government should spend money so that teachers could do even less work'.

This was not the end of the POLY 1. Some people used the machines for gaming, and Progeni won a contract with the Australian army. The POLY was adapted with Chinese characters and was sold to the Beijing Institute of Aeronautics and Astronautics. Then in 1989 the financial difficulties of the Bank of New Zealand resulted in receivers forcing the closure of Progeni. After that year, no more POLYs were produced, although it is said that one Massey academic found the machine so reliable that he continued to run aerospace software on his POLY until 2000.

The death of the POLY did not signal the end of personal computers in New Zealand. Far from it. By 1985 some 5 per cent of New Zealand households possessed a PC, usually in homes where the householder was young and relatively affluent. The most popular use was gaming, and this explained the popularity of the Commodore 64 where users could program games themselves. However, the introduction that year of Microsoft's Windows operating system and their provision of MS Word for Apple machines signalled the growing importance of word-processing. Laser printers came onto the market and by the end of the 1980s there had been a major transformation in the way documents were prepared. Typists lost their jobs; and in organisations ranging from universities to businesses, personal computers began to sit on every desk. As early as 1985 the University of Otago had 400 micro-computers.

Elsewhere the computerisation of society saw the introduction of EFTPOS banking services and the gradual disappearance of cash and cheques. So important was computing to society that when Jim Bolger's government took office in 1990, he appointed Maurice Williamson as the first Minister of Information Technology. By then the country's first experiments with the internet had begun within universities and the DSIR (Department of Scientific and Industrial Research), and within five years the internet was a crucial driver of business communications and user interests. In 1995 *Internet: A New Zealand Users' Guide* was published; and there was a surge in the sales of personal computers as New Zealanders both at home and at work become interested in surfing the web.

By the mid-1990s, of course, computers from United States firms, Apple's Macintosh and Microsoft-based PCs like Compaq were dominating this booming market. The dream of a New Zealand-based home computer industry had long died; and while New Zealanders would soon reveal abilities to use the new technology for creative purposes — from digital film-making to accountancy packages — the possibility that once existed in those few crucial years when personal computers were just emerging, that New Zealand might become a home for a significant personal computer industry, had died. But the memory of the creativity and potential opportunity of the POLY 1 should not be forgotten.

88.

Barry Brickell's memorial post to Ralph Hotere's father

OTAGO MUSEUM

Against the background of political challenges and protests, economic transformations and new family structures, the half-century after World War II saw a remarkable outpouring of cultural creativity in New Zealand. It is not easy finding one object to tell that story, but this beautiful stoneware post by Barry Brickell is a good place to start.

Exactly when and where Barry Brickell met Ralph Hotere is not clear. They were both sociable individuals — so it was probably in Auckland over glasses of wine. They grew closer in 1975 when Brickell went south to Dunedin while his pottery studio was being built at Driving Creek near Coromandel. Hotere turned up one day with a pile of bricks, suggesting Brickell make a kiln. He did so and the two spent many convivial evenings around the hot kiln. Seven years later, in 1982, when Hotere's father, Tangirau Hotere, died, Brickell contacted Ralph and offered to make a memorial. His father was a major influence on Ralph's life, so he readily agreed. Hotere's family of 15 children lived at Mitimiti, in a distinctive Māori Catholic area on the west coast of Northland, just north of the Hokianga Harbour. Because Tangirau was a hard-working dairy farmer, Ralph suggested the memorial take the form of a fence strainer, since 'his father had the reputation of installing the mightiest tōtara strainer posts in the district'. Accordingly, Brickell made an upright fence strainer, complete with holes for the wires. That did not quite do it. Both men enjoyed using words and poetry in their craft. Asked by Brickell for a poem, Hotere offered a Muriwhenua chant which he had learned from his father:

He kuaka	A godwit
He kuaka mārangaranga	A hovering godwit
Kotahi manu	One single bird
I tau ki te tāhuna	Has landed — on the sandbank
Tau atu	It has settled there
Tau atu	Others are landing
Kua tau mai	Now they have all landed here

Hotere had used the chant previously in extended form — in the magnificent 18-metre mural which he completed in 1977 for the arrival hall of Auckland International Airport. In the airport the words were a fitting welcome to travellers landing in a new country. They were equally appropriate for the man who had given them to his son.

The journey of Ralph Hotere and Barry Brickell to the memorial post tells us much about the emergence of a creative culture in New Zealand. Both were born in the 1930s — Hotere at Mitimiti in 1931; Brickell the son of a New Plymouth civil engineer in 1935. In these years of economic depression a cultural nationalist movement slowly emerged, especially in Christchurch among young poets like Allen Curnow and Denis Glover, and painters like Rita Angus, Toss Woollaston and Colin McCahon. The post-war years began with Allen Curnow's plea: 'Strictly speaking, New Zealand doesn't exist yet . . . It remains to be created — should I say invented — by writers, musicians, artists, architects, publishers; even a politician might help'! Two years later in 1947 Charles Brasch launched his path-breaking journal *Landfall* as an outlet for new voices.

In the 1950s the cultural nationalist leadership moved to Auckland, as artists like McCahon and poets like Curnow relocated, and the art gallery held lively exhibitions under the leadership of an innovative director, Peter Tomory. During these years both Brickell and Hotere spent much time in the city. Brickell moved to Devonport with his family at the age of seven, and discovered a fascination with fire, steam and brick. He built a kiln at his parents' house, and met an inventive local potter, Len Castle, in 1950 who taught him to make pots on a wheel. He also took Thursday-evening painting classes with Colin McCahon. Hotere came to Auckland from 1946 to 1952 to study, first at the Catholic school Hato Petera College, and then at the Teachers' Training College. He became an arts adviser for the Education Department in Northland but continued to frequent Auckland, mixing with poets, sculptors, architects, academics, and of course painters. At the Kiwi Tavern he socialised with McCahon, and encountered other intellectuals such as the poets Hone Tuwhare and Allen Curnow, and the novelist Bill Pearson.

A lively intellectual community was one factor encouraging creative artists like Hotere and Brickell; another was an audience to buy their works. Some of this came from the private sector, where in the affluent years after World War II wealthy patrons wished to purchase art as conspicuous consumption. The expansion of tertiary education heightened appreciation for modernist works of art. In Brickell's case, handmade pottery became affordable because, until the 1980s, locally made pottery was protected by steep tariffs. Some New Zealanders could afford Crown Lynn tea sets, others chose Barry Brickell mugs. The sale of works by artists like Hotere and Brickell was aided by the emergence of private galleries, such as the New Vision and Barry Lett galleries in Auckland, the Peter McLeavey Gallery in Wellington, and the Bosshard Gallery in Dunedin which held a joint showing of works by Hotere and Brickell in 1977.

Public agencies also gave support. From the Labour government of the 1940s, governments encouraged local creative artists. The Queen Elizabeth II Arts Council was established, and both our artists received patronage from the council. The $15,000 paid to Hotere for his Auckland airport *Godwit/Kuaka* mural came from the Arts Council and the Ministry of Works; and in 1978 the council gave him a scholarship to travel overseas. Brickell was the recipient of Arts Council grants in 1963, 1974 (to build the country's first wood-fired stoneware kiln), 1978 and 1986. He was commissioned by the government to create works for the 1992 Seville Expo and for the National Library. Other organisations offered support. In 1961 Hotere studied at the Central School of Art in London on a New Zealand Art Societies Fellowship, and then he went south to Dunedin in 1969 on a Frances Hodgkins Fellowship from the University of Otago. There was also patronage from public organisations. Hotere was commissioned for a mural at the Founders Theatre in Hamilton in 1973; while Brickell completed murals for the Coromandel and Devonport libraries, and commercial outfits such as Shell BP and Waitaki Refrigerating Company. The emergence of a supportive cultural infrastructure was crucial in allowing creative artists to work full-time in the later twentieth century.

A third major influence in facilitating creative high culture was overcoming the cultural cringe — the assumption that overseas art was always better than local work. Not that Hotere and Brickell were hostile to overseas ideas. Hotere's four years in London and travelling through Europe in the early 1960s were hugely influential. He saw both the historical masters like Goya and Rembrandt and contemporary creators such as the American expressionists. Other young New Zealand artists, such as Bill Culbert, Pat Hanly, Don Peebles and John Drawbridge similarly fed off London's cultural buzz at that time. For Brickell, the Englishman Bernard Leach's *A Potter's Book* and the work of the Japanese master Shōji Hamada were highly influential. Brickell also found inspiration in Pacific masks and New Guinea Sepik pottery. But both men were determined their work would reflect the character of this land. Always a devotee for native plants, and an early explorer of New Zealand clays, Brickell abandoned a career as a teacher in 1961 to buy land at Driving Creek near Coromandel township which had bush remnants and available clay. He built a railway to ferry clay uphill to his potter's wheel. The clay was not the refined imported substance or highly processed material used by outfits such as Crown Lynn. His local clay had lumps and 'warty surprises', to produce what he said were 'Rough, crude but vigorous, solid pots'. He drew on local glazes for his distinctive brown stoneware. Carrying visitors on his railway into the bush was also central to his mission.

Hotere too drew on local materials. One reason he stayed in Dunedin after the Hodgkins Fellowship was because Dunedin was a good source of old used objects. He turned unused window frames or abandoned bits of corrugated iron into works of art. Hotere once came upon the burnt-out hulk of a trawler, which he then transformed into one of his finest works, *Black Phoenix*. He portrayed the power and mythology of local landscapes — whether the beach at Mitimiti where he grew up, or Aramoana Harbour just along from his home from the 1970s at Careys Bay, which was a favourite site for collecting kai moana or spiritual reflection. When Aramoana was threatened by a smelter as one of Robert Muldoon's 'Think Big' projects in the late 1970s, Hotere produced a series of paintings using the word 'Aramoana'.

In both *Black Phoenix* and *Godwit/Kuaka*, Hotere drew on whakataukī (sayings) learnt from his father. In this way Hotere expressed his Māori origins. For many creative artists in the late twentieth century, Māori culture was a major source. In the 1950s Hotere had been one of those Māori artists employed by Gordon Tovey in the Department of Education to teach art drawn from Māori traditions in the schools. They included, besides Hotere, Para Matchitt, John Bevan Ford, Cliff Whiting and Marilynn Webb. Unlike other Māori artists, Hotere did not draw on the distinctive motifs of traditional Māori art, saying, 'I am Maori by birth and upbringing. As far as my works are concerned this is coincidental.' But he undoubtedly received from his father both a sense of a spiritual relationship with the land and a store of ancient whakataukī. It is arguable that Hotere's fascination with blackness has roots in his own tribal background. Te Aupōuri take their name from a famous incident when they escaped under the dark cover of smoke (au) and darkness (pōuri).

It is not easy summing up the emergence of a creative high culture in New Zealand through one object, a memorial post, and two artists, Barry Brickell and Ralph Hotere. But both men had extensive links within the creative community, finding a kinship with the work of others who in turn recognised their contributions. Hotere interacted extensively with poets, such as Bill Manhire, Ian Wedde, Hone Tuwhare, James K. Baxter, and his wife from 1974 to 1986, Cilla McQueen. His friend the novelist, poet and academic Vincent O'Sullivan wrote his biography. Brickell was equally rich in his cultural acquaintances. They included the writers A. R. D. Fairburn, and painters like Robert Ellis, Gordon Brown, Toss Woollaston, Michael Illingworth, Nigel Brown, Tony Fomison, not to forget of course Colin McCahon, and Hotere. Hotere and Brickell were both powerfully original artists, but their concerns and achievements provide one window into several generations of New Zealand creative culture.

89.

The New Zealand AIDS quilt

MUSEUM OF NEW ZEALAND TE PAPA TONGAREWA

It was the best of times; it was the worst of times. For gay New Zealanders the late 1980s and early 1990s were a moment when they could finally enjoy their freedom from arrest and promote gay liberation; but they had also to deal with a terrible disease, AIDS (acquired immune deficiency syndrome), which struck their community in disproportionate ways. This remarkable quilt, consisting of 16 blocks, all but one with eight individual segments, displays both the grief and the pride at this moment in gay history.

Although it seems likely that homosexual people were accepted in Māori society, European settlement brought public hostility. The missionary William Yate who engaged in oral sex and mutual masturbation with local men was a target of strong disapproval, and in 1858, following English law, male homosexual acts became illegal. Eleven years later, sodomy and forcible assault became criminal offences with punishment of between 10 years and life imprisonment. Despite this sanction, there was considerable sexual activity between men, especially in the heavily male communities found in mining areas or on pastoral stations. On Canterbury wool stations homosexual sex was known as 'ram'. There were few prosecutions.

Things changed in the late nineteenth century. In 1893 hard labour and flogging were added as punishments and consent removed as a defence. The trial of Oscar Wilde in Britain in 1895 heightened public fears; and

PHOTOGRAPH BY GARETH WATKINS. THE NZ AIDS MEMORIAL QUILT IS IN THE COLLECTION OF TE PAPA.

in the first half of the twentieth century there were several notorious incidents — including the shooting by the Mayor of Wanganui, Charles Mackay, of D'Arcy Cresswell in 1920 for an attempted blackmail; and the arrest of Norris Davey (later the famous author Frank Sargeson) in 1929. But the police brought few cases and a secret homoerotic world existed in certain city places, such as public baths or wharves. The biggest issue for homosexuals was less police harassment than the sense of moral shame.

In the 1950s and 1960s these 'queer worlds' expanded, and legal punishments were eased — hard labour and flogging ceased. Most men convicted were given three to six months' imprisonment. But cases of police entrapping men in public toilets increased, and there was more use of 'treatments' — electroconvulsive and aversion therapies and hypnotism. Gay men suffered fear of prison and public disgrace. There was an appalling case in 1964 when six youths who had set out to 'bash a queer' in Christchurch's Hagley Park, killed Charles Aberhart, but the jury returned a verdict of not guilty.

The fightback started soon after. In 1967 the New Zealand Homosexual Law Reform Society was established following the UK decision to legalise consenting sexual activity between men. The society pushed for a similar change here. Five years later, gay liberation groups emerged in Auckland, Wellington, Christchurch and Hamilton. Modelled on a United States movement which had been sparked by the raid on the Stonewall Inn in Greenwich Village, gay liberation was a classic countercultural movement. A leader, Lindsay Taylor, said, 'All the people had backgrounds in the women's liberation movement or the anti-war movement or Maoist politics or whatever.' Gay liberation emerged from the universities and was as much about changing attitudes as changing laws. It aimed to replace the shame of being gay with openness and pride; and it built strong links between male homosexuals and lesbian women. There was an emphasis on 'coming out' visibly.

In July 1986, after a sixteen-month campaign which was the most intense public debate in the country since the Springbok tour, homosexual acts for those aged 16 and over were legalised. For gay people that was just the beginning. A drive to outlaw discrimination on the basis of sexual orientation was temporarily defeated but finally achieved in 1993. Other legal protections such as civil union and equal marriage rights were accomplished in 2004 and 2013 respectively.

Sadly, just as these steps were being considered, and even before law reform had been achieved, a new menace arrived. In 1983 scientists announced that AIDS was the result of the human immunodeficiency virus (HIV), and the first New Zealander died. The virus was thought to have spread from chimpanzees and attacked the immune system, making the body vulnerable to infections and cancers, which eventually caused death. It rapidly became clear that homosexual men were disproportionately at risk from the virus which was spread mainly through sexual contact. Several gay men recalled the impact — Peter Wells said, 'Gay men were being made into victims again. It was like watching the gains of the last 15 to 20 years being washed away in an afternoon.' Michael Stevens remembered, 'How could something so wonderful, so powerful, so joyous, and fun as fucking now be the cause of our death?' Sadly, death came. In 1987, 14 New Zealanders died; and from 1989 until 1997 when anti-retroviral drugs were found to suppress the virus, the average mortality was over 50 people a year, most, though not all, gay men.

So, during the years when gays fought for legal rights, they also confronted the epidemic. Some opponents of legalisation even suggested that AIDS was God's penalty for unnatural acts and legalisation would spread the disease. Gay activists replied that the disease could never be fought without open discussion of safe sex. They established a support network and the New Zealand AIDS Foundation.

Making the AIDS quilt was another response by the gay community. The idea of a quilt as a memorial to individual AIDS victims was born in San Francisco in 1978 when a gay activist made a panel in memory of his friend. Others joined in making quilt panels for friends and lovers who had died of AIDS, and within two years the AIDS memorial quilt had been nominated for a Nobel Peace Prize. By 1996 the US quilt was so large that it covered the whole National Mall in Washington. Forty-two countries developed quilt projects. The New Zealand project began in December 1988. Each panel was 1.8m x 900cm, and eight panels were then joined together into blocks 3.6m x 3.6m. There were 15 such blocks, with 120 separate panels, plus one large international panel. Designed initially as a way of paying tribute to each person it featured, the quilt also educated New Zealanders about AIDS and the gay experience. It was taken to 50 high schools and viewed by 800,000 people before being handed over to Te Papa for safe keeping.

The quilt panels display great beauty of design and needlework; but they also repay closer study for their content. Most panels are devoted to one individual, with the sewing and design carried out by mothers or sisters, and in a few cases by lovers and friends. 'Making Mike's quilt,' said Michael Jelicich's mother, 'was a tangible expression of feelings of love, pride and sadness.' Several quilts are memorials to groups. One shows a yellow flower with petals and names 39 men who received touch massage from Darren Horn before they died. The last two petals are for Darren and his partner, also lost to AIDS. Another panel gives the Christian names of nine Waikato people who died, while the huge international block answers the question, 'But who do I know that has died of AIDS?' with 54 names, 42 from overseas. The names include creative artists like Anthony Perkins, Liberace, Rudolf Nureyev and Miles Davis, writers like Bruce Chatwin, and Arthur Ashe the tennis player. Eve van Grafhorst, who died aged 11, has a block of eight panels to herself, paying tribute to her remarkable life as a child afflicted by AIDS through a blood transfusion and who came to New Zealand to escape social discrimination in Australia.

The international block highlights the gay community's international character. Many spent parts of their lives overseas — some in New York or London, more in Sydney, a favourite refuge for gay men facing family and community prejudice back home. The Sydney Harbour Bridge features on several panels. Symbols representing individuals' interests are also revealing. Several speak of the man's love of rugby league football, while many highlight the cultural interests of the gay community. Numerous panels feature musical instruments; others show paint brushes; and cameras are common.

The most striking message of the quilt is the extent to which the battle for visibility and pride remained an ongoing process when these people died. To accept that a man died of AIDS was to accept that he was gay, and many families could not cope with that. There are a number where only a Christian name, or even initials, is given, because families did not want the full name revealed. 'Oh for the day when the fullness of their lives and accomplishments can be proudly proclaimed' writes the maker of the Waikato panel; while Robin Murie's sister explained that she created a panel in his memory, 'To help deal with anger directed at the extended family who weren't there for him on his "coming out", who wasn't there to help and support him through all his AIDS-related conditions, but were all at his funeral.' Others write of the pain of avoiding a person's name — 'what does this say about the status and acceptance of HIV and AIDS in our community' was one comment. Instead we get a panel to 'A son whose name we cannot reveal'.

Rainbows are everywhere on the quilt, the symbol for gay pride born in the United States in the 1970s. But the most powerful message of the quilt is that pride and public acceptance of being gay, while hugely better than in the early twentieth century, still had a way to go in the 1990s. If the beauty of the New Zealand AIDS quilt leads New Zealanders to a more tolerant future, then it will have accomplished its most important role.

90.

Mike Smith's chainsaw

MUSEUM OF NEW ZEALAND TE PAPA TONGAREWA

On the evening of 28 October 1994, the 159th anniversary of the signing of the Declaration of Independence by northern Māori, a 36-year-old man from Ngāti Kahu and Ngāpuhi took this recently repaired chainsaw, borrowed from a friend, and proceeded to the top of One Tree Hill in Auckland. He politely warned bystanders to beware of a falling tree, cut the security fence and began sawing the trunk of the century-old pine tree. Neighbours saw the action and called the police who came with dogs and firearms. Smith immediately stopped the engine, leaving the saw in the tree, and upon arrest stated that he 'wanted people to understand the outrage the Maori people were feeling . . . We're feeling outrage not just about one tree but about all our forests, fishing and land.'

While Mike Smith acted alone, he was correct that there were widespread feelings of outrage among Māori. In the years since the 1975 Land March important steps had been taken — most significantly the establishment in that year of the Waitangi Tribunal and its extension a decade later to examine historical Treaty infringements. But there were continuing grievances about failures to honour the Treaty. In Auckland on 5 January 1977 an occupation had begun of Takaparawhau (Bastion Point), traditional Ngāti Whātua land which the Crown had taken for defence purposes. Robert Muldoon's government intended to sell the land for housing. After 506 days, on 25 May 1978, more than 600 police and army forcibly removed the protesters; 222 were arrested.

The same year, Eva Rickard led an occupation of the Raglan golf course, which had been taken for defence purposes and not returned. From the early 1970s there were annual protests at Waitangi on 6 February, initially with the claim that the Treaty was a fraud, then with the plea that the Treaty promises be honoured. These protests peaked in 1984 with a hīkoi from Ngāruawāhia to Waitangi led by Eva Rickard. At the 1990 commemorations a young Māori woman threw a wet T-shirt at the Queen, and the Bishop of Aotearoa, Whakahuihui Vercoe, spoke passionately about the Crown's failure to honour the Treaty.

Following these two decades of unrest, protest was reignited by the proposal of Jim Bolger's National government to impose a 'fiscal cap' on Waitangi settlements. National had campaigned in 1990 on completing all Waitangi settlements within 10 years, and they emphasised the Crown's fiscal constraints. By 1994, two major settlements were close to completion — with Waikato-Tainui and Ngāi Tahu. Neither would sign unless a total settlement figure had been determined. A fiscal cap of $1 billion was proposed. While the Crown did consult, the fiscal cap was non-negotiable and Māori saw it as a fundamental attack on tino rangatiratanga — a unilateral interpretation of the Tribunal's rulings. Protests included a thousand-strong hui at Hīrangi marae in Tūrangi. Mike Smith's action came at the start of the unrest, when the 'fiscal envelope' was known but not announced.

Another influence on Mike Smith was the Rogernomics reforms of the 1980s. The selling off or corporatisation of state assets such as forestry disproportionately affected Māori employment. Further, in 1991, Ruth Richardson's 'mother of all budgets' slashed benefits. She believed people should be squeezed financially into seeking a job. But jobs were not available; and the cuts produced the most serious growth in inequality since the Second World War. Māori were among those paying the price. As Jane Kelsey commented later, the annual cost savings of the benefit cuts, $1.3 billion, were larger than the proposed payout for all historic Treaty breaches.

Treaty breaches, benefits cuts and a fiscal cap — no wonder Mike Smith felt outraged. There was an irony in his choice of weapon, a chainsaw. One of the enduring effects of the transfer of Māori land to Pākehā ownership was the cutting of the bush to provide pastures for dairy cows and sheep. Historically, the tool used in this process was the axe, not the chainsaw. While chainsaws originated in eighteenth-century Europe where a small hand-turned saw with linked chains was used for bone surgery, it was not until the 1950s that aluminium, forged steel and the development of a two-stroke engine made possible light portable wood saws. They were quickly adopted in New Zealand, for clearing bush and cutting pine trees.

CHAINSAW, CIRCA 1985, ITALY, OLEO-MAC. PURCHASED 2009. TE PAPA (GH012615)

Mike Smith's target was also carefully considered. The pine tree on One Tree Hill was planted by the British on an ancient volcanic cone with a rich Māori history. Maungakiekie, mountain of the kiekie vine, had erupted 60,000 years ago and was the second largest volcano after Rangitoto in Auckland. Following Māori arrival it became an important pā. Rising 183 metres, it offered superb views of any approaching enemy and was positioned between the Waitematā and Manukau harbours. Its volcanic soil fed productive kūmara gardens. By 1600, it had been occupied by Māori — at first by Ngāti Awa en route from Northland to the Bay of Plenty, and soon after by Te Wai-o-Hua. In the eighteenth century their famous chief, Kiwi Tāmaki, established a complex fortification on its slopes, probably the world's largest earth fort at that time. The gardens covered 1000 hectares, supporting over 4000 inhabitants. About 1740, Kiwi Tāmaki was defeated at Parutoa on the Manukau by Ngāti Whātua forces who had come south and became Auckland's dominant tribe. Led by the chief Tuperiri, Ngāti Whātua took over Maungakiekie and established Hikurangi pā on its slopes. So for several hundred years the hill was an important Māori site.

The tree, too, had a history. Oral tradition records that when Korokī was born to a chief in the sixteenth century, his umbilical cord was buried beneath a tōtara seedling on the summit. Eventually tōtara flourished and for a time the hill was known as 'te tōtara i ahua' — the tōtara that stands alone. At some point the tōtara died and was replaced by a pōhutukawa. About 1852, after the hill passed into Pākehā hands under the English name of One Tree Hill, some Pākehā workmen allegedly chopped down the tree, perhaps for firewood, perhaps as an act of vandalism when a parcel of meat failed to arrive. John Logan Campbell, who with his fellow Scot William Brown had become the town's first trading merchants, purchased land around the hill in 1853, and planted native trees and radiata pines on the summit in the 1870s. But the natives died and only two pines took root. One was cut down in 1960, leaving a lone pine to fulfil the English name. The pine, introduced by British settlers, must have seemed a perfect symbol of the takeover of a sacred Māori place. When Eva Rickard supported Smith the day after his action she said, 'the pine tree was a symbol of Maori oppression because it was planted by Pakehas to replace a 300-year-old totara felled last century'.

There was another layer of meaning. Beside the pine tree a soaring stone obelisk had been unveiled in 1948. This was another initiative of John Logan Campbell. When he died in 1912 he was buried on its summit. In his will he gave £5000 for an obelisk which might express his admiration for Māori of the past. At that time many Europeans considered Māori 'a dying race', and in a paternalistic way Campbell wanted their historic feats acknowledged. His ambition took time to achieve fruition. In 1933 Richard Gross was commissioned to prepare a model to be cast in bronze 'of a Maori chief of heroic mould'. The project became a 1940 centennial project, but delays in construction, and Māori views that a dedication was not possible during wartime 'while blood is being spilt', deferred the unveiling until 1948. While some Māori supported the monument, others such as Apirana Ngata were cool on the idea. After all, the obelisk was a European monument, and the motivation was paternalistic. The monument did not acknowledge the hill's history or the Māori communities who had lived there. It was easily regarded as another act of colonisation.

Mike Smith's felling of the tree sparked immediate reactions. He anticipated an outcry focused on the treatment of Māori. Instead this was drowned out by 'outpouring of grief towards the tree'. For Pākehā, One Tree Hill had 'come to represent the city as a whole', as one correspondent wrote; and this was recently highlighted when the Irish band U2 recorded a much-played song 'One Tree Hill' after the accidental death of a Kiwi friend in 1986. Floral tributes appeared at its base. A few newspaper letters revealed some Pākehā with different perspectives: Helen Yensen and Tim McCreanor thought Mike Smith's act offered a wonderful opportunity to replace a pine with a tōtara 'as a symbol that the pain and suffering of Maori people as a result of more than 150 years of colonial oppression is beginning to be heard'. Activist Māori expressed their sympathy for Smith's action. At Pipitea marae in Wellington a hui gave full support for his act and some argued that other colonial monuments should come down — the obelisk, and also statues to Queen Victoria. Long before the international 'Black Lives Matter' movement, Māori were inspired by Mike Smith's act to call for attacks on symbols of colonialism.

In the long term, despite efforts to bind the tree, another attack in 2000, reportedly by Mike Smith's whānau, spelled its end. The pine was felled and in 2016 nine tōtara and pōhutukawa were planted. The chainsaw was offered for sale on Trade Me in 2007, and purchased by Te Papa where it now resides. As for Mike Smith, he has become a powerful voice for environmentalism, working closely with Greenpeace. Protecting trees, albeit native trees, is now part of his mission.

91.

Peter Blake's red socks

NEW ZEALAND MARITIME MUSEUM | HUI TE ANANUI A TANGAROA

These are not Sir Peter Blake's own red socks, the ones he wore on the successful America's Cup challenge at San Diego in April and May 1995. Those socks were placed on his coffin after his tragic shooting by pirates on the Amazon in 2001, and buried with the man at Emsworth in Hampshire on England's south coast. These socks were one pair of thousands sold to New Zealanders in 1995 and 2000 in support of the country's America's Cup efforts.

The origin of the red socks campaign is well known. Peter Blake liked a good-luck charm on boat races, and during the 1988 Whitbread Round the World race, when Blake's *Steinlager 2* was in a fierce battle with Grant Dalton's *Fisher & Paykel*, he wore fluorescent green socks at crucial moments. The crew accepted that their bringing good luck overruled their smell. One afternoon in December 1994 Peter Blake and his English-born wife Pippa were in a boat on San Diego Harbour with a French photographer, Christian Février. They were greeting the America's Cup yachts after training. Christian was wearing dark pink socks. Blake teased Christian about them, but Pippa liked them and determined to get some for Peter for Christmas. She went to a San Diego store, could only find bright red golfing socks as her Christmas present. Blake wore the red socks in Team New Zealand's first race. The team won; and Blake wore them subsequently. But in one race against *oneAustralia* in the challenger final, Blake developed tendonitis and pulled out. The boat lost; and Peter

Montgomery, television commentator, noted that 'the man with the lucky red socks was missing'. The crew also noticed. They too began to wear red socks and won again.

Back home as the country celebrated Team New Zealand's advance to the America's Cup final, a young TVNZ staff member suggested that the country wear red socks in support. Full-page ads appeared in newspapers, 'Sock it to 'em'. Socks were sold for $10 from Lotto outlets and supermarkets with the proceeds injecting money into the team's empty coffers. Wearing red socks became a national expression of support, worn by everyone from the Governor-General and Prime Minister to the elephant in Auckland's zoo. The *New Zealand Herald* had a cartoon of spectators on a launch and one comments, '3 days in the same socks . . . Phew!' — an unfair cut because Pippa washed the socks each night!

The adoption of red as a colour for New Zealand sporting was unusual. From the All Blacks to Olympic uniforms, black was the nation's sporting colour, and indeed *NZL32*, the winning boat at San Diego in 1995, was called *Black Magic*. Yet in other ways the America's Cup success had traditional elements. Peter Blake was very much the conventional Kiwi male hero. By comparison with his 'loud-mouthed' competitor at San Diego, Dennis Conner, Blake was seen as modest, 'an ordinary do-it-yourself Kiwi' who had taught himself to sail in a P-Class yacht aged eight. Success had not come easy. He had competed five Whitbread Round the World races before finally in *Steinlager 2* pipping his compatriot Dalton's *Fisher & Paykel* in an epic race in 1989–90. There was huge public interest as the two Kiwi-led yachts battled down the Northland coast to Auckland Harbour in the halfway leg. Over 300,000 people and 9000 boats saw the fleet depart Auckland, the day after the Commonwealth Games there ended. When he reached the pinnacle of success at San Diego, comparisons with Edmund Hillary flowed. Blake was described as scaling 'the Everest of sailing'.

Further, like traditional Kiwi male heroes, Peter Blake emphasised that he was only one of a team. Pippa recalled, 'Teamwork was what Peter really loved.' Loyalty and mateship were core values. And his teams were traditional in character. At a time when the country, Auckland especially, was becoming increasingly multicultural, Blake's teams were white male Kiwis.

As for yachting, the sport had roots in the country's European traditions. Early European settlement was close to harbours, and from the 1840s every major New Zealand port held sailing regattas, often on their anniversary days. Initially the races were between working sailing boats, but from the 1870s decked craft were used. Specialist yacht builders emerged, especially in Auckland, where Robert Logan's and

GIFTED BY PHILIPPA WERE. NEW ZEALAND MARITIME MUSEUM HUI TE ANANUI A TANGAROA (2009.213.14)

Charles Bailey's firms became prominent. They used kauri which was strong and durable. From this tradition, internationally successful sailors emerged. At Melbourne in 1956, Tokyo in 1964 and Los Angeles in 1984, New Zealand yachtsmen won Olympic gold medals. Yachting became recognised as an area of sporting success. Blake's *Ceramco* famously won the Sydney to Hobart race; Blake's *ENZA* won the Jules Verne Trophy and both Peter Blake and Grant Dalton won the Whitbread. Public interest rose with the America's Cup challenges of merchant banker Michael Fay in Fremantle in 1987 and San Diego in 1992. In both cases the New Zealand boats reached the finals of the challenger series.

So Peter Blake's red socks success at San Diego in 1995 emerged from powerful traditions. But it also implied some major changes. For a start, Peter Blake was an Aucklander whose success reflected that city's yachting culture. The America's Cup represented the increasing dominance of Auckland in New Zealand life. In population terms, Auckland had become by far the largest urban area. By 1996 its population had rocketed to one million. New Zealand was no longer a country of four main centres; but of two centres, Wellington and Christchurch, and one large international metropolis. Auckland's dominance had other facets. In 1989 Television New Zealand followed the banks and large businesses in relocating to Auckland, from Wellington; and major sporting events took place in the 'City of Sails' — the final of the 1987 Rugby World Cup, and the 1990 Commonwealth Games.

Reflecting its Auckland base, Peter Blake's America's Cup show presented a new image of New Zealand culture and economy. Instead of the 'number 8 fencing wire' society with strong rural roots and a land-based export economy, the America's Cup presented an image of sophisticated technological know-how. For the 1987 Fremantle challenge New Zealand's boats were not made of kauri but were 'Plastic Fantastics'. For the 1995 challenge Peter Blake called together a group of technicians and scientists with impressive academic credentials. There were computer flow analysts and wind tunnel technicians. A lecturer in management studies, Peter Mazany, drew up a mission and vision statement. New Zealand was seen, not just as a producer of milk and sheep meat, but of sophisticated luxury boats. Team New Zealand's success, editorialised the *New Zealand Herald*, 'underlines the fact that New Zealand has become a global centre of competitive excellence'. Such expertise expressed economic ambitions. The mission stated: 'This challenge will become synonymous with, and in significant ways, could be a catalyst in New Zealand's fight back to economic and social well being.' The spectre of Britain's entry into Europe and the 1980s ambitions to reshape the New Zealand economy are clearly visible.

Peter Blake's career, and the America's Cup challenge, also replaced the amateur sporting tradition with a big business alliance. In his Whitbread days, Blake turned to first Tom Clark, the leading spirit behind Crown Lynn and Ceramco; then to Doug Myers, the managing director of Lion Breweries. Myers' largesse was rewarded in yacht names, *Lion New Zealand* followed by several *Steinlagers*. When *Steinlager 2* was battling *Fisher & Paykel* an observer might have imagined a sharemarket competition rather than a yacht race. Our pair of red socks proudly displays the sponsors' names — Steinlager, Television One, Telecom, Lotto and Toyota. The red socks challenge reflected the new Auckland-based economy — a set of thriving business entrepreneurs.

Yachting was not the only sport witnessing professionalism. In the 1980s both New Zealand cricketers and athletes began competing for money; and in 1995, the year of the America's Cup success, a professional rugby league team, the Warriors, entered a trans-Tasman competition. The next year rugby union also became professional and before long football and basketball followed. Sponsorship, privileged membership, advertising on team jerseys appeared and amateur traditions disappeared.

Much can be read into Peter Blake's red socks of the economic, cultural and sporting revolutions of the 1990s. A world of Auckland-based urban expertise and professional sport had replaced the old rural mythology in the country's sporting triumphs. But the red socks were also a tribute to Peter Blake's inspiring leadership. In 2000 he had the vision to turn his back on his sporting and corporate success and lead a larger and more significant campaign to protect the environment of the world's oceans and waterways. After he was cruelly killed in December 2001, and buried at a churchyard close to his home, visitors to the site quite often left behind a pair of red socks — a suitable tribute to a man widely regarded as a New Zealand hero.

92.

Dorene Robinson's Swanndri

PUKE ARIKI

This bush shirt or Swanndri, in striking orange and brown tartan, and displaying some 20 badges, belonged to a New Plymouth woman. It is a favourite item at Puke Ariki Museum. The 'Swanni' was worn by Dorene Robinson on many tramps, usually with the New Plymouth Women's Institute, in the 1990s. This distinctive shirt points to the transformation in the meaning of the great outdoors as New Zealand became a predominantly urban society.

Throughout the nineteenth century, New Zealand had been mostly a country of farms and small towns. But in 1911 the census announced that there were more New Zealanders living in 'urban places' (locations of over 2500 people) than in rural areas. By 1926, urban numbers were two thirds of the population; by 1971, over four fifths.

City life brought excitement such as movies, cafes and big sporting fixtures. But it also brought noise and pollution; and the nation's rural origins were still treasured. In the last half of the century the great national hero was the pre-eminent outdoorsman and mountaineer Edmund Hillary, while Barry Crump's tales of deer hunting were bestsellers. The pioneering heritage and the allure of the bush and mountains for city people explain Dorene Robinson's Swanndri.

The Swanndri was invented, and trademarked in 1913, by an English-born tailor, William Broome, 'a tall, dapper gentleman', with a clothing business in New Plymouth. Made of a heavy woollen fabric, and often with a dark tartan design, the bush shirt was dipped in a secret mixture that made it warm and showerproof. Broome claimed the water ran off the garment as off a swan's back — hence the name. In 1955, 15 years after Broome's death, the family sold the business to John McKendrick, a tailor with a clothing factory in Waitara. He added an olive-green colour option. The market was primarily Taranaki dairy farmers needing protection from the area's showers and chilly westerlies. The Swanni was also used by hunters with a desire to keep warm and dry. In 1975 Alliance Textiles bought the trademark and in turn sold it in 1994 to Swanndri New Zealand who made the items in Asia. By then the market had broadened. Still a favourite among farmers and hunters, the Swanni attracted urban consumers. The colours became flamboyant; and city people began wearing them on bush walks or tramps.

Dorene Robinson was a city-dweller, married to a policeman, and she wore the garment on tramps in the 1990s. By then New Zealand produced a remarkable number of sophisticated items for trampers — from packs and sleeping bags to merino tops and polypropylene jackets. But Swannies continued to appeal to city-dwellers. Arguably this was because, as cities grew, the outdoors took on a romantic quality as the site of the real 'Kiwi' identity. Swannies evoked a homespun rural origin, a cowshed heritage.

The interplay of city and country also lies behind the Women's Institute which organised Dorene's tramps. The institutes began in Hawke's Bay in 1921 to provide community life for rural women. Crafts and adult education were emphasised, to reduce the attractions of city life so that 'New Zealanders may continue to be a race of contented country dwellers'. In 1952 the word 'Country' was added to the name and a rule confined institutes to places of under 4000. Despite this urban clubs were formed by women keen to retain rural links. By 1990, urban members outnumbered their country cousins. Dorene's New Plymouth club organised many day walks and tramps in the surrounding districts. The walkers normally started their walks with 'banana time' when they all ate a banana as an energy boost.

In putting on her Swanni and heading into the bush, Dorene followed a common recreation for urbanites, but it had not always been the case. For much of the nineteenth century Pākehā New Zealanders regarded the bush as an obstacle to progress that needed to be burnt and cleared for farming. The bush evoked fear and discomfort. Signs of change came at the end of the century when overseas tourists praised the sublime character of the alpine regions, and some genteel city-dwellers, such as William Pember Reeves, expressed nostalgia about the passing of the bush. A New Zealand Alpine Club was set up for gentlemen climbers;

and national parks were established around the central North Island volcanoes. Huts and accommodation were erected on their flanks. But on the whole, those who went into such areas for recreation were regarded as unusual — harmless but misguided.

From about 1920 things changed. New Zealand became urbanised, and those wanting to reconnect with their pioneering heritage put on their Swannies and explored the mountains and bush. Several developments facilitated this. One was the coming of automobiles, and roads to back-country areas. Another was the emergence of tramping clubs. The first was the Tararua Tramping Club in 1919 which attracted middle-aged city people, and opened up tramping to women. Initially men wore worsted suits and waistcoats and women riding breeches, but by the 1950s both wore shorts and Swannies. By 1965, there were 99 such clubs. Dorene's involvement with several is signalled by two badges on her Swanndri — the New Plymouth Tramping Club, which expanded facilities on Mt Taranaki, and the North Taranaki CWI Tramping Club. Another badge recalls her devotion to the Royal Forest and Bird Protection Society established in 1923. The clubs built tracks and huts. The back country was further opened up by the Forest Service which, to assist deer cullers, built 644 huts, many in the distinctive orange colour, and 4000 kilometres of track. By the 1990s when Dorene was tramping, there were 1400 back-country huts, most managed by the Department of Conservation (DOC).

Another development was the creation of new national parks in addition to Tongariro and Egmont/Taranaki. A National Park Act in 1952 saw their number increase to 14 by the 1990s, especially along the South Island's west coast. Facilities for trampers in the form of huts, tracks and bridges followed. Dorene's badges reflect this. There are badges to Doubtful Sound, Lake Te Anau, Chateau Tongariro, Queenstown, Franz Josef Glacier, Fox Glacier, Lake Manapōuri, and Hollyford, all of which sit in national parks. In addition, from 1954 forest parks were set up, and by the 1990s there were 19. In the 2000s, under Helen Clark's government, conservation parks opened up alpine areas of the South Island.

The creation of DOC in 1987 led to the Great Walks, with comfortable lodges, catering especially for walkers from overseas. There were private initiatives — farm walks, and in 2011 the opening of a tramping route, Te Araroa, from Cape Rēinga to Bluff. Dorene anticipated this by joining 16 others from the New Plymouth YWCA in walking from Bluff to Cape Rēinga in 1990. Walking on roads, rather than in bush, and in all weathers, the group took 84 days on the journey. Dorene went through four pairs of footwear! There is a Y badge on her Swanni which she wore throughout.

The rise of long-distance tramping in the great outdoors was driven initially by Kiwis' desire to escape the city, for a sense of unspoilt New Zealand. But from the 1990s the experience developed an international appeal. Cheaper airfares and promotion by tourism authorities and guidebooks attracted younger 'hikers' from Australia and the northern hemisphere. Certain tracks such as the Tongariro Crossing became crowded with overseas visitors. There were as many visitors as locals on the Great Walks in 2019, and indeed on the more challenging walks such as the Kepler and Routeburn a majority were from overseas. Yet it was New Zealanders who had opened up the facilities and equipment for tramping. In 2019 about a quarter of all people in the country aged 18 or over had tramped or day-walked in the previous year; and if the numbers were greater among Pākehā than among Māori, Asian or Pasifika people, bush walking had become a national characteristic. New Zealanders aspired to leave the city for a direct encounter with the bush and the mountains. Dorene Robinson's Swanni expresses that story.

PA2016.019 JACKET, SWANNDRI, DORENE ROBINSON, COLLECTION OF PUKE ARIKI, NEW PLYMOUTH

93.

Helen Clark's trousers

AUCKLAND WAR MEMORIAL MUSEUM

It was late summer, 25 February 2002. Queen Elizabeth II and the Duke of Edinburgh were on a Commonwealth tour to celebrate the Queen's golden jubilee. They were on a five-day visit to New Zealand and the high point was a state banquet in Parliament. The hostess was the Prime Minister Helen Clark. The Queen wore a crown, and a white lace pearl-embroidered evening dress. The Prime Minister wore an elegant silk jacket and these black trousers. Immediately after the banquet the New Zealand media did not comment on the Prime Minister's clothes. They focused on other issues — the fact that Helen Clark had almost broken protocol when she began taking her seat before the Queen, that she had not been in the country to greet the monarch when she arrived three days earlier, and was reported telling the London School of Economics that it was 'absurd' for New Zealand to be ruled by a head of state 20,000 kilometres away.

Then the British press got into the act. It was claimed that some in the press contingent 'believed that Helen Clark was inappropriately dressed, at the state banquet'. With the British giving a lead, the locals chimed in. Samantha Browne of Thorndon wrote to the *Herald*, 'I was horrified to see our Prime Minister wearing pants'; Jackie McCabe was grateful that 'Mabel Howard and her bloomers are a thing of Labour's past'; and P. K. Ellwood was not surprised at Clark's wardrobe — 'Is it not well known that Miss Clark wears the pants?'. It was mostly light-hearted stuff but the ACT MP Stephen Franks took the issue more seriously: 'The women were glittering, the men bemedalled. But the women at my table, and others, were angrily focussed on one woman who was dressed casually — the host, the Prime Minister . . . If the Prime Minister was using the state dinner just to make a political point of her republicanism, her behaviour was shameful.'

There is no evidence that Helen Clark was making a point about republicanism; and it is revealing that the talking point was the trousers not republicanism. Admittedly, compared with the extraordinary scenes on the Queen's first visit in 1953, New Zealanders in 2002 showed little passion about royalty in their midst. Only 1000 people turned up to see her visit St Paul's Cathedral in Thorndon. This was understandable: 2002 was the Queen's tenth visit and royalty was no novelty. In many ways New Zealand had turned away from the United Kingdom. Entry into the EEC reduced Britain's share of our exports by value from two thirds in 1953 to under 5 per cent in 2002. Immigrants were increasingly from China and India, not the old home country. We no longer listened to BBC news. Further, the royal family had experienced some tawdry days since 1953. But there was no great push for republicanism. A T-shirt declaring New Zealand a 'monarchy-free zone' was seen in Thorndon but no large populist movement; 58 per cent believed monarchy had little relevance to their lives, but the same number rejected republicanism. To the question whether the tour advanced republicanism, Helen Clark responded, 'the short answer is "no". The long answer is there's very little public interest in the subject, one way or another.' Not that the British link was irrelevant. The Prime Minister had returned home with an assurance from Tony Blair that the working holiday scheme for young New Zealanders in Britain would remain.

If trousers, not republicanism, were the talking point in 2002, it is worth asking why. Historically, trousers were associated with men, and it was women with feminist aspirations who considered wearing them. In the United States, nineteenth-century feminists promoted 'bloomers', loose trousers gathered at the ankle, but they never became popular. Even an energetic woman like Freda Du Faur became the first woman to climb Aoraki/Mt Cook in 1910 wearing a skirt. During World War II, women in men's roles, either civilian or military, did wear trousers; but not until a feminist movement re-emerged from the early 1970s were trousers once again commonly worn by women.

It seems likely that Helen Clark was aware of the feminist implications of her clothing, for her political career reflected a major change in the country's sexual politics. Although New Zealand prided itself upon leading the world towards women's suffrage, the challenge to male dominance was not maintained. After the 1893 suffrage act, no woman sat in Parliament until 1931, and from then until 1968, there were 11 women MPs, but 950 men. Born on a Waikato farm as the oldest of four daughters, Helen Clark went to a single-sex school, Epsom Girls' Grammar. When she arrived at Auckland University in 1968 the counterculture of alternative politics was in full flood. She recalled getting 'swept along with all of that', becoming active in the Auckland committee for the mobilisation against the Vietnam War; an early member of HART when it was set up in 1970 to oppose South African sporting contact; and an enthusiast for Norman Kirk's anti-nuclear stance. But her major focus became the party rather than identity politics. Joining the Labour Party in 1971, she threw herself into helping at election campaigns, membership drives and preparing policy. The back room, not the streets, took her energy.

Those years also saw a second women's movement. Baby-boomer women held women's conventions, campaigned for equal pay and more childcare, demanded improved access to abortion, and raised awareness about domestic violence. One issue raised was women's lowly representations in positions of power. The Women's Electoral Lobby (WEL) was set up in 1975 to improve the representation of women in Parliament and on public boards, quickly attracting 2000 members. Among those active in WEL were two National Party women who achieved major breakthroughs — Ruth Richardson, the first woman Minister of Finance, gained notoriety in the 1990s by cutting benefits and presenting the 'mother of all budgets'; and Jenny Shipley, like Helen Clark one of four daughters, who succeeded Jim Bolger in 1997 as the country's first woman Prime Minister.

Helen Clark was fully aware of the women's movement but was not active in it and not involved in WEL. She threw herself into the Labour Party where she found support from other women, such as Margaret Shields, Margaret Wilson and Ruth Dyson. She lived for several years with Cath and Judith Tizard. Within the party, women faced some derision — Norman Kirk had opposed a stronger women's council on the grounds that 'he had a wife to advise him'. So Clark recognised that her selection for the winnable seat of Mt Albert in 1981 expressed 'a decade of feminism . . . starting to have its effect'. Arriving at Parliament, one of only eight woman MPs, she faced 'some really entrenched dislike of women. You were dealing with old-style Labour men who just didn't see you . . . the male MPs were used to women playing a subordinate and service role in their lives.' It was not just Labour men. The government was lorded over by Robert Muldoon whose style was aggressively macho. Clark remembered a generational clash, but even more a gender clash.

TRISH GREGORY TROUSERS WORN BY HELEN CLARK AT STATE BANQUET FOR QUEEN ELIZABETH II, 2002. AUCKLAND WAR MEMORIAL MUSEUM TĀMAKI PAENGA HIRA. 2010.1.16.

It was a male workplace where returned servicemen were prominent on both sides of Parliament. Their recreations — billiards, alcohol and poker — were not hers.

She discovered things were different for women politicians. Being unmarried or childless was never an issue for male politicians — it had not stopped Mickey Savage becoming the most loved leader in New Zealand history. Despite being happily married to Peter Davis since 1981, Helen Clark was constantly suspected of being lesbian. When she challenged Mike Moore successfully for the Labour Party leadership in 1993, her Mt Albert house was picketed with 'Mike vs the Dyke' placards; and when she challenged Jenny Shipley in the 1999 election, Shipley called her 'the spinster of New Zealand politics'.

Clark was also suspect because of her 'mannish' voice and 'unfeminine' hairstyle. This became an issue as Clark's position rose. In the tumultuous second term of Lange's Labour government she avoided the internecine war to become a successful minister. Then in 1993 when she succeeded Mike Moore as leader, her appearance and image became significant. She responded. The Tizards assisted her with her hairstyle; and Brian Edwards and Judy Callingham gave instructions on media presentation.

From 1993 Helen Clark gave greater attention to her clothes and paid visits, two to four times a year, to an Auckland designer, Jane Daniels. British-born Daniels grew up in Auckland, becoming interested in fashion design and winning a competition to design the suits for medal bearers at the Christchurch Commonwealth Games in 1974. After study in London she started her own label in 1986. Her clothes were elegantly styled using fine materials, and expressing urban sophistication. Daniels joined others, mostly in Auckland, making a name for fashion design. Until World War II, most fashionable clothes were imported; but from the 1970s, creative designers found a market with the affluent urban middle class. This development climaxed in 1999 when Karen Walker, NOM*d, WORLD and Zambesi showed to widespread plaudits at the London Fashion Week. In patronising a local designer Helen Clark was supporting local creativity — she was also Minister for Arts, Culture and Heritage.

That was only part of the meaning of her clothes. The state banquet outfit was stylish and elegant, fitting an important public occasion. The jacket, designed by Jane Daniels, was in pure silk and laced with gold sequins. It was definitely not 'down market'. But the trousers, manufactured, not custom-made, by another innovative local designer, Trish Gregory, did carry a quiet message that a woman was now wearing the trousers and that women had achieved significant positions in New Zealand political life. In 2002 the eight women MPs of 1981 had become 35. The Governor-General was Sylvia Cartwright, the second woman to fill the position; the Chief Justice was Sean Elias, the first woman in that role; Margaret Wilson was Attorney-General and soon to become the first female Speaker of the House. The top four constitutional positions were held by women and Theresa Gattung headed the country's largest business.

Before she retired, Helen Clark could claim many successes — the introduction with Michael Cullen of KiwiSaver, the Superannuation Fund, a Working for Families scheme to assist poorer New Zealanders, civil union for gay people, the acceptance of prostitution, the assertion of an independent foreign policy on Iraq, and leading Labour to three electoral victories despite the wounds caused to her party by Rogernomics. But another contribution was signalling that women belonged at the centre of political power. Helen Clark's clothes at the 2002 banquet subtly carried that message. Twenty years later, it was not worthy of comment when once again women occupied the positions of Prime Minister, Leader of the Opposition and Governor-General.

94.

King Théoden's armour

WĒTĀ CAVE

You arrive at an unprepossessing house in suburban Miramar and are greeted by a guide with an American accent. You enter a small room and your eyes are drawn to this extraordinary suit of armour. The breastplate, the arm protectors and the fierce helmet are made from brass with delicate acid-etched designs and covered in handsome red leather. Beneath is a skirt of leaf mail with some scales decorated in horse heads. Even the inside of the breastplate, seen only by the wearer, is lined with embossed leather. There is a long sharp sword with an intricately designed handle ready for action. The outfit weighs 22 kilograms. What is this masterwork of medieval Europe doing in Wellington, New Zealand?

It turns out this is no ancient heirloom brought across the seas. It was made in Wellington by Wētā Workshop in 1998 as a prop to be worn by Bernard Hill playing King Théoden in two films of Peter Jackson's *Lord of the Rings* (*LOTR*) trilogy: *The Two Towers* (2002) and *The Return of the King* (2003). King Théoden is King of the Rohan people, who after being saved by Gandalf from the influence of the wicked wizard Saruman, regains youthful energy and leads his people successfully against Saruman's army of orcs in the battle of Helm's Deep. In the last film Théoden dies heroically in the arms of his niece before the forces of the evil Sauron are defeated, and the magic ring, the source of Sauron's power, falls with Gollum into Mt Doom. Good triumphs over evil, with Théoden playing a starring role.

The question remains — does this armour and its story have anything to do with the history of New Zealand? True, the film was directed by a New Zealander, Peter Jackson, and the armour was made in Wellington by Richard Taylor's Wētā Workshop. The film featured New Zealand landscape — King Théoden's Rohan capital, Edoras, lies amid the snowy mountains of Canterbury's Rangitātā Valley; the battle of Helm's Deep was filmed around Twizel, further south; while the final battle takes place on the Rangipō Desert in the central North Island. But the plot was based on books by an English Anglo-Saxon scholar, J. R. R. Tolkien, and the culture is straight out of medieval Europe. Théoden's armour could have been worn by Crusaders in the thirteenth century. Further, the film was produced and funded by a Hollywood studio, New Line, with leading actors from the USA or the UK. Yet the films are of real significance to this country.

To understand their impact, we need a brief history of New Zealand film in the years since Rudall Hayward's pioneering efforts. Hayward had few immediate successors, and in the 1950s and 1960s, although there were over twice as many cinemas per head as in the USA, New Zealanders watched almost exclusively Hollywood's latest plus a few British comedies. There were only three local feature films, and none very successful. In 1977 Roger Donaldson's *Sleeping Dogs* was the first 35mm feature produced here. To hear New Zealand accents on the big screen was a shock. The film was a huge success. Within a year the New Zealand Film Commission was established to support feature film. A tax shelter was set up to encourage investment in movies. The result was a relative flood of local features — 19 from 1978 to 1983; and 14 in 1984 alone. Then the tax concession was closed and the output diminished.

Many local film-makers of the 1980s and 1990s explored New Zealand environment and culture. Some, such as *Goodbye Pork Pie*, *Smash Palace*, Vincent Ward's *Vigil* and Jane Campion's *The Piano*, drew on the distinctive landscape. A number, including *Pork Pie*, *Smash Palace* and the animation *Footrot Flats*, evoked Kiwi male culture. Others drew on Māori experience such as Geoff Murphy's tale of the New Zealand Wars, *Utu*, and several features by Māori directors Barry Barclay, Merata Mita, and Lee Tamahori with his commercially successful *Once Were Warriors*. Women directors explored the lives of significant local women — Jane Campion's *An Angel at My Table* recounted Janet Frame's early life; and Gaylene Preston's *Bread and Roses* told the story of the labour activist Sonja Davies.

But such cultural nationalism failed to attract international audiences. Some directors moved to Hollywood; others used the landscape as a backdrop for overseas stories such as the television series *Xena: Warrior Princess* and the feature film *The Last Samurai* featuring Tom Cruise with Mt Taranaki playing Mt Fuji.

In this context, where did Peter Jackson, director of *The Lord of the Rings*, sit? His approach certainly drew on distinctive New Zealand elements. His first film, *Bad Taste*, made part-time on weekends while Jackson worked as a photo-engraver for the *Evening Post* newspaper, emerged from do-it-yourself traditions. Determined not to follow the flight to Hollywood, Jackson made films in his home town of Wellington. When he desired technical assistance he set up his own facilities there — his own studio, production house and technical support. He wanted to show that 'you can actually stay in your own country and make really good top-quality films'.

However, Jackson's work was pitched at international audiences. His early films, *Bad Taste* (1988), *Meet the Feebles* (1989) and *Braindead* (1992), were a combination of horror and comedy — 'splatstick', with nothing local about the content; and while his next feature, *Heavenly Creatures* (1994), had a New Zealand setting, Jackson saw the story of two young girls murdering a mother as of international interest. *The Frighteners* (1996), a horror movie made for Hollywood's Universal Studios and featuring the well-known Hollywood actor Michael J. Fox, had a set which looked like an American Midwest town.

Unsurprisingly, Jackson happily took on the old-world fantasy *The Lord of the Rings*. Originally commissioned by Harvey Weinstein's Miramax to develop a one-film script of the book, Jackson had ambitions for several films and was given a month to find a new funder. He did — New Line Cinema, owned by Time Warner, agreed in August 1998 to three films and a budget of US$320 million. The three parts appeared, one a year, between 2001 and 2003, and were a huge success. They gathered an extraordinary 17 Oscars, the most ever by any series; and the third film, *The Return of the King*, won all 11 Academy Awards for which it was entered. The series grossed over US$2.9 billion in box-office sales, and about U$6 billion in total earnings — not bad given the initial investment.

New Zealanders bathed in the films' success. The high point came on 1 December 2003 when the world premiere of *The Return of the King* was held on a warm, early summer day in Wellington. An estimated 100,000 people lined the route as the cast travelled in open-top classic cars from Parliament to Courtenay Place where they walked the red carpet to a refurbished Embassy Theatre. The *Dominion Post* described the day as 'Wellington's biggest-ever party'. Why did New Zealanders react with such enthusiasm? It was partly that the local boy had seemingly conquered the world — New Zealanders love overseas approval. Pete Hodgson, 'the *LOTR* Minister', got closer to the films' significance with his claim that they showed New Zealand was 'more than a green mountainous country with lots of sheep'. It boosted the tourist industry, and 'the information technology that went into the trilogy was the best in the world'.

The *LOTR* phenomenon was not about presenting distinctive local traditions to the world. Rather, it represented a major economic transformation. At a time when New Zealand was trying to recover from the downturn in agricultural exports, the films showcased two alternative sources of wealth. The first was international tourism. The natural settings which were the backdrop to the films' fantasy might attract visitors from around the world. New Zealand was presented as a stunning but empty landscape, stripped of its people and culture and repopulated as Middle Earth. The tourist industry promptly got into the act. Air New Zealand embellished its planes with *LOTR* imagery, promoting itself as 'the airline to Middle-earth'. Over 40 operators offered *LOTR* tours of the country; a *LOTR* location guidebook sold over

200,000 copies. A popular destination was Hobbiton in Matamata where the landscape was transformed into an old-world shire. The campaign had impact. In 1995 there were over two million overseas visitors; but in the next decade as *LOTR* had its effect the numbers rose to over 4.2 million in 2005, and 5.4 million in 2015. Total spending of international tourists increased from under $6000 million in 1999 to over $9750 million in 2005. By then international tourism comprised over one fifth of overseas receipts and was the country's most valuable export earner. Of course *LOTR* was not the only factor, but its impact was profound in those years when the tourist industry took off.

The second meaning of *LOTR* highlighted by Pete Hodgson was 'information technology' and technical talent. The reference was primarily to Wētā Workshop and Wētā Digital. Their origins go back to 1988 when Peter Jackson called upon Richard Taylor and his partner Tania Rodger to produce models for his first feature *Bad Taste*. They worked with Jackson on his later films and in 1993 formally established Wētā Workshop to provide props for Jackson's films. *Heavenly Creatures* in 1994 required digital effects so Wētā Digital was established to produce computer enhancements and graphics. Both outfits blossomed. The Workshop, making physical items like weapons, armour and vehicles, and Wētā Digital, making digital special effects, attracted business from Hollywood studios. The Workshop has won five Academy Awards, Wētā Digital six. The celebration of the *LOTR* success in 2003 was as much about the creativity of Wētā as the genius of Peter Jackson; and the idea of 'Wellywood' as a world centre for creative film-making took off. The work of Wētā Workshop was seen around Wellington — in the huge figures of the 2015 *Gallipoli* exhibition at Te Papa, in the giant eagles greeting arrivals at Wellington Airport, in the tripod sculpture in Courtenay Place, and the UK Memorial at Pukeahu Memorial Park.

While Wētā was the major creative industry hailed in 2003, other digital industries emerged, providing jobs and overseas earnings. By the 2000s Xero was selling its accounting systems internationally; New Zealand was providing weather forecast graphics for use in overseas television; and Ian Taylor's Animation Research was providing computer graphics for sporting events such as the America's Cup. By 2015 the tech sector was contributing $6.3 billion of export income and was the third largest export earner after tourism and dairy exports.

When we look at King Theoden's armour, then, we are reminded how a country that once earned its living solely through wool, sheep meat and dairy products was earning its way through attracting tourists from abroad and selling smart programs and computer graphics to the world. *LOTR* represented, not a cultural revolution, but a major economic one.

95.

The world's first iPhone 3G

JOCK PHILLIPS

Jonny Gladwell began queueing outside the Vodafone shop on Auckland's Queen Street about 5 pm on Tuesday 8 July 2008. For 55 hours he waited. It was mid-winter. There were periods of drizzle and temperatures were chilly. Jonny got 45 minutes' sleep on Tuesday, about six hours on Wednesday. His friends supplied an exercycle and a masseuse! Finally, at 12.01 am on Friday 11 July, Jonny Gladwell purchased the world's first iPhone 3G. By then 300 people were in the queue. Apple had announced that sales would begin on 11 July, and because of the international dateline New Zealanders, notably Jonny Gladwell, were the first purchasers. At $979, it was not a bargain. But over the next three days New Zealanders bought some 3000. Within a week Vodafone's initial shipment had sold out.

The smartphone had reached New Zealand, bringing a portable phone, a sound system, access to the web, and an ability to communicate with others through texts, emails and photos. It heralded a social revolution. Let's look closely at the iPhone 3G. A mere 11 millimetres thick, 10.4 centimetres tall, and weighing only 135 grams, it is small and light enough to fit comfortably in a pocket. You can explore the phone's functions by touching or swiping the screen icons; and when you type, a keyboard appears. Nearly all previous cellphones had possessed a hard keyboard, and the idea that the screen could become the workplace was the inspired hunch of Steve Jobs, Apple's entrepreneurial genius. The firm had been developing a tablet and Jobs suggested a multi-touch display for finger-typing. Within six months the engineers returned with a workable model. Immediately Jobs realised this would work even better for a phone. The tablet was put aside and the iPhone developed. The first model, the iPhone 2, appeared in June 2007; but with teething problems was never sold here. The 3G was the improved version.

Looking at the screen, the first symbol you notice is a green square with a white phone. You press it to access the phone. Phones had a long history in New Zealand. The first telephones were demonstrated in 1878. They were large wooden devices with a separate hanging ear trumpet and a handle used for calling. Major growth came in the interwar years under the Post and Telegraph Department. Businesses benefited hugely. 'Party' lines (where a group of households shared a line), broke down rural isolation. 'Working? Working?' became the catch-cry of farmers' wives. Post-war affluence brought phones into most homes. They became smaller and built of plastic, not wood, with a circular dial.

In 1987 the break-up of the Post Office created Telecom which was privatised within three years. Competitors promoted mobile phones. Internationally the first mobile phone, the Motorola, appeared in 1973 but did not reach New Zealand until the 1980s. These mobile phones used analogue, not digital, communication. Handsets were bulky, known colloquially as 'bricks', and required an aerial. They were commonly used by tradesmen out on jobs. In the late 1990s when mobile phones became digital, they became smaller and cheaper. Nokia was the popular brand. New functions emerged. One was texting, a feature accessed on the iPhone 3G by pressing the green square with a fat white quote mark. The first text message was sent in New Zealand in 1998.

While teenagers texted, older New Zealanders used cellphones for emails. Emailing took off on computers in the 1990s; but quickly mobile phones were used. In the 2000s the BlackBerry was especially fashionable for emailing. On the iPhone 3G, email was accessed from the blue square with a white envelope. As emails increased, physical letters declined. By 2013 New Zealand Post was carrying 265 million fewer items than 10 years before.

Several other functions on the smart iPhone were not on earlier phones. One was instant access to music. For much of the twentieth century, people listened to music on record players. People on the move listened to the transistor radio. From the 1970s, cassette tapes were used in cars, and in the next decade Sony introduced a Walkman with earphones which offered a choice of music while you moved. In the late 1990s portable MP3 players using the internet appeared, but Apple considered they were either big and clunky or small and useless, and in 2001 launched the iPod. This was light and sleek and and able to play music from a digital

IMAGE BY JOCK PHILLIPS.

library or Apple's iTunes store. While iPods were spectacularly successful, Steve Jobs realised that people wanting both to listen to music and text had to carry two easily lost items, a phone and an iPod. His original inspiration in developing the iPhone was to combine the two. The iPhone was as much an iPod with a phone as the reverse. Revealingly, the relevant symbol on the 3G screen has an image of an iPod.

As music entered the digital age, so did photos. In the 1990s small digital cameras were widely used. But again, people did not want to carry another small device. Although in 2000 Samsung had introduced a camera phone, it was really the iPhone that successfully combined a high-quality camera and phone. On the screen the camera was accessed from a grey square with a camera lens, and the resulting photos from a blue square with a yellow flower. Video recording appeared on the 3G's successor, the 3GS.

A third feature that was new on the 3G was a mapping facility powered by GPS data. The phone allowed you to discover your location and direct you to a chosen destination.

Many iPhone functions were dependent on the World Wide Web; and the 3G allowed users to surf the net directly by pressing the blue square with a compass dial. There were even links to YouTube and weather information. Web access was the biggest revolution of the smartphone. The roots of this change also dated from the late twentieth century.

Although in the 1980s there had been experimental digital user groups here, it was not until Tim Berners-Lee launched the World Wide Web in 1991 that the internet became viable. At that time few had personal computers and in 1996 only one in five knew of the internet. Then it all changed. Websites exploded — to over half a million sites in New Zealand by 2013 — and users discovered the web. At first most surfing was on computers, but in the new century cellphones, such as Nokia, offered web access. By 2005, 62 per cent of our mobile phones did so, but reaching the web was often clunky. The iPhone was the first mobile phone to offer easy access and simple browsing. The process was aided by the arrival of 3G (3rd Generation) networks in 2009. The iPhone 3G was specifically designed for this advance. By 2013 over two thirds of users accessed the internet via phones, compared with 7 per cent before the iPhone. New Zealanders explored the web in cafés, on buses, even 'double-screening' while watching television.

Having the internet on smartphones was revolutionary. Daily life was affected in multiple ways. Shopping changed. In 1999, 24-year-old Sam Morgan founded Trade Me, to sell second-hand goods, and it became hugely popular. Increasingly, new purchases were also online — electrical goods, clothes, entertainments and travel. People bought houses and sought jobs through the web. Retail outlets closed stores in main streets. Banking was conducted on the web, and paper-based cheques disappeared.

The web affected the way people learnt about the world. The advertising revenue of printed newspapers collapsed as they moved online. The Stuff site, set up in 2000, was by 2020 the most popular local site; while web-only sites such as Newsroom and The Spinoff took off. Flashes on smartphone home-screens provided instant news. For other information, students turned to Google where they were directed to international sites such as Wikipedia or local sources such as Te Ara. Even television and radio streamed on demand.

Perhaps the most profound impact of cellphone access to the web was social networking. 'Web 2.0' sites allowed users to contribute and comment appeared about 2002. But it was with the iPhone and other smartphones that their use exploded. By 2012, 76 per cent of New Zealanders aged 17 or over were on Facebook. Twitter with its short messages, and Instagram with its photos, were also forums where younger New Zealanders picked up social gossip.

It was impossible to predict the long-term effect of a world in which relationships were conducted on a smartphone screen. But there was little doubt that the economic, social and psychological changes unleashed by the iPhone 3G were among the most profound changes ever experienced by New Zealanders.

96.

'Thunder Down Under' Christchurch portable toilet

CANTERBURY MUSEUM

It had been a warm February morning, not unlike another February morning in Napier eight decades earlier. Many were eating their lunch. Then there was a rumble. Everyone assumed it was another minor aftershock from the 7.1-magnitude earthquake at Darfield, west of Christchurch, the year before (4 September 2010). Philip Willis remembered: 'But within less than a second the vibrations had undergone a massive crescendo, and the house began roaring around me . . . It wasn't like riding big waves, or being blown around in a high wind. It was the sharpest, most violent kind of shaking; as though the house sat on some giant mechanism of limitless force that was snapping it back and forth, up and down, however it liked.'

The quake in Christchurch at 12.51 on 22 February 2011 was 6.3 in magnitude — not as big as the Darfield shake. But it was shallow and the epicentre was only 10 kilometres from the city centre. The destruction was terrifying. Some high buildings were flattened, like the offices of Canterbury Television where 115 people died and of Pyne Gould Corporation where 18 were killed. The Forsyth Barr building lost its stairwell, and office workers were rescued by crane. In the streets old masonry collapsed, killing many beneath. In all, 185 people died, and some 300 were treated for serious injuries in the hospital over the next hours. A quarter of the buildings in the CBD had to be demolished, and over 16,000 were severely damaged. Some famous heritage buildings became ugly ruins, such as the Canterbury Provincial Chambers, and the Anglican and Catholic cathedrals. With houses damaged and roads impassable, Christchurch people experienced an unforgettable and horrifying time that afternoon and evening.

The drama of 22 February was but the beginning for Christchurch's citizens. Even where homes were still standing, the community faced a major loss of infrastructure. Landlines did not work, even cellphones were erratic. Most houses had lost electricity. None in the eastern suburbs had running water, which meant no showers, no washing machines, no toilets. Two weeks later a fifth of all households still lacked water. Roads were impassable, covered in rubble and liquefied mud. For many the greatest challenge was the damaged sewerage system. The quake had broken many underground pipes that carried sewage; others were blocked when liquefaction forced silt inside through joints; 41 per cent were damaged; 136 pumps were disabled. For those in the eastern suburbs, it would be months before the sewers and running water were restored and people could again enjoy the simple convenience of a private lavatory. Three weeks after the big quake, half of Christchurch's homes, 80,000 residences, had no sewer connection.

Christchurch's sewerage system was about a century old — too old to be reliable, but old enough that citizens regarded it as an essential of life. This had not always been the case. When Europeans first arrived and established towns, they accepted a high degree of dirt and excrement. Streets were unpaved and became seas of mud after rain, their surfaces covered in horse dung. Dunedin became known as 'Mud-edin'. In most places, maids hurled slops, scraps and night-soil into the street at night. In Auckland the Ligar Canal was an open drain carrying garbage and sewage down Queen Street to the harbour. For most of the nineteenth century few people had a clean internal toilet. Often they dug garden long-drop 'dunnies', which would regularly overflow. Then they would dig another pit, or hire a night-soil man to collect the full pails of sewage. Little wonder that, typhoid, cholera and dysentery were endemic.

The colony was supposed to be a healthy place free of old-world diseases, so such problems raised concern. As the germ theory of disease replaced the old idea that bad odours, or 'miasmas', were the problem, a public health movement emerged. An 1872 Public Health Act recommended improved drainage and a water supply in cities. Even then the high costs of importing iron pipes and a scepticism about the need delayed progress. Christchurch was actually the first major city to complete a sewer system when pumping began in 1882; and all cesspits were ended. Even so, only 639 houses were connected to the system and the rest relied on night-soil men. Realistically, it was not until the First World War that most urban centres had a piped sewerage system and running water to provide for toilets. By 1914 Christchurch had 12,844 houses connected and over 1000 miles of sewers. Even then most places poured the sewage direct into rivers or harbours.

So, in 2011, Christchurch people accepted that a water closet attached to sewerage was a necessity of life; but the system was about a century old and vulnerable to the earthquake. How did people cope in the months after the quake? Undoubtedly pooing was a major source of discussion. Eric Cummins was driven to poetry:

> We've got the water on at last but we are not amused,
> They tell us we will have to wait, the toilets can't be used.
> Without a proper toilet how on earth will we survive,
> We can't keep watching toilet paper floating down the drive!

Going to a public toilet down the road was no solution because they too were out of commission and instead became the targets of inventive graffiti. People had to fend for themselves. Some drew on their bach experience and dug a long drop in the garden with a bucket carved to form a comfortable seat. Often these were decorated — creating such interest that a website, 'showusyourlongdrop', became a favourite site. Others used sheds and pulled out a Thunder Down Under, our chosen object. This was an Australian product, originally produced for camping in the bush. But it worked well in the earthquake emergency. Relatively light and 35 centimetres high, it could hold 20 litres and came with a comfortable snap-on, hinged plastic seat. If you were lucky, you might have a supply of plastic liners. The only problem was that you had to have one in your shed or garage because camping stores quickly ran out.

The most common alternative to the do-it-yourself loo was a portaloo distributed by Civil Defence. Within the first fortnight almost 1200 had been placed in affected areas. It was not enough. There were reports of people trying to wave down trucks carrying portaloos and beseeching them to be dropped in their street. Then came news of 'portaloo wars' as people in poorer

'THUNDER DOWN UNDER / PORTABLE TOILET 2010' ELEMENTAL MANUFACTURER CANTERBURY MUSEUM 2012.1.43

areas of the eastern suburbs, alleging that richer localities were getting priority, sent vigilantes to steal them. The alleged discrimination became such an issue by early March that the Prime Minister, John Key, originally a Christchurch boy, was forced to answer on the issue. Civil Defence felt under so much pressure that they displayed a toilet in their Christchurch Art Gallery headquarters with a sign, 'We know that 80,000 people need loos . . . we're number one with number twos!'

The next official solution was chemical toilets: 37,000 were ordered and distributed by the city council, allegedly exhausting the world's supply! The Mayor of Christchurch, Bob Parker, was called on to illustrate their use. The chemical toilets were accompanied by large tanks placed at convenient points where residents could dispose of their waste. But these toilets, made of plastic, were squat objects and many people, especially the elderly and disabled, found them awkward, while some large people were reported to have broken them. It proved difficult to carry the full containers down the street to be emptied.

Fortunately, the loo crisis did not last forever. Two months after the quake, only 20 per cent of the population had sewage issues and they had the use of no fewer than 2800 portaloos and 31,000 chemical toilets. Three months later Sumner saw portaloos removed from its streets; and by the end of the year sewerage and running water were sufficiently restored that most inhabitants of Christchurch could safely flush. There were only 30 portaloos left. The temporary crisis was over.

But the long-term crisis was far from over. The Christchurch earthquake pointed up major problems with the infrastructures of New Zealand cities. In nearly every city the sewers were over a century old. Old clay pipes were liable to cracking; joints were loose; and as the need for new housing became more urgent in the late 2010s, the ability of the sewerage system to cope became an issue. The distribution of fresh water, also highlighted by the earthquake, was another concern — especially in Auckland which faced growing consumption and diminishing rainfall. Further, the earthquake emphasised the dangers of buildings that were not able to withstand such movement. When five years later, on 14 November 2016, a magnitude 7.8 quake hit Kaikōura, both Wellington and Christchurch received another warning about the dangers of earthquake-prone buildings. The quakes also added urgency for alternative transport routes to get people out of the major cities in the event of a major disaster. These experiences and warnings signalled for New Zealanders the need for major investments in the infrastructures of cities and communities. This would become a major policy and political challenge for 2020s New Zealand. The saga of Christchurch's loo problem would take a long time to flush away.

97.

Austin Wang's language cards

MUSEUM OF NEW ZEALAND TE PAPA TONGAREWA

This set of 100 language cards was used by the parents of four-year-old Austin Wang in about 2012 to teach him the Mandarin language. On one side of each card are Chinese characters, and on the other is an image which represents the correct answer, such as 'tree' or 'run' or 'lip' or 'dog'.

Austin's mother, Li Yingjia, known by the New Zealand community as Kilihelsey or Kili, first came over in 2002 to study at Auckland University of Technology (AUT). She had grown up in Shenyang City in north-east China, some 200 kilometres from the border with North Korea. A city of over seven million people, Shenyang suffers from extreme cold winters and hot humid summers, and is a centre of heavy industry which, along with coal-fired heating, creates much pollution. It is home to some fine universities, but the competition is tough. Kili had an aunty and cousins already in Auckland, so she decided to follow them and enrol.

Kili was part of a major outflow of young Chinese to western countries for higher education at that time. Although the number of Chinese universities was rising, competition to enter was intense. The economic reforms in China from the late 1970s introduced the profit motive and created a heightened competition for skilled jobs. People from the urban middle classes, such as Kili, realised the need for specialised education and saw in overseas universities training in valuable language skills and western ways of doing business. The entrance to Chinese universities, the Gaokao, was based on a stressful one-off exam which many preferred to avoid. In 1998 New Zealand became one of the first western countries offering open access to student visas for Chinese nationals. Young Chinese quickly voted with their feet. In 1999 there had been only 173 Chinese students in New Zealand and most of these originated from Malaysia. A year later the number had risen close to 1000. By 2002, when Kili arrived, there were over 8000 students from China, more young women than men. The following year saw over 12,000. For the next few years the numbers stabilised before falling somewhat, but in 2012 international education was bringing in $2.5 billion to the New Zealand economy, the country's fifth most valuable export. Auckland was consistently the most popular destination — either at the University of Auckland or AUT.

For many Chinese, education in New Zealand was a pathway to settlement, and so it proved for Kili. At first she found the adjustment difficult. She was 22 years old, and lived in a homestay with a young Kiwi family. The food was simple Kiwi fare, very different from Chinese cooking. She was still learning English and found communicating with the family difficult. There was no internet, and phone calls home were expensive. Understandably, Kili was lonely. She welcomed her trips to see her aunty and cousins, and on one such visit met Wilson Wang who was also studying at AUT. Wilson and Kili decided to get married and settle in Auckland. Their first-born, Austin, followed in 2008, and four years later came a daughter, Yolanda. Obviously, Wilson and Kili missed their homeland and its culture. They returned for visits, but enjoyed the opportunities, lack of pollution and mild climate of Auckland. One major issue remained. Like all Chinese of their generation, Wilson and Kili came from one-child families, and there was a traditional Chinese obligation for the young to care for the old. Having decided to settle, both brought their parents to Auckland.

Kili and Wilson thus became part of a major new movement of Chinese people to New Zealand. Chinese men first came to New Zealand in the 1860s when invited by the Otago province to search for gold, and by 1881 there were 5000 Chinese, only nine of whom were women. They quickly became the target of racist discrimination in the form of a poll tax, restrictions on naturalisation, and exclusion from state support such as the old-age pension and family allowances. Some Chinese returned home, others drifted to the city where they worked in fruit shops, laundries and market gardens, and found solace from public hostility in gambling. During World War II the Chinese fight against the Japanese in Asia, and the contribution of their market gardens to the war effort raised their public reputation and acceptance. Finally, some Chinese women were allowed in as wives of the men settled here. But as late as

SET OF MANDARIN LANGUAGE FLASH CARDS, CIRCA 2008, UNKNOWN MAKER. GIFT OF AUSTIN WANG, 2012. TE PAPA (GH023305)

1986 there were under 5000 people born in China living in New Zealand and under 20,000 of Chinese ethnicity — tiny proportions.

Then in 1987 New Zealand changed its criteria for admitting immigrants, ending the preference for those from the United Kingdom and Ireland, and instead basing entry on educational qualifications, skills, age and assets. This opened up possibilities for those from Asia, and within 10 years the Chinese numbers had quadrupled four times to over 80,000. At first, many people came from Chinese communities in different parts of Asia, but in the 2000s, like Kili and Wilson, migrants arrived from mainland China. By 2018 there were almost a quarter of a million people of Chinese ethnicity living here. Many had come from China itself. Most were from an educated middle-class urban background in the north, quite different from the southern peasants who had arrived in search for gold.

As Kili, Wilson and their family made a life in New Zealand, a continuing issue was how far they would integrate, and how far they would keep Chinese culture alive. Obviously, they had to integrate to some extent just to make a living. For a time they ran two massage parlours in Newmarket, but the outbreak of Covid made such enterprises unviable. They both went back to school to learn real estate. Language was another primary issue. Both Kili and Wilson had to speak English simply to operate successful businesses. But what about the children? From television and school the two kids would learn English easily. But both Kili and Wilson had their parents with them who did not speak English. If Austin and Yolanda were going to communicate with their grandparents they would need to speak and understand Mandarin. Hence the language cards became important to the family. Both children succeeded in learning some Mandarin, which they used with their parents and grandparents, and Austin had a weekly language class. Interestingly Austin and Yolanda speak English when talking together.

Another crucial issue was where the family would live. A majority of recent migrants settled in Auckland. In 2018, 171,309 or 69.1 per cent of the Chinese people in New Zealand were in the Auckland region. Ten per cent of Aucklanders were Chinese by ethnicity, over twice their representation in New Zealand generally. Over a quarter of Aucklanders had an Asian ethnicity. So if Kili and Wilson wanted to be close to other Chinese then Auckland it would be. Further, in the Pakuranga, Howick, Botany Downs area of East Auckland the numbers of Chinese were much higher. In the Howick local board area in 2018 there were more people living of Asian background (46.5 per cent) than of European ethnicity. It was not surprising that Kili and Wilson chose to live in that area. They set up home in Dannemora where in 2018 almost two thirds were Asian. Many were from Korea or other Asian countries such as the Philippines; but in the Howick area 11 per cent of the population could speak Mandarin in 2018 and another 11 per cent another Chinese language. So when Austin and Yolanda went to school they would find other children whose families had recently migrated and who knew some Mandarin.

The existence of a Chinese community in East Auckland opened up other opportunities for Austin and Yolanda to keep alive Chinese interests. The Pakuranga community offered traditional dancing, singing and opportunities to play the great Chinese sport of table tennis. Chinese New Year remained a significant community event. Austin also had a cellphone from the age of 11 and used the Chinese social app WeChat which allowed members of the community to send short messages and photos. The Chinese app TikTok is another of his favourites. Alongside his Chinese language class, Austin also learned to use the traditional Chinese abacus for his maths, and his parents are ambitious for him to succeed in that area.

Yet despite the continuing relevance of Chinese culture, Austin absorbed much from wider New Zealand. He was keen on volleyball, he had ambitions to be a professional footballer. He ate McDonald's Happy Meals alongside Chinese food. When, on a family visit back to see relatives in China, he fell ill with a bad stomach bug, Austin told his parents that he did not want to die in China. For Austin, and one suspects Yolanda, Kili and Wilson, New Zealand became their long-term home, and the rich Chinese culture they brought with them was forging a different New Zealand of the twenty-first century.

98.

Rā Maumahara pouaka petihana (New Zealand Wars Day petition box)

AUCKLAND WAR MEMORIAL MUSEUM

This handsome cardboard box, one of three originally bought from The Warehouse for $6, then decorated, contained some of the 13,000 signatures in support of a petition for a national day to remember the New Zealand Wars (Rā Maumahara). The three boxes, plus a fourth elaborately carved wooden box also holding signatures, were presented to Parliament by pupils of Ōtorohanga College in November 2015.

The petition grew out of the 150th anniversary of the Waikato War in 2014. A major anniversary event was planned at the site of the battle of Ōrākau but local kuia and kaumātua thought the commemorations would be crowded. Instead they offered to take students from Ōtorohanga College to both Ōrākau and Rangiaowhia and tell them about the events there. The college was about 60 per cent Māori and valued te ao Māori (the Māori world) with frequent pōwhiri, kapa haka and a week-long Matariki festival. 186 pupils set off in six buses to visit the battle sites. It was Rangiaowhia that really affected them. The elders described the events at that place 150 years before — how having promised to leave the township as a refuge during the wars, government forces then attacked Rangiaowhia and, in one shocking incident, a whare protecting women, children and old people was set on fire, and several were allegedly shot at as they fled for their lives. One of the students, Leah Bell, remembered that as the stories flowed, 'We were looking at the ground in shame . . .

It was awful to think that people had died there, that history had been covered up. There was this shared sense of horror that this offence had occurred in our own country in our great-grandparents' time.' At school the pupils had been learning about atrocities in distant countries — in the United States in their civil war, in Adolf Hitler's Germany. The realisation dawned that horrors had occurred no more than half an hour up the highway.

It was not that Leah knew nothing about history or Māori experience. She was descended through her father from William Colenso, the pioneering missionary. Although her father was a Pākehā farmer, he had been kicked out of a shearing gang in the early 1980s for opposing the Springbok tour, while her mother worked in a Pākehā anti-racist environment supporting groups such as Māori Women's Refuge. Leah learnt her alphabet painting letters on placards opposing the Foreshore and Seabed Bill. She had gone to Waitomo Caves primary school and become aware of how little benefit local Māori earned from the millions of tourist dollars pumped into the caves. At primary school she became good friends with Waimarama Anderson, who was also shocked by the visit to Ōrākau and Rangiaowhia. Independently both Leah and Waimarama asked the principal of the college, Timoti Harris, why they had not heard about these events before and why they were not taught about them in class. The idea emerged of organising a petition to Parliament, asking for a raised awareness of the New Zealand Wars, the introduction of local histories into the school curriculum, and a day of recognition for the wars.

Over the next few months with the support of the community and teachers, especially Mariana Papa and Linda Campbell, the Ōtorohanga pupils, led by Leah, Waimarama and their friends, set out to attract signatures. They had the support of a local MP, Nanaia Mahuta, who helped them develop the correct wording, and also Mariana's husband, Rahui Papa, a distinguished Tainui kaumātua and historian. The pupils sold their message at shopping centres, town halls and festivals like Polyfest, Matatini, the Tūrangawaewae Koroneihana and at Waka Ama regatta. They spoke on radio. It was not easy. Every signature 'was hard won, face-to-face'. The conversation was as important as the signing. By November 2015 they had 13,000 signatures.

In November 2015 on Parliament's forecourt Waimarama Anderson, Tai Te Ariki Jones, Rhiannon Magee and Leah Bell with 1500 supporters presented the petition, in its boxes, whose poutama pattern signalled the upward journey of education, on Parliament's forecourt. In place of their normal school garb of polyfleece tops, shorts and jandals, Leah wore a college blazer, while Waimarama, Tai and Rhiannon wore ancestral Ngāti

PETITION BOX, FOR NATIONAL DAY COMMEMORATING NEW ZEALAND WARS. AUCKLAND WAR MEMORIAL MUSEUM TĀMAKI PAENGA HIRA. 2018.12.1.

Maniapoto kākahu — a conscious decision by kaumatua to represent their respective positions and ancestries in this shared movement. Five months later, in March 2016, the girls gave evidence before the Māori Select Committee — the youngest ever to appear before a parliamentary committee. The MPs listened. In August 2016 the government announced that a national day of commemoration, Rā Maumahara, would be established. The first was held at Kororāreka in March 2018, and thereafter 28 September, the anniversary of the Declaration of Independence, became the annual Rā Maumahara. Then in September 2019 Prime Minister Jacinda Ardern announced that from 2022 the teaching of New Zealand history would become compulsory in all schools and kura. This was a remarkable achievement for a movement that, as Leah Bell recalled, 'grew from the passion of small-town gumboot-wearing, hormone-driven teenagers into a nationwide petition'. As for the boxes, Nanaia Mahuta was presented with the carved wooden book which remains in parliament, and her office returned the other three boxes. One was presented to Rahui Papa, and one to the school principal, Timoti Harris. He in turn gave it to the Auckland Museum where it sits at the entrance to the New Zealand Wars gallery.

In one sense the campaign for Rā Maumahara represented the voice of a young generation. They spoke in a similar way to the youthful movement which campaigned soon after for greater action on climate change. On 15 March 2019 about 20,000 New Zealand school students marched in support of the School Strike movement sparked off by Sweden's Greta Thunberg; and in April 2021 once again thousands carried beautifully illustrated placards about climate change in the main centres.

In another sense the petition for a greater awareness of the New Zealand Wars followed a half-century of activism. Since the 1970s there had been a remarkable movement by Māori to reinvigorate their own culture. It had begun with the language. As Māori moved to the city in the years during and after World War II, many no longer used te reo even in their own homes. Then in 1972 Nga Tamatoa based at Auckland University and Te Reo Maori Society based at Victoria collected over 30,000 signatures on a petition for the teaching of Māori language in schools. Teaching the coming generations was the key. In 1982 the first kōhanga reo began at Wainuiomata and this became a national movement as kaumātua taught te reo and Māori traditions to preschoolers. Graduates of kōhanga reo needed to continue their language and the first full-immersion Māori language kura kaupapa started in Auckland in 1985. A tertiary institution, Te Wānanga o Raukawa, began at Ōtaki in 1981 and others followed. Then in 1986 the Waitangi Tribunal found that te reo was a taonga and should be recognised as an official language. This occurred in 1987. That year the first Māori radio station was set up, and by the 2000s there were 21 such stations around the country. After a striking increase in Māori programmes on New Zealand television in the 1990s the Māori Television Service was launched in 2004 with government support, and by 2008 there were two channels, one broadcasting entirely in te reo. Alongside this revival of the language and media came a new energy in kapa haka. Again the beginnings were in the 1970s with the establishment of a national festival of performing arts in 1972 which became in 2004 Te Matatini, a major showcase of waiata and haka.

This impressive movement over 40 years helped maintain and enrich Māori culture for the growing Māori population. The next challenge was to involve all New Zealanders, Pākehā and migrants, in an understanding of Māori history and language. Until the 1970s most Pākehā had little sense of their country's history. They believed New Zealand was 'a young country' — real history lay overseas among older civilisations. Ignorance of Māori history was woeful. But then New Zealand history, especially the historical experiences of Māori, began to attract skilled writers and presenters. Māori film-makers like Barry Barclay and Tainui Stephens worked with Pākehā historians Michael King and James Belich to bring Māori history to television screens, and those historians joined others such as Ranginui Walker, Claudia Orange and Vincent O'Malley in describing the pains caused by the New Zealand Wars and failures to honour the Treaty. Their work was supplemented by the impeccable scholarship of the Waitangi Tribunal. For some 50 years scholars had been uncovering the history which came to the Ōtorohanga students as such a bombshell. What was new was the students' insistence that these perspectives should not be lost in scholarly tomes but become part of daily understandings for both Māori and also Pākehā.

Because this was a movement that involved all New Zealanders, it was important that it was led together by the Pākehā Leah Bell and the Māori Waimarama Anderson working as Treaty partners. Their efforts came when there were other signs of responsiveness by non-Māori to the indigenous language and traditions. There was a steady movement to restore traditional names — in 1986 Mt Egmont was given the alternative name of Taranaki and from 2020 just Taranaki Maunga; Mt Cook became Aoraki/Mt Cook in 1998. Many places had macrons restored to their names. Newsreaders on television and radio made efforts to pronounce place names correctly and began using phrases of te reo. The successful movement led by the Ōtorohanga students for a day to remember the New Zealand Wars was supplemented by a growing interest in commemorating Matariki, the arrival of the Pleiades each winter. In 2020 the government announced that from 2022 Matariki would be formally recognised as a Friday public holiday in late June or early July. Matariki fireworks looked set to replace Guy Fawkes Day of British tradition. At local schools throughout the country Pākehā children were being taught correct pronunciation of te reo and learning simple phrases. Pōwhiri had become appropriate ritual when welcoming parents and visitors to the school.

No one could pretend that by 2021 New Zealand was a bicultural country; and alongside signs of Pākehā interest in Māori culture, newspaper correspondence columns showed some redneck resentment. In 2018 only 1.9 per cent of people of European heritage spoke Māori, while, equally concerning, only 20.6 per cent of Māori did so. There remained stubborn continuities. In 2018 there were also huge discrepancies in jobs — a much higher proportion of Māori than other groups were labourers or manual operators, far fewer were managers or professionals. The prison population was 53.1 per cent Māori, far higher than their 16.5 per cent in the population. New Zealand remained a long way from a fair bicultural Aotearoa. Yet significant change had occurred and this petition box expressed an important moment in that change. Whether it represented the beginnings of a cultural revolution, or just another false start, the next half-century would reveal.

99.

Tariq Omar's football

CANTERBURY MUSEUM

This yellow, blue and red football was laid down amid the football-field length of tributes which sprang up just outside the Botanic Gardens in Christchurch in the days after the mosque shootings on 15 March 2019. On the football, inscribed in white paint or corrector fluid, are tributes from his team-mates to Tariq Omar. This was one of the items chosen by the Muslim community for permanent keeping in Canterbury Museum.

Tariq Omar was a smiling 24-year-old with long hair and a beard. He was the son of Rashid Bin Omar and Rosemary Omar. Rosemary came from Timaru and had met and fallen in love with Rashid in Singapore. Rashid migrated to Christchurch in 1990 when they were married. Rosemary converted to Islam. Tariq was a devoted Muslim who would never miss Friday prayers. On this Friday, Rosemary had intended to wait for him outside the Al Noor Mosque on Deans Avenue; but Tariq joked that she might get a parking ticket so she drove round the back to Brockworth Place. At 1.40 pm she heard shots and immediately texted Tariq telling him to hide in the mosque because someone was firing in the neighbourhood. Then she realised that the shots were coming from inside the mosque.

The shooter, a troubled Australian temporarily living in Dunedin, had arrived at the mosque on a mission to kill Muslims as an example to his white nationalist colleagues around the world of the alleged threat of Islam to western civilisation. He brought with him five weapons, two of them semi-automatic assault rifles, and he livestreamed his actions on several websites. As the gunman approached the mosque entrance a worshipper greeted him with the words 'Hello, brother' and became the shooter's first victim. The individual entered the mosque and fired wildly at women at worship in the women's hall, then at men in the men's hall. He returned to his car, picked up another weapon and returned to shoot once more in the men's hall. Five minutes after arriving he left. There were over 40 dead and many wounded. Then he headed eastwards in his car at speeds of 150 kilometres an hour and soon after 1.50 pm arrived at the Linwood Islamic Centre where 100 people were at worship. Unable to find the main door, he began shooting through a window. The carnage would have been worse had not one of the brotherhood, Abdul Aziz Wahabzada, grabbed a credit card reader and thrown it at the attacker. When the shooter returned to his car for another weapon, Abdul Aziz picked up an abandoned weapon and threw it at the car, shattering the rear window. The assailant drove away and some minutes later was stopped and arrested by the police. At Linwood, seven people died, bringing the eventual total including two who died later to 51 victims: 47 men and four women. There were over 40 wounded.

Rashid Omar said that after the attack the family texted and rang Tariq's phone but it was not answered. They made a desperate plea for

'FOOTBALL TRIBUTE', CANTERBURY MUSEUM, 2019.60.2

information on television. Eventually, after midnight on Saturday, the Imam read out the names of the victims and Tariq's was among them. 'My body felt completely weak and everything went silent,' Rashid recalled. Identifying his son's body was the hardest thing he had ever done. It was months before he could return to his work as an electrical supervisor. Sleep was difficult. Tariq was a handsome young man and was widely praised for his gentle, helpful nature. His death was hugely painful for his parents and siblings — his older sister Qariah had told him the date of her forthcoming wedding the night before his death.

Tariq would not be forgotten by his family; nor by the community of which he was a part. In the days after the horrific 15 March shooting, there was an extraordinary level of outrage and sympathy for the victims from the Christchurch community. Outside the Botanic Gardens were laid heaps of flowers and cards, and also objects — painted rocks, balloons, Buzzy Bees, teddy bears, stuffed kiwis, photographs, paintings, origami, sewn hearts, flags, candles, poems, pāua shells in a kete, skateboards — anything that could carry a message. Tariq would be remembered appropriately by this football. For Tariq Omar, like many young people the world over, was a lover of the game. He played for FC Twenty 11 in the Christchurch competition and had also coached the under-nine and under-12 teams for the Christchurch United club. Among the messages inscribed on the ball are the words: 'Tariq. All any kid wants is the ball at their feet. May this one stay at yours forever.' The director of the club's academy and junior teams wrote on the club's website that Omar was 'a beautiful human being with a tremendous heart and love of coaching'. Others said he was one of the hardest-working players on his team who brought a smile to every week's training. His own team wore black armbands in his memory and hung his photo in their changing rooms. Another message on the football reads: 'There will be a huge gap in everyone's life without you.' In November 2019, some seven months after his death, the Tariq Omar five-a-side competition for a memorial cup was held, with the proceeds going to families affected by the mosque shooting.

Tariq was not the only footballer remembered from 15 March. Also laid on the gardens' fence was a bronze medal from the Christchurch Under-15s International Cup plus a football jersey which was probably a tribute to Sayyad Milne, a fourteen-year-old who had been chosen to play in the tournament. Like Tariq, Sayyad had one parent from Singapore and the other was New Zealand-born. In addition, Atta Elayyan, the 33-year-old goalkeeper for the indoor New Zealand football team, the Futsal Whites, was among the victims of the shooting. He had played 19 games for the team.

It was unsurprising that football was a passion for three of the mosque victims, for football, the pre-eminent international game, has long been an interest of New Zealand immigrants. While rugby union and league took off among the Pākehā and Polynesian communities, football remained an interest for many migrants from Britain and Europe in the years after World War II. 'Ten-pound Poms' or their children dominated the New Zealand soccer team which won its way to the World Cup finals in 1982. European migrants organised soccer clubs such as the Olympic Club for Greek migrants and the Hungaria club set up by refugees from the 1956 Hungarian uprising, both in Wellington.

Many of the Muslims who were at prayer on 15 March had arrived fairly recently. The Muslim population in the country had risen from just over 20,000 at the turn of the century to over 60,000 by 2018. Of those, almost two thirds were from Asian countries. About half of these were Indian in background, with other smaller groups coming from places such as Pakistan, Sri Lanka and Singapore, like Tariq's father Rashid. Despite popular assumptions, only about a fifth of the nation's Muslims in 2018 were from the Middle East. Among those killed in the mosque shootings, however, those from the Middle East and Africa were overrepresented, comprising 22 of the 51 killed. People of Asian background numbered 28.

Despite the increase in the Muslim community they still represented only about 1 per cent of the New Zealand population — a tiny minority — and there is little evidence that the Muslim presence had sparked much hostility. The Human Rights Commission reported that from 2002 to 2009 they had received a few complaints about the wearing of Muslim women's headscarves (hijabs) and some women in the Muslim community reported street harassment about the hijab. Of course, historical conflicts with the Muslims, especially the Crusades, were part of the British heritage and recalled by the name of the Canterbury Super Rugby team. But hostility to Islam was never a strong impulse for most New Zealanders. Rather, the shootings represented the impact of global forces facilitated by international air travel and the World Wide Web.

Internationally the alleged Islamic threat was sparked by the attack on the Twin Towers in New York in September 2001. The event was replayed endlessly on television, producing heightened concerns about 'Muslim terrorism'. New Zealand stood at some distance from this fear and Helen Clark refused New Zealand's participation in the invasion of Iraq which followed 9/11. But the country was unable to avoid international paranoia. The Christchurch shooter found a community of interest online. He soaked up anti-Muslim ethno-nationalist ideas and 'the Great Replacement' theory which viewed Muslim migration as the major threat to western Christian civilisation.

On the day of the Christchurch attack he uploaded the anti-Muslim manifesto of the Oslo terrorist Anders Breivik, who had killed 77 people; and he followed Breivik in claiming to be associated with the Knights Templar, a Christian order active in the Crusades. He plastered his weapons with international white supremacist symbols including references to Hitler's *Mein Kampf*, played a Serbian nationalist song in his car as he drove to the mosque, posted his manifesto online and began a livestream on Facebook of his 'attack against the invaders'. In other words, influenced by overseas ideas imbibed from the web, he saw his action as contributing to an international white supremacist movement. He chose Christchurch as the site of his horrific attack because levels of security were so low; and New Zealand because semi-automatic rifles were obtainable.

In this way the Christchurch shootings represented the end of New Zealand's isolation and innocence. Further, the shooting attracted world headlines. Jacinda Ardern's actions immediately following the event — her wearing of a hijab and comforting the families of victims, her speedy calls for a ban on semi-automatic weapons and international measures to restrict racist ideas on the internet — established her standing in the world's eyes. Few New Zealand leaders — or indeed events — have ever become so well-known overseas.

Arguably, 15 March 2019 was the moment when New Zealand found itself firmly part of the international community; and it is fitting that the object we have chosen, a football, represents the pre-eminent international game, one enjoyed the world over by people of all faiths.

100.

Ailys Tewnion's crocheted bears

THE NELSON PROVINCIAL MUSEUM, PUPURI TAONGA O TE TAI AO

Our final object consists of two crocheted teddy bears, portraying the Prime Minister, Jacinda Ardern, and the Director-General of Health, Dr Ashley Bloomfield. They were created during the total lockdown experienced by all New Zealanders in March and April 2020. The cause of the lockdown was the outbreak of a highly contagious respiratory coronavirus, SARS-CoV-2, which emerged in Asia at the end of 2019. Within months Covid-19 had spread. around the world; and with open borders New Zealand could not remain untouched. On 28 February 2020 the country's first case appeared, a recent traveller from Iran. On 19 March the borders were closed to non-residents and non-citizens, and a week later, when cases were over 500, the country went into full lockdown. All but essential businesses like supermarkets were closed. Schools were shut. People were instructed to remain within their family 'bubble', not to mix face-to-face with friends or wider family. The country remained shut like this for just over a month, and even then many restrictions remained. It was not until 9 June that all limitations on domestic life ceased — although periodically over the next year there were three short reversions to lockdowns in the Auckland region, followed by another, even lengthier, lockdown in Auckland when the Delta variant arrived in August 2021.

So how, cut off from neighbours, their community and the world, did New Zealanders cope with the lockdown? They explored the media to link to the outside world. They surfed the net and called family, friends and work colleagues on Zoom calls. They listened to the radio, and watched streamed content on televisions or computers. They followed news of the growing tolls overseas, gaining comfort from the low totals of cases and deaths here. Many tuned in each day to a 1 pm news conference when Prime Minister Jacinda Ardern and Director-General of Health Ashley Bloomfield presented news about the number of cases or deaths in the country over the previous 24 hours and discussed the possibility of changing alert levels. The two appeared direct, honest and sympathetic and won great plaudits in the community, which partly explained the landslide electoral victory for Ardern's Labour Party in the election at the end of 2020.

As a central ritual of the lockdown it was not surprising that the 1 pm news conference inspired artistic responses. An Auckland couple, Scott Savage and Colleen Pugh, put together models of Ardern and Bloomfield from recycled household rubbish — chip and coffee packets, pill containers, toilet rolls and juice lids. They showed the PM and the D-G standing behind lecterns with the distinctive yellow and white stripes of the government's Covid response placed between them and two New Zealand flags behind them. The models received such publicity that the national museum Te Papa promptly acquired them, while the museum's shop offered for sale laser-cut MDF kitsets of the two leaders.

Our chosen object is a variation on this theme. It also presents the 'Jacinda and Ashley' show. Bloomfield, with his distinctive black-rimmed glasses and dark suit, is standing at a lectern, while Ardern stands beside him in a skirt, her characteristic long loose hair and earrings. Behind Bloomfield a clock shows the time, 1 pm, on the Nelson clock tower. For this was a Nelson creation. The artist was Ailys Tewnion. With her husband and two children, Ailys had moved to Nelson from Christchurch in 1986, and developed many domestic interests from cooking and gardening to crocheting, sewing, knitting and floral art. She spent much time with her children, her grandchildren and her friends; but lockdown cut off those options. She was isolated at home and had time to fill. Many people throughout the country responded to this situation by developing their creative instincts. Some wrote poems, some redeveloped their gardens, some baked sourdough bread, while many sewed masks which became necessary wear on the occasional visits to the supermarket. Ailys's response was to get out her sewing kit and crochet relevant scenes. She placed them on the top of the fence in front of her house so that passers-by saw a new scene each time they passed.

In this way Ailys was contributing to the 'great New Zealand bear

DIORAMA, JACINDA AND ASHLEY BEARS. NELSON PROVINCIAL MUSEUM COLLECTION: NPM2020.57.2

hunt'. These two figures may be Jacinda and Ashley, but they are also recognisable as crocheted teddy bears. Displaying teddy bears in the windows of houses became a feature of the lockdown. It became a way that people could send a supportive message to others suffering through the isolation. People, especially those with young children, were encouraged to walk around their neighbourhood spotting teddy bears. The teddy bear hunt, although directed at a local audience, again showed the importance of international media. The idea of people locked inside sending positive messages to the outside world spread via the internet. In Italy people drew rainbows and posted them in their windows with the slogan 'andrà tutto bene' (everything will be alright). Then in Britain, the United States, Iceland, Australia, the Netherlands, Canada and many other countries teddy bears began appearing. Where it started is not clear but the habit became more widespread in New Zealand than anywhere and was initially promoted by a Facebook group set up by a Hamilton mother, Annelee Scott. Before long the group 'We are going on a bear hunt — NZ' had over 18,000 members. Jacinda Ardern gave support noting that she and her daughter Neve would post a teddy bear in the windows of Premier House.

Despite the effort to give the movement a local aspect by adding a long kiwi beak to the face of a teddy bear on the website home page, the teddy bear was an expression of international children's culture. Teddy bears date from 1902 when the US President, Theodore Roosevelt, commonly known as 'Teddy', went on a hunt for a black bear with the Governor of Mississippi whose recent criticisms of lynchings in the state had won Roosevelt's support. Roosevelt was impatient to get a trophy; and on the second day the guide, a former slave, located a 235-pound black bear, smashed it over the head and tied it to a tree to allow the President to claim the kill. When Roosevelt arrived he saw a bloodied panting old bear tied to a tree, but he refused to shoot it on the grounds that this would be unsportsmanlike. He ordered instead that it be put out of its misery. Roosevelt's gesture received wide publicity as an act of compassion and a few days later a *Washington Star* cartoon showed Roosevelt turning his back on a cuddly bear as a metaphor for his rejection of lynching. The cartoon was picked up by a Brooklyn candy store owner who decided to make a stuffed bear and call it 'Teddy's bear'. The toy bear was so well received that the owners abandoned selling candies for selling stuffed bears. Before long the teddy bear had spread round the western world.

Its popularity was heightened in New Zealand by the appearance in the 1920s of stories and poems by A. A. Milne about Winnie-the-Pooh, a teddy bear which was a favourite of Milne's son, Christopher Robin.

Pooh was a friendly bear 'of very little brain', who loved 'hunny' and became a childhood favourite. A third item of international children's culture which became popular among New Zealand children was a picture book, *We're Going on a Bear Hunt*, written by Michael Rosen in 1989. At the very time when New Zealand was in lockdown, Rosen was in intensive care in Britain with Covid; but his book became a central symbol of the New Zealand campaign. The first lines: 'We're going on a bear hunt / we're going to catch a big one / what a beautiful day! / we're not scared' became the virtual motto of the movement. The last line seemed perfect reassurance at a time of anxiety. Indeed, another Facebook page was called 'We're Not Scared — NZ Bear Hunt'.

So Ailys Tewnion's figures were part of the New Zealand bear hunt. She added delightful variations to the theme. After the Jacinda and Ashley bears came a teddy bears' picnic with grandpa holding a dog, grandson on a skateboard and the family about to polish off cakes at 'COVID CRESCENT RESERVE'. If big family parties were not allowed under lockdown at least passers-by could contemplate a virtual picnic. When the alerts dropped to level 2 in May 2020 and Kiwis were encouraged to explore their own country and give suffering tourist operators some custom, Ailys put on her fence a crocheted car pulling a 'COVID CRUISER' caravan. There was also a post with AA signs topped by the words 'Teds on tour' and pointing to various attractions such as 'hot pools', beaches', and 'Abel Tas'.

We began this book with Aotearoa as a lonely island in the Pacific, uninhabited and touched only by sea creatures swimming from afar or birds bringing seeds from distant continents. In one sense the history since then has been a progressive increase in exposure to the outside world. The sea brought settlers, first from the Pacific; then from Europe from the eighteenth century. Shipping opened the country to international trade, and from the mid-nineteenth century New Zealand's economy became dependent on the sea lanes to the world. The twentieth century brought air travel, bringing a new influx of migrants from more diverse places — Europe and the Pacific once again, but also from Asia, the Middle East and the Americas. Aircraft brought cabin-loads of tourists from afar to explore the landscapes and culture of these islands. International trade and tourism became the drivers of the New Zealand economy in the twenty-first century. The outside world invaded people's minds through the media — the telegraph in the nineteenth century, radio and movies in the first half of the twentieth century, television in the second half, and in the twenty-first century the internet brought the world to the screens within people's pockets.

It is interesting therefore that we end this book with a strange, if temporary, retreat into family isolation. But the virus came from overseas and what helped to make the lockdown survivable was international culture mediated through children's literature, television, and the internet. Ailys Tewnion's brilliant works are inconceivable without these international influences. In the twenty-first century, even in lockdown, New Zealand was emphatically not adrift in a lonely sea.

Further reading

Detailed references available at penguin.co.nz/a-history-of-new-zealand-in-100-objects-9781761047213

2. Te Arawa taumata atua

George Grey, *Polynesian Mythology and Ancient Traditional History of the New Zealand Race*, H. Brett, Auckland, 1885, first published 1855

Vincent O'Malley and David Armstrong, *The Beating Heart: A Political and Socio-economic History of Te Arawa*, Huia, Wellington, 2008

Don Stafford, *Te Arawa: A History of the Arawa People*, Oratia Books, Auckland, 2016, first published 1967

3. Tairua pearl shell lure

Louise Furey, 'Tairua trolling lure', Auckland War Memorial Museum — Tāmaki Paenga Hira, first published 2 October 2015, updated 10 March 2016, www.aucklandmuseum.com/discover/collections/topics/tairua-trolling-lure

R. C. Green, 'Sources of New Zealand's East Polynesian culture: The evidence of a pearl shell lure hank', *Archaeology and Physical Anthropology in Oceania*, vol. 2 (1967), pp. 81–90

Matthew Schmidt and Thomas Higham, 'Sources of New Zealand's East Polynesian culture revisited: The radiocarbon chronology of the Tairua archaeological site, New Zealand', *Journal of the Polynesian Society*, vol. 107, no. 4 (December 1998), pp. 395–403

https://teara.govt.nz/en/te-hi-ika-maori-fishing

4. Wairau Bar necklace

Atholl Anderson, *Prodigious Birds: Moas and Moa-hunting in Prehistoric New Zealand*, Cambridge University Press, Cambridge, 1989

Janet Davidson, *The Prehistory of New Zealand*, Longman Paul, Auckland, 1984

Roger Duff, *The Moa-Hunter Period of Maori Culture*, Government Printer, Wellington, 1977, first published 1950

James R. Eyles, *Wairau Bar Moa Hunter: The Jim Eyles Story*, River Press, Dunedin, 2007

Chris Jacomb, Richard N. Holdaway, Morten E. Allentoft, Michael Bunce, Charlotte L. Oskam, Richard Walter and Emma Brooks, 'High-precision dating and ancient DNA profiling of moa (Aves: Dinornithiformes) eggshell documents a complex feature at Wairau Bar and refines the chronology of New Zealand settlement by Polynesians', *Journal of Archaeological Science*, vol. 50 (2014), pp. 24–30

5. Monck's Cave kurī

W. Colenso, 'Notes, chiefly historical, on the ancient dog of the New Zealanders', *Transactions of the New Zealand Institute*, vol. 10 (1877), pp. 135–155

Karen Greig, James Boocock, Melinda S. Allen, Elizabeth Matisoo-Smith and Richard Walter, 'Ancient DNA evidence for the introduction and dispersal of dogs (*Canis familiaris*) in New Zealand', *Journal of Pacific Archaeology*, vol. 9, no. 1 (2018), pp. 1–10

Basil Keane, 'Kurī — Polynesian dogs — Traditional accounts of kurī', Te Ara — the Encyclopedia of New Zealand, http://www.TeAra.govt.nz/en/kuri-polynesian-dogs

Michael Trotter and Beverley McCulloch, *Unearthing New Zealand*, GP Books, Wellington, 1989, p. 54

6. Kahungunu hei tiki, Te Arawhiti

Atholl Anderson, Judith Binney and Aroha Harris, *Tangata Whenua: An Illustrated History*, Bridget Williams Books, Wellington, 2015

Dougal Austin, *Te Hei Tiki: An Enduring Treasure in a Cultural Continuum*, Te Papa Press, Wellington, 2019

Mere Whaanga, *A Carved Cloak for Tahu*, Auckland University Press, Auckland, 2004, pp. 87–88

Mere Whaanga, 'Ngāti Kahungunu', Te Ara — the Encyclopedia of New Zealand, http://www.TeAra.govt.nz/en/ngati-kahungunu/pages

John White, *The Ancient History of the Maori, his Mythology and Traditions*, Government Printer, Wellington, 1887

7. Puketoi kete

Atholl Anderson, Moira White and Fiona Petchey, 'Interior lives: The age and interpretation of perishable artefacts from Māori rockshelter sites in inland Otago, New Zealand', *Journal of Pacific Archaeology*, vol. 6, no. 2 (2015), pp. 41–48

Augustus Hamilton, 'Discovery of a Maori kete at upper Taieri', *Transactions of the New Zealand Institute*, vol. 29 (1896), pp. 174–175
Mick Pendergrast, *Te Aho Tapu: The Sacred Thread — Traditional Maori Weaving*, Reed, Auckland, 1987
Awhina Tamarapa (ed.), *Whatu Kākahu: Māori Cloaks*, Te Papa Press, Wellington, 2011
Moira White, Catherine Ann Smith and Kahu Te Kanawa, 'Māori textiles from Puketoi Station, Central Otago, New Zealand', *Textile History*, vol. 46, no. 2 (2015), pp. 213–234

8. Taiaha kura

Gilbert Archey, 'Evolution of certain Maori carving patterns', *Journal of the Polynesian Society*, vol. 42, no. 3 (September 1933), pp. 171–190
Hirini Moko Mead, *Te Toi Whakairo: The Art of Māori Carving*, Reed, Auckland, 1986
Matthew Schmidt, 'The commencement of pa construction in New Zealand pre-history', *Journal of the Polynesian Society*, vol. 105, no. 4 (1996), pp. 441–460
Te Rangi Hiroa, *The Coming of the Maori*, Maori Purposes Fund Board / Whitcombe & Tombs, Wellington, 1949

9. Māori war trumpet or pūkāea

Brian Flintoff, 'Māori musical instruments — taonga puoro', Te Ara — the Encyclopedia of New Zealand, http://www.TeAra.govt.nz/en/maori-musical-instruments-taonga-puoro
Brian Flintoff, *Taonga Pūoro: Singing Treasures — The Musical Instruments of the Māori* Craig Potton, Nelson, 2004
Hilary and John Mitchell, *Te Tau Iho o Te Waka, vol. 1*, Huia, Wellington, 2004
Richard Nunns, *Te Ara Puoro: A Journey into the World of Māori Music*, Craig Potton, Nelson, 2014
Anne Salmond, *Two Worlds: First Meetings between Maori and Europeans, 1642–1772*, Viking, Auckland, 1991
Andrew Sharp, *The Voyages of Abel Janzsoon Tasman*, Clarendon Press, Oxford, 1968
Te Rangi Hiroa, *The Coming of the Maori*, Maori Purposes Fund Board / Whitcombe & Tombs, Wellington, 1949

10. De Surville's anchor

John Dunmore, *The Fateful Voyage of the St Jean Baptiste: A True Account of M. de Surville's Expedition to New Zealand & the Unknown South Seas in the Years 1769–70*, Pegasus Press, Christchurch, 1969
Anne Salmond, *Two Worlds: First Meetings between Maori and Europeans, 1642–1772*, Viking, Auckland, 1991

11. Joseph Banks' or Daniel Solander's kōwhai specimen

J. C. Beaglehole (ed.), *The* Endeavour *Journal of Joseph Banks, 1768–1771, vol. 1*, Trustees of the Public Library of NSW in association with Angus & Robertson, Sydney, 1962
J. C. Beaglehole (ed.), *The Journals of Captain James Cook: 1. The Voyage of the* Endeavour, *1768–1771*, Hakluyt Society, Cambridge, 1968
David Mabberley (ed.), *Joseph Banks' Florilegium: Botanical Treasures from Cook's First Voyage*, Thames & Hudson, London, 2017
Anne Salmond, *Two Worlds: First Meetings between Maori and Europeans, 1642–1772*, Viking, Auckland, 1991

12. James Cook's cannon

J. C. Beaglehole (ed.), *The* Endeavour *Journal of Joseph Banks, 1768–1771, vol. 1*, Trustees of the Public Library of NSW in association with Angus & Robertson, Sydney, 1962
J. C. Beaglehole (ed.), *The Journals of Captain James Cook: The Voyage of the* Endeavour, *1768–1771*, Hakluyt Society, Cambridge, 1968
Anne Salmond, *Two Worlds: First Meetings between Maori and Europeans, 1642–1772*, Viking, Auckland, 1991

13. Solander Island sealskin purse

Robert McNab, *Murihuku: A History of the South Island of New Zealand and the Islands Adjacent and Lying to the South, from 1642 to 1835*, Whitcombe & Tombs, Wellington, 1909
Rhys Richards, *Sealing in the Southern Oceans, 1788–1833*, Paremata Press, Wellington, 2010
June Starke (ed.), *Journal of a Rambler: The Journal of John Boultbee*, Oxford University Press, Auckland, 1986

14. Te Pahi's medal

Angela Ballara, 'Te Pahi', Dictionary of New Zealand Biography, first published in 1990. Te Ara — the Encyclopedia of New Zealand, https://teara.govt.nz/en/biographies/1t53/te-pahi

Deidre Brown, 'Te Pahi's whare: The first European house in New Zealand', in *Fabulation: Myth, Nature, Heritage — Proceedings of the Society of Architectural Historians Australia and New Zealand, vol. 29*, Launceston, Tasmania, 2012, pp. 181–183

Vincent O'Malley, *The Meeting Place: Māori and Pākehā encounters, 1642–1840*, Auckland University Press, Auckland, 2012

Anne Salmond, *Between Worlds: Early Exchanges between Maori and Europeans, 1773–1815*, Viking, Auckland, 1970

Anne Salmond, *Tears of Rangi: Experiments Across Worlds*, Auckland University Press, Auckland, 2017

Mark Stocker, 'A silver slice of Māori history: the Te Pahi medal', *Tuhinga*, vol. 26 (2015), pp. 31–48

15. Hannah King's chair

Judith Binney, *The Legacy of Guilt: A Life of Thomas Kendall*, Oxford University Press, Christchurch, 1968

Barbara Brookes, *A History of New Zealand Women*, Bridget Williams Books, Wellington, 2016

Vivien Caughey, 'The King chest of drawers and the early NZ missionaries', Auckland War Memorial Museum — Tāmaki Paenga Hira, first published 4 June 2015, updated 16 June 2015, www.aucklandmuseum.com/discover/topics/king-chest-of-drawers-and-early-nz-missionaries

John L. Nicholas, *Narrative of a Voyage to New Zealand*, 1817, http://www.enzb.auckland.ac.nz/document/?wid=521&action=null

Kathryn Rountree, 'Re-making the Maori female body: Marianne Williams's mission in the Bay of Islands', *Journal of Pacific History*, vol. 35, no. 1 (2000), pp. 49–66

16. Hongi Hika's gun

Angela Ballara, *Taua: 'Musket Wars', 'Land Wars' or Tikanga? Warfare in Māori Society in the Early Nineteenth Century*, Penguin, Auckland, 2003

Dorothy Urlich Cloher, *Hongi Hika: Warrior Chief*, Viking, Auckland, 2003

R. D. Crosby, *The Musket Wars: A History of Inter-iwi Conflict*, Libro International, Auckland, new edition 2012

Paul D'Arcy, 'Maori and muskets from a pan-Polynesian perspective', *New Zealand Journal of History*, vol. 34, no. 1 (2000), pp. 117–132

Basil Keane, 'Musket wars', Te Ara — the Encyclopedia of New Zealand, http://www.TeAra.govt.nz/en/musket-wars

Anne Salmond, *Tears of Rangi: Experiments Across Worlds*, Auckland University Press, Auckland, 2017

17. Te Rauparaha's mere, Tuhiwai

Angela Ballara, *Taua: 'Musket Wars', 'Land Wars' or Tikanga? Warfare in Māori Society in the Early Nineteenth Century*, Penguin, Auckland, 2003

Patricia Burns, *Te Rauparaha: A New Perspective*, Reed, Wellington, 1980

Ross Calman (trans. and ed.), *He Pukapuka Tātaku i ngā Mahi a Te Rauparaha Nui nā Tamihana Te Rauparaha*, Auckland University Press, 2020

Steven Oliver, 'Taiaroa, Te Mātenga', Dictionary of New Zealand Biography, first published in 1990, updated June 2018. Te Ara — the Encyclopedia of New Zealand, https://teara.govt.nz/en/biographies/1t2/taiaroa-te-matenga

Steven Oliver, 'Te Rauparaha', Dictionary of New Zealand Biography, first published in 1990. Te Ara — the Encyclopedia of New Zealand, https://teara.govt.nz/en/biographies/1t74/te-rauparaha

18. Betty Guard's comb

Harry C. Evison, *Te Wai Pounamu: The Greenstone Island*, Aoraki Press, Wellington & Christchurch, 1993

Don Grady, 'Guard, Elizabeth', Dictionary of New Zealand Biography, first published in 1990. Te Ara — the Encyclopedia of New Zealand, https://teara.govt.nz/en/biographies/1g23/guard-elizabeth

Don Grady, *Guards of the Sea*, Whitcoulls, Christchurch, 1978

Robert McNab, *The Old Whaling Days*, Whitcombe & Tombs, Christchurch, 1913

Jock Phillips, 'Our history, our selves: The historian and national identity', *New Zealand Journal of History*, vol. 30, no. 2 (1996), pp. 119–123

19. Bishop Pompallier's printing press

Robert Glen (ed.), *Mission and Moko: The Church Missionary Society in New Zealand, 1814–1882*, Latimer Fellowship, Christchurch, 1992

R. A. McKay (ed.), *A History of Printing in New Zealand, 1830–1940*, Wellington Club of Printing House Craftsmen, Wellington, 1940

Paul Moon, *Ka Ngaro Te Reo: Māori Language Under Siege in the Nineteenth Century*, Otago University Press, Dunedin, 2016

R. M. Ross, *A Guide to Pompallier House*, Government Printer, Wellington, 1970

E. R. Simmons, 'Pompallier, Jean Baptiste François', Dictionary of New Zealand Biography, first published in 1990, updated November 2010. Te Ara — the Encyclopedia of New Zealand, https://teara.govt.nz/en/biographies/1p23/pompallier-jean-baptiste-francois

20. Flag of the United Tribes

Basil Keane, 'He Whakaputanga — Declaration of Independence', Te Ara — the Encyclopedia of New Zealand, http://www.TeAra.govt.nz/en/he-whakaputanga-declaration-of-independence

Malcolm Mullholland, 'Ngā haki — Māori and flags', Te Ara — the Encyclopedia of New Zealand, http://www.TeAra.govt.nz/en/nga-haki-maori-and-flags

Museum of New Zealand Te Papa Tongarewa, *Icons Ngā Taonga*, Te Papa Press, Wellington, 2004, p. 194

Waitangi Tribunal, *He Whakaputanga me te Tiriti: The Declaration and the Treaty — Report on Stage 1 of the Te Paparahi o Te Rahi Inquiry*, Wai 1040, Wellington, 2014

21. William Wakefield's epaulets

Miles Fairburn, 'Wakefield, Edward Gibbon', Dictionary of New Zealand Biography, first published in 1990. Te Ara — the Encyclopedia of New Zealand, https://teara.govt.nz/en/biographies/1w4/wakefield-edward-gibbon

Philip Temple, *A Sort of Conscience: The Wakefields*, Auckland University Press, Auckland, 2002

Waitangi Tribunal, *Te Whanganui a Tara Report*, Wai 145, Wellington, 2003

22. The Waitangi sheet of the Treaty of Waitangi

William Colenso, *The Authentic and Genuine History of the Signing of the Treaty of Waitangi*, Government Printer, Wellington, 1890

NZHistory, 'Treaty sheets and signing locations', https://nzhistory.govt.nz/politics/treaty/making-the-treaty/treaty-of-waitangi-signing-locations (Ministry for Culture and Heritage)

Claudia Orange, *The Treaty of Waitangi*, Allen & Unwin, Wellington, 1987

Waitangi Tribunal, *He Whakaputanga me te Tiriti: The Declaration and the Treaty — Report on Stage 1 of the Te Paparahi o Te Rahi Inquiry*, Wai 1040, Wellington, 2014

23. Ruapekapeka waka huia

James Belich, *The New Zealand Wars and the Victorian Interpretation of Racial Conflict*, Auckland University Press, Auckland, 1986

James Cowan, *The New Zealand Wars, vol. 1, 1845–64*, Government Printer, Wellington, 1922

Vincent O'Malley, *The New Zealand Wars / Ngā Pakanga o Aotearoa*, Bridget Williams Books, Wellington, 2019

24. Millstones from Pātea

Hazel Petrie, *Chiefs of Industry: Māori Tribal Enterprise in Early Colonial New Zealand*, Auckland University Press, Auckland, 2006

Tony Sole, *Ngāti Ruanui: A History*, Huia, Wellington, 2004

25. John Buchanan's table

Erik Olssen, *A History of Otago*, John McIndoe, Dunedin, 1984

Brad Patterson, Tom Brooking and Jim McAloon, *Unpacking the Kists: The Scots in New Zealand*, McGill-Queen's University Press / Otago University Press, Montreal & Dunedin, 2013

Jock Phillips and Terry Hearn, *Settlers*, Auckland University Press, Auckland, 2008

https://collections.toituosm.com/objects/51247

26. Emma Barker's sewing box

Barbara Brookes, *A History of New Zealand Women*, Bridget Williams Books, Wellington, 2016

Canterbury Centennial Committee, *A History of Canterbury, vol. 1*, Whitcombe & Tombs, Christchurch, 1957

Raewyn Dalziel, 'The colonial helpmeet: Women's role and the vote in nineteenth-century New Zealand', *New Zealand Journal of History*, vol. 11, no. 2 (1977), pp. 112–123

Frances Porter, Charlotte Macdonald and Tui Macdonald (eds), *'My Hand will Write what my Heart Dictates': The Unsettled Lives of Women in Nineteenth-century New Zealand as Revealed to Sisters, Family and Friends*, Auckland University Press / Bridget Williams Books, Auckland & Wellington, 1996

27. Te Reko's cast-iron pot

Roger Frazer, 'Chalmers, Nathanael', Dictionary of New Zealand Biography, first published in 1990. Te Ara — the Encyclopedia of New Zealand, https://teara.govt.nz/en/biographies/1c11/chalmers-nathanael

W. G. McClymont, *The Exploration of New Zealand*, Department of Internal Affairs, Wellington, 1940

Philip Temple, *New Zealand Explorers: Great Journeys of Discovery*, Whitcoulls, Christchurch, 1985

Waitangi Tribunal, *Ngai Tahu Land Report*, Wellington, 1991

Alan Ward, 'A Report on the Historical Evidence: The Ngai Tahu Claim Wai 27', May 1989

28. Colonel Robert Wynyard's epergne

Kelly Dix, 'The Wynyard Testimonial', Auckland War Memorial Museum — Tāmaki Paenga Hira, first published 21 June 2016, updated 1 July 2016, http://www.aucklandmuseum.com/discover/collections/topics/the-wynyard-testimonial

W. David McIntyre, 'Self-government and independence', Te Ara — the Encyclopedia of New Zealand, http://www.TeAra.govt.nz/en/self-government-and-independence

Frank Rogers, 'Wynyard, Robert Henry', Dictionary of New Zealand Biography, first published in 1990. Te Ara — the Encyclopedia of New Zealand, https://teara.govt.nz/en/biographies/1w40/wynyard-robert-henry

Russell Stone, 'Auckland party politics in the early years of the provincial system, 1853–58', *New Zealand Journal of History*, vol. 14, no. 2 (1980), pp. 153–178

General Robert Henry Wynyard in *The Cyclopedia of New Zealand [Auckland Provincial District]*, The Cyclopedia Company Limited, Christchurch, 1902

29. Wiremu Kīngi's tauihu

James Cowan, *The New Zealand Wars and the Pioneering Period, vol. I*, Government Printer, Wellington, 1922

Kelvin Day, 'Voyaging Taonga: The Kīngi Tauihi', in Annabel Cooper, Lachy Paterson and Angela Wanhalla (eds), *The Lives of Colonial Objects*, Otago University Press, Dunedin, 2015, pp. 35–39

Andrew Moffat, *Flashback: Tales and Treasures of Taranaki*, Huia, Wellington, 2010, pp. 209–12

Vincent O'Malley, *The New Zealand Wars / Ngā Pakanga o Aotearoa*, Bridget Williams Books, Wellington, 2019

Ann Parsonson, 'Te Rangitake, Wiremu Kingi', *Dictionary of New Zealand Biography*, first published in 1990. Te Ara — the Encyclopedia of New Zealand, https://teara.govt.nz/en/biographies/1t70/te-rangitake-wiremu-kingi

30. James Quedley's red coatee

Michael Barthorp, *British Infantry Uniforms since 1660*, Blandford Press, Poole, Dorset, 1982

James Belich, *The New Zealand Wars and the Victorian Interpretation of Racial Conflict*, Auckland University Press, Auckland, 1986

James Cowan, *The New Zealand Wars and the Pioneering Period, vol. I*, Government Printer, Wellington, 1922

Richard Holmes, *Redcoat: The British Soldier in the Age of Horse and Musket*, HarperCollins, London, 2001

Vincent O'Malley, *The Great War for New Zealand: Waikato 1800–2000*, Bridget Williams Books, Wellington, 2016

Keith Sinclair (ed.), *A Soldier's View of Empire: The Reminiscences of James Bodell, 1831–92*, Bodley Head, London, 1982

31. Moutoa flag

Judith Binney, 'Māori prophetic movements — Ngā Poropiti — Te Ua Haumēne — Pai Mārire and Hauhau', Te Ara — the Encyclopedia of New Zealand, http://www.TeAra.govt.nz/en/maori-prophetic-movements-nga-poropiti

Paul Clark, *'Hauhau': The Pai Marire search for Maori Identity*, Auckland University Press / Oxford University Press, Auckland & Wellington, 1975

James Cowan, *The New Zealand Wars and the Pioneering Period, vol. II*, Government Printer, Wellington, 1922
Jock Phillips, *To the Memory: New Zealand War Memorials*, Potton & Burton, Nelson, 2016
Waitangi Tribunal, *Te Kāhui Maunga: The National Park District Inquiry Report*, Wai 1130, Wellington, 2013

32. Kereopa's bailer

Paul Clark, *'Hauhau': The Pai Marire search for Maori Identity*, Auckland University Press / Oxford University Press, Auckland & Wellington, 1975
Steven Oliver, 'Te Rau, Kereopa', Dictionary of New Zealand Biography, first published in 1990, updated June 2014. Te Ara — the Encyclopedia of New Zealand, https://teara.govt.nz/en/biographies/1t72/te-rau-kereopa
Vincent O'Malley, 'Frontier justice?: The trial and execution of Kereopa Te Rau', *Journal of the Polynesian Society*, vol. 120, no. 2 (2007), pp. 183–191
Evelyn Stokes, 'Völkner, Carl Sylvius', Dictionary of New Zealand Biography, first published in 1990. Te Ara — the Encyclopedia of New Zealand, https://teara.govt.nz/en/biographies/1v5/volkner-carl-sylvius
Ranginui Walker, *Ōpōtiki-Mai-Tawhiti: Capital of Whakatōhea*, Penguin Books, Auckland, 2007
Peter Wells, *Journey to a Hanging*, Random House, Auckland, 2014

33. Paddy Galvin's pipes

Stevan Eldred-Grigg, *Diggers, Hatters & Whores: The Story of the New Zealand Gold Rushes*, Random House, Auckland, 2008
Tom Field and Erik Olssen, *Relics of the Goldfields: Central Otago*, John McIndoe, Dunedin, 1976
John Hall-Jones, *Goldfields of Otago: An Illustrated History*, Craig Printing, Invercargill, 2005
Jock Phillips and Terry Hearn, *Settlers: New Zealand Immigrants from England, Ireland and Scotland, 1800–1945*, Auckland University Press, Auckland, 2008

34. Chinese li-ding scales

Stevan Eldred-Grigg with Zeng Dazheng, *White Ghosts, Yellow Peril: China and New Zealand 1790–1950*, Otago University Press, Dunedin, 2014
Manying Ip, 'Chinese', Te Ara — the Encyclopedia of New Zealand, http://www.TeAra.govt.nz/en/chinese
James Ng, *Windows on a Chinese Past*, Otago Heritage Books, Dunedin, vol. 1, 1993; vol. 2, 1995

35. Timaru life-saving rocket

Gerard Hutching, 'Shipwrecks', Te Ara — the Encyclopedia of New Zealand, http://www.TeAra.govt.nz/en/shipwrecks
C. W. N. Ingram, *New Zealand Shipwrecks: 195 Years of Disasters at Sea*, 7th rev. ed., Beckett Books, Auckland, 1990
Gavin McLean, *Shipwrecks and Maritime Disasters*, Grantham House, Wellington, 1991

36. The locomotive *Josephine*

Neill Atkinson, *Trainland: How Railways Made New Zealand*, Random House, Auckland, 2007
Robin Bromby, *Rails that Built a Nation: An Encyclopedia of New Zealand Railways*, Grantham House, Wellington, 2003
Geoffrey Churchman and Tony Hurst, *The Railways of New Zealand: A Journey through History*, Transpress, Wellington, 2001
David Leitch and Bob Stott, *New Zealand Railways: The First 125 Years*, Heinemann Reed, Auckland, 1988.
Euan McQueen, *Rails in the Hinterland: New Zealand's Vanishing Railway Landscape*, Grantham House, Wellington, 2005
A. N. Palmer and W. W. Stewart, *Cavalcade of New Zealand Locomotives*, A. H. & A. W. Reed, Wellington, first published 1956, revised 1965

37. Elisabet Engebretsdatter's hair embroidery

Val A. Burr, *Mosquitoes & Sawdust: A History of Scandinavians in Early Palmerston North and Surrounding Districts*, Scandinavian Club of Manawatu, Palmerston North, 1995
G. C. Petersen, *Forest Homes*, A. H. & A. W. Reed, Wellington, 1956
Carl Walrond, 'Scandinavians', Te Ara — the Encyclopedia of New Zealand, http://www.TeAra.govt.nz/en/scandinavians

38. Ōkaramio blacksmith bellows

Tom Brooking, 'The equine factor: the powerhouse of the colonisation of New Zealand to 1945', in Lily Baker (ed.), *On the Horse's Back: Proceedings of the 2004 Conference of the New Zealand Society of Genealogists*, Auckland: New Zealand Society of Genealogists, Auckland, 2004, pp. 53–60

Emma Meyer, 'Horses', Te Ara — the Encyclopedia of New Zealand, http://www.TeAra.govt.nz/en/horses

Carolyn Mincham, *Attitude and Heart: The Horse in New Zealand*, David Bateman, Auckland, 2011

Steve Waterman, *Blacksmith*, Price Milburn, Wellington, 1981

39. Parihaka plough/parau

Rachel Buchanan, *The Parihaka Album*, Huia, Wellington, 2009

Danny Keenan, 'Te Whiti-o-Rongomai III, Erueti', Dictionary of New Zealand Biography, first published in 1993, updated November 2012. Te Ara — the Encyclopedia of New Zealand, https://teara.govt.nz/en/biographies/2t34/te-whiti-o-rongomai-iii-erueti

Hazel Riseborough, *Days of Darkness: Taranaki, 1878–1884*, Allen & Unwin, Wellington, 1989

Dick Scott, *Ask That Mountain: The Parihaka Story*, Reed, Auckland, 1975

Ailsa Smith, 'Tohu Kākahi', Dictionary of New Zealand Biography, first published in 1993. Te Ara — the Encyclopedia of New Zealand, https://teara.govt.nz/en/biographies/2t44/tohu-kakahi

Waitangi Tribunal, *The Taranaki Report: Kaupapa Tuatahi*, Wellington, 1996

40. Totara Estate killing knives

Mervyn Palmer, 'Brydone, Thomas', *Dictionary of New Zealand Biography*, first published in 1993. Te Ara — the Encyclopedia of New Zealand, https://teara.govt.nz/en/biographies/2b45/brydone-thomas

Mervyn Palmer, 'William Soltau Davidson: A pioneer of New Zealand estate management', *New Zealand Journal of History*, vol. 7, no. 2 (1973), pp. 148–164

Rebecca J. H. Woods, 'Breed, culture, and economy: The New Zealand frozen meat trade, 1880–1914', *Agricultural History Review*, vol. 60, no. 2 (2012), pp. 288–308

41. Gaelic Society targe

Tom Brooking, *Lands for the People? The Highland Clearances and the Colonisation of New Zealand: A Biography of John McKenzie*, Otago University Press, Dunedin, 1996

Alison Clarke, *Holiday Seasons: Christmas, New Year and Easter in Nineteenth-Century New Zealand*, Auckland University Press, Auckland, 2007

Brad Patterson, Tom Brooking and Jim McAloon, *Unpacking the Kists: The Scots in New Zealand*, McGill-Queen's University Press / Otago University Press, Montreal & Dunedin, 2013

Jock Phillips and Terry Hearn, *Settlers: New Zealand Immigrants from England, Ireland and Scotland, 1800–1945*, Auckland University Press, Auckland, 2008

42. Amelia Haszard's sewing machine

Geoff Conly, *Tarawera: The Destruction of the Pink and White Terraces*, Grantham House, Wellington, 1985

Jenifer Curnow, 'Hinerangi, Sophia', Dictionary of New Zealand Biography, first published in 1993, updated July 2015. Te Ara — the Encyclopedia of New Zealand, https://teara.govt.nz/en/biographies/2h37/hinerangi-sophia

R. F. Keam, *Tarawera: The Volcanic Eruption of 10 June 1886*, R. F. Keam, Auckland, 1988

Kerryn Pollock, 'Sewing, knitting and textile crafts — Home sewing', Te Ara — the Encyclopedia of New Zealand, http://www.TeAra.govt.nz/en/sewing-knitting-and-textile-crafts

43. 1893 Women's suffrage petition

Judith Devaliant, *Kate Sheppard: A Biography*, Penguin Books, Auckland, 1992

Patricia Grimshaw, *Women's Suffrage in New Zealand*, Auckland University Press, Auckland, first published 1972, 2nd edition 1987

Jock Phillips, *A Man's Country? The Image of the Pakeha Male — A History*, Penguin Books, Auckland, 1987

The Women's Suffrage Petition — Te Petihana Whakamana Pōti Wahine, Bridget Williams Books, Wellington, 2017

44. Meri Te Tai Mangakāhia's parliamentary chest

Angela Ballara, 'Mangakahia, Meri Te Tai', *Dictionary of New Zealand Biography*, first published in 1993. Te Ara — the Encyclopedia of New Zealand, https://teara.govt.nz/en/biographies/2m30/mangakahia-meri-te-tai

Angela Ballara, 'Wāhine rangatira: Māori women of rank and their role in the women's Kotahitanga movement of the 1890s', *New Zealand Journal of History*, vol. 27, no. 2 (October 1993), pp. 127–139
https://www.aucklandmuseum.com/discover/stories/history/meris-parliamentary-chest-and-mary-anns-chair
Tania Rei, *Māori Women and the Vote*, Huia, Wellington, 1993
John A. Williams, *Politics of the New Zealand Maori: Protest and Cooperation, 1891–1909*, Oxford University Press for the University of Auckland, 1969

45. H. B. Lusk's cricket bat

Dan Reese, *Was it All Cricket?*, Allen & Unwin, London, 1948
Greg Ryan, *The Making of New Zealand Cricket, 1832–1914*, Frank Cass, London, 2004

46. Northland gum cathedral

Alfred Reed, *The Kauri Gumdiggers*, Bush Press, Auckland, 3rd edition, 2006
Carl Walrond, 'Kauri gum and gum digging — Gum digging methods', Te Ara — the Encyclopedia of New Zealand, http://www.TeAra.govt.nz/en/kauri-gum-and-gum-digging

47. Westport District Gold Miners' Union banner

https://collections.tepapa.govt.nz/object/953428
Len Richardson, *Coal, Class & Community: The United Mineworkers of New Zealand, 1880–1960*, Auckland University Press, Auckland, 1995
H. Roth, *Trade Unions in New Zealand: Past and Present*, Reed, Wellington, 1973

48. George Bradford's bandolier

John Crawford with Ellen Ellis, *To Fight for the Empire: An Illustrated History of New Zealand and the South African War*, Reed, Auckland, 1999
Jock Phillips, *A Man's Country? The Image of the Pakeha Male — A History*, Penguin, Auckland, 1987
Jock Phillips, 'South African War', Te Ara — the Encyclopedia of New Zealand, http://www.TeAra.govt.nz/en/south-african-war
Nigel Robson, *Our First Foreign War: The Impact of the South African War 1899–1902 on New Zealand*, Massey University Press, Auckland, 2021

49. Richard Seddon's coronation coatee

Tom Brooking, *Richard Seddon: King of God's Own*, Penguin Books, Auckland, 2014
David Hamer, 'Seddon, Richard John', Dictionary of New Zealand Biography, first published in 1993. Te Ara — the Encyclopedia of New Zealand, https://teara.govt.nz/en/biographies/2s11/seddon-richard-john
https://collections.tepapa.govt.nz/topic/1580

50. Christchurch 1906 exhibition waharoa

Roger Blackley, *Galleries of Maoriland: Artists, Collectors and the Māori World, 1880–1910*, Auckland University Press, Auckland, 2018
Paul Greenhalgh, *Ephemeral Vistas: The Expositions Universelles, Great Exhibitions, and World's Fairs, 1851–1939*, Manchester University Press, Manchester, 1988
Roger Neich, *Carved histories: Rotorua Ngāti Tarāwhai Woodcarving*, Auckland University Press, Auckland, 2008
Conal McCarthy, *Exhibiting Māori: A History of Colonial Cultures of Display*, Berg, Oxford & New York, 2007
John Mansfield Thomson (ed.), *Farewell Colonialism: The New Zealand International Exhibition Christchurch, 1906–07*, Dunmore Press, Palmerston North, 1998

51. Jimmy Hunter's All Blacks jersey

Erik Dunning and Kenneth Sheard, *Barbarians, Gentlemen and Players: A Sociological Study of the Development of Rugby Football*, Martin Robertson, Oxford, 1979
Ron Palenski, *Century in Black: 100 years of All Black Test Rugby*, Hodder Moa Beckett, Auckland, 2003
Jock Phillips, *A Man's Country? The Image of the Pakeha Male — A History*, Penguin, Auckland, 1987
Greg Ryan, *Forerunners of the All Blacks*, Canterbury University Press, Christchurch, 1993
Greg Ryan (ed.), *Tackling Rugby Myths: Rugby and New Zealand Society, 1854–2004*, Otago University Press, Dunedin, 2005
Jill White, 'Women at work in the Manawatu Knitting Mills', *Manawatu Journal of History*, vol. 3 (2007), pp. 20–29

52. The *Dundonald* coracle

Herbert Escott-Inman, *The Castaways of Disappointment Island*, S. W. Partridge and Co., London, 1933

Conon Fraser, *Beyond the Roaring Forties: New Zealand's Subantarctic Islands*, Government Printer, Wellington, 1986

Jock Phillips, 'Subantarctic islands', Te Ara — the Encyclopedia of New Zealand, http://www.TeAra.govt.nz/en/subantarctic-islands

53. Dannevirke Plunket scales

Barbara Brookes. 'King, Frederic Truby', Dictionary of New Zealand Biography, first published in 1993, updated October 2011. Te Ara — the Encyclopedia of New Zealand, https://teara.govt.nz/en/biographies/2k8/king-frederic-truby

Linda Bryder, *A Voice for Mothers: The Plunket Society and Infant Welfare*, Auckland University Press, Auckland, 2003

Lloyd Chapman, *In a Strange Garden: The Life and Times of Truby King*, Penguin, Auckland, 2003

Mary King, *Truby King the Man: A Biography*, G. Allen & Unwin, London, 1948

Erik Olssen, 'Truby King and the Plunket Society: An analysis of a prescriptive ideology', *New Zealand Journal of History*, vol. 15, no. 1 (1981), pp. 3–23

Jim Sullivan, *I was a Plunket Baby: 100 years of the Royal New Zealand Plunket Society (Inc)*, Random House, Auckland, 2007

54. Hurleyville dairy cooperative cheese crate

Gordon McLauchlan, *The Farming of New Zealand: An Illustrated History of New Zealand Agriculture*, Australia and New Zealand Book Company, Auckland, 1981

Hugh Stringleman and Frank Scrimgeour, 'Dairying and dairy products', Te Ara — the Encyclopedia of New Zealand, http://www.TeAra.govt.nz/en/dairying-and-dairy-products

Eric Warr, *From Bush-burn to Butter*, Butterworths, Wellington, 1988

55. Leslie Adkin's baton

Anthony Dreaver, *An Eye for Country: The Life and Work of Leslie Adkin*, Victoria University Press, Wellington, 1997

Melanie Nolan (ed.), *Revolution: The 1913 Great Strike in New Zealand*, Canterbury University Press with Trade Union History Project, Christchurch, 2005

Erik Olssen, *The Red Feds: Revolutionary Industrial Unionism and the New Zealand Federation of Labour, 1908–1913*, Oxford University Press, Auckland, 1988

Herbert Roth, *Trade Unions in New Zealand: Past and Present*, A. H. & A. W. Reed, Wellington, 1973

56. Edmond Malone's lemon-squeezer

John Crawford, *No Better Death: The Great War Diaries and Letters of William G. Malone*, Exisle Publishing, Auckland, 2014

Glyn Harper, *Johnny Enzed: The New Zealand Soldier in the First World War*, Exisle Publishing, Auckland, 2015

Ian McGibbon, *New Zealand's Western Front Campaign*, Bateman, Auckland, 2016

Jock Phillips, Nicholas Boyack and E. P. Malone (eds), *The Great Adventure: New Zealand Soldiers Describe the First World War*, Allen & Unwin, Wellington, 1988

Christopher Pugsley, *Gallipoli: The New Zealand Story*, Hodder & Stoughton, Auckland, 1984

57. Margaret Cruickshank's gold watch

Sandra Coney, *Standing in the Sunshine: A History of New Zealand Women Since they Won the Vote*, Viking, Auckland, 1993

Beryl Hughes, 'Cruickshank, Margaret Barnet', Dictionary of New Zealand Biography, first published in 1996. Te Ara — the Encyclopedia of New Zealand, https://teara.govt.nz/en/biographies/3c41/cruickshank-margaret-barnet

Jock Phillips, 'Timekeeping', Te Ara — the Encyclopedia of New Zealand, http://www.TeAra.govt.nz/en/timekeeping

Geoffrey Rice, *Black November: The 1918 Influenza Epidemic in New Zealand*, Allen & Unwin / Department of Internal Affairs Historical Branch, Wellington, 1988

58. Tainui mere and the Prince of Wales

Michael Belgrave, *Dancing with the King: The Rise and Fall of the King Country, 1864–1885*, Auckland University Press, Auckland, 2017

Sarah Carter and Maria Nugent (eds), *Mistress of Everything: Queen Victoria in Indigenous Worlds*, Manchester University Press, Manchester, 2016
Rahui Papa and Paul Meredith, 'Kīngitanga — the Māori King movement', Te Ara — the Encyclopedia of New Zealand, http://www.TeAra.govt.nz/en/kingitanga-the-maori-king-movement
Jock Phillips, 'Māori and royal visits, 1869–2015: From Rotorua to Waitangi', *Royal Studies Journal*, vol. 5, no. 1 (2018), pp. 33–54
Guy H. Scholefield, *Visit of His Royal Highness the Prince of Wales to the Dominion of New Zealand, April–May 1920*, Government Printer, Wellington, 1920

59. ATCO motor mower

Mark Derby, 'Suburbs', Te Ara — the Encyclopedia of New Zealand, http://www.TeAra.govt.nz/en/suburbs
Kenneth T. Jackson, *Crabgrass Frontier: The Suburbanization of the United States*, Oxford University Press, New York & Oxford, 1985
Malcolm McKinnon (ed.), with Barry Bradley and Russell Kirkpatrick, *New Zealand Historical Atlas*, David Bateman, Auckland, 1997, plate 72
Ben Schrader, *The Big Smoke: New Zealand Cities, 1840–1920*, Bridget Williams Books, Wellington, 2016

60. Rudall Hayward's camera

Bruce Babbington, *A History of the New Zealand Feature Film*, Manchester University Press, Manchester, 2007
Geoffrey B. Churchman (ed.), *Celluloid Dreams: A Century of Film in New Zealand*, IPL Books, Wellington, 1997
Annabel Cooper, *Filming the Colonial Past: The New Zealand Wars on Screen*, Otago University Press, Dunedin, 2018
Sam Edwards and Stuart Murray, 'A rough island story: The films of Rudall Charles Hayward', in Ian Conrick and Stuart Murray (eds), *New Zealand Filmmakers*, Wayne State University Press, Detroit, 2007
Diane Pivac (ed.) with Frank Stark and Lawrence McDonald, *New Zealand Film: An Illustrated History*, Te Papa Press, Wellington, 2011

61. The thermette

Graham Hawkes, *On the Road: The Car in New Zealand*, GP Books, Wellington, 1990
Eric Pawson, 'Cars and the motor industry', Te Ara — the Encyclopedia of New Zealand, http://www.TeAra.govt.nz/en/cars-and-the-motor-industry

62. Harold Pond's Napier Technical School uniform

J. C. Dakin, *Education in New Zealand*, David and Charles, Plymouth, 1973
Helen McConnochie, *Afterwords: Interviews and Letters from Survivors of the 1931 Hawke's Bay Earthquake*, Friends of Hawke's Bay Cultural Trust, Napier, 2004
Robert McGregor, *The Hawke's Bay Earthquake: New Zealand's Greatest Natural Disaster*, Art Deco Trust, Napier, 1998
Anna Rogers, *New Zealand Tragedies: Earthquakes*, Grantham House, Wellington, 1996
Matthew Wright, *Quake: Hawke's Bay 1931*, Reed, Auckland, 2001

63. Eva Bowes' flour bag bloomers

Francis Bennett, A *Canterbury Tale*, Oxford University Press, Wellington, 1980
Gerard Hindmarsh, 'United we stand: Blackball and the working-class struggle', *New Zealand Geographic*, vol. 47 (July–September 2000), https://www.nzgeo.com/stories/united-we-stand-blackball-and-the-working-class-struggle/
Malcolm McKinnon, *The Broken Decade: Prosperity, Depression and Recovery in New Zealand, 1928–39*, Otago University Press, Dunedin, 2016
Tony Simpson, *The Sugarbag Years: An Oral History of the 1930s Depression in New Zealand*, Alister Taylor, Wellington, 1974

64. Photo of Michael Joseph Savage

Peter Franks and Jim McAloon, *Labour: The New Zealand Labour Party 1916–2016*, Victoria University Press, Wellington, 2016
Barry Gustafson, *From the Cradle to the Grave: A Biography of Michael Joseph Savage*, Penguin, Auckland, 1986

William Main, 'Digby, Spencer Harry Gilbee', Dictionary of New Zealand Biography, first published in 2000. Te Ara — the Encyclopedia of New Zealand, https://teara.govt.nz/en/biographies/5d19/digby-spencer-harry-gilbee

65. Pacific Burleigh radio

Patrick Day, *The Radio Years: A History of Broadcasting in New Zealand, vol. 1*, Auckland University Press with Broadcasting History Trust, Auckland, 1994
Peter Downes and Peter Harcourt, *Voices in the Air: Radio Broadcasting in New Zealand — A Documentary*, Methuen / Radio New Zealand, Wellington, 1976
John W. Stokes, *The Golden Age of Radio in the Home*, 2nd edition, Craigs Printers and Publishers, Invercargill, 1998

66. *Te Winika* waka

Michael King, *Te Puea: A Biography*, Hodder & Stoughton, Auckland, 1977
Ann Parsonson, 'Hērangi, Te Kirihaehae Te Puea', Dictionary of New Zealand Biography, first published in 1996. Te Ara — the Encyclopedia of New Zealand, https://teara.govt.nz/en/biographies/3h17/herangi-te-kirihaehae-te-puea
Beryl Te Wiata, 'Te Wīata, Īnia Mōrehu Tauhia Wātene Iarahi Waihuihia — Te Wīata, Īnia Mōrehu Tauhia Wātene Iarahi Waihuihia', Dictionary of New Zealand Biography, first published in 2000. Te Ara — the Encyclopedia of New Zealand, https://teara.govt.nz/en/biographies/5t12/te-wiata-inia-morehu-tauhia-watene-iarahi-waihurihia

67. Centennial exhibition souvenir ashtray

N. B. Palethorpe, *Official History of the New Zealand Centennial Exhibition*, New Zealand Centennial Exhibition Company, Wellington, 1940
Jock Phillips, 'Smoking', Te Ara — the Encyclopedia of New Zealand, http://www.TeAra.govt.nz/en/smoking
William Renwick (ed.), *Creating a National Spirit: Celebrating New Zealand's Centennial*, Victoria University Press, Wellington, 2004

68. The Pope's accordion

W. A. Glue and D. J. C. Pringle, *20th Battalion and Armoured Regiment*, Historical Publications, Wellington, 1957
Megan Hutching (ed.), *Inside Stories: New Zealand Prisoners of War Remember*, HarperCollins, Auckland, 2002
W. Wynne Mason, *Prisoners of War*, Historical Publications, Wellington, 1954
I. C. McGibbon, *New Zealand and the Second World War: The People, the Battles and the Legacy*, Hodder Moa Beckett, Auckland, 2004
Christopher Pugsley, *A Bloody Road Home: World War Two and New Zealand's Heroic Second Division*, Penguin, Auckland, 2014

69. Victory cot blanket

John Crawford (ed.), *Kia Kaha: New Zealand in the Second World War*, Oxford University Press, South Melbourne, 2002
Eve Ebbett, *When the Boys Were Away*, Reed, Wellington, 1984
Heather Nicholson, *The Loving Stitch: A History of Knitting and Spinning in New Zealand*, Auckland University Press, Auckland, 1998
Nancy Taylor, *The Home Front, vol. II*, Historical Publications, Wellington, 1986

70. Chip and Rona Bailey's typewriter

Michael Bassett, *Confrontation '51: The 1951 Waterfront Dispute*, A. H. & A. W. Reed, Wellington, 1972
Peter Franks, 'Bailey, Chip', Dictionary of New Zealand Biography, first published in 2000. Te Ara — the Encyclopedia of New Zealand, https://teara.govt.nz/en/biographies/5b2/bailey-chip
Peter Franks, 'Bailey, Rona', Dictionary of New Zealand Biography, first published in 2018. Te Ara — the Encyclopedia of New Zealand, https://teara.govt.nz/en/biographies/6b3/bailey-rona
David Grant (ed.), *The Big Blue: Snapshots of the 1951 Waterfront Lockout*, Canterbury University Press, Christchurch, 2004
Dick Scott, *151 Days: History of the Great Waterfront Lockout and Supporting Strikes, February 15–July 15, 1951*, Auckland Waterside Workers Union, Auckland, 1952

71. Jim Bradley's flagon case

Michael Donaldson, *Beer Nation: The Art and Heart of Kiwi Beer*, Penguin, Auckland, 2012

Jock Phillips, 'Alcohol', Te Ara — the Encyclopedia of New Zealand, http://www.TeAra.govt.nz/en/alcohol
Bernard Teahan, *A Great Social Experiment: The Story of Licensing Trusts in New Zealand*, Fraser Books, Masterton, 2017

72. Godfrey Bowen's handpiece

Ron Palenski, 'Bowen, Walter Godfrey', Dictionary of New Zealand Biography, first published in 2000. Te Ara — the Encyclopedia of New Zealand, https://teara.govt.nz/en/biographies/5b37/bowen-walter-godfrey
Des Williams, *Don't Forget the Sweat Towel: Great Days in New Zealand Shearing*, Last Side Publishing Ltd, Hamilton, 2020
Des Williams, *Top Class Wool Cutters*, Shearing Heritage Publications, Hamilton, 1996

73. Crown Lynn coronation mug

Valerie Ringer Monk, *Crown Lynn: A New Zealand Icon*, Penguin Books, Auckland, 2006
Jock Phillips, *Royal Summer: The Visit of Queen Elizabeth II and Prince Philip to New Zealand, 1953–54*, Daphne Brasell Associates Press, Wellington, 1993

74. Happy Families card game

Barbara Brookes, *A History of New Zealand Women*, Bridget Williams Books, Wellington, 2016
Bronwyn Labrum, *Real Modern: Everyday New Zealand in the 1950s and 1960s*, Te Papa Press, Wellington, 2017

75. Winston Reynolds' television set

R. Boyd Bell, *New Zealand Television: The First 25 Years*, Reed Methuen, Auckland, 1985
Roger Horrocks and Nick Perry (eds), *Television in New Zealand: Programming the Nation*, Oxford University Press, South Melbourne, 2004
Television New Zealand, *50 Years of New Zealand Television*, https://www.nzonscreen.com/title/50-years-of-nz-tv-2010/series

76. Mt Eden gallows

Mark Derby, *Rock College: An Unofficial History of Mount Eden Prison*, Massey University Press, Auckland, 2020
Donald F. MacKenzie, *While We Have Prisons*, Methuen, Auckland, 1980
Robert Muldoon, *The Rise and Fall of a Young Turk*, A. H. & A. W. Reed, Wellington, 1974
Sherwood Young, *Guilty on the Gallows: Famous Capital Crimes of New Zealand*, Grantham House, Wellington, 1998

77. Margaret Sparrow's contraceptive pills

Barbara Brookes, Claire Gooder and Nancy De Castro, '"Feminine as her handbag, modern as her hairstyle": The uptake of the contraceptive pill in New Zealand', New Zealand Journal *of History*, vol. 47, no. 2 (2013), pp. 208–231
Barbara Brookes, *A History of New Zealand Women*, Bridget Williams Books, Wellington, 2016
Sandra Coney, *Standing in the Sunshine: A History of New Zealand Women Since they Won the Vote*, Viking, Auckland, 1993
Helen Smyth, *Rocking the Cradle: Contraception, Sex and Politics in New Zealand*, Steele Roberts, Wellington, 2000

78. Dunedin Committee on Vietnam protest banner

Stephanie Gibson, Matariki Williams and Puawai Cairns, *Protest Tautohetohe: Objects of Resistance, Persistence and Defiance*, Te Papa Press, Wellington, 2019
David McCraw, 'Reluctant ally: New Zealand's entry into the Vietnam War', *New Zealand Journal of History*, vol. 15 (April 1981), pp. 49–60
Roberto Rabel, *New Zealand and the Vietnam War: Politics and Diplomacy*, Auckland University Press, Auckland, 2005

79. Save Manapouri Campaign share certificate

John E. Martin (ed.), *People, Politics and Power Stations: Electric Power Generation in New Zealand, 1880–1998*, 2nd edition, Electricity Corporation, Wellington, 1998 (first published 1991)
Neville Peat, *Manapouri Saved: New Zealand's First Great Conservation Success Story*, Longacre Press, Dunedin, 1995
Roger Wilson, *From Manapouri to Aramoana: The Battle for New Zealand's Environment*, Earthworks Press, Auckland, 1982

David Young, *Our Islands, Our Selves: A History of Conservation in New Zealand*, Otago University Press, Dunedin, 2004

80. Ngati Poneke record Aku Mahi

Patricia Grace, Irihapeti Ramsden and Jonathan Dennis, *The Silent Migration: Ngāti Pōneke Young Māori Club 1937–1948*, Huia, Wellington, 2001
Bradford Haami, *Urban Māori: The Second Great Migration*, Oratia Books, Auckland, 2018

81. Te Rōpū o te Matakite (Māori Land March) pou whenua

Aroha Harris, *Hīkoi: Forty Years of Māori Protest*, Huia, Wellington, 2004
Michael King, *Whina*, Hodder & Stoughton, Auckland, 1983
Geoff Steven (director), *Te Matakite o Aotearoa — The Maori Land March*, 1975 documentary, https://www.nzonscreen.com/title/te-matakite-o-Aotearoa-1975

82. Tepaeru Tereora's tīvaevae

Sandra Kailahi, *Pasifika Women: Our Stories in New Zealand*, Reed, Wellington, 2007
Lynnsay Rongokea, *The Art of Tīvaevae: Traditional Cook Islands Quilting*, Random House, Auckland, 2001
Jeffrey Sissons, *Nation and Destination: Creating Cook Islands Identity*, Institute of Pacific Studies and the University of the South Pacific Centre in the Cook Islands, Suva & Rarotonga, 1999
Carl Walrond, 'Cook Islanders', Te Ara — the Encyclopedia of New Zealand, http://www.TeAra.govt.nz/en/cook-islanders

83. Montana Blenheimer wine cask

Bronwyn Dalley, 'Wine', Te Ara — the Encyclopedia of New Zealand, http://www.TeAra.govt.nz/en/wine
Warren Moran, *New Zealand Wines: The Land, the Vines, the People*, Hardie Grant Books, Melbourne, 2016
Keith Stewart, *Chancers and Visionaries: A History of New Zealand Wine*, Godwit, Auckland, 2010

84. Biko shield

Geoff Chapple, *1981: The Tour*, A. H. & A. W. Reed, Wellington, 1984
Tom Newnham, *By Batons and Barbed Wire*, Graphic Publications, Auckland, 2003, first published 1981
Trevor Richards, *Dancing on Our Bones: New Zealand, South Africa, Rugby and Racism*, Bridget Williams Books, Wellington, 1999

85. Nuclear-free badges

Philip Attwood, *Badges*, British Museum Press, London, 2004
Kevin Clements, *Back from the Brink: The Creation of a Nuclear-free New Zealand*, Allen & Unwin, Wellington, 1988
Elsie Locke, *Peace People: A History of Peace Activities in New Zealand*, Hazard Press, Christchurch, 1992

86. Save Our Post Office poster

Michael Bassett, *Working with David: Inside the Lange Cabinet*, Hodder Moa, Auckland, 2008
Mark Derby, Jennifer Rouse and Ian Wedde (eds), *We Will Work With You: Wellington Media Collective 1978–1998*, Victoria University Press, Wellington, 2013
Brian Easton, *Not in Narrow Seas: The Economic History of Aotearoa New Zealand*, Victoria University Press, Wellington, 2020

87. POLY 1 personal computer

Reuben Schwarz, 'How New Zealand took on IBM and Apple, and lost', *Dominion Post*, 24 April 2006, p. C7
Michael Smythe, 'Poly 1 Educational computer', http://www.creationz.co.nz/kiwinuggets/2007/03/poly-1-educational-computer_07.html
Janet Toland (ed.), *Return to Tomorrow: 50 Years of Computing in New Zealand*, New Zealand Computer Society, Wellington, 2010
W. R. Williams (ed.), *Looking Back to Tomorrow*, New Zealand Computer Society, Wellington, 2010

88. Barry Brickell's memorial post to Ralph Hotere's father

David Craig and Gregory O'Brien, *His Own Steam: The Work of Barry Brickell*, Auckland University Press, Auckland, 2013

Gregory O'Brien, *Hotere: Out the Black Window — Ralph Hotere's Work with New Zealand Poets*, Godwit, Auckland, 1997
Vincent O'Sullivan, *The Dark is Light Enough: Ralph Hotere — A Biographical Portrait*, Penguin, Auckland, 2020
Peter Simpson, *Bloomsbury South: The Arts in Christchurch 1933–1953*, Auckland University Press, Auckland, 2016

89. The New Zealand AIDS quilt

Lynne Alice and Lynne Star (eds), *Queer in Aotearoa New Zealand*, Dunmore Press, Palmerston North, 2004
Chris Brickell, *Mates and Lovers: A History of Gay New Zealand*, Random House, Auckland, 2008
Chris Brickell and Judith Collard (eds), *Queer Objects*, Otago University Press, Dunedin, 2019
Brent Coutts and Nicholas Fitness, *Protest in New Zealand*, Pearson, Auckland, 2013

90. Mike Smith's chainsaw

Aroha Harris, *Hīkoi: Forty Years of Māori Protest*, Huia, Wellington, 2004
Richard S. Hill, *Maori and the State: Crown–Maori relations in New Zealand/Aotearoa, 1950–2000*, Victoria University Press, Wellington, 2009
The Spinoff 'Object' series, https://www.youtube.com/watch?v=MGycxYHIVV4

91. Peter Blake's red socks

Harold Kidd and others, *Southern Breeze: A History of Yachting in New Zealand*, Viking, Auckland, 1999
Harold Kidd, 'Sailing and windsurfing', Te Ara — the Encyclopedia of New Zealand, http://www.TeAra.govt.nz/en/sailing-and-windsurfing
Alan Sefton, *Sir Peter Blake: An Amazing Life*, Viking, Auckland, 2004

92. Dorene Robinson's Swanndri

Shaun Barnett and Chris Maclean, *Tramping: A New Zealand History*, Craig Potton, Nelson, 2014
Carl Walrond, 'Tramping', Te Ara — the Encyclopedia of New Zealand, http://www.TeAra.govt.nz/en/tramping

93. Helen Clark's trousers

Claudia Pond Eyley and Dan Salmon, *Helen Clark: Inside Stories*, Auckland University Press, Auckland, 2015
Kerryn Pollock, 'Fashion and textile design', Te Ara — the Encyclopedia of New Zealand, http://www.TeAra.govt.nz/en/fashion-and-textile-design
Elspeth Preddy, *The WEL Herstory: The Women's Electoral Lobby in New Zealand, 1975–2002*, WEL New Zealand, Auckland, 2003
Denis Welch, *Helen Clark: A Political Life*, Penguin, Auckland, 2009

94. King Théoden's armour

Ian Conrich and Stuart Murray (eds), *New Zealand Filmmakers*, Wayne State University Press, Detroit, 2007
Jonathan Dennis and Jan Bieringa (eds), *Film in Aotearoa New Zealand*, 2nd edition, Victoria University Press, Wellington, 1996
Alfio Leotta, *Touring the Screen: Tourism and New Zealand Film Geographies*, Intellect, Bristol, 2011
Diane Pivac (ed.) with Frank Stark and Lawrence McDonald, *New Zealand Film: An Illustrated History*, Te Papa Press, Wellington, 2011

95. The World's first iPhone 3G

Russell Brown, 'Digital media and the internet — Social media', Te Ara — the Encyclopedia of New Zealand, http://www.TeAra.govt.nz/en/digital-media-and-the-internet
Walter Isaacson, *Steve Jobs*, Simon & Schuster, New York, 2011
Matthew Jones, 'iPhone History: every Generation in Timeline Order 2007–2021', *History Cooperative*, 14 September 2014, https://historycooperative.org/the-history-of-the-iphone/
A. C. Wilson, 'Telecommunications', Te Ara — the Encyclopedia of New Zealand, http://www.TeAra.govt.nz/en/telecommunications

96. 'Thunder Down Under' Christchurch portable toilet

https://quakestudies.canterbury.ac.nz

R. Potangaroa, S. Wilkinson, M. Zare and P. Steinfort, 'The management of portable toilets in the eastern suburbs of Christchurch after the February 22, 2011 earthquake', *Australasian Journal of Disaster and Trauma Studies*, vol. 2 (2011), pp. 35–48

Geoffrey W. Rice, *Christchurch Changing: An Illustrated History*, Canterbury University Press, Christchurch, 1999

Pamela Wood, *Dirt: Filth and Decay in a New World Arcadia*, Auckland University Press, Auckland, 2005

98. Rā Maumahara pouaka petihana (New Zealand Wars Day petition box)

Auckland Museum Blog, 'Nation changers', https://www.aucklandmuseum.com/discover/stories/blog/2017/nation-changers

Rawinia Higgins and Basil Keane, 'Te reo Māori — the Māori language', Te Ara — the Encyclopedia of New Zealand, http://www.TeAra.govt.nz/en/te-reo-maori-the-maori-language

Tainui Stephens, 'Māori and television — whakaata', Te Ara — the Encyclopedia of New Zealand, http://www.TeAra.govt.nz/en/maori-and-television-whakaata

99. Tariq Omar's football

Royal Commission of Inquiry into the terrorist attack on Christchurch masjidain on 15 March 2019, Department of Internal Affairs, Wellington, 2020

100. Ailys Tewnion's crocheted bears

https://www.smithsonianmag.com/history/the-history-of-the-teddy-bear-from-wet-and-angry-to-soft-and-cuddly-170275899

Utsa Mukherjee, 'Rainbows, teddy bears and "others": The cultural politics of children's leisure amid the COVID-19 pandemic', https://www.tandfonline.com/doi/full/10.1080/01490400.2020.1773978

Index

Bold page numbers indicate photographs.

Acknowledgements

A work of this length and range of subject-matter would never have been completed, let alone begun, without the support and generosity of many people, and I would like to express my huge indebtedness and gratitude to some of those individuals.

As for the beginning, a history of New Zealand in 100 objects was suggested to me by Jeremy Sherlock. Initially I was sceptical and not confident that I would be able to do justice to the full span of the history. But Jeremy's gentle encouragement won me to the task and once I began, his wise counsel and enthusiastic response to drafts kept me working and greatly improved the final product. In the later stages Jeremy handed me over to the professional team at Penguin Random House, and I owe a great debt to Grace Thomas, the senior editor, Mike Wagg and Claire Gummer whose eagle eyes picked up my inconsistencies and tightened up my loose text, Rachel Eadie who organised publicity and Faith Wilson and Carla Sy.

The book was only possible because of the magnificent collections in the museums of New Zealand. I was able to visit most of the major museums through a generous Research Grant in 2018 from Copyright Licensing New Zealand; and I also was fortunate to receive interested support from Philippa Tocker at Museums Aotearoa and from three major museum directors – Anthony Wright at Canterbury, David Gaimster at Auckland and Ian Griffin at Otago. Their staff were also very helpful indeed including Chanel Clarke and Lucy Mackintosh at Auckland, Sarah Murray and Emma Brooks at Canterbury and Moira White at Otago. Once on the road I received enthusiastic encouragement from many curators who walked me around their wonderful collections, answered queries on particular objects and responded quickly and professionally to my endless questions by email. I am especially grateful to Katie Cooper, Grace Hutton, and Bronwyn Labrum at Te Papa, Nikolas Brocklehurst at Wellington Museum, Sandi Black and Patricia Nugent-Lyne at the Whanganui Museum, Gary Bastin, Megan Wells and Trudi Taepa at Puke Ariki in New Plymouth, David Dudfield at the Southland Museum in Invercargill, David Clarke at the Lakes District Museum in Arrowtown, Chloe Searle at the North Otago Museum in Oamaru, Sue Asplin at the Hokitika Museum, Amanda McGrath at the Buried Village Museum, Sherri Murphy at Shantytown, Gail Pope at MTG Hawkes Bay, Caitlin Timmer-Arends from Te Kōhangu Museum of Waitangi, Diana Stidolph from the Museum of Childhood in Masterton and Ian Stewart from Masterton's The Woolshed: National Museum of Sheep and Shearing. Sean Brosnahan spent a morning exciting me with Te Toitū's astonishing holdings and making my choices exceedingly difficult. I must note too Richie Gould whose infectious enthusiasm for the history of shearing led him to compile a magnificent museum of hand-pieces and other shearing equipment on his Pleasant Point property in South Canterbury.

Once I had chosen the objects I was excited to find people who knew their origins personally and could enrich their meaning with family stories. I am hugely grateful to Gary Pond who gave me a fascinatingly detailed account of his father's ordeal in the 1931 Napier Earthquake, to Rosalie Sugrue who recounted the extraordinary story of her father's homemade television set in Hokitika, to Tepaeru Tereora and her daughter Joy who told me the personal history behind Tepaeru's beautiful tivaevae, to Bill Nathan who was welcoming and informative about Ngāti Poneke, to Ailys Tewnion who told me about her clever Jacinda and Ashley bears, to the Wang family (Kilihelsey, Austin, Yolanda and Wilson) who were wonderfully honest about the challenges of migrating from China to New Zealand, and Leah Bell who recalled in rich detail the origins of the Rā Maumahara petition. Without their generous willingness to share their passions and deeply personal information, this book would be much diminished.

Then there were many individuals who provided advice along the way. They include Dougal Austin, Annabel Cooper, Anthony Dreaver, Brian Easton, Marguerite Hill, John Hine, Az James, Margaret Lovell-Smith, Stuart Park, Mark Stocker, Clive Rivers and Alan Tennyson. I received constant encouragement and suggestions from a Wellington history writers group which included Barbara Brookes, Ross Calman, Peter Clayworth, Elizabeth Cox, Mark Derby, Paul Diamond, Grace Hutton, Emma Jean Kelly, Ewan Morris, Ben Schrader, Tim Shoebridge, Jane Tolerton, and Ross Webb. In addition this group read drafts towards the end which was immensely helpful.

Two book clubs, of which I was a member, also read draft stories and provided valuable feedback from the perspective of the general reader. One group included Rebecca Denford, John and Pauline Hannah, Alison Neville, Heather and Nigel Roberts, Lew Skinner, and Jiff Stewart. The other group was Kevin Clark, Loretta Desourdy, Anne Gilbert, Ken Howell, Stanley Marshall, Clare Murray, Wendy Nelson, Lynette Squire, and Pat Webster. I am very grateful to the energy and commitment these two groups put into critiquing the drafts. Chris Maclean also gave the final manuscript a helpfully critical read. Inevitably in a work of this scale there will be errors – and for those I hold no-one responsible but myself.

The final acknowledgement conventionally goes to the author's family which is fully justified because any author knows that writing and research can make you unreasonably grumpy or annoyingly loquacious. So I must thank the Harper family, of all generations, who have followed the project from the start and whose interest, especially that of Philip Harper, has extended to asking searching questions in a tramping hut late at night. I must also thank my sister, Elizabeth Caffin, for her consistently wise counsel, and my Phillips children and grandchildren who have willingly gone along with my strange passions. The greatest acknowledgement is to my wife and partner-in-life Frida Harper, who has been the most consistent champion of the whole project and who has worn its challenges no less than me. To all of you, I trust, the travails have been worth it.